A Greater Prize Than Gold

Augustus Oldfield, 19th century botanical collector and ethnographer in Australia

M. Helen and William G. (Bill) Henderson

Published: 16 August 2018 by
The Book Reality Experience

ISBN: 978-0-6482222-3-1
Paperback Edition

For Noel, Georgina, Ully, Elena,
Olivia, Lucia, Will and Alex

Contents

PART 2

CHAPTER 6

CHAPTER 7

CHAPTER 8

CHAPTER 9

CHAPTER 10

PART 3

CHAPTER 11

CHAPTER 12

CHAPTER 13

PART 4

CHAPTER 14

CHAPTER 15

CHAPTER 16

CHAPTER 17

CHAPTER 18

CHAPTER 19

CHAPTER 20

PART 5

CHAPTER 21

CHAPTER 22

APPENDICES

Maps

Abbreviations

ADBO	Australian Dictionary of Biography Online. National Centre of Biography,Australian National University.
ABRS	Australian Biological Resources Study.
AMANI	Australian Marine Algal Name Index.
ANHBiog	Australian National Herbarium, *Biography: plant collectors and illustrators or others relevant to Australian botany.*
ALH	*Checklist of Australian Liverworts and Hornworts.*
ABRS	Australian Biological Resources Study, Canberra.
AO	Augustus Oldfield.
APC	Australian Plant Census, Council of Heads of Australasian Herbaria.
APNI	Australian Plant Name Index, Australian National Herbarium.
AusMoss	*Catalogue of Australian Mosses*, National Herbarium of Victoria.
Botanic Garden Edinburgh	Royal Botanic Garden Edinburgh Herbarium.
Botanic Gardens Kew	Royal Botanic Gardens Kew.
Botanic Gardens Melbourne	Royal Botanic Gardens Melbourne.
Botanic Gardens Victori	Royal Botanic Gardens Victoria
Botanic Gardens, Sydney	Royal Botanic Gardens, Sydney.
BFSS	British Foreign School Society.
CKB	Court of King's Bench.
CSD	Colonial Secretary's Department.
CSO	Colonial Secretary's Office.
CWA	City of Westminster Archives Centre.
DAN, NLA	Digitised Australian Newspapers 1803–1954, National Library of Australia.
DECNSW	Department of Environment and Conservation (NSW).
EDTA	Education Department Tasmania Archives.
FloraBase	*FloraBase: the Western Australian Flora,* Western Australian Herbarium specimen catalogue online.
GAPNA	Geoscience Australia Place Name Search Australia, Australian Government.
GB	George Bentham.
GL&A	Guildhall Library & Archives, London.

HCWA	Heritage Council of Western Australia.
Herb.	Oldfield herbarium specimen record
MHH	M. Helen Henderson.
HOA	Henderson Oldfield Archive.
HAODb	The M.H. & W.G. Henderson Augustus Oldfield digitised database of botanical collections.
JSBL	J.S. Battye Library of West Australian History, State Library of Western Australia.
JDH	Joseph Dalton Hooker.
K&CLSL	Kensington and Chelsea Local Studies Library.
KCLA	Kings College London Archives.
LAIT	*Checklist of the Lichens of Australia and its Island Territories,* Australian Biological Resources Study, Canberra..
LMA	London Metropolitan Archives
LSD	Lands & Survey Department, State
FM	Ferdinand Mueller.
MELISR	National Herbarium of Victoria, Royal Botanic Gardens Melbourne MELISR database.
ML	Mitchell Library, New South Wales.
NAL	National Archives, Kew, London.
NHNSW	National Herbarium of New South Wales, Royal Botanic Gardens Sydney.
NHV	National Herbarium of Victoria, Royal Botanic Gardens Melbourne.
NLA	National Library of Australia.
Natural History Museum	Natural History Museum, London.
NRA	National Register of Archives.
PFA	Pimblett Family Archive.
PlantNet	*PlantNET: New South Wales Flora Online,* The Royal Botanic Gardens [Sydney] and Domain Trust.
PROV	Public Records Office of Victoria.
Publ.	Published Oldfield specimen record.
RBGKA	Royal Botanic Gardens Kew Archives.
Kew, RBGK	Royal Botanic Gardens Kew Herbarium.
RBGML	Royal Botanic Gardens Melbourne Library.
RBGSL	Royal Botanic Gardens Sydney Library.
STLV	State Library of Victoria.
SLWA	State Library of Western Australia.
SRONSW	State Records Office of New South Wales.
SROWA	State Records Office of Western Australia.
TAHO	Tasmanian Archive and Heritage Office.
Tasmanian Herbarium	Tasmanian Herbarium, Tasmanian Museum and Art Gallery.
The Times	*The Times*, London.
TTDA	*The Times* Digital Archive.
UCLA	University College London Archives.
ULA	University of London Archives.

UWAL	University of Western Australia Library.
UWAP	University of Western Australia Press.
VDL	Van Diemen's Land.
WH	William George Henderson.
WHH	William H. Harvey
WJH	William Jackson Hooker.
WLHSL	Wandsworth Local History Service Library.

International codes for herbaria referred to in footnotes and main text of this work

B	Botanischer Garten und Botanisches Museum Berlin-Dahlem, Berlin, Germany.
BM	National History Museum, London, England.
BR	Jardin Botanique National de Belgique, Meise, Belgium.
BRI	Queensland Herbarium, Toowong, Queensland.
CANB	Australian National Herbarium, Canberra, Australian Capital Territory.
CGE	Department of Plant Sciences Herbarium, University of Cambridge, Cambridge, England.
E	Royal Botanic Garden Herbarium, Edinburgh, Scotland.
G	Conservatoire et Jardin Botaniques de la Ville de Genève, Switzerland.
GH	Harvard University Herbaria, Cambridge, Massachusetts, USA.
H	Viktor Ferdinand Brotherus Herbarium, University of Helsinki, Helsinki, Finland.
HBG	Biozentrum Klein Flottbeck und Botanischer Garten der Universita Hamburg, Germany.
HO	Tasmanian Herbarium, University of Tasmania, Hobart, Tasmania.
JE	Friedrich-Schiller Herbarium, University of Jena, Jena, Germany.
K	Royal Botanic Gardens, Kew Herbarium & Economic Botany Collection, Kew, England.
LD	Agardh Herbarium, Botanical Museum, University of Lund, Lund, Sweden
M	Botanische Staatssammlung, München, Germany.
MANCH	University of Manchester Herbarium, Manchester, England.
MEL	National Herbarium of Victoria, Royal Botanic Gardens, Melbourne, Victoria
MICH	University of Michigan Herbarium, Ann Arbor, Michigan, USA.
MO	Missouri Botanical Garden Herbarium, St Louis, Missouri, USA.
NBG	Compton Herbarium. South African Biodiversity Institute, Cape Town, South Africa.
NSW	National Herbarium of New South Wales, Royal Botanic Gardens, Sydney, New South Wales.
NY	New York Botanical Gardens Herbarium, New York, USA.
OXF	Fielding-Druce Herbarium and Boswell's Collection, Oxford University, Oxford, England.
PC	Muséum Nationale d'Histoire Naturelle, Paris, France.
PERTH	Western Australian Herbarium, Biodiversity Conservation Science Centre. Kensington, Western Australia.
PH	Philadelphia Herbarium, the Academy of Natural Sciences?[1]

[1] A search of the Academy's online database found no records for 'Oldfield'. 'A F Oldfield', 'A Oldfield', 'Augustus Oldfield' or 'Augustus Frederick Oldfield'. However, James A. Mears, 'Guide to plant collectors represented in the Herbarium of the Academy of Natural Sciences of Philadelphia', *Proceedings of the Academy of Natural Sciences of Philadelphia*, vol. 133 (1981), pp. 141–165 includes an Oldfield. Also, W.L. Dix, 'The lichen collection of the Academy of Natural Sciences of Philadelphia, *The Bryologist*, vol. 62, no. 1, 1959, pp. 65–66 states that there are 53 Australian lichen specimens in the herbarium collected by 'E. Oldfield'. An unsuccessful search of the Academy's database was made for this name. It is probably Augustus Oldfield since he is the only collector with that surname listed in Alex. S. George, *Australian Botanist's Companion*, Four Gables Press, Kardinya, Western Australia, 2009.

S	Naturhistoriska Riksmuseet, Stockholm, Sweden.
TCD	School of Botany, Trinity College, Dublin, Ireland.
U	National Herbarium Nederland, Utrecht University branch, Utrecht, Netherlands.
US	Smithsonian Museum of Natural History Herbarium, Washington DC, USA.
W	Naturhistorisches Mueum Wien, Vienna, Austria.

Acknowledgements

The research for this work covered the period 1995 to 2014 and it is not possible to express here the full sense of our indebtedness to the archivists, colleagues, Oldfield descendants, family and friends who have contributed to the making of this book. Nevertheless, the following acknowledgements are given as a mark of our gratitude.

A very special thank you is due to Dr Michael O'Connor who gave so generously of his time and expertise to making our Oldfield specimen records database more workable, for developing the index and for re-working the maps when it became obvious that the maps Bill had prepared could not be completed due to difficulties with mapping software. We also thank Michael for his tremendous support since 2005, for giving us access to his PhD thesis and many other publications in his library and for always being on the lookout for references relevant to our research.

We are indebted to The Board of Trustees of the Royal Botanic Gardens Kew for granting permission for copies of Oldfield's and related official correspondence, his digitised specimen records and digitised historical images to be made available to us. We also thank the Royal Botanic Gardens Victoria Board for making available downloads of all digitised Oldfield specimens in the National Herbarium of Victoria. Our gratitude is also extended to the Royal Botanic Gardens Sydney Library for allowing us to photocopy Augustus Oldfield's unpublished manuscript on natural philosophy. To descendants of Augustus's father, Frederick Oldfield, we owe special thanks: to Jeff and Anne Pimblett, Cumbria, England, Desmond and the late Lyn Williams, Albert Park, Victoria and Elaine Taylor of Perth, Western Australia for so generously making family archival and genealogical material available to us and for all the support they have given over the many years of the project.

Grateful thanks are due to our readers who gave so much of their time and expertise to improving the quality of aspects of the text. Early drafts of the entire manuscript were read by Frederick Oldfield descendant, Dr Murray Williams, Narrabundah, ACT and Jeff Pimblett. Specialist readers were: Dr Alison Alexander, historian, Tasmania; Dr Jane Balmer, Department of Primary Industries, Parks, Water and Environment, Tasmania; Alex S. George, School of Veterinary and Life Sciences, Murdoch University, Perth, W.A.; Professor Stephen Hopper, Centre of Excellence in Natural Resource Management and School of Biological Sciences, University of Western Australia, Albany, W.A.; and Dr Sara Maroske, editor, Mueller Correspondence Project, Royal Botanic Gardens Victoria, and editor, *Historical Records of Australian Science*.

We are greatly indebted to the staff of herbaria and botanic gardens who provided details of Oldfield's specimens held in their collections: ***Australia:*** Australian Biological Resources Study, Department of Environment, Canberra; Australian National Herbarium, Integrated Botanical Information System (IBIS), Australian National Botanic Gardens, Canberra; Centre for Plant Biodiversity, CSIRO, Canberra, Jo Palmer; John Ray Herbarium, School of Life and Environmental Sciences, University of Sydney, Associate Professor Murray Henwood; National Herbarium of New South Wales, Dr Katherine Downs; Royal Botanic Gardens, Sydney, New South Wales, Anna Hallett; National Herbarium of Victoria, Royal Botanic Gardens, Melbourne, Dr Alison Vaughan, Dr Catherine Gallagher and Dr Pina Milne; King's Park and Botanic Garden, Perth, Roger Fryer; Queensland Herbarium, Queensland Department of Environment; Tasmanian Herbarium, Tasmanian Museum and Art Gallery, Alex Buchanan, Dr Gintaras Kantvilas, Dr Jean Jarman and Patrick (Paddy) Dalton; Western Australian

Herbarium, Department of Parks and Wildlife, Dr Kevin Thiele, Neville Marchant, Karina Knight and Albany Regional Herbarium, Ruth Moir and Libby Sandiford; ***Austria:*** Naturhistorisches Museum Wien, Wien, Dr Ernst Vitek; ***England & Scotland:*** Cambridge University Herbarium, Dr Gina Murrell; Fielding-Druce Herbarium, Oxford University Herbaria, Department of Plant Sciences, Serena Marner; Bryology Department, Natural History Museum London, Dr David Sutton, and Dr Len T. Ellis; Royal Botanic Garden Edinburgh, Adele Smith; Royal Botanic Gardens Kew, including the Museum of Economic Botany; Dr Peter Bostock, Dr Roberta Cowan, Dr Juliet Wege, Dr Jenny Tonkin, Dr Daniela Zappi and Dr Mark Nesbitt. ***Germany:*** Botanische Staatssammlung München, Dr Hans Hertel. ***Finland:*** Viktor Ferdinand Brotherus Herbarium, University of Helsinki Herbarium, Helsinki, Finland, Dr Sinikka Piippo. ***Sweden:*** Agardh Herbarium, Botanical Museum, Lund University, Lund; Swedish Museum of Natural History, Stockholm, Dr Per Lassen; Departments of Cryptogamic Botany and Phanerogamic Botany, Swedish Museum of Natural History, Dr Anders Tehler & Dr Arne A. Anderberg, Stockholm.***United States of America*:** Bailey Hortorium Herbarium, Cornell University, Robert Dirig. We also offer thanks to all those curators and directors of herbaria included in our survey who informed us they either held Oldfield specimens but could not provide information or had none of his specimens.

Much appreciation is extended to librarians, archivists and officers of official registries who have so readily responded to our requests for assistance, especially the staff of the following: ***Australia:*** Augusta Museum, Western Australia, particularly Graham Davis; Australian Academy of Science Library, Canberra; Australian Institute of Aboriginal and Torrest Strait Islander Studies Library, Canberra; Cambridge University Archives, J. Fox; Correspondence of Ferdinand von Mueller Project, Department of History & Philosophy of Science, University of Melbourne, Dr Sara Maroske and Dr Monika Wells; Department of Indigenous Affairs, Western Australia; Education Department Tasmania Archives, Pat Lloyd; J.S. Battye Library of West Australian History, State Library of Western Australia. Perth; Nomenclature Section, Landgate, Department of Lands Administration (Western Australia), Brian Goodchild; Office of the Surveyor General, Department of Primary Industries, Water and Environment, Hobart, John Vander Niet; Mitchell Library, Sydney; National Library of Australia, Canberra; Royal Botanic Gardens Victoria Library, Helen Cohn, Jill Thurlow and Sally Stewart; Office of Surveyor-General Victoria, Geographic Names, Anita Davids; Royal Botanic Gardens Sydney Library; South Australian Museum Archives, Adelaide; State Library of Victoria, Melbourne, Margaret. McCormack; State Records Office of New South Wales, Sydney; State Records Office of Victoria, Melbourne; State Records Office of Western Australia, Perth; Tasmanian Archive and Heritage Office, Hobart, Jennifer R. Jerome and Fiona MacFarlane; ***England:*** Birkbeck College Library, London; British & Foreign School Society Archives, London, Brian York; Brunel University, London; Cambridge University Archives; Chipping Norton Historical Society, Eian Caws, John Mann; City of Westminster Archives Centre, London; Crown Estate Office, London, Andrew Thomas and attached private researcher E.R. Talbot-Rice; General Register Office, England, London; Guildhall Library & Archives, London; Institute of Education, London; Kensington and Chelsea Local Studies Library, Hazel Cook; Kings College London Archives; London Metropolitan Archives; National Archives, Kew, attached private researcher R.W. O'Hara; Probate Registry, London, Manchester, York; Oxford University Archives, Oxford, Alice Blackford; Royal Botanic Garden Edinburgh Library; Royal Botanic Gardens Kew Archives, Pat Wilson, Kate Pickard, Helen Broderick and Sara Cox; Royal Society, London; Royal Parks, Brompton Cemetery, Laurence Epstein; Royal Society London, Rupert Baker; St Mary's Church, Rotherhithe; Church of St Mary Abbots, Kensington, Pat Wilson;University College London Archives; University of London Archives; Wandsworth Heritage Service, Julie Gregson, Ruth MacLeod; ***France*:** Gûines Town Museum, Société Historique de Guînes et des Environs, Jacques Louf.

A number of individuals have provided us with contextual information and images relevant to the botany and history of various places through which Oldfield travelled and contributed in other ways for which we are grateful: Special mention must be made of the late Dr Robin Adamson for her assistance in locating the information about Monsieur Liborel's schools in Guînes, France and for translating our letters and information received; Dr Jayne Balmer for so patiently providing information of the flora relevant to Oldfield's Tasmanian collecting places; Don and Barbara Bellairs who showed us live flowering plants of most of the species in the Kalbarri National Park named in Oldfield's honour; Stephen Carrier for digitising all our 35mm slides of Oldfield's collecting localities; Alex. S. George for kindly providing a list of Oldfield's Proteaceae specimens that he located at the Royal Botanic Gardens, Kew; Peter Grant who prepared a description of the ascent of Mt La Perouse which we have incorporated in the text; Dr Patrick M. McCarthy for checking the taxonomic status of many of Oldfield's lichen specimens; Dr Richard Selleck for advice on teaching during the Kay-Shuttleworth period, England; Robin Simkin for making special efforts to photograph live plants of additional species named in Oldfield's honour in the vicinity of the Murchison River. Others who generously contributed with images or information were Chris Bassham, Dr Anne Brearley, Bradley Commins, Gillian Craig, Craig Dickman, Stan Gratte, Noël Henderson. Greg Jordan, Lenore Layman, Dr Jim Leavesley, Ruth Moir, Dr Barry Mendelawitz, the late Dr Philip Playford, Diana Scott; and Robert Wiltshire. Thanks also to Jenny Eaton, Bowes Homestead, near Northampton and Calum Carruth, Murchison House Station for permission to photograph places on their properties relevant to this study.

We also thank Mary Elgar, of Bridgetown, WA, for her conscientious attention to detail in editing and proofreading the manuscript and with whom it was such a pleasure to work. We also greatly appreciate the never-ending patience and sound advice that our publisher, Ian Hooper of The Book Reality Experience, has given in getting this work ready for publication.

For their never-failing encouragement we owe a huge debt of gratitude to our beloved children Noël and Georgina and their families for their support and patience in the face of an apparently endless project!! We are indebted to our dear friend the late Professor Patricia Crawford for her invaluable advice and support throughout many years of the project. To other close friends and relatives—the late Dr Robin Adamson, Chris and Paul Bassham, Penny Blumann, Stephen and Janice Carrier, Yvonne Coote, Dr Ian Crawford, Lee and Geoff Fernie, Dr Jenni Ibrahim, Anne Harse, Lennie McCall AM, Ruth Marchant-James, Margaret Medcalf, Dr Barry and Margaret Mendelawitz, Ruth Moir, Gillian Flecker, Diana Scott and Diana Wald—we thank for all the encouragement and time they spent listening to Helen airing her ideas about aspects of Augustus Oldfield's life and the difficulties we faced with the research.

Preface

Among the items that were Helen's legacy from the estate of her paternal grandmother, Grace Henrica Williams (née Oldfield), there were three documents that aroused the authors' curiosity. One was a testimonial, dated 2 June 1875, from Dr Joseph Dalton Hooker, President of the Royal Society, London and Director of the Royal [later: Botanic] Gardens, Kew, England (hereafter Botanic Gardens, Kew, and Kew Gardens), certifying that he had

> known Mr Augustus Oldfield for nearly twenty years as a most active, able, industrious & trustworthy Naturalist, and especially Botanist whose disinterested labors & collections have thrown great light on the Flora of many distant, and some of them previously wholly unexplored districts of Australia.[2]

The other documents were two newspaper clippings. A rather lengthy and illuminating obituary from a Melbourne newspaper, *The Argus*, of 13 July 1887, which noted, among other things, Augustus' 'wonderful feats of pedestrianism in the pursuit of his favourite study, botany'.[3] The second was a report of a monthly meeting of the Royal Society of Tasmania from the *Hobart Mercury,* 16 August 1887, which recorded Augustus' death and his 'disinterested scientific service in Tasmania' as a botanist and collector.

Perhaps it would be surprising to some that Helen had never heard of Augustus. Grace Henrica, Augustus' niece and the daughter of his younger brother, Edwin Douglas, was born in Hobart in 1863, a year after Augustus left Australia forever. As far as we know, Grace never travelled to England where her uncle resided. Nevertheless, she must have known of him since there is evidence that her father was well acquainted with his brother's circumstances during the time he was in England. If Grace had told her children about Augustus and his botanical pursuits, her youngest son, Noël, did not pass on the information to his daughter, Helen. In any event, prior to Noël's death in 1955 she was not interested enough in family history to persistently quiz her parents for details of their forebears. On the few occasions that she and Grace were in each other's company, Augustus was never discussed.

Once aroused, curiosity led us to the State Library of Western Australia in search of more information. These initial pickings were lean indeed. A brief entry in F.A. Sharr's *Western Australian Plant Names and Their Meanings*[4] noted that Augustus was an

> English botanist and zoologist who collected in the south-west and in the Murchison (WA) in the 1850s and 1860s, and earlier extensively in Tasmania, his collections being sent to the Melbourne botanical gardens and to Kew (~ 13 species).[5]

Rica Erickson's *Bicentennial Dictionary of Western Australians* added to Sharr's information

> About 1863 he was in England & offered his collections to Kew Herbarium for use in compiling Bentham's *Flora Australiensis*. Lived alone for many years at Devonport, Tasmania.[6]

---

[2] JDH, Augustus Oldfield testimonial, 2 June 1875, from the estate of Grace Henrica Williams, Henderson Oldfield Archive (hereafter HOA).

[3] The author of the obituary was only referred to as 'a correspondent' but it was almost certainly Oldfield's brother, Edwin Douglas, *The Argus*, 13 July 1887, p. 5.

[4] F.A. Sharr, 1988, *Western Australian Plant Names and their Meanings: a Glossary*, 2nd edn, University of Western Australia Press, Perth, p. 171.

[5] It will become apparent that this figure is incorrect.

[6] R. Erickson, 1979, *Dictionary of Western Australians 1829–1914*, vol. 3, *Free 1850-1868*, University of Western Australia Press (hereafter UWAP), Nedlands, p. 644.

There is no evidence that he ever lived in Devonport, either in Tasmania or New Zealand.

Subsequent discussions with botanists in Western Australia and elsewhere revealed that although known to them through his specimens and brief biographical entries in directories of plant collectors, little was known about him personally or his botanical pursuits. We decided to try and fill this gap in the historiography of Australian botany.

Aims

While it is hoped that this biographical exploration will have wider appeal, the study has been undertaken with the object of assisting those professional and amateur botanists, ethnographers and historians who require an understanding of Oldfield's life and his contribution to science beyond the brief, occasionally incorrect, published information about him and his expeditions in Australia.

Oldfield was one of several dedicated botanical collectors in Australia who provided specimens to the successive Directors of the Royal Botanic Gardens Kew, Sir William Jackson Hooker and his son Dr Joseph Dalton Hooker.

The Director of the Botanic Gardens Melbourne and from 1858, Royal Botanic Gardens Melbourne (now Royal Botanic Gardens Victoria from 2015 but hereafter Botanic Gardens Melbourne when referring to the name as it applied in Oldfield's time) and Victorian Government Botanist, Dr Ferdinand Jakob Heinrich Mueller, was also a major recipient of Oldfield's specimens after January 1856. Mueller did not receive any honours until 1867 and, as a result his various titles will only be used in this work at the point in the narrative when each honour was conferred.[7] Furthermore, in 1849, soon after his arrival in South Australia, he took out Australian citizenship. A short time later he anglicised his name from the German 'Müller' to 'Mueller'.[8]

An important aim of this study is to examine Oldfield's relationship with Joseph Hooker to test whether it was in accord with any of the proposals put forward in the literature concerning the way in which he, and other 'men of science' in Britain, related to their colonial botanical collectors. Oldfield's relationship with Ferdinand Mueller will also be considered.

Method

This work embraces Oldfield's family background, his formative years and his life in the countries of birth and adoption. Emphasis has been placed on his dealings with the taxonomists to whom he sent his specimens, his income-earning occupations, his collecting and other natural history activities, and the geographical, economic and socio-cultural contexts in which he was operating. Along the way we discovered that his interests extended beyond collecting plants to plant biology, zoology, anthropology and contemporary natural philosophy.

The search for information about Oldfield's life was conducted over sixteen years, by postal and email communications to herbaria and archives in Great Britain, Europe and some other countries around the world. We visited selected archives in England and Australia and undertook online searches of British archival databases.[9] Footnote reference citations to these websites include the dates

[7] In 1871 Mueller was appointed a hereditary baron by the King of Württemberg, having been granted his 'von' in 1867. He was made Companion of the Order of St Michael and St George C.M.G. in 1869 and Knight Commander of the Order of St Michael and St George (K.C.M.G.) in 1879, Diedre Morris, 'Mueller, Sir Ferdinand Jakob Heinrich von (1825–1896)', Australian Dictionary of Biography (hereafter ADBO), viewed 28 July 2011at adb.anu.edu.au.

[8] R.W. Home, A.M. Lucas, Sara Maroske, D.M. Sinkora and J.H. Voigt (eds), 1998, *Regardfully Yours: Selected Correspondence of Ferdinand von Mueller* vol. 1, 1840–1859, 1998, Peter Lang, Bern, pp. 9 & 22.

[9] Historical Manuscripts Commission database, Historical Manuscripts Commission National Register of Archives, personal name search combining the Personal Index and the Diaries and Papers Index; and the English Strand of the UK Archives Network Database, viewed 9 April 2014 at www.nationalarchives.gov.uk.

viewed and their URLs except for citations from *The Times Digital Archive, 1785 – 2006* (hereafter TTDA), and *Digitized Australian Newspapers 1803–1954* (hereafter DAN, NLA). It became apparent during our investigation that the archival databases we searched online may not have reflected the totality of the respective archives' manuscript holdings and that not all archives had made their catalogues accessible via the internet. Therefore, other manuscript material of relevance to Oldfield's life may emerge in future.

Oldfield's letters and his only published work bear testimony to him having kept field notes but efforts to locate them were unsuccessful.[10] Augustus never married and as far as we are aware he had no descendants. Nevertheless, some of his manuscript material pertinent to this study has survived. Copies were obtained of 44 items of correspondence comprising Augustus' and others' correspondence, including official internal memoranda, held in the Botanic Gardens Kew Archives. These embraced the years 1850–1890. Unfortunately, none of the correspondence from the Hookers to Oldfield has survived.[11] Also, copies were obtained of Augustus' application for employment with the New Zealand Company (1841)[12] and his unpublished, untitled manuscript on natural philosophy (1865a).[13] This was transcribed by Bill Henderson. Apart from Oldfield's wide–ranging discussion of contemporary natural philosophy issues, this work contains personal accounts of his observations in mainland Australia and Tasmania, which he employed at times to support theoretical claims.

Between c.1853 and 1862 Oldfield periodically collected for, and corresponded with, Dr Ferdinand Mueller. Unfortunately, none of their letters have survived.[14] It seems that they may have been destroyed, along with other early records of the Botanic Gardens Melbourne, in the 1920s or 1930s.[15] However, Mueller's correspondence with Sir William and Dr Joseph Hooker and others, his official reports published in volume 1 (1840–1859) and volume 2 (1860–1875) of *Regardfully Yours: Selected Correspondence of Ferdinand von Mueller*[16] and our personal communications with one of the editors, Dr Sara Maroske, have shed light on aspects of Oldfield's and Mueller's relationship.

Augustus' lengthy paper *On the Australian Aborigines*, published in 1865[17], is a major primary resource, particularly about his relationship with the *Watchandie* (*Nanda*) people of the lower Murchison River area in Western Australia and of the theoretical approach which underpinned the interpretation of his observations of their socio-cultural lives. This paper and the manuscript material that we have been able to obtain provide important clues as to Oldfield's character, the breadth of his scientific interests, his theoretical stance on many scientific issues, his observational and analytical skills, methods, motivations, and his frustrations as a botanical collector.

Mr Jeff Pimblett and his wife, Anne, of Kendall, Cumbria, England, have kindly provided us with copies of Oldfield family documents covering the date range 1828–1916 from their private family archive (hereafter Pimblett Family Archive or PFA). They include 97 pieces of correspondence (letters, notes on postcards and visiting cards), seven images and 39 miscellaneous items, among

[10] Following Oldfield's death in 1887 his private herbarium was donated to the Botanic Gardens Kew, but the Royal Botanic Gardens Kew Archive (hereafter RBGKA) has assured us that it holds none of his diaries and field notes (personal communication Dr Juliet Wege, Australian Botanical Liaison Officer, Kew, email 25 November 2005). An extensive search of online archival databases in England was conducted and personal approaches were made during visits in October 2007 to the archive of the Royal Botanic Gardens Victoria Library (RBGML) and the Royal Botanical Gardens Sydney Library (RBGSL). The disposal of Oldfield's private herbarium following his death in 1887 is discussed in Chapter 22.

[11] Apparently, copies of outgoing correspondence were rarely prepared in the mid-19th century since the institution did not have the resources to employ clerical staff, personal communication, Librarian, RBGKA, November 2002.

[12] AO to the Directors, New Zealand Company, London, 3 September 1841, PRO CO208 /18/1841, 41/4605, p. 249, NAL.

[13] AO 1865a, Untitled ms., 860 pp., Species Manuscripts, RBGSL. In M. Helen Henderson (MHH), 2005, 'Introduction' to Augustus Oldfield 'On the Aborigines of Australia', *Transactions of the Ethnological Society of London*, n.s., vol. 3, 1865, pp. 215–298, reprint Hesperian Press, Victoria Park, Western Australia, untitled manuscript c.1861. Subsequently Jim Endersby, 2008, (*Imperial Nature, Joseph Hooker and the Practices of Victorian Science*, University of Chicago Press, London, fn. 54, p. 360) has given the date as '1865' because many of its pages bear the paper manufacturer's watermark: 'E Towgood 1865'. In this work, 1865 is adopted.

[14] R.W. Home *et al.*, 1998, vol. 1, p. 551.

[15] ibid., 'Introduction', pp. 43-45.

[16] R W Home et al., 1998 & 2002, vols. 1, and 2.

[17] AO 1865b, 'On the Aborigines of Australia', *Transactions of the Ethnological Society of London*, n.s., vol. 3, pp. 215–298.

which were a catalogue of a private sale, receipts, cheques, accounts, addressed envelopes, certificates & lists. They also kindly forwarded copies of 25 letters they had located at the National Archives Kew, London. Anne is a direct descendant of Augustus' youngest brother Vincent Howard Oldfield. Although it is disappointing that there are only a few references to Augustus in the Pimblett Family Archive, it did provide a window into the latter part of his life and valuable information about his parents and siblings during that time. Mrs Elaine Taylor, Perth, Western Australia, a descendant of Edwin Oldfield through his son Frederick Augustus, assisted with some relevant genealogical documents. Another descendant of Edwin Oldfield, Mr Desmond Williams, provided copies of family photographs and an image of an 1825 sampler created by Lavinia Oldfield.

As far as we can ascertain, Oldfield's only other writings that survived are his herbarium specimen labels.

In the absence of personal detailed accounts of all his collecting expeditions our only option for discovering where he collected was to compile, within the limits of our resources, a dataset of as many of his specimen collection records as possible. A relational electronic database of these records has been compiled entitled 'The Henderson Augustus Oldfield database' (Oldfield database/HAODb). These records variously contain his determinations, the localities where he gathered his specimens and brief notes on the habitats and habits of the plants from which he took them. It is rare to find a specimen label that includes the date of collection.[18]

Published and herbarium specimen records are separated in the dataset. Each record in the dataset includes the source, herbarium specimen number if applicable, the species' name noted in the source reference, current taxonomic and type[19] status and, if available, collection location, any plant habitat, and other field information provided by Oldfield. To facilitate searches, collection localities have been standardised with the major locality placed first in the field (for example, if the locality 'near the Blackwood River' is given in the source herbarium's database, it is presented as such in our database under 'Notes' and standardised to 'Blackwood River, near the' for each specimen record under 'Locality'. While as far as we are aware there is no duplication of individual records within the herbarium specimen dataset, there is some duplication between the herbarium and published records, and within the latter. No attempt has been made to identify such duplications as it was a complicated task beyond our resources. Many specimen labels include what appear to be field numbers and have been entered as 'Collector No.' in various publications and herbarium databases. There are sound reasons for claiming that Oldfield's specimen numbers given as 'Coll. No.' in herbarium records are not his field numbers (see below).

Where it is appropriate to refer to the number of herbarium and published specimen locality records they are separately cited. Thus, for example, 'Near Bunbury (3 publ., 12 herb.), HAODb'. In some cases, the source reference for specimens is given in addition to the name of the species, date and locality collected and if relevant, the relevant herbarium accession number. International herbarium codes are confined to the citing of specimens. For example, MEL 2278624 refers to a specimen in

---

[18] For example, of 3992 records of Oldfield specimens obtained from the Botanic Gardens Victoria National Herbarium of Victoria (MEL, prefix when citing a specimen number, or NHV when citing the herbarium as the repository), only eight specimens had the full date descriptor (i.e. day, month, year), nine had month and year and there were 18 with just the year on the label. Some of these last were stated to have been approximate datings by taxonomists and others who subsequently handled the specimens. There are also collection dates on MEL (international abbreviation for the National Herbarium of Victoria, Royal Botanic Gardens, Melbourne) specimen labels that are incorrect, namely MEL10584, 1850, Western Australia, Kalgan River; MEL705706, 1867, Upper Kalgan River; MEL 2073762, 1875, Western Australia, Esperance Bay; and MEL 1610975 & 1610974, Kalgan River; MEL 706642, Locn: Western Australia, Lake Logue; and MEL 291015, Western Australia, Manjinup [not Manjimup] were all stated to have been obtained in 1860 (MELISR database 8 March 2007, NHV, Royal Botanic Gardens Victoria (NHV). Oldfield was not in Western Australia in any of those years.

[19] A nomenclatural type (typus) is that element to which the name of a taxon is permanently attached, whether as the correct name or as a synonym. The nomenclatural type is not necessarily the most typical or representative element of a taxon, J. McNeill, F.R. Barrie, W.R. Buck, V. Demoulin, W. Greuter, D.L. Hawksworth, P.S. Herendeen, S. Knapp, K. Marhold, J. Prado, W.F. Prud'homme van Reine, G.F. Smith, J.H. Wiersema & N.J. Turland 2012, *International Code of Nomenclature for algae, fungi, and plants (Melbourne Code) adopted by the Eighteenth International Botanical Congress Melbourne, Australia, July 2011*, Koeltz Scientific Books, Kőnigstein, Germany.

the National Herbarium of Victoria, Royal Botanic Gardens Victoria. The codes referred to are separately listed as Part B of the Abbreviations.

With no detailed chronological accounts of his experiences in Australia, an effort has been made to fill some of the gaps from a variety of published historiographical sources. In the first place, brief descriptions of the places he visited and the floristic associations through which he travelled are provided. When first mentioned in the text of each chapter, the common names of plants are given in association with their respective scientific names, thereafter in the chapter they are referred to only by their common names.[20] Those plants without common names are always referred to by their scientific names. Every endeavour has been made to give contextual accounts, which pertain to the time of his visits. It is hoped this will give the reader some appreciation of the conditions under which he worked and the difficulties he must have faced as he moved through remote parts of Australia that could best only be described as at the very forefront of European settlement. Secondly, when we encountered gaps in the historical record about what Oldfield was doing during particular periods, we have offered some suggestions, drawing upon his specimen record localities, what he has revealed of his skills and interests, and what we have discovered of contemporary activities in which he might have been engaged. Thirdly, some publications, his letters and specimen labels have revealed information about his botanical acquaintances and colleagues. Unless we found that they were an important influence on his work or were co-collectors only relevant biographical details have been provided.

To locate the whereabouts of Oldfield's specimens we tested the comprehensiveness of the list of 16 herbaria given as housing Oldfield's specimens in I. Hettie Vegter, *Regnum Vegetabile*.[21] From online herbarium directories, 128 major herbaria, including those in the *Regnum Vegetabile* list were identified and emails sent to them in 2001 requesting information about their Oldfield specimen holdings. These institutions were not randomly selected but on our judgement, they might have been repositories of his material. They embraced 29 countries namely: England, Scotland, Republic of Ireland, Sweden, Finland, Norway, Denmark, Germany, Austria, France, Belgium, Netherlands, Switzerland, Spain, Italy, Turkey, Estonia, Slovenia, Croatia, Lithuania, South Africa, Israel, India. United States of America (including Hawaii), Canada, Australia, New Zealand, Japan, Taiwan, Russia.[22] Of these, 75 herbaria responded, 14 were returned as address unknown and 39 did not reply, including several institutions that were listed in *Regnum Vegetabile*, 12 stated that they held Oldfield's specimens, but with one exception, details were not accessible as digitised catalogues of their collections had not been prepared. The exception was the Western Australian Herbarium which provided a hard copy list of Oldfield specimen records in its collection.[23] A few herbarium respondents, however, kindly gave details of specimens of which they were aware. Since then (from 2005), some of those institutions, which had previously indicated they held Oldfield's specimens and were intending to digitise their collection catalogues, were successfully approached with a request for electronic downloads of his specimen records that had been digitised. These were: the Fielding-Druce Herbarium and Boswell's Collection, Oxford University[24], Botanic Gardens, Kew Herbarium[25], National Herbarium of Victoria, Royal Botanic Garden Victoria (hereafter National Herbarium of Victoria)[26], Tasmanian

[20] Common names used in the text were drawn from the relevant state government online floras and other databases.

[21] I. Hettie Vegter, 1983, *Index herbariorum: a guide to the location and contents of the world's public herbaria*, Part 2(5). Collectors N–R, International Association for Plant Taxonomy, vol. 109, and F.A. Stafleu (ed), **1981,** *Index Herbariorum: a Guide to the Location and Contents of the World's Public Herbaria* Pt 2 (5), Scheltema & Holkema, Bohn.

[22] Copies of all emails sent and received are in the Henderson Oldfield Archive, HOA.

[23] Kindly provided by Dr Neville Marchant, then Curator of the WAH, 9 June 1999, HOA.

[24] Dr Serena Marner, Manager of the Fielding-Druce Herbarium, Oxford University Herbaria, Department of Plant Sciences (OXF), personal communication, email 27 January 2006, HOA. In footnotes OXF prefix will be used when citing the number of a specimen from this herbarium and OXF when citing the herbarium as the repository.

[25] Dr Daniela Zappi, Herbarium, Royal Botanic Gardens, Kew, personal communication, email 22 May 2007, HOA. In footnotes K prefix will be used when citing the number of a specimen from this herbarium and RBGK when citing the herbarium as the repository.

[26] Catherine Gallagher, coordinator, Curation, (NHV), personal communications, emails, 18 November 2005, 8 March 2007, 1 December 2010 and 22 December 2011, HOA. In footnotes NHV prefix will be used when citing the number of a specimen from this herbarium and NHV when citing the herbarium as the repository.

Herbarium, Tasmanian Museum and Art Gallery[27], New York Botanical Gardens Herbarium[28], National Herbarium of New South Wales, Royal Botanic Garden Sydney[29], and Queensland Herbarium.[30] Of the remainder who stated they held Oldfield material, internet searches were undertaken of their websites with the object of discovering any online catalogues where his specimens were located. All details of his specimen records discovered during this investigation were entered into the Oldfield database. We ceased searching for specimen records in December 2012 to enable the finalisation of the manuscript.

The National Herbarium of Victoria, Royal Botanic Gardens, Kew, and the Natural History Museum, London (Natural History Museum) hold the bulk of Oldfield's collections. The Natural History Museum received all his cryptogams[31] from Kew Gardens under the Moreton Agreement, 1961. Unfortunately, comprehensive information on Oldfield's collections at Kew Gardens and at the National History Museum could not be obtained as they were still in the process of being digitised when searches for specimen records ceased.

Oldfield's published specimen records were derived mainly from major 19th century taxonomic works, but other references to his specimens in books, journal articles and monographs for the period 1850–2010 have been included. The main sources were Joseph Dalton Hooker, *The Botany of the Antarctic Voyage of H.M. Discovery Ships Erebus and Terror, in the Years 1839–1843 under the Command of Captain Sir James Clark Ross, Kt., R.N., F.R.S. & L.S, etc.*, Part 1 *Flora Novae-Zelandiae*, 2 vols, 1853, Part 3, *Flora Tasmaniae,* 2 vols, 1860[32] (hereafter *Flora Novae-Zelandiae* and *Flora Tasmaniae* respectively); George Bentham, assisted by Ferdinand Mueller, *Flora Australiensis: A Description of the Plants of the Australian Territory*, 7 vols, 1863–1878[33] (hereafter *Flora Australiensis*); Ferdinand Mueller, *Fragmenta Phytographiae Australiae*, vols 1–9 , 1858–1875 (hereafter *Fragmenta*).[34]

In all, Oldfield's specimens were found in 32 herbaria across 13 countries (see Appendix 1). Since the survey did not cover all the world's herbaria, and more complete digitised herbarium catalogues of the future may reveal many more of his herbarium specimen records, the Oldfield database cannot claim to be comprehensive in that regard. Furthermore, it is doubtful that the information regarding his published records is complete. Nevertheless, although there are 503 herbarium and 71 published records without locality information, the database contains 5,120 herbarium and 2,648 published specimen records and we are confident that the data provide a sound basis for the claims we make regarding where he collected and the routes he took on each of his expeditions.

Since the original taxonomic determinations were made, there have been revisions of many, but by no means all, of the genera collected by Oldfield and many scientific names given by original

---

[27] Alex Buchanan, Tasmanian Herbarium, Tasmanian Museum and Art Gallery, personal communication, email 28 September 2007, HOA. List compiled by Dr Gintaras Kantvilas, Head Curator, in footnotes HO prefix will be used when citing the number of a specimen and Tasmanian Herbarium when citing the herbarium as the repository.

[28] New York Botanical Gardens, Herbarium Catalogue online, viewed 31 May 2007 at www.sciweb.nybg.org. In footnotes NY prefix will be used when citing the number of a specimen from this herbarium and NYBG when citing the herbarium as the repository.

[29] Karen L. Wilson, Acting Manager, Plant Diversity Section, National Herbarium of New South Wales (NHNSW or NSW), personal communication, email 23rd October 2006, HOA. In footnotes NSW prefix will be used when citing the number of a specimen from this herbarium and RBGS when citing the herbarium as the repository.

[30] Paul Robins, Queensland Herbarium, Queensland Department of Environment (BRI), personal communication, email, 14 December 2000, HOA. In footnotes BRI prefix will be used when citing the number of a specimen from this herbarium and Queensland Herbarium when citing the herbarium as the repository.

[31] *Cryptogamia*, an old primary division of plants comprising those without true flowers and seeds and including the orders *Filicales* (ferns), *Bryophyta* (mosses, liverworts and hornworts), *Thallophyta* (lichens, algae, fungi), Joseph Dalton Hooker, 1878, *The Student's Flora of the British Islands*, 2nd edition, Macmillan & Co., p. 487.

[32] Joseph Dalton Hooker, 1847-1860, *The Botany of the Antarctic Voyage of H.M. Discovery Ships Erebus and Terror, in the Years 1839–1843 under the Command of Captain Sir James Clark Ross, Kt., R.N., F.R.S. & L.S, etc.*, 1844–60, Pt. 1, *Flora Antarctica*, 2 vols, Pt. 2, 1847, *Flora Novae-Zelandiae*, 1855, 2 vols, Pt. 3, *Flora Tasmaniae*, 2 vols, 1860, Reeve, London.

[33] George Bentham (GB), assisted by Ferdinand Mueller, 1863–1878, *Flora Australiensis: A Description of the Plants of the Australian Territory*, 7 vols, Lovell Reeve & Co., London, 1870, reprint, A. Asher & Co, Amsterdam & L. Reeve & Co., Brook, Near Ashford, Great Britain, 1967.

[34] Ferdinand Mueller (FM), 1858–1881, *Fragmenta Phytographiae Australiae* 11 vols., Auctoritate Guberni, Coloniae Victoriae. ex officina Joannis Ferres, Melbourne.

taxonomists have changed as a result. These revisions reveal that some original determinations of specimens were incorrectly assigned to genera or, if to the correct Australian genus, to the wrong species or subspecies. Furthermore, some of Oldfield's specimens were originally assigned to taxa[35] which have since been reported as not being indigenous Australian species. Type status of these specimens is as given in the respective herbarium catalogue specimen record downloads from which they were obtained. The published specimen record dataset includes a list of references to Oldfield's specimen records extracted from a CSIRO Forest Research database printout of records of *Eucalyptus* species from within 160 kilometres of Albany, Western Australia. This list was compiled from material housed in various herbaria in Australia and Britain[36] and although they duplicate many herbarium records provided by various institutions, this was not always the case. It is for this reason that the records in this list have been included in the database. All the species named in Oldfield's honour illustrated in this present work were photographed live on the plant and professionally identified. Specimens of all the Western Australian species were identified by staff at the Western Australian Herbarium and incorporated into its collection.

Scientific names given in publications and current names are included in the database. These last were derived from the *Australian Marine Algal Name Index*[37], *Australian Plant Name Index*[38], *AusMoss*[39], *FloraBase: the Western Australian Flora*[40], *PlantNET: New South Wales Flora Online*[41], *Natural Values Atlas, Tasmania*[42] and *Checklist of Australian Liverworts and Hornworts*.[43] Dr P.M. McCarthy, Australian Biological Resources Study, Canberra, kindly provided us with the then (c.September 2000) current names of lichens collected by Oldfield, which were found in published works.[44] The *Checklist of Australian Lichens*[45] has since been consulted (October 2011) and name changes updated. The *Australian Plant Name Index* was an important source of information regarding published references to the specimens which have subsequently been designated to particular type categories.

Electronic copies of the Oldfield database have been deposited at the National Herbarium of Victoria, Tasmanian Herbarium, Royal Botanic Gardens, Kew, and the Western Australian Herbarium and will be freely available to those who wish to search it. Not all the species represented in Oldfield's collections were referred to in the major nineteenth century taxonomic works. One reason for this is provided by Joseph Hooker in his *Flora Novae-Zealandiae*:

---

[35] Taxa (noun, singular: taxon): 'Every individual organism is treated as belonging to an indefinite number of taxa of consecutively subordinate rank [for example: phylum, order, family, genus or species] of which the rank of species is basic', J. McNeill *et al.*, *International Code of Nomenclature for algae, fungi, and plants* (Melbourne Code)…, 2012, pp. 4–5.

[36] This list was kindly made available by the Albany Herbarium who received it from Marjorie M. Free per A.J. Boland, Curator, Forest Research, CSIRO, 11 May 1984, see the Henderson Augustus Oldfield Database (HAODb).

[37] R.A. Cowan, 2006, *Australian Marine Algal Name Index* (AMANI), ABRS, and Murdoch University, Perth, viewed 2006–2013, www.anbg.gov.au.

[38] Australian National Herbarium, Australian National Botanic Gardens, *Australian Plant Name Index* (APNI), part of the Integrated Biodiversity Information System (IBIS) in collaboration with the Centre for Australian National Biodiversity and the ABRS, viewed 2000–2013, www.cpbr.gov.au.

[39] National Herbarium of Victoria, *AusMoss: Catalogue of Australian Mosses*, Royal Botanic Gardens Victoria. Viewed 2011–2013, www.rbg.vic.gov.au.

[40] Western Australian Herbarium (PERTH), *FloraBase*: the Western Australian Flora, Western Australian Herbarium catalogue, viewed 2000-2013, www.florabase.dec.wa.gov.au.

[41] National Herbarium of New South Wales (NHNSW), PlantNET: New South Wales Flora Online, The Royal Botanic Gardens [Sydney] and Domain Trust, viewed 2010–2013 www.plantnet.rbgsyd.nsw.gov.au.

[42] Department of Primary Industries, Parks, Water and Environment, *Natural Values Atlas*, viewed 2017, www.naturalvaluesatlas.tas.gov.au.

[43] P.M. McCarthy, 2006a, *Checklist of Australian Liverworts and Hornworts* (ALH), Australian Biological Resources Study, Canberra (ABRS), viewed 2006–2013, www.abrs.gov.au.

[44] P.M. McCarthy, ABRS, personal communication, email circa September 2000. This email and its attachment have been lost.

[45] P.M. McCarthy, 2006b, *Checklist of the Lichens of Australia and its Island Territories (LAIT)*, ABRS, viewed 2008–2013, www.anbg.gov.au.

> Only discoverers named of the common plants (discoverers namely: Banks, Solander, Forster). I name more recent collectors only when they are discoverers, or detectors of rare plants in new or remarkable localities.[46]

Although it cannot be confirmed, it seems that the numbers Oldfield applied to his specimen labels were not field numbers linked to records in his field notes. He retained in his private herbarium duplicates of specimens he provided to various institutions. On at least two occasions he offered specimens to Kew in return for identifications. During 1850 he made several requests to Sir William Hooker to identify cryptogamic specimens from New Zealand and Australia. In one such letter he referred to an attached list of numbers: 'My own specimens are marked with corresponding numbers, so that the name after a number will be sufficient'.47 What appears to have been that list (numbered 1–384) is in the Botanic Gardens Kew Archives but only 19 blanks were filled in. In 1855 when Oldfield sent a collection to Sir William he stated:

> As in my former collection I have numbered most of the specimens in this, hoping that you would favour me with the names of some, for I am so far in arrears in that respect that I fear that otherwise I shall never be able to arrange my specimens properly.[48]

'Collector Number[s]' were assigned to 1682 of the 3,992 Oldfield specimens in the download provided by the National Herbarium of Victoria in 2007.[49] There are many instances of specimens gathered from a locality, which have an assigned 'Collector Number' in official databases. In some cases, the specimens were taken on the same expedition (for example, this was the case for the specimens on the trips he did to Mount La Perouse in Tasmania in 1857 and to Bald Island in 1862).[50] National Herbarium of Victoria data show quite clearly that Oldfield commenced numbering from #1 collections obtained in vastly different locations. Thus, for example, listed are five different specimens each with the Collector Number 5, four of which have different locality information.[51] There are instances, however, of specimens of the same species with the same Collector Number which have been gathered at the same locality.[52] Of the specimens with a Collector Number which were taken from Bald Island the numbering appears to have been sequential although some numbers are missing and there is one instance of two specimens bearing the same Collector Number.[53] The interrupted sequencing of specimen Collector Number(s) from similar localities may be due to specimens being traded with other herbaria either by von Mueller, Oldfield himself or later RNHV, RBGM directors.

We believe the evidence supports the view that all the numbers assigned by him to specimens were intended to facilitate the communication of determinations to him by the Hookers, and this may well have obtained for material collected for Mueller.

Comparison of the number of plant records we have found that were accorded the same collection locality (or close to it) is of little use in allowing us to form a firm opinion as to how long he spent at the various places he collected. In the first place, our data set is not complete. Secondly, his selection

---

[46] JDH, 1855, *Vol. II, Flora Novae-Zelandiae, Part 1,* p. 4.

[47] AO to W.J. Hooker (WJH), [?April] 1850, Australian Letters, vol. 73, 1850, fol. 257 (488), RBGKA. This was apparently the first letter Oldfield wrote to William Hooker.

[48] AO to WJH, 26 December 1855, Australian Letters, vol. 74, 1851–1858, fol. 191, RBGKA.

[49] MELISR database, download 8 March 2007, NHV, RBGM.

[50] Of the MEL specimens in HAODb, there are 24 from Bald Island and 44 from Mount La Perouse of which 13 and 17 respectively have 'Collector's Numbers'.

[51] MEL 2175221, *Grevillea* sp., locality not given; MEL 2202506, *Lepidosperma inops*, Browns River, Tasmania; MEL 2077893, *Acacia latipes* subsp. *latipes*, Darling Range, Western Australia; MEL 20679, *Lomandra pauciflora*, Mount Elphinstone, Western Australia; MEL 232290, *Eucalyptus urnigera*, Mount Wellington, Tasmania (MELISR download 8 March 2007, NHV, RBGM).

[52] For example, three specimens of *Waitzia* (MEL 1585028, MEL 1585153, MEL 1585162) with the Locn 'Blackwood River' bore the same 'Collector Number' 55 (ibid.).

[53] #881, MEL 226912, *Callitris preissii*; #882, MEL 2097946 *Boronia alata*; #884; MEL 271289, *Ixiolaena viscosa*; #885, MEL 12389, *Solanum symonii;* #886, MEL87788 & MEL 87790, *Rhagodia baccata* subsp. *baccata*; #887, MEL 602468, *Melaleuca lanceolata*; #888, MEL 605543, *Threlkeldia diffusa*; #889, MEL 2252846, *Apium prostratum* subsp. *prostratum*; #890, MEL 50559, *Pimelea clavata*; #894, ME L87914, *Rhagodia crassifolia*; #896, MEL 74949, *Disphyma crassifolium* subsp. *clavellatum*; #898, MEL 2227120, *Hibbertia cuneiformis,* ibid.

criteria may have been an important contributing factor. Thus, for example, it will become apparent that during his first decade in Australia at least his primary motivation was to discover species new to science and to build his own herbarium. Thirdly, there may have been a disparity between the number of species of interest to him occurring in the various places he visited and those that were flowering when he was there. Nevertheless, some suggestions will be made as to what he might have been doing at various places on the basis of the number of records in our database of specimens obtained there.

Except where otherwise stated, Geoscience Australia's gazetteer, Place names of Australia54, was the source of information regarding the location and current spelling of places given on Oldfield's plant specimen labels. Many place names, however, were not found in the gazetteer and with varying success attempts were made to establish their location in other ways. For example, for Western Australian localities, advice was sought from Mr Brian Goodchild, Secretary, Nomenclature Committee, Department of Land Administration, Western Australia.

'Van Diemen's Land' was the official name for what on 1 January 1856 officially became Tasmania. Except where Van Diemen's Land is used in publication titles and quotations, throughout this work Tasmania will be used since it grew in common usage from 1803-1804 when the botanist, Robert Brown, visiting the island, named a plant *Tasmannia lanceolata*.[55] It was also frequently used in 19th century taxonomic works that were published prior to 1856.

Birth and death dates were derived from many sources, including online biographies and, for Oldfield's family members, from official birth, baptismal and death records. The main sources for other people were *Australian Dictionary of Biography Online*[56] (ADBO), *Biography: Biographical notes on plant collectors and illustrators*[57], and for Western Australia, Rica Erickson, (ed), 1979, *Dictionary of Western Australians 1829–1914*, vol. 3, *Free 1850–1868*[58], University of Western Australia Press, Nedlands. It should be noted that in the case of the *Australian Dictionary of Biography Online*, authors' dates of publication relate to when the entry was first published in hard copy. Furthermore, to save space for entries in footnotes and the bibliography, all online subject and biographical databases references (e.g. *Wikipedia*, *Australian Dictionary of Biography Online* and Australian National Herbarium *Biography: plant collectors and illustrators or others relevant to Australian botany*), where searches for subjects and individuals are based on the names given in footnotes and the bibliography only, the URLs for the databases' search pages are provided.

One difficulty relating to the material Oldfield collected in the Murchison River district deserves special mention. His paper *On the Aborigines of Australia* has a section on the language of the indigenous people of the area, the *Watchandie*.[59] This includes a vocabulary list, which contains names of places within the *Watchandie* territory that appear to be descriptions of particular types of country, '*Wan-yu*, a grassy flat' for example, and not identifiers of unique places.[60] Some of these terms are similar to the names of specific places along the Murchison River that either appear on modern maps or are known to people who have lived in the area. These include: Bettie Crossing, Murchison River[61], and Wittacarra Gully stemming from '*Whih-ta-ra-ra* (perhaps more properly *Whit-to-ka-ra*)'.[62] Other

[54] Geoscience Australia, *Gazetteer of Australia, Place Names of Australia* (GAPNA), Australian Government, viewed 2001-2014, www.ga.gov.au.

[55] Terry Newman, 2005, 'Tasmania, the name' in *The Companion to Tasmanian History*, Centre for Tasmanian Historical Studies, University of Tasmania, Hobart, p. 353.

[56] National Centre of Biography, *Australian Dictionary of Biography Online* (ADBO), Australian National University, viewed 2001–2014, www.adb.anu.edu.au.

[57] Australian National Herbarium, *Biography: Biographical notes on plant collectors and illustrators* (ANHBiog), Council of Heads of Australasian Herbaria, viewed 2001–2014, www.anbg.gov.au.

[58] Rica Erickson, 1979, vol. 3.

[59] AO, 1865b, pp. 289–298.

[60] ibid., p. 293.

[61] *Bet-thie* is *Watchandie* for a 'large pool'(ibid.). Given on specimen labels as 'Bettie' or 'Betthie' or 'Betty', Murchison River and may refer to the gazetted 'Bettie Crossing' on the Murchison River or simply to any of the large pools on that stream.

[62] *Whih-ta-ra-ra* ('perhaps more properly *Whit-to-ka-ra* from its abounding in that species of acacia, the seeds of which are called *Whit-to* by the *Watch-an-dies*'), where there are numerous springs', ibid., p. 238.

place names listed in Oldfield's *Watchandie* vocabulary and used by him to signify specimen collection localities have not been found in any gazetteer or on any maps.

Further difficulties have arisen regarding identifying the localities of some of Oldfield's specimens referred to in *Flora Tasmaniae*, *Flora Australiensis* and to a lesser extent, Mueller's *Fragmenta.* The first related to Hooker's tendency to list the localities where specimens were collected followed by a list of the collectors. In many cases where this occurred it related to cryptogams that were originally sent to the Botanic Gardens, Kew, and are now housed at the Natural History Museum. In 2002, we visited that Museum and personally checked the labels of the relevant Oldfield cryptogamic specimens we could locate and, in most cases determined where they had been obtained.

The second problem was that Bentham quite frequently summarized the locality of different specimens of the same species provided by two or more collectors into one broad statement indicative of its apparent range. Dr Juliet Wege, Australian Botanical Liaison Officer, Royal Botanic Gardens, Kew, kindly confirmed this by inspecting a selection of the relevant species' records. For example, *Rhagodia baccata* was stated by Bentham as occurring from 'King George's Sound to Swan & Murchison Rivers, along the coast'.[63] Oldfield's specimen contribution to this apparent range was from 'Wet places, near the sea, King George Sound'.[64] Mueller occasionally adopted this practice. Except where Oldfield's specimen localities were identified, locality ranges have been included in the database but ignored in the specimen tallies from localities in this book.

Thirdly, where W. Wilson[65] cited Oldfield's specimen label numbers for more than one location for a particular species it is impossible to associate the location with its label number without examining the label itself.

The absence of collection dates on all but a very few field labels prevented us from preparing a comprehensive chronological account of Oldfield's movements within New Zealand, mainland Australia and Tasmania and, as explained above, his numbering on specimen labels is of no help in separating collections made on each of his various expeditions. This has posed limitations on the validity of some aspects of the narrative, particularly where he apparently traversed the same geographical area on more than one occasion. Nevertheless, the narrative has been presented chronologically using various sources, some of which did permit us to tie his activities to specific dates or more broadly within particular months or years with some degree of certainty and others which allowed us to make reasonable but tentative claims as to what he might have been doing within a particular period. Overall, his travels in Tasmania are the most problematic regarding their timing. For example, we do not know when or how often he visited his collecting localities within easy reach of Hobart, Browns River and some further afield to the west, east and north of the city. We have good reason to believe that most of the visits to these places were undertaken between 1845 and 1856. With these limitations and the efficient presentation of the narrative in mind, we have systematically discussed his visits to those sites in accordance with their geographical position in relation to Hobart and Browns River but with no attempt to assign any finite chronological order.

Western Australian species' flowering times are provided in the species' 'taxon profiles' of *FloraBase* but do not appear to be available online for other States and Territories. Given that, in the past, collectors usually collected specimens with flowers to facilitate the identification of species, we hoped that flowering status would enable the separation of those collections from areas Oldfield visited more than once. To this end we commenced recording flowering times as given in *FloraBase* for each of the Western Australian species represented in the Oldfield published specimen records that we located. However, this was discontinued because species' flowering time data were too problemat-

[63] GB, 1870, vol. 5, reprint, 1967, p. 152.

[64] Dr Juliet Wege, Australian Botanical Liaison Officer, Royal Botanic Gardens, Kew, personal communication to MHH, email 2 May 2006, HOA.

[65] See for example, W. Wilson, 1860, in J.D. Hooker, 1860, vol. 2, p. 194: *Bartramia crassa*, 'Near springs: Top of Western Mountains; Mount Wellington; Creek above Wellington Falls (Gunn) (Oldfield 101, 104, 106, 111).

ic. In the first place, there were many specimens in *FloraBase* which did not have flowering status recorded, particularly prior to the 1990s, and we did not have the resources to check all the specimens for each species in the Western Australian Herbarium. Secondly, not all the collection dates of flowering specimens for particular species in that herbarium have been taken into account in working out the flowering times given in *FloraBase* taxon profiles (for example, *Taxandria linearifolia*).[66] Thirdly, subsequent research has revealed that the vast majority of the flowering times of the species collected by Oldfield in particular areas overlapped to a varying extent both times when we have good reason to believe he was in certain localities. Fourthly, as far as we are aware no account is taken of the flowering times for particular species, which occur across a wide geographical range in which climatic and geophysical conditions vary quite markedly. Thus, for example, if a species occurs to the north of Perth it may flower earlier than it does in the lower south-west where it remains colder for longer. However, for some purposes species' flowering times have been helpful when used in conjunction with other information, allowing us to make tentative assertions as to when he was in some places. In each of these instances, the relevant data for all the specimens listed in *FloraBase* on each of the respective taxa have been checked. It should be noted that many records did not specify whether the specimens were flowering and again we did not have the resources to undertake the searches to check this on the actual specimens.

All the settlement, road and track locations, were identified by Bill using 19th and 20th century official plans and maps. He then prepared drafts of all the maps. Unfortunately, the mapping software program he used for this purpose proved too problematic for him to perfect each map. Dr Michael O'Connor has kindly produced the final maps drawing upon Bill's drafts. Each map covers Oldfield's collecting localities and places mentioned in relation to contextual and collecting places of people who visited the various areas prior to Oldfield's visits.

Except where otherwise stated, all photographs of species named eponymously for Augustus and some of his collecting places were shot by Helen. We have endeavoured to visit as many of these places as possible. Photographs were taken only where the bushland appeared reasonably intact. However, exotic incursions may be present in the images. Also, species' loss may have occurred since Oldfield's time. No effort was made to include in the photographs plant species collected by Augustus because of our very limited understanding of the flora in each locality. Rather, the images are provided to give the reader some appreciation of the varied environments in which he worked.

A complete set of the photographs and biographical documents and those documents relating to Oldfield's collections that we gathered for this study have been deposited in the Library Archives of the Royal Botanic Gardens, Victoria.

Despite the apparently meagre offerings from the historical record and the problematic nature of many of our sources, we believe we have managed to expand considerably on what is known of Oldfield's life, including his motivations and consequent actions, and to bring together a view of his scientific contribution which hitherto could not be obtained in published sources. To create a coherent and meaningful narrative we have endeavoured to describe the socio-cultural and bio-geographical contexts in which he operated in England and Australia.

Oldfield was born in England into a socio-cultural milieu that was changing and continued to do so throughout his life. Furthermore, during the two decades he was in Australia, the changes that were taking place in his homeland were strongly influencing, although by no means totally, the way in which socio-cultural life was evolving in the country to which he had emigrated. The impact of what was happening in Britain on the direction Oldfield's life took was considerable and because of this the first chapter provides a broad overview of those developments.

[66] The flowering times of individual Western Australian Herbarium (PER) specimens for some species listed in *FloraBase* were compared with the flowering times given in the relevant Taxon Profiles. The latter were found to be incomplete.

Chapter 1

The changing socio-cultural environment in England 1800–1887

A brief overview

Augustus Oldfield was born in 1821, a year after George IV ascended the throne, and died in 1887 toward the end of the reign of Queen Victoria. He lived in an era when internal and external influences were changing the nature of England's earlier social structure and its power relationships within the global context. These developments began in the 18th century, during which enlightened sectors of the British population increasingly acted to replace 'the Providence-dominated world of early modern and medieval Europe by the world of expanding knowledge and science, discoverable nature and rational exploration'.[1] English social structure changed as people variously promoted the processes of industrialisation, colonisation and the practices of the established Anglican and non-conformist churches in their own interests, or acted to ameliorate their adverse effects, to improve the quality of life across the social spectrum. Augustus' life took its course under the influence of these developments, particularly from the changing interests, aspirations and values of the evolving middle-class and its relationship with the elite upper class.

The aristocracy and landed gentry held to the ideal of a gentleman based on a lifestyle that required neither mental nor manual labour[2], and grounded in a conviction of their moral superiority and dominance over all others in society, together with a chivalrous code of honour towards each other.[3] This elite society was an open aristocracy based on property and patronage[4] that enabled them to maintain their social exclusivity and their dominance in the nation's parliament and the civil service during most of the eighteenth century.[5] Nevertheless, societal changes during the 19th century weakened their politically powerful position and from the latter half of the 18th century, this 'old society', claimed Harold Perkin, evolved to a new structure, arising from 'vertical class antagonisms' and the development of a powerful middle class.[6]

The pursuit of pleasure dominated the lives of many in the aristocracy and gentry during the late 18th and well into the next century, with gambling for high stakes in their homes, clubs and high-class gaming houses being one of the most popular pastimes during most of that period.[7] The creative arts, particularly visual arts and literature, were also popular among members of the upper class. There were those, however, who developed an interest 'in exploring both the world of Nature and of humankind' and this increased the emphasis on the possibilities of social and technological improvement.[8]

---

[1] Francis Sheppard, 1998, *London: A History*, Oxford University Press, pp. 171–173.
[2] Leonore Davidoff & Catherine Hall, 1987, *Family Fortunes: Men and Women of the Middle Class 1780–1850*, Hutchinson, London, p. 73.
[3] Harold Perkin, 1972, *The Origins of Modern English Society 1780–1880*, Routledge & Kegan Paul, London, pp. 273, 274.
[4] ibid., p. 17.
[5] Linda Colley, 1992, *Britons: Forging the Nation, 1707–1837*, Yale University Press, New Haven, pp. 147–193.
[6] ibid., pp. 176–195.
[7] S. Margetson, 1971, *Regency London*, Cassell, London, pp. 112, 113.
[8] John Gascoigne, 1994, *Joseph Banks and the English Enlightenment: Useful Knowledge and Polite Culture*, Cambridge University Press, Cambridge, p.185.

Advances in agricultural technology over the 17th and 18th centuries were proof of the benefits of improvement as an outcome of human ingenuity and application. As a result, members of the aristocracy became increasingly involved in applying their wealth to geographical and scientific discovery, including the collection and naming of 'new' botanical and zoological species, the application of science to deal with the challenges of everyday life, tackling the problem of mass pauperism through poor relief, and the establishment of charity schools and hospitals.[9]

Other technological inventions were important contributions to the massive expansion in industrial output in Britain and to population drift to the centres of production, population growth and increasing local and overseas trade.[10] This motivated the expansion and consolidation of British colonial rule over throughout the 19th century facilitated by the 'ever shrinking world' brought about by the application of scientific discoveries to major engineering works, which made for safer and faster transport s and to inventions that improved communication.[11]

Over the 19th century other significant government reforms came about gradually. The male franchise was increased[12] and the Civil Service unified[13] with the system of patronage recruitment replaced with competitive entry examinations.[14] A more effective system of policing was established.[15] Subsequent reforms of the criminal law included the massive reduction of capital crimes from 200[16] to four for civilians.[17] Imprisonment for debt, however, 'remained an absurd and cruel anomaly'[18] until 1869 when it was abolished.[19] 'Reform by industry' and local administration underpinned the passing of the *Gaols Act 1823* which also did away with privately run prisons. The debtors' prisons were left untouched in that regard.[20] Reform of the court system commenced in 1830, which included the simplification of the procedure of the high courts.[21] The ongoing reform of the public health system was assisted by scientific discoveries relating to environmental causes of transmissible and non-communicable diseases.[22]

As a condition of pardon from a death sentence, the transportation of convicts to British colonies for the terms of their natural lives had been practiced since 1614 but later the courts had the power to sentence those convicted of lesser felonies and crimes to terms of transportation (seven or 14 years). By 1821 transportation had ceased in all British colonies except New South Wales and Van Diemen's Land, where it continued until the 1840s. It was taken up by Western Australia in 1851, but ceased in 1868.[23]

The changes in the character, scope, and extension of elementary and secondary schools and universities throughout Britain between 1815 and 1870 were, according to E.L. Woodward, the central feature of what he refers to as the 'era of reform'.[24] The methods employed in the majority of schools to inculcate school curriculum content usually involved rote learning and severe discipline, fortified by the mainstream Christian doctrine that infants were 'beings who bring into the world a corrupt nature, evil disposition, which it should be the great end of education to rectify'.[25] Since the 17th century

[9] ibid., pp.185–236.
[10] Juliet Gardiner & Neil Wenborn, 1995, *The History Today Companion to British History*, Collins & Brown, London, pp. 410—411.
[11] ibid., p. 735.
[12] ibid, p. 584.
[13] Harold Perkin 1972, pp. 267–279.
[14] Juliet Gardiner & Neil Wenborn, 1995, p. 166 and Harold Perkin, 1972, pp. 268–269.
[15] P.T. Smith 1985, *Policing Victoria London: Political Policing, Public Order, and the London Metropolitan Police*, Contributions in Criminology and Penology, no. 7, Greenwood Press, Westport, p. 17.
[16] ibid., p. 17.
[17] Juliet Gardiner & Neil Wenborn, 1995, p.130.
[18] E.L. Woodward 1954, *The Age of Reform 1815–1870*, Oxford at Clarendon Press, London, p. 453.
[19] ibid., pp. 438–454.
[20] ibid., pp. 448–450.
[21] ibid., pp. 438–454.
[22] ibid., pp. 444–446. See also: G. Melvyn Howe, 1997, *People, Environment, Disease and Death: A Medical Geography of Britain Throughout the Ages*, University of Wales Press, Cardiff, pp. 167–170.
[23] Juliet Gardiner & Neil Wenborn, 1995, p. 757.
[24] E L Woodward, 1954, p. 455.
[25] Roy Porter, 2000, *Enlightenment: Britain and the Creation of the Modern World*, Allen Lane, London, England, pp. 339–340.

growing numbers of enlightened philosophers, importantly including John Locke[26], and educators believed that children should be trained to use their powers of observation to know the world at hand, and by appealing to their powers of reason to learn and obtain conformity.[27] This change in approach increasingly underpinned teaching methods in middle-class homes and schools, and in some working class schools, particularly those run by non-conformist sects.[28]

The late 18th and early 19th centuries saw real incomes rise for the upper and 'middling ranks' of the 'old society'.[29] For the latter, this was the result of an increase in demand for the services of professionals such as doctors, lawyers, engineers and the goods of merchants, shopkeepers and self-employed craftsmen. As they consolidated their position, they created a new social pattern of sobriety between the extravagant lives of the very rich and the desperate lives of the very poor.[30]

The second major change emanated from the recognition that British industry needed an increased supply of administrators and technicians.[31] Diversification and expansion of the professions from the beginning of the 19th century created demand for well-defined preparation and formal credentials. This, in turn, created a demand for improved preparation for tertiary education through expansion of the secondary school education system.[32] As a consequence a different kind of secondary private school, for example, the University College School in London offering a more humane supervision of students and a greater readiness to incorporate subjects such as natural history, astronomy, physics and commercial subjects into the curriculum, started to come into existence in the late 1820s.[33]

Until the 1820s 'higher education' in England had been the preserve of the strictly male Oxford and Cambridge Universities. In the late 1840s and early 1850s they began to offer a broader range of courses in the sciences, law and history.[34] Scotland had long had four universities, of which the Universities of Edinburgh and Glasgow had specialised professorships since the mid-1700s, which embraced 'a galaxy of talent, philosophical, scientific, including natural history, and medical'.[35] Commencing in 1826, with the opening of the non-denominational University College, London, and in 1829, Anglican King's College, London, other 'civic' universities were established in London and other centres in England, providing an increase in degree courses for established professions.[36] The secularisation of Oxford and Cambridge Universities came about with the passing of the *University Tests Act 1871*.[37] Britain's first Polytechnic opened in London in 1838[38], and from 1853 the government took an active role in applied science through the activities of the Science and Art Department of the Board of Trade and later the Education Department.[39]

Through self-education, facilitated by participation in informal improvement societies and, from 1823, the emergence of mechanics' institutes across England, working class political leaders believed that the power of workers would be enhanced through knowledge attained by intelligent reading, especially in the fields of politics, science and economics.[40]

---

[26] John Locke put forward his ideas regarding education in his *Some Thoughts Concerning Education*, London, 1693 and his *Essay Concerning Human Understanding*, London, 1690.
[27] ibid., pp. 339–341.
[28] Roy Porter, 2000, pp. 346–347.
[29] Harold Perkin, 1972, pp. 134–135.
[30] ibid., pp. 134–135 & 252–255.
[31] ibid. and W.B. Stephens, 1998, *Education in Britain*, St Martin's Press, New York, pp. 56–76.) and William Boyd & Edmund J. King, *The History of Western Education*, A. & C. Black, London, 1975, p. 367.
[32] W.B. Stephens, 1998, p.116.
[33] E.L. Woodward, 1954, p. 466.
[34] ibid., pp. 472–473.
[35] Roy Porter, 2000, p.246 and E.L. Woodward, 1954, p. 474.
[36] Juliet Gardiner & Neil Wenborn, 1995, p. 772.
[37] ibid.
[38] *University of Westminster*, undated, 'History of the University of Westminster', viewed 24 December 2014 at www.westminster.ac.uk.
[39] H. C. Barnard, 1955, *A Short History of English Education from 1760 to 1944*, University of London Press, London, pp. 157–159.
[40] ibid., pp. 304–307.

The general weight of opinion within the 'old society' was that women were socially and mentally inferior to men.[41] Women had no legal *persona*.[42] Their location in the social structure, therefore, was tied to that of their fathers, and later to that of their husbands. However, growing numbers of enlightened men and women from the upper- and middle-classes insisted that both sexes have equality before the law and be regarded the equal of each other in mental capability and educability[43], although many still held to women's old status[44], and as the century progressed, the view hardened within the middle-class that female relatives should be dependents.[45] Nevertheless, during the 1830s, the 'frustrated energies of growing numbers of able and intelligent [upper- and middle-class] women ...burst out in a concerted movement of protest'.[46] That movement grew and was largely responsible for the passage of legislation that improved, but did not wholly rectify, the situation of married women in relation to divorce, custody and maintenance of, and access to, children following divorce or separation, protection from a cruel husband, and to retention of earnings and property rights.[47] As the century wore on, single women's choice of occupations increased but salaries remained meagre and working conditions were often poor.[48] By 1850 more single women were establishing their own schools or becoming hospital and workhouse administrators.[49]

During the 19th century, the path to reform was not easy. Furthermore, the gains were not always to the satisfaction of those who had pushed for betterment. In the fight for reform, the social classes competed on an uneven political 'playing field'. Although parliamentary *Reform Acts* weakened the power of the aristocrats in the House of Commons with increasing members of the middle- and working-classes winning seats, the aristocrats nevertheless retained their political dominance through the House of Lords. However, as the century progressed the middle-class, particularly its highly placed professionals in the Civil Service and professional associations, became more politically influential than the working-class in developing reform policies and implementing them. They became wealthier, better educated and were professional in their approach to solving problems, thereby enabling them to get into positions of power and influence. This played an important role in the consolidation of the middle-class in the newly forming emerging social structure.[50]

Against this background of social reform, science and the way it was practised in the 19th century changed. These developments played an important role in the scientific career that Oldfield chose to follow.

The British public's fascination with botany and other branches of natural history mushroomed and spread across layers of society. European interest in botany during the 18th and 19th centuries was boosted by the stream of new plants brought in from abroad by professional collectors, maritime explorers, and members of the British Navy, traders and travellers. As the desire for previously unknown exotic plants grew, plant collecting around the world became more organised and was financed by nurserymen, botanical gardens or syndicates of private gardeners, and there developed an urgent need for the new plants to be named.[51] By 1840 there were at least fifteen botanical gardens in Britain.[52]

---

[41] Amanda Vickery, 1998, *The Gentleman's Daughter: Women's Lives in Georgian England*, Yale University Press, London, p. 72.
[42] Juliet Gardiner & Neil Wenborn, 1995, p. 818.
[43] Roy Porter, 2000, pp. 320–338.
[44] ibid., pp. 253–255.
[45] ibid., pp. 238–239.
[46] Harold Perkin, 1972, p. 160.
[47] ibid., pp. 159–160, and Juliet Gardiner & Neil Wenborn, 1995, p. 818
[48] Martha Vicinus, 1985, *Independent Women: Work and Community for Single Women 1850–1920*, Virago Press, London, pp. 5–6.
[49] ibid., p. 15.
[50] See for example, Harold Perkin, 1972, pp.271–339, and Juliet Gardiner & Neil Wenborn (eds), 1995, p. 818.
[51] See for example, Richard P Aitken, 2006, *Botanical Riches: Stores of Botanical Exploration*, Melbourne University at The Miegunyah Press, Carlton with State Library of Victoria, Chapters 1–24, and David Elliston Allen, 1978, *The Naturalist in Britain: A Social History*, Penguin Books, Harmondsworth, England, pp. 30–31.
[52] Alexandra Hasluck, 1955, *Portrait with Background: A Life of Georgiana Molloy*, Appendix E, Oxford University Press, Melbourne, p. 260.

Previously, botany held an important place among expanding, learned, London clubs and societies.[53] Under the presidency of Sir Joseph Banks between 1778 and 1820, the Royal Society (which was formed in 1660 and received its Royal Charter in 1662) was particularly influential in the development of the British Government's scientific and trade policies.[54]

English men of science were wealthy members of the upper class with few focusing their endeavour on a single branch (for example, botany); their activities were at times referred to as 'gentlemanly science'. They were amateurs, and patronage was crucial if they needed a salary.[55] According to Jim Endersby, 'the ways in which men of science imagined and planned their working lives were shaped by shifting notions of gentility and by the accompanying discussions of conduct, character, and etiquette'.[56]

European plant taxonomists struggled to make sense of the flood of new plants; many attempts were made to find practical ways to classify them. From c.1750, the binomial sexual system of plant classification developed by the Swedish naturalist Carl von Linné, or Linnaeus, was adopted by taxonomists in Europe and Britain. According to Linnaeus there were two kinds of natural groups in the plant kingdom. The first, 'species' and 'genera', were divinely created as discrete entities. That is, each one contained plants that were clearly related to each another, and not to plants in other similarly ranked groups. In the second kind of natural group — families, classes and so on — plants were also related to each other, but not in a clear-cut way. Linnaeus described the pattern they made as 'continuous'.[57] The simplicity of the Linnaean sexual system played a large role in making botany popular both in Britain and its colonies. This was strengthened in Britain by the publication of popular works on this system of classification and an upsurge in the involvement of the new middle class and gentry in horticulture and gardening.[58] However, among some European and British naturalists there grew recognition that a new system based on natural relationships between entities in the animal and plant kingdoms was required, and a number of competing systems were developed.[59] Eventually, a system developed by Antoine de Jussieu, proposed a pattern of continuity that applied to genera as well as supra-generic groups, leaving only species as discrete entities. He employed this system in his *Genera plantarum* (1789).[60] Unlike Linnaeus and Jussieu, the Swiss botanist Augustin Pyramus de Candolle did not believe that nature was continuous but like them, he placed his groups in a ranked hierarchy, including a series of 'sub-classes' that were more or less the same as Jussieu's 'secondary classes' and Linnaeus's 'classes' and remained artificial.[61] Their joint efforts created a binomial classificatory system that became known in the 19th century as the 'natural system'. It was based on natural relationships hierarchically organised according to seven main ranks (kingdom, phylum/division, class, order, family, genus, species) derived from characteristics inherited from common ancestors. Plant classification using this system entailed a detailed knowledge of plant structure and development

---

53 Juliet Gardiner & Neil Wenborn, 1995, p. 175.

54 See for example, John Gascoigne 1994, and John Gascoigne 1998, *Science in the Service of Empire: Joseph Banks, the British State and the Uses of Science in the Age of Revolution*, Cambridge University Press, Cambridge.

55 Jim Endersby, 2008, *Imperial Nature: Joseph Hooker and the Practices of Victorian Science*, The University of Chicago Press, London, p. 9.

56 ibid., p. 9.

57 Sara Maroske, 2006. 'Ferdinand Mueller and the Shape of Nature: Nineteenth-century Systems of Plant Classification', *Historical Records of Australian Science*, 2006, vol., 17, p. 148.

58 A.G. Morton, 1981, *History of Botanical Science: An Account of the Development of Botany from Ancient Times to the Present*, Academic Press, London, p. 234, and David Elliston Allen, 1978, p.42.

59 Jim Endersby, 2008, pp. 180–183. According to Endersby, prominent among these early attempts were Edward Newman's septenary system which proposed that 'all living things could be classified in groups of seven, with affinities joining the members of the groups and analogies connecting the groups to one another'. Others included William Sharp MacLeay's and William Swainson's quinarian system and Thomas Baskerville's *Arrangement of the System of Vegetable Affinities, on the Principle of a Sphere.*

60 Antoine Laurent de Jussieu, 1789, *Genera Plantarum, secundum ordines naturales disposita juxta methodum in Horto Regio Parisiensi exaratam*, Paris.

61 Sara Maroske, 2006, pp. 148–149.

requiring long study and microscopic analysis. Although de Jussieu's system had been introduced into Britain by Robert Brown in 1810, it took many years to become established in that country.[62]

Classification was, according to Jim Endersby, a basic tool of Victorian natural history, 'the key to providing stable names that facilitated scientific communication and the efficient exploitation of the world's natural resources'. Accumulating specimens and expertise at the British Empire's centre (London) was a strategy for building the power and authority of those at the centre and extending their imperial reach. However, to undertake this work, they relied on collectors in the colonies to provide the specimens and descriptions of the plants as they grew in their natural surroundings. Classifications played a key role in this accumulation; concepts were devised, implemented and enforced by those at the centre.[63]

By the mid-nineteenth century taxonomy was the dominant force in the botanical sciences, although there had been advances in other branches through microscopical study and experimentation.[64] Despite the huge public interest in botany, opportunities for formal training were few but nevertheless on the increase.

Although by no means the first to postulate the transmutation of species, Charles Darwin published his great work *On the Origin of Species by Means of Natural Selection, or the Preservation of Favoured Races in the Struggle for Life* in 1859, in which he put forward the theory, backed by his extensive empirical observations, that populations evolved by a slow process of change over long ages through a process of natural selection. Darwin did not offer any suggestions as to the actual physiological process by which the transmutation of species happens. The concept of natural selection was originally developed in the absence of a valid theory of heredity; at the time of Darwin's writing nothing was known of modern genetics.[65] Even so, his theory completely revolutionised the sciences of botany and zoology. Roy MacLeod has asserted that

> developments in science over the previous centuries led to the rise of the agnostic ideology of 'scientific naturalism', which by the 1860s and 1870s enjoyed considerable acceptance among many eminent French, German and English scientists. Scientific naturalism was 'based upon the assumption that the universe was ultimately material, mechanical, and uniform. The two great symbolic syntheses—the evolution of the species and the conservation of energy—were to be understood through a secular catechism of unification, simplification, correlation and universality'.[66]

While adherents held that the dogma of scientific naturalism was 'not necessarily irreconcilable with religion' they 'accepted in principle the possibility of reducing the physical world to material causes'. They retained as their *modus operandi* the scientific method, with its emphasis on the iteration and reiteration of hypothesis testing through observation, quantification, measurement, controlled experiment and accurate recording.[67] Nevertheless, scientific naturalism, claimed MacLeod, 'ultimately derived its strength from acknowledgement of a Supreme power—the "Divine mind in Nature"'.[68] Proponents of such a supreme power included the biologist T.H. Huxley, the mineralogist W. Whewell and the astronomer J.F. Herschel.[69] 'Warfare' between religion and science arose over difficulties with

[62] Jim Endersby, 2008, pp. 172–191.
[63] ibid., pp. 137–138.
[64] See Charles Singer, 1997, *A Short History of Science to the Nineteenth Century*, republication of the 1943 edition, Dover Publications, London, pp. 327–379 and A.G. Morton 1981, Chapters 7 & 8.
[65] Charles Singer, 1997, pp. 379–387.
[66] Roy M. MacLeod, 2000, 'The "Bankruptcy of Science" Debate: the Creed of Science and its critics', in Roy M. MacLeod (ed), 2000. *The 'Creed of Science' in Victorian England*, Varium Collected Studies Series, Ashgate, Aldershot, III, pp. 4–5.
[67] See for example, W.S. Fowler, c.1962, *The Development of Scientific Method*, Pergamon Press, London.
[68] Roy M. MacLeod, 2000, p. 6.
[69] ibid.

reconciling the creed of 'scientific naturalism' and Christian beliefs with their doctrines of revealed truth and biblical liberalism.[70]

During the Enlightenment, the notion had developed that humankind was in the process of progressing socially.[71] In 1824 Auguste Comte argued that there were laws governing human social evolution 'as for the fall of a stone' but he was content only to deduce those laws and leave the verification and development to others. Between 1830 and 1842, Comte published a six-volume work, *The System of Positive Philosophy*, in which he proposed his law of three stages through which knowledge developed. The third and final stage, which he referred to as the positive stage of human knowledge, is reached when people come to rely on empirical data, reason and the development of scientific laws to explain phenomena.[72] The belief that human societies were progressing was dominant by the 1860s. In 1857 Herbert Spencer asserted that socio-cultural progress, like 'organic progress consists in a change from the homogeneous to the heterogeneous'.[73] Later writers have referred to Spencer's ideological approach as 'Social Darwinism', which they claimed was based on a naïve interpretation of Charles Darwin's *On The Origin of Species* (1859).[74] Now discredited, Social Darwinism enjoyed widespread popularity, especially in Britain and North America, during the late 19th and early 20th centuries, and its influence covered the entire political spectrum. Not only did it serve to justify European colonialism but Social Darwinism was also used by some early anthropologists to describe and hierarchically classify the anthropomorphic characteristics, mental capacities and differing cultures of humans across the globe, against the yardsticks of Europeans and their industrialising cultures.[75]

By the 1820s the French set the pace in all the sciences. They also led the way in the applications of science.[76] Spurred on by patriotism and flamed by the Napoleonic Wars, rivalry with the French dominated British science until about 1875 when leadership in science, at least quantitatively, had passed to Germany.[77]

Unlike the situation in France, opportunities for formal training in Britain were limited[78] and there was no professional scientific community.[79] Nevertheless, before and during that period English scientists contributed to some notable advances in the physical and biological pure and applied sciences.[80] The British Museum, instituted in 1759, the British Museum of Natural History (now the Natural History Museum), which split from the British Museum in 1880, the Museum of Economic Botany, 1848, and the Herbarium at the Botanic Gardens, Kew, 1852, were among the important institutions of science founded in England during the 18th and 19th centuries and are still in operation today.

Concomitant with the development of scientific societies among the elite and the publication of their researches in learned journals, scientific activity started to take on a life over and above that of its individual adherents.[81] An important outcome was the emergence of occupations that relied for their very existence and advancement on scientific discoveries. It led to the recognition of the need for practitioners to band together in associations to ensure compliance with ethical standards of behaviour,

[70] W.H. Brock and Roy M. Macleod, 2000, 'The Scientists' Declaration: reflection on science and belief in the wake of *Essays and Reviews*, 1864–5', in Roy M. MacLeod, 2000, pp. 39–66.

[71] John Gribbin, 2002, *Science: a History*, Penguin Books, London, p. 241.

[72] H. Russell Bernard, 2001, *Research Methods in Anthropology: Qualitative and Quantitative Methods*, 3rd edition, Altamira Press, Oxford, pp. 12–13.

[73] Herbert Spencer, April 1857, 'Progress: its law and causes', *The Westminster Review*, pp. 445–465.

[74] Pierre van den Berghe, 1985, 'Social Darwinism', in Adam and Jessica Kuper (eds), 1985, *The Social Science Encyclopaedia*, 2nd edn, Routledge, London, p. 783.

[75] ibid., p. 783.

[76] Ian Inkster, 1991, *Science and Technology in History: An Approach to Industrial Development*, MacMillan, London, pp. 43–45.

[77] J.D. Bernal, 1953, *Science and Industry in the Nineteenth Century*, Indiana University Press, London, p. 149.

[78] The Cambridge University mathematical tripos had long had a high reputation, and at Oxford University Mathematics and Physics became a separate school in 1807 but study in other branches of science by these elitist and religiously exclusive institutions was not offered until later in the century (E.L. Woodward, 1954, p. 471). After 1750 a swelling tide of English students attended Edinburgh University where philosophy, 'moral and natural science' subjects taught by eminent scholars were on offer Roy Porter, 2000, p. 243.

[79] David Knight, 1986, *The Age of Science: The Scientific World-view in the Nineteenth Century*, Basil Blackwell, Oxford, pp. 11–31.

[80] See for example, David M. Knight, c.1998, *Science in the Romantic Era*, Ashgate, Brookfield, USA: and John Gribbin, 2001.

[81] David Elliston Allen, 1976, p. 5.

and to promote their own professional interests.[82] These began to exercise greater control over the recruitment, competence and conduct of their professions; for example, the Provincial Medical and Surgical Association (1834) that evolved into the British Medical Association (1856).[83] As the century advanced, the physical and biological sciences became more professional as governments and universities began to demand that their scientists were trained. Degree courses in the various branches of science were gradually introduced into universities and although patronage remained important throughout the century, it slowly gave way to individual expertise and intellectual merit.[84]

****

These, then, were the major socio-cultural developments that took place in England during the 19th century and which, through extensive migration and British rule, were transferred in large measure to the Australian colonies, thereby importantly influencing the direction of Augustus Oldfield's life in England and in his adopted country.

[82] ibid.
[83] Harold Perkin, 1972, p.120.
[84] Jim Endersby, 2008, p. 9.

Chapter 2

Augustus Oldfield's contribution to science

Augustus Oldfield was a naturalist with an eclectic interest in the world around him and a reputation for being an 'industrious & trustworthy Naturalist and especially Botanist'[1], a 'careful and acute observer'[2], an 'intelligent collector'[3] and 'extremely well informed'.[4] He was not merely one who gathered plants without an eye to their biology. He noted their habits of growth, physical structure and habitats, taking into account geological features and the plant communities which surrounded them. He was also interested in the socio-cultural lives of the indigenous people and the fauna in the areas through which he travelled but it was the flora that claimed the greater part of his attention and his passion. It is indeed unfortunate that, although some of this information was recorded on many of his plant labels, most of his botanical, zoological and geological observations were set down in field notebooks that are apparently lost to the historical record.[5] Nevertheless, some observations have survived in his unpublished treatise on natural philosophy[6] and his lengthy article *On the Aborigines of Australia*.[7]

In this chapter, we endeavour to provide an overview of Oldfield's contribution to science in the fields of botany and anthropology, and to place it within the context of the way in which science was practiced in the mid-nineteenth century. The evidence to support claims made here relating to Oldfield's scientific activities and his relationships with the botanists with whom he dealt in Australia and England is presented in the following chapters.

Oldfield's scientific approach

Before considering Oldfield's contribution to Australian botany and anthropology some of the major theories which guided his thinking about the natural world will be outlined briefly. Oldfield discussed these in his unpublished manuscript on natural philosophy[8], completed c.1865. He devoted a considerable amount of his time and intellectual energy to its production.[9] It is a very long work, some 860 handwritten pages plus 56 footnotes. The work demonstrates his broad reading of the scientific literature, his wide-ranging knowledge of, and deep interest in, the natural world. It provides an insight into the theoretical underpinning of his scientific approach and the way he grappled with the issues that claimed the attention of scientists of his day.

---

[1] JDH, Augustus Oldfield testimonial, 2 June 1875, HOA.
[2] JDH, 1860, vol. 1, pt. 1, p. cxxvii.
[3] GB, 1863, vol. 1, p. 14.
[4] JDH to FM, December 1865, RBGKA. Transcription and copy kindly provided by Dr Sara Maroske, emails 4 March 2012 & 16 March 2012 respectively from the database of The Correspondence of Ferdinand von Mueller Project, Department of History, Department of History & Philosophy of Science, The University of Melbourne.
[5] It is known that he kept such records from his RBGK correspondence.
[6] AO, 1865a, RBGSL.
[7] AO, 1865b, pp. 215–298.
[8] AO, 1865a.
[9] See Chapter 20.

He discussed, systematically, theories that attempted to understand the nature of the organic world, embracing a consideration of the nature of matter, 'species', their 'genesis' (origin), 'transformation' (evolution), similarities and differences in appearance and behaviour and the vast array of reproductive and developmental processes across the organic world, as well as the distribution of species and their classification. It is beyond the scope of this work to provide more than a brief insight into his thinking on some aspects of the topics he covered. While he agreed with some of the theories being debated at the time, he took issue with others, including an important aspect of Charles Darwin's theory on natural selection (1859)[10] (see below). He assessed the theories of other scientists according to whether they were deduced from systematic empirical investigations. Oldfield's apparently original speculations were not so derived but were either inferred from the findings of others or from his own *ad hoc* accumulated observations, mostly made in Australia, of human cultures, indigenous plants and animals in the field, animals in captivity or in the service of settlers and introduced cultivated plants.

Although he sought Joseph Hooker's advice as to who might comment on the manuscript there is no evidence that Hooker ever responded to the request.[11] Hooker supported Darwin's theory of natural selection, so if he (Hooker) did read Oldfield's manuscript he very likely distanced himself because of Oldfield's criticism.[12] There were other reasons why Hooker might not have supported the publication of the work (see below). Not having been published may have meant that the treatise was never critically considered by scientists of his day and, as far as we are aware, it was only subsequently referred to in 1911 by Joseph Maiden, then Government Botanist in New South Wales[13], and in 2008 by Jim Endersby.[14]

Implicit was Oldfield's acceptance of what Roy MacLeod termed the 'articles' of scientific naturalism.[15] Oldfield argued that, 'All through nature we observe that every object is in a state of continual change'[16] but 'there is order in all its operations [that is] amenable to strict and eternal laws.'[17] He considered speculation to be an important part of the scientific process, but distinguished sharply between the validity of claims about the origin and processes of the natural world that were purely hypothetical and those based on what he considered to be a systematic view of nature. The latter were discoverable through direct observation and experiment and, at least as far as the physical sciences were concerned, were generally describable through mathematics. He believed that scientific theories should be based on 'long and extended experience derived from actual observation'. He went on to say that all such observations 'require verification, and their accuracy established, they must be carefully compared *inter se* ere the deductions drawn from their discussion can be fully admitted'.[18] He believed that nature was created for a purpose but it was a 'dark enigma whose purport will never be fully revealed to man'. This suggests, but does not confirm, that Oldfield agreed with some of his contemporaries, among them the eminent agnostic proponents of scientific naturalism Thomas Henry Huxley, William Whewell and John Frederick Herschel, who adhered to the notion of the 'Divine mind in Nature' rather than holding to 'a cause theory of matter operating according to the rules of chance'.[19]

Science, Oldfield stated, as yet knew 'nothing of the prime causes—those hidden springs which move only at nature's command—that regulate the phenomena of life'.[20] Such knowledge, he argued, would be incomplete and not gained in the foreseeable future. He believed that the accumulation of

[10] Charles Darwin, *On the Origin of Species by Means of Natural Selection*, John Murray, London, 1859.
[11] See Chapter 20.
[12] ibid.
[13] J.H. Maiden, 'Records of Australian botanists', 1st Supplement, Proceedings of Section D, *Report of the Thirteenth Meeting of the Australasian Association for the Advancement of Science*, Sydney, 1911, pp. 233–234.
[14] Jim Endersby, 2008.
[15] Roy MacLeod, 2000, pp. 4–5.
[16] AO, 1865a, p 158.
[17] ibid., pp. 151.
[18] ibid., p. 10.
[19] MHH, 2005, pp. xii–xiv.
[20] AO, 1865a, p. xviii.

scientific knowledge would only ever enable scientists to guess at how nature and its processes originated.[21]

He thought that scientists would have to look to the physical sciences (chemistry, mechanics, geology, mineralogy, astronomy), 'for all that we can hope to know of those unknown conditions which regulate the phenomena of life'.[22] Each of the physical sciences, he claimed, 'are strictly particular cases of the science of Chemistry for one and all consider the peculiar properties of the several elements either as agents or subjects, or in other words, they take account of the phenomena resulting from the chemical union of some of those elements'.[23]

He was aware of the atomic and molecular theories that had currency at the time and was clearly under the influence of John Dalton's theory. Through experimentation Dalton concluded that matter was composed of a large number of elementary, homogeneous and distinct substances, themselves composed of indivisible, indestructible, uncreatable atoms and it must also be assumed that all the atoms of any particular element are like each other but different from the atoms of other elements.[24] Thus, Oldfield stated that

> the existence of any solid, coherent mass of matter...is composed of some two or more of the prime elements [i.e. atoms, for example O, N, H] in chemical union and that all the composing atoms of any proximate element [i.e. molecules, for example, oxygen (O_2), sodium hydroxide (NaOH)] are of the same definite figure, but different from the form of any other proximate element.[25]

He went on

> Matter (as apparent to our senses) may be regarded as some special cumulation of forces, these being united by other forces generated by their mutual actions and reactions; and reciprocally, forces may be considered as free, *uncombined* [sic] or disintegrated matter.[26]

He inferred that the laws regulating the chemical union of the prime, as well as the proximate, elements are similar to, if not identical with, those that have been efficient in grouping the several members of the solar system.

> [T]hat is to say, those principles which by their combination originate the greatest intensity of physical force, occupy the central position, while the other compounds of the same or different principles, arrange themselves at distances proportionable to the amount of physical force generated by the mutual actions and reaction of their component elements, each particular atom or planetule...moving in a definite orbit about the central atom which in turn moves in a certain trajectory related to the same of associated and similarly composed atoms.[27]

Here he attributes the forces that combine atoms in a molecule as akin to the forces of rotation, whereas the Swedish chemist Jöns Jakob Berzelius concluded based on experimentation that electrochemical reactions were responsible.[28]

At the time, atomic theory was in its infancy. It was not until 1895 that the first sub-atomic particle, the electron, was discovered by British physicist Joseph John Thomson.[29] A later discovery early in the 20th century by Ernest Rutherford led him to propose a planetary model in which a cloud of

---

[21] ibid., p. xxix.
[22] ibid. pp. xix.
[23] ibid. pp. xlix.
[24] Charles Singer, 1997 edn, p. 293.
[25] AO, 1865a, pp. 111—112 & 123.
[26] ibid., p. 95.
[27] ibid., p. 140—141.
[28] Charles Singer, 1997 edn, p. 295.
[29] Luke Mastin 2009, 'Quantum theory and the uncertainty principle: early development in atomic theory', in *The Physics of the Universe*, viewed 14 June 2011 at www.physicsoftheuniverse.com.

electrons surrounded a small, compact nucleus of positive charge.[30] The purely speculative proposal which Oldfield discussed is entirely different from that of Rutherford, which focuses on the structure of atoms, and not, as in the case of the former, on the structure of molecules. Rutherford was on roughly the right path but there followed over the course of the next century much greater insight into the nature of sub-atomic particles and the formation of molecules.[31]

Oldfield entered the debate on the transformation of species (evolution). Thus, he stated: '[T]he more reasonable of the theories that had been put forward to account for the various forms of life, express the existence of other forms of life anterior to the origin of man'.[32] He considered Darwin's theory on the origin of the species through a process of natural selection to be inadequate.

> [F]or applying that theory to any isolated country—as New Holland—it requires us to admit that some forms are stable, while others are variable under the same contingencies, a supposition at variance with the spirit of that theory which assumes that all forms are equally susceptible of variation under the same conditions.[33]

Oldfield thought, along with some of his contemporaries including the French zoologist Alfred Moquin-Tandon[34], that the creation of new species was effected in sexually reproducing species

> [by] the blending of embryos; that is to say, the assemblage of a certain number of prime utricles [? cells formed during the early phases of blastula formation and gastrulation] of a particular nature, those utricles by their combination under certain conditions, originating even the most complex forms of organic life.[35]

The outcome leading to physically modified individuals, 'monsters' in some cases, changed individuals that would survive to sexual maturity and be able to reproduce their like and, in turn, produce viable offspring.[36] Unlike Darwin, at no time did he address the important question of how he thought a distinct aggregate of new organisms would come about that fitted his definition of 'species', namely,

> an assemblage of individuals [or 'organised beings'] in time and space, whose progeny, for a definite period, oscillates in attributes about a certain permanent, but arbitrary type that are 'nearly alike in all their more important characteristics.[37]

This definition, he said, 'must be regarded as excluding the probability of breeding between members of distinct species'.[38] He attributed this latter criterion of a 'species' to the French physician and anatomist Marie-Jean-Pierre Flourens who, as Oldfield remarked, 'justly regards as the most essential characteristic of a species...continuous fecundity'.[39]

According to the eminent palaeontologist, Ernst Mayr, until the second half of the nineteenth century it was virtually universal for species to be recognised by their differences; 'they were kinds, they were types'. This concept was usually referred to as 'the morphological or typological species concept'. The definition employed by Oldfield is somewhat similar to what has been referred to as the 'biological species concept (BSC)', which in essence is widely accepted today and defined by Mayr as

---

[30] John Gribbin, 2003, pp. 504—505.
[31] ibid., pp. 505–528.
[32] AO, 1865a, p. viii.
[33] ibid., pp. 161–163.
[34] ibid., pp. 207–8.
[35] ibid., p. 155.
[36] The theory of 'hopeful monsters' which is currently being debated by some scientists and not yet confirmed, is different to and perhaps does not suffer the same theoretic difficulties as Oldfield's theory. Thus, according to Günter Theißen, 'hopeful monsters' are organisms with a profound mutant phenotype [not formed by the blending of embryos] that have the potential to establish a new evolutionary lineage [presumably because they are able to pass on their macro–changes through being able to continue to reproduce with those of their own species who are not so changed]. The Synthetic Theory of evolutionary biology has rejected the evolutionary relevance of 'hopeful monsters', but could not fully explain the mechanism and mode of macroevolution. On the other hand, several lines of evidence suggest that 'hopeful monsters' played an important role during the origin of key innovations and novel body plans by saltational rather than gradual evolution. *Homeotic mutants* [A mutant in which one body part, organ or tissue, is transformed into another part normally associated with another segment] are identified as an especially promising class of hopeful monsters', Günter Theißen, 'The proper place of hopeful monsters in evolutionary biology', *Theory in Biosciences,* vol. 124, 2006, pp. 349–369.
[37] AO, 1865a, p. 50.
[38] ibid.
[39] ibid., p. 11.

> '[t]he segregation of the total genetic variability of nature into discrete packages, so called species, which are separated from each other by reproductive barriers, prevents the production of too great a number of disharmonious incompatible gene combinations'.[40]

Oldfield discussed at length what was understood of embryonic development in animals and plants. He agreed with the theory of recapitulation[41] (often referred to as the 'biogenetic law'[42]). This asserted that embryonic developmental stages (ontogeny) recapitulate the evolutionary transitions of adult forms of an organism's past (phylogeny). First formulated by Étienne Serres in 1824–1826, apparently drawing upon the ideas of Étienne Geoffroy Saint-Hilaire and his German colleagues of the *Naturphilosophie* school,[43] with some modification by Ernst Heinrich Philipp August Haeckel and others, this theory had credibility in various quarters for over a century. However, it was attacked as early as 1828 by the Estonian zoologist and founding father of embryology, Karl Ernst von Baer. Von Baer's empirical observations in embryology led him to conclude that recapitulation cannot possibly occur since

> The embryonic vertebrate, at every stage, is an undeveloped and imperfect vertebrate; it can represent no adult animal whatever. Furthermore, there can be no transformation of any kind between types, either in ontogeny or phylogeny. The four great groups cannot be arrayed in a progressive sequence; they are simply different from each other, not higher or lower.[44]

There is nothing in Oldfield's writings to suggest that he was aware of this attack on recapitulation. Under the weight of subsequent developments in embryology and genetics the theory finally collapsed. Nevertheless, according to American palaeontologist, evolutionary biologist, and historian of science Stephen Gould, even though the whole subject of parallels between ontogeny and phylogeny fell into disrepute, the temporal displacement of characters during the ontogeny of a descendant, 'heterochrony'[45], is still one of the important themes of evolutionary biology.[46]

The evolution, geographical origin, distribution of the genus *Homo* and the development and decline of civilisations were briefly discussed by Oldfield.[47] He thought the 'various races (or species) of "man"'

> were probably of independent origin, both as regards epoch and locality, but amidst the conflicting views on this subject, it is impossible to determine the precise number of those races since no traces of primitive man, or of his immediate predecessors have been 'preserved for us'. Considering however, the necessity for the concurrence of

[40] Ernst Mayr, June 1996, 'What is a Species, and What is Not?' originally published in *Philosophy of Science*, vol. 63, pp. 262–277, viewed online 13 June 2011 at www.darwiniana.org.

[41] AO, 1865a, pp. 189–190.

[42] Stephen Jay Gould, 1977, *Ontogeny and Phylogeny*, The Belknap Press, Harvard University Press, Cambridge, Mass., pp. 47–49.

[43] Initially propounded by Johann Gottlieb Fichte and Friedrich Wilhelm Joseph Schelling, the *Naturphilosophie* School adhered to a view of nature that flourished in Germany during the Romantic era. It criticised the Newtonian scientific view that treated nature as mechanistic and meaningless and suggested that nature undergoes a process of self–development culminating in a state of self–preservation. In opposition to the scientific method of exploring nature through external observation and experiment, it sought to understand nature's own language through intuition and contemplation. In assessing the relationship of *Naturphilosophie* to science, Timothy Lenoir has argued that no single system of natural philosophy is characteristic of the entire Romantic period in Germany. From its first appearance and throughout its stormy career, for instance, the *Naturphilosophie* of Schelling and his school was severely criticized. Rather than a single approach to understanding nature, Lenoir says it seems more appropriate to regard the German science of this era as having been formed from a common fund of scientific concepts and methods, metaphysical predispositions and epistemological concerns, which received differing emphases in the various approaches to natural philosophy of the period. The Gottingen School, for example, were empiricists who established and tested hypotheses on the basis of observations of the real world, Timothy Lenoir, 'The Gottingen School and the Development of Transcendental *Naturphilosophie* in the Romantic Era', *Studies in the History Biology*, vol. 5, 1981. This paper is part (pp. 111—205) of a larger work the details of which could not be determined, viewed 25 May 2012 at www.jhfc.duke.edu.

[44] Stephen Jay Gould, 1977, pp. 52—57.

[45] Gould employed von Baer's definition of 'heterochrony' which refers to phyletic change in the onset or timing of development, so that the appearance or rate of development of a feature in a descendant ontogeny is either accelerated or retarded relative to the appearance or rate of development of the same feature in an ancestor's ontogeny (ibid., pp. 221 and 482).

[46] ibid., pp. 47–114, 167–206, 405–409.

[47] AO, 1865a, pp. 620–650.

> many favourable conditions for his development and subsequent existence, we may infer that localities so favoured would be few, and consequently the number of distinct races could not be many, probably two or three in Asia, one or two in Africa, and one in America.[48]

In making these claims, apparently Oldfield did not recognise that interbreeding he had observed between his so-called races had over the years resulted in 'continuous fecundity' across generations.

Oldfield's contribution to the science of the Australian flora.

Oldfield was in Australia from either 1844 or 1845 until the latter half of 1862 with about a year's sojourn in England in 1849–50. During his time in Australia he periodically collected in Tasmania, mostly from 1844 or 1845 until 1857, in Western Australia 1856, 1858–59 and 1861–62, New South Wales 1848 and 1861, Victoria c.1848 and c.1852–53 and South Australia for one or more short periods possibly in 1850, 1856 and/or 1859. During his time in Western Australia he collected over vast areas, including remote, virtually unsettled areas that had been the object of very little or no botanical collecting on the part of Europeans.

Jim Endersby, in his *Imperial Nature: Joseph Hooker and the Practices of Victorian Science* (2008), referred to Oldfield's relationship with Hooker both as a collector and theoretician.[49] Apart from this and references to his specimens, Oldfield's contribution has received little attention in the published historiography of Australian botany, being mainly confined to one or two paragraphs in biographical dictionaries of Australian plant collectors.[50]

Botanists in Britain and Europe who were working on the floras of their respective nations' colonies were almost wholly dependent upon colonial collectors for the specimens they needed for their taxonomic work. Shortly after Oldfield began botanising in Australia in the latter half of the 1840s, the then Director of the Botanic Gardens, Kew, William Jackson Hooker, ceased sending collectors abroad due to insufficient funds and poor results from previous expeditions, although he was involved until 1861 in the selection of some naturalists aboard Royal Navy vessels who collected for Kew Gardens. He and his successor, Joseph Dalton Hooker, who was appointed Assistant Director at Kew Gardens in 1855, respectively overcame the paucity of official collectors by establishing wide networks of colonial collectors who supplied them with specimens, usually without remuneration.[51] Oldfield was one such collector. However, it is apparent that it was not William who sought his cooperation in the first place; it was Oldfield who made the initial overtures.

Although Bentham and Hooker spoke highly of Oldfield in their major Australian floras, the almost total lack of recognition of his contribution in the subsequent historiography of Australian botany must be viewed in terms of the way that science was practised from the mid-nineteenth century.

R.M. and W.R. Barker, in their paper 'Botanical contributions overlooked: the role and recognition of collectors, horticulturalists, explorers and others in the early [prior to 1850] documentation of the Australian flora' discussed the absence of acknowledgement of the contributions of many collectors. They included Aborigines, convicts, subordinate members of exploratory expeditions and a few

---

[48] ibid., pp. 626–627.

[49] Jim Endersby, 2008, pp. 184–185 & 229-230.

[50] See for example: A.E. Orchard, 1999, 'Oldfield, Augustus F. (1821—1887)', *Biography: Biographical notes on plant collectors and illustrators*, Australian National Herbarium, Council of Australian Herbaria (hereafter ANHBiog), viewed 2 June 2012 at www.anbg.gov.au; J.H. Maiden 1909a, 'Records of Tasmanian botanists', *Papers and Proceedings of the Royal Society of Tasmania for the year* 1909, pp. 24–25 and 'Records of Western Australian botanists and Australian botanists who have dealt with the Flora of Western Australia', *Journal of the Western Australian Natural History Society* for 1909b, p. 22; Norman Hall 1978, *Botanists of the Eucalypts*, Commonwealth Scientific & Industrial Research Organisation, Melbourne, p. 107; Ray Desmond 1977, *Dictionary of British and Irish Botanists and Horticulturalists including Plant Collectors and Botanical Artists*, Taylor & Francis, London, p. 471.

[51] See Chapter 8.

resident collectors who often provided information about the habits and habitats of the plants they gathered.[52]

Mid-nineteenth century major taxonomic works on the Australian flora include the names of the respective collectors of the actual specimens (often the 'type' specimens) that were used to describe each of the species and where they were gathered. Furthermore, in Britain, Europe and Australia collectors' names were frequently applied to newly discovered genera or species. See for example, Joseph Hooker, *Flora Tasmaniae,* vols. 1 & 2, 1860, George Bentham assisted by Ferdinand Mueller, *Flora Australiensis*, vols. 1–7, 1863–1878, and F. Mueller, *Fragmenta*, vols.1–12, 1858–1882. While these practices afforded collectors some recognition they by no means provided a true picture of the extent and nature of their various contributions.

The Barkers offered several factors contributing to this situation, most of which, to a greater or lesser extent, are applicable to Oldfield. In the first place, the identities of many of those collectors of Australian plants have been obscured because principal British and European botanists on expeditions, such as Robert Brown on Matthew Flinders' voyage or Jacques-Julien Houtou de La Billardière who accompanied Bruny D'Entrecasteaux on a French expedition to the Pacific Ocean, are remembered and fêted because they published their journals of exploration and taxonomies of the plants they collected. They made little mention of those who accompanied them on their journeys and made important plant discoveries.[53] Although the obscurity of other collectors could be attributed to this, it was not the case with Oldfield; he never participated in a major expedition.

Another reason put forward by the Barkers relates to the tendency for details of plant habits and habitats provided by collectors not to be attributed to them by many late 18th and early 19th century taxonomists who included the information in their publications. Thus, educated resident collectors 'often provided short descriptions of the plants, commented on their novelty, and sometimes named them':

> In the present day it is likely that the principal recipients of their collections [in Britain], Robert Brown, John Lindley, Sir William Hooker and Dr Joseph Hooker would have published the many new plants they described and named under joint authorship.[54]

Ferdinand Mueller did, however, publish descriptions of two new species under joint authorship with Oldfield— (*Eremophila clarkei* and *Halgania bebrana*)[55] and a new variety (*Patersonia umbrosa* var. *xanthina*)[56] —based on Oldfield's specimens because, as Mueller explained: 'I liked to give him every encouragement in my power'. While this may seem to some to have been a token gesture, Mueller was more generous than the Hookers and Bentham in that regard since it was a recognition they never accorded Oldfield. Each of them did, however, commemorate him by naming species for him (see below).

Across ten volumes of Mueller's *Fragmenta* there are species descriptions based solely on Oldfield's specimens.[57] These descriptions frequently include brief details of the habitats in which they were found and which only Oldfield could have provided but he was never specifically recognised for these contributions in those volumes. For example, *Eremophila oldfieldii*: '*In loco Yattoo collium arenoso-rupestrium calcaria tectorum*' [In the locality of Yattoo hill hidden in limey sandstone].[58]

[52] R.M. Barker & W.R. Barker, 1990, 'Botanical contributions overlooked: the role and recognition of collectors, horticulturists, explorers and others in the early documentation of the Australian flora', in P.S. Short 1990, *History of Systematic Botany in Australasia: Proceedings of a Symposium Held at the University of Melbourne, 25–27 May 1988*, Australian Systematic Botany Society Inc., South Yarra, Victoria, pp. 37–85.

[53] ibid., pp. 39–73.

[54] ibid., p. 37.

[55] FM, 1858–59, vol. 1, pp. 208–209.

[56] ibid., p. 214.

[57] ibid.

[58] ibid., p. 208.

Similarly, among Mueller's plant descriptions are unattributed particulars as to their habit of growth, height etc. which must have come from Oldfield. For example, *Anthocercis arborea* (=*Anthocercis intricata*): '*Arborescens, ramis intricata...Arbuscula altituden 12' attines, coma intricato trunco 4–6' longo...species eximae staturae*' [Becoming tree-like, tangled branches...Small tree, attaining 12 feet in height, tangled leafy crown trunk 4–6 feet in height...species exceedingly statuesque].[59] Perhaps, however, Mueller assumed such acknowledgement by citing Oldfield as the collector but this is not apparent to the reader. To be fair to Mueller he did occasionally cite Oldfield and his other collectors as the authors of botanical observations in some other publications, particularly in newspaper articles.

Although by no means a common practice among British taxonomists, from the mid-nineteenth century onwards acknowledgement of this kind was not unknown. In some cases, Bentham in his *Flora Australiensis* attributed to Oldfield particulars he had provided on the habits and habitats of the plants from which he had taken the specimens. For example, *Eucalyptus patens*: 'A tree attaining a great height, with a rough half-fibrous, persistent bark (Oldfield)'.[60] However, he appears to have been inconsistent in that regard. Thus, it is unlikely that he would have been in a position based solely on Oldfield's herbarium specimens to provide the unattributed details of the form of *Calothamnus blepharospermus*, namely: 'A rigid straggling shrub hirsute with spreading hairs the foliage scarcely becoming glabrous with age'.[61]

Thirdly, continued the Barkers, was the failure of most 19th century resident colonial collectors, such as Allan Cunningham, Charles Fraser, James Backhouse, James Drummond and Ludwig Leichhardt, to publish their own taxonomic descriptions. This led to them losing credit for much of their original work.[62] There were many reasons why early Australian colonial collectors did not publish. Among them, according to the Barkers, were conditions of employment, little access to the botanical works of the day, lack of an authentically named Australian reference herbarium and want of an accessible means of publication.[63] No evidence has emerged that Oldfield desired to publish plant descriptions *per se*. After he returned to England in 1862 he yearned to undertake research but was thwarted because of the failure of his eyesight[64], and his attempt to publish his manuscript on natural philosophy was prevented by his lack of financial resources.[65]

Oldfield was highly educated, claimed to be 'a gentleman' and was recognised to be such by Ferdinand Mueller. He was not high-born, wealthy, a member of the clergy, or a professional of standing in the community, nor did he occupy a senior position in the colonial civil service. There is no evidence that Oldfield was a member of the Tasmanian Society of Natural Science, Agriculture, Statistics etc., which was still operating at the time of his arrival in the Colony. While there is the likelihood that this might have been because of its elitist nature this could not be confirmed. However, it seems that eventually he did become a member of the Royal Society of Tasmania, but *The Papers and Proceedings of the Royal Society of Tasmania* contain none of his writings.[66] It has not become apparent as to why this was the case. Apart from the possibility that he had no inclination to publish taxonomic descriptions, he may have been actively discouraged.

Joseph Hooker argued that species could be named only by naturalists like himself who commanded global specimen collections.[67] Thus, while he discouraged collectors from publishing their taxonomic descriptions, at the same time he encouraged them, by various means, to classify the plants they gathered. His motives are explained by Endersby:

---

[59] ibid., p. 212.
[60] GB, 1866, vol. 3, p. 247.
[61] ibid., p. 175.
[62] R.M. Barker & W.R. Barker, 1990, p. 37.
[63] ibid., pp. 65–66.
[64] Chapter 21.
[65] Chapter 20.
[66] Chapter 7.
[67] Jim Endersby, 2008, p. 137.

> Hooker regularly bemoaned the "chaos of synonymy which has been accumulated by thoughtless aspirants to the questionable honour of being the first to name a species"...Encouraging collectors to learn plant identification, and thus the rudiments of classification, evidently encouraged some of them to try naming species themselves; from Hooker's perspective that was preferable to the "dreadful work" of sorting through badly labeled [sic] duplicates...Yet, even if it had been possible to entirely separate collecting and classifying, it is unlikely that Hooker would have done so, because teaching collectors how to identify plants was also his best opportunity to persuade them he was right about classification.[68]

Oldfield was classifying plants before he made his initial contact with William in 1850.[69] Although it is known that his (Oldfield's) interest in natural history began when he was a boy[70] extant records do not reveal how he acquired classification skills or from whence he derived the notion, also expressed in 1850, that the scientific worth of his plant specimens depended on his ability to name the plants he gathered.[71] While it was to William Hooker that Oldfield attributed 'the chief part of what little [he] knew of Botany'[72], in addition to his own self-directed reading he may have received some training through botanical courses at King's College London or some other institution in England.[73] If so, he probably used Linnaeus' sexual system of classification and referred to his *Philosophia Botanica* for instructions on how to describe, measure, and preserve plants.[74] Since Oldfield was educated in France[75] it cannot be ruled out that initially he had been taught, or at least read of, the natural system classification, which he advocated and which had been the preferred system in that country for many years before it was adopted in Britain.[76]

At times during the early to mid-1850s Oldfield expressed concern that his efforts at classifying his specimens might not meet with William Hooker's approval but gave no reasons other than the unavailability of suitable taxonomic reference texts in Tasmania.[77] No specific mention has emerged of the views held by the Hookers, Bentham or Mueller as to the accuracy of Oldfield's plant classifications or the quality of his plant preservation, label data and packaging for transport to their institutions for taxonomic investigation. Perhaps the epithet 'intelligent collector' meant that he embraced the requirements as outlined by Endersby:

> The craft aspects of collecting remind us that natural history specimens—botanical, zoological, and geological—are not natural objects but artifacts, things that were made according to exacting standards. Simply picking a plant did not produce a botanical specimen: in addition to knowing which plants to pick...collectors had to gather specific parts of the plant, label them correctly, and preserve and mount them properly...Learning to do all this was part of a collector's training...Precisely because collecting was so popular, it was a largely voluntary activity. As a result, local collectors had to be encouraged to comply with expert judgments (a standard that no one complied with would be worthless), so teaching someone how to collect was part of a complex process in which botanical standards were simultaneously being set and enforced.[78]

---

[68] ibid., pp. 162–163.
[69] Chapter 7.
[70] Chapter 5.
[71] ibid.
[72] Chapter 20.
[73] Chapter 5.
[74] Jim Endersby, 2008, p. 143.
[75] Chapter 5.
[76] Jim Endersby, 2008, p. 171.
[77] Chapter 7.
[78] Jim Endersby, 2008, p. 55.

William Hooker made available copies of his instructions to botanical collectors that were contained in *A Manual of Scientific Enquiry: Prepared for the Use of Her Majesty's Navy and Adapted for Travellers in General*,[79] to Oldfield in 1850 for distribution in Australia. These may have helped him to hone his skills in accordance with the 'expert judgments' that prevailed at the time among the taxonomists at Kew Gardens.[80] The focus of Hooker's published 'Instructions', as revealed in a later version[81], was on the selection, preservation and preparation for transport of plant specimens. Nothing is mentioned about recording details of the habits or habitats of the plants from which they were obtained, place and date of collection, or a requirement for collectors to classify their plants. This may account for the absence of date of collection on the vast majority of Oldfield's specimens. It seems, however, that it was Hooker's practice from at least 1842, through personal communications to his collectors, of emphasising the importance of recording on 'a ticket' basic details of each plant from which the specimen was taken.[82] If Oldfield received instructions to that effect the historical record has remained silent. Although by no means consistently provided, specimen labels on plants he gathered in Tasmania, Western Australia and New South Wales bear ecological information on the geographical locations in which plants grew.[83] Bentham's instructions to collectors were not published until after Oldfield left Australia permanently in 1862, which probably explains why his specimen labels did not comply with all of Bentham's requirements.[84]

Endersby went on to explain that Oldfield's ideas on plant classification would have been particularly problematic for Hooker because 'he [Oldfield] used his wide reading and knowledge of the sciences to blend widely held opinions with unorthodox ones, creating a mixture that might have appealed to colonial naturalists'. For example, according to Endersby, Oldfield's 'view of the Linnaean system was one that many in the metropolis would have echoed', that is, the sexual classification's 'very simplicity' had turned botany into a 'mere jargon of names', resulting in 'the total neglect of the important study of the physiology of plants':

> Oldfield condemned the "lax notions" of naturalists who named new species on spurious grounds, adding that their "unphilosophical and groundless assumption that nature has plainly labelled all her various and multifarious productions" leads such men into "continually registering new species." He also argued that splitters—motivated by "their inordinate ambition to leave their names deeply engraven on the tablets of science" —were turning botany "into a perfect chaos of synonymy", to such an extent that "people begin to doubt the existence of species".[85]

William Hooker employed both the Linnaean and natural systems in his books, with 'the former as an introductory key to the latter'. He used the simpler Linnaean system in the *Botanical Magazine* until his death in 1865.[86] Oldfield was a self-professed adherent of natural system taxonomy. First proposed by Georges-Louis Leclerc, Comte de Buffon and later extended by the Göttingen School, its supporters argued that classification should be not only on multiple morphological criteria but also on systemic physiological criteria as well.[87] It is clear from his letters that Oldfield was using available taxonomic works to identify his plants including works by the Hookers. Perhaps initially, Oldfield followed the sequence of families proposed by Augustin Pyramus de Candolle (1819) or the works of

---

[79] These instructions were originally published in *A Manual of Scientific Enquiry: Prepared for the Use of Her Majesty's Navy and Adapted for Travellers in General*, 1849, a copy of which could not be found.
[80] AO to WJH, 5 July 1850, Australian Letters, vol. 73, fol. 258, RBGKA
[81] Although it cannot be confirmed, Sir William Hooker's 'Instructions for collecting and preserving and sending home plants and their useful products', in 'Botany', *Admiralty Manual*, 1856, pp. 1–8, RBGKA were probably essentially the same as the document provided to Oldfield in 1850 by Helen Broderick to MHH, email 29 January 2007, HOA.
[82] Jim Endersby, 2008, p. 139.
[83] See for example, MELISR downloads, 18 November 2005, 8 March 2007, 1 December 2010, NHV, RBGM.
[84] GB, 1863, *'Collection, Preservation and Determination of Plants', Flora Australiensis...*, vol. 1, *Ranunculaceae to Anacardiaceae*, Lovell Reeve & Co., London, pp. xxxii.
[85] Jim Endersby, 2008, p. 184.
[86] ibid., p.178.
[87] Timothy Lenoir, 1981, p. 121.

the Swiss botanist, Carl Daniel Friedrich Meisner and the Austrian botanist, Stephan Ladislaus Endlicher. He stopped collecting just before Joseph Hooker's and George Bentham's three volume *Genera plantarum* (1862–1883) was published. This work arose because the authors recognised the need for a new plant classification to replace the outdated ones of Endlicher and Meisner. While not a phylogenetic system, *Genera plantarum* is a natural one following, with modifications, the sequence of families proposed by de Candolle in 1819.

Oldfield's views on classification, expressed in his unpublished manuscript, give, according to Endersby, a clear sense of why Joseph Hooker attempted to dissuade colonial naturalists from speculating about classification. Thus, 'if similar opinions had become widely known and adopted in the colonies, they could have led colonial naturalists to use different classifications and names from those in use at Kew'. Such speculations posed an even more serious threat to the prestige and authority of imperial botany, and thus to Hooker's fragile status.[88] Hooker would have particularly approved of Oldfield's claim that the real issue was the validity, 'not of species, but of the methods by which we attempt to discriminate them'.[89] It was there, however, that Endersby considered Hooker's agreement with Oldfield's views would have ended because alongside these comments were others that showed the influence of Lorenz Oken and the German *Naturphilosophie* school, which Hooker would have found totally unacceptable. For example, Oldfield argued 'that there were no genuinely intermediate forms in nature' a claim, 'that would have undermined Hooker's classificatory method'.[90] Oldfield's manuscript was never published so, continued Endersby,

> there is no way of knowing how colonial botanists would have reacted to it, but many of his ideas were not noticeably stranger than others in circulation in the mid-nineteenth century; indeed, its immediate inspirations appear to have been German *Naturphilosophie* and Adrian Haworth's binary system. Seen in the context of binary, quinary, septenary, and "Baskervillean" classifications, Oldfield's speculations—far from being bizarre—are clearly typical of the classificatory debates, not least because of his opposition to artificial systems and his focus on the search for affinities between plants. Had his ideas been published, there is every chance they would have been taken seriously [? by many taxonomists in Britain, Europe and the colonies].[91]

Mueller's few surviving references in the 1850s suggest that he thought nature was continuous, the position taken by Joseph Hooker, but in the 1860s, 'he reaffirmed his view of species as "special creations" — discrete and unalterable — above which the pattern of relationships in nature was continuous'. Nevertheless, Mueller admitted that 'in its execution we would lose the important points of reference,which support particularly our memory',and that therefore it was probably still worthwhile combining species into groups of 'never precisely definable genera', and so on. Mueller followed Candolle's system in the 1860s but became increasingly dissatisfied with it. [92] Nevertheless, there was common ground between Mueller and Oldfield in that there was no continuity in nature.

Following his reading of Joseph Hooker's *Introductory Essay to the Flora of New Zealand*,[93] Oldfield wrote that he had a 'desire to throw some light on the distribution of Tasmanian plants'[94] which, according to Endersby, had unwelcome consequences,

> since it led Oldfield to read widely on distribution issues [this is evidenced in his unpublished natural philosophy manuscript].[95] For example, 'he discussed—and criti-

[88] Jim Endersby, 2008, p. 184.
[89] ibid.
[90] ibid., p. 185.
[91] ibid.
[92] Sara Maroske, 2006, p.154.
[93] JDH, 1853, *Introductory Essay to the Flora of New Zealand*, pamphlet, Lovell Reeve, London. This essay was reprinted as the introduction to JDH 1864, *The Handbook of the New Zealand Flora: a Systematic Description of the Native Plants of New Zealand and the Chatham, Kermadec's, Lord Auckland's, Campbell's, and Macquarrie's* [sic] *Islands*, Lovell Reeve, London.
[94] See chapter 10 of this work.

> cized—several of Darwin's [theories], including his claims about waterbirds spreading plant seeds via the mud adhering to their feet. Oldfield thought the latter distinctly implausible, whereas, Ronald Gunn was more confident about the effects of accidental migrations…From Hooker's perspective, the immediate problem presented by the colonial botanists who took too close an interest in distribution was that, like Gunn and Oldfield, they might arrive at different conclusions, which might encourage different collecting practices…However, that was a minor problem compared to what would happen once their speculations led them to dispute Hooker's theories and authority, as Oldfield's did not.[96]

It is not known whether Hooker read Oldfield's manuscript.[97] However, according to Endersby, had he done so, it

> would have reinforced his desire to dissuade his colonial correspondents from speculating, since they contain ideas that were typical of the current of speculation in Britain. Once again, botany's more philosophical topics created a dilemma for Hooker: as in the case of the natural system itself, persuading his colonial collectors of the importance of distribution—because he needed their data—could also lead to speculation.[98]

In 1911 Joseph Maiden included a quite lengthy excerpt from Oldfield's unpublished manuscript in his paper, 'Records of Australian botanists' which appeared in the *Report of the Thirteenth Meeting of the Australasian Association for the Advancement of Science*. Maiden used this piece to support his conclusion that it was 'a pity that [Oldfield] published nothing on Australian flora'.[99] It is perhaps unfortunate that Maiden, as one of the senior professional botanists in Australia, did not explain why he thought this.

Endersby, while acknowledging that metropolitan 'scientific men', such as the Hookers in England, were dependent upon colonial collectors to obtain the specimens they required for their taxonomic work, stated that there were probably 'several Oldfields, who failed to advance either science or themselves'.[100] Oldfield's passion for botanical collecting and his strong desire to collect for lengthy periods in remote parts of Australia certainly interfered with his ability to obtain a financially secure working life and retirement. That he failed to advance science is questionable, unless of course such advancement is viewed wholly in terms of published taxonomic descriptions and scientific theories derived from empirical experiments and systematic observation of the physical world. We do not agree that Oldfield failed to make a scientific contribution. Oldfield's scientific legacy to Australian botany lies not in published botanical works authored by him. His collections, and the observations he made regarding the form and habitat of many of the plants from which his specimens were obtained, are his contribution to the advancement of the taxonomy of the Australian flora, a contribution that we believe is significant and worthy of his receiving far higher recognition than he has been accorded in the past. The evidence in support of this claim is derived from published taxonomic works and records of herbarium specimens that we have accumulated in our Oldfield database. He also contributed to the science of anthropology that will be discussed below.

In all we found references to 2,648 of Oldfield's specimens in publications. Most of these were derived from major 19th and early 20th century taxonomic works on Australian flora in which many of Oldfield's specimens were used to describe new species and to extend the ranges of previously

---

95 AO, 1865a.
96 Jim Endersby, 2008, p. 229–231.
97 See Chapter 21 of this work. Endersby has referred to Oldfield's unpublished natural philosophy as a collection of manuscripts, but it is our opinion that it is a single ms which he hoped to publish as a book.
98 Jim Endersby, 2008, p. 231.
99 J.H. Maiden, 1911, pp.233–234.
100 Jim Endersby, 2008, p. 100.

named taxa (see Table 1). Embracing the period 1855 to 2010, articles citing Oldfield's specimens were found in 32 botanical journals and stand-alone monographs published in England, France, Australia, New Zealand, Belgium, Netherlands, Sweden, Norway and Switzerland. This count, which does not pretend to be comprehensive, does not include all the individual papers across the volumes of those publications in which there were references to his specimens.[101] The majority of these papers and monographs were revisions of genera and/or related to the designation of various specimens into type categories.

Author	**Publication**	**No. of Oldfield's specimens mentioned**
J.D. Hooker	*Flora Tasmaniae,* vols. 1 & 2[102]	332
G. Bentham	*Flora Australiensis,* vols. 1-7[103]	1,692
F. von Mueller	*Fragmenta Phytographiae Australiae,* vols. 1–11[104]	350
F. von Mueller	*Eucalyptographia: A Descriptive Atlas…,* decades I–X[105]	8
F. von Mueller	*Systematic Census of Australian Plants…,* pt. 1[106]	15
J.H. Maiden	*Critical Revision of the Genus Eucalyptus,* vols. 1–6[107]	189
	Other	62
	TOTAL	2,648

Table 1: Number of Oldfield's specimen records referred to in publications between 1855 and 2010 on Australian flora in the Oldfield database.

In most cases types will be among the earliest collected specimens of a species and are extremely important to botanists who are attempting to determine the correct application of a name. 'A nomenclatural type (typus) is that element to which the name of a taxon is permanently attached, whether as the correct name or as a synonym'.[108] They are often specially curated in herbaria. In all we located 5,121 of his herbarium specimens of which 715 are type specimens[109] embracing 315 species and 45 infraspecies. These fall into several different categories (see Table 2).

The bulk of his type material comes from Champion Bay to the north of the Murchison River in Western Australia (421 types of various categories, with 334 of them from the vicinity of the Murchison River). His journeys through other parts of Australia resulted in further types—south west of Western Australia 209, Tasmania 49, Victoria 2, New South Wales 2, locality in Australia not known 31, and New Zealand 1. Among the type specimens collected by Oldfield is the magnificent Karri (*Eucalyptus diversicolor*) of the lower south west of Western Australia and the popular Western Australian park and garden tree, Illyarrie (*E. erythrocorys*). In Western Australia he collected in

[101] See HAODb for details of all the publications located.
[102] JDH, 1860, Part 3, vols. 1 & 2. See also HAODb.
[103] GB assisted by FM, 1863–1878, vols. 1–7. See also HAODb.
[104] FM, 1878–1882, vols.1–10, which includes 75 in C.M. Gottsche's 1878-1881 'Musci Hepatici (Adans., Hedw.) Sive Lichenastri (Dill., Wallr.) Australiani' Supplement iv, vol. 11, 1878-1881. See also HAODb.
[105] FM, 1879–1884, *Eucalyptographia: A descriptive Atlas of the Eucalypts of Australia and the Adjoining Island*, Decades I–X, J. Ferres, Government Printer, Melbourne.
[106] FM, 1882a, *Systematic Census of Australian Plants with Chronologic, Literary and Geographic Annotations, Part 1–Vasculares*, Victorian Government, Melbourne.
[107] J.H. Maiden, 1909–1933, *Critical Revision of the Genus Eucalyptus*, vols. 1–6, William Applegate Gullick, Sydney, PDF electronic edition, University of Sydney Library, Sydney, 2002.
[108] J. McNeill *et al.*, 2012, Article 7.2.
[109] See HAODb for details of all the herbarium specimens we located.

particularly botanically rich areas of the world, boosting his opportunities to make a significant contribution to botany.

Type category	No.
Holotypes[110]	65
Isotypes[111]	62
Syntypes[112]	247
Isosyntypes[113]	10
Paratypes[114]	2
Lectotypes[115]	61
Isolectotypes[116]	61
Residual syn-types/paralectotypes[117]	78
Types	129
Total	**715**

Table 2: Type specimen records in the Henderson Oldfield database from the sandplains to the north of Hill River in Western Australia[118]

Another indication of Oldfield's importance as a collector can be derived from the number of taxa that bear his name. Our research indicates that originally, he was commemorated in 29 genera but with subsequent revisions of genera and new taxa identifications this number now stands at 21. These are: *Acacia, Actinobole, Calothamnus, Calytrix, Chthonocephalus*, *Darwinia, Encephalartos (=Macrozamia fraseri), Eragrostis (=E. speciosa), Eremophila, Eucalyptus, Hakea, Helichrysum (=Rhodanthe citrina), Herbertus, Isolepis, Lasiopetalum, Lepidosperma, Leucopogon, Luzula (=L. australasica* subsp. *australasica), Melaleuca, Monotaxis (=M. bracteata), Myriocephalus, Phebalium (=Leionema oldfieldii), Polypogon tenellus* var. *oldfieldii (=Polypogon tenellus), Quoya, Scaevola, Senecio, Solanum* and *Trichostomum* (*=Ditrichum difficile*).[119]

The distributed collections of plants that were made by collectors in regional and remote areas of Australia prior to, or in the very early stages of, their colonisation by Europeans could, perhaps, be used as baselines from which to judge changes in the flora, in particular areas since that time for the

[110] *Holotype*: The type (holotype, lectotype, or neotype) of a name of a species or infraspecific taxon is either a single specimen conserved in one herbarium or other collection or institution, or an illustration (Article 8.1). A holotype…is the one specimen or illustration…used by an author, or designated by the author as the nomenclatural type. As long as the holotype is extant, it fixes the application of the name concerned Article 9.1, J. McNeill et al. 2012, pp. 13—14 and 16.

[111] *Isotype* is any duplicate of the holotype, it is always a specimen. Article 9.4, ibid., p. 17.

[112] *Syntype* is any specimen cited in the protologue when there is no holotype, or any one of two or more specimens simultaneously designated in the protologue as types, Article 9.5, ibid., p. 17.

[113] *Isosyntype*: a duplicate of a syntype, Article 9.12, ibid., p.19.

[114] *Paratype* is any specimen cited in the protologue that is neither the holotype nor an isotype, nor one of the syntypes if in the protologue two or more specimens were simultaneously designated as types, Article 9.6, ibid., p. 18.

[115] *Lectotype* is a specimen or illustration designated from the original material as the nomenclatural type, Article 9.2, ibid., p. 17.

[116] *Isolectotype* is a duplicate specimen of a lectotype. Recommendation 9C, ibid., p. 22.

[117] *Residual syntype/paralectotype*: each of the remaining syntypes when a lectotype has been selected from among syntypes, MELISR Data Entry Manual, NHV, RBGM, viewed 30 October 2009 at www.rbg.vic.gov.au.

[118] HAODb.

[119] ibid.

Figs. 2.1a, b, c and d Herbarium sheets of Oldfield's specimens

purpose of reconstructing ecosystems.[120] Thus, taken together, James Drummond's[121], Pemberton Walcott's[122] and particularly Oldfield's collections from Champion Bay to the Murchison River may provide a useful reference point for this purpose undertaken in the Murchison River district. Unlike those of the other collectors, Oldfield's specimens from Port Gregory and the Murchison River areas were gathered over extended periods—from late September to circa December 1856 and from December 1858 to November of the following year. The latter expedition embraced all the seasons and because of this he was able to gather specimens of a wide range of flowering species suitable for identification and classification. The explorers Augustus and Frank Gregory visited the area, but we have found only one specimen and that was collected by Augustus Gregory at the Murchison River.[123]

From the foregoing, it seems clear that Oldfield's contribution to the science of plant taxonomy as a collector was considerable. Although many of his specimens were used in conjunction with those of other collectors to describe new species, many descriptions of new species were based on his specimens alone. This is particularly true of his collections from Champion Bay north to the vicinity of Murchison River. Without his, together with Pemberton Walcott's and the commercial botanist James Drummond's, collecting efforts the major 19th century taxonomic floras of Australia would have been almost totally deficient in descriptions of species from that part of Western Australia.

Contribution to Australian anthropology

Oldfield's only published work was his lengthy article 'On the Aborigines of Australia' in the *Transactions of the Ethnological Society of London* (1865).[124] This paper is a scholarly work, part of which is devoted to a consideration of some prevailing theories on Aboriginal origins. Whenever he could he recruited the services of Aboriginal people to assist him with his collecting activities and while he was in the Murchison River district in 1858–1859 he lived with a *Watchandie* group, as they moved around their territory on the lower Murchison River. This afforded him opportunities to observe and record aspects of their way of life. Although he makes some reference to other Australian Aboriginal groups, the major focus of his paper is an account of his observations of the *Watchandie*. At times, he also cited pre- and post-settlement incidents involving Europeans and Aboriginal people and occasionally bio-climatological phenomena that had come to his attention. Toward the end of the paper is a description of grammatical and semantic aspects of the *Watchandie* language and a supplementary word list of the 'Champion Bay Tribe' contributed by Dr Robert T. Foley, medical practitioner at Geraldton and Medical Attendant at the Lynton Convict Depot.[125]

Norman Tindale, writing in 1974, stated that '*Watjandi* (Oldfield: *Watchandie*) is a *Nokaan* term for the *Nanda* and means westerners and [f]rom the evidence of Oldfield, the name *Watjandi* probably was the one by which they were known also to the natives of Geraldton'.[126] William Burges who was appointed the first Resident Magistrate of the Victoria District in 1851, recorded during the 1850s that

[120] An example of the use to which the plant collections of early explorers and naturalists is being undertaken jointly by the Harden Murrumburrah Landcare Group and the Centre for Plant Biodiversity Research, Canberra. One component of this project, 'Greening the Grainbelt', has been the development of a database which provides a floristic view of the original vegetation of the Shire. There is the intention to revegetate where possible various ecosystems across the district using these data as a guide, CSIRO Publishing, 'Herbarium lends landcarers the vision of Hume', *ECOS: Science for sustainability*, 108, July–September 2001, pp. 30–31, viewed 29 May 2012 at www.ecosmagazine.com.

[121] James Drummond briefly collected in the area during 1850 and early 1851 while accompanying his son, First Constable John Nichol Drummond on his patrols within that part of the Victoria District (see Chapter 11 of this work).

[122] Pemberton Walcott and Oldfield collected together from the vicinity of the Greenough River, south of Geraldton to the Murchison River in 1856–1857 (see Chapters 11 & 12).

[123] MELISR database download, 02 March 2012, NHV, RBGM.

[124] AO, 1865b, pp. 215–298.

[125] Chapter 11.

[126] Norman B. Tindale, 1974, *Aboriginal Tribes of Australia: Their Terrain, Environmental Controls, Distribution, Limits & Proper Names*, Australian University Press, Canberra, pp. 249-250.

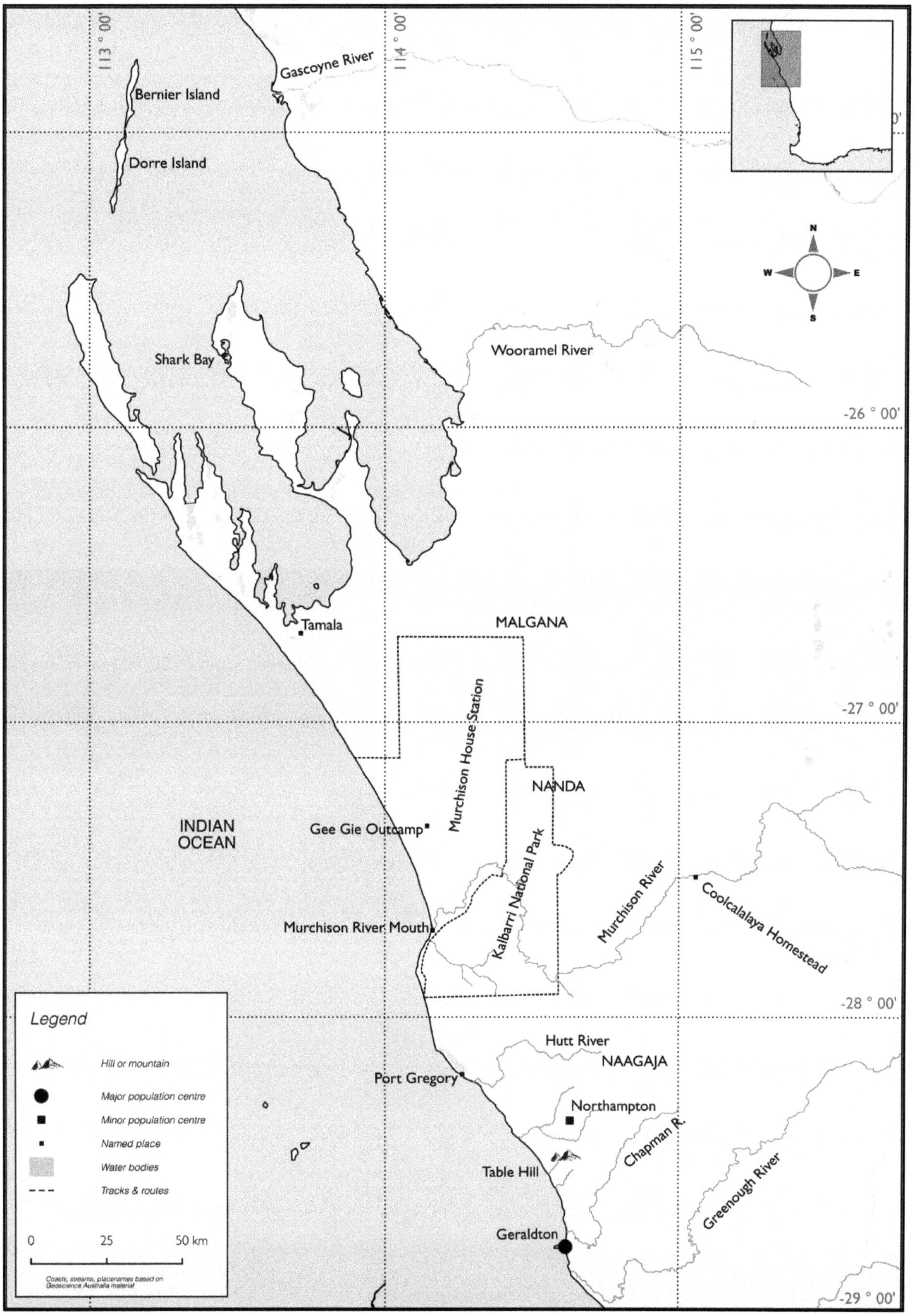

Map 1: Aboriginal tribal territories north of Geraldton to Shark Bay

the Aboriginal tribe to the north of Table Hill, approximately 10 kilometres south of Northampton (see Map 1 p. 25), were the *Waten-dé or Water-dé.*[127] If W*aten-dé* is correct perhaps it was a synonym of W*atchandie* and meant westerners. The *Watchandie* were a non-circumcising group that practiced cicatrisation of the chest at male initiation. According to Tindale's map, the western boundary of the *Nanda* territory stretches from Willigabi on the coast near Northampton northward to Shark Bay, Hamelin Pool, and Yaringa to the south of the Wooramel River and includes the islands of Shark Bay. Inland the territory varies in width from the coast from four kilometres in the south, 60 kilometres near the Murchison River to about 112 kilometres at the southern end of Shark Bay. The tribal area of the *Nokaan,* a circumcising group, lies immediately to the east of about the southern third of the *Nanda* territory and the *Malgana*, immediately to the north of the *Nokaan*. The *Malgana* did not practice circumcision.[128]

Although the location of the *Nanda* territory boundary has not found general agreement among anthropologists, it was, for a time, officially in accordance with the Western Australian Department of Indigenous Affairs Gascoyne/Murchison Tribal Boundaries Map (11 February 1999). The Murchison/Gascoyne tribal elders, who determined the tribal territories on that map, included within one boundary with no lines separating the *Malgana*, *Nanda,* and *Naagaja* (currently *Naaguja*), the latter according to Tindale, to be a synonym of *Nokaan.*[129]

According to the map, the *Nanda* language group is centrally located and embraces the Murchison River, extending to the north possibly as far as Tamala Station, inland for about 150 kilometres, extending south to the point where the southern boundary runs due west, and to the coast just to north of Hutt River, possibly including Port Gregory; the western boundary includes coastal waters to about 50 kilometres west of the mainland.[130] It appears that the *Watchandie* occupied the territory from just north of the lower Murchison River and south of east to about 100 kilometres from the coast to just east of Coolcalalaya Station. Today, Murchison House Station and the Kalbarri National Park lie within the territory of the *Watchandie.*[131] The *Malgana,* whose southern boundary is in the vicinity of Tamala Station, are now officially recognised as being a quite distinctive socio-cultural grouping to the *Nanda*.[132]

As far as can be ascertained, Oldfield was the only 19th century naturalist in Western Australia who undertook serious botanical fieldwork while living and moving around the bush for an extended period with an Aboriginal group following their traditional nomadic way of life. He could speak the *Watchandie* language. For them to tolerate his presence for such a lengthy period must be indicative of someone who had learnt to relate to them in a friendly, respectful and accommodating way. Although this speaks well of his interpersonal skills, there were times when he tested their patience and good humour.[133]

The question arises as to why he, an Englishman, undertook the most unusual course by participating in the nomadic life of Aboriginal people. Unfortunately, he shed little light on his motives. He must have been aware of Aborigines' deep knowledge of their natural surroundings. This realisation was probably a driving force. Although there is evidence that he inveigled them into his service as collectors, he would have been aware that, with suitable coaxing, they could draw his attention to ecological aspects and ethnographic usage of plants, which as a short-term visitor he might have missed. Added to this, Oldfield had an eclectic scientific interest in his surroundings. His detailed

---

[127] William Burges, 9 June 1851, CSR Inward Correspondence, vol. 218/291-293, SROWA. Difficulty in deciphering this word poses these as possible alternatives.
[128] Norman Tindale, 1974, accompanying map.
[129] ibid., pp. 249-250.
[130] Department of Indigenous Affairs Western Australia, Gascoyne/Murchison DIA Region Tribal Boundaries, determined by the Gascoyne/Murchison/ traditional tribal elders, 11 February 1999.
[131] MHH, 2005, pp. xvi–xix.
[132] Dr Philip Playford, personal communication, April 2016, HOA.
[133] MHH, 2005, p. xvi and Chapters 12 & 18.

ethnographic account of *Watchandie* language and life-ways suggests that he was aware virtually nothing had been published about these particular people and this motivated him to observe and record their customs and associated day-to-day behaviour. What better way could he achieve this than by developing their trust to the point where they allowed him to accompany them as they moved through their territory? His thirst for knowledge and perhaps scientific recognition must have helped him to overcome any challenges he might have faced regarding the experience of entering and living with a group whose way of life was so vastly different from his own.

In 2005 Hesperian Press published a reprint of Oldfield's paper and an Introduction in which there is a consideration of some of the more general theories he considered credible and of aspects of his observations of the way of life of the *Watchandie*.[134] This present discussion draws heavily upon that work. For details of the discussions leading to these proposals the reader is directed to the reprint of Oldfield's paper.[135]

Oldfield's theoretical speculations on aspects of Australian Aboriginal origins, migrations and psycho-physical makeup

Oldfield's discussion of the origins and migrations of Australian Aborigines and on aspects of their psycho-physical lives is a consideration from his perspective of the credibility of related theories that were current at the time. The theories he addressed were highly speculative. They were inferred almost entirely from observations, usually *ad hoc* in nature, of extant indigenous groups within Australia and elsewhere.[136] It is unfortunate that Oldfield only occasionally cited his published sources.

He proposed that Australia was peopled from South-East Asia and New Guinea who colonised the shores bordering the Torres Strait and from thence spread southward. He thought this likely because in his time Malay proas sometimes visited the northern shores of the continent. This was slated as 'humbug' in 1864 by a *Brisbane Courier* reader[137] although such theories were not new by the 1860s. Some had suggested entry via New Guinea and others from other points to the north of the continent through Timor. Recent systematic archaeological and oceanographical research has tended to affirm an entry from South-East Asia, which it is now thought was achieved by island hopping via Sulawesi across an island chain to New Guinea and via Timor onto the north-western shelf of the Australian mainland. In 1977, Sandra Bowdler suggested that during the Pleistocene Epoch, Australia was colonised by people adapted to a coastal way of life and that initial colonising routes were around the coasts and thence up the river systems.[138] This supports Oldfield's position.

Oldfield referred to Australian Aborigines by the Portuguese word *Alfoura* meaning 'free men' but gave no explanation as to his reasons for using it. The Portuguese assigned the name *Alfoura* to the inland tribes of the Indonesian Aru Islands that lie to the south of Irian Jaya in the Arafura Sea. 'Arafura' is derived from *Alfoura*, which suggests that Oldfield's use of the term had something to do with his belief that the peopling of Australia commenced with migrations of people of Malay origin, principally via New Guinea.[139]

One theory being debated at the time, and which Oldfield considered credible, has since been refuted; thus, he thought it probable that Australia was inhabited prior to its colonisation by the *Alfoura* from evidence of 'the existence of relics of a civilisation far higher than can be claimed by any tribes of the Malay family'. Although no source was cited, there is little doubt that this notion was derived

---

[134] ibid., pp. xii–xiii.
[135] ibid., pp. iii–lviii.
[136] ibid., p. xv.
[137] 'A Sydney Man', News and Notes, *The Brisbane Courier*, Saturday 30 April 1864, p. 3.
[138] MHH, 2005, p. xv.
[139] ibid., p. iii.

from the writings of George Grey who, in 1838, discovered and sketched details of paintings of figures and an engraved head on rocks in the Kimberley in the region of the Glenelg River.[140] In his journal, Grey opted for a cautious approach by not speculating on the origins of these images. However, later he thought them to be the work of 'Asiatic' artists and evidence that a superior race once dwelt there.[141] In the 1960s, then Curator of Anthropology and Archaeology at the Western Australian Museum, Dr Ian M. Crawford, conducted a detailed investigation of the rock art in the area visited by Grey. He found no trace of the engraved head and concluded that the other figures that Grey saw were *Wandjina* paintings of local Aboriginal religious tradition.[142]

Occasionally, Oldfield added some ideas of his own. For example, he suggested that human migration across the country may have followed the lines marking the distribution of plants. This perhaps foreshadowed recent debate over the role that ecological factors played in early migrations to and within Australia.[143]

He also discussed the issue of the survival of the *Alfoura* in the face of European colonisation and paid some attention to their intelligence and educability as far as settlers' efforts to 'civilise' them were concerned. For the most part, however, he made general claims about aspects of their culture and social organisation which he supported with empirical examples, drawn from his own observations of the Aboriginal people who had assisted his botanical endeavours, principally the *Watchandie*.

Some late 19th century anthropologists and other writers were strongly critical of one or more of Oldfield's theories.[144] Thus, for example, his assertion, along with that of Count Paul Edmund de Strzelecki, that Aboriginal women 'never' bore children to an Aboriginal man once they had born children by Europeans was discredited, with some qualification, by James Bonwick in 1870.[145] Sir John Lubbock[146] and the District Surveyor at Ballarat, Philip Chauncy[147], were sceptical of Oldfield's proposals relating to the intelligence of Aborigines. Edward M. Curr was damning in his criticism when he said, 'Mr Oldfield's paper abounds with important misstatements'.[148] On the other hand, there were others, for example, the Director of the Anthropological Institute in London and prominent anthropologist, C. Staniland Wake, who supported Oldfield's argument relating to the primitiveness of Australian Aborigines with some of Oldfield's observations about *Alfoura* behaviour and mental characteristics.[149]

There were times when Oldfield, like some of his critics and supporters, interpreted his observations of the *Alfoura* from the viewpoint of a Eurocentric Social Darwinist. For example, although deploring the situation, he claimed their extinction was inevitable. This, he said, was because they lacked the necessary degree of intelligence to adapt to colonial society, which was increasingly taking and clearing their land and thereby depleting their traditional food sources giving rise to conflicts with settlers, at times, causing fatalities, particularly among the Aborigines. This was further exacerbated by the infectious diseases the settlers introduced into the country, and to which the Aborigines had no

[140] George Grey, 1841, *Journals of Two Expeditions of Discovery in North–West and Western Australia during the Years 1837, 38, and 39*, vol. 1, T. & W. Boone, London, pp. 201–216.

[141] George Grey, 1838, 'A brief outline of the recent expedition to the north–west coast of Australia…,' *Journal of the Royal Geographical Society*, viii, pp. 454–9 cited in I.M. Crawford 1968, *The Art of the Wandjina: Aboriginal Cave Paintings In Kimberley, Western Australia*, Oxford University Press, Melbourne, p. 66.

[142] ibid., pp. 62–68.

[143] MHH, 2005, p. xv.

[144] ibid., pp. vii–viii.

[145] James Bonwick, 1870, *The Last of the Tasmanians: or The Black War of Van Diemen's Land*, Sampson Low, Son & Marston, London, facsimile edition, 1969, pp. 386–388.

[146] John Lubbock, 1st Baron Avebury 1865, *Prehistoric Times as Illustrated by Ancient Remains and the Manners and Customs of Modern Savages*, facsimile 7th edn, 1913, Anthropological Publications, Oosterhout, Netherlands, 1969, p. 442.

[147] Philip Chauncy, 1878, 'Notes and anecdotes of the Aborigines of Australia', in R. Brough Smyth, *The Aborigines of Victoria: With Notes Relating to the Habits of the Natives of Other Parts of Australia and Tasmania*, Appendix A, vol. 2, Government Printer, Melbourne, p. 257.

[148] Edward M. Curr, 1886–1887, *The Australian Race: Its Origin, Languages, Customs, Place of Landing in Australia, and the Routes by Which it Spread Itself Over that Continent*, vol. 1, Trubner & Co., London, pp. 310–311.

[149] C. Staniland Wake, 1870–1871, 'The mental characteristics of primitive man, as exemplified by the Australian Aborigines', *Journal of the Royal Anthropological Institute*, vol. 1, pp. 75,78 & 80.

resistance. His conclusions regarding the Aborigines' low intelligence are based on naïve judgements about their educability, anthropomorphic characteristics and their resistance to settlers' efforts to civilise them.[150] Unlike the explorer, George Grey, Oldfield apparently did not recognise that the Aborigines often rejected colonial society and its 'civilising' efforts for sound pragmatic reasons. Among these was their need to meet their tribal obligations and their lack of trust in the invaders who showed them little or no respect, often physically abused them and inadequately rewarded them for their services.[151]

Oldfield's approach to understanding Aboriginal society was not just influenced by Spencerian (Herbert Spencer) thought but also by the tenets of science that guided his botanical labours. It has become apparent that in his anthropological fieldwork he put into practice his avowed belief that careful observation and recording was a vital part of the heuristic process. However, some of his explanations of aspects of *Watchandie* behaviour were not based on sound application of the 'articles' of scientific naturalism. Thus, at times, he drew conclusions without considering other contributing factors that might have been influencing the behaviour which he observed. For example, occasionally these were inferred from ingenuous experiments that he conducted with the object of ascertaining his subjects' intelligence or eliciting their emotional reactions.[152] There are instances where Oldfield made sweeping moral judgements with little or no supporting evidence. For example, 'Egotists as these savages [all *Alfoura*] naturally are from their mode of existence'. Notwithstanding this he occasionally indicated the speculative nature of his theories by prefacing them with statements like: 'From these considerations, it seems reasonable to infer…'[153]

Oldfield's ethnographic record of the *Watchandie*

Oldfield's ethnographic record, which is his most important contribution to Australian anthropology, embraces aspects of *Watchandie* language, religion (beliefs and rituals), sorcery (beneficial and harmful), social organisation (male and female initiation, marriage, division of labour, relative status of men and women), camp life routine including amusements (play, fighting, songs and story-telling), the exploitation of the environment for food and water, arms and other tools shelter, clothing and body adornments and the associated technology. He gave some indication of social relationships with outsiders, primarily the people living at or towards Shark Bay, that included trade, social gatherings, and conflicts over land and revenge. He also made observations regarding the number of acres from which each *Watchandie* drew his/her subsistence and the causes of death, including introduced diseases. For the most part, Oldfield appears to have reported, without interpretive embellishment, what he observed or was told. His accounts are often lively and provide the reader with a colourful insight into *Watchandie* life as he experienced it. The paper, particularly the section on language, includes references to the ethno-geography, ethno-botany and ethno-zoology of the *Watchandie* territory.

The credibility of some of Oldfield's *Watchandie* ethnography was tested against what anthropologists have reported from other Australian Aboriginal groups and what *Nanda* descendants and others have said specifically in relation to that tribal group. As yet, however, his entire *Watchandie* ethnography has not been investigated in those terms. So far, current support for the plausibility of the record tends to be confined to language, land ownership and accessibility, subsistence technology, interment of the dead and details of some other rituals, religious beliefs, including beliefs in malign spirits of the dead and sorcery, and aspects of their social organisation, including marriage.[154]

---

[150] MHH, 2005, p. xii.
[151] ibid., p. xi–xii. See also George Grey 1841, vol. 2, Australiana Facsimile edn, 1964, pp. 367–370.
[152] MHH, 2005, pp. xi–xiv.
[153] ibid., p. xiv.
[154] ibid., pp. xix–xxxv.

Oldfield specifically rejected the view prevailing among anthropologists and others well into the 20th century that traditionally oriented Australian Aborigines were not a religious people 'although [they do] not, like most barbarous people, impersonify [their] deities by visible representations, adoring graven images the work of [their] own hands'. Oldfield considered the *Alfoura* to be a religious people because they attempted to rationalise and control, through superstitious beliefs and practices, phenomena that were inaccessible and inexplicable; a view that is now held widely by social anthropologists.[155] His notion of *Alfoura* religion seems to have accorded somewhat with Herbert Spencer's view that religious beliefs were human creations to explain the 'unknowable'.[156] Oldfield supported his claim with an account of what he had found, particularly among the *Watchandie*, of the constellation of beliefs in the supernatural, which he asserted were held by *Alfoura* to explain their 'cosmogony', the advent of catastrophic events, such as fire, flood, violent storms, disease, death and spiritual life after death. Nevertheless, more modern anthropological studies of traditional Aboriginal societies indicate that Oldfield's record on *Watchandie* religion and social organisation is far from complete.[157]

Apparently, Oldfield did not recognise that there was a secret/sacred aspect to *Alfoura* religion, probably because it was deliberately kept from him. Although he observed that there existed a special relationship between individuals and specific animals[158], there is no indication that he discovered that the *Watchandie* were part of the wider *Nanda* totemic group which, according to the early 20th century anthropologist A.R. Radcliffe-Brown, was organised around patrilineal totemic groups.[159] This is not surprising, since it was not until 1915 with the publication of Durkheim's *The Elementary Forms of the Religious Life* that the importance of totemism in Australian Aboriginal religious life began to be understood.[160] Also, Oldfield did not recognise that Aboriginal religion was, as the eminent Australian anthropologist Professor W.E.H. Stanner claimed later, 'a living faith, something quite inseparable from the pattern of everyday life and thought', the connection being so intimate that 'there is no sharp demarcation between secular and sacred life'.[161]

Oldfield was aware that laws controlled *Watchandie* social relationships but he offered no details as to how they governed consanguinity and affinity. It is now known that, although there are different arrangements between tribal groups, it is a feature of Aboriginal social organisation across the continent that the social behaviour of members of those groups is determined by the operation of complex, supernaturally sanctioned laws. These laws assign individuals to particular groupings within a tribe (for example, the clan, the horde and the religious unit) with such assignment governing their kinship relationships, and their associated economic, social and religious rights and obligations.[162] An Aboriginal person coming into a strange group is always allocated a kinship position[163] and this has been assigned to some non-indigenous men and women in other parts of the continent. Had Oldfield been given such a position he would have been taught his associated rights and obligations.[164] That his ethnographic record provides so little information about *Watchandie* kinship suggests that Oldfield was never accorded *Watchandie* membership.

---

[155] ibid. p. xxi.
[156] Herbert Spencer, 1857, 'Progress: its law and causes', *The Westminster Review*, April 1857.1862, *First Principles*, Williams & Norgate, London, England,Ch. 1.
[157] MHH, 2005, p, xxi–xxxiii.
[158] ibid. pp. xxiv–xxvi.
[159] A.R. Radcliffe–Brown, 1931, 'The social organization of Australian tribes', *The Oceania Monographs*, no. 1, The Australian National Research Council, Sydney, pp. 215–216.
[160] Emile Durkheim, 1915, *The Elementary Forms of the Religious Life*, G. Allen & Unwin, London.
[161] W.E.H. Stanner, 1965, 'Religion, totemism and symbolism', reprinted from Ronald M. Berndt (ed) *Aboriginal Man in Australia: Essays in Honour of Emeritus Professor A.P. Elkin*, Angus & Robertson, Sydney, in Max Charlesworth, Howard Morphy, Diane Bell and Kenneth Maddock (eds), 1984, *Religion in Aboriginal Australia: An Anthology*, University of Queensland Press, St Lucia, p. 145.
[162] Ronald M. & Catherine H. Berndt, 1965, *The World of the First Australians*, Ure Smith, Sydney, pp. 42–91.
[163] ibid., p. 69.
[164] see, for example, Diane Bell, 1993, *Daughters of the Dreaming*, 2nd edn, University of Minnesota Press, Minneapolis, pp. 18–39.

Recent studies have made various uses of Oldfield's paper.[165] Robert Cross, in his ethno-history of Western Australia's Midwest (1995), within which the Murchison River–Geraldton area is located, acknowledged that Oldfield's paper contained a wealth of ethnographic information concerning *Watchandie* subsistence settlement and social landscapes. Cross noted that 'many of [Oldfield's] observations seem reasonable and are congruent with other ethno-historical records'. Cross referred extensively to Oldfield's accounts of *Watchandie* subsistence technology, exploitation of the environment, corroborees, burials and social organisation.[166] In the late 1990s Rupert Gerritsen attempted to locate the sites that he termed '*Nhanda* villages'. He undertook a visit to the Midwest in his endeavour to locate these supposed sites using explorers' journals, particularly George Grey's and the writings of settlers and others including Oldfield's.[167] More recently, in 2001, the linguist Professor Juliette Blevins, a highly experienced linguist who has undertaken extensive study of the *Nanda* language, noted that Oldfield's article 'is the only source portraying *Nanda* language and customs as they might have been in pre-European contact times'.[168] In his doctoral thesis on the historical ecology of the Greenough Flats (2001), Michael O'Connor referred to plant species reported by Oldfield as being important in the diet of Aboriginal people of the west coast region.[169] Prior to this in 1994, Gerritsen claimed that there has been a Dutch influence on the *Nanda* language.[170] Among the data he employed to support his claim were several items from Oldfield's *Watchandie* language record. Professor Blevins disputes Gerritsen's claim on several grounds and concludes that there is no sign of Dutch sounds or Dutch words in the *Nanda* language.[171] Lullfitz et al. in their paper 'Human Niche Construction:[172] Noongar Evidence in Pre-colonial Southwestern Australia' have referred to Oldfield's paper in connection with his documentation of land and resource rights as well as trading among Noongar groups, in-situ management and protection of specific plant taxa, and restrictions and territorial markers.[173] According to Stephen Hopper, Oldfield's insight on Talyerak (*Eucalyptus pleurocarpa*) as a territorial marker is a long overlooked but highly significant contribution on how Noongars of the south-coast perceive their territorial limits.[174]

Oldfield's *Watchandie* language vocabulary includes lists of terms the *Watchandie* used for particular plants, animals (mammals, birds, reptiles, fish and insects) and places in their tribal area. The place names refer to descriptions of vegetation and geological formations. Some place names have survived and appear on today's maps; for a few others that are not found on maps he described where they were located. It is possible, therefore, that this contribution to ethno-botany, ethno-zoology and ethno-geography could be useful to those investigating the derivation of the common names of plants, animals and places which have survived to this day in contemporary *Nanda* and European languages.

---

[165] MHH, 2005, p. ix.

[166] Robert Cross, 1995, The Social Landscape of the Midwest Aboriginal People (1840–1860) as Reflected Through Ethnohistory, Unpublished B.A. (Hons) thesis, Centre for Archaeology, University of Western Australia, Nedlands, pp. 23 & 37.

[167] Rupert Gerritsen, 2002, Nhanda villages of the Victoria District, Western Australia, Intellectual Property Publications, Canberra, mimeo., JSBL.

[168] Juliette Blevins 2001, '*Nhanda*: an Aboriginal language of Western Australia', *Oceanic Linguistics Special Publication*, no. 30, University of Hawai'i Press, Hawai'i, p.9.

[169] M.H. O'Connor, 2001, The Historical Ecology of the Greenough Flats, Western Australia, Unpublished PhD, Curtin University of Technology, Bentley, table 7.2, pp. 226–228.

[170] Rupert Gerritsen, 1994, *And Their Ghosts May Be Heard…*, Fremantle Arts Centre Press, Fremantle, Ch. 8.

[171] Juliette Blevins, 1998, 'A Dutch influence on Nhanda? Wanyjidaga innga!' *Australian Aboriginal Studies*, no. 1, pp. 43–45.

[172] Niche construction is'the process whereby organisms, through their metabolism, their activities, and their choices, modify their own and/or each other's niches [environments]', F.J. Odling-Smee. 2003, *Niche Construction: the Neglected Process in Evolution*, Princeton University Press, Princeton, J.J. p. 419.

[173] Alison Lullfitz, Joe Dortch, Stephen D. Hopper, Carol Pettersen, Ron (Doc) Reynolds, and David Guilfoyle, 2017, 'Human niche construction; Noongar evidence in pre-colonial southwestern Australia', *Conservation and Society*, vol. 52 (2), pp. 203, 209, and 212. We are grateful to Stephen Hopper for drawing our attention to this recent paper, personal communication, email attachment, 22 December 2017, HOA.

[174] See Chapter 12.

Conclusion

Oldfield's contribution to Australian botany and anthropology was important. The credibility of much of his paper on Australian Aborigines has been recognised recently but insofar as his botanical contribution is concerned, he and his achievements have received little attention in the published historiography of Australian botany. His specimens are housed in herbaria in at least 13 countries, with most of his collections at the Royal Botanic Gardens Kew, London, the Natural History Museum, London, and the National Herbarium of Victoria at the Royal Botanic Gardens Victoria. They have been widely referred to in taxonomic publications, particularly major 19th century works authored by Joseph Hooker, George Bentham and Ferdinand Mueller in which many of his specimens, especially those from the vicinity of the Murchison River, were used to describe species new to science and to extend the ranges of others previously known; a famous example is *Nuytsia floribunda*.[175] He operated in remote areas of the Australian continent where only a few dedicated collectors of his day had created opportunities to work. In addition, he provided information on the forms and habitats of many plants from which he took samples and which were incorporated in published species descriptions, particularly by Ferdinand Mueller. Unfortunately, this information was only occasionally attributed to him. His contribution was, however, recognised eponymously in at least 31 genera, but with subsequent revisions of genera and re-identifications this number has been whittled to 21.[176]

Although Oldfield wrote at length on major issues in natural philosophy being debated in the mid-nineteenth century, and in so doing demonstrated his broad reading and deep interest in the world around him, he never published his treatise. Endersby considered that his speculations in botanical classification were typical of the associated debates of his time, 'not least because of his opposition to artificial systems and his focus on the search for affinities between plants', and that had his ideas been published, there is every chance they would have been taken seriously by colonial botanists. Joseph Maiden thought it a pity that Oldfield never published on the Australian flora. Some reasons for this have been discussed, including the possibility that Joseph Hooker discouraged him. Whatever the reason/s, it is clear, on the testimony of William Harvey[177], that Oldfield had a strong desire to be a natural history collector and traveller. Furthermore, as shown by the great effort he expended on his unpublished manuscript, he apparently had a deeply held ambition to be a natural philosopher.[178]

His more general discussion on Australian Aborigines shows that, through his broad reading of the anthropological literature and his understanding of the Australian flora, geology and climatology, his interests in the world around him went far beyond botany. Oldfield's record, with all its inadequacies, of the *Watchandies'* way of life, is possibly the most important primary historical source concerning the *Nanda* people who lived near the lower Murchison River at the onset of European settlement in the area. It is certainly the most detailed account of their socio-cultural lives from the period when colonial disruption to traditional ways of life had only just begun.

The text, written in the first person, is a colourful and animated account of Oldfield's direct observations and of what he was told by *Watchandie* people and other informants. It gives the reader some insight into the day-to-day lives of the *Watchandie* and how they coped with the vicissitudes of living in an environment that seemed harsh to Europeans. Further, written as it was by someone who was imbued with the idea of presenting accurate records of what he saw and heard, it is perhaps not surprising that recent investigators have come to accept the credibility of much of his ethnography.

[175] S. D. Hopper, 2010. '660. Nuytsia floribunda. Loranthaceae', Curtis's Botanical Magazine, vol. 26, p.343. We are grateful to Professor Hopper for providing this information, personal communication to MHH, email attachment, 17 December 2017.
[176] HAODb.
[177] See Chapter 10.
[178] See Chapter 21.

Unlike other writers of his day, he acknowledged that the *Watchandies* and other Australian Aborigines were a religious people but his description of their belief system is far from complete. He apparently did not understand that there was a secret/sacred aspect to their religious life and that there was no sharp demarcation between their secular and sacred lives. He realised that there were rules governing consanguineal and affinal relationships but there is no evidence that he had worked out the structure of their mode of social organisation. Perhaps it would have been remarkable if he had solved these intricacies since anthropologists did not begin to comprehend them until well into the 20th century.

Oldfield appears to have been unique in the early history of botanical collecting in Western Australia and possibly in Australia, in that he pursued his botanical and ethnographical endeavours, for lengthy periods while living and moving around the country with tribal Aborigines who had had little contact with Europeans. Furthermore, that his observations of *Watchandie* socio-cultural life took into account its ethno-ecological aspects were rare, if not unique among contemporary ethnographers. It is a mark of his ability to adapt to widely differing socio-cultural and environmental conditions, an ability which no doubt greatly assisted him to work effectively in remote areas and under circumstances few, if any, European scientists in the nineteenth century would have contemplated.

Bury Street, St James's showing Georgian houses, c. 1850

Part 1

Family background, birth and growing up

1765–1841

Fig. 3.0 Lasiopetalum oldfieldii

Chapter 3

Family background 1765–1820

Thirty-two-year-old William Oldfield[1], Augustus' paternal grandfather, was indicted on 3 June 1797[2] and found guilty on 1 July of keeping a public/common gaming house at 26 Oxendon Street, Westminster[3] and 'suffering persons of all descriptions to resort there, for the purpose of playing unlawful games'.[4] Not all gaming was illegal at that time.[5] Lotteries had been declared to 'be *public nuisances* in whatever way they may be drawn or arranged' and also unlawful were a number of card and dice games where wagers were made against a bank; these included hazard, ace of hearts, faro, basset and *rouge et noir*. Common gaming houses were places where unlawful games of this nature were played, or where a lawful game was so arranged that the chances of winning were in favour of the table.[6] As had happened to William, a keeper of one of these establishments was 'indictable and punishable as for a misdemeanour'.[7]

It was said of hells [unlawful gaming houses] in general, 'that they are *infernally* productive, since M. T----l---r finds that the banking business is nothing compared to these money mills, and...so many fortunes have been made from them'.[8] The location, décor and services of common gaming houses varied considerably according to the social status of the clientele for which they catered. According to an article in *Fraser's Magazine* (August 1833): 'On an average, during the last twenty years, about thirty hells were regularly open in London for the accommodation of the lowest and most vile set of hazard players'[9] and most were kept by 'prize-fighters and other desperate characters, who bullied and hectored the more timid out of their money by deciding that bets had been lost, when in fact, they had been won'. These were known as 'copper hells', many of which were located in Covent Garden.[10] The game of hazard was the principal one played at these places, which, stated one writer, was 'the most desperate and ruinous of all games'. To loosen inhibitions, bread, cheese and beer were supplied to the players, and, when called for, a glass of gin was handed to them *gratis*.[11] In contrast

---

1 Anonymous (? Juliana Oldfield), handwritten note giving birth and death details of William Oldfield, Sally (Sarah) Carruthers and of Frederic[k] Oldfield, November 1860, Pimblett Family Archive (PFA).

2 The King v. Oldfield, Westminster, 1 June 1797, *The Times*, 3 June 1797, p. 3. This hearing was before Lord Lloyd Kenyon but *The Times* report does not state the Court in which it occurred. This and all subsequent references to *The Times*, London were sourced from *The Times of London Digital Archive, 1785 –2006,* viewed at the Reid Library, University of Western Australia and will be referred to by the acronym TTDA.

3 The King v. Oldfield, Court of King's Bench (CKB), 1 July 1797, *The Times*, 3 July 1797, p. 3, TTDA.

4 The King v. Oldfield, CKB, 25 January 1798, *The Times*, 26 January 1798, p. 4, TTDA.

5 During the first half of the 19th Century gambling, 'by playing at Cards, Dice, or any other Games, and all exercises, when practiced innocently and as a recreation' were 'not offences at common law' G.H.H. Oliphant 1847, *The Law Concerning Horses, Racing Wagers and Gaming, with an Appendix Containing Recent Cases, Statutes, &c.,* T. & J.W. Johnson, Philadelphia, p. 216.

6 ibid., pp. 211 & 216.

7 ibid., p. 216.

8 Quoted from an unsighted pamphlet in John Ashton 1898, *The History of Gambling in England,* Duckworth & Co., London, microform, Reid Library, University of Western Australia, p. 108.

9 From an excerpt of an article in *Fraser's Magazine* for August 1833 quoted in John Ashton, 1898, pp.135–39.

10 Roger Munting 1996, *An Economic and Social History of Gambling in Britain and the USA*, Manchester University Press, Manchester, p. 21.

11 John Ashton, 1898, p. 135.

The Court of King's Bench, 1805
Thomas Rowlandson and Augustus Pugin

Fig. 3.1 Court of King's Bench, 1805

were the fashionable gaming houses of St James Street and Pall Mall, known as 'gold and silver hells'.[12] Perhaps the most famous of these was Crockford's. His premises in St James Street were the most palatial in London and the suppers, prepared by his famous chef, Ude, were provided free of charge with the best wines and every luxury in season.[13] All gaming houses had their strategies for avoiding indictments for offences against the gambling laws.[14] The police were far more likely to act on copper hells, which could not afford to match the bribes offered by the 'gold and silver hells'.[15]

William Oldfield's criminal record gives his occupation as labourer.[16] His affidavit when he was brought to the Court of King's Bench on 1 July 1797 to receive sentence for his guilty verdict of 1 June 1897, stated that he had for many years been in His Majesty's Navy and while so serving had lost the use of his right hand, which had curtailed his naval career. On this occasion, he pleaded not to be sent to prison since he had a mother, a wife and five children depending on him for support who would be 'in a very bad state [if his] gaming house was entirely put down'.[17] In consequence of it being represented to the court that he had left off gaming, and could support himself and family creditably, by his honest industry and labour, the Court, in its humanity, postponed passing judgment on him, on condition that he would give bail for his good conduct in future, and that he would come up, when called upon to receive the Judgment of the Court'.[18]

On 24 November 1797, bail was offered but the Counsel for the Prosecution moved for the judgment of the Court against him. This was on the ground that 'on the very first search after he was

[12] Roger Munting, 1996, p. 21.
[13] Stella Margetson, 1971, p. 67.
[14] John Ashton, 1898, pp. 135–139.
[15] Roger Munting, 1996, p. 21.
[16] Record for William Oldfield, December Sessions, 1797, *England and Wales Criminal Registers, 1791–1892*, viewed 28 January 2012 at www.ancestry.com. We are grateful to Jeff Pimblett for drawing our attention to this record.
[17] *The Times* 1 July 1797, p. 3, & 3 July 1797, p. 3, TTDA.
[18] The King v. Oldfield, CKB, 8 & 24 November 1797, *The Times*, 9 November 1797, pp. 3 & 27 November 1797, p. 3, TTDA.

discharged' he had been found 'in a gaming house between 11 and 12 o'clock at night at No 5 under the Piazzas [Covent Garden]'.[19]

On 25 January 1798, William was before the Court to receive the judgement of the Court; he was found guilty of breaking his good behaviour bond. In his defence, he explained that he had been at the gaming house to recover money owed to him by the owner. Not able to pay the fine, again he sought the mercy of the Court, this time claiming that he was affected 'with a very dangerous complaint' of his lungs 'which would endanger' his life if he were to be 'sentenced to any confinement'. The Judge was not sympathetic:

> Of that prison, to which you will probably go, it has been reported, that prisoners who had gone into in a sickly state, had after the expiration of their confinement, come out much healthier than they went there.[20]

William was sentenced to imprisonment in the House of Correction for the County of Middlesex on Clerkenwell Green for a period of one year and after the expiration of that period he had to give security for his good behaviour for the term of seven years, in his recognizance of £500.[21]

At the time of these court appearances William was married to Sarah (Sally) Carruthers.[22] The names of only four of his other dependent family members are known for certain, namely, his mother, Elizabeth Oldfield (née Newton)[23], his ten-year-old son, Frederick[24], and two daughters, Louisa Camilla[25] and Juliana Sarah.[26]

Frederick, Augustus' father, was baptised at St Mary's Rotherhithe on 25 May 1787, '6 days after he was born' on 20 May 1787 to William and Mary Oldfield.[27] Sarah Carruthers, therefore, seems to have been William's second wife and Louisa Camilla and Juliana Sarah were their children. Nothing is known of either Mary or Sarah's family backgrounds, when and where their respective marriages to William took place or the reason why his first marriage ended. William and Mary may have had two other sons, Jonathan and Henry (see below), who were still William's dependents in 1797.

Since no record could be located of William having been a naval officer[28] he was probably an ordinary seaman but details of his naval career, including the ships on which he served, remain a mystery.[29] His income as an ordinary seaman would have been meagre, less than that of a Warrant Officer's salary of £22 10s per year.[30] Under such circumstances it must have been extremely difficult for him to provide adequate shelter, food and clothing for the members of his family unless he was one of the few lucky sailors who had their incomes regularly bolstered by prize money through serving on vessels that had opportunities to capture and plunder enemy ships.[31]

William's inability to pay the fine suggests that his gaming house had not been particularly successful. If, as he argued, his imprisonment left his family in very straitened circumstances they may

---

[19] *The Times* 24 November 1797, p. 3, 27 November 1797, p. 3, TTDA. The Piazzas were within the Covent Garden area *Covent Garden Life* undated, 'Covent Garden: A brief history', viewed 23 March 2014 at www.coventgardenlife.com.

[20] *The Times* 25 January 1798, p. 4, 26 January 1798, p. 4, TTDA.

[21] ibid.

[22] Anonymous (? Juliana Oldfield), handwritten note, November 1860, (PFA). No official record of their marriage has been found.

[23] Record of baptism William Oldfield, 20 October 1765, Parish Register, St Mary Magdalen, Bermondsey, Surrey. His parents given as Jonathan and Elizabeth Oldfield, *London, England, Baptisms, Marriages and Burials, 1538–1812*, viewed 30 January 2012 at www.ancestry.com. Their marriage recorded as 30 September 1764, Parish of Mile End New Town St Olave in the Borough of Town Hamlets, *London, England, Marriages & Banns, 1754–1921*, viewed 15 December 2011, www.ancestry.com.

[24] Record of baptism Frederick Oldfield, 25 May 1787, Parish of St Mary, Rotherhithe, County of Surrey, parish register, viewed 15 December 2011 at www.ancestry.com.

[25] Record of baptism Louisa Camilla Oldfield, 1793, born 1792, Soho, Middlesex, St Anne's, Soho, England, Pallots Baptism Index for England, 1780–1837, viewed 16 June 2014, www.ancestry.com.

[26] Record of baptism of Julia Ann Sarah Oldfield, 1793, born? Soho, Middlesex, baptised St Anne's, Soho, ibid.

[27] Record of baptism Frederick Oldfield, 25 May 1787, Parish of St Mary, Rotherhithe, County of Surrey, parish register, ibid.

[28] Relevant Royal Navy service records at the National Archives London (NAL) were searched in 2002 by researcher Robert O'Hara.

[29] Tracing naval seamen prior to 1853 is mainly based upon the: (a) muster and pay books for the individual ships they served upon; (b) pension applications which include certificates of service. Warrant officers were more likely to receive pensions around 1800 than ordinary seamen, Stuart C Blank, undated, 'Royal Navy Service Records', *Military Archive Research* viewed 9 January 2012 at www.militaryarchiveresearch.com.

[30] Dudley Pope 1981, *Life in Nelson's Navy*, Chatham Publishing, London, p. 87.

[31] ibid. p. 87.

have sought and gained financial support from William's brothers, Jonathan and Henry Oldfield[32], but no evidence of this has been found.

William did not serve out his sentence at the House of Correction on Clerkenwell Green. At some stage, he was transferred to the House of Correction in Coldbath Fields. On 5 June 1800, 18 months after his release from prison, an action was brought against him by one Ballad, the cook at the Coldbath Fields prison, to recover £5 1s 8d 'for some provisions which [he] had furnished to the Defendant beyond the gaol allowance'. The case was heard at the The Mayor's and City of London Court within the precincts of the Guildhall. Lord Lloyd Kenyon, the 'Noble and Learned Judge', ruled the plaintiff nonsuited because by providing such items, which included porter, Ballad's action was unlawful and, therefore, no claim could be made against the defendant.[33] Nothing more is known of William except that he died on 6 June 1808.[34]

It was reported in *The Times*, 2 July 1806, that three Oldfield brothers had been indicted at the Bow Street Court with having 'no visible estate, calling, or profession to maintain themselves by, but gaming'. Of these young men, the names of only two, Jonathan and Henry, were mentioned in the article; the other brother was not in attendance.[35] The case did not proceed; the witness had previously been convicted of conspiracy and, therefore, his testimony was judged to be inadmissible. Frederick, who was about sixteen years of age at the time, was probably the missing brother but this could not be affirmed. Jonathan and Henry's link with gaming, their names being the same as William's brothers and Jonathan, the name of their paternal grandfather, lend some support that these young men were also William and Mary's sons.

Of Frederick and his siblings' childhoods and educational backgrounds information is scanty. If we are to believe one Mr Adolphus, defence counsel in the civil court case Oldfield v. Miller (1827), Frederick's family background would not have been conducive to his sound moral development. Thus, Adolphus contended:

> There was no reliance to be placed on the statement of such a man as the prosecutor [Frederick Oldfield] who had been reared up from his very infancy in a gambling house, where scenes of the most infamous description were daily, nay hourly, witnessed.[36]

There is the inference in Adolphus' claim that the establishment run by Frederick's father was one of the many copper hells that abounded in London's West End Quadrant toward the end of the 18th century.[37] However, this must be treated with caution coming as the allegation did from a defence counsel intent upon obtaining a favourable verdict for his client and in a climate where members of the judiciary often, if not always,

> entirely concurred with the policy which made gaming an offence, and thought it the sacred duty of the legislature, as far as possible, to suppress a practice which led to ruin and a course of folly and of crime.[38]

Frederick was both literate and numerate. His letters[39] and the nature of the business in which he was involved bear witness to this. How he obtained these skills and how far he went with his education is not known. He may have attended a common day (or 'private venture') elementary school or a local parish school. As a gaming house keeper William was probably in a position to pay the small fee

---

[32] Apart from William, his parents, Jonathan and Elizabeth Oldfield, had five children: Henry, born 10 September 1768; Jonathan, born 13 May 1770; Elizabeth, born 6 August 1773; Mary Anne, born 23 May 1775; and Margaret, born 11 February 1778, Baptisms, Parish Register, St Mary Rotherhithe (personal communication, Jeff Pimblett to MHH, email 27 December 2010, HOA.

[33] Ballad v. Oldfield, 5 June 1800, Guildhall, London, *The Times*, 6 June 1800, p.3, TTDA.

[34] Anon.? Juliana Oldfield, handwritten note, November 1860, PFA.

[35] 'Qui Tam Informations', *The Times,* London, 2 July 1806, p. 3, TTDA.

[36] Oldfield v. Miller, Surrey Sessions, 23 October 1827, *The Times*, 24 October 1827, p. 3, TTDA.

[37] John Ashton, 1898, pp.133–135.

[38] The Lord Chief Justice, The King v. Oldfield and Bennett, CKB, Westminster, 24 May 1823, *The Times*, 25 May 1823, p. 3, TTDA.

[39] See for example: Letters from Frederick Oldfield to Vincent Oldfield written in 1859, PFA.

Fig. 3.2 St Mary Abbots, the old church, c. 1810

demanded by such schools. Otherwise, a local charity school was an option.[40] At about the time William was incarcerated, Frederick was of an age to commence his secondary schooling. Unless he qualified for entry to a free grammar school or his parents had the means to send him to a fee-paying grammar school, his education may have terminated at that point.[41] If, on the other hand, his parents had required him to enter the workforce at a very early age, as commonly happened in those days, he may have attended one of the many Sunday Schools which had emerged in the 1780s as an aspect of the evangelical revival. At any one of these schools he would have received religious instruction, and some grounding in the 'Three-Rs' and perhaps, if he was lucky, in grammar, geography and other subjects as he got older.[42]

Frederick was 23-years-old when he married spinster Mary Howard, three years his junior, at the Anglican Church of St Mary Abbots, Kensington on 23 December 1810.[43] When Frederick signed their Marriage Licence Allegation, two days before they married, he swore that they were both 'twenty-one years and above'[44], although Mary had about a month to go before she came 'of age'. Frederick was given as a resident of the Parish of Kensington and Mary, the Parish of St Marylebone.

The daughter of John and Elizabeth Howard, Mary was born on 25 January 1790[45] in or close to the ancient market town of Crediton, Devon, or in nearby Sandford[46] and was christened on the same

---

[40] Anne Digby and Peter Searby, 1981, *Children, School and Society in Nineteenth-Century England*, Macmillan Press, London, p. 5.

[41] W.B. Stephens, 1998, p. 2.

[42] ibid., pp. 1–2.

[43] Certified Copy of Church Register Entry No. 361, Marriage Solemnized at Church of St Mary Abbots, Registration District: Parish of Kensington in the County of Middlesex, in the presence of Thomas Johnson, Rebecca Brown and Henry Taylor, M.Cert./Church D/13 112305, St Mary Abbott Archives.

[44] Record for Frederick Oldfield and Mary Howard, 21 December 1810, *Faculty Office Marriage Licence Allegations (1701–1850)*, viewed 29 January 2003 at www.englishorigins.com.

[45] Marian Oldfield to Mary Oldfield, 24 for 25 January 1869, files 0102–0104, PFA.

[46] Mary's birthplace is given as Crediton, Devon, *1851 England Census*, Registration district: Chelsea; Sub-registration district: Chelsea North West; Household schedule no: 35; Piece: 1473; Folio: 10; Page: 13, and Sandford, Devon, *1861 England Census*, Registration

day in Sandford.[47] Nothing is known about her parents' social background or her early life but her only two surviving letters indicate that she was educated and appears to have come from a middle-class family.[48] St Marylebone was home to the terraced houses of many members of the middle-class.[49] In what capacity she resided there—dependent daughter, house servant, governess or lodger—remains an unanswered question.

Frederick and Mary's first child, Marrian (hereafter Marian) Howard, was born on 4 January, 1812[50] in St James, Westminster[51] and was baptised on the thirty-first of that month at the Anglican St James Church Piccadilly.[52] She was followed by Frederica Juliana (otherwise referred to as 'Freddie' by her siblings), born on 30 November 1813[53] which event, she later stated, had occurred at 'St George's Pimlico, Middlesex'.[54] At the time, Pimlico was an area of public gardens and was not opened up for residential development until 1825. The first keys of houses surrounding St George's Square in Pimlico were not handed to residents until 1854.[55] Looking back over the intervening years Frederica may have mistakenly referred to her birthplace as Pimlico which was more likely to have been at St George's Hospital on Hyde Park Corner in the adjacent district of Belgravia; this could not be confirmed.[56] From the outset, 1733, this institution operated as a training institution for medical practitioners and was a general voluntary hospital.[57] Frederick and Mary's third child, another daughter, Lavinia Sarah, was born on 19 January 1815 but no record has been found of where this event took place.[58] Frederica and Lavinia were christened at St James Church Piccadilly on 15 February 1816 at which time the family were residing at '[28?] Bury Street'.[59]

Unfortunately, the nature of Frederick and Mary's relationship remains largely shrouded in mystery. What, then, has the historical record revealed about their life together in those early years of their marriage that might shed some light on the kind of social environment into which Augustus was born in 1821 and why lowly-born Frederick thought of himself as a gentleman?

Where the couple resided between the time they married and 1815 is not known for certain. They may have started out together at Frederick's residence somewhere in Kensington and shifted to Westminster prior to Marian's birth in 1812.

---

district: Wandsworth; Sub-registration district: Wandsworth, Household schedule no: 84; Piece: 372; Folio: 50; Page no: 15, viewed: 4 January 2012 at www.ancestry.com.

[47] Record of baptism of Mary Howard, 25 January 1790, The Church of Jesus Christ of the Latter-Day Saints, *Family Search International Genealogical Index*, v5.0, British Isles, Batch no.: CO52651, 1603–1837, Source Call no.: 0916931, Type: Film, Call no.: 0933498, Type Film, Sheet: 00, viewed 14 January 2006 at www.familysearch.org. An attempt was made to access this record again on 22 December 2011 but it appears to have been deleted.

[48] Mary Oldfield to Sir C. Abbott, Chief Justice of the King's Bench, undated, c.1824, Home Office: Criminal Petitions: Series 1, Item no: Ho34, Piece no: 49, NAL (copies of 2 letters kindly provided by Mr Jeff Pimblett).

[49] Francis Sheppard 1998, pp. 171–173.

[50] Record for Marrian Howard Oldfield, Register of Baptisms, Church of St James Piccadilly, Parish of St James, Westminster, 1812, microfilm 6, Parish Records, City of Westminster Archives (CWA). Marian's date of birth was given in this register, viewed 9 January 2012 at www.ancestry.com. 'Marrian' and 'Marian' are variously used in records relating to this individual. The latter will be employed in this work since it was the spelling consistently used by members of her family.

[51] Record for Marian Oldfield, *1891 England Census*, Source information: RG12/598; Registration District: Croydon; Sub–registration district: Croydon; Folio: 127; p. 29. Last viewed 9 February 2014 at www.ancestry.com.

[52] Register of Baptisms, Church of St James Piccadilly, Parish of St James, Westminster, 1812.

[53] Record for Frederica Juliana Oldfield, no. 137, 1816, microfilm 6, parish records, Register of Baptisms, St James's Church, Piccadilly, Parish of St James, Westminster, CWA.

[54] Record for Frederica Oldfield, *1871 England Census*, Source information: RG10/1641; Registration district: Epping; Sub-registration district: Harlow; Folio: 67; Household schedule no.: 65, GSU no.: 829943, p. 12, viewed 13 January 2012 at www.ancestry.com.

[55] St George's Square is a long narrow garden that lies between Grosvenor Road and Lupus Street in Pimlico, Wikipedia, 'St George's Square [Pimlico]', viewed 23 February 2012 at www.wikipedia.org.

[56] 'St George's Hospital, London', *Hospital Records Database*, NAL, viewed 4 January 2012, www.nationalarchives.gov.uk. Hospital admissions and discharges records for the 1800s are not available for this institution.

[57] Humanities Research Unit, University of Sheffield 2012, 'Hospitals', *London Lives, 1690–1800,* viewed 4 January 2012 at www.londonlives.org.

[58] Record for Lavinia Sarah Oldfield, no. 137, Register of Baptisms, St James's Church, Piccadilly, Parish of St James, Westminster, 1815, microfilm 6, parish records, CWA.

[59] ibid. and record for Frederica Juliana Oldfield, no. 136, Register of Baptisms, St James's Church, Piccadilly, Parish of St James, Westminster, 1816, microfilm 6, parish records, CWA.

Fig. 3.3 St James's Church, Piccadilly, 1814

Her baptismal record does not mention a place of residence but she later gave her birthplace as St James, Westminster.[60] Frederica and Lavinia were given as residing in 'Bury Street' in their christening records.[61] By 1819, Frederick and Mary had produced two sons—Frederick Henry Howard (hereafter Frederick Jnr), born on 30 January 1817 and christened in the Church of the Parish of St James Piccadilly on 15 July 1817[62] and Alfred Bennett, born on 6 March 1819 and baptised at the same church on 1 April 1819.[63] They too resided in 'Bury Street'. The family's residence from the time of Marian's birth was probably 28 Bury Street where it was stated to be in January 1821.[64]

Mary Smith was the occupier/lessee of 28 Bury Street from 1809 to 1821. Frederick is not recorded as the subsequent lessee.[65] The Oldfields must have lived there under a sub-tenancy arrangement. It is not known how much of the house at 28 Bury Street Frederick rented at any time during his tenancy, or whether the family and Mary Smith ever co-resided there.

60 Record for Marian H. Oldfield, *1861 England Census*.

61 Records for Frederica and Lavinia, nos. 136 & 137, Register of Baptisms, St James's Church, Piccadilly.

62 Record for Frederick Henry Howard Oldfield, no 755, Register of Baptisms, Church of St James's Piccadilly, Parish of St James Westminster, 25 January 1817, microfilm 9, parish records, CWA.

63 Record for Alfred Bennett Oldfield, 2 April 1819, no 393, Register of Baptisms, Church of St James's Piccadilly, Parish of St James Westminster, 1819, microfilm 9, parish records, CWA.

64 Record for Augustus Frederick Oldfield, no 122, Register of Baptisms, St James's Church, Piccadilly, Parish of St James, Westminster, 1821, microfilm 10, parish records, CWA.

65 Records for 28 Bury Street, Parish Records Rate Books for 1814–1822, Pall Mall Ward, Parish of St James, Westminster, CWA, L Thomas, Local Studies Librarian, C.W.A, email to MHH, 10 March 2003, HOA.

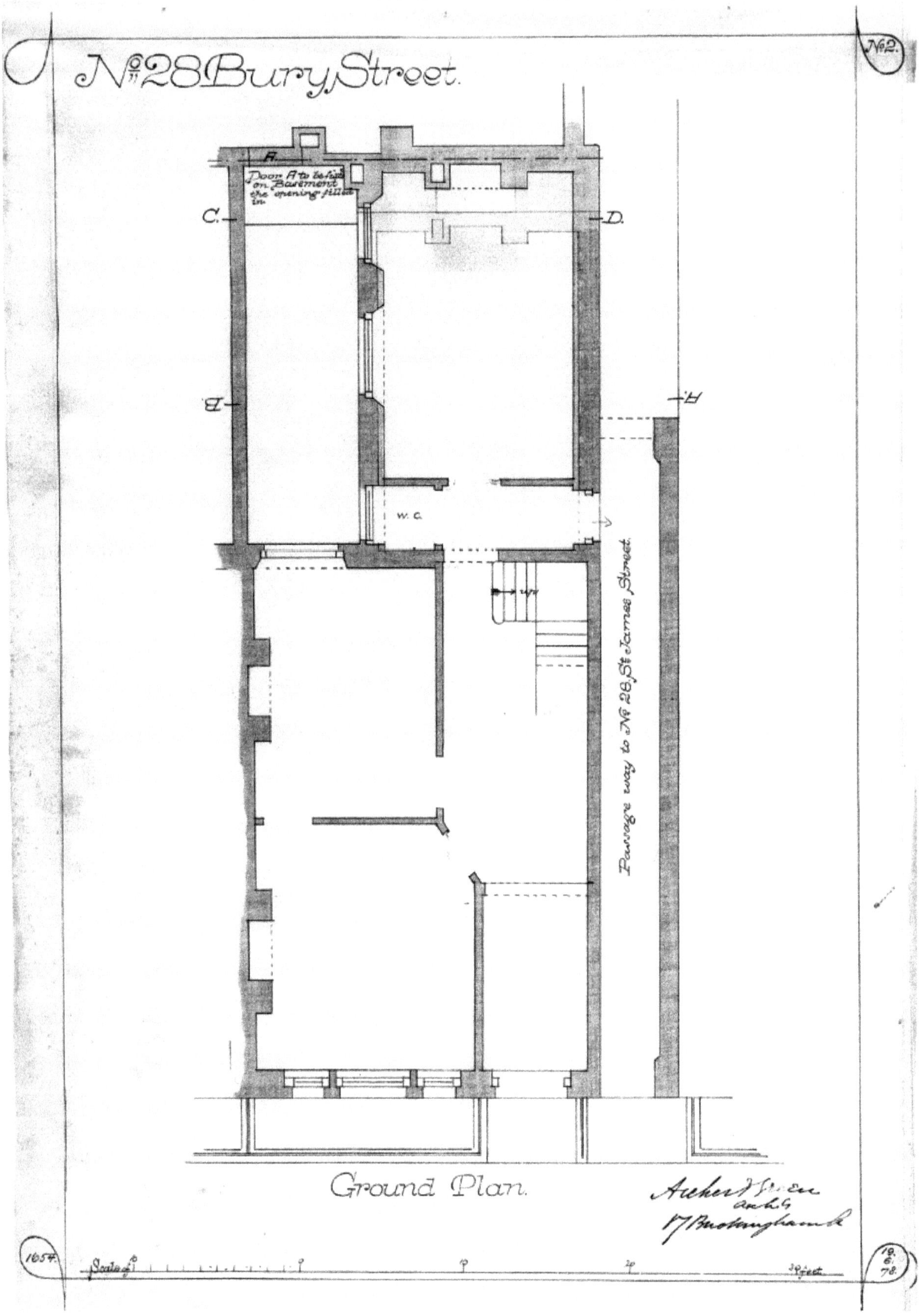

Fig. 3.4 28 Bury Street, plan for renovations, 1878

The freehold of the whole of Bury Street belonged, as it still does, to the Crown but all the existing buildings have been erected since the late 19th Century.[66] A sketch based on a photograph, c. 1880, in the Crown Estate Office shows that the buildings in Bury Street consisted for the most part of late 17th or early 18th century, single-fronted terrace-houses, generally four storeys high.[67] There is little doubt, therefore, that much the same streetscape was there when the Oldfields were in residence.

28 Bury Street was located on the west side of the road between Ryder and Jermyn Streets.[68] Renovation plans for this house, drawn up by the architects Archer & Green on 6 June 1878,[69] delineate between the building's original layout and the proposed alterations. Similar to other Georgian houses, the original building was a large gentleman's residence comprising a basement, ground floor and four storeys. In addition to stairwells, landings, water closets, of which there was one on every level, and a basement storeroom, the house had nineteen rooms, varying in number on each level from two in the basement to four on the second and third floors. A 'bath' is shown as being in the basement backroom but it is not clear whether this was part of the proposed 1878 additions. Apparently, bathrooms were not included in 18th century gentlemen's houses; it was considered dangerous to wash too much.[70]

Apart from the water closets, the plans provide no indication of the use to which each room was intended. The front basement room, the largest one in the house, may have been the kitchen since it is likely that two juxtaposed fireplaces shown on the south wall of the room were cooking stoves. The second largest was the first floor, front room that was probably a reception room. The main entrance to the building was from Bury Street and at about the centre of the ground floor, behind the stairwell on the north side, there was a doorway to the 'passage way leading to No. 28 St James Street'.[71] Although this lane way was not represented in Horwood's 1813 map, probably because it was covered by the overhang of the first and subsequent storeys of 28 Bury Street[72], it separated 27 and 28 Bury Street which Horwood's 1813[73] map of Westminster shows to be directly behind 28 St James Street, as was also the case in a later undated map Ordnance map of the area.[74]

Frederick Oldfield, gaming house keeper

Following in his father's footsteps, Frederick became a common gaming house keeper sometime between 1811 and 1813 but the location of his business during those years is not known.[75] It is not hard to imagine that Frederick's desire to become wealthy was sown early in his life, particularly if his father's absence in prison when he was about ten years old brought great hardship upon his family. As a young adult facing a lifetime of financial adversity through not possessing the means to obtain a lucrative lawful occupation, his options for achieving wealth were extremely limited. The most obvious to him, regardless of the risks, must have been to operate his own gaming establishment catering for the upper classes. The example provided by his father and uncles would have allowed him to follow that occupational path free of the hindrance of conscience.

---

[66] F.H.W. Sheppard (ed), 1960, 'Bury Street', *Survey of London: The Parish of St James Westminster, Pt 1. South of Piccadilly,* vol. 29, for the London City Council, The Athlone Press, University of London, London, p. 313 and confirmed by Andrew Thomas, Crown Estate Office, email to MHH, 15 May 2003, HOA.

[67] ibid., Fig. 55, p. 314. This is supported by a comment on the buildings on p. 313.

[68] Horwood, Richard, 1813, *The A to Z of Regency London: Plan of the Cities of London & Westminster with the Borough of Southwark',* reduced reprint, H. Margary, 1985 in association with Guildhall Library, London.

[69] Archer and Green, 19 June 1878, Architects, plan drawings for renovations at 28 Bury Street, Westminster, LRRO/2348 109860, NAL.

[70] S.E. Rasmussen (ed), 1934, *London: The Unique City,* Faber & Faber, London, 1982 edition, MIT Press, Massachusetts, p. 294.

[71] Archer and Green, Architects, 19 June 1878.

[72] ibid. and F.H.W Sheppard, (ed), 1960, 'Map of the Southern part of the Parish of St James, Westminster...' *Survey of London...*, Pocket D Index Map to vols. 29 & 30.

[73] Richard Horwood, 1813.

[74] F H W Sheppard, (ed), 1960, 'Map of the Southern part of the Parish of St James Westminster, Pocket D Index Map to vols. 29 & 30)

[75] *The Times* 23 October 1827, p. 3, 24 October 1827, p. 3, 23 May 1823, 24 May 1823, p. 3, TTDA. This puts his age at commencing the keeping of gaming houses at between 24 and 26 years.

Wealth from gaming could not be achieved without first learning how stylish hells were managed for maximising profit and avoiding the attention of the police. Moving around the gaming houses of Westminster as a very young man and observing their operations must have provided the necessary training, albeit informal, with his winnings allowing him to assume the outward trappings of a gentleman. Of course, this does not necessarily mean that he was not a loving and compassionate husband and father or that he did not take his obligations to support and protect his wife and children seriously.[76] In the chapters that follow consideration is given to the way he played out his spousal and parental roles and how this may have affected his children.

For varying periods between 1814 and 1821, when finally he gave away the 'play world'[77], Frederick, always in partnership with others, operated gaming houses at 75 St James Street, 6 Pall Mall, 71 Pall Mall as well as at 27 and 28 Bury Street.[78] By the early 1820s, his partner at 6 Pall Mall, 71 Pall Mall, and 27 - 28 Bury Street was Richard Bennett[79], who was also one of Frederick's partners at 75 St James's Street together with John Phillips and Thomas Carlos.[80] It was also alleged in 1823 that Oldfield, Bennett, Carlos and Phillips had 'branch establishments' in the country, at Brighton, Preston, Bath, and many other places.[81] This has not been confirmed. No light has been shed on the sequence of events that led to the formation of these businesses. Apparently, Frederick and Richard Bennett did not become involved in the house at 71 Pall Mall until sometime during or after 1817; 'Taylor, Phillips, Lowe and Fielder' were reported as being the proprietors in 1817.[82] At least one of the establishments, 27 Bury Street, was open for play on Sundays.[83]

The Oldfield home and all of Frederick's gaming establishments were in London's West End where gambling was well-entrenched in the way of life of the wealthy and poor, although they practised it in markedly different ways and under varying conditions. St James Street, a favoured social rendezvous of male elite society in London, had long been famous for its coffee and chocolate houses.[84] Gentlemen's clubs in and around St James Street and Pall Mall had superseded the popularity of the coffee houses of the 18th century. These clubs, of which White's, Boodle's, the predominantly Tory Cocoa Tree, and the Whig Brooks's were but a few, provided 'luxurious rendezvous for groups of gentlemen united by common tastes and social background'.[85] Here gentlemen in their skin-tight trousers, waist pinchers, full cravats and tail coats[86] gossiped, schemed, discoursed on issues of the day, drank and dined. Many of these venues, including White's and Boodle's, provided opportunities for gambling, often for high stakes.[87] Although less popular with their female counterparts than it had been in the latter part of the 18th century, gambling was still a feature of the lavish entertainments in grand houses. Elegantly coifed and gowned in the classical lines of the Empire age or in the more fashionable style of the hour-glassed silhouette and tapering 'leg-of-mutton' sleeves[88], upper-class women gambled, often for considerable amounts.[89]

The district's many other privately-operated gaming houses provided opportunities for gamblers from across the social spectrum to wager anything from a few pence to thousands of pounds on the roll

---

[76] Leonore Davidoff & Catherine Hall, 1987, pp. 329–330.
[77] The King v. Oldfield, Bennett, Carlos and Phillips, CKB, Westminster, 11 February 1823, *The Times*, 13 February 1823, p. 3, TTDA.
[78] The King v. Bennett and Oldfield, CKB, 24 January 1824, *The Times*, 26 January 1824, p. 3, TTDA.
[79] Carter v. Oldfield and Bennett, CKB, 26 October 1822, *The Times*, 28 October 1822, p. 3, TTDA. Bennett's first name is given in another court case, The King v. Oldfield and Others, Court of King's Bench, 18 December 1822, p.3, 19 December 1822, p. 3, *The Times*, London, TTDA.
[80] The King v. Oldfield, Bennett, Carlos, and Phillips, CKB, 12 February 1823, *The Times*, p. 13 February 1823, p. 4, TTDA.
[81] ibid.
[82] John Ashton, 1898, p.115.
[83] ibid., 1898, p. 117.
[84] F.H.W. Sheppard, (ed) 1960, vol. 30, p. 432.
[85] F.H.W. Sheppard, 1998, pp. 244–247.
[86] Tatiana de Fircks, 1992, *History of Costume and Style*, Pitman, Sydney, pp. 208–209.
[87] Andrew Steinmetz, 1870, *The Gaming Table: Its Votaries and Victims: In All Times and Countries, Especially in England and in France*, vol. 1, Tinsley Brothers, London, Microform, University of Western Australia, pp. 131–133.
[88] Tatiana de Fircks, 1992, pp. 209–210.
[89] Andrew Steinmetz, 1870, pp. 273–278.

of the dice or the turn of cards. While their clientele was primarily male, it was not uncommon for women to try their luck at the gaming tables.[90]

Oldfield and Bennett were conducting their gaming house at 27 Bury Street in 1817[91] and probably had being doing so since 1814 when Frederick became the official occupier and continued to be until at least 1822. Rate books for 27 Bury Street from 1814-1822 give Frederick as the lessee.[92] This establishment was included in a list of the most fashionable gaming houses in existence in 1817 but a description of the premises was not provided.[93] The list was compiled by a dissatisfied gambler who believed that:

> Upon the whole, the French first-rate gaming-houses beat our hells hollow, and they are carried on upon a much more extensive, handsome, and attractive plan.[94]

As a stylish gaming house, the proprietors of 27 Bury Street aimed to attract wealthy, genteel gamblers. Accordingly, its interior apartments would have been 'furnished with uncommon elegance' as was reputedly the case with at least two of Oldfield and Bennett's other establishments, 71 Pall Mall[95] and 75 St James Street[96], where 'every accommodation and luxury was offered which could heighten the fascination of the scene'.[97] At 75 St James Street, for example, 'Wine, spirits, and every species of refreshment, were provided *gratis* for the visitors'.[98] The décor and customer services at the other house, fashionably located under the Colonnades in Pall Mall, were probably of a similar nature.

The partners were reported on several occasions during 1819 and 1820 to be operating a gaming house at 28 Bury Street[99]; but no evidence has been found of this prior to 1819. It is perhaps no coincidence that they had businesses at 27 and 28 Bury Street; the covered passage between these buildings terminated at the rear of 28 St James Street where Boodle's gentlemen's club was located. This highly exclusive and palatial association was privately owned; the proprietors at the time Oldfield and Bennett were in Bury Street were Richard Cuddington (1811 to 1817), then Cuddington and John George Fuller (1818 to 1830).[100] The proximity of 27 and 28 Bury Street to Boodle's and the ease with which the club's members could have moved unobtrusively between each of the establishments must have served Oldfield and Bennett's business interests well. For example, perhaps gaming rules at Boodle's were like those at the Almack's Club where

> 40. ...every person playing at the new guinea table do keep fifty guineas before him.
>
> 41. ...every person playing at the twenty-guinea table do not keep less that twenty guineas before him.

If a member, for whatever reason, was not able to abide by these rules he had the option of visiting one of the many nearby gaming houses where, if considered safe, he could play after having borrowed money from the proprietors.[101] This almost certainly would have given Oldfield and Bennett a decided financial advantage over other, less opportunely placed, gaming houses in the district.

The partners apparently had an interest in another gaming establishment. There is on record an indenture of lease, dated 19 April 1821, whereby a Thomas Le Blanc of Lincolns Inn Fields did 'demise' the lease of a property in Jermyn Street to Frederick Oldfield who was stated in that docu-

---

[90] ibid., pp. 267–270.
[91] John Ashton, 1898, p. 117.
[92] Records for 27 Bury Street, Parish Records Rate Books for 1814–1822, Pall Mall Ward, Parish of St James, Westminster, CWA, L Thomas, Local Studies Librarian, C.W.A, to MHH, email 10 March 2003, HOA.
[93] John Ashton, 1898, p. 117.
[94] ibid.
[95] *The Times*, 26 October 1822, p. 3, 28 October 1822, p. 3, TTDA.
[96] *The Times*, 18 December 1822, p.3, 19 December 1822, p. 3, TTDA.
[97] *The Times*, 24 May 1823, p. 3, TTDA.
[98] *The Times*, 12 February 1823, p. 4, TTDA.
[99] The King vs. Oldfield, CKB, 3 July 1820, *The Times*, p. 3; Carter v. Oldfield and Bennett, CKB, *The Times* 27 October 1822, p.3, 28 October 1822, p. 3, and The King v. Oldfield and Bennett, CKB 24 May 1823, p. 3, *The Times*, 25 May 1823, p. 3, TTDA.
[100] F.H.W. Sheppard, (ed), 1960, vol. 30, p. 443.
[101] John Ashton, 1898, pp. 144–146.

ment as 'then in occupation' of the premises.[102] Although the street number is not given, its location is fully described relative to other adjacent buildings. From this, we deduced that its address was 60 Jermyn Street and that the northerly part of its western wall abutted onto part of the rear wall of White's Club.[103] According to Captain Rees Howell Gronow, who wrote of this very exclusive club at 37-38 St James Street as it was in 1814:

> White's is decidedly the most difficult of entry: its list of members comprised nearly all the noble names of Great Britain...It was here that play was carried on to an extent which made many ravages in large fortunes, the traces of which have not disappeared at the present day.[104]

Given the strategic location of the building at 60 Jermyn Street in relation to White's and that the writer who listed London's fashionable hells in 1817 referred to 'Bennet [stet.] of Jermyn Street'[105], it is supposed that Bennett and Oldfield also operated a house at that address under some kind of tenancy arrangement. For reasons that will be explained presently, Oldfield started to divest himself of his gaming houses in 1821. It was probably as part of this process that on the very same day he finalised the indenture of lease with Thomas Le Blanc, he entered into another signing over the lease of the property to one George Elphick, a breeches maker of St James Westminster[106], perhaps making a sizeable profit on the deal.

Admission to at least some of the establishments in which Oldfield had an interest was dependent upon gamblers meeting certain criteria; these appear to have varied from house to house according to clients' perceived socio-economic status and, no doubt, whether they could be trusted not to inform the police. Thus, a Lieutenant Rennie stated of 71 Pall Mall:

> No-one could go to the house in question unless he had the appearance of a gentleman. Mr Grant never had that appearance, he was always shabby. The other houses where I saw him allowed play from half a crown up to £30 or £40; but at 71, Pall-mall [sic], the play was from 5s. to £100. [107]

Not all of Oldfield's gaming houses catered only for the upper classes. A Mr Woodroffe was reported to have observed that at 75 St James Street:

> Play was carried on in these rooms to a great extent, and was shared by young men and lads, apparently the clerks of merchants and tradesmen. Mr Woodroffe remembered one of these 'stung to madness by his losses, declaring that the money he had lost was the property of his employers'.[108]

There is no certainty as to the exact location of the gaming house at 75 St James Street. An 1839 profile of the west side of that street suggests, but does not confirm, that it was within a four-storey building situated on the southerly corner of St James and a laneway. However, the sequential street numbers shown in that profile include only Nos. 73 and 76 on the south and north corners of the laneway respectively. The lane was too narrow for there to have ever been two buildings. The 1839 illustration of the facade of No. 76 shows two entrances, both opening onto St James Street. In 1839 the building may have been under a single tenancy but sometime prior to that it may have been occupied by two tenants, with No. 75 being on the corner of the laneway. [109]

---

[102] Middlesex Deeds Registers MDR 1821/3/634, 19 April 1821, LMA.

[103] cf. 'Ordnance Map of the Southern Part of the Parish of St. James, Westminster' in F.H.W. Sheppard, (ed) 1960, Plan Pocket D, Index Map, vols. 29 & 30.

[104] R.H. Gronow, 1889, *The Reminiscences and Recollections of Captain Gronow: Being Anecdotes of the Camp, Court, Clubs, and Society, 1810–1860*, Scribner & Welford, New York, 1969 edition, abridged, Bodley Head, London, 1969, p. 58.

[105] John Ashton, 1898, p. 116.

[106] Middlesex Deeds Registers, MDR1821/3/635, 19 April 1821, LMA.

[107] *The Times* 24 May 1823, p. 325 May 1823, p. 3, TTDA.

[108] *The Times* 12 February 1823, p. 4, 13 February 1823, p. 4, TTDA.

[109] F.H.W. Sheppard (ed), 1960, Pocket Drawing C to vols. 29 & 30.

71 Pall Mall was situated on the south side of Pall Mall and about five buildings west of St James Palace.[110] This house had been in the possession of Sir Edward Walpole, second son of Sir Robert Walpole, from 1740 until 1778, when he was succeeded by Sir Edward's natural daughter Laura, widow of the Hon. Frederick Keppel, Bishop of Exeter. She and subsequently her son, Frederick Keppel, remained there until 1830.[111] This means that either Laura's, or her son's, occupancy was contemporaneous with the gaming house operated at that address by Oldfield and Bennett from at least 1820[112] until 1823.[113]

The establishment at 6 Pall Mall was operated in one of nine separate apartments opening, either directly or indirectly, onto the Doric colonnades that ran along the southerly wall of the now demolished Haymarket Opera Theatre. This building was situated on the north-easterly corner of the Haymarket and Pall Mall. Oldfield and Bennett's occupancy probably commenced soon after the improvements to the theatre's interior and exterior (designed by John Nash and George Repton) were completed c.1818.

The partners were tenants of a Thomas Holloway[114] who held the long-term lease of the entire building from 1815 with expiration due in 1912.[115]

Like other common gaming houses in London, the entrances to each of the establishments in which Oldfield was a partner 'were strongly barricaded, like the entrance of a prison'.[116] The door at 71 Pall Mall, for example, was heavily guarded. The door was secured with chains[117] and there was 'a hole in it with a small glass, through which the porter peeped, and admitted no faces but such as were familiar to him. When any doubt existed, the door was opened with the chain up'.[118] There was an inner door with a round window, through which another porter watched and observed the parties entering.[119]

The players at these houses were often intoxicated, but the proprietors were sober at all times.[120] *Rouge et noir* and hazard were the games on offer at their establishments. At No. 71 *Rouge et noir* was 'played at a large oblong table, covered with a cloth marked in red and black partitions, with six packs of cards'.[121] The game was described by one Mr Rotch:

> The players staked their money on which colour they pleased from 5s. to £100; the cards were turned up successively for each colour, and that won for which the pips on the cards most nearly approached 31. The bank paid all [its] losings and received all [its] winnings. When both colours were 31, the bank called an *après*, and won half the stakes. I have seen £1,200 on the table at once, £600 of which the defendants have received from an *après*. These *après* happened often in the course of an evening.[122]

The proprietors, always the bankers, reputedly took steps to minimise or avoid losses. Thus, one of their clients, a Mr Abitbol, observed:

> Sometimes when the bank was broken, a fresh supply of notes was procured; at other times, the bankers ceased to play, and they often ceased to play when they were winners, before the hour. Mr Oldfield often subtracted large notes from the bank, and re-

110 ibid.
111 ibid., vol. 30, 1960, pp. 378–9 and vols. 29 & 30, Pall Mall, Pocket Drawing B. This building was one of those that were later demolished to make way for the Oxford and Cambridge University Club.
112 The King v. Oldfield, CKB, 3 July 1820, *The Times*, 4 July 1820, p. 3, TTDA.
113 *The Times* 24 November 1823, p. 3, London, 25 November 1823, p. 3. TTDA.
114 This may have been Thomas Holloway the English philanthropist and manufacturer, Magnus Magnusson, (ed), 1990, *Chambers Biographical Dictionary*, 5th edn, Chambers, New York, p. 725.
115 F.H.W. Sheppard (ed), 1960, vol. 30, p. 241–243. Similar in design to the Nash-Repton Doric colonnades facing Pall Mall, the later-built Royal Opera Arcade, along the east wall of the Theatre and facing Market Lane is the only part of this building to survive to this day.
116 *The Times* 18 December 1822, p. 3, 19 December 1822, p. 3, TTDA.
117 *The Times* 24 May 1823, p. 3, 26 May 1823, p. 3, TTDA.
118 The Times, 3 July 1820, p. 3, 4 July 1823, p. 3, TTDA.
119 The King v. Oldfield and Bennett, CKB, 24 May 1823, *The Times*, 25 May 1823, p. 3, TTDA.
120 The King v. Oldfield and others, CKB, 18 December 1822, *The Times*, 19 December 1822, p. 3, TTDA.
121 *The Times*, 25 May 1823, p. 3, TTDA.
122 ibid.

placed them with small ones, to enable him to secure something as winner, which often drew loud expostulations of the witness.[123]

Despite their best efforts to avoid detection, Frederick was indicted on the Saturday prior to 24 October 1814 at the Bow Street Court for keeping an illegal gaming house in Bury Street, St James.[124] Frederick, who was about 27 years of age at the time, failed to appear in Court as required on 22 November 1814 due to an indisposition. The magistrate Mr Birnie, on receiving 'certificates from Dr Hooper and Mr Brand', adjourned the case; the outcome of this indictment is not known. Had he been found guilty, and he probably was, he would have been liable for a penalty of £100.[125]

There is no indication in *The Times* that Frederick was indicted for any other offences until almost a year before Augustus was born. On 'Hilary' (13 January) 1820 Oldfield and Bennett were indicted for certain 'nuisances and misdemeanours' related to keeping gaming houses.[126] On the information of the gambler, John Cox, Frederick was found guilty on 3 July 1820, at the Court of King's Bench, Westminster, of keeping two common gaming houses, one at 28 Bury Street, St. James and the other at 71 Pall Mall.[127] In the Michaelmas term of 1820, Frederick, along with Richard Bennett, was again indicted for certain nuisances and misdemeanours, this time on the information of a gambler, Thomas Erskine Grant. It would not be until 24 January 1824 that the defendants would be brought into court for sentencing on these judgements.[128]

By 1820 social acceptance of the gambling craze that had gripped England from about 1777 began to wane. There was a widespread belief, particularly among the middle-classes, that common gaming houses promoted cheating and other corrupt practices, and incited idleness and avaricious ways of gaining property and were also apt 'to draw together great numbers of disorderly persons'.[129] Gambling on the part of men and women also led to family breakdown, child neglect and abject poverty.[130] Members of the clergy railed at their congregations about the evils of gambling, not the least of which, according to some, was that it kept 'men from the public furtherance of religion' on the Sabbath. These sentiments were supported by a growing number of industrial employers since gambling, as well as drink, was judged to be in conflict with the work ethic. From 1802 the Society for the Suppression of Vice campaigned against Sabbath-breaking, licentious publications, private theatricals, fairs, brothels, dram houses, illegal lotteries and gaming houses.[131]

The question of why gaming was so popular among both the rich and the poor at that time does not appear to have attracted much attention from historians after the 19th century. J.H. Plumb has explained it in terms of a response to the uncertainty of survival: 'In the midst of death, the people sought palliatives and found them in drink, gambling and violence.'[132] While this may partly explain the phenomenon, there are other factors that should be considered: gambling by the rich as a diversion from boredom or as a way of increasing their wealth, the poor seeing it as means of becoming wealthy or at least releasing them from the grip of grinding poverty. For some it had become an addiction that saw the wealthy lose their fortunes and the poor driven deeper into poverty.

[123] *The Times*, 26 October 1822, p. 3, 28 October 1822, p. 3, TTDA.
[124] 'Gaming Houses', Bow Street Court, *The Times*, 24 October 1814, p. 3, TTDA.
[125] ibid.
[126] *King's (Queen's) Bench Prison and Queen's Prison Commitment Books 1719–1862*, PRO Pris4 v. 35, November 1822 – November 1823, 24 January 1824, NAL.
[127] ibid. and *The Times* 3 July 1820, p. 3, 4 July 1820, p. 3TTDA.
[128] The King v. Bennett and Oldfield, *The Times*, p. 3, TTDA.
[129] G.H.H. Oliphant, 1847, pp. 209–210.
[129] ibid.
[130] Andrew Steinmetz, 1870, pp. 112–115.
[131] Roger Munting, 1996, pp. 21–22.
[132] J.H. Plumb, quoted in ibid., p. 21.

Where did Frederick Oldfield's family fit into English society?

One of the baptismal records of Augustus's older siblings gives their father's occupation as 'Gentleman'.[133] Consistent with what is known of his father's occupational background, Frederick's stated genteel status was assumed rather than derived from his family of origin and this was recognised on two 1821 indentures of lease, where he was referred to as 'ass^{d} Genl^{n}'.[134]

Did Frederick satisfy contemporary requirements for social acceptance as a gentleman? By 1823 he had a reputation for being 'possessed of great wealth' and living 'in a costly style'.[135] There is little doubt, given the dress code enforced at his more stylish establishments that he was attired in the manner of an upper-class wealthy gentleman and that Mary, too, was fashionably clad, and their children clothed in accordance with their counterparts in the social class to which their father aspired. Would this have been enough? Probably not. Aristocrats still clung to the old association of gentility with an income and style of life requiring neither mental nor manual labour regardless of how they behaved in terms of their code of conduct:

> The typical vices of the Georgian gentleman—gambling, drunkenness, sexual indulgence—were private and confined chiefly to his own class, 'whereas his influence for good was public, and spread to all his neighbours.

Even his vices were less vicious because 'they were indulged in with decorum and good taste thus, according to some, 'losing half their evil by losing all their grossness'.[136] With the growth of the middle class and its culture of sobriety this old view of gentlemanliness gave way to one that was about fulfilling expectations that conferred the epithets 'genteel' and 'respectable'[137] and above the mercenary money-grabbing pursuits of the commercial classes.[138]

Frederick's occupation certainly did not fall within either of these realms. The nature of his business located him squarely among the most mercenary in London. The illegal nature of his operations and the prevailing anti-gaming sentiments being widely and vehemently expressed among the emerging middle-classes is unlikely to have earned him an invitation to any of their social events.

Middle-class social recognition as a gentleman almost certainly escaped Frederick if one observer's characterisation of him in 1817 was anywhere near correct:

> Oldfield is not a well-proportioned man. He has red hair, and soon betrays his dunghill origin. He is a pragmatical, bloated, officious, flippant coxcomb, with the *tout-ensemble* of a waiter.[139]

Then there is the matter of whether he abided by the upper-class gentlemanly code of honour or the middle-class notion that a genteel person is honest in all his dealings with others. Frederick apparently had no compunction about dishonestly reducing the value of banks in his gaming houses when it suited him. The resulting cries of outrage from his genteel customers suggest that he would not have earned the epithet of 'gentleman' in those quarters.

Middle-class gentlemanliness demanded a Christian way of life with church or chapel attendance and familiarity with religious texts. Records of Mary and Frederick's marriage and their children's baptisms bear testimony to their allegiance with the Anglican Church but this does not mean that the couple was religious or that they attended church regularly. Operating his gaming houses on Sundays indicates that Frederick was by no means a devout Christian. For the sake of respectability, however, perhaps he occasionally appeared at Sunday services.

---

133 Record for Frederick Henry Howard Oldfield, no. 755, Register of Baptisms, Church of St James's Piccadilly, 1817.
134 Middlesex Deeds Registers MDR 1821/3/634, 19 April 1821; MDR 1821/3/635, 19 April 1821.
135 *The Times*, 13 February 1823, p. 4, TTDA.
136 Harold Perkin, 1972, p. 275.
137 Leonore Davidoff & Catherine Hall, 1987, p. 73.
138 Harold Perkin, 1972, p. 275.
139 John Ashton, 1898, p. 117.

His involvement all his life in 'the play world' in and around St James must have meant that as a young man he socialised with some of the upper-class bucks that frequented hells in the district. Later appeals for his release from prison indicate that he and Mary had respectable contacts who were willing to speak on his behalf.[140] Aside from his family, Frederick's intimates were probably largely confined to his business partners, their wives and perhaps some other men and women whom he befriended through his gaming houses. His association with his partner Richard Bennett, of whom one writer said, 'is tall and robust, with black hair and eyes, and a rather blue beard' [141], seems to have been one of a trusted friend, at least prior to their imprisonment in the 1820s. This is borne out by Frederick and Mary naming their second son Bennett.

Throughout the fifty years of her marriage, Mary remains largely a shadowy figure in the historical record. However, it will become apparent in the next chapter that she considered Frederick to be an affectionate husband and father, and that she was not backward in coming forward to do all she could to secure his release from prison. If, at the time of their marriage, Frederick was the 'not well-proportioned', 'bloated' man portrayed in the only description of him to have surfaced, how he attracted such a young woman is hard to comprehend, unless, of course, his appearance had changed radically by 1817. As a twenty-three-year-old bridegroom in love, he may have presented himself to Mary as an amiable, witty and attractive gentleman able to provide her with a comfortable and stylish way of life. As those early years rolled by, however, the ravages of excessive indulgence in food and drink, apparently always outside the hours of play[142], could have taken its toll on his physique. The very negative image of Frederick painted by the unknown, angry author of extremely disparaging comparisons of London's upper-class hells with their counterparts in France must be treated cautiously in the absence of other descriptions.

Mary must have known of Frederick's strong involvement with gaming before they married and that from the beginning their family income would be derived from this type of occupation. If that was the case, she probably was aware that she and their children could experience social stigma and perhaps rejection from some quarters. Sources have disclosed very little of the nature of Mary's relationships with her own and Frederick's immediate families. If, as seems to have been the case, she had a middle-class background, regardless of how dear her husband was to her and how much she had believed at the outset of their marriage that she would be able to cope with social stigma, any rejection by much loved relatives and friends or by others whom she hoped to cultivate as intimates, meant that her feelings of loneliness and unhappiness may have been considerable. On the other hand, this may not have occurred since there is some evidence that some respected her apparently and well-connected members of the middle-class and received support from her mother during trying times.[143]

Frederick's assumption of gentlemanly status meant that adopting a stylish lifestyle was of paramount importance to him, since 'in no part of the world [was] genteel visiting founded on esteem, in the absence of suitable furniture and complete dinner-service'.[144] Furthermore, it is likely that Mary befitted his social ambitions and as such was educated with all the attributes of a gentlewoman. She would have kept well away from becoming involved in the day-to-day operations of the gaming establishments as to do so branded her unladylike.[145] Regardless of her social background and aspirations and any acts of rejection she experienced, she must have found it most agreeable to live in comfortable and fashionable circumstances. Nevertheless, the demands of the gaming houses on her husband's time may have been somewhat irksome to her and she may have insisted on restricted hours

[140] See Chapter 4.
[141] John Ashton, 1898, p. 117.
[142] 'The players were often intoxicated, but the masters of the house [Oldfield and Bennett] were always sober', King v. Oldfield and Others, CKB, 18 December 1822, *The Times*, London, 19 December 1822, p. 3, TTDA.
[143] See Chapter 5.
[144] George Eliot, 1994, *Middlemarch*, Wordsworth Classics edition, Wordsworth Editions Limited, Ware, Hertfordshire, p. 191.
[145] Catherine Hall, 1992, *White, Male and Middle Class: Explorations in Feminism and History*, Polity Press, Oxford, p. 110.

at 28 Bury Street—the hours at 28 Bury Street were from 'three to six [pm]' whereas at 71 Pall Mall they were 'from nine until two in the morning' and often prolonged after that hour.[146]

Frederick, it seems, was a caring and apparently loving parent. By keeping their residence at 28 Bury Street separate from his businesses until about 1820, and after that by confining the hours of play at that address from three pm until six pm he possibly sought to protect the family's privacy and separate his domestic and business affairs. Of course, this may have been Mary's doing by successfully acting to prevent her husband from opening up the house to much longer hours of play. The location of their residence meant that he was near his other gaming houses in Westminster, particularly 27 Bury Street, and in a position to become readily involved in the lives of his wife and children; whether or not he did remains an unanswered question.

Letters written from the 1850s to 1870s show that all their daughters as older adults were deeply religious.[147] Perhaps this stemmed from Mary's beliefs, practices and teachings. If, as we believe she was, she was a devout Christian she must have come to terms with the conflict between her life as Frederick's wife and what the Church taught regarding gambling, particularly with his gaming houses being open for business on the Sabbath. Nevertheless, church attendance and piety may have afforded Mary personal comfort and security in the unstable and unsafe world in which her husband's occupation had placed her and their children. It is unlikely she found acceptance among middle-class parishioners once she became known as the wife of the proprietor of a hell. Furthermore, some of the congregation who were established members of the gentry may have condemned her for betraying her gentility.[148] However, given the number of men and women who were gaming house keepers and gamblers around St James perhaps she found some satisfying social relationships in that quarter.

While the historical record suggests that by 1820 the Oldfields were enjoying a lifestyle only considerable wealth could provide, their hold on it was tenuous due to the illegal nature of Frederick's income earning occupation. As for his claim to be a gentleman, there is considerable doubt that it was a quality in him that was recognised by those of his associates, who were firmly entrenched in the upper- and middle-classes, howsoever they respectively defined it. Neither Frederick, Mary nor any of their children left any first-hand details of how, if at all, this apparent incongruence affected them.

[146] *The Times*, 26 October 1822, p. 328, October 1822, p. 3, 23 May 1823, 24 May 1823, p. 3, TTDA.
[147] See Chapter 22.
[148] Catherine Hall, 1992, p.110.

Fig. 4.0: Scaevola oldfieldii

Chapter 4

Birth and early childhood 1821–1829

Augustus Frederick was safely delivered on 12 January 1821.[1] His entry into the world took place in St James, London, probably at their family home, 28 Bury Street.[2] Perhaps the ministrations of a trusted, local medical practitioner or man-midwife were preferred to those of a female midwife; increasing numbers of genteel women were choosing male *accoucheurs* as they 'could deliver a living child in an obstructed delivery'.[3] If so and in common with many husbands of the time, it is unlikely that Frederick would have raised any objections[4] or that he was at his wife's bedside during the birth.[5]

Frederick's wealth almost certainly ensured that the family had several domestic servants to carry on as usual with the cooking, cleaning and caring for the other children during Mary's confinement, as well as assisting with the care of baby Augustus during his mother's approximately four-week lying-in period. During that time, her favoured female kin and friends may have been relied upon for comfort and counsel.[6] To what extent this was wanted or needed may have depended on how difficult parturition had been, whether Mary and Augustus suffered any setbacks during the post-natal period and to what extent Frederick was meeting her emotional needs at the time.

From what is known of Mary's concern for the welfare of her children, she probably breastfed her babies. Perhaps, like many other British women, she was dissatisfied with the alternatives to mother's own breast-milk. Human milk provided by a diseased wet nurse or milk from non-human sources such as a cow, goat, sheep, ass or horse had proved risky to infants' health and even fatal in many instances.[7] The advice offered in medical texts and child care manuals promoting mother's milk as the safest and most nutritious food for the infant may have been a sufficient reason for Mary to choose to breastfeed her babies.[8] However, other influences could have been at work. Concern for her own reputation might have been a factor; there was at the time a widely and long-held view that it was a mother's 'natural duty' to breastfeed her babies.[9] Belief in the contraceptive effect of lactation, possi-

[1] Record for Augustus Frederick Oldfield, Register of Baptisms Solemnised at St James's Church, Piccadilly, Parish of St James, Westminster, 1821, no. 122, microfilm 10, Parish Records, WCA. Some authors have erroneously stated his date of birth to have been 12 January 1820. For example, J.H. Maiden, 1909, 'Records of Tasmanian botanists', *Papers and Proceedings of the Royal Society of Tasmania for the year 1909*, p. 24 and R. Desmond, *Dictionary of British and Irish botanists and horticulturists including plant collectors and botanical artists*, Taylor & Francis, London: 1977, p. 63.

[2] ibid.

[3] Amanda Vickery, 1998, pp. 94–101.

[4] ibid., pp. 101–102.

[5] Tim Blackshaw, 2003, 'Fathers and childbirth' in Caroline Squire (ed), *The Social Context of Birth*, Radcliffe Publishing, Oxford, pp. 215–219.

[6] Amanda Vickery, 1998, pp. 94–95.

[7] If lactation was problematic, a woman had no option but to accept the risks associated with artificial feeding or employing a wet nurse. However, if such a course was taken it was usual for middle-class couples to make every effort to obtain the services of a healthy wet nurse and ensure that she did not mistreat their offspring, Valerie Fildes, 1988, *Wet Nursing, A History from Antiquity to the Present*, Basil Blackwell, Oxford, pp. 191–196.

[8] ibid., 1988, p. 191.

[9] Amanda Vickery, 1998, p. 107.

bly reinforced by the conservative prohibition against sexual intercourse while breastfeeding also could have been powerful driving forces.[10]

If Mary were in accord with contemporary family management guides popular with middle-class mothers, she would have had little or no time 'for the pleasures and [social] activities of spinsterhood or the honeymoon years' in which many wealthy women were wont to indulge, and leave their children's upbringing in the hands of a nurse.[11] Preferring to devote her days to amusing, educating and disciplining the very young children 'in such a manner as to generate that warm and steady affection which forms the truest joys of married life'[12] and by so doing aim to provide for her children a 'nursery of virtue'.[13] Plausibly the very high regard and affection her adult daughters had for their mother[14] reflected a situation where she did play an important and loving role in her children's upbringing. Nevertheless, as a well-to-do mother she probably engaged a nurse to take charge of the infant Augustus for much of the time, tend her when she was breast-feeding and assist her in the management of all the children.[15] A nursery maid may have performed 'menial tasks such as cleaning, washing, making children's meals and sewing'.[16] These servants very likely 'lived in' with their bedrooms on the top floor of the house.

Mary and Frederick may have adhered to contemporary advice and expected any nurse in their employ to monitor the children's health and report to them signs of sickness immediately, and take appropriate safety precautions when bathing the children to make sure they did not suffer over-exposure to the cold, heat or to the water itself. Another requirement was to 'assist nature' in bringing the children's bodies 'to a proper shape' by ensuring they had plenty of exercise and 'the utmost freedom of motion and expansion' by allowing them to develop their motor skills naturally without the aid of contrivances. She was to ensure that her charges had bland, nourishing food at body temperature and encourage sleep through gentle motion, soft lullabies and comfortable, well-aerated cots. Also, she was obliged to avoid totally the giving of 'various things of a stupefactive quality' to encourage sleep, make sure that her charges did not suffer much from 'the rapid variations of the English climate', and teach them the dangers of fire and scalding hot water. Although it was Mary's duty to improve and cultivate her children's minds, the nurse

> must always speak truth to children, always perform her promise to them, and never terrify them with idle fears. She should never teach them to resent injuries, or to be proud of their clothes. She should neither be churlish, nor too fond, but always treat them with that cheerfulness and kindness, which they will soon learn to imitate.[17]

Augustus was baptised on 9 February 1821. Like his elder siblings, he received this sacrament at the elegant Grinling Gibbons (1686) font[18] in the Anglican Church of St James, Piccadilly, Westminster[19], that

> barbarous brick-cased and ill-shapen pile which stands on the South side of Piccadilly...[enclosing] one of the choicest and most elegantly formed interiors...which dis-

[10] ibid., p. 109.
[11] ibid., p. 110.
[12] See for example, Watkin Poole (ed), 1826, pp. 511–541.
[13] Leonore Davidoff & Catherine Hall, 1987, pp. 149–192.
[14] See Chapter 22.
[15] Watkin Poole, (ed), 1826, *The Female's Best Friend; or, The Young Woman's Guide to Virtue, Economy, and Happiness: Containing A Complete Modern System of Cookery Formed upon Principles of Economy for Private Families*, New Edition, J. Cleave & Son, Manchester, England, pp. 84–94.
[16] Leonore Davidoff & Catherine Hall, 1987, p. 110.
[17] Watkin Poole (ed), 1826, pp. 84–94.
[18] This font has stood in the church since 1686. Its shaft 'represents the Tree of Life, with a serpent coiled around it. The figures of Adam and Eve stand on either side. On the bowl are three bas-reliefs showing the Ark of Noah with the dove bearing an olive-branch, the Baptism of our Lord, and the Baptism of the Treasurer of Candace by St Philip', *St James's Church Piccadilly Guide*, pamphlet, undated.
[19] Record for Augustus Frederick Oldfield, Register of Baptisms Solemnised at St James's Church, Piccadilly, 1821.

plays, in the highest degree, the extraordinary talents of our great architect, Sir Christopher Wren.[20]

While their home itself might have been a reasonably healthy place, the surrounding city was not. The air was ever thick with coal smoke from the fires of factories and homes. It is doubtful that the well-to-do Oldfields depended on the polluted water of the Thames for drinking and other domestic purposes[21], instead relying on rainwater collected in tanks at their home, or, if not, on water drawn from springs miles away and supplied by a private company.[22]

Although in many respects London was not a pleasant place, to young Augustus and his siblings growing up in its midst must have been wondrous and exciting. Integral to the life of the city was the incessant clamour of vendors' voices 'crying' their wares and services. In streets, alleys and courts they competed for attention to sell merchandise of almost every description.[23] Adding to the din of the vendors was the clatter of more than a thousand hackney-carriages plying the streets.[24] He may have delighted in watching the steam yachts that travelled up and down the Thames between London and Gravesend 'at a degree of velocity which [was] imminently perilous to those who may be afloat in small boats and wherries at the same time'.[25] Perhaps on some mornings he accompanied his mother and siblings on shopping expeditions to Leicester Square, Covent Garden, Oxford Street or Mayfair and gazed with interest at stylishly dressed 'high-bred ladies', attended by their maids and liveried footmen or pages, intent upon the purchase of the 'whim-whams of fashion'.[26] Along Oxford Street they probably lingered to gaze at the brightly-lit displays of goods behind the plate-glassed bow windows of shops or were sometimes permitted to explore toy shops. On occasions they may have attended such popular entertainments as the panoramas in Leicester Square and Regent's Park, the menagerie at Exeter Change in the Strand[27] or enjoyed splendid fireworks and other entertainments at one of London's pleasure gardens, the most notable of which was at Vauxhall off the South Lambeth Road.[28] They may also have been afforded opportunities to explore the natural history and other displays at the British Museum and the exotic plants at the private royal garden at Kew Gardens, where limited access was afforded to those members of the public 'who could afford the cost of getting to Kew Gardens by the Brentford coach to Kew Bridge or, during the summer months, by hired boats'.[29] It is likely that the children were firmly encouraged to shun the gin-sodden drunks staggering along the streets and the many near-starving, ragged adults and children who roamed the city desperately trying to eke out a living by whatever means presented themselves.

A gambling business in the family home

By the time Augustus was born, Frederick, in partnership with Richard Bennett, was operating a gaming business at 28 Bury Street on a part-time basis. This must have necessitated a division of the building during the opening hours of 3 pm to 6 pm to safeguard the family's privacy and protect them from any noise and other disturbances caused by rowdy, abusive drunken gamblers. Just how this

[20] A.C. Pugin, W.H. Leeds & J. Britton, 1838, *Illustrations of the Public buildings of London with Historical and Descriptive Accounts of Each Edifice,* 2nd edn, London, p. 82.
[21] ibid.
[22] Marjorie and C.H.B. Quennell, 1938, *A History of Everyday Things in England. Done in Four Parts of Which This is the Third. The Rise of Industrialism, 1733–1851*, B.T. Batsford Ltd., London, p. 94.
[23] C. Hibbert, 1987, *The English: A Social History 1066–1945*, Guild Publishing, London, pp. 528–529.
[24] F. Barker & P. Jackson, 1974, *London: 2000 Years of a City and its People*, Cassell, London, pp. 252, 253.
[25] R.P. Cruden, 1831, *An Account of the Origin of the Steam-boats in Spain, Great Britain and America and of Their Introduction and Employment Upon the River Thames Between London and Gravesend to the Present Time*, Effingham Wilson, London, pp. 43–44.
[26] Stella Margetson, 1971, pp. 68–69.
[27] F. Sheppard, 1998, p. 245.
[28] Ben Weinreb & Christopher Hibbert, (eds), 1983, *The London Encyclopaedia*, 2nd edn, 1993, Macmillan London Ltd., London, p. 937.
[29] Ray Desmond, 1995, *The History of the Royal Botanic Gardens Kew*, The Harvill Press with the Royal Botanic Gardens Kew, London, p. 167.

Fig. 4.1 Interior of a gambling establishment

division was arranged is a matter for conjecture. Since the separation was only required for three hours per day the children may have spent the time in the nursery or school room, almost certainly located above the first floor, where under supervision they played or were taught until their evening meals were brought up from the kitchen in the basement and possibly remaining there until bedtime.[30]

If, as has been suggested[31], the gaming establishment at 28 Bury Street aimed to attract upper class clients, in addition to stylish furnishings, the presentation of food and beverages and the gaming services were of a high standard. To achieve this and an environment safe from the prying eyes of the police and intrusions from unwanted gamblers, the proprietors employed servants to keep the premises, table linen, eating and drinking utensils and food platters clean, prepare and serve food and drink, and manage the stores, as well as those who served to maintain security and act as croupiers. Some of these retainers, such as the cook, housemaids, laundry and scullery maids and possibly a housekeeper, would have serviced the family. Any such servants would have delivered food prepared in the basement kitchen to the family and the gaming house while laundry emanating from these sources was washed, dried and ironed behind the kitchen in the basement and then delivered to storage cupboards. Such a situation may have required Mary and Frederick to come to some agreement regarding the management of the servants operating in their respective domains of responsibility. Perhaps Mary acted as overall manager for those jointly serving the house and the gaming house or a housekeeper was engaged to oversee all the domestic employees, except perhaps the nurse and nursery maid.

Newspaper reports describing the use of space in other Bennett and Oldfield gaming houses suggest that the three rooms on the first floor of 28 Bury Street were opened for play; the large front room, which was most likely Mary and Frederick's main reception room, for playing *rouge et noir* and the others for hazard and faro. Each room would have had a large oblong table covered in a cloth, possibly of velvet, and in the main room, another table for refreshments.[32] The ground floor had sufficient rooms to accommodate the security guards required to watch out for the police and monitor those others who sought entry via the Bury Street or laneway entrances.

Whatever the precise arrangements, prior to the hours of gaming, the house must have been a bustling hive of industry every day. Servants going about their chores probably chided children who

[30] Kathryn Hughes, 1993, *The Victorian Governess*, The Hambledon Press, London, p. 70.
[31] See Chapter 3.
[32] See for example, *The Times,* 23 May 1823, 24 May 1823, p. 3, TTDA.

interfered with their work or gossiped to one another as the occasion arose. Meanwhile, perhaps Bennett and Oldfield, the demands of other establishments permitting, were busying themselves with administration of the gaming business somewhere in the residence. Upstairs for part of the day, some of the older children would attend to their schooling under the guidance of Mary or a governess and afterwards add to the din of the younger children's play and incessant chatter. When the gaming house was operating, play would have been fierce, fast and noisy with one, or both, of the partners keeping watchful eyes out for cheating or for opportunities to take advantage of one of the less experienced players.

Difficult times for the family

The year of Augustus' birth marked the beginning of a new way of life and nearly a decade of adversity for the family with Frederick becoming 'a person of considerable notoriety in the neighbourhood of St James'.[33]

During 1821, on his own word, Frederick 'embarked his capital in trade' and withdrew from 'all concern and interest' in the gaming house at 75 St James Street in June 1821. By 'September of the same year the nuisance was abated'.[34] According to an affidavit given in 1826 by one Isaac Tooke, Frederick was planning to quit the gambling industry as early as 1820 because he believed it 'was now beginning to fail'.[35] Perhaps he was experiencing the beginning of the fading of 'the flush of success' as valued upper-class clients ran out of funds they could afford to lose at the tables. This was said to be happening by 1844[36] due in some extent to the effectiveness of the anti-gaming lobby in changing public opinion but fuelled more by expediency than high moral concern. 'Disputes over gambling debts had been clogging the courts and a number of judges expressed displeasure at tiresome cases'.[37] It is clear from newspaper court reports that at least some of the judiciary had been intolerant of gambling since the early 1820s. For instance, Mr Justice Bayley, before passing sentence on Frederick and some of his partners in 1823 for keeping common gaming houses, stated

> the offence of the defendants...was a very serious offence; highly prejudicial to public morals, and, tending not only to destroy the character and the Fortune of those who were led astray, but to blast their integrity, and leave them destitute of virtue and [full] of shame.[38]

More than foreshadowing the loss of wealthy clients, it may have been Frederick's assessment of the seriousness with which some judges were viewing gaming offences and the cost of attendant adverse verdicts—repayment of gamblers' losses, fines, imprisonment—which could put at risk his ability to support himself and his family. Affidavits support this conclusion he presented to the Court on two occasions in which he stated that his wife and children 'would be reduced to poverty if any heavy fine should be inflicted'.[39]

The sale of his partnership in the silver and gold hell in St James Street and the proceeds of the sale of the indenture of lease on the property in Jermyn Street, probably provided him with funds to finance the establishment of his Westminster Wine Company at 51-2 Pall Mall under the Colonnades of the Opera House [King's Theatre], Haymarket.[40,41] By 1822, Frederick was an excise wine merchant. Not properly understanding the wine trade, he obtained the assistance of the aforementioned

[33] For example, Jarman v. Oldfield, CKB, 18 January 1828, p. 3, *The Times*, 19 January 1828, p. 3, TTDA.
[34] *The Times*, 12 February 1823, 13 February 1823, p. 4, TTDA.
[35] The Attorney-General v. Oldfield, Court of Exchequer, 30 November 1826, *The Times*, 1 December 1823, p. 3, TTDA.
[36] Roger Munting, 1996, pp. 21–24.
[37] ibid.
[38] *The Times*, 12 February 1823, p.4, 13 February 1823, p. 4, TTDA.
[39] ibid.
[40] *The Times*, 30 November 1826, 1 December 1826, p. 3, TTDA.

Fig. 4.2 King's Theatre, Haymarket from Pall Mall east, 1822

Tooke, a long-time London wine merchant, who for about eight or nine years had served wine to Oldfield's gaming houses. It was Tooke's understanding that there had been a verbal agreement that he would manage the business as a partner while Oldfield provided all the necessary capital. Despite efforts on Tooke's part to finalise a deed of partnership, Oldfield refused. Tooke took legal action in 1826 to obtain £500 for teaching Oldfield the secrets of the trade and for managing the concern but his case was unsuccessful.[42]

Frederick and Mary's fourth son, Edwin Douglas, was born on 22 November 1822.[43] By this time, Augustus was nearly two years old and the family had left Bury Street. They probably moved directly to 54 Norton [now Bolsover] Street, St Marylebone, where they were living at the time Edwin was baptised at the Church of Saint Mary, St Marylebone Road on 1 January 1823. On this occasion Frederick gave his occupation as 'trade'.[44] Following the middle-class trend of that time, the Norton Street residence was quite separate and distanced from any of Frederick's business premises.[45]

On this property, which has since been demolished, rates were charged on annual rentals that ranged from £60 to £70 over the period the family was in residence. These rentals indicate it was a substantial dwelling. It stood on the western side of the street near its northern end and opposite to what is now the Royal Orthopaedic Hospital for Children.

This marked change in their lifestyle suggest that the couple were harbouring dreams that they and their children would be fully accepted as respectably upper middle-class. Perhaps they were

---

[41] *Robson's Commercial Directory, Street Key & Conveyance List for 1828*, Part II, Robson, Blades & Co., London, *London Directories from the Guildhall Library 1677–1900*, Group II, excluding Post Office directories, vol. 179, microfilm, NLA. The exact location of these premises within the Colonnades has not been established. The only indication of Pall Mall street numbering for the Colonnades is from Horwood's map of 1792 (in Joseph Wisdom, *The A to Z of Regency London*, Map 23, Harry Margary, Lympne Castle, Kent in association with Guildhall Library, London, 1985). However, we cannot reconcile this numbering with G.S. Repton's floor plan of the J. Nash and Repton designed renovations to the theatre precinct that were carried out from 1816–1818, F.H.W. Sheppard (ed), 1960, vol. 30, Plate 33.

[42] *The Times*, 30 November 1826, p. 3, TTDA.

[43] Record for Edwin Douglas Oldfield, no. 2, *Register of Baptisms Solemnised at Church of Saint Mary – St Marylebone Road*, 1823, microfilm 10, parish records, WCA.

[44] ibid.

[45] Leonore Davidoff and Catherine Hall, 1987, p. 181.

disappointed. It was not long after they took up residence in Norton Street that Frederick's previous occupation and subsequent indictments and convictions were so conscientiously reported in *The Times* that his notoriety must have been common knowledge. This may have resulted in the family being shunned by their neighbours. One general article in *The Times* referring to a judgement requiring Oldfield and Bennett to repay a gambling loss[46] was particularly damning:

> We this day publish the reports of two trials, which, so far as they go, ought to be viewed with no little satisfaction. They are of the nature of vermin-hunting; and the noxious animals have not escaped their pursuers. Some of these miscreants against whom we have lately been employed in rousing the public vigilance and the arm of justice, viz., the keepers of common gaming tables, have been attacked in a way which it is probable they will feel more severely than any police prosecution...and have been forced to disgorge, in one or two instances, a portion of their spoil. Three of them, named Abbott, Holdsworth, and Davis, have kept the "Hell" at 5, King-street, St. James's...The other vagabonds (for in law their occupation affixes upon them the designation of "rogues and vagabonds") were called Oldfield and Bennett, and kept two houses – 71 Pall-mall [sic], and 28 Bury-street.[47]

During 1822 and 1823 two gamblers, Joseph Carter and John Woodroffe, were awarded repayment of losses incurred at one or other of the gaming houses run by Oldfield and his partners. The total amount awarded was £912 but how the debt was apportioned among the partners is not known.[48]

As a direct result of these court decisions Oldfield, Bennett, Carlos and Phillips were first indicted on 18 December 1822 and found guilty of keeping a common gaming house at 75 St James Street on the information of John Woodroffe, a gambler seeking to recoup his losses.[49] The defendants were at the Court of King's Bench, Westminster on 12 February 1823 to receive judgement for this offence. Oldfield, in addition to pleading the dire consequences to his family if a heavy fine were imposed, said he 'was subject to a disease in the head, arising from a determination of blood to that part, which protracted imprisonment would dangerously heighten'.[50] Unsuccessful efforts were made, through consultation with a medical historian,[51] to identify the disease that Frederick claimed he suffered, and which was confirmed by the affidavit of his medical attendant. 'Determination of the blood' according to the *Concise Oxford English Dictionary,* refers to a tendency of blood to flow to some part.[52] Since this is a normal aspect of the body's functioning, perhaps Oldfield colluded with his doctor to 'pull wool over the eyes' of the judge and jury, blinding them with spurious medical terminology. However, further investigation might reveal that the term was used at that time to refer to hypertension or some other disorder of the circulatory system.

Although Carlos and Bennett also appealed for leniency on much the same grounds, Oldfield and Bennett were each fined £1,000 and sentenced to one year in Marshalsea Prison on 13 February 1823, and to remain there until the said fines were paid. They were also required, at the end of their terms of imprisonment, to 'give Security for their good behaviour for five years, themselves in £2,000, and two sureties in £1,000 each'.[53] They escaped 'being kept at hard labour' as the relevant legislation had not been given Royal assent at the time of the commission of the offences.[54]

[46] *The Times,* 27 October 1822, p. 328 October 1822, p. 3, TTDA.
[47] Anon., *The Times*, 28 October 1822, p. 2, TTDA.
[48] *The Times*, 27 October 1822 28 October, p. 3, and Woodroffe v. Oldfield and others, CKB, Guildhall, 28 February 1823, *The Times*, p. 3, TTDA.
[49] *The Times*, 18 December 1822, p. 3, 19 December 1822, p. 3.
[50] *The Times*, 12 February 1823, p. 4, 13 February 1823, p. 4, TTDA.
[51] Dr Jim Leavesley, personal communication to MHH, c.2006.
[52] H.W. & F.G. Fowler, 1964, *The Concise Oxford Dictionary of Current English*, Clarendon Press, Oxford.
[53] Entry for Frederick Oldfield and Richard Bennett, no. 769, *King's (Queen's) Bench Prison and Queen's Prison Commitment Books 1719–1862*, PRO Pris 4/35, November 1822–November 1823, NAL.
[54] *The Times*, 12 February 1823, p. 4,13 February 1823, p. 4, TTDA.

The Marshalsea Prison, primarily a gaol for debtors of all social ranks, was situated 'a few doors short of the church of Saint George, in the borough of Southwark, on the left-hand side going southward'.[55] Notoriously cramped for inmates, it was an oblong barrack pile divided into eight houses, each with three floors, containing 56 rooms and standing back to back, surrounded by a narrow, paved yard, which was hemmed in by high walls duly spiked at top. In addition to barracks, the debtors' section consisted of a kitchen, a public room, and a 'tap room', where debtors could drink as much beer as they wanted, at about five pence a pot. The rooms were accessed directly from the outside via eight narrow wooden staircases. [56]

The prison was a privately run, profit-making institution where to ensure a reasonably comfortable internment, Oldfield and Bennett needed to ingratiate themselves with the Marshal, keepers, turnkeys, tipstaffs,[57] watchmen, chaplains and surgeons who presided over and participated directly in the ceaseless trading of essential and illicit goods, including tea, coffee, tobacco and alcohol beyond the limits allowed, provided by local tradesmen to inmates.[58] Despite their concerns at the prospect of incarceration, Oldfield and Bennett probably entered quickly into the internal economy of the Marshalsea and worked to maximise its operations to their own benefit. Based on what is known of Oldfield's later confinement in the King's Bench Prison, they rented their own rooms from the Marshal. In common with other well-to-do inmates, they either paid to have their meals prepared for them by the prison cook or sent in from nearby restaurants.[59] Mary was probably among the many frequent visitors to the gaol, perhaps at times accompanied by the children. Like other wives of inmates, there is little doubt that she kept Frederick supplied with information about his business affairs, the children's activities and any goods that could not be obtained easily through the prison networks. Because they were held on civil and not criminal charges, Oldfield and Bennett as paying prisoners may have been afforded the privilege of day trips outside the prison wall but there is no evidence that that they 'lived within the Rules', that is, were permitted to reside within three square miles of the prison.[60]

If Bennett and Oldfield did rent their rooms, they qualified to belong to 'The Master's Side', one of two spatial divisions within the gaol that served to distribute inmates according to their social and legal status; the other division, 'the Common Side'. This division underpinned a ranked array of distinctions, privileges and amenities that demarcated the inmate population according to wealth and social pretension. Oldfield and Bennett may have asserted their status and recruited Common Side debtors to act as their servants but who 'otherwise frequented the back ports of the prison, where walls obscured the daylight, open sewers ran from the privies and where inmates deposited the prison slops'.[61]

In addition to government regulations administered by prison staff, debtor self-government operated in Marshalsea and served a variety of pragmatic functions; it arbitrated in disputes, regulated prison hygiene and the behaviour of inmates. This was conventional and highly formalised, modeling their structures of authority on the legal and political institutions that maintained order in the outside world.[62]

The Oldfields were said to have lived in Norton Street in a 'costly style'.[63] Presumably, therefore, the Norton Street residence prior to Frederick's detention at Marshalsea was well attended by a range

---

[55] Charles Dickens, *Little Dorrit*, 1857 edn., digitally scanned, Amazon kindle, location 1259.
[56] This was the price in 1815, *Wikipedia*, 'Marshalsea', viewed, 27 March 2012 at www.wikipedia.org.
[57] 'Tipstaff': A constable, bailiff, or sheriff's officer; so-called because he carried a staff tipped with a bull's horn or with metal, *Brewer's Dictionary of Phrase & Fable*, 1952, revised edition, Cassell, London, p. 906.
[58] Margot C. Finn, 2003, *The Character of Credit: Personal Debt in English Culture, 1740-1914*, Cambridge University Press, Cambridge, pp. 129–132.
[59] ibid., pp. 135–136.
[60] J. Gardiner & N. Wenborn, (eds), 1995, p. 403.
[61] Margot C. Finn, 2003, p. 121.
[62] ibid.
[63] *The Times*, 12 February 1823, p. 4, 13 February 1823, p. 4, TTDA.

Fig. 4.3 Social life in a debtors' prison, 1821

of domestic servants.[64] In addition to being responsible for the children's care and education, Mary must have assumed full responsibility for the management of their domestic staff now that the family residence was at some distance from Frederick's place of business. For Mary, the extra burden brought about by Frederick's imprisonment probably centred on visiting him in prison or at some nearby meeting place within 'the Rules', taking over Frederick's role of employing and discharging domestic servants and coping with any stigma their neighbours imposed upon her and the children. If she regularly visited Frederick, however, there were plenty of opportunities to make decisions relating to major expenditures and resolving any serious problems with the children, provided, of course, that he was amenable to such involvement. Any concerns Mary might have had regarding Frederick's business affairs could have been allayed since court reports indicate that he continued to direct his business all the time he was in prison.[65] He did this with the assistance of his trusted clerk, William Porter[66], who managed the business, and at least three other servants, George Bedell, Samuel Mole and James Garvan.[67] Nevertheless, Mary's actions to retrieve the money from the swindler Thomas Miller in 1828 (see below) suggests that she was involved to some extent in Frederick's business affairs, even if it meant compromising her gentility.[68]

If Frederick's children visited him in Marshalsea (and from at least November 1825, the King's Bench Prison[69] where he was later held), they may have been quite upset at finding him in such dreary confines surrounded by all manner of humanity. As time went by, however, they may have realised that Frederick was, in fact, reasonably comfortable and came to enjoy the exciting life of the courtyards and possibly the company of some their father's inmate acquaintances of The Master's Side.

---

[64] June Purvis, 1989, *Hard Lessons: the Lives and Education of Working-class Women in nineteenth-century England, Polity, Cambridge, England*, p. 61.
[65] *The Times*, 23 October 1827, p.3 24 October 1823, p. 3, TTDA.
[66] Oldfield v. Miller, Guildford, 2 August 1828, *The Times*, 4 August 1828, p. 3, TTDA.
[67] *The Times* 30 November 1826, p.3, 1 December 1826, p. 3, TTDA.
[68] From an affidavit given by Mrs Oldfield and read at the case: Oldfield a Bankrupt, Vice-Chancellors Court, 27 February 1828, *The Times*, 28 February, p. 6, TTDA. TTDA.
[69] The King and Oldfield, Court of King's Bench, 29 November 1825, p. 3, TTDA.

Apparently because of a 'falling out' between Carlos and Oldfield, the latter was indicted on the information of Carlos on 30 April 1823 for 'wilful and corrupt perjury'.

> The perjury was assigned on his [Oldfield's] affidavit offered in mitigation of the punishment, in which he swore that he ceased to be connected with the house in question [75 St James Street]; and also, that he was destitute of pecuniary resources.[70]

He was found guilty but later acquitted on the grounds of insufficient evidence.[71] There was a suggestion that Carlos had been paid off by Oldfield but later this was strongly denied.[72]

Once more, on 23 May 1823, Bennett, Oldfield and Phillips were charged and found guilty of keeping a common gaming house at 75 St James Street.[73] The witness, Mr Andrews, denied that his evidence related to the same period for which the defendants had been previously convicted and imprisoned. Sentencing was held over until a later date. Oldfield and Bennett were back in the same court next day to stand trial for keeping a gaming house at 71 Pall Mall during the months of September, October and November 1819. They were found guilty and again sentencing was held over.[74]

Oldfield and Bennett were indicted on 3 November 1823 for 'certain nuisances and misdemeanours', following a series of affidavits from one William Roubel, gentleman. This conviction apparently aroused Oldfield and Bennett's ire to such an extent that they retaliated with violence and on 3 and 4 November 1823 they were respectively charged with

> riotously and tumultuously assaulting with force and arms William Roubel and pumping divers large quantities of water upon the said William Roubel on the Twenty fourth day of September last within the King's Bench Prison in the Parish aforesaid [St George the Martyr] and also with intent him the said William Roubel then and there feloniously to kill and murder against the Peace.

They were each fined £1,500 with sureties of £2,000 for good behaviour.[75]

On 24 January 1824, they were before Mr Justice Bayley for judgements relating to keeping gaming houses at 75 St James Street, 71 Pall Mall and 6 Pall Mall, and Oldfield, for a similar offence 'in Bury Street, St James's' (reports of the last two original indictments were not found). Oldfield was fined £1,500 and Bennett, $2,000 with each giving security of £10,000 with two sureties for each of them in £2,000 for his good behavior for five years and to remain in custody until the fine was paid and the securities given.[76] During this court appearance they denied on oath an accusation by Thomas Erskine Grant that they kept a gaming table in the Prison. It seems from the newspaper report of the case that the charge was dropped.[77]

At the trial Oldfield produced affidavits from many wine merchants testifying that he had engaged in the wine-trade, and conducted himself fairly in the way of his business. Mr Adolphus, counsel for the defendants, pleaded:

> These men were already suffering for their fault; prosecution after prosecution had been heaped on them by those who had no moral claim to attack them, and for the basest purposes; they had already given up the practice for which they were indicted; and therefore, had perhaps been done for justice, and for prevention. They were broken men; broken in health and in fortune; and it could not be necessary,

---

[70] King v. Frederick Oldfield, Court of King's Bench, 30 April 1823, *The Times*, p. 4, TTDA.

[71] Oldfield v. Miller, Surrey Sessions, 23 October 1827, *The Times*, p. 3, TTDA.

[72] Oldfield v. Miller, Surrey Sessions, 23 October 1827, *The Times*, 24 October 1827, p.3, TTDA.

[73] The King v. Bennett, Oldfield and Phillips, CKB, 24 May 1823, *The Times*, 24 May 1823, p. 3, TTDA.

[74] The King v. Oldfield and Bennett, CKB, 26 May 1823, *The Times*, 25 May 1823, p. 3, TTDA.

[75] The King v. Richard Bennett and Frederick Oldfield and further charge against Frederick Oldfield, 24 January 1824, *King's (Queen's) Bench Prison and Queen's Prison Commitment Books 1719–1862*, PRO Pris 4/35, November 1822 – November 1823, p. 257, NAL.

[76] *King's (Queen's) Bench Prison and Queen's Prison Commitment Books 1719–1862*, 4 November 1823 and 24 January 1824, PRO Pris4 v. 35, November 1822–November 1823, NAL.

[77] The King v. Bennett & Oldfield, CKB, 24 January 1824, *The Times*, 26 January 1824, p. 3, TTDA.

Fig. 4.4 King's Bench Prison, London, c. 1825

> for any pur pose, that they should be utterly crushed into the earth, and deprived of the means of providing for their families by honourable industry.[78]

Following deliberation by the judges, it was reported that Mr Justice Bailey

> after expounding on the pernicious consequences of gaming, and observing that the ill-health of the defendants, though no doubt brought on by their course of life, had prevented the Court from inflicting additional imprisonment…passed sentence.[79]

Oldfield was fined £500 for the indictment on which he alone was convicted. On the second conviction Oldfield was fined £1,500 and Bennett, £2,000 and on the third, they were each required to pay £1,500. Each was ordered to give security by his own recognizance of £10,000 and sureties of £2,000 each.[80] This appears to have been the last of Frederick's indictments for illegal gaming and claims for the return of monies lost during such activities but it was not the end of his court appearances. In all, Oldfield's fines amounted to £4,500.

The degree of distress Mary experienced throughout Frederick's imprisonment appears to have been considerable. Sometime during September 1824 Mary appealed to the then Home Secretary, Hon. Robert Peel, for Frederick's release on the grounds of family hardship and again three months later[81] but to no avail, even though Richard Bennett had been released with a remission of his fines.[82] She referred to Frederick as her 'dear husband' and stated that she

> now had the much wished-for satisfaction to well know he never will enter a [gaming house] again, and that several very respectable merchants in the city are willing to give security for his future good conduct.[83]

In February 1825 Frederick appealed for his release because he was unable to pay the fine.[84] Again it fell on deaf ears, as did several similar submissions from people outside the family. David

---

[78] ibid.
[79] ibid.
[80] ibid.
[81] Mary Oldfield to Right Honorable, Robert Peel undated (c. September 1824) & undated (c. December 1824), Series: HO17, Piece no: 49, Item no. HK34, Home Office: Criminal Petitions: Series 1, Record set: Crime, Prisons & Punishment, NAL.
[82] Mary Oldfield to Right Honorable, Robert Peel, undated (c. December 1824*),* ibid.
[83] ibid.
[84] Frederick Oldfield to Right Honorable Robert Peel, 9 February 1825, ibid.

Manley, Frederick's most persistent appellant, wrote to Robert Peel on 3 October 1825, urging for Frederick's release, saying that Mary had declared that her husband was 'a most tender and affectionate Husband and Father' and she and her mother were 'quite disconsolate' that Bennett had been released but not Frederick. In a further submission, 8 December 1825, David Manley stated that Mary was far advanced in pregnancy with her 'eighth child',

> and is in a bad state of health through constant unhappiness of mind she seems to think and fear that she gave offence to Mrs Peel at the time she intruded herself into that Lady's presence in the country with a letter she had received and owing to her Husband often falling in a dejected state of mind and spirits she has a dread that he will destroy himself.[85]

Manley also noted that Frederick's wine business was suffering.[86] Augustus and those of his siblings who were at home at the time were probably very much aware of Mary's state of mental and physical health but no details of how they reacted have survived. On 21 November 1826 Frederick appealed again stating that:

> I have at this moment a child lying dead, and my wife in consequence of that loss, and her attendance on it during its sickness, is now alarmingly ill, and that my health is so much impaired of late, that I am seldom able to leave my room.[87]

Although born a short time after Manley's letter to Peel of 8 December 1825, no record of the birth or baptism of this child has been found. He or she appears to have been conceived in the prison.

In both her letters to Peel, Mary gave her address as '6 Grove Street, London', yet court and St Marylebone rate records show their residence during Frederick's imprisonment to have been '54 Norton Street' (see below). Perhaps this was her mother's address; Mary and the children may have been staying with her temporarily.

On 29 May 1825, the Marshal of the King's Bench [Prison] was summoned to show cause why Oldfield should not be restrained within the walls of that prison as he had been seen 'out of the rules of the prison at a sale in the city'. Since the Marshal could not provide a satisfactory explanation, the Court ordered that the rule be taken and served on both Oldfield and the Marshal.[88] Just when Oldfield was transferred to the King's Bench Prison or the circumstances under which the change occurred is not known. Frederick had his own room and therefore would have served his sentence on the master's side.[89]

The King's Bench Prison, also located in the Parish of St George the Martyr in the Borough of Southwark, was so-called because it was 'linked' to the Court of King's Bench. It was a large and cheerless place with 224 rooms and an open courtyard surrounded by a high wall.[90] It would not have taken Oldfield long to become familiar with the operations of the prison and to enter its internal economy since the prison was conducted in a very similar manner to that of Marshalsea Prison. Rules permitted debtors day trips, 'living within the Rules'[91], buying the 'freedom of the Rules[92] and visitors at will. Oldfield was a 'ruler' of the Prison.[93]

By the early 19th century the prison was notorious for the laxity of its rules and no doubt Frederick, who had his own room[94], took full advantage of it being 'the most desirable place of incarceration in London'. The courtyard thronged with life: there were tailors, barbers, hatters, piano-makers,

---

85 David Manley to Right Honorable Robert Peel, 8 December 1825, ibid.
86 ibid.
87 Frederick Oldfield to Right Honorable Robert Peel, 10 November 1826, ibid.
88 The King & Oldfield, Court of King's Bench, 26 November 1825, *The Times*, 29 November 1825, 3, TTDA.
89 Oldfield v. Miller, Surrey Sessions, 23 October 1827, *The Times*, 14 October 1827, p. 3, TTDA.
90 Ben Weinreb & Christopher Hibbert, 1983, p. 445.
91 J. Gardiner & N. Wenborn, (eds.), 1995, p. 403.
92 Ben Weinreb & Christopher Hibbert, 1983, p. 446.
93 *The Times*, 23 October 1827, p. 3,14 October 1827, p. 3, TTDA.
94 ibid.

chandlers and oyster-sellers. Alcoholic beverages flowed freely: there were no fewer than 30 gin shops and 120 gallons sold weekly. Affluent inmates dined in style with a regular cook to prepare their meals; poorer prisoners 'took their turns with a begging box at the gate' or waited for their wives or other members of their families to bring them food.[95]

On 30 November 1826, whilst still in prison, Oldfield was indicted on several counts relating to wrongful use of liquor permits by not returning them to the Excise Office, including one count in which his mother, Sarah, was implicated. A quantity of wine had been had been assigned to her by Oldfield and she held one of the permits in question. At the time, she was residing in Knightsbridge and was stated to have 'always been in indigent circumstances'. Formerly supported by her son-in-law, Murray Hollis, the husband of Frederick's sister, Juliana, Sarah was at the time of the court case dependent on Frederick.[96] He was also accused of adulterating wines by mixing Cape wine with sherry for selling the mixture as pure sherry. The jury returned a guilty verdict for the Crown on one permit and Oldfield was fined £500.[97] Tooke was the chief witness for the Crown in the indictment relating to the adulteration of wine. His credibility as a witness was so seriously questioned that the Attorney-General dropped the charges.[98]

Frederick was again in court on 23 October 1827, this time with an action to retrieve monies that allegedly had been fraudulently obtained from him by Mr Thomas Miller. This case excited considerable interest and 'the court was crowded to excess during the trial'. It came about because Miller, whom Oldfield did not know at that time, visited him in prison in July 1824 with the offer that 'he had no doubt he could procure Oldfield's liberty, if he would consent to pay £1,000 as a douceur for the purpose to a certain well-known nobleman.' Oldfield, at Miller's dictation, gave him an IOU for £1,000. Oldfield mortgaged property to get the money and on 30 July, gave Miller a cheque for that amount and at the same time requested the return of the IOU. Despite repeated requests Miller failed to comply nor did Oldfield secure his release from prison. The jury found in Miller's favour because, as Oldfield explained later, of the absence of a material witness.[99] On the day after his acquittal Miller, having retained the original IOU, bought an action against Oldfield for £1,000 on a claim for money laid out and expended in his behalf.[100] Oldfield refused to pay and on 27 December 1827, Miller, as a petitioning creditor, sued and won a commission of bankruptcy against Oldfield. As a result, his stock of wines was seized by the Commissioners, thereby putting him out of business.[101]

Two months later, on 27 February 1828, Oldfield entered a petition to the Vice-Chancellor's Court in which he prayed that the commission of bankruptcy be superseded and his business restored, 'or, if that prayer was not granted that an inquiry might be directed to be taken before the commissioners as to the validity of the petitioning creditor's debt'.[102] According to his counsel, at this time all that remained of Oldfield's dues to the Crown was £60 and that should be immediately paid, and security given for the payment of the debt if it were declared valid. Oldfield ultimately preferred a bill of indictment against Miller for obtaining money under false pretenses but on this occasion, he had a credible witness who could testify that Miller had confessed to receiving the £1,000. That witness was Mary who, on 23 May 1827 with her daughter, 'Mary Anne' [Marian], called on Miller at his home for the purpose of 'desiring him to refund the money which he had obtained from her husband'. During the ensuing conversation, which occurred after Miller insisted that Marian leave the room, he acknowledged that he had received the £1,000 but had given it to the nobleman who could not return it

[95] Ben Weinreb & Christopher Hibbert, 1983, p. 446.
[96] *The Times*, 30 November 1826, p.3,1 December 1826, p. 3, TTDA.
[97] Oldfield v. Miller, Guildford, 2 August 1828, *The Times*, 4 August 1828, p. 3, TTDA.
[98] *The Times*, 30 November 1826, p. 3,1 December 1826, p. 3, TTDA.
[99] Oldfield v. Miller, Surrey Sessions, 23 October 1827, *The Times*, 24 October 1827, p. 3, TTDA.
[100] Re Oldfield, a bankrupt, Vice-Chancellor's Court, 27 February 1828, *The Times*, 28 February, p. 6, TTDA.
[101] *The Times*, 2 August 1828, p.3, 4 August 1828, p. 3, TTDA.
[102] Re Oldfield, a bankrupt, Vice-Chancellor's Court, 27 February 1828, *The Times*, 28 February, p. 6, TTDA.

as he was short of money.[103] The case was heard at Kingston where the jury found that Oldfield did not owe Miller £1,000 and the commission of bankruptcy was superseded on 2 August 1828.[104] In the meantime, Oldfield had sustained 'serious injuries [loss of income] from the conduct of Miller'. On 2 August 1828, Oldfield indicted Miller for compensation for monies lost during the period of the commission of bankruptcy. The judgement was in the plaintiff's favour and he was awarded damages of £225 16s 1d.[105]

In 1828, another daughter, Juliana Maria Louisa, conceived in prison was born to Frederick and Mary.[106] Frederick was declared to be the owner of much property in 1828.[107] He had his business at 51/2 Pall Mall and another at Tottenham Court Road and to the west of which, in Norton Street, his rented family residence.[108] Nevertheless, on the 28 January 1829, just after Augustus' eighth birthday, Frederick was once again declared bankrupt.[109] At the time, the family was still at the Norton Street address[110] but they must have moved out soon afterwards as the building was empty at each of the quarterly rate collections for 1829.[111] Apparently their next residence was on Bishops Walk, Lambeth, where it seems they were residing in 1833.[112]

Frederick was finally released from prison on 14 July 1829 after having given the securities required by the Court and paid his fines and court-ordered gambling loss repayments.[113] How he came to be declared bankrupt in January 1829 and yet six months later was able to clear his final debt to the court is a mystery.

Added to his substantial fines was his share of the £912 court-ordered repayment of gambling losses, the debt incurred in 1824 by mortgaging property to provide £1,000 for the *douceur* to secure his release from prison and, no doubt, there were also legal fees. The profit from his wine business was reported in 1828 to have been £800 per year less the book-keeper's salary, the rent on his warehouse and some other incidental expenses, probably leaving approximately £700. The actual profit could have been much less in 1827 since Porter had not included in his reckonings £500 drawn from the business to pay the fine for the false permit.[114] If this payment is taken into account, over the 6 years and 5 months Oldfield was in prison the total profit from the wine business would have been approximately £4,050.[115] This amount alone would not have provided the means to pay his debts and fines as well as affording a comfortable, middle-class standard of living for his mother in Knightsbridge, his family at 54 Norton Street and for himself in the King's Bench Prison as well as keeping five of the children at expensive boarding schools in France.[116]

Why he remained in gaol for almost six and a half years when apparently he could have cleared his debt to the Crown quite quickly through the sale of assets was a question raised by the Court but seems never to have been answered by Frederick.[117] Perhaps he chose to remain in prison so that he

---

[103] From an affidavit given by Mrs Oldfield and read to the Court, ibid.
[104] *The Times*, 4 August 1828, p. 3, TTDA. A report of this case could not be found.
[105] ibid.
[106] Record for Juliana Oldfield, 1841 England Census, County of Middlesex (PRO Ref: 107/680/2, File: 10105) gives Juliana's age as 10 which would have put the year of her birth as 1831. However, her records for all subsequent *England Census* returns and her death certificate, Certified copy of the entry of death for Frederica Juliana Hold, no. 273 for 1901, Registration District: Doncaster; Sub-district: Doncaster in the County of York, DYC993484, given at the General Register Office, England all give 1828 as her year of birth, HOA, viewed 12 May 2005 at www.ancestry.com.
[107] Re Oldfield a Bankrupt, Vice-Chancellor's Court, 27 February 1828, *The Times*, 28 February 1828, p. 6, TTDA.
[108] ibid. Sources have not shed any light on the whereabouts of the Tottenham Court Road business; it may have been the warehouse referred to in the 1826 court case relating to liquor permits and adulteration of wine.
[109] Frederick Oldfield, bankrupt notice, *The Times*, 28 January 1829, p. 2, TTDA.
[110] Record for 54 Norton Street, St Marylebone, Parish of St Marylebone Rate Book, 1828, Occupier: Frederick Oldfield, Reel 55, Part 2, WCA.
[111] Record for 54 Norton Street, Parish of St Marylebone Rate Book. 1829, Occupier: Nil, Microfilm Reel 56, Part 1, WCA.
[112] Record for Lavinia Sara Oldfield, London, England, Deaths and Burials, 1813–1980, viewed 29 March 2010 at www.ancestry.com.
[113] *The King's Bench Prison and Queen's Prison: Discharges for 1828, PRO PRIS 7/47* 1776–1862 and *King's Bench Prison and Queen's Prison Commitment Books 1719–1862*, PRO Pris4 v. 35, November 1822 – November 1823, NAL.
[114] *The Times*, 2 August 1828, 4 August 1828, p. 3, TTDA.
[115] During the 1820s income tax was not applied in England, Juliet Gardiner & Neil Wenborn, (eds), 1995, p. 404.
[116] See Chapter 5.
[117] *The Times* 27 & 28 February 1828, p. 6, TTDA.

could pace the clearing of the fines and debt repayments over a longer period to allow income-earning investments, besides the wine business, to realise sufficient cash funds to meet those commitments. By this strategy he would have maintained his investments and ensured the continuance of his wine business upon the profits of which the family depended for its income.

The family's financial position resulting from Frederick's imprisonment appears to have been made somewhat precarious, not only because David Manley's statements in his submissions to Peel but also by the declaration of bankruptcy in 1827. Nevertheless, Mary and the children seem to have maintained a way of life that was both respectable and comfortable as evidenced by their continuing residence at 54 Norton Street until about seven months before Frederick's release[118], and their ability to keep some of the children at a boarding school in France (see next chapter). It is, however, hard to imagine that Mary did not find it necessary to exercise restraints in her spending.[119] If so, perhaps she reduced her complement of domestic servants, retaining only a housemaid to do the heavy cleaning and laundry. Since 1827 Marian, then a fifteen-year-old, had accompanied her mother on her visits to Thomas Miller, and it seems likely that by that time the girl's education had terminated. Under these circumstances, Marian may have assisted with other domestic chores and with the care and education of the younger children. Whether Frederick and Mary's mothers played any part in the day-to-day care of the children is not known.

In other ways, the period of Frederick's incarceration must have been a most difficult period for the family but the historical record is silent on their reactions to the situation. Although during his detention he apparently continued to play a pivotal role in running his businesses, no matter how liberal his visiting privileges, the outcomes of the various court cases, the vilification from the bench and the press, the William Roubel incident and Frederick's long absence from home must have placed an enormous strain on his and Mary's marriage. Nevertheless, it seems that their relationship was affectionate and their physical separation painful. Apart from their anxiety over obtaining sufficient funds to meet their commitments, there must have been other stressors such as the ever-present threat of more gamblers suing her husband for return of losses at the tables prior to 1823 or of further attempts to fraudulently obtain their assets.

At the time Frederick was imprisoned, Marian was about eleven years old, Frederica, nine years, Lavinia, almost eight, Frederick Jnr, six and Alfred Bennett, four. For these children, the loss and associated stigma could have been devastating. Two-year-old Augustus' social and emotional development may have been adversely affected by feelings of insecurity and loss, particularly if his father had been in the habit of interacting with him in an affectionate and playful way. Although there is no evidence that any of the children in their teen or adult years reacted to Frederick's incarceration by delinquent behaviour, as children they may have experienced feelings of abandonment, embarrassment, shame, guilt and excessive sadness resulting in withdrawal, difficulties with their schooling and with their relationships with others in the short and long-term.[120] Since Edwin's father was not living at home for most of his first six to seven years of life, he may not have been conscious of the loss of nurture but nevertheless suffered from paternal deprivation in other ways.[121] Any of the children's

---

[118] The January 1829 bankruptcy notice in *The Times* gives Frederick's address as 'Norton Street'. The St Marylebone Rate Books, Ward Berniers (Berniers St), 1828, for 54 Norton Street show that Oldfield was the occupier during 1828 with paid up rates at the Michaelmas rate collection but unreceived rates for the Christmas collection Microfilm Reel 55, Part 2, WCA. The Division B Parish of St Marylebone Rate Book. 1829, for 54 Norton Street shows the building to have been empty at each rate collection during the year, Microfilm Reel 56, 1829, Part 1, Ward B, WCA.

[119] According to Leonore Davidoff & Catherine Hall most historians agree that incomes from £200 to £300 secured a place within the middle-class for an average family in the early part of the nineteenth century, 1987, p. 23.

[120] Lois E. Wright & Cynthia B. Seymour, 2002, 'Effects of parental incarceration on children and families' and Creasie Finney Hairston, 2002, 'Fathers in prisons: responsible fatherhood and responsible public policies' in Eileen Trzcinski, (ed), 2002, *What About Me? Children with Incarcerated Parents*, Michigan Family Impact Seminars, Briefing Report no, 2002–1, pp. 2 & 9, viewed 12 May 2005 at www.ancestry.com.

[121] Through their interaction and stimulation with their infants, fathers encourage babies to distinguish different sounds, faces, and touches at an earlier age. This tends to make infants more receptive to change, less fearful of new experiences, and better able—having established close ties with more than one person—to form trusting relationships with others, T. Ballard, 1995, *Statement of Travis Ballard, Esq.*,

negative reactions to their father's absence or associated social stigma was probably a matter of great concern to Mary.

Frederick's performance as a husband and father did not match the expectations applied to those roles by the middle-class of his era. Given Frederick's low-class background, his involvement with gaming houses had enabled him to attain a degree of wealth that potentially allowed him to assume a higher status way of life. However, whatever gains were made by wealth, his notoriety and imprisonment put at risk his ability to fulfill his role as an acceptable middle-class role model and as the family's caring protector and advisor, which deflected any improved social standing his family might have enjoyed.[122] Nevertheless, he sacrificed his own freedom to achieve this. It would, however, be unwise to draw any firm conclusions as to Frederick's failure or success as a father. As M.E. Lamb emphasises: '[H]igh paternal involvement may have positive effects in some circumstances and negative effects in others. The same is true of low paternal involvement'.[123]

We can only speculate as to the way Frederick's interpersonal relationships with the young children and their mother were played out. We know nothing of the kind of contact he had with his children while he was incarcerated. It is possible, for example, that family visits to their father, particularly if they took place outside the prison walls, were so positive that they felt secure in his affection and support.

President of National Congress for Fathers and Children, At the White House, 27 November 1995, viewed 16 April 2012 at www.ancestry.com.

[122] See for example, *The Times*, 12 February 1823, 13 February 1823, p. 4, p.3, 2 August 1828, 4 August 1828, p. 3, TTDA.

[123] M.E. Lamb, 1997, *The Role of the Father in Child Development*, 3rd edition, Wiley, New York, p. 13.

Fig. 5.0 Melaleuca oldfieldii

Chapter 5

Education, entry into a profession and decision to emigrate 1826–1841

In common with other middle-class children, Augustus probably commenced his schooling c.1826 when he was about five.[1] His father was still in prison at the time. His elementary education was almost certainly undertaken at home by either Mary or a governess or both[2], and conducted in a room specially set aside for the purpose with instruction organised as formally as the local elementary school. That is, their school year was divided into terms, their lessons ran according to timetables, the room furnished with one or more tables and chairs, a blackboard and possibly globes (celestial and earth) and their instruction undertaken for five and a half days a week, from Monday to Saturday lunchtime. Holidays were rare and the children may have worked after dinner, and on occasions until bedtime, finishing what had been left undone from the day's lessons. There is little doubt that music took a central place in their curriculum—piano, violin and singing, probably practised daily, with lessons from visiting masters once or twice a week. Physical exercise also played a large part in their schooling—running, jumping and playing cricket. For the girls, however, as they got older their daily exercise was a long walk with their governess. Even then the time may have been put to good use, observing and identifying flora and fauna.[3]

Augustus' initial instruction would have centred on reading, writing, spelling and arithmetic. As he grew older and became a reasonably fluent reader and writer, he would have been exposed to a variety of specialist books such as *Mangnall's Questions*—row after row of general knowledge questions and answers to be learnt by rote—as well as classic and contemporary works by writers such as Jacques-Bénigne Bossuet, Edward Hyde, 1st Earl of Clarendon, Samuel Taylor Coleridge, Jean Racine, and Thomas Macaulay.[4] He may have received basic instruction in history, geography, Latin and German[5] and 'object lessons focusing on topics of general interest...often of a scientific nature'.[6] One of his early teachers may have awakened his curiosity and love for botany and the broad field of 'natural philosophy' through exposing him to the scientific writings of women like Jane Marcet's *Conversations on Chemistry*[7], *Conversations on Political Economy and Adapted to the Comprehension of Young Pupils*[8], and *Conversations on Natural Philosophy*[9], and Priscilla Wakefield's *An Introduction to Botany, in a Series of Familiar Letters*. [10]

---

[1] J. Walvin, 1982, *A Child's World: A Social History of English Childhood 1800–1914*, Penguin Books, Harmondsworth, England, p. 112.
[2] J. Purvis, 1989, p. 71.
[3] Kathryn Hughes, 1993, pp. 70–73.
[4] ibid., p. 73.
[5] J. Walvin, 1982, p. 112.
[6] A. Digby & P. Searby, 1981, pp. 31–32.
[7] Jane Marcet, 1825, *Conversations in Chemistry; in which the Elements of that Science Are Familiarly Explained and Illustrated by Experiments*, 2 vols., Longman, Hurst, Rees, Orme, Brown and Green, London.
[8] Jane Marcet, 1816, *Conversations in Political Economy; in which the Elements of that Science are Familiarly Explained*, Longman, Hurst, Rees, Orme, Brown and Green, London.
[9] Jane Marcet, 1822, *Conversations in Natural Philosophy; in which the Elements of that Science are Familiarly Explained and Adapted to the Comprehension of Young Pupils*, Longman, Hurst, Rees, Orme, Brown and Green, London.
[10] Priscilla Wakefield, 1807, *An Introduction to Botany in a Series of Familiar Letters*, Darton & Harvey, London.

Wakefield's *An Introduction to Botany* may have provided Augustus' first insight into Linnaean plant classification. However, from it he would not have derived a clear idea of that system since Wakefield sidestepped the topic of using sexual parts of plants in classifying them[11] due, no doubt, to prevailing opinion that it was improper for women and children.[12] But, of course, Augustus' passionate interest in botany might have been kindled later in his life. His brother Edwin was, however, convinced that Augustus' 'love of science was innate, and while yet a boy he had picked up at the book stalls in London, and carefully studied the best works of the time in science and literature'.[13]

A letter in *The Times,* dated 8 July 1841, stated that 'at different periods, from the year 1820 to the present time' children of all the signatories, of which Frederick Oldfield was one, had been educated at the Liborel family's schools in Guînes, 6 miles south of Calais, France.[14] Frederick and Mary left no details as to why they decided on that course of action. Perhaps the move was driven by a belief that it would ensure the children had the best start in life they could provide—a belief probably born of the excellent reputation of French schools[15]—and Mary and Frederick's desire to provide opportunities for them to escape the negative effects of their father's incarceration and notoriety.

Liborel's academy for young gentlemen had, according to an advertisement in *The Times* in 1847, 'for the last 26 years enjoyed a continued and increasing patronage by most respectable families in England'; this puts its establishment at 1821.[16] Unfortunately, available sources do not provide particulars of management and staffing arrangements at the school, where it was located or a description of the building prior to 1828, nor do they reveal any indication that the Liborels were running a seminary for young English ladies prior to 1834.[17] According to Jacques Louf, in his article on boarding schools in Guînes from 1700, Louis Joseph Berthe Eugène Liborel, known as Eugène Liborel, was the headmaster of the local state boys' school until 1838 and was the 'original Liborel' who had purchased his first property on which to establish a school in 1828.[18] What then of the original Academy? A feasible scenario is that he and his wife, Dorothée Laude, established it in a rented building in 1821 with Eugène as Principal, assisted by Dorothée in the role of part-time school mistress and manager of the boarding house, its retinue of domestic staff and the school's additional teachers recruited locally and possibly from England. He and Dorothée had five children and four of them became teachers in schools their father established in Guînes. Their son, Louis-Joseph Liborel probably taught at this school from its inauguration.[19]

This Academy may have been where the Oldfield boys, Frederick Jnr and Alfred Bennett, were educated until about 1828 and then completed their education at Eugène's recently purchased Beaulieu House in the Rue Calvaire.[20] Eugène purchased land in the Rue du Parcage in 1832 and built a small boarding school, 'no doubt intending it to serve as a school for girls to be run by one of his children'.[21] By that time, it is likely that Marian, Frederica and Lavinia had completed their education; very few

[11] ibid. p. 85.
[12] David Elliston Allen, 1976, p. 48.
[13] E.D. Oldfield, 'An Australian Naturalist', *Bairnsdale Advertiser and Tambo and Omeo Chronicle*, 16 July 1887, p. 3, DAN, NLA. This obituary could not have been written by Alfred because by the time of Augustus' death, he had been deceased for over a decade.
[14] 'French Schools', Letter to the Editor, *The Times*, 8 July 1841, p. 8, TTDA.
[15] Chantal Courbot, 2005, *Education des Jeunes Britanniques: Une Eclosion de Pensionnaits*, Introduction, Museum of Calais Exhibition, 2005 catalogue, p. 1 (excerpts kindly provided by Monsieur Jacques Louf, Société Historique de Guînes et Des Environs, 18 January 2006). Apart from the families of British Army officers stationed in France after the Battle of Waterloo, other Britishers residing in France and in Great Britain 'did not hesitate to put the education of their children in the hands of French teachers, because French schools had an excellent reputation', ibid.
[16] Advertisement, 'Education in France', *The Times*, 12 January 1847, p. 10, TTDA.
[17] The first advertisement we found calling for enrolments at a Liborel seminary for young ladies appeared in *The Times* 17 July 1840, p. 2, TTDA.
[18] Jacques Louf, 1987, 'L'enseignement à Guînes et les éducateurs due XVIe siècle à nos jours', *Bulletin Historique et artistique du Calais*, no. 111–112, Sept.–Dec., p. 186.
[19] Chantal Courbot, 2005, exhibit no. 100. This is surmised because the text associated with this exhibit stated that Louis had been teaching for 24 years in 1844.
[20] Advertisements in *The Times* give no indication that Eugène was operating two academies for boys at that time.
[21] According to Louf, Liborel bought this land in 1832 but 'it is possible that it was his home and his first school', Sept.–Dec. 1987, p. 187. Until then there may have been a building on the site which housed Eugene's first academy for young gentlemen.

girls continued with their schooling beyond the age of fifteen. It is a matter for speculation as to whether the girls attended a seminary for young ladies run by the Liborels; no evidence was found that such existed at that time. There were other schools with links to private boarding houses catering for British girls in Calais from 1825[22], which they may have attended. If the three girls were at a school in Calais, Marian could not have been there before she was thirteen years old, Frederica, eleven years and Lavinia, ten years. From what we know of Frederica she was fluent in French and German and sufficiently proficient musically to have later established herself as a music teacher. If, as may well have been the case, she gained these skills in France, there is little doubt that Marian and Lavinia were given the same opportunity.

Augustus was sent to Monsieur Liborel's academy to complete his schooling.[23] Although it is not known at what age he went to France, the school he attended would have been the Beaulieu House Academy. Boys under eleven years were accepted[24] but no indication of the minimum age has been found. If he was younger than nine years when he was sent to Guînes his father was still in prison or, if he had his freedom, the boy was a year or two older. The school may have been under the day-to-day direction of Louis Liborel and his wife Horténse Garenaux; it seems that Eugène and Dorothée retained an interest in it until 1838 when they sold it to Louis.[25]

As was the case in Britain, teacher training in France was rudimentary; teachers learned through experience in the classroom but to begin practising they were required to show competence at one of three levels of scholarship and to obtain a certificate of good conduct signed by the mayor and the parish priest. Eugène and Louis were holders of a second level teaching qualification, which meant that in addition to being numerate, literate and able to write they had demonstrable knowledge of spelling, arithmetic and calligraphy. Private schools like the Liborel's, did, however, provide quality education. They offered a variety of subjects at levels above the minimum standards required by law by recruiting more highly qualified teaching staff locally and from England.[26]

In an undated Beaulieu House Academy prospectus[27] the school was promoted as ranking among the first in the north of France and, according to Jacques Louf, it was one of two boarding schools that 'had the biggest influence on Guînes' local history in the nineteenth century.[28] Liborel claimed to provide pupils with 'a sound education, enabling them to become honourable members of society' and enhancing their chances of 'obtaining a good situation, whether in government or mercantile office', and for which 'of late years, the knowledge of the French Language, taught where it is the language of the Country, has become absolutely necessary'. Pupils' health was stated to be of the highest importance 'for where there is neither health of the body nor contentment of mind, the intellectual faculties cannot so fully develop themselves'. A physician visited daily to attend to the students. Their 'religious duties' were 'not neglected' —there was a Protestant Church and an English minister had been engaged to perform Divine Service every Sunday. The fee per annum for boys under eleven years of age was 15 guineas, for those under fourteen, 30 guineas and for those who were older, 35 guineas. Apart from optional extra subject tuition, additional charges included: the 'sitting' at the

---

[22] Chantal Courbot, 2005, p. 1.

[23] Frederick Oldfield, *The Times*, 8 July 1841, p. 8 TTDA. In this letter support was expressed by 17 parents of past and present pupils for the education and care provided to their children by the Liborels 'at different periods, from the year 1820 to the present time'. One of the signatories was J. Halford, 14 Down Street, Piccadilly, father of George Britton Halford who later became Professor of Medicine at the University of Melbourne and whom Mueller stated was a school fellow of Augustus, F. Mueller to G. Bentham, 5 February 1866, Kew Correspondence, Australia, Mueller, 1858–70, fols. 196–201, RBGKA, cited in R. W. Home *et al*, (eds), 2002, vol. 2, 1860–1875, p. 337.

[24] Louis Liborel (?) *Beaulieu House Acadėmy for Young Gentlemen Under the Direction of M[onsieur] L. Liborel at Guînes, near Calais*, Prospectus, Calais, undated, c.1840, copy kindly provided by Jacques Louf, 24 November 2005, HOA. The original is in the Guînes Town Museum, France.

[25] Jacques Louf, Sept.–Dec. 1987, p. 186.

[26] Jacques Louf to MHH, 18 January 2006, HOA.

[27] Louis Liborel, (?), c.1840.

[28] Jacques Louf, Sept.–Dec. 1987, p. 184.

Protestant Church, 4s 0d; medical attendance, 5s 0d; repairing clothes, linen, stockings, etc., 5s 0d and cricket club, 2s 6d.[29]

It seems likely that Augustus' journey to Guînes commenced at a London hotel or some other pre-arranged meeting place where Monsieur Liborel or one of his representatives received children ready to escort them to France.[30] It is not difficult to imagine that if Augustus was under eleven years of age, and he probably was, he was upset and apprehensive at the prospect of leaving his parents and all that was familiar to him, particularly as he must have been aware that he would be away from them for the long haul; there were no vacations from one year's end to another.[31] Frederick Jnr and Alfred Bennett were probably still at the school. If so, Augustus' fears may have ameliorated to some extent at the thought of reuniting with his brothers. For Mary, too, the parting must have been difficult no matter how much she thought it was for the best or was reassured by the claims made about the education and care provided by the Liborels. She was almost certainly there to bid him good-bye. Including the clothes he was wearing, the remainder of his luggage comprised of the required two pairs of trousers, two jackets and two waistcoats, six day shirts, three night shirts, six pairs of stockings, twelve handkerchiefs, twelve collars, two or three pairs of boots, brushes and combs, 6 towels, a knife, fork, spoon, a Bible and prayer book. One or two favourite books and other items may have been included, but nothing valuable—watches and jewellery were prohibited.[32]

Following their farewells, Augustus, along with the rest of Monsieur Liborel's charges, travelled by coach to Dover where they boarded a ferry for the short voyage across the English Channel to Calais, and then into another coach for the six-mile ride south to Guînes. At times, the Channel crossing can be very rough; if Augustus and his companions experienced such a voyage and accompanying sea sickness, they must have craved for their mothers.

If the prospectus reflected the actual situation, on arrival and subsequently 'the care and attention of Madame Liborel [were] most maternal, and by whose strenuous exertion and kindness towards the Pupils, the loss of home [was] scarcely felt by them'.[33] Augustus' new residence comprised a basement, two stories and an attic. His dormitory was 'spacious, lofty and perfectly ventilated' and his 'bedstead', exclusively for his own use, was of iron—'experience having shown their superiority over those of wood'.[34] If the Liborels' claims are correct, from the outset the boys enjoyed the meals that were dished up to him; food was apparently 'of the best quality, bountifully and unrestrictedly supplied to Pupils'.[35] The school's facilities included a gymnasium and 'a large field surrounded with trees, in which there was a reservoir expressly for the pupils to bathe, skate or slide', [and] every other recreation 'that is consistent with their health and amusement'.[36]

Augustus and his fellow students were required to wear the uniform of the Academy which was apparently provided after they arrived; it cost from £2 5s 0d to £3 10s 0d. This outfit comprised a tunic, trousers, cap and belt and was one 'generally approved of by everyone for its simplicity and elegance'. The illustration of Beaulieu House Academy shows students in uniform marching out of the building; the tunic reached to just above the knees and seems to have flared from the waist and the hat, round in the crown and wide-brimmed.[37]

[29] Louis Liborel (?), c.1840.
[30] See for example, Advertisement, 'Education in France', *The Times*, 2 October 1833, p. 1, TTDA.
[31] Louis Liborel (?), c.1840.
[32] ibid.
[33] ibid.
[34] ibid.
[35] ibid.
[36] ibid.
[37] ibid.

Fig. 5.1 Beaulieu House Académy for Young Gentlemen, Guînes, near Calais, c. 1840

The normal curriculum at the Academy embraced French and English languages, geometry, algebra, arithmetic, book-keeping, writing, history and geography. The cost of stationery was included in the fees but text books were not included. Optional extras were German and Latin, £1 1s 0d each; music (any instrument), £1 0d; use of the piano, 5s 0d; and drawing and dancing, 15s 0d each. Augustus' teachers were Monsieur Liborel and 'competent' French, English, and if he studied the language, German masters. There is no evidence that one of Augustus' subjects was German but Latin may have been since he was able, later on, to use words where descriptions of plant species were in that language.

Neither Augustus nor his siblings left any details of their experiences at the Liborels' Academies nor how long they were there. It seems, however, that their father and the other signatories to the letter to *The Times*, London dated 8 July 1841, 'having confided the instruction of [their] children' to the Liborels were 'happy hereby to testify to the particular care and kind treatment they have there received, and to the rapid progress they have made in every branch of their education'.[38] It is difficult to imagine that during their time at the Academy the boys had no face-to-face contact with their parents. Perhaps they were taken out of school to return to London for a week or two every year running the risk of them getting behind in their studies, or Frederick and/or Mary may have visited them periodically in Guînes.

Apparently, Augustus did not receive formal instruction in any of the sciences at the Academy but he may have been able to advance his interest in them in other ways: through accessing books on scientific subjects at the school, or elsewhere in Guînes, and by exploring the local terrain. On the other hand, perhaps one of his masters or the Anglican minister was an amateur botanist, and recognising the boy's interest, encouraged him by taking him, and possibly others, on field excursions and teaching him to collect, preserve and name local plants. During this time, he may have been exposed to the natural systems of classification promoted in the late 18th and early 19th centuries by the French botanists, Bernard and Antoine-Laurent de Jussieu.[39]

[38] Frederick Oldfield, 8 July 1841, *The Times*, p. 8.
[39] Jim Endersby, 2008, p. 174.

Edwin, it seems, was also educated at Beaulieu House, joining Augustus and his other brothers there sometime between 1830 and 1833. Early in 1833 they would have received the sad news of Lavinia's death. The historical record has revealed nothing of the cause of her demise at Bishop's Walk, Lambeth, in January 1833 at the very young age of seventeen. She was buried on 23 January of that year at the Anglican Church of St Mary at Lambeth.[40] Apart from this and her baptismal record the only other evidence of her existence to have survived is an embroidery sampler which she created when she was about ten years old; it is dated 22 November 1825.[41]

After Frederick's release from prison he and Mary had one more child, Vincent Leopold Howard, who was born c.1834 in Pimlico, Middlesex, London.[42] This event possibly occurred at St George's Hospital where, it has been suggested, Frederica came into the world.[43]

At what age Augustus left Beaulieu House Academy is not known. We believe he might have been there until c.1836 when he was about fifteen. He later stated that he had capability in bookkeeping and the French language, proficiencies which he would have acquired at the Academy, and 'surveying small tracts' indicating that he could serve as an Assistant Surveyor.[44] J. H. Maiden, after reading Oldfield's unpublished manuscript, considered him a competent mathematician and linguist, 'but especially biologist in both branches of zoology and botany'.[45]

It was not uncommon for biologists of that era to gain their scientific expertise from their own reading; one was William Hooker, Director of the Royal Botanic Gardens, Kew from 1841 to 1865. He was educated at an academy for young gentlemen in England but did not attend university.[46] Oldfield's writings, including his letters, his paper on Australian Aborigines and his natural philosophy manuscript, indicate a broad general education that may have included some time at a university or college where university-level academic courses were provided in biological sciences and cadastral survey.

During the mid- to late-1830s, when Oldfield was most likely to have undergone survey training, it could only be obtained through an apprenticeship under a master surveyor or via courses of instruction within engineering curricula at either King's College, London, or the University of Edinburgh, or Trinity College, Dublin or Durham University.[47] No evidence has emerged, however, that he was a fully qualified surveyor. Augustus' name did not appear on any of the, often incomplete, alumni records of any of the universities in England, Scotland or Ireland.[48] Although student records for the late 1830s are incomplete[49], he could have attended the Anglican King's College London since, as far as we have been able to ascertain, it was the only institution where both surveying and the biological sciences were taught toward the end of the 1830s. Scientific subjects taught at King's included natural philosophy and astronomy, experimental philosophy, chemistry, botany and zoology. Engineering, arts

---

[40] Record for Lavinia Sarah Oldfield, London, England, Deaths and Burials, 1813–1980, viewed 29 March 2010 at www.ancestry.com.

[41] In the possession of Mr & Mrs Desmond Williams, Albert Park, Melbourne. Photo by kind permission of the owners.

[42] Record for Vincent Oldfield, 1851 England Census, Source information: HO107.1473; Registration district: Chelsea; Sub-registration district: Chelsea North West; folio: 10; p. 13, viewed 29 March 2010 at www.ancestry.com. The 1841 England Census, County of Middlesex record for the Oldfield residence, 14 Dorset Street, Portman Square (PRO Ref: 107/680/2, File: 10105, viewed 29 March 2010, at www.ancestry.com, gives Vincent's age as 7 years which tends to confirm that he was born in 1834.

[43] see Chapter 2.

[44] AO to the Directors, New Zealand Company, 3 September 1841, New Zealand Company General Correspondence, 41/4605, PRO CO208/18/1841, NAL.

[45] J.H. Maiden 1911, p. 232–233.

[46] Mea Allan, 1972, 'Hooker, William Jackson', in C.C. Gillispie (ed), 1970–1980, *Dictionary of Scientific Biography*, vol. 6, Charles Scribner's Sons, New York, pp. 492-495.

[47] M.R. Bailey, M.M. Chrimes, R.C. Cox, P.S.M. Cross-Rudkin, B.L. Hurst, R.C. McWilliam, R.W. Rennison, E.C. Ruddock, R.J.M. Sutherland, T. Swailes (eds), 2008, *Biographical Dictionary of Civil Engineers in Great Britain and Ireland*, vol. 2, 1830–1890, The Institution of Civil Engineers, London, p. 21, viewed 25 April 2012 at www.thomastelford.com.

[48] Augustus was not among the students listed in available alumni records for the late 1830s. Universities and colleges known to be operating at the time, namely, King's College London, Birkbeck College (formerly the London Mechanics Institute), University College London, Oxford University, Cambridge University, Glasgow University, Edinburgh University, Trinity College, Dublin, and Durham University were contacted by email to ascertain whether Oldfield was ever an enrolled student; the responses were all negative.

[49] King's College student registers for the period 1828–1848, which are incomplete for 1835–1840, and college calendars for 1829–1844, which included prize winners, King's College London Archives (KCLA). The authors are grateful to Ms Katy Thomas, Records Assistant, for her help in identifying the relevant KCLA records.

and manufacturing courses were apparently established in 1838.[50] These included, among other subjects, land surveying, mineralogy, mathematics and geology, all subjects where Oldfield's writings show he had more than a passing knowledge.[51]

Family developments 1833 to 1841

Augustus may have been present when Juliana and Vincent, respectively eleven- and five-year-olds, were christened on 24 July 1839 at the Anglican Holy Trinity Church, Brompton Road, in the Parish of Kensington. The family was residing at 47 Hans Place, Kensington. Frederick recorded his occupation as 'gentleman' in the baptismal register which suggests he was no longer actively involved in the wine trade.[52] Just why he and Mary waited so many years to have the children baptised is puzzling. Perhaps the delay reflected their disenchantment with the Church and they were goaded into action to ensure that Juliana's and Vincent's religious status was clear when they entered the schools in Calais.

Although there is no firm evidence, Frederick probably continued to run his business in Pall Mall when the family was residing at Bishop's Walk. When they moved to Hans Place is not known; possibly not long after Lavinia's death. Frederick Jnr operated a wine retail business at 7 Queen's Buildings, Brompton Road, Brompton, which traded under the name 'Frederick Howard Oldfield, wine merchant'.[53] Alfred Bennett was employed by his brother as a salesman.[54] The business was operating in September 1839 by which time young Frederick Jnr had married Mary Jane Myers[55] and they with their two daughters, two-year-old Mary Jane and Lavinia Howard aged about three months, resided at 35 Jura Street, Chelsea.[56] Frederick Jnr appears to have been a published composer at this stage for in 1836 his composition 'Four Waltzes, for the Piano Forte' was published.[57]

Queen's Buildings, Brompton, was part of a series of terraces fronting the south side of the turnpike road to Fulham, now Brompton Road. The terrace previously called Queen's Row was, by 1800, referred to as Queen's Buildings, Brompton. It ran from Hooper's Court to what was then New Street, now Hans Road.[58] The exact location of Frederick's premises can be seen on Horwood's map published in 1813.[59] None of the original terrace houses were large, 'although most were at first decently inhabited [from c.1766]' and enjoyed substantial front gardens to shield them from the traffic. After about 1830, however, the district began to deteriorate. Single-storey shops began to take over the gardens on Brompton Road while the back-courts and streets 'slipped into multi-occupation'. As fashionable development moved westwards the area gained a reputation for its 'insalubriousness'.[60]

Frederick Jnr and his family had moved from Jura Street by the time of the *1841 England Census* (6 June 1841); they were residing at a more stylish address, 8 Montpelier Row, Knightsbridge. His

---

[50] Anon., Statement of the Arrangements for Conducting the Various Departments of King's College London, August 1838, KCLA

[51] King's College London, Reports from the Council to the Annual General Court of Governors and Proprietors for the years 1840 –1842, KCLA.

[52] Register of Baptisms Solemnised in the Holy Trinity Church, Brompton Road in the Parish of Kensington, Records Nos 291 (Juliana) and 292 (Vincent), London Metropolitan Archives, Arch. P84/TR12, microfilm 089/007.

[53] *Pigot's & Co. Directory of London for 1839*, vol. 197, 1839, CEP Microform Ltd, Research Publications Group 2 1800–1855, Directories, vols. 118–205, Film 96917, reel 53, NLA. See also indictment of Samuel Matthews for a misdemeanor on 16 September 1839, Eleventh Session, 1839, Ref. no: t18390916–1656, *The Proceedings of the Old Bailey*, viewed 24 November 2011 at www.oldbailey.org. In this case the witnesses were Frederick Henry Howard and Alfred Benjamin [sic] Oldfield. The defendant was alleged to have fraudulently obtained wine on two occasions from Oldfield's shop. The Oldfield brothers quite clearly indicated in their respective statements to the Court that Frederick Henry Howard owned the business.

[54] ibid.

[55] No record of their marriage has been found.

[56] Record for Mary Jane Oldfield, Register of Baptisms, Chelsea Ranelagh Chapel, 5 July 1837 LMA, lma/4404/b, Item: 007, viewed 17 October 2012 at www.ancestry.com. Jura Street, Chelsea could not be located on modern or historical maps of the area.

[57] Frederick Henry Howard Oldfield, 'Four Waltzes, for the Piano Forte', 1836, viewed 9 December 2013 at www.amazon.co.uk.

[58] F.H.W. Sheppard (ed), 1983, vol. 41, pp. 9–10 & 12. Name changes to the terraces originally called Queen's Buildings are described in this publication. The turnpike road to Fulham was frequently referred to in the 1830s and 1840s as Brompton Road, but not officially until 1863.

[59] Richard Horwood, 1813, Map. 21.

[60] F.H.W. Sheppard (ed), 1983, vol. 41, p. 12.

occupation was given as 'wine merchant'; no live-in servant was recorded.[61] He was still operating his business in Queen's Buildings.[62]

Frederica, at the time of the *Census*, was a governess residing at Collinia House, Bathwick, Bath Forum, Somerset. She was not living with her employers because no residents under the age of 20 were recorded as part of the household.[63] Alfred and Edwin were not mentioned in the *1841 England Census*.[64] Frederick and Mary, Marian, Juliana and Vincent had moved to 14 Dorset Street, Portman Square. The *Census* recorded a live-in family servant, twenty-year-old Elizabeth Watson and Frederick's occupation was given as 'wine merchant'.[65] Frederick Jnr apparently transferred his business to his parents' residence or, as events unfurled, it is probable that his father took over his company and retained the name. Thus, by at least 1843 'Frederick H. Oldfield & Co.' was located at 14 Dorset Street, Portman Square.[66]

Originally covering an area of 270 acres stretching from Oxford Street to Regents Canal, the estate on which the Oldfields were living was acquired by Sir William Portman of Somerset, Lord Chief Justice to Henry VIII, in 1532. During the late 18th century, Henry William Portman started to develop the area by building large houses for the wealthy around the perimeter of the Square, the central part of which was a landscaped garden. Although the estate dictated the lines of streets and the open spaces, it leased the land to private and speculative builders who were also responsible for providing services.[67] By 1820 the road system existed as it is today, and surrounding the original building development were blocks of terraces for residential and commercial purposes. The Oldfield residence was on the north side of Dorset Street and the third terrace house to the east of Baker Street. It was adjacent to the south entrance to the mews which served all the terraces that bounded Spring, Crawford, Baker and Dorset Streets[68]; it has since been demolished and replaced with modern commercial multi-storey buildings.[69]

Augustus enters the teaching profession

Augustus, as a 21-year-old, was School Assistant at the Chipping Norton Free Grammar School in Oxfordshire under School Master Edward Hartley (aged 25 years) on 6 June 1841.[70] Allowing for the possibility of two or three years at King's College, Augustus may have commenced at the grammar school in 1838 or 1839. In common with his contemporaries, he was unlikely to have undergone any formal teacher training.[71] Rather, armed with what prior education he had received he would have entered the teaching profession expecting to acquire his teaching skills through experience on the job.[72]

[61] England Census, 6 June 1841, Civil Parish: Knightsbridge; Hundred: Westminster; County/Island: Middlesex: Registration District: Westminster; Sub-registration District: St Margaret; Source information: HO107/738/17; Folio: 43; Page: 3; GSU Number: 438839, viewed 18 March 2010 at www.ancestry.com.

[62] *The Post Office London Directory 1841*, Frederick Kelly, London, p. 502: Oldfield, Fras. [sic] Howard, wine merchant, 7 Queens Buildings, Brompton, microfilm, NLA.

[63] Record for Frederica Oldfield, *1841 England Census*, Registration district: Bath; Sub-registration district: Bathwick, Book: 6, Folio: 6, Page no: 4. Last view 11 May 2012 www.ancestry.com.

[64] There were no records for them in the *1841 England Census*.

[65] *England Census 1841*, PRO Ref: 107/680/2, certificate of search provided by CENSEARCH, Morley, Western Australia, File: 10105, Rec. no. 1929–1934.

[66] *Robson's London Directory: Street Key & Royal Court Guide for 1843*, 24th edition, Bowtell & Co., London Directory and Court Guide Office, Cheapside, microfilm, NLA. This Directory gave 'Dorset Street, Manchester Square. Manchester Square is part of the Portman Estate as is Dorset Street (see Richard Horwood 1792, Map 11, from Joseph Wisdom, *The A to Z of Regency London*, Harry Margary, Lympne Castle, Kent in association with Guildhall Library, London, 1985, p. 23).

[67] 'Portman Estate', viewed 22 March 2010 at www.wikipedia.org.

[68] Richard Horwood, 1792, Map. 11.

[69] MHH and WH visited the street in 2002 and noted the modern buildings on the site.

[70] Record for Augustus Oldfield, *1841 England Census*, Source information: HO107/879/2, Registration District: Chipping Norton, Sub-registration district: Chipping Norton; Folio 27, p. 18, GSU No.: 474573, viewed 15 February 2012 at www.ancestry.com.

[71] See Chapter 1.

[72] R.J.W. Selleck, personal communication to MHH, email, 25 October 2006, HOA.

Fig. 5.2 Grammar School, Chipping Norton, Oxfordshire, 1823

The school, which is still in existence, dates back to 1450 when the vicar of St. Mary's Church and certain parishioners were licensed to found the Gild [sic] of the Holy Trinity. In 1818 Carlisle's account of the school shows that it was in a prosperous state, having forty to seventy pupils. By 1824 there was a large school building and residence for the master. In the same year the Government commissioners also reported very favourably. The free scholars were still being regularly appointed and a good classical education was given. At the time Augustus was at the school there were, apart from day pupils, eighteen students in residence, as were principal Hartley, his mother, Ann, two of his siblings and two servants.[73] Unfortunately, no details of the school's management, curriculum and status as an educational institution in the 1830s appear to have survived[74] nor did Augustus leave any comments on his stay there.

Three brothers emigrate

Augustus, however, was not the first of the Oldfields to leave England. Frederick Jnr gave up his wine business and emigrated to New York, USA, aboard the *Mediator*, arriving there on 26 April 1842.[75] He was joined by Mary Jane and their two young daughters on 11 August of that year having made the Atlantic crossing on the *Montreal*.[76]

Nineteen-year-old Edwin, who had previously served as an Ordinary Seaman on the *Royal Admiral*, signed on in the same capacity to the barque *Kinnear* on 1 June 1842.[77] Under the command of William Lidderdale, the *Kinnear* left London on 10 July bound for Sydney, New South Wales, with

[73] Record for Augustus Oldfield, *1841 England Census*. There is no indication in the record as to how the Hartleys were related. Ann was 45 years and the other two 23 and 14 years so it is assumed Ann was Edward's mother and the younger Hartleys his siblings.

[74] L.F. Salzman (ed), 1939, 'Schools', *A history of the County of Oxford*, vol. 1, Victoria County History (sponsor), pp. 457–490, at www.british-history.ac.uk.

[75] Record for Frederick H. Oaldfrel, age 24, New York Passenger Lists, 1820–1957, viewed 8 May 2012 at www.ancestry.com. The spelling of the surname in the original record is 'Oaldfiel [last letter indecipherable]' and although it has been translated as 'Oaldfrel' the database search engine considered it to be a variant of Oldfield. His occupation is given as 'mason' on the passenger list. Perhaps he gave this occupation to obtain assisted travel under a scheme to encourage skilled migrants.

[76] Record for Mary Jane Oldfield, age 21, New York Passenger Lists, 1820–1957, viewed 8 May 2012 at www.ancestry.com. The two children are listed on the original record.

[77] Agreement with William Lidderdale, Master of the Ship *Kinnear*, 1 June 1842, Admiralty and Board of Trade: General Register and Record Office of Seamen: Agreements and Crew Lists, Series 1, Port of Registry: London Ships Names, JO–K: BT 98/370, 1835-1844, NAL.

180 male convicts on board. It arrived in Hobart Town on the 23 October 1843[78] where Edwin was discharged.[79] From his very meagre wage of one shilling per month[80] it is evident that he worked his passage to Australia.

It is not known on which ship Alfred travelled to Australia, but he was in Van Diemens Land (hereafter Tasmania) employed as a police constable in 1842, attached to the Hobart Town Police Force; he, too, may have worked his passage. On 9 August 1842, the Chief Constable notified the public through the pages of the *Colonial Times* that he had distributed the sum of £10 to Constable Oldfield and eight others 'given by Mr Whitby, of the Whaler's Return, New Wharf, for the discovery of Wilson and Werry, who committed a burglary at his house'.[81]

Edwin and Alfred had landed in Tasmania during a barrage of resentment and criticism, which, fuelled by 'a mischievous' local press, had been aimed at the Lieutenant-Governor, Sir John Franklin, and his wife Lady Jane very soon after their arrival in the Colony in January 1837 and which led to their recall in 1843. Despite the carping and fault-finding of both Franklin and his wife, emanating from certain sectors of the local community[82], and Franklin's problems with the Colonial Office, not all the press and society were hostile, and a large section of society admired and supported the couple.[83]

Alfred left the police force some time before 29 August 1843, by which time he and Edwin had established a private school in Hobart[84], probably in Roxburgh House on the south-west corner of Elizabeth and Brisbane Streets.[85] On 2 September 1843 the *Launceston Examiner* reported that Alfred was a tutor at 'Cannaway's grammar school'[86]; this may have been the name of the brothers' school, or it may have been another establishment where Alfred tutored part-time when not teaching at Roxburgh House.

On 29 August Alfred gained some notoriety when he was involved in the apprehension of the armed runaway convict, Martin Cash, who had shot Constable Peter Winstanley in Brisbane Street, Hobart Town that night.[87] Subsequently, Alfred's 'exertions' in capturing Cash were deprecated by a reader in the Advertiser (1 September 1843) and this prompted him to write to the Editor of the Colonial Times:

> Last Tuesday evening, while engaged at home I heard the cry of 'stop thief,' and on going out saw several persons running in the direction of Trinity Church. I also joined in the pursuit; and on reaching the junction of Brisbane and Argyle streets, heard the report of a pistol. I then quickened my pace, and almost immediately after saw a man on the ground struggling with two others. Having heard that he had committed murder, and the two men requesting some one to hold his legs, I proceeded to render im-

---

78 'Ship News', *Colonial Times*, Hobart, 25 October 1842, p. 2, DAN, NLA.

79 List of the crew of the trading barque *Kinnear* at the period of her sailing from the Port of London in the United Kingdom and of the men who joined the ship subsequent to such departure and until her return to the Port of London being her Port of Destination in the United Kingdom, 7 July 1843, BT 98/370, Admiralty and Board of Trade (General Register and Record Office of Seamen: Agreements and Crew Lists, Series 1: Port of Registry: London Ships Names, JO–K, 1835–1844, NAL).

80 ibid.

81 *Colonial Times* (Hobart), 9 August 1842, p. 4, DAN, NLA.

82 Lloyd Robson, 1983, *A History of Tasmania*, vol. 1, *Van Diemen's Land from the Earliest Times to 1855*, Oxford University Press, Melbourne, pp. 336–338 and Jenny Parrott, 'Franklin, Sir John and Jane, Lady Franklin', in Alexander, (ed), 2005, *The Companion to Tasmanian History*, Centre for Tasmanian Historical Studies, University of Tasmania, Hobart, p. 145.

83 Lloyd Robson, 1983, vol. 1 pp. 370–385. Also see George MacKaness, 'Some Private Correspondence of Sir John and Lady Jane Franklin; Tasmania, 1837–1845, Part 2, *Australian Historical Monographs*, Vol. 18, New Series, 1947.

84 Edwin D. Oldfield, 'Bushranger Martin Cash', Letter to the Editor, *The Argus*, 30 September 1905, p. 14, DAN, NLA.

85 Alfred Bennett Oldfield, Letter to the Editor, *Colonial Times*, 5 September 1843, p. 3, DAN, NLA. In this letter, which bore the address, 'Roxboro' House' Alfred stated that on the night of the apprehension of Martin Cash 'while engaged at home' he heard the cry of 'Stop thief' and as a result went to find out what was happening. It is on this basis that the suggestion is made that his and Edwin's school was at Roxburgh House. The location of Roxburgh House was according to William Graham Robertson, Hobart Streets, 1 January 1952, (unpublished ms, NM59, c.1919, p.6., Tasmanian Archive and Heritage Office, TAHO.

86 *Launceston Examiner*, 2 September 1843, p. 4, DAN, NLA.

87 'Apprehension and committal of Martin Cash. Coroner's Inquest', *Colonial Times* (Hobart), 5 September 1843, p. 3, DAN, NLA. See also T.O. Beedham, Letter to the Editor, 'The late Mrs Oldfield', *The Mercury*, 6 March 1917, p. 7, DAN, NLA.

> mediate assistance. While thus engaged, someone exclaimed, "he has a pistol—take it from him!" when one of the parties abandoned his hold, and I endeavoured to secure the weapon; but before I could effect my intention, I received the contents in my face. Bleeding and confused from the effects of the wound, I was unable to afford any further assistance, and retired to procure surgical aid.[88]

The Lieutenant-Governor, Sir John Eardley-Wilmot, subsequently approved a reward of £10 each to Alfred and the constables involved in Cash's capture.[89] According to Edwin, reporting some six decades later

> [Alfred] was not much upset by his narrow escape from death and soon recovered from the effects of the wound but he carried to his grave the scars it left and the powder stains on his face.[90]

By 1841 Augustus had decided to emigrate. He wrote to the London office of the New Zealand Company from his parent's home in Dorset Street on 3 September of that year stating his willingness

> to be employed in any capacity where Accounts, French, or surveying of small tracts may be of service.
>
> I should prefer a situation as Surveyor's Assistant but should not object to be employed in all three of these branches. I am 24 years of age, and would be happy to sign articles for a few years.[91]

His readiness to sign articles 'for a few years' suggests that he had plans other than to follow a career in the survey profession. Unfortunately, a Company advertisement for applicants to fill a number of Assistant Surveyor positions in New Zealand escaped Augustus's attention because his enquiry was 'too late; 7 Sept'; the appointments had been settled.[92] Augustus inflated his age in this letter but why he did this has not become apparent. Although he had made up his mind to emigrate by 1841, he did not join Edwin and Alfred in Hobart until c.1845. In the meantime, he probably remained at the school in Chipping Norton.

The historical record is silent as to why the teaching profession was chosen by Augustus, Alfred, Edwin, Frederica and Juliana, or why the four brothers emigrated. The curriculum taught at Monsieur Liborel's school would have prepared them for direct entry into the teaching profession and the Civil Service, either at home or abroad, but there is no evidence that any of them attempted to pursue the latter path. Why Frederick Jnr backed away from his wine venture is a mystery. Perhaps it was because his father interfered too much in the running of the business, or he had another ambition which he wished to satisfy unfettered by parental expectations. It seems that when he emigrated Alfred left behind a business in London. What kind of an operation it was, where it was conducted and when it was established are not known but it was still operating in 1860.[93]

For the brothers, leaving England afforded them the chance of escaping any social stigma they might have experienced as the result of their father's notorious past and which might have dogged their careers in their home country. They were all young men in their early to mid- twenties when they decided to migrate. It is on record that Augustus was eager to travel and botanise[94] so this might have been his primary motivation for leaving England. If so, school teaching was a respectable vocation in the event of his not being able to engage in surveying or to earn a living in botanical exploration. Alfred and Edwin, on the other hand, established a school soon after arriving in Hobart, which indicates an ambition on their part to carve out reasonably lucrative careers in the teaching profession.

---

[88] Alfred Bennett Oldfield, 5 September 1843, *Colonial Times* (Hobart), p. 3, DAN, NLA. Article in the *Advertiser* not sighted.
[89] Government Notice no 259, *The Gazette*, *The Courier*, Tasmania, 20 October 1843, p. 2, TAHO, DAN, NLA.
[90] Edwin D. Oldfield, 30 September 1905, *The Argus*, p. 14, DAN, NLA.
[91] AO to New Zealand Company, London, 1841, p. 249, NAL.
[92] New Zealand Company, Alphabetical List of Surveyors, *1839–1850:* PRO CO 208 281, NAL.
[93] see Chapter 17.
[94] see Chapter 9.

Unfortunately, however, their school was discontinued prior to 1846.[95] What went wrong with that arrangement could not be determined. Perhaps the brothers simply needed to develop sound reputations as teachers, worthy of parents' committing children to their tutelage before they could attract enough students for the school to be economically viable, or their family background may have become known and their school was shunned as a result.

By 1841 Marian was twenty-nine and Frederica twenty-seven. Any education that the sisters had received in France was designed to prepare them for marriage but by 1841 they seemed to be well and truly on the way to life-long spinsterhood. There is no evidence to support an assertion that marriage held no attraction for them, as was the case with some women. On the other hand, men whom they considered to be suitable may have shied away because of Frederick's background or because he was not in a position to enable them to bring any money into the union. Marian remained at home but whether it was because her parents expected her to continue to live there to look after them in their old age is not known. Nor is there any information as to why Frederica entered the teaching profession. Perhaps Frederick could not support her or she may have yearned independence to create a more fulfilling life for herself? Employment choices for well-educated, middle-class women in the early 1840s were very limited. Lowly-paid teaching positions that offered live-in accommodation in genteel families and girls' schools were perhaps the most sought after by those seeking independence or avoiding destitution. Such situations enabled them to retain their respectability, however marginal; as single women, their lives continued to be in those newly adopted social spheres.[96] Apparently Frederica was not a live-in governess, which suggests that she enjoyed her independence. What is known of the way Edwin and Frederick Jnr travelled to their new homeland, indicates that by the 1840s Frederick could not offer them and their two other brothers financial support that would have established them as upper middle-class members of society in Tasmania and the USA respectively.

[95] see Chapter 6.
[96] Anon., 'Governess', *Wikipedia,* viewed 16 January 2018 at www. en.wikipedia.org.

Mt. Wellington, Tasmania

Part 2

Plant collecting becomes a passion

1845–1856

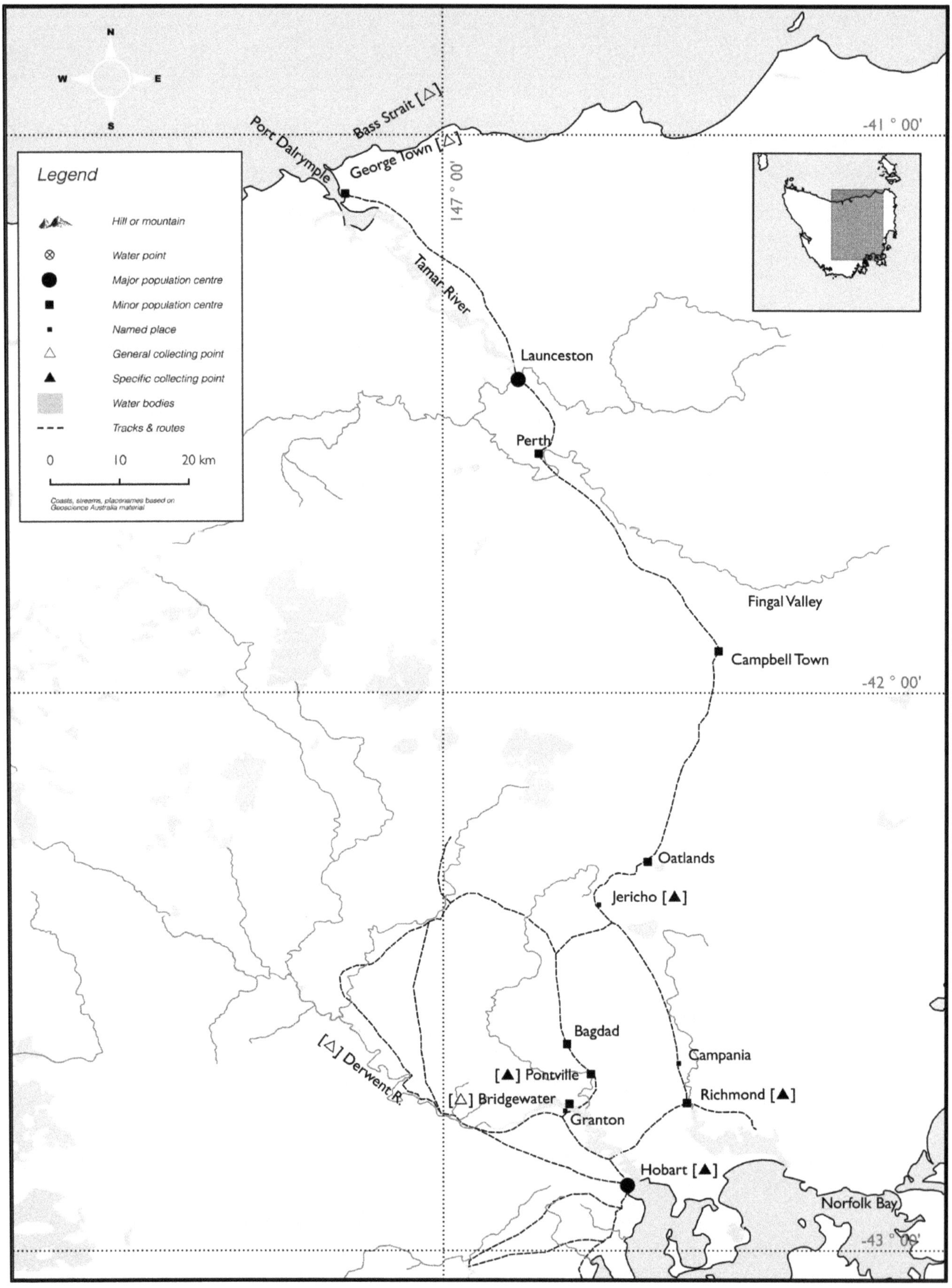

Map 2: Launceston To Hobart

Fig. 6.0: Solanum oldfieldii

Chapter 6

Early years in Australia 1845–1849

Van Diemen's Land and Australian mainland

Augustus' 'Ship to Colony [Van Diemen's Land]' was the barque *Winchester*[1] either owned by Henty & Co. of Portland Bay (then in New South Wales), or Damien Dunbar and Henry Thomas Buckle of London in partnership.[2] Prior to June 1846, when Augustus was already in Hobart Town[3], this vessel transported passengers and cargo from London to Van Diemen's Land three times.[4] Oldfield was not among the passengers and crew on its first voyage to Hobart Town in 1843.[5] Taking four and a half months to complete the journey, the second voyage from London commenced on 15 August 1844 with all passengers bound for Launceston.[6] It arrived on 23 December 1844[7] travelling from London via Cape of Good Hope.[8] The third voyage, under the command of Captain William Milligan, departed London on 25 July 1845[9] and reached Launceston on 2 December 1845.[10] For both these journeys Oldfield was not listed among the passengers and the crew lists are missing.[11] It seems, therefore, that, like Edwin, he worked his passage on either the 1844 or 1845 voyage.

Prior to 1850 Oldfield obtained botanical specimens in New Zealand[12] of which we have located nine undated records, but only three give any indication of where specifically they were collected; these were from Mount Wellington (Auckland) and Wellington.[13] Although it could not be confirmed that the *Winchester* travelled via Cape Horn on its 1845 journey to Launceston, the New Zealand plants were probably obtained during his voyage to Van Diemen's Land since no evidence has emerged to suggest that he visited that country at any other time. Captain Milligan did on occasion use that route when travelling from England to Australia; he returned that way to London in 1845.[14] At this stage, therefore, it is proposed that Oldfield came to Van Diemen's Land in 1845 and made landfall in New Zealand at Wellington and Auckland on the way.

---

1 In an entry for 'A. Oldfield', travelling steerage aboard the *Yarra Yarra* from Launceston to Melbourne on 20 October 1852, he identified the *Winchester* as being his 'Ship to Colony' (Index of Departures 1817–1867, ref: POL2201.2, p. 205, TAHO, viewed 15 April 2008 at portal.archives.tas.gov.au. This passenger was not Alfred Bennett because he was already in the Colony by the time the *Winchester* made its first voyage from London in 1843.

2 'Shipping intelligence' *Hobart Town Courier*, 31 December 1844, p. 2, DAN, NLA. According to *Lloyd's Register of British and Foreign Shipping 1 July 1845 to 30 June 1846*, London, 1845, SLOWA, the *Winchester* was owned by 'Dunbar & Buckles'.

3 'Shipping intelligence' *The Maitland Mercury & Hunter River General Advertiser*, 27 June 1846, p. 2, DAN, NLA.

4 *Lloyd's Register of British Shipping*, 1840–1845, SLOWA.

5 'Shipping intelligence', *Hobart Town Courier*, 29 September 1843, p. 2, DAN, NLA. An Agreement and Crew List for this voyage was found but Augustus' name was not on it, Robert O'Hara, Researcher, NAL, to MHH, email 12 November 2008, HOA.

6 'Shipping intelligence', *Colonial Times*, 19 October 1844, p. 2, DAN, NLA.

7 *Hobart Town Courier*, 31 December 1844, p. 2, DAN, NLA.

8 ibid., 4 December 1844, p. 2. In this issue, it was stated that a valuable horse being transported from England on the *Winchester* died off the Cape of Good Hope.

9 'Shipping intelligence' *Sydney Morning Herald*, 22 November 1845, p. 2, DAN, NLA.

10 *Hobart Town Courier*, 31 December 1844 and 25 July 1845, p. 2.

11 'Shipping intelligence', *Hobart Town Courier*, 6 December 1845, p. 2, DAN, NLA, and Robert O'Hara, Researcher, NAL, to MHH, email 12 November 2008, HOA.

12 AO, to WJH, 15 April 1850, English Letters, vol. 30, I–Z, 1850, fol. 154, RBGKA.

13 (HAODb). With one exception, a *Hypnum crinitum* (now *Wijkia extenuata*) collected on North Island, all the four Oldfield published records we located were given as 'New Zealand'. Of the five herbarium specimens in HAODb three have a specific location. These are: NY236774, NY236775, *Lepidozia procera*, Locn: Mt Wellington, New Zealand, digital download, nybg_vh_specimen list. 20070531–035221– MS Excel, New York Botanical Garden Herbarium (NYBG), and K000772085, *Leptostigma setulosum* Locn: 'Wellington, New Zealand', Kew Herbarium Catalogue, viewed 9 September 2012 at www.apps.kew.org.

14 It was reported that 'The *Winchester* spoke the *Eleanor*, for Sydney, on 15th March [1845] in lat. 56° S, long. 39°W', 'Shipping intelligence', *Hobart Town Courier* 24 September 1845, p. 2, DAN, NLA.

He left no details as to how he weathered the journey. Passengers varied greatly in the way they coped with the voyage regardless of whether they travelled cabin or steerage. Commenting generally about conditions aboard immigrant sailing vessels, Robin Haines concluded:

> The ship, in reality, constituted a tenement block of two to five hundred people or more, with a similar mix of ages, the main difference being that the seafarers were under the strict supervision of an authority to whom submission was mandatory. And there was no escape.[15]

Most were prone to seasickness, particularly in the early stages of the voyage and under certain weather conditions, but as time went by there were those who adapted well to shipboard life. They did what they could to maintain personal care and shipboard sanitation routines with only basic facilities at their disposal and most steerage passengers came to terms with their overcrowded, stuffy accommodation below deck. They variously passed the time organising and participating in entertainments, religious activities, schooling children, reading and writing, and taking in their stride the 'incessant din, uproar, movement and vile odours, which...accompanied the emigrants as they rolled, lurched or sped their way to their new lives in Australia'.[16]

In the process, many formed friendships that continued and thrived in their adopted country. Crew members and male passengers often became firm friends but relationships between seamen and female emigrants were actively, but not always successfully, discouraged. Others, however, had great difficulty in coping with strange food, the lack of privacy, boredom and unpleasant living conditions, possibly exacerbated by anxiety about what their future would bring and chronic poor health due to deprivations experienced in Britain.[17]

Since the 1820s mortality and morbidity rates among passengers aboard convict and migrant ships had greatly decreased due to improved sanitation. However, all passengers, particularly children under five years of age, still faced the risk of serious illness and death from infectious disease, malnutrition, accidents on board, shipwreck and, for some women, birthing complications. The comfort and health of passengers, however, depended much on the medical skills, sanitation and nutrition management by ships' surgeons and the extent to which ships' commanders sympathetically supported them.[18]

Augustus may have followed the practice of many young male emigrants and kept a diary throughout the voyage containing weather observations, descriptions of birds and fish studied or caught by passengers and crew[19], and how he spent his time when the *Winchester* stopped at ports.

What feelings and thoughts he experienced as the ship came within sight of Van Diemen's Land we will never know. Perhaps, as David Burn wrote in the early 1840s, Oldfield along with many migrants when their long and weary journey was drawing to a close,

> thought…of what may be their fortune in the land that lies sleeping in such death like repose before them, casts an awe upon the heart, imbued (as are those of all new emigrants) with an undigested impression of the Botany-bay character of the place, with a superadded [sic] and correspondingly increased estimate of their own immaculate and superior worth. These, and the fact that a certain and not uninteresting portion of life has for ever passed; that, in a few hours more, they who have for months formed as it were members of one common family, will be scattered and dispersed, never to meet again.20

---

[15] Robin Haines 2003, *Life and Death in the Age of Sail: The Passage to Australia*, University of New South Wales Press, Sydney, p. 80.
[16] ibid.
[17] ibid., pp. 32–74 and pp. 123–165.
[18] ibid.
[19] ibid., p. 33.
[20] David Burn 1973, *A Picture of Van Diemen's Land: A Facsimile of a Work Published in 'The Colonial Magazine' 1840–41*, facsimile, Cat & Fiddle Press, Hobart, p. 37.

Van Diemen's Land as Oldfield might have first experienced it

Map 2 p. 84 - Launceston To Hobart

Augustus' initial experience of the Colony was George Town, the clearing port at the mouth of the Tamar River, and from there, journeying up the river before disembarking at the port of Launceston.[21] In common with other contemporary observers, he would have found much that was familiar in his new surroundings. It is unlikely that his first forays into the town and surrounding countryside were made without discovering that the settlers had worked hard to create built landscapes reminiscent of Britain. The climate had allowed them to replicate English patterns of land use and plant cultivation[22] and to retain the modes of dress of their home country. Parks and gardens were planted out with species from the homeland—oak, ash, elm, birch, cedars, spruce and larch, and lanes and hedgerows with sweet-briar, hawthorn or poplars, river banks with willow.[23]

Soon after his arrival at the port of Launceston, Augustus set out for Hobart Town to find Edwin and Alfred. Nothing is known of their circumstances or whereabouts at that time but there is no reason to suggest they had left the Colony's capital. They may still have been operating their school in Hobart Town. If it was at Roxburgh House on the corner of Brisbane and Elizabeth Streets, Augustus would have had little trouble finding them; it was within easy walking distance of the wharf. Wherever it took place, the reunion must have been a happy one with much exchange of news and relief that Augustus had arrived safely.

It is doubtful that Oldfield travelled from Launceston to Hobart Town on a coastal ship. Keen to see his adopted country and his brothers he probably opted for the faster journey by coach along the main road which connected Launceston with the city. At that time, a daily stage coach and mail service to and from Launceston was operating and he had the choice of that conveyance and the rival private carriers, Hyrons' 'Comet' and Mrs Cox's 'larger diligence'. By 1845, with the labour of 1,000 convicts, work had progressed on improving the 'execrable state' of the main road and some more solidly constructed deviations and river crossings had been opened to traffic. While all carriers completed the journey in one day—Mrs Cox's diligence was the fastest, taking fifteen hours—there was still much work to be done to lessen significantly travellers' discomfort, particularly as rival day coaches raced along, sometimes neck and neck, putting passengers' lives at risk. 'Though slowed on the hills, they dashed along at eight or nine miles an hour, to the sound of cracking whips, clattering hoof-beats and the rumble of wheels'; accidents were quite common. The situation was exacerbated when the interior and roof of a coach were overcrowded and relief only came when it stopped for refreshments at the various staging inns along the way. Coach travel along the main road was not without other dangers; the unwanted attentions of bushrangers, convict road gangs and other felons intent upon relieving passengers of their valuables were common occurrences.[24] Oldfield must have been quite exhausted when he arrived in Hobart Town, although fascination with the unfamiliar countryside may have taken his mind off the perils of travel.

Not far away from the towns perhaps the bush at first sight seemed to him sombre and monotonous, as the eminent Irish botanist William Harvey found when he encountered it for the first time near George Town: 'When the light is uniformly diffused the dull green of the gum trees & the sombre

---

[21] Peter Cox, 'Tamar Valley', in Alison Alexander, (ed), 2005a, Alexander, Alison (ed), 2005a, *The Companion to Tasmanian History*, Centre for Tasmanian Historical Studies, University of Tasmania, Hobart, p. 351.

[22] Peter Boyce, 2005, 'Britishness', in ibid., p. 403.

[23] ibid.

[24] G. Hawley Stancombe, 1969, *Highway in Van Diemen's Land*, G. Hawley Stancombe, Glendessary, Western Junction, Tasmania, abridged edition, pp. 33–41.

Fig. 6.1 Launceston, V.D.L. from the Westbury Road, 1844

She Oaks (*Casuarinae*) have nothing to diversify them'.[25] Any such thoughts were probably quickly dispelled when, over the coming months, Oldfield discovered the diversity and the beauty of the indigenous flora adorning the still uncleared forests, mountains, plains, coast, and river banks of his adopted homeland.

Further confirmation of the colonists' British background was manifest in town and rural housing along the way. In settlements from George Town and Launceston in the north through the Midlands, south along the main road, to Bridgewater and Hobart Town, there were brick and stone churches, inns, shops, mills, government buildings and houses that were predominantly Georgian.[26] Many were Palladian Georgian structures to which were added 'the clip-ons' of architectural style.[27] The gardens of free standing cottages, villas and great houses in the city and elsewhere were filled with plants from the homeland. Holly trees and beds of roses, vines, geraniums and fuchsias were surrounded by closely clipped mint borders, all enclosed with hedges of sweet-brier.[28]

During his first journey along the main road, Augustus could not have avoided observing the great disparity in the wealth of the settlers that was evident in town and rural housing. Perhaps a source of concern to the young immigrant was the presence of impoverished men and women and their families, but possibly by no means as strange to the urbanite fresh from England as their wattle-and-daub and slab huts on the edges of towns and in rural areas.[29] At the other end of the spectrum were the magnificent country houses of wealthy sheep farmers of the Midlands, south of Launceston and, for him yet to be discovered, along the spurs of sheep country running east to the Fingal Valley and

[25] William Henry Harvey to Jane L. Gray, 6 February 1855 cited in Sophie C. Ducker (ed), 1988, *The Contented Botanist: Letters of W. H. Harvey* [hereafter WHH] *about Australia and the Pacific*, Melbourne University Press at the Miegunyah Press, Melbourne, p. 179.
[26] G. Hawley Stancombe, 1969, map following p. 64.
[27] Clive Lucas, 1992, 'The Architecture of Van Diemen's Land to 1856', in Gillian Winter (ed) *Tasmanian Insights: Essays in Honour of Geoffrey Thomas Stilwell*, State Library of Tasmania, Hobart.
[28] G.C. Mundy, 1852, *A Record of Observations in Van Diemen's Land*, 1986 reprint, Sullivan's Cove, Adelaide. This is a reprint of the Tasmanian section of the author's *Our Antipodes*, Richard Bentley 1852, London.
[29] Lloyd Robson, 1983, *A History of Tasmania*, vol. 1, Oxford University Press, Melbourne, p. 364.

towards Oyster Bay, and west and inland up towards the central highlands, in the very rich agricultural and pastoral land that lay in the Norfolk Plains area and in the region of Pittwater and Richmond in the south.

Oldfield collected a few specimens along the main road[30] but we doubt that they would have been gathered during this journey. It is unlikely he went to the trouble of unpacking his luggage to equip himself for the task and at each stop he was probably eager to partake of the refreshments on offer.

To Augustus, as with many first-time visitors, Hobart Town, the seat of government, might have excited his 'unqualified admiration'. In the early 1840s David Burn waxed lyrical:

> Seated in the sheltered bight called Sullivan's Cove—reared, like Rome, upon (more than) seven hills, it rises amphitheatrically from the water's edge; its edifices towering one above the other in proud and pleasing succession-the snow—capt [stet.] Mount Wellington, girdling them as it were with a rampart of living rock. The port is a magnificent basin, with great depth of water towards its western side, the whole being formed by the natural and already oft-mentioned indentation, Sullivan's Cove.[31]
>
> Hobart Town, continued Burn, is remarkably clean, well lit, and well conducted. Its streets are wide, airy, and diverge from one another at right angles. Its numberless eminences, rising and falling like the wavelets of a summer sea, each pinnacle presenting its crescent, place, or terrace, give a peculiarity and picturesque character to the embryo city; and when it is added that many of [the] houses have been erected at a cost varying from £2 to £12,000, and that many of the shops would not discredit the most respectable parts of London, the reader may, perchance, wonder at the persevering assiduity of his distant fellow-countrymen.[32]

There were many substantial buildings, among them churches embracing the Anglican, Presbyterian and Catholic faiths, military and prisoners' barracks in Campbell Street, schools, the Supreme Court, Government House, the colonial hospital in Liverpool Street, commissariat stores at the lower end of Macquarie Street and fashionable residences in certain quarters, including Battery Point. Cornmills, a sugar-bakery, pottery, several tanneries, two foundries, coach-makers, and 'every description of cabinet and other furniture makers' contributed to the diverse economy, which had developed in the Colony.[33] Hobart Town possessed deep, sheltered anchorages at Sullivans Cove almost in the heart of the city, which since the 1830s was surrounded on three sides by busy wharves facilitating the City's trade through the ease of loading and unloading goods.[34] While there were inns and hotels catering for travellers, for the lower classes, many taverns and beer houses provided the only kind of public building, 'where their thoughts and actions were not being in some ways arranged for them'.[35]

Although the streets were well laid out, everything in Hobart Town was not as rosy as the picture Burn had painted. According to Peter Bolger, 'the grandeur of mountain dominance' created a 'harmony of scale' that went 'a long way towards obscuring the poverties and banalities of colonial improvisation in a raw town of the frontier'—the rough plank boozing dens, and the shabby little shops and the wooden shanties that were the homes of the poor, clustered along the creek banks too close to the water that was used for drinking.[36] The resulting polluted state of the city's water supply contributed to a spate of epidemics, which had given rise to a call for the incorporation of the city in 1843.[37]

---

30 See Chapter 9.
31 David Burn, 1973, pp. 38–39.
32 ibid., p. 43.
33 ibid., pp. 41–48.
34 Michael Cannon, 1975, *Life in the Cities: Australia in the Victorian Age*, vol. 3, Thomas Nelson, Melbourne, p. 187.
35 Kim Pearce, 'Public houses', in Alison Alexander (ed), 2005a, p. 294. See also Peter Bolger 1973, *Hobart Town*, Australian National University Press, Canberra, pp. 57–8.
36 ibid., pp. 55–64.
37 Margaret Mason-Cox, 'Water', in Alison Alexander (ed), 2005a, pp. 380–381.

Fig. 6.2 Hobart Town and Harbour, [184-]-[185-]

There was resentment among the settlers at the continuing presence of the convicts who they believed were 'the cause of most of their ills and blamed them for their own impediments'.

Free labourers resented the competition from cheap convict labourers. The settlers were hostile at the high incidence of crime perpetrated by assigned convicts and ticket-of-leavers, and rankled at the responsibility of being called upon as supervisors of bonded labour.[38] They were becoming impatient to be free of the 'despotic condition'—the push for local democratic self-government had begun.[39]

Augustus, Edwin and Alfred arrived in Van Diemen's Land at a time when, like the other Australian Colonies, it was in the grip of severe economic depression. This was the first major interruption to the rapid growth of Hobart Town as a flourishing mercantile centre and the Colony's prosperity, due in large measure to the booming wool-growing industry.[40] The effects of the Depression[41] on the people of Van Diemen's Land were harsh—banks restricted credit, the influx of British capital ceased and lower earning power led to reduced spending. Following the cessation of transportation to New South Wales in 1840, all convicts sentenced to transportation in Britain were forwarded to Van Diemen's Land. At the same time, free immigration was encouraged on the basis that newcomers would bring money and stimulate consumption and investment. Rather, the labour market was glutted as free workers competed with ticket-of-leave men and probation passholders, causing mass unemployment.[42]

As he became familiar with his new surroundings, Augustus would have discovered that 'class was a core ideal' of Van Diemen's Land's 'free' immigrant society and its structure largely modelled on the one he had left behind in England. 'Nowhere,' claimed Louisa Anne Meredith c.1844, 'were the decent and becoming observances of social and domestic life more strictly maintained'.[43] The Lieutenant Governor and some other senior officials were men from the higher ranks within British society. Along with a few of the powerful local landholders, merchants and churchmen, they formed the

[38] Peter Bolger, 1973, p. 44.
[39] ibid., p. 43.
[40] Lloyd Robson, 1983, p. 413.
[41] For details of the causes of this depression see for example, Wendy Rimon, 'The Depression of the 1840s' and Bruce Felmingham, 2005, 'Economy', in Alison Alexander (ed), 2005a, pp.103 & 421 respectively.
[42] Wendy Rimon, 2005, p. 103.
[43] Peter Bolger, 1973, p. 25.

appointed governing elite, the Legislative Council. Combined with the system of convict assignment, the land grant system had laid the basis for the creation in the 1830s of a three-tiered rural class structure of landlord, tenant farmer and itinerant farm labourer, mainly in the Colony's central north, where wetter, more fertile country encouraged mixed agriculture. Urban class structure included merchants and bankers, professionals and self-employed, the waged middle-class, and the labouring-class. The 'usual distinctions of rank' existed within the military and the navy. Each class in the Colony's society was broken down into sub-classes and the structure was ordered top-down according to social background and income in much the same way as it had in England.[44] Furthermore, during the 1840s the 'ideological power' of the Colony's urban middle-class of small producers, professionals and civil servants was increasing. Temperance and education movements were 'expressions of this power, as was middle-class control of outdoor charitable relief for the Colony's destitute'.[45] By that time, too, the Colony's major towns, which played administrative and trading roles, were governed by the local landed elites whose members, as magistrates, controlled law enforcement in rural districts. Magistrates also administered landlord and tenant laws that were heavily weighted in favour of landlords. Nevertheless, the ruling elites still had substantial economic and legal power.[46]

Where then might Augustus have stood within the Colony's social structure? He was well educated under circumstances that were designed to prepare him for English upper-middle-class membership. It seems that neither he nor his brothers brought substantial material assets to the Colony to assist in establishing any kind of business enterprise or otherwise allowing them to outwardly demonstrate materially their worthiness for upper-middle- or higher-class membership.

His assignment to a low sub-class within the middle-class was therefore inevitable. He may, however, have found acceptance in somewhat more highly placed families where there were shared scientific interests but no evidence of this has emerged. Of course, the question arises as to how he perceived his position within this socio-economic context and what kind of bearing it had on his career as a botanist. Some indication of this will become apparent in the coming chapters.

From the time Augustus landed in the Colony, and until 1846, the historical record will only allow us to speculate about how he occupied his time.[47] Despite the depression and the attendant severe unemployment within Tasmania, soon after his arrival it seems he was involved in an outdoors occupation, which suggests he was a member of a survey team. Thus, writing some eighteen years later:

> When I first visited Tasmania [1845], chance and my vocation led me into a part of the 'bush' where I found a hut recently deserted, evidently for the reason that all the available timber in its immediate vicinity had been felled.[48]

It is highly unlikely that he was gainfully employed as a botanical collector—his activities in that regard until 1857 appear to have been altruistic; for the advancement of science without any expectation or receipt of financial recompense.

Initially Augustus may have joined the staff of Edwin and Alfred's school but it is not known whether it was still operating in 1845. If not, his educational background and interests may have caused him to seek and obtain a position in a private school or a tutorship outside the city. If what David Burn claims was true of the early 1840s, Augustus' chances of finding employment in one of these capacities were quite considerable, even in those economically depressed times:

> The people of Van Diemen's Land are sensibly alive to the paramount importance of education. The proof of this is the intense anxiety they evince to attract and encourage

---

[44] Shayne Breen, 'Class', in Alison Alexander (ed), 2005a, pp. 408–409.

[45] ibid., p. 409.

[46] ibid., pp. 409–410.

[47] A search of *The Annual Official, Financial and Statistical Reports*, 'Blue Books' Tasmania, 1841–1847, CSO50/16 – CSO50/22, Acc. nos 3333, 3334, 3335, 3336, 3337, 318/12, 3338, TAHO, failed to reveal any indication that Augustus had been employed by any government department during that period.

[48] AO, c.1865a, pp. 827–829.

> teachers and masters of every description in every quarter of the island. Not only are respectable and capable seminaries, male and female, numerous in Hobart Town and Launceston, but they have also struck root in the less important villages, whilst many of the settlers have had highly accomplished tutors and governesses sent out to them from the parent land.[49]

Fear of moral corruption through association with the children of convicts, and dissatisfaction with the monitorial system, kept the sons and daughters of respectable people away from local government subsidised schools.[50] Isolation and the cost of boarding children away from home were probably also factors contributing to the demand for home tutors.

A rural placement as a teacher would have afforded Augustus opportunities, when he was not working at his vocation, to traverse the local countryside collecting botanical specimens. We doubt it. Augustus kept 'kangaroo dogs' which assisted him in his quest for food while he was in the bush.[51] This is more indicative of someone, like a surveyor, who spent lengthy periods in the countryside away from settlements and not of one who was engaged in teaching with opportunities for botanising confined to Saturday afternoons and Sundays. At the time, private surveyors were operating in the Colony.[52] Apparently Augustus' name does not appear on any survey plans for the period[53], so it is unlikely that he was either a private principal or assistant surveyor in charge of a team but he may have worked for one of them as a junior surveyor or survey hand.[54] Nevertheless, it will become apparent in the next chapter that his patterns of collecting localities suggest that he was with a survey team at various times.

Augustus left Hobart Town on 13 June 1846 for Sydney aboard the barque *Augustus*, arriving there on 23 June but the purpose of this journey remains untold.[55] Employment may have been his quest. When he returned is uncertain but he may have been the 'Mr Oldfield' who sailed from Sydney for Hobart Town on 29 August 1846 on the schooner *Rebecca*.[56]

On 28 October 1846 the Tasmanian Board of Education recommended Edwin for the appointment of Master[57] of the Public Day School, Campbell Street, Hobart Town for a three month trial. His conditions of employment included a salary of £90 per annum plus parents' fees and accommodation in the school house.[58] This prompted the Colonial Secretary, James Ebenezer Bicheno, to reply two days later that the 'desired or implied' directions from the Secretary of State in London 'that the Govt. schools are to be abridged in number as occasion offers' seemed 'a proper time to consider how far this arrangement may be forwarded by appointing some Master who may lose his position on [closure] of this school'.[59] Charles Bradbury, Secretary to the Board of Education, responded by arguing against the reduction in the number of the government schools and added:

> The Board further directs me to state that the appointment of Mr Oldfield was made by them, and he himself at once placed in charge of the school, under circumstances which render any difficulty in the way of his confirmation especially inconvenient and undesirable. The former Master left the school under pressing necessity connected

---

[49] David Burn, 1973, p. 31.
[50] Michael Sprod, 'Education', in Alison Alexander, (ed), 2005a, p. 115.
[51] AO, c.1865a, p. 574.
[52] See, for example, P.L. Lette, Letter of application for employment in the Lands & Surveys Department, Victoria, 19 April 1848, VPRS97/P/001, Unit 000002, DSC 1, Applications and Appointments, Locn. V/AK/162/03/02, PROV.
[53] It was beyond our resources to undertake a search of plans for the period. The Office of the Surveyor General, Department of Primary Industries, Water and Environment, Hobart, advised that the records of the Surveyors' Board (now defunct) do not stretch back that far; nor do staff recall any survey plans from that period with his name on them', John Vander Niet to William (Bill) Henderson (hereafter BH), 1 May 2003, email, HOA.
[54] Searches conducted at TAHO failed to reveal any lists of employees of private surveyors undertaking work for private persons.
[55] 'Shipping News', *The Maitland Mercury & Hunter River General Advertiser*, 27 June 1846, p. 2, DAN, NLA.
[56] ibid., 2 September 1846, p. 2.
[57] As Master, Edwin was the principal of the school.
[58] C. Bradbury, Chief Inspector of Schools, Board of Education Office to Colonial Secretary, 28 October 1846, CSO 13/ 678. Education Board Correspondence, TAHO.
[59] ibid., note on reverse side by James E. Bicheno, Colonial Secretary.

> with his impaired health, and with much difficulty the Board succeeded in obtaining Mr Oldfield's services at a short notice, so as to avert the disadvantages of closing even for a day, a school which from the large number of children attending it is probably one of the most important in the Colony.[60]

Bicheno subsequently approved the appointment.[61]

Edwin experienced some consternation when, on 13 November 1847, he returned to his home in Campbell Street at about 7 pm to observe a light shining through the window of one of the bedrooms. 'This excited his suspicions' as no candle was left burning when he left the house. He went over to the Penitentiary opposite the house for a constable, and on their return the burglar, one William Edward Harper, a convict 'pass-holder', was apprehended on the premises before any goods could be removed. Harper was found guilty and 'sentenced to ten years transportation'.[62]

Augustus was in Hobart Town as a 'Person appointed by Commissioners to take Population Account' in the *1848 Tasmanian Census* conducted on 3–5 January to determine the size and nature of the population on the night of 1 January 1848. He was the census collector for Bathurst, Elizabeth, Argyle, Campbell, Melville, Collins and Brisbane Streets within Parish No. 2, Hobart Town.[63] He must have been a last-minute replacement collector since his name was not published in the official notice of appointees in the *Hobart Town Gazette*.[64]

The *Census* record for 51 Campbell Street, completed by Augustus and dated 4 January 1848, shows that the head of the household was Edwin Oldfield. The residence, the stone school house, was also occupied by another man on *Census* night. Both were single in the aged category '21 and under 45' and their religious affiliation was Church of England. They had 'arrived free' and were in the occupational category 'Land Proprietors, Merchants, Bankers, & Professional Persons'.[65] Augustus was almost certainly the other resident; at the time Alfred was principal master at a government-funded school in the town of Melbourne.[66]

Why Augustus put himself forward to assist with the *Census* can only be conjectured.

[60] C. Bradbury, Chief Inspector of Schools to The Auditor, 24 November 1846, CSO11/1/160, Education Board Correspondence, TAHO.
[61] ibid.
[62] Hobart Quarter Sessions 24 November 1847, *The Hobart Town Courier and Gazette*, 1 December 1847, p. 4, DAN, NLA.
[63] This is evidenced by his signature as Recorder/Collector on 1848 Tasmanian Census records for those streets in Parish No. 2, Microfilm CEN 1/81 79, 80, 81,82, TAHO. Record details by kind permission TAHO.
[64] *Hobart Town Gazette*, 7 December 1847, p. 1193, TAHO.
[65] Record for E. Oldfield, Census VDL 1848, CEN1/81, pp. 73–74, microfilm, TAHO. Record details by kind permission TAHO.
[66] 'Port Phillip, Annual Statistics', *The South Australian*, Adelaide, 28 January 1848, p. 3, DAN, NLA. There were 38 males and 15 females at this school.

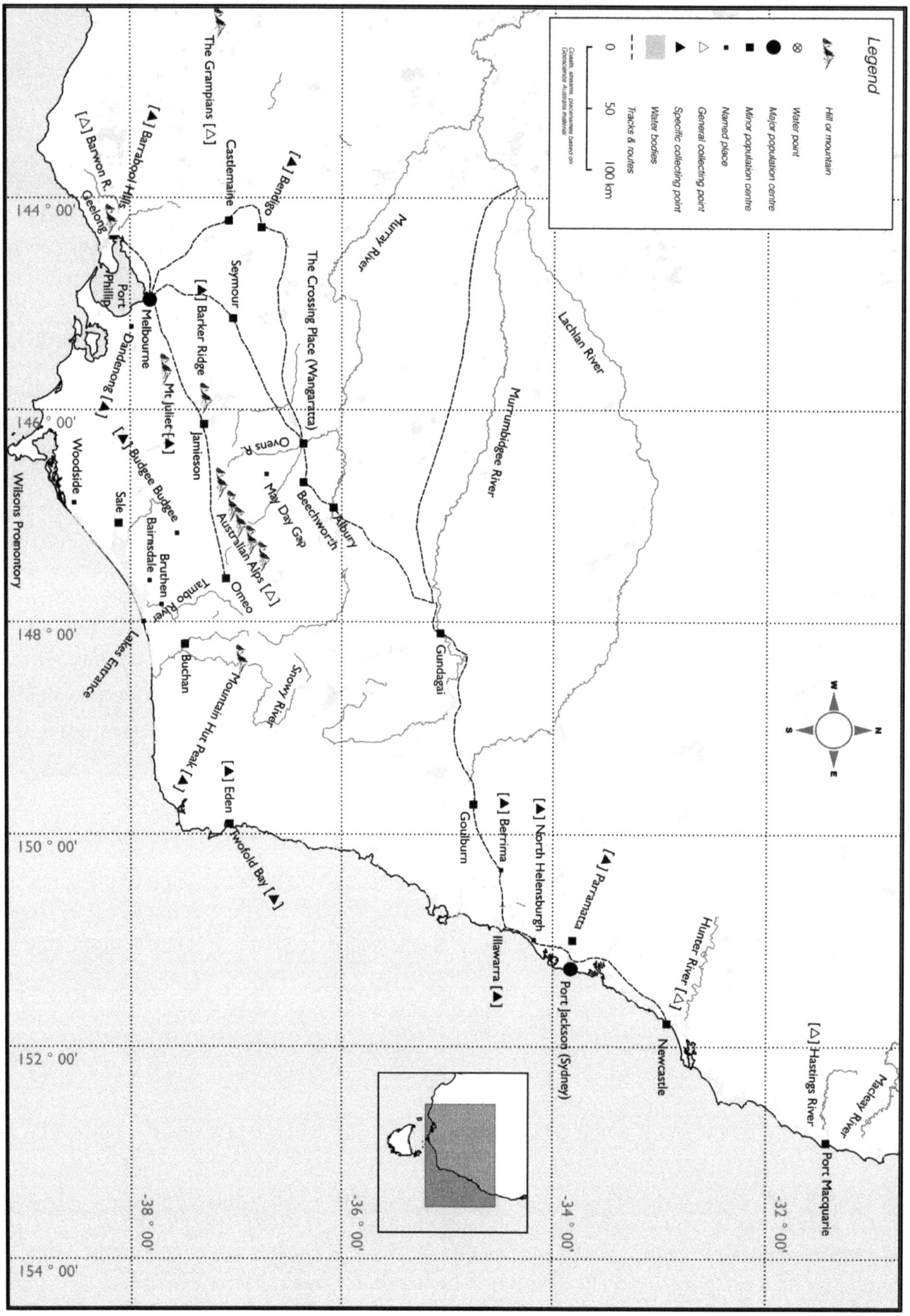

Map 3: Sydney to Melbourne

A long walk: Sydney to Melbourne

Map 3 p. 94 – Sydney to Melbourne

Around 1848, Augustus walked from Sydney to Melbourne, taking a month to complete the journey, 'the greatest part of the road being at that time mere bullock tracks'.[67] Few other details of that trip have survived. He collected some plants along the way, which 'for the most part' he 'mounted on larger paper [than was his usual practice]'. Although in 1850 he informed William Hooker that he would send them to him, it is not known whether he ever did.[68] There is no reference to any of the specimens in the volumes of Bentham's *Flora Australiensis* or in any other publication that we have been able to locate. This, however, cannot be taken as proof that they are not at Kew Gardens; they may have been specimens of species that were not new to science or did not extend the ranges of any earlier collections referred to in that publication.[69]

His route to Melbourne from Sydney probably followed the old road that linked the two cities[70] and which approximated today's Hume Highway, about 900 kilometres. We have, however, only located records of two plants collected along that road, one from Berrima and the other Parramatta, New South Wales.[71] There is the possibility Oldfield gathered it during his 1848 visit, kept it in his personal herbarium and presented it much later to Mueller. In one of his 1850 letters to William Hooker, Oldfield mentioned a plant that he collected on the Murrumbidgee and which he took to be 'a *Lythrum*'.[72] He also visited Gundagai, a town on the Murrumbidgee River in New South Wales through which the old Sydney to Melbourne road passed.[73]

This personal account shows that he was already familiar with the signs of flooding in the Australian landscape and was by no means afraid to speak his mind if he thought the circumstances warranted it:

> In 1848, when near the Murrumbidgee, I was particularly struck by the evidences that the plains around the township of Gundigai [sic] had been under water at no remote period. The land had that even aspect and hardened surface peculiar to soil deposited by water, while in the forks of many trees I remarked collections of brush-wood which had evidently been lodged there by the same agency. I called the attention of some residents to these facts, assuring them that the town was built in a very insecure situation, and that sooner or later, it would suffer from floods. Of course this warning was unheeded, but unfortunately the prediction was verified, for a few years since [1852 and 1853][74] Gundigai [sic] was swept away by an inundation, with great loss of life and property.'[75]

It is perhaps understandable that the inhabitants took no notice. They probably thought they were secure as the settlement had been shifted to higher ground from the opposite (southern) bank of the Murrumbidgee River soon after a devastating flood in 1844.[76]

In one of his obituaries, Edwin recalled the journeys undertaken by Augustus

[67] Edwin Oldfield, ?16 July 1887, p. 3.
[68] AO to WJH, undated letter probably written some time during April–May 1850, Australian Letters, vol. 73, p. 256 (487), RBGKA.
[69] See the Preface of this work relating to the citing of collectors in nineteenth century taxonomic works on the Australian flora.
[70] J. Jones, *Map of New South Wales and Victoria Showing the Course of the Mountains, Rivers and Roads, the boundaries of Old, New, and Proposed Counties, the limits of the settled Intermediate and Unsettled Districts and the Situation of the Principal Towns and Villages from a Map by Arrowsmith,* J Jones, Sydney, 1850, Mitchell Library Sydney (ML).
[71] MEL 201654, *Platylobium obtusangulum,* Berrima, New South Wales, (MELISR database, 18 November 2005, NHV, RBGM) and *Eucalyptus siderophloia*, Parramatta, New South Wales, J.H. Maiden, 1909, vol. 1, pt. 1, p. 363.
[72] AO to WJH 15 April 1850, English Letters, vol. 30, I – Z, 1850, folio 154 (155), RBGKA.
[73] J. Jones, 1850.
[74] Department of Environment and Conservation (NSW) (DECNSW), 2011 'South Western Slopes – regional history', viewed 24 December 2014 at www.environment.nsw.gov.au.
[75] AO, 1865a, pp. 215–298.
[76] DECNSW, 2011, viewed 24 December 2014 at www.environment.nsw.gov.au.

> in his scientific investigations, alone and afoot, were not unattended with danger at a time when many aborigines were still in possession of their native home—the bush. However, by tact, he in some way placated them, although sometimes threatened, he was never in any way harmed. This was particularly the case while walking from Sydney to Melbourne some 40 years ago.[77]

Edwin did not elaborate. Augustus, however, left an undated account of an incident of this nature:

> Once, when travelling near the junction of the rivers Lachlan and Murrumbidgee, darkness overtook us ere we had pitched on a suitable locality for camping. Arrived on the side of a hill where there was plenty of grass and water, we were disconcerted at perceiving, on the face of an opposite hill, very many fires, indicating the presence of a great number of natives. As the blacks in those parts had the character of being very fierce (they had slain many of the neighbouring settlers), and our party was but small and badly armed, we were about to retire, when a shepherd proposed the following plan to scare the savages. Under his direction we each made several heaps of leaves, brushwood, &c. at certain distances apart, and at a given signal these were all lighted rapidly as possible. No sooner were they all well a-light, when, as if by magic, all the other fires disappeared, for the natives imagining that some foreign—and therefore hostile—tribe was near to them, quickly decamped, not however going very far, as we afterwards learned to our cost.[78]

Oldfield left no details of the group he was with, what they were doing and what they learned afterwards to their 'cost'. In 1848, he may have taken a side-trip with drovers driving stock to a pastoral run. At the time, there were no boats or barges carrying people, stock and goods along the river.[79] It is perhaps more likely that he was there sometime over the next five or six years with a party of surveyors although it doubtful that they would have been 'poorly armed'. In referring to the shepherd and camping where there was 'plenty of grass and water', Oldfield could have been alluding to animals being driven to one of the station properties or to surveyors' horses and stock for slaughter along the way.

What transpired during the remainder of his journey to the south could not be determined. He probably returned to Hobart Town via Port Phillip.

Before giving an account of where Oldfield stood as someone entering the quest for understanding the flora of Van Diemen's Land in the mid- to late-1840s, and what we believe his movements were in that Colony prior to 1850, the remainder of this chapter is devoted to giving some initial indication of his connectedness with the fauna and flora he encountered during his travels.

Some insights into Oldfield's relationship with nature

In his unpublished manuscript on natural philosophy, Augustus chronicled some of his experiences in Tasmania and New South Wales, which demonstrate his powers of observation and his interest in, and feelings for, nature. Unfortunately, there is no way of knowing precisely when or where the events he described took place.

The following account, which he gave as an illustrative example when discussing language in 'Man' and other animals, is a delightful example that shows that it was not only Australia's flora that claimed his attention:

> I was once travelling in Tasmania, when my attention was drawn to a flock of white cockatoos by their unusually clamorous proceedings, there being something quite new

[77] E.D. Oldfield (?), 13 July 1887, p. 5.
[78] AO, 1865a, p. 288.
[79] Gwenda Painter, 1987, *The River Trade: Wool and Steamers*, Turton & Armstrong with The Pioneer Settlement Press, Swan Hill, Victoria.

> to me in their notes. Stealthily approaching their place of congress, I beheld one whose squalid appearance proclaimed him to be an escaped domestic pet, occupying the highest branch of a dead tree, while the rest of the flock were grouped about him, in the manner of an auditory. 'Polly put the kettle on' sang the educated bird, whereupon the rest of the flock flapped their wings, and uttered the peculiar shriek before referred to, which probably was their laugh. For above half an hour did the travelled bird go through its repertory of acquired sounds, much to the amusements of its companions, the performance being brought to an apparently untimely close, by the unwary approach of a wayfarer which caused an immediate dispersion of the assembly.
>
> The different cries of the birds just referred to, constitute a very expressive language, a knowledge of which is sometimes serviceable to the hungry traveller in search of a meal. The sentinel's cry of security – which may be translated – All's well! – is very different from the cry of caution – which may be rendered 'beware!' – uttered on the approach of any suspicious object, and this again from that which warns the flock of the presence of real danger, while the shriek of alarm which causes the whole flock to take to immediate flight, is quite different from either of the preceding cries. Then there is the note of endearment of the male to the female, the cry of defiance of the former to his rival; the grunt of satisfaction when enjoying a good meal obtained with little trouble; the imploring cry of the wounded bird to his fellows for assistance, and the shriek of despair when he finds himself deserted by them, and so forth to the number of above fifty distinct cries, which I have noticed to be uttered under different occasions.[80]

This interest in animals was also accompanied by a concern for their welfare which is attested by an anecdote he gave when discussing domestication of animals by Man:

> I once slightly wounded a black magpie (or piping crow) [probably the Forest Raven, *Corvus tasmanicus* race *tasmanicus*[81]] of Tasmania, but repenting me of the evil I had done, I took the bird home with me, carefully tended it until it was quite recovered, and then set it at liberty near the spot where I had shot it. About a month after, chance again led me to that place: gun in hand, I was anxiously watching my dogs which were striving to round up a Kangaroo towards me, when I was startled by a bird perching on my shoulder. It was my friend the magpie which had recognised me, and by flapping its wings, and uttering its peculiar cry, now seemed to welcome me to the spot where I had acted so cruelly towards it. For several hours did it continue these demonstrations, only quitting me when I neared the few scattered houses to be found in those parts, and on several subsequent occasions, it acted in like manner.[82]

The following serves to further illustrate his ability to gain the trust of certain indigenous species:

> Sometimes when travelling through the unsettled districts[83], I have had the company of one of these birds [Native Companion or Brolga (*Grus rubicunda*)] for several days, but they always forsook me when I neared the abodes of the white man, for, to his shame be it said, he acts less honourably towards these confiding birds, than does the black man, who never harms them, even though he be suffering extremes of hunger. When travelling the Native Companion walks behind, and rather to one side of

---

[80] AO, c.1865a, unnumbered fn., p. 633.

[81] 'Black crow' may have been a local name for the Tasmanian Magpie (*Gymnorhina tibicen* race *hypoleuca*) but it is strongly black and white. The bird in question may have been what is now known as the Forest Raven (*Corvus tasmanicus* race *tasmanicus)* which is all black. There appear to be no true crows in Tasmania and only one raven, Peter, Pat and Raoul Slater, 1992, *The Slater Field Guide to Australian Birds*, Weldon, Smithfield, New South Wales, pp. 324–325 & 334–335.

[82] AO, c.1865a, unnumbered fn., p. 573.

[83] Oldfield probably encountered this bird when he was in Victoria or New South Wales, Peter & Pat Slater & Raoul Slater, 1986, p. 50.

> me, so that our heads were side by side, and on all occasions, seemed to court, rather than share, familiarities.[84]

In a discussion in which he stated that previously cultivated deserted areas that had become overrun with introduced species eventually reverted to indigenous vegetation when deserted by humans he gave many examples to support his claim. One of these related to the aforementioned hut which he discovered c.1845 and around which

> was about half an acre of cleared ground which had evidently been used as a kitchen garden, and not far distant from it, was an enclosure of about an acre, in which wheat had been grown…the whole of the cleared space was one tangled mass of introduced weeds. I observed *Carduus lanceolata*[85] growing luxuriously, 8–12 feet high, and quite gay with its purple blossoms: there was *Urtica urens* [Stinging Nettle] — esteemed in the Australian colonies the infallible indication of the vicinity of women or sheep—in rank profusion, while in many places two species of *Polygonum* (*P. aviculare* [Wireweed] and *P. convolvulus* [Black Bindweed]) carpeted the soil, besides which I noticed two species of *Rumex*, two of *Plantago*, one of *Chenopodium,* some common and worthless European grapes and a few other weeds. The sweet-briar—as is usual in cultivated and waste localities in Tasmania—was plentiful, attaining a height of 12–18 feet, an altitude it seldom reaches in its native country.[86] Some years after, chance again led me to that spot, and I was much struck with its altered aspect. Nature had begun to resume her rights, and the lawful possessors of the soil were fast expelling the invaders. The nettles had quite disappeared and the thistles—weak and straggling—were vainly contending with the native shrubs for air and light. The rest of the weeds were nearly smothered by the native burr (*Acaena sanguisorba*) and the sweet-briar looked very sickly. On two subsequent visits - both made to satisfy my curiosity as to the progress of events in that secluded locality, where nevertheless a great battle was being fought—I found that the strangers were fast losing ground, while at my last visit—18 years after my first—I had the gratification of beholding the total rout of the intruders, for even the sweet briar, and the lively insects had been subdued.[87]

This suggests that by the time he reached Van Diemen's Land he had accumulated a sufficient knowledge of exotic species[88] to be able to distinguish them from indigenous plants. However, it cannot be ruled out that his recollection of the names of species around the hut on his various visits was influenced by botanical knowledge gained over the 18 years or so he was in Australia. While there may well be instances where indigenous species choke out exotic plants, today there would be many observers who would seriously question Oldfield's assertion by pointing to at least one introduced species, Blackberry (*Rubus fruticosus* aggregate). This is regarded as one of the worst weeds in Australia because of its invasiveness, potential for spread, and economic and environmental impacts. Blackberry has invaded the banks of watercourses, roadsides, pastures, orchards, plantations, forests and bushland throughout temperate Australia.[89]

---

[84] AO, c.1865a, fn. 40, p. 571.

[85] Species name not found in APNI or the Australian Plant Census (APC), Council of Heads of Australasian Herbaria, viewed 20 June 2012 at www.cpbr.gov.au with link to the APC.

[86] AO, 1865a, pp. 827–829. Oldfield's reference to the presence of *Urtica urens* as being 'the infallible indication of women and sheep' probably reflected his observation of women's preference for the plant in their gardens because of its value as a food and farmers' belief that it was a nutritious feed for sheep, more than to some derogatory association of women and sheep.

[87] AO, c.1865a, pp. 827–829.

[88] For example, *Biosecurity, Queensland Division,*2013, '*Weeds of Australia', P. aviculare'* probably originated in Europe, viewed 18 June 2013 at www.daff.qld.gov.au, and *Wikipedia,* '*Fallopia convolvulus*' (synonym of *P. convolvulus*) is native throughout Europe, Asia and northern Africa, viewed 18 June 2012 at www.wikipedia.org.

[89] Department of Environment, Australian Government, 'Blackberry (*Rubus fruticosus* aggregate)', *Weeds in Australia,* viewed 23 March 2013 at www.environment.gov.au.

Augustus' deep interest in and passion for the natural environment is revealed not only in the above quotations from his unpublished manuscript but also in the observations he made on the labels of the plants he collected and wrote about in his letters and unpublished manuscript—their habitats, forms and in some cases their physiology. This will become apparent in the chapters that follow.

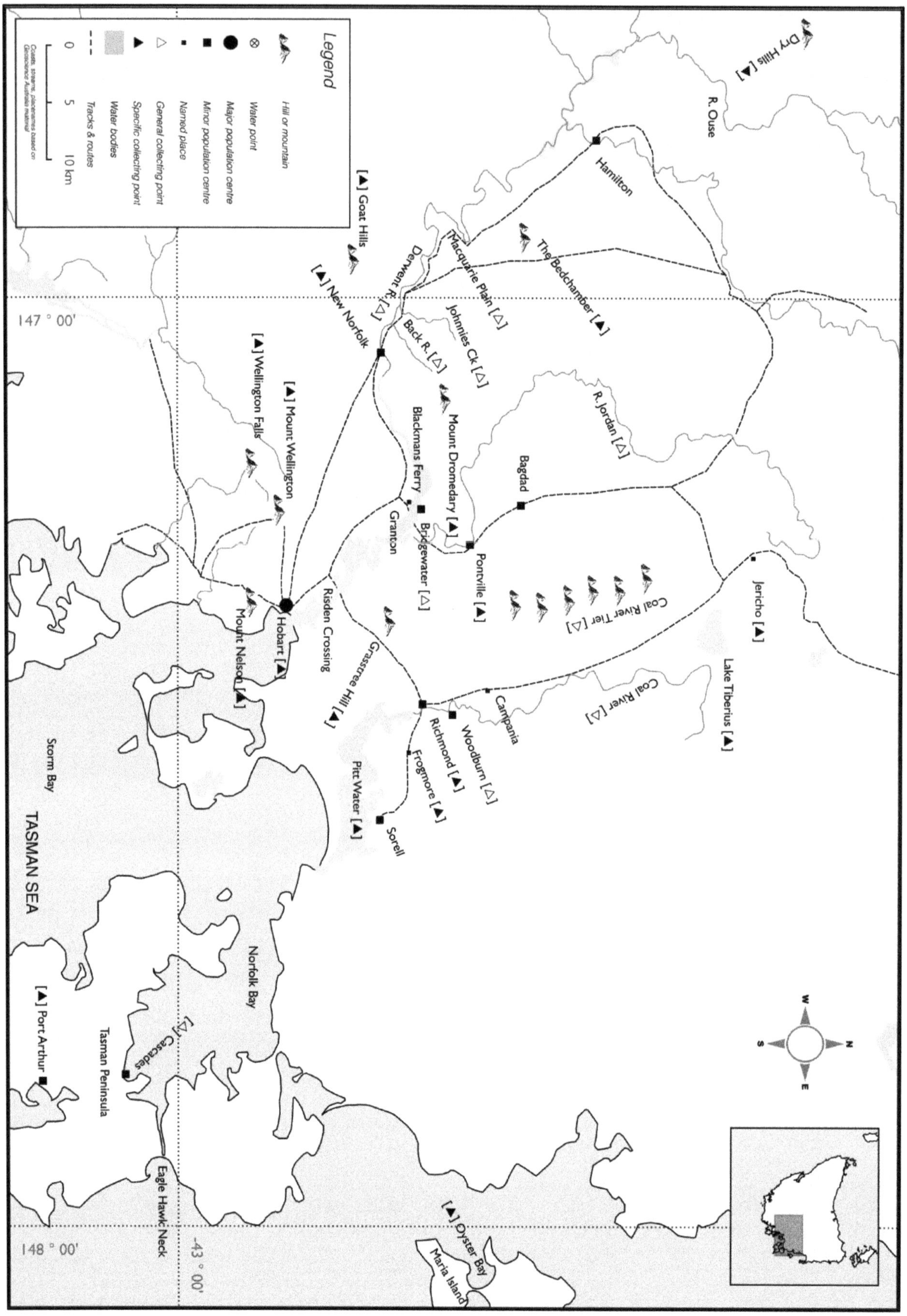

Map 4: Central and Eastern Tasmania

Fig. 7.0: Lepidosperma oldfieldii

Chapter 7

Early years in Australia 1845–1849 Oldfield enters the field of Australian botanical endeavour

Tasmania was by no means culturally or scientifically arid when Augustus first set foot there. Despite its somewhat turbulent internal politics and shaky existence, a Mechanics Institute in Hobart had long been offering periodical and gratuitous lectures on a wide array of art and science subjects.[1] There was a Book Society and a Debating Society but a theatre group established in 1833 had not survived. The elite Union Club, modelled along the lines of those establishments in London that catered for the upper-classes, was opened in 1834.[2] Not overly keen on large social occasions, the Governor Sir John Franklin and his wife, Jane, Lady Franklin, much preferred 'getting out and about'[3] and entertaining visiting dignitaries and distinguished savants.[4]

Scientific interest in Tasmania's 'natural productions' to 1845

Map 2 p. 84, Map 3 p. 94, Map 4 p. 100, Map 5 p. 132

Local interest in the Colony's 'natural productions' was considerable, particularly from the perspective of their potential economic value, and the Colony supported a number of societies which aimed to address that and other scientific interests. The Tasmania Scientific Society was founded in 1829 under the patronage of Lieutenant-Governor Arthur but was short-lived.[5] The Tasmanian Society of Natural Science, Agriculture, Statistics & etc. (Tasmanian Society of Natural Science) was established in 1839.[6] Franklin is generally credited with founding the Society. It was Jane Franklin 'who initiated [the Society], set it up and kept it going...' However, to fulfil her self-imposed role of boosting Sir John, she wanted credit for any achievement to him.[7] Meetings were held in the Governor's parlour and papers delivered there 'were often drawn from original discoveries'.[8] By 1842, the Society had 32 resident members and 39 corresponding members. These last included His Excellency Sir George Gipps, Governor of New South Wales, His Excellency Captain George Grey, Lieutenant-Governor of South Australia, and distinguished scientists and explorers from as far a-field as Russia, England, France and New Zealand.[9] The list of 35 resident members for 1846 indicates that it was an exclusive

[1] For information on the Mechanics Institute and its internal conflicts see Colin Finney, 1993, *Paradise Revealed: Natural History in Nineteenth-century Australia*, Museum of Victoria, Melbourne, pp. 32–35.
[2] David Burn, 1973, pp. 32–33.
[3] Lloyd Robson, 1983, vol. 1, p. 336.
[4] *The Tasmanian Journal of Natural Science, Agriculture, Statistics & etc.*, vol. 2, no. 6, 1846, p. 80.
[5] Gillian Winter, 1993, 'The foundation years of the Royal Society of Tasmania, 1843–86', *Papers and Proceedings of the Royal Society of Tasmania*, vol. 127, p. 62.
[6] T.E. Burns, 1966, 'Gunn, Ronald Campbell (1808–1881)', ADBO, viewed 23 August 2012 at www.adb.anu.edu.au.
[7] Alison Alexander, 2013, *The Ambitions of Jane Franklin, Victorian Lady Adventurer*, Allen & Unwin, Crows Nest, New South Wales, p. 109.
[8] Peter Bolger, 1973, p. 27.
[9] *The Tasmanian Journal of Natural Science, Agriculture, Statistics & etc.* 1842, vol. 1, p. 424.

group, being confined to men and women of the upper echelons of society[10], following the lead of the major scientific societies in Britain. Nevertheless, a 'naturalist' was listed among the members as 'M. Verreaux'. He was almost certainly Monsieur Jules Pierre Verreaux who came to Australia in 1842 and collected in Tasmania and New South Wales for the Paris Museum of Natural History.[11] At times the Society's meetings were attended by visiting British scientists.[12] The first volume of its scientific journal publication, *The Tasmanian Journal of Natural Science, Agriculture, Statistics, & etc.*[13], was edited by Ronald Gunn and 'laid before the public' in 1842. The volume comprised five issues that were all produced during that year and included serious contributions relating to the Colony's flora, fauna and geological formations from local and corresponding members, including Joseph Hooker and the ornithologist and bird artist John Gould.[14]

When John Gould and his wife Elizabeth were in Tasmania, between 1838–1840, collecting and preparing drawings for John's major illustrated work on Australian birds, the Franklins assisted him in many ways in his quest for birds. Lady Jane took the homesick Elizabeth under her wing and invited her to stay at Government House while John was away in the bush.[15]

Joseph Hooker was received at Government House in 1840 and 1841. His visits occurred during the Royal Navy's Antarctic expedition of H.M.S. *Erebus* and *Terror* under the command of Captain Sir James Ross.[16] Hooker was Assistant Surgeon on the *Erebus* and the expedition made two extended visits to Tasmania. In August, September, and October 1840, he and Dr David Lyall, Assistant Surgeon on the *Terror*, collected extensively around the Derwent River, the Great Lakes district and at Port Arthur. After the ships returned to Hobart in 1841 to refit, Hooker made collections in the Huon River and Richmond districts during March, April and May of that year. In his quest for plants, Ronald Gunn assisted Hooker. Of his acquaintance with Gunn, Hooker wrote later

> [I] am indebted to him for nearly all I know of the vegetation of the districts I then visited...I can recall no happier weeks of my various wanderings over the globe, than those spent with Mr Gunn, collecting in the Tasmanian mountains and forests, or studying our plants in his library, with the works of our predecessors Labillardière and Brown.[17]

Gunn was instrumental in the establishment of both the Launceston Horticultural Society (1838) and the Gardeners and Amateurs Horticultural Society, Hobart Town (1845). The objects of the Launceston Society included the introduction and acclimatising of plants[18], and experimental research on the properties of the native plants of the island. The Hobart organisation aimed to encourage the introduction, growth and improvement of horticultural productions, manufacture of articles arising out of them and to ascertain the capabilities of the soil and the climate of Tasmania, with a view to increase the Colony's exports and reduce its imports.[19]

---

[10] ibid. 1846, vol. 2, no. 6, p. 80. ADBO biographical information on the members listed in this issue indicated that most of the members were highly placed civil servants, military and naval officers, clergymen, medical and legal practitioners, rural landowners and the wives of two leading citizens, Mrs Joseph Allport and Mrs Whitford Smith (ADBO, viewed 7 October 2006 at www.adb.anu.edu.au). However, biographical information on all the members could not be traced so we cannot assert with certainty that the Society was as exclusive as we think.

[11] Alex S. George, 2009, *Australian Botanist's Companion*, Four Gables Press, Kardinya, Western Australia, p. 583.

[12] Peter Bolger, 1973, p. 27.

[13] By 1851 this journal and the Tasmanian Society of Natural Science were defunct. Its members had joined The Royal Society of Van Diemens Land which published its journal as *Papers and Proceedings of the Royal Society of Van Diemen's Land 1851 to 1859* and from then to present continued as *Papers and Proceedings of the Royal Society of Tasmania*.

[14] *The Tasmanian Journal of Natural Science, Agriculture, Statistics & etc.*, vol. 1, 1842.

[15] Isabella Tree, 1991, *The Bird Man: The Extraordinary Story of John Gould*, Ebury Press, London, pp. 75–91.

[16] Lloyd Robson, 1983, vol. 1, p. 357.

[17] JDH, vol. 1, 1860, pp. cxvii, cxxv.

[18] The importance accorded the acclimatization of exotic species across the British Empire is reflected in an order made by Her Majesty in Council, dated 23 August 1843, which allowed specimens illustrative of natural history to be admitted free of duty into any of her Majesty's colonial possessions, James Wood (compiler), 1849, *Woods Tasmanian Almanack for 1849*, J W H Walch, Hobart, p. 39, TAHO.

[19] ibid., pp. 110–111.

In 1843, Franklin's replacement, Lieutenant-Governor Sir John Eardley-Wilmott set up the Van Diemens Land Horticultural and Botanical Society which shortly became the Royal Society of Van Diemens Land Horticultural and Botanical Society (Horticultural and Botanical Society). In the same year Eardley-Wilmott transferred to the Society the responsibility for the Colonial Gardens in Hobart (now the Royal Tasmanian Botanical Gardens). The Society soon set about forming its own museum and library.[20] The members of the Tasmanian Society of Natural Science were entitled to be among the first fifty foundation Fellows of the Society but most of them refused to pay the entrance fee. Although no formal amalgamation of the two societies took place, pressure from Eardley-Wilmot's successor, Lieutenant-Governor Sir William Denison, made it easier from 1848 for members of the Tasmanian Society of Natural Science to join the Horticultural and Botanical Society and within 18 months all active resident members had transferred to that organisation.[21]

The young Oldfield may have eagerly purchased issues of *The Tasmanian Journal of Natural Science, Agriculture, Statistics, &c* and yearned to join such a group. The membership of the Horticultural and Botanical Society was open to anyone interested in science, 'open in fact to anyone who cared to belong'.[22] He is not included in surviving membership records of that organisation. There is some evidence that he was a member in the 1850s[23] but this cannot be confirmed due to gaps in the membership records.[24] Nevertheless, despite membership being open to anyone interested in science, the published list of Fellows (apparently the entire membership) showed the organisation to be elitist in practice, possibly kept that way by high subscriptions that may have precluded him from joining when he first came to Tasmania. Oldfield's response to a question from William Hooker in 1850, asking whether he knew Ronald Gunn, tends to support a claim that he did not belong:

> I am acquainted with him by name only in fact few of the inhabitants of any place where I have sojourned know that such a person as myself exists, my pursuits not permitting me to join in much society. Mr Gunn's name is chiefly known to me from his connexion with you.[25]

Given the very active role that Gunn was playing in the Society during the 1840s, it is difficult to imagine that any member would not have known him personally. As far as Oldfield was concerned, this seems more certain because of the deep interest in botany he and Gunn shared.

The founding of the TSNSAS and the visit of Joseph Hooker did not, however, mark the beginning of the accumulation of European knowledge of the flora of Tasmania. This had commenced with the visit to Adventure Bay of *Resolution* during James Cook's third voyage to the Pacific Ocean in 1776. According to Helen Hewson, the natural history legacy from this stopover does not seem to have been particularly significant, with one notable exception. William Anderson, surgeon and naturalist, together with David Nelson, a gardener at the Royal Botanic Gardens, Kew, and one of Banks' collectors, gathered a specimen of the Stringy Bark (*Eucalyptus obliqua*), the type species of the genus *Eucalyptus.*[26]

Other collectors followed. Antoine Raymond Joseph de Bruni d'Entrecasteaux's expedition in *Recherche* and *Espérance* visited Recherche Bay in the autumn of 1792 and January 1793. The naturalists accompanying the expedition—Jacques Julien Houtou de Labillardière, Louis Ventenat, Louis-Auguste Deschamps and Claude-Antoine-Gaspard Riche—made important collections in the

---

[20] Gillian Winter, 1993, pp. 62–63.
[21] ibid., p. 63.
[22] ibid., p. 61.
[23] See Chapter 21.
[24] The Reports of the Royal Society of Van Diemen's Land and subsequently the Royal Society of Tasmania for the Years 1847 – 1861 only list Fellows, Hon. Members and Corresponding Members, TAHO.
[25] AO to WJH, undated (probably written sometime in April–May 1850), Australian Letters, vol. 73, fol.. 256 (487), RBGKA.
[26] Helen Hewson, 1999, *Australia: 300 Years of Botanical Illustration*, CSIRO Publishing, Collingwood, Victoria, p. 29. Hewson refers to David Nelson as a midshipman but Ray Desmond states he was a Kew gardener and one of Banks' collectors, *The History of the Royal Botanic Gardens Kew*, Harvill Press, London, 1995, p. 434.

region of Recherche Bay and Bruny Island while seeking traces of the lost expedition of Jean François de Galaup, Comte de La Pérouse. Between 1804 and 1807 Labillardière published his *Novae Hollandiae Plantarum Specimen*[27], which was the most comprehensive account of the Australian flora to that time and was largely based on his own collections during the expedition.[28]

In 1805 George Caley, trained at Kew Gardens, made collections in Tasmania for Joseph Banks.[29] A list of Caley's plants is included in Brown's unpublished manuscripts in the Natural History Museum in London.[30] He was not only the first of the early naturalists to obtain the help of Aborigines but he made an important contribution to the development of Australian bushcraft. He was the first to recognise that 'the essential elements of travelling in New South Wales were lightness, speed and packhorse transport' although he 'failed to develop the capacity to stay in the field for extended periods of time'.[31]

Robert Brown's visit to Tasmania, between 1803–1804, was much more productive botanically than those of his predecessors. Taking up a proposal put to him by Lieutenant Matthew Flinders in 1800, Banks successfully encouraged the British Admiralty to commission Flinders in 1801 to survey Australia's southern and western coast in HMS *Investigator*. Brown, selected and financially supported by Banks, was the naturalist with the expedition. Banks also recruited Peter Good, a gardener from Kew Gardens, the Austrian botanical illustrator Ferdinand Bauer and the landscape artist William Westall to accompany the expedition.[32] Due to the unseaworthiness of the *Investigator*, Flinders set off for England from Port Jackson to obtain another vessel. While he was awaiting Flinders' return Brown travelled three times south to Tasmania aboard the *Lady Nelson* under the command of Lieutenant Colonel William Paterson, who had been instructed to found a British colony in Tasmania at Risdon Cove[33] on the Derwent River. The first expedition, between November 1803 and January 1804, was to the Kent Group in Bass Strait and to Port Dalrymple, where Paterson had been instructed to establish a British settlement. He explored the upper Tamar River and eventually settled on the present site of Launceston in 1806. The second voyage, from February to April 1804, was to the settlement at Risdon Cove and Sullivans Cove, on the shores of which Hobart Town was founded on 19 February. From Hobart, Brown ascended Mount Wellington three times, undertook two boat and walking trips up the Derwent and to the river south of Hobart that was later to be named 'Brown's River' (later Browns River) in his honour. The vessel visited the Derwent River again between May and August 1804, during which time he explored overland to the south and reached the Huon River, spent time near Risdon and made a journey to the north of Mount Wellington and ascended Table Mountain. He also visited Storm Bay, the mouth of the Huon River and Esperance River.[34] T.G. Vallance, D.T. Moore and E.W. Groves have located 733 'higher plants', and many lichens and 'lower' plants collected by Brown in Tasmania.[35]

After returning to England in 1805, Brown became the first librarian of the Linnean Society in 1806, and in 1810 was appointed Joseph Banks' librarian. During that year he published his *Prodro-*

[27] J.J.H. Labillardière, 1804–1806, *Novae Hollandiae Plantarum Specimen…*, 2 vols., Paris.

[28] Orchard, A.E., 'Labillardière, Jacques J.H. de (1755-1834)', *Biography: Biographical notes on plant collectors and illustrators*, Australian National Herbarium, Council of Heads of Australasian Herbaria (hereafter ANHBiog), viewed 15 June 2010, at www.anbg.gov.au. For a detailed account of Labillardière's writing and publications see Edward Duyker, 2003, *Citizen Labillardière: A Naturalist's Life in Revolution and Exploration (1755–1834)*, The Miegunyah Press, Melbourne, pp. 223–241.

[29] Helen Hewson, 1999, p. 80.

[30] R.M. Barker & W.R. Barker, 1990, 'George Caley (1770-1829)', in P. S. Short (ed), 1990, p. 54.

[31] Glen McLaren, 1996, *Beyond Leichhardt: Bushcraft and the Exploration of Australia*, Fremantle Arts Centre Press, Fremantle, pp. 23, 24, 38, 39.

[32] T.G. Vallance, D.K.T. Moore & E.W. Groves, 2001, *Nature's Investigator: The Diary of Robert Brown in Australia, 1801–1805*, ABRS, p. 7.

[33] The first settlement was at Risdon Cove on the Derwent River in September 1803.

[34] For a detailed account of Brown's visit to Tasmania see T.G. Vallance *et al.*, 2001, chapters 20–22, pp. 459–526.

[35] ibid., 2001, p. 459.

mus Florae Novae Hollandiae et Insulae Van-Diemen…[36] This was followed in 1814 by his 'General remarks, geographical and systematical on the botany of Terra Australis'.[37] After Banks' death in 1820, Brown took up residence in Banks' home in Soho Square.[38]

Motivated by the political situation in Europe, in 1800 the French sent an expedition under the command of Nicolas Baudin to establish a French presence in New Holland and the Pacific. Baudin sailed in the *Géographe*, accompanied by the *Naturaliste* under the command of Jacques Félix Emmanuel Hamelin. The expedition set out from France with a large scientific team, There were two head botanists (Jean-Baptiste Leschenault on the *Géographe* and André Michaux on the Naturaliste), a head gardener (Anselme Riedlé on the *Géographe*), an apprentice botanist (Jacques Delisse on the *Naturaliste*) and four assistant gardeners (Antoine Guichenot and Antoine Sautier attached to Riedlé; Jean-François Cagnet and the young slave, Merlot who accompanied Michaux). In addition there were, two mineralogists and two astronomers. When the ships reached Ile de France (Mauritius) in March 1801, forty men abandoned the expedition, some due to sickness and others not prepared to endure Baudin's authoritarian leadership. Those deserting included André Michaux and Jacques Delisse, two zoologists and two artists—approximately a third of the scientific team], with the result that there were only four left on arrival in the Terres australes (Leschenault, Riedlé and the assistant gardeners, Guichenot and Sautier). Two of these would die during the voyage (Riedlé and Sautier), a third, Leschenault, was disembarked in Timor during the expedition's second stay there because of illness. Baudin died from tuberculosis at Ile de France on the return journey in 1803.[39] Only three scientists, among them Leschenault de La Tour, returned to France, Antoine Guichenot, who cared for live plants on the *Géographe*'s return voyage, was the only gardener to complete the voyage. [40]

From January to March 1802, Baudin explored and mapped the east coast of Tasmania before sailing north into Bass Strait and, after a visit to Sydney, returned to explore further the west coast of the continent. Botanical and zoological specimens were collected at various places around Tasmania, including Partridge Island, in D'Entrecasteaux Channel off the south of Bruny Island, North West Bay, Norfolk and Blackmans Bays, the Tasman Peninsula and Maria Island.[41] Over 200,000 specimens, were deposited in Muséum National d'Histoire Naturelle and Labillardière was entrusted with studying them but he did not take the opportunity although he is believed to have used some of the material in his *Novae Hollandiae Plantarum Specimen* (1804-1806).[42] They were integrated by Desfontaines, Professor of Botany at the Museum, into what was to become the general herbarium of the National Museum of Natural History.[43] Live plants and animals were also sent to Jardin des Plantes, Malmaison and Chateau de Navarre. According to Jean Fornasiero *et al.*, the scope and variety of the Expedition's collections 'were impressive by any standards'. However, even though Leschenault published a report on the botanical collections in the second volume of the official account of the voyage (1816), 'the timing meant that Brown's work [on his collections from Terra Australis] was already well known' and 'Leschenault's contribution far too late to secure for French botanists the recognition that counted—in scientific nomenclature and publication'.[44]

Another of Banks' collectors, Allan Cunningham, accompanied Phillip Parker King in the *Mermaid* to survey Macquarie Harbour on the south-west coast of Tasmania from 25 December 1818 to 14

[36] Robert Brown, 1810, *Prodromus Florae Novae Hollandiae et Insulae Van-Diemen, Exhibens Characteres Plantarum quas Annis 1802–1805 per Oras utriusque Insulae Collegit et Descripsit Robertus Brown…*vol. 1, J. Johnson & Co., London.

[37] R. Brown, 1814, 'General remarks, geographical and systematical on the botany of Terra Australis' in Matthew Flinders, *A Voyage to Terra Australis*, vol. 2, G. & W. Nichol, London, pp. 533–612.

[38] T.G. Vallance *et al.*, 2001, p. 10.

[39] 'Nicolas Baudin', *Wikipedia*. Accessed 27 July 2017 at www.wikipedia.org.

[40] Michel Jangoux, ? date, The Herbarium of the Terres Australes (Baudin Expedition, 1800-1804), Marine Biology Laboratories, Free University of Brussels (ULB) & University of Mons, viewed 27 December 2017 at http://sydney.edu.au, p. 1.

[41] Jean Fornasiero, Peter Monteath & John West-Sooby, 2004, *Encountering Terra Australis: The Voyages of Nicolas Baudin and Matthew Flinders*, Wakefield Press, Kent Town, South Australia, pp. 81–121.

[42] Helen Hewson, 1999, pp. 62, 63.

[43] Michel Jangoux, ? date, p. 3.

[44] Jean Fornasiero *et al.*, 2004, p. 148.

February 1819. During this voyage, Cunningham collected near Hobart and Mount Wellington and at various coastal locations around Macquarie Harbour, and nearby Kelly Channel and Pilot Bay.[45] Many taxa have been based on his collections, and many of his manuscript names have since been taken up, but he published little himself.[46]

Prior to 1845 some residents, including Robert Lawrence and Ronald Gunn, made collections in various places in the Colony and amassed their own herbaria, although parcels of dried specimens were sent for identification to William Hooker at Kew Gardens.[47] Ronald Gunn was perhaps the most prolific and knowledgeable of these individuals. So valuable was his private collection considered to be that it was later presented to the Royal Society of Tasmania.[48] One expedition of note was Lady Franklin's exploring expedition of the south and west coasts of Tasmania. Ronald Gunn and the ornithologist John Gould were members of the party. They left Hobart on 10 December 1838 on a government schooner bound for Port Davey on the south coast but storms prevented them going further than Recherche Bay.[49] During this expedition Gunn collected plants at Recherche Bay, Whale Head, Catamaran River, and towards and at South Cape Bay but we have found no evidence that he ever penetrated as far west as Mount Laperouse or any other mountains in the vicinity.[50]

While Gunn himself collected quite extensively in northern and central Tasmania, c.1842 he engaged the services of the commercial collector Charles Stuart.[51] Stuart was to become one of Oldfield's colleagues.[52]

Prepared for surviving and collecting in the Australian bush

Apart from a notebook in which to record certain characteristics of plants—the general habitat, locality where it was gathered, growth habit, height and the colour of flowers and leaves[53]—from which Oldfield took specimens, he probably carried one of several types of hand-held lens available at the time[54] to assist in observing the very small characteristics of plants. He also would have needed to carry quite a lot of other equipment in the field. Important were a small axe and spade to assist in the collection of plants with roots intact, for breaking through heavy undergrowth and for camping purposes, and secateurs (invented in 1789)[55] to ensure plant stems were cut cleanly.

Necessary too were the means to convey the specimens to ensure that they remained separate and undamaged for a few days or until he could commence the drying process. For cuttings or entire specimens of flowering plants and cryptogams, he probably took a vasculum or two depending on how long he intended to be out in the field. A vasculum was a painted or lacquered ovoid-shaped cylindrical case, usually made of tin or thin tinned iron, 'the size of an octavo volume, or larger' with a generous opening covered by a hinged lid along its side and an attached leather strap to enable it to be slung across the body so that the vessel hung in a horizontal position. The case was 'calculated to preserve plants in a state of freshness... by confining their exhausting evaporation'.

---

[45] S. Curry, B.R. Maslin & J.A. Maslin, 2002, *Allan Cunningham: Australian Collecting Localities*, ABRS, pp. 25-27.
[46] Australian National Herbarium, 'Allan Cunningham (1791–1839), ANHBiog, viewed 3 July 2010, at www.anbg.gov.au.
[47] See for example, J.H. Maiden, 1909a, pp. 9–29, and Colin Finney, 1993, pp. 39–44.
[48] This collection has had a chequered history with most of it ending up in the National Herbarium of New South Wales and the remainder in the Tasmanian Herbarium, Hobart, A.M. Buchanan, 1990, 'Ronald Campbell Gunn (1808–1881)' in P.S. Short (ed), 1990, pp. 179–192. See this reference also for details of Gunn's collecting localities and a list of literature sources where his collecting trips are described.
[49] Isabella Tree, pp. 77–82.
[50] A.M. Buchanan, 1990, 1990, p. 183.
[51] C. Daley, 1935, 'Charles Stuart, An early Australian botanist', *The Victorian Naturalist*, vol. 52, October, p. 106. Stuart's collecting activities in Tasmania and elsewhere are outlined in Chapters 10 & 12.
[52] See Chapters 9 and 13.
[53] AO to JDH 16 March 1866, English Letters, vol. 97, 1859–1900, NAT–OXC, fol. 289 (296), RBGKA.
[54] These included a lens mounted on a stand, a hinged set of lenses of different magnification, or single pocket lenses, Jim Endersby, 2008, p. 76.
[55] Silas Clifford-Smith, 2013, 'The reflective gardener', viewed 2 July 2014 at www.reflectivegardener.blogspot.com.au.

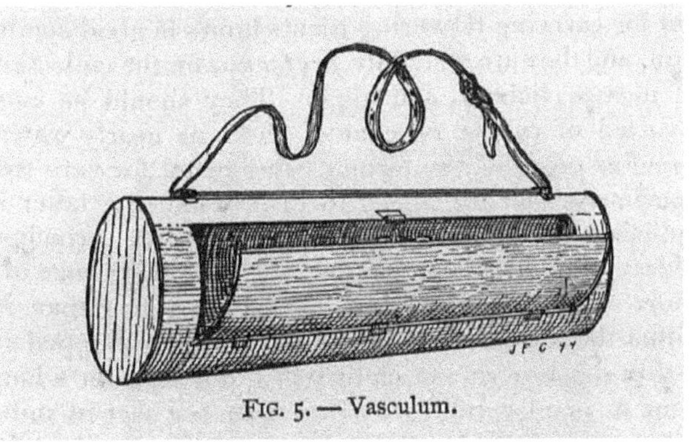

Fig. 7.1 Vasculum, from W.W. Bailey

A gentle sprinkling of water or lining the box with damp newspaper helped to maintain the plants' freshness.[56]

Alternatively, or probably additionally if the weather was hot or he was in the field in a cooler clime for a week or more, he may have carried one or two light portfolios of pasteboard, covered with calico or leather. This would have contained sheets of absorbent 'blotting' paper for initial drying, coarse 'brown' paper for late stage drying and furnished with straps and buckles, one for closing the press tightly and another for slinging on the shoulder.

The specimens were placed between the sheets of paper and could be crowded together if they were not left long without sorting and transferring to proper drying paper and presses.[57] Bentham considered that the press was superior to the 'the old-fashioned tin box [? vasculum] (except, perhaps, for stiff prickly plants and a few others)'[58] because the drying process would have commenced in the field and the damage to loosely packed specimens, due to movement, was minimised. Oldfield may have preferred to use a bag of heavy material to carry large woody or prickly specimens. We have come across herbarium sheets of Oldfield's specimens which include a packet of seeds. It seems, therefore, that he would also have carried a pouch with small papers for folding, keeping and carrying seeds. Of course, if collecting was not the main reason for him being in the bush, he may have had to be selective about the equipment he carried for that purpose. No evidence has emerged of Oldfield having travelled anywhere by horseback in Tasmania although he most certainly would have done so if he was with a survey team.

The rough terrain and dense undergrowth necessitated tough clothing. He probably wore items favoured by gold diggers and other outdoors workers of his time—a serge shirt, moleskin trousers, long leather hobnail boots, thick woollen socks for warmth and to prevent blisters, and a wide-brimmed hat made of leather or soft felt[59] possibly with a fly-net. Protection from Tasmania's often cold, windy and wet weather required a sturdy warm woollen coat and an oiled cotton cape. It is on record that he carried a gun[60], and there is little doubt that he took with him a compass, a small water canteen, which could be replenished in one of the streams that abound in the island, and the means of starting a fire—flint, steel and tinderbox or Lucifer matches.[61] His portable food was probably confined to tea, sugar, salt, flour for damper and for convenience he may have included dried salted meat.

[56] D.E. Allen, 1959, 'The history of the vasculum', *Proceedings of the Botanical Society of the British Isles*, vol. 3, pp. 135–150, viewed 30 August 2012 at www.archive.bsbi.org.uk.

[57] GB, 1863, vol. 1, p. xxxii.

[58] ibid.

[59] See, for example, Frances Jones, 2007, 'Life on the goldfields: living there', *Primefacts*, New South Wales Department of Primary Industries, no. 575, February, p. 3 and Jennifer A. Jones, 2010, *Old Age in a Young Colony: Image and Experience in South Australia in the Nineteenth Century*, PhD thesis, School of History & Politics, University of Adelaide, p. 172.

[60] See Chapter 6.

[61] Geoffrey Blainey, 2003, *Black Kettle and Full Moon*, Penguin Books Australia, Camberwell, Victoria, pp. 60–62.

If not, there was no shortage of edible fauna in the bush and in the waterways that could be obtained with the aid of a gun, fishing line or trap and native fruits and roots could be sourced at various times of the year.[62] He would have carried his gun, a musket, rifle or shot gun, either by hand or across his back by means of an attached strap and his other equipment, dry food and spare clothes in canvas or leather bags toted either on his back or by his sides by means of straps attached to the bags. If in the field for a few days his bed roll of a blanket and light canvas would have been fastened to the back pack or directly to his back by means of shoulder straps.

Important among game animals in the Colony were the Tasmanian or Red-bellied Pademelon (*Thylogale billardierii*), the Tasmanian Kangaroo (or Forester) (*Macropus giganteus*) and the Wombat (*Vombatus ursinus*)[63], which, Oldfield said, is

> very easily domesticated, and...seeks to be the companion of man, for it will follow its owner, and utter piercing cries, if he walks so fast that his tardy companion [the Wombat] cannot keep-pace with him. Indeed, the domestication of the wombat would be doubly advantageous to man, for not only would he secure a most faithful companion, but also an occasional and dainty dish, for the flesh of that animal is far more savoury than that of any other Marsupial, thus confirming the assertion of the French abbé, that Providence has compensated certain animals for the slowness of their motions, by the succulency of their flesh.[64]

He leaves us wondering whether he would have eaten a pet wombat!

In a discussion on the 'aptitude of [animals'] instincts for receiving education', Oldfield indicated that he had been hunting wild sheep for food when he came across two small flocks in Tasmania:

> These had assumed all the habits of the wild goat—which also exist in some parts of that island—having sentinels posted, which gave the alarm as I and my dogs approached, on hearing which, the whole flock took to the rocks, running so swiftly that my Kangaroo dogs were not able to overtake them before they had reached that place of safety, although they had no greater start than is considered a fair one for capturing the Kangaroo, that is, about 600 Yards. Arrived [sic] at the rocks (basalt), they sprang from pinnacle to pinnacle with all the agility of goats, so that after all, I had to return without what I was in quest of—mutton.[65]

It would have been imperative for Oldfield to become aware of, and how to identify and avoid, the perils of venturing into the Australian bush, waterways and ocean. There are species of plants and animals—snakes, spiders, insects, ticks, fish and jellyfish—that transmit disease-causing organisms or toxins, which can be fatal.[66] If from his own reading or fellow settlers' warnings he understood these dangers, consultations with local bushmen, medical practitioners and Aborigines may have helped him to recognise these hazards and how to treat poisoning. According to John Pearn and Kenneth Winkel, the medical management of those poisoned by snakes, spiders and fish in Australia featured extensively in the writings of 19th century doctors, expeditioners and anthropologists. Against the background of this introduced medical doctrine there already existed an extensive tradition of Aboriginal medical lore; techniques of heat treatment, suction, incision and the application of plant-derived pharmacological substances featured extensively in the management of poisoned victims.[67]

---

[62] See for example, Glen McLaren, 1996, pp. 126–128, for a discussion of the explorers' food sources in mid-nineteenth century.

[63] Ellis Troughton, 1957, *Furred Animals of Australia*, 6th edition, Angus and Robertson, Sydney, pp. 194 and 224, and W.D.L. Ride, 1970, *A Guide to the Native Mammals of Australia*, Oxford University Press, Melbourne, pp. 44, 48. *Macropus tasmaniensis* is a synonym of *Macropus giganteus*, Troughton, 1957, p. 224.

[64] AO, c.1865a, fn. 40. p. 571.

[65] ibid., pp. 574–575.

[66] John Pearn & Kenneth D. Winkel, 2006, 'Toxicology in Australia's colonial era: A chronology and perspective of human envenomation in 19th century Australia', *Toxicon*, vol. 48, pp. 726–737.

[67] ibid.

Mounting and classifying specimens

After returning home from the field, other commitments permitting and when the specimens were sufficiently dry, the task would begin of mounting dried specimens on high quality paper, bottling in spirits those that were too fragile to dry and securing packets of seeds. It was imperative that each specimen, including seeds, was clearly labelled—Oldfield's labels usually included his taxonomic determination, a collection locality, very rarely the date he obtained the specimen and quite often a brief description of the plant and its habitat. He left no comments on where he sourced his initial drying and mounting paper; suitable paper was expensive and often difficult to find in the colonies.[68] There is, however, no indication that his specimens were the object of the adverse criticism that the Hookers and others levelled at those James Drummond sent to England from Western Australia. One such complaint came from William Harvey:

> Drummond's plants are splendid—lovely things—but put up most piggishly. I would subscribe to buy the man a screw press, if one could induce him to use it. It is a thousand pities that he has not such a contrivance—and a better stock of paper. Fortunately, he lives in an oven or his plants would all be rotten as well as crumpled.[69]

Even though he and Drummond spent some time together in Western Australia in 1854[70], Harvey does not seem to have appreciated the impediments to proper plant preservation faced by early Australian collectors, due to difficulties in obtaining the correct equipment and consumables. However, by the time Oldfield reached the Colony in Tasmania, it was much better served by trading ships from England and elsewhere than the Swan River colony had been during the 1830s and 1840s when Drummond was actively collecting. It is likely, therefore, that supplies of paper and screw presses were more readily available to Oldfield. Nevertheless, like Drummond, he may have used newspapers for initial drying, replacing the paper when it became damp and drying it for re-use.[71]

The plant identifications on his labels indicated that Oldfield had the necessary equipment for classification. This task usually required a microscope for dissection of the floral parts to reveal some, or all, of the key characteristics of the species such as the ovaries. In common with many botanists, he probably used a simple dissecting single lens microscope although, being interested in plant physiology[72], he might have preferred a more expensive compound microscope with two or more achromatic lenses which provided higher magnification.[73] For dissecting a range of instruments was used, including fine forceps, handled blades and thick needles, dividers[74] and a scale for linear measurement. A means of dampening dried specimens to facilitate dissection was also important.

Taxonomic works were vital to ensure classifications were acceptable to Joseph Hooker and, as was the case with many colonial collectors, Oldfield was eager to obtain them. As this narrative progresses it will become apparent that he found it difficult to get appropriate published taxonomic material in Tasmania but nevertheless seems to have had access to some.

---

[68] Jim Endersby, 2008, p. 67.
[69] WHH quoted in Jim Endersby, 2008, p. 67.
[70] Sophie C. Ducker (ed), 1988, pp. 109–110.
[71] Rica Erickson,1969, *The Drummonds of Hawthornden*, Lamb Paterson, Osborne Park, Western Australia, p. 36.
[72] AO, c.1865a, fn. 40. pp. 748–749.
[73] Jim Endersby, 2008, p. 71.
[74] ibid., p. 73.

Fig. 7.2 Joseph Hooker's dissecting microscope

Oldfield's botanising from Hobart before 1850

Map 4 p. 100

Plants collected by Oldfield in the Colony prior to 1850 have already been alluded to and the absence of dated and systematically numbered specimens precludes any claims as to exactly when he was in the areas recorded on his specimen labels. Joseph Hooker's *Flora Tasmaniae*[75] was not published until 1860 and includes many of Oldfield's plant collections, most of which we believe were obtained during the 1840s.[76] Oldfield took a collection of 350 cryptogams to England in 1849 which, with the exception of those that were obtained in New Zealand, were most likely to have been gathered from a variety of vegetation types close to Hobart Town or within two days' walk from the city in four directions—to the north west, north, east, north east and south-east. This, of course, does not negate the possibility that he collected in these localities after 1850 and in other places relatively close to Hobart Town, where for a time he resided with Edwin. Nevertheless, for reasons that will become apparent in subsequent chapters it seems that the majority of his collections to the south of Hobart were probably not obtained until after 1850.

Oldfield collected at Hobart Town and on the hills around and 'at the back' of the capital.[77] The locations he attributed to some of the plants he took from the mountain show that he explored it from its base to the summit, covering the southern and northern aspects extending as far as Wellington Falls, nestled between Mount Wellington, Mount Montagu and other mountains, 13 kilometres south

[75] JDH, 1860, vol. 2, reprint 1860. For a complete listing of Oldfield's specimens referred to in this publication see HAODb.
[76] See Chapter 8.
[77] HAODb.

Fig. 7.3 Wellington Falls track, Mount Wellington

east of Hobart, and North West Bay River above the Wellington Falls.[78] Our specimen records suggest that he focused a lot of attention on the cryptogams in the bogs, springs, creeks and gullies on Mount Wellington.[79] At other localities during his early years in the Colony he seems to have directed his attention more to cryptogams than to angiosperms but the numbers are too small to present meaningful statistics.

The journey from Hobart to the summit of Mount Wellington was not an easy one although for the young and energetic Oldfield it may not have posed much of a challenge. It was a one-day pedestrian return expedition frequently undertaken by men and women of the capital.[80] When Oldfield was residing in Hobart and was free of a weekend he may have remained on the mountain overnight.

According to Butler Stoney, who described his climb to the summit in 1851, it was 'arduous but rewarding'. With his description to guide us, Oldfield's route probably commenced from the edge of the city just beyond the western ends of Macquarie and Davey Streets after which the path skirted along the valley by the Town Rivulet and the Cascades past the Female Penitentiary. The last establishment on this road was 'the extensive brewery, house and grounds of Mr de Graves, one of the wealthiest gentlemen in the Colony'. From there the track soon started to wind up the hillside, which at first sight seemed almost perpendicular. The steep slopes and ridges on mudstone are dominated by the Silver Peppermint (*Eucalyptus tenuiramis*) and form dry open forests.[81] 'Surmounting the first ascent' the collector would have 'come on a level platform; and again, crossing it, ascend[ed] another; but a dense wood'[82] dominated by Messmate Stringybark (*Eucalyptus obliqua*), which is dominant in the gullies and wetter south facing areas—usually on dolerite substrate. This forest had a wet sclero-

---

[78] ibid.

[79] Of all the records of specimens collected on Mount Wellington, Wellington Springs and North West Bay River above Wellington Falls (69 publ., 64 herb.), the great majority were cryptogams (82% publ. and 73% herb.). All the remainder were angiosperms, ibid.

[80] H. Butler Stoney, 1856, *A Residence in Tasmania: With a Descriptive Tour Through the Island from Macquarie Harbour to Circular Head*, Smith, Elder & Co., London, Tasmaniana Facsimile Editions no. 6, Melanie Publications, Hobart, 1982, p. 22.

[81] Information on plants occurring in Oldfield's collecting localities in Tasmania was kindly provided to us by Dr Jayne Balmer, Senior Ecologist, Vegetation Conservation Section, Biodiversity Conservation Branch, Department of Primary Industries & Water, Hobart, personal communication, emails to MHH, c.6 October 2008, HOA.

[82] H. Butler Stoney, 1856, pp. 22-23.

phyll understorey of Musk (*Olearia argophylla*), Silver Wattle (*Acacia dealbata*), Satinwood (*Nematolepis squamea*) and Dogwood (*Pomaderris apetala*)[83], which blocked his view until after he reached 'the first spur of the hills called the Springs. The way runs here along a cool and gushing streamlet for some distance' and he may have been 'gladly refreshed by its sweet water'. From this resting-place he had a 'very fine view of the Derwent into Storm Bay. D'Entrecasteaux Channel, Brown's River and the Huon seeming like silver threads amid the dense mass of foliage around. But [he was] only now half-way'.[84] Here in the wet forest on south facing slopes at mid-elevation on a sandstone plateau he encountered eucalypts such as Tasmanian Yellow Gum (*Eucalyptus johnstonii*) and Gum-topped Stringy Bark (*E. delegatensis*) and in some places, Urn Gum (*E. urnigera*).[85]

Oldfield may have shared Stoney's experience and found 'the ascent high up…still more laborious' but 'the view from this is so very grand that [he] gain[ed] fresh courage [to] hurry up the towering hill above' with 'some level places afford [ing relief] to [his] weary feet'.[86] Here, below the treeless alpine vegetation of the summit and high plateau, he passed through Tasmanian Snow Gum (*E. coccifera*) woodlands and forests. [A]s he approached the top, 'the air [became] rarefied, cool, and refreshing'. At length, like Stoney, he probably threw his 'exhausted frame on the highest rock', and rejoiced that his challenging task was completed.[87] At the summit he would have been greeted with a scattering of cushion moorlands interspersed within eastern alpine heathland and sedgelands.[88] There are extensive rocky boulder scree slopes and boulder fields on the higher slopes of the mountain also providing habitat for many lichens as well as a diverse alpine heath flora.[89]

Along the Derwent River, at least as far as New Norfolk, Oldfield gathered specimens of cryptogams among which marine algae appear to have predominated.[90] Apart from those stated to have been obtained near New Norfolk, the other algae, mosses and liverworts could have been taken from anywhere along the length of the estuary. It is likely, however, that those collected before 1850 came from between Hobart and New Norfolk. A few miles up-stream from Hobart were The Flats where Oldfield may have needed to wade across what David Burn described in 1840, as 'oozy slime of prodigious depth'[91] to reach water where the algae occurred. The Derwent estuary contains large areas of tidal flats (13.2 km^2) with mudflats predominating in the upper estuary and sand flats.

Oldfield may have enjoyed observing the waterfowl on these flats and on the adjacent wetlands and saltmarshes, which occur mainly along the river above Granton.[92] Between the Bridgewater Bridge and New Norfolk these areas are vegetated with freshwater aquatic sedgelands and rushlands dominated by species such as the Common Reed (*Phragmites australis*), Shiny Bogsedge (*Schoenus nitens*), Slender Twine-rush (*Leptocarpus tenax*), and Broadleaf Cumbungi (*Typha orientalis*) and rushes (*Juncus* spp.).[93] If Oldfield collected the algae from water flowing through mudflats, perhaps he obtained them during a day's or weekend's walking expedition from Hobart Town. The muddy and time-consuming activity of searching for these plants is hardly something he would have undertaken during a short break in a coach journey to the hinterland.

While there is no confirmatory evidence, he may have botanised along the overland track which ran through uncleared bush from Hobart Town to New Norfolk. In 1840, according to Burn, this was the most direct route but was 'only known to pedestrians' and which traversed 'among the inferior

83 Jayne Balmer, 6 October 2008.
84 H. Butler Stoney, 1856, pp. 23-24.
85 Jayne Balmer, c.6 October 2008.
86 H. Butler Stoney, 1856, p. 24.
87 ibid.
88 Jayne Balmer, c. 6 October 2008.
89 ibid.
90 HAODb.
91 David Burn, 1973, p. 88.
92 Department of Primary Industries, Parks, Water and Environment (Tasmania), 2010, 'Derwent Estuary Habitats', *State of the Derwent Estuary*, 2009, Hobart, pp. 133–158, viewed 10 July 2012 at www.derwentestuary.org.au.
93 Jayne Balmer, c.6 October 2008.

*Fig. 7.**4** Flats on Derwent River estuary, Tasmania*

hills that skirt the base of Mount Wellington'. It had not been opened for vehicular traffic due to 'the great want of labour, and the perfect conditions of the present route' which ran along the south bank of the Derwent.[94]

Perhaps, however, there were occasions when Oldfield found it more expedient to travel by coach or on horseback on the macadamised road along the river through New Town, crossing Humphries Rivulet at O'Brien's Bridge, the Islet Rivulet near Undine and then along the southern bank, crossing a few other rivulets, past Austin's Ferry, the Hestercombe Inn and Ferry to the Black Snake Inn located just to the south of the river of that name.[95] From Bridgewater, the road continued along the south bank to New Norfolk. He may have concurred with Burn that the road was one of surpassing beauty, 'following the river entirely' and that the several stage coaches plying their way along the route were

> a public means of conveyance, [were] in no respects inferior either in horses or other appointments to the generality of English carriages—the journey being performed in about two hours and a half.[96]

It is likely that there was little to interest the young botanist immediately to the south of this road. By the late 1830s and early 1840s the forests along it had largely disappeared:

> [T]he axe, the plough, and the spade have done their duty—noble roads have been formed—goodly dwellings have arisen—and smiling fields with blooming gardens glad the eye and repay the toil. In fact, this picture may be said to hold good in a greater or less degree through the whole two and twenty miles that divide New Norfolk from Hobart Town, which places are connected by the noble river (the Derwent) on whose bosom, after several failures, steam has at length commenced its beneficial influence.[97]

---

[94] David Burn, 1973, p. 85.
[95] ibid., pp. 84–87.
[96] ibid., p. 84.
[97] ibid.

Fig. 7.5 New-Norfolck, Ile Van Diemen, [1841-]

The settlements that lined the route did not penetrate the hinterland for any distance and the country to the south of New Norfolk was very little known; of this, stated Burn, penetration on horseback was prevented by the 'matted and impervious…jungle'.[98]

We can only guess at Oldfield's first impressions of the thriving town of New Norfolk, originally called Elizabeth Town. It was a busy agricultural and timber centre and an important stopping place for carts and coaches transporting passengers, farm and timber products and other goods to and from Hobart Town and settlements along the river road to Hamilton and beyond. The town was also a terminal for row boats and steam boats providing similar services to and from the capital and settlements along the way. Navigation above New Norfolk was entirely barred by a ridge of rocks 'called in local parlance The Falls'.[99]

In addition to the slab huts of the labouring poor on the fringes of the settlement, there were some well-constructed public and private buildings in the town—an Anglican Church with an associated school room, a Wesleyan Chapel, a court of police and quarter sessions, gaol, hospital, flour mill, shops and offices, three or four comfortable inns, including the Bush Inn, one or two academies for young ladies and gentlemen, and the residences of a few officials, shopkeepers, and professionals. Sited opposite the town on the northern side of the river was the 'pretty little village upon the stream, dignified by the style of Back River' around which 'agriculture, floriculture, and horticulture contend[ed] for supremacy'.[100]

Oldfield collected near the town of New Norfolk[101] and in quite a few places in the surrounding district where land was increasingly being taken up for farming. The wooden bridge spanning the Derwent at New Norfolk[102] provided him with convenient access to the road that ran to the north-west

[98] ibid., p. 89.

[99] ibid., p. 94. 'The Falls' mentioned here by Burn were probably the same as 'Major's Falls, New Norfolk' and 'The Falls, New Norfolk' which was one of Oldfield's collecting places but not located on any gazetteer (see HAODb for details).

[100] ibid., p. 94–96.

[101] HAODb.

[102] Joan Goodrick 1977, *Life in Old Van Diemens* [sic] *Land*, Rigby, Sydney, p. 136. This bridge was completed in 1841.

Fig. 7.6 Rivulet, Back River valley, near New Norfolk

along the north bank of the Derwent to Hamilton and on to more northerly settlements in the valley of the River Clyde. By following this road, he would have reached his collecting localities at Back River, Johnny's Creek and the Macquarie Plains in a short time on foot from New Norfolk.[103] Some of Oldfield's specimens[104] indicate that at the time of his visit, as is the case with some sections of the river today, the banks of the Back River were ferny with wet forest species at the river margins, although introduced plants such as the Blackberry (*Rubus fruticosus*) and the Weeping Willow (*Salix babylonica*) may have become well established in some parts. What remained of the surrounding bush after clearing was probably dry eucalypt forest and woodland containing Tasmanian Blue Gum (*Eucalyptus globulus*) and White Gum (*E. viminalis*) with an understorey dominated by native grasses with a sparse cover of tall shrubs that may have included species such as Prickly Box (*Bursaria spinosa*) and Silver Wattle (*Acacia dealbata*). [105] If today's forest to the east of Johnny's Creek is any guide, in Oldfield's time uncleared land in the vicinity of the Creek was dry White Peppermint forest and woodland including such other eucalypts as Black Peppermint (*E. amygdalina*), White Gum and Tasmanian Blue Gum over a variable understorey of grasses and Sagg (*Lomandra longifolia*) and sparse layers of shrubs and small trees.[106]

Goat Hills, south of west of New Norfolk, was another area where Oldfield collected.[107] What native vegetation remains on these hills today suggests that when he visited them they were clothed in a wet sclerophyll forest of Gum-topped Stringy Bark and Messmate Stringy Bark under which grew a variety of *Acacia* species including Silver Wattle.[108] The understorey was typically composed of broad-leaved shrubs, most commonly Satinwood, Dogwood and Musk with a high proportion of ground ferns.[109]

---

[103] HAODb.
[104] For example, MEL 54041, *Blechnum chambersii*, Back River, New Norfolk (MELISR database, 8 March 2007, NHV, RBGM).
[105] S. Harris & A. Kitchener (eds) 2005, p. 251.
[106] Jayne Balmer circa 6 October 2008).
[107] HAODb.
[108] Jayne Balmer circa 6 October 2008.
[109] S. Harris & A. Kitchener (eds) 2005, p. 371.

Fig. 7.7 Eastern slopes, Mt. Dromedary

He apparently traversed the New Norfolk to Hamilton road across the fertile, grassy Macquarie Plains to collect in the more distant dry Stringy Bark forest and Swamp Gum (*Eucalyptus ovata*) woodland[110] of Dry Hills[111] to the north-east of Hamilton. Prior to 1850 there does not appear to have been a track to that locality from Hamilton. Perhaps the easiest route was to follow the northern bank of the Derwent through the bush from just to the west of Hamilton and, in turn, the eastern bank of the Ouse River, a tributary of the Derwent, that runs along the eastern side of the Dry Hills. Other localities where he gathered a few plants were The Bedchamber, which lay among hills to the north-east of New Norfolk, and on the Jordan River.[112]

Oldfield also obtained plants on Mount Dromedary[113] which lies five kilometres north of the Derwent and is clearly visible from the Causeway at Bridgewater. It is clothed in dry sclerophyll forests with several canopy species including Gum-topped Stringy Bark, Snow Gum, Silver Peppermint (*Eucalyptus tenuiramis*), Tasmanian Blue Gum and White Gum over a shrubby understorey.[114] He may have visited there from the vicinity of the Causeway or by travelling cross-country near farms and through bush from New Norfolk.

To the north-east of the capital Oldfield collected at the base of, and on, Grass Tree Hill.[115] This prominence, named for the many Southern Grass Trees (*Xanthorrhoea australis*) that occur there, lies in the path of the old road from Risdon Crossing to Richmond. He also obtained specimens at Richmond[116], which is about 20 kilometres to the north-east of Hobart and is situated on Coal River Flats on the banks of the Coal River. Traversing these Flats, the dominant native vegetation he would have encountered in uncleared areas were populations of Swamp Gum (*Eucalyptus regnans*), White Gum

---

[110] Jayne Balmer, c.6 October 2008.
[111] HAODb.
[112] ibid.
[113] ibid.
[114] Forestry Tasmania 1999, *Derwent Forest District Management Plan*, Forestry Tasmania, Hobart, p. 56, viewed 15 July 2012 at www.forestrytas.com.au and Jayne Balmer, personal communication emails 6 October 2008.
[115] HAODb.
[116] ibid.

Fig. 7.8 Grass Tree Hill, Tasmania

and Silver Wattle, and the 'upper terraces', which fringed the Flats, supported woodlands predominantly White Gum, Black Peppermint and an understorey of Silver and Black Wattle (*Acacia mearnsii*)[117] with a dense tall shrub layer and commonly occurring ground ferns.[118]

Richmond's apparent prosperity may have surprised Oldfield. It was the headquarters of Richmond police district and the thriving centre of the wheat, sheep, cattle and hop producing areas of the Coal River Valley, with a gaol, courthouse 'where also divine service is held every Sunday by a clergyman of the Established Church…and in the immediate neighbourhood are several well-built storehouses'.[119] Businesses had been established—several inns, blacksmiths, saddlers, stockyards, tanneries, a market place, brick and lime kilns, a windmill and general stores. Since 1825 a convict-built bridge over the Coal River had facilitated travel to settlements to the south- and north-east.[120]

Oldfield obtained mosses, lichens and a grass along the Coal River,[121] among the tree ferns (*Dicksonia antarctica*), Musk (*Olearia argophylla*), Dogwood (*Pomaderris apetala*), Silver Wattle, Black Wood (*Acacia melanoxylon*), below the canopy of Stringybark and White Gum of the riparian vegetation and on the alluvial flats.[122]

Apart from gatherings close to the river at Richmond and at nearby Woodburn [Farm][123], these specimens could have been taken anywhere along the stream, which rises to the north in the hills

[117] Department of Infrastructure, Energy and Resources, *Richmond Bridge Conservation Management Plan*, Department of Infrastructure, Energy and Resource, Hobart, 2010, p. 2, viewed 16 July 2012 at www.transport.tas.gov.au.

[118] S. Harris & A. Kitchener (eds), 2005, pp. 374, 375.

[119] J Syme, 1848, *Nine Years in Van Diemen's Land Comprising an Account of its Discovery, Possession, Settlement, Progress, Population, Value of Land, Herds, Flocks etc.; an Essay on Prison Discipline and the Results of the Working of the Probation System; With Anecdotes of Bushrangers*, J Syme, Dundee, p. 49. See also Henry Melville 1851, *The present state of Australasia, including New South Wales, Western Australia, South Australia, Victoria and New Zealand, with practical hints on emigration: also, Remarks on prison discipline: with suggestions for obviating the difficulties attending the transportation of convict; to which are added the land regulations, and description of the aborigines and their habit*, G. Willis, London, p. 235.

[120] Dianne Snowden, 'Richmond', in Alison Alexander (ed), 2005, p. 308. The convict-built bridge at Richmond is still in use today.

[121] HAODb.

[122] K.A. Johnson and J.B. Kirkpatrick, 1996, 'The vegetation and flora of the Coal River Gorge Nature Reserve', *Papers and Proceedings of the Royal Society of Tasmania*, vol. 130, pp. 15-23.

[123] HAODb. Today bed and breakfast accommodation is provided in the residence now called Hatchers Manor on 'Woodburn Farm', Anon., 'Hatchers Manor', viewed 22 August 2012 at www.laterooms.com.au.

south-east of Tunnack at some 580m altitude and discharges into Pittwater to the south of Richmond.[124] Lake Tiberias, Pittwater and rocky hills near Frogmore were places where Oldfield obtained a few specimens and also Jericho which is only about 10 kilometres to the north-west of Lake Tiberias. It is feasible that he visited there from the lake. Plants he gathered at Pontville may have been taken when he was on his way to or from the dry forest and woodland on the Coal River Tier to the south-west of Campania.[125] However, Jericho and Pontville were stopping places for coaches on the Hobart to Launceston Road so there is the possibility he did some botanising while he waited to continue his journey to Launceston a few years later.[126]

Oldfield's visits to Tasman Peninsula & Maria Island—pre-1850

Map 4 p. 100 – Central and Eastern Tasmania

Oldfield apparently only collected a few specimens on the Tasman Peninsula.[127] Some were from among the *Allocasuarina* and Honeysuckle Banksia of the Black Peppermint dry coastal forest around the Cascades Convict Station and amidst the ground ferns and other understorey species such as Satinwood, Dogwood and Musk of the tall Messmate Stringybark wet forest in the vicinity of 'Port Arthur'.[128] The eminent Irish botanist William Harvey stated that the tree fern (*Dicksonia antarctica*) is 'so abundant [around Port Arthur] that they [convicts] use the split logs (if so you can call a fern-trunk) for making corduroy roads through the forest; and very pleasant roads they are to walk on'.[129] Oyster Bay on Maria Island, off the east coast, was also one of Oldfield's plant gathering places.

Although the penal settlement at Port Arthur operated until 1877, the convict station on Maria Island was closed in 1832, but was reopened as a convict probation station in 1842. It was shut down again in 1850 and, following its abandonment, the island was used to run sheep.[130] The convict station at The Cascades, located about 12 kilometres west of Eaglehawk Neck and about the same distance to the north of Port Arthur, was established to cater for convicts from Norfolk Island operating in the local timber industry. A military station had been at Eaglehawk Neck since the early 1830s to prevent escapees leaving the convict establishments on the peninsula. Provisions and personnel from Hobart were delivered by a weekly government steamer to settlements on Forestier Peninsula, Eaglehawk Neck station and the convict stations at Port Arthur, Maria Island and The Cascades on Tasman Peninsula.[131] At Eaglehawk Neck there was a jetty 300 metres long which allowed supplies to be brought into the shallow Eaglehawk Bay.[132] A network of tracks connecting the settlements on the Tasman Peninsula was developed in the 1830s and 1840s.[133] The records of specimens we have located that were collected in these places suggest that Oldfield travelled by ship. This seems feasible since in 1855 William Harvey had gone to Port Arthur, to collect algae, on a government ship carrying 'a large party of chained convicts…guarded by soldiers'.[134]

In contrast to Back River, Johnny's Creek and Goat Hills, the very small number of records we have located of Oldfield's specimens from each of the other localities away from Hobart Town and

[124] David Horner, 2003, *Index of River Condition for the Coal River Catchment*, Department of Primary Industries, Environment and Water, Tasmania, p. 3, viewed 15 August 2012 at *www.dpiw.tas.gov.au*
[125] HAODb.
[126] Chapter 9.
[127] HAODb.
[128] S. Harris & A. Kitchener, 2005, pp. 224–25, 372–373.
[129] WHH to WJH, 31 March 1855, cited in Sophie C. Ducker (ed), 1988, p. 210.
[130] Hamish Maxwell-Stewart, 2005, 'Maria Island' in Alexander (ed), 2005, p. 227.
[131] H. Butler Stoney, 1856, pp. 34–53 & pp. 74–75. Stoney incorrectly refers to 'The Cascades' as 'Cascades Bay'
[132] Parks and Wildlife Service 2008, 'Eaglehawk Neck History', Department of Primary Industries, Parks, Water and Environment, Hobart, viewed at www.parks.tas.gov.au.
[133] Jennifer R. Jerome, Librarian, Heritage and Tasmanian Information Services, State Library of Tasmania, personal communication to MHH, email, 22 February 2008, HOA.
[134] Sophie C. Ducker (ed), 1988, pp. 202–204 & 210–211.

Mount Wellington, and the geographical spread between those places, suggests that, since they were apparently floristically diverse at the time, botanising was not the prime reason for his visits. He may have been working with a survey team with little time for collecting or on a tour dependent upon others for transport, as may well have been the circumstance of his Tasman Peninsula visit.

Overall, cryptogams[135] appear to have predominated his pre-1850 collections with an emphasis on mosses and liverworts[136] but as we do not have a comprehensive dataset of his specimen records this suggestion must remain tentative. If that focus was the case the question arises as to why. One explanation could be that he had discovered that they were relatively neglected by his predecessors and contemporaries who had concentrated on the showier angiosperms and ferns desired by nurserymen and curators of botanical gardens in England and Europe. This is supported to some extent by a comment made by William Hooker in a letter to Mueller that Oldfield

> is not neglectful of other plants but has a most extraordinary affection for the Acotyledons: & we are very thankful for anyone who will be at the pains to collect them: so, I give him all the encouragement I can, & there are several charming things among them.[137]

On the other hand, as Hooker's statement also suggests, his fascination for them may have been greater than for angiosperms. He left no details.

[135] Percentage of cryptogams and angiosperms in HAODb which we believe were obtained prior to 1850 in Tasmania: cryptogams (publ. 85%., herb. 75%), angiosperms (publ. 15%, herb. 25%) where N = 207 publ. and 113 herb., HAODb.

[136] *Bryophyta* (51% publ., 82% herb.); *Hepaticae* (41% publ., 12% herb.); Lichens (3% publ., 0% herb.); Algae (publ. 5%., herb. 6%) where N = 175 publ. and 85 herb., ibid.

[137] WJH to FM, 9 April 1854, RB MSS M2, RBGML, quoted in R.W. Home *et al.* 1998, vol. 1, *1840–1859*, p. 175.

Fig. 8.0: Quoya oldfieldii

Chapter 8

Sojourn in England 1849–1850

Oldfield was in London in 1850.[1] He was probably the 'Mr Oldfield' listed among the passengers on the barque Wellington which left Hobart for London on 8 February 1849.[2] Meeting up with his family may have been an important motive for making the voyage. Another intention, however, was to obtain identifications of a collection of 300 cryptogamic specimens he had brought with him from Tasmania.[3]

If they had advance knowledge of his ship's arrival, one or more members of the family may have met him at the dock and taken him about 10.5 kilometres across the city in his father's carriage to 14 Dorset Street, Portman Square, where, apparently, he was to stay with his parents while he was in London.[4] Otherwise, another option was to hire a Clarence carriage, commonly called a "growler", for his journey to Dorset Street. This was the most comfortable and convenient mode of public transport for a recently arrived voyager with a great deal of luggage.[5]

Frederick, Augustus' father, was still operating as a wine merchant from the Dorset Street address.[6] There is, however, no certainty as to which, if any, of Augustus' siblings were living there at the time. Marian, Frederica, whose occupation was given as teacher of French, German and Music, and 17-year-old Vincent were residing there with Frederick and Mary on the night of the England Census, 31 March 1851.[7] It is reasonable to suppose, therefore, that they were living with their parents during Augustus's visit. At the time of the Census the 23-year-old Juliana was a governess in the home of Mrs Mary Mammett at Ashby de la Zouch, Leicestershire, and may have been employed in that capacity when her brother arrived home.[8] Frederick Jnr was still in America with his family. During 1849–1850 he was in New York City earning his living as a musician.[9] On 13 March 1850 he petitioned the Common Pleas court in New York County for naturalisation as an American citizen.[10] Later in the year, 17 August 1850, he and his family were residing in Hackensack, Bergen, New Jersey, where he was an auctioneer.[11]

---

[1] AO to WJH, April 1850, Australian Letters, vol. 73, fol. 257 (488) and AO to WJH, 15 April 1850, English Letters I–Z, 1850, vol. 30, fol. 154, RBGKA. It is clear from these letters that the one dated 'April 1850' preceded that of 15 April 1850.

[2] 'Shipping News', *The Courier* (Hobart Town), 10 February 1849, p. 2, DAN, NLA.

[3] AO to WJH, April 1850, Australian Letters, vol 73, fol. 267 (488), and AO to WJH, 15 April 1850, English Letters, 1850, fol. 154, RBGK..

[4] All the letters he wrote to WJH while he was in London during 1850 were written from the Dorset Street address.

[5] Cindy Crank, 2011, 'Transport and carriages in the Victorian era (1837–1901)', *Horses and History Throughout the Ages*, 2 May 2011, viewed 4 September 2012 at www.horsesandhistory.wordpress.com.

[6] Entry for Oldfield, Frederick H. & Co, wine merchant, 14 Dorset Street, Portman Square, *Post Office London Directory, 1850*, Frederick Kelly, London, p. 222, English Census Directories Project (ECDP), vol. 94, 1993, NLA.

[7] Record for Frederick Oldfield, *1851 England Census*, Source information: Class: HO107; Piece: 1473; Fol.: 10; P.:13, viewed 4 September 2012 at www.ancestry.com. At the time of this Census the family had moved from Dorset Street to 1 Somerset Villas, Chelsea.

[8] Record for J.M. Oldfield, *1851 England Census*, Source information: Class: HO107; Piece: 2084; Fol.: 443; P.:33, 4 September 2012, www.ancestry.com.

[9] *Doggett's New York City Directory for 1849–1850*, p. 322, New York Genealogical Records, 1675–1920 database, viewed 14 May 2005 at www.ancestry.com.

[10] Record for Frederick Henry Howard Oldfield, no. 0431; vol. or bundle no: 100; copy of record no: 229, 13 March 1850, New York Petitions for Naturalization, Common Pleas Court, New York County, US Immigration Records Database, viewed 17 November 2004 at www.ancestry.com. The address of petitioner: New York City and former nationality: English. Spaces provided for occupation, birth date or age, and the port and date of arrival in America were all left blank. Witness to Naturalization: Samuel Drury, Astoria, L. 1.

[11] Record for Frederick H. H. Oldfield, *1850 United States Federal Census*, M532, 1009 rolls, National Archives and Records Administration, Washington D.C., viewed 3 November 2005 at www.ancestry.com.

Sir William Hooker and the Royal Botanic Gardens—background to 1850

It was during this visit to London that Augustus appears to have commenced his association with Sir William Hooker. Hooker, a self-taught and widely published taxonomist, had been knighted in 1836 for his services to botany.

Established as a scientific garden in 1759, Sir Joseph Banks used his considerable influence and entrepreneurial skills to advance the Royal Gardens at Kew. Following his death in 1820 the gardens suffered from neglect. As a result of political action by London's scientific elite, a working party led by John Lindley, Professor of Botany at the UnivesityCollege of London and reporting to a parliamentary committee, investigated the future of Kew Gardens in February 1838. It recommended that a national botanic garden funded by Treasury, with a paid Director, be established at the site of the Royal Botanic Gardens. In 1840, this proposal was adopted by the government.[12] Under this arrangement the gardens were to be administered by the Office of Woods and Forests. Hooker became the first Director in 1841. During that year William Andrew Nesfield began landscaping the great avenues, grasslands and vistas of 'modern Kew' while, at the same time, Hooker set about replanning and rebuilding the ten major glasshouses, which were heated by fires producing deleterious soot and in a deplorable state.[13] 'His crowning achievement was the Palm House, "an exquisite winged structure" completed in 1848 to the design of Decimus Burton'. By 1846 Hooker had increased the gardens to their present size of nearly 300 acres and, for the first time, they were opened to the public.[14]

Hooker's commitment to supporting the British horticultural industry was evident for many years before he became Director. Thus, from 1827 until 1845 he was wholly author and editor, and for nearly a decade, the illustrator of *Curtis's Botanical Magazine,* a gardening and botanical journal started by Sir William Curtis circa 1790. Now well over two hundred years old this magazine has been published continuously since its inception although there have been several series within the overall sequence. It is the longest-running botanical periodical featuring colour illustrations of British and exotic plants.[15]

There were no collectors from Kew Gardens working abroad when Sir William became Director. During the years 1842 and 1843, Hooker entered into cost-share arrangements with wealthy members of the aristocracy, which enabled the full-time employment of Joseph Burke to collect in California and William Purdie in Jamaica and Columbia. Although some interesting new species were brought back to Kew Gardens the overall result to the end of the decade was extremely disappointing.[16]

During the 1840s Sir William was involved in the selection of naturalists for service on some Royal Navy vessels. The first, 1847, was Berthold Seeman, naturalist aboard HMS *Herald*, then engaged in a hydrographic survey of the Pacific. The Botany of the Voyage', published between 1852 and 1857, embraced the floras of Panama, north-west Mexico, West Esquimauxland, and Hongkong.[17] Over the next decade, Seeman was followed successively by William Milne to the South Seas and Charles Wilford and Richard Oldham to Japan in 1861, the latter being Kew's last official collector.[18] Sir William's son, Dr Joseph Hooker, who in 1848 was still on the payroll of the Royal Navy, was authorised by the First Lord of the Admiralty and former Governor General of India, George Eden, 1st

---

[12] Stephen D. Hopper, April 2015, 'Royal Botanic Gardens Kew', *ELs: Essential Life for Science*, John Wiley & Sons, Chichester, viewed 28 December 2017 at http://www.els.net, pp. 1-3.

[13] Mea Allan, 1972, 'Hooker, William Jackson', in C.C. Gillispie (ed), 1970–1980, *Dictionary of Scientific Biography*, vol. 6, Charles Scribner's Sons, New York, p. 494. This article contains a bibliography of Hooker's books and a list of the journals to which he contributed. A list of secondary sources relating to his life and scientific contribution is also presented. See also Ray Desmond 1995, pp. 169–189.

[14] Mea Allan, 1972, vol. 6, p. 492.

[15] Ibid.

[16] Ray Desmond, 1995, pp. 206–213 and 432–434.

[17] George Simonds Boulger, c.2011, Dictionary of National Biography, 1885-1900, vol. 51, Viewed 28 December 2017 at www.en.wikisources.org.

[18] Ray Desmond, 1995., p. 434.

Fig. 8.1 Sir William Jackson Hooker, 1855?

Earl of Auckland, to collect in India for Kew for two years on reduced naval pay on the proviso that he spent a third year on the island of Labuan, North Borneo, assessing its agricultural potential. Between 1848 and 1851 he collected plants in Sikkim, Nepal, Bengal and the Khasia mountains. Hooker had spent three years in India when the Labuan trip was cancelled but obtained permission to remain in Greater India for a further year..[19] Hooker made substantial collections during his time there and produced a number of important publications, among them his *Himalayan Journals: or notes of a naturalist in Bengal, the Sikkim and Nepal Himalayas, the Khasia mountains, &c* (1854 & 1855).[20]

At the time of Sir William's appointment, there was no herbarium or library at the Botanic Gardens but he allowed botanists to access his private plant and book collections. These occupied thirteen rooms of his residence, West Park, a plain three-storey Georgian house ten minutes' walk from the Gardens. His herbarium was the largest and most valuable in private hands and attracted scientists from all over the world. During the late summer of 1846 he tried to transfer the expense of maintaining the herbarium to the Garden's budget held by the Office of Woods and Forests but was only able to obtain some mounting paper and presses. For the remainder of the decade he worked to obtain alternative housing for the collection, which was expanding year-by-year because of his growing international network of correspondents. Specimens were obtained primarily by donation but some

[19] ibid. p. 208.
[20] Vols 1 & 2 London.

were purchased from his own pocket. By the time Oldfield arrived back in England, Sir William had not been successful in his quest for more suitable herbarium accommodation.[21]

Under Joseph Banks' long presidency (1778–1820) of the Royal Society, the government came to depend on it for 'scientific advice in a period when the British State was having to expand its activities to cope with the challenges posed by increasing foreign rivalry and rapid economic growth'. These challenges prompted the establishment in distant colonies of plantations to provide food for the British population following the agricultural crisis at home in 1795, to produce cotton to meet the increasing demand for raw product from the industrialised cotton mills of England, and to meet the British public's demand for exotic products such as Chinese teas and plants. It was against the backdrop of such changes in the character and activities of the British State, then, that Banks was to integrate scientific policy into the fabric of government. Thus, for example, as President of the Royal Society, Banks served on special committees of enquiry requiring scientific expertise. The Board of Ordnance frequently called on the Society for scientific advice and he became unofficial scientific adviser to the government's Committee for Trade and Plantations following its establishment in 1774.[22] He was elevated to Privy Councillor in 1797, chiefly on account of his work on the Committee for Trade, and in this position, he was able to influence government policy across a number of sectors.[23]

Sir William acted on a number of fronts to position the Gardens as a vital source of information having implications for the industrial and commercial exploitation of exotic plants, action which he probably hoped would strengthen the Gardens' position with the government. He founded the Museum of Economic Botany, the first of its kind, in 1847. Here specimens of vegetable products and materials from around the world were displayed for the benefit of manufacturers, tradesmen, and craftsmen.[24] In 1849 Hooker reported that he was endeavouring to introduce plants useful 'in commerce, in medicine, in agriculture, in domestic economy'. A year later he announced that it was his 'especial object to cultivate what may be useful and valuable for our colonies'. Exactly what he was doing in the latter part of the 1840s to address these objectives is not clear. In 1851, however, Wardian cases, which had been in use since 1835, were filled with live 'useful' plants and sent to New Zealand, Tasmania, India, Trinidad, Jamaica, Sierra Leone and British Honduras.[25]

Fig. 8.2 Wardian case, 1852

Wardian cases were tightly sealed boxes with a glass roof which, if there was sufficient light, allowed photosynthesis to take place. The transpiration of the plants saturated the air in the container with water vapour, which condensed on the glass; the droplets moistened the soil in a continuous cycle. Even though more plants survived long sea voyages to England from Australia and other colonies in these containers than by the earlier method of conveying exposed plants in pots housed on the ships' decks, losses were still incurred. Modifications to packaging procedures and to the positioning of the cases on board ship gradually reduced the failure rate.[26] Following these initial consignments from Kew Gardens, a growing two-way traffic of live plants and parcels of seeds developed with manufactured goods of vegetable origin also being sent from colonies for inclusion in the Museum of Economic Botany.[27]

Many botanists and others recognised that the precise identification of plants is an essential preliminary to their use in commerce, medicine and agriculture. This prompted the pioneering work at

---

[21] ibid., pp. 197, 198.
[22] John Gascoigne, 1998, pp. 22–33.
[23] ibid., pp 111–146.
[24] Mea Allan, 1972, vol. 6, p. 494, and Ray Desmond, 1995, p. 191
[25] ibid., pp. 213, 258
[26] ibid.
[27] ibid., p. 213.

Kew Gardens, initiated by Hooker, in publishing colonial floras. This could not have proceeded without the resources of its herbarium. Thus, the discipline of taxonomy requires comprehensive reference collections of plants embracing all the taxa being investigated. Furthermore, the plants must be preserved and recorded in such a way that, as far as possible, they retain the morphological and physiological characteristics required for taxonomic study and can be referred to almost indefinitely.[28]

Oldfield's first communications with Sir William Hooker

Oldfield wrote to Sir William early in April 1850 stating that he had 'taken the liberty of sending the enclosed specimens in hopes' he would 'be so kind as to name them' as he, Oldfield, did not have 'access to any works in which [he] could find them'. To this end he enclosed a list of numbers (#1 to #384) with space beside each one for the plant identification. He stated, 'My own specimens are marked with corresponding numbers, so that the name after a number will suffice'. From this, it is apparent that the specimens he sent to Sir William on this occasion were all duplicates of those held in his own herbarium. Against 19 of the numbers in the list are the names of plants which appear to have been written by Oldfield.[29]

In the same letter, Oldfield offered Hooker seedlings of some flowering plants: '*Acacia lophantha* [= *Paraserianthes lophantha*], *Culcitium salicinum* [= *Bedfordia salicina*] and *Xanthorrhoea asphodelea* [name not in APNI]'. How he had transported these live plants from Australia is not evident in the letter but they probably travelled in a Wardian case as later correspondence shows he possessed one.[30]

In this letter, Oldfield did not introduce himself, provide information as to his background, his interest in botany or where he obtained the specimens, which suggests that the pair had corresponded previously. If so, there is no evidence of it at Kew Gardens, nor that they had met in person already.[31] Perhaps Oldfield's association with Kew Gardens was initiated by a visit to the Gardens when he may have spoken to John Smith, the then Curator[32], or another member of staff who encouraged him to forward his plants to Hooker for identification and perhaps later briefed the Director on Oldfield's background and botanical interests.

It is clear from subsequent correspondence that the specimens were all cryptogams from New Zealand and Tasmania (see below). Sir William had published widely on the taxonomy of the *Cryptogamia* of which he was a recognised authority.[33] Oldfield was probably aware of this from his own reading, although someone else may have suggested that Sir William was the appropriate person to undertake the identifications.

Oldfield wrote again on 15 April 1850, thanking Sir William for his kindness in answering his letter, expressing awareness that his, Hooker's, duties left him 'no time to spare', and regretting 'that the late severe weather destroyed' the *Xanthorrhoeas*. The letter continued:

> I am particularly anxious that my collection of cryptogamous plants (more than 300 in number) from Australia and New Zealand should be examined by a Botanist, for I think them likely to include something new, having had opportunities of collecting which few persons can hope for…If you would permit me to send some more specimens to be named, I shall esteem it a great favour, or should your time permit I could bring any given number to Kew for your inspection…[S]hould there be any of my

---

[28] Ray Desmond, 1995, pp. 195–197

[29] AO to WJH, April 1850, Australian Letters, vol. 73, fol. 257 (488), RBGKA.

[30] See Chapter 9.

[31] AO to JDH, 16 March 1866, English Letters, vol. 97, 1859–1900, NAT–OXL, fol. 289 (296), RBGKA.

[32] John Smith was Curator from 1841–1864. Prior to this (since 1822) he had been foreman in the propagating department, Ray Desmond, 1995, p. 430.

[33] Mea Allan, 1972, vol. 6, p. 493.

> specimens you particularly desire, you are welcome to them, as they would be more useful in your hands than in mine: besides I shall be able to replace them.[34]

It seems from this letter that initially Oldfield only sent part of his collection of 384 specimens to Sir William.

Apparent in this letter is Oldfield's desire to be the discoverer of species new to science and to assist in the taxonomic endeavours of scientists such as Sir William, an aspiration he shared with many other colonial collectors.[35] It is also evident from Oldfield's earlier communication that he possessed the knowledge and skill to identify plants provided he had access to the appropriate literature. Nevertheless, he did not look upon himself as being a 'Botanist'. Underlying this is the implication that he reserved the term for taxonomists and others who undertook botanical research and publication, a view that is reinforced when he said that his specimens 'would be more useful' in Sir William's 'hands' than in his.[36] Furthermore, although Oldfield may have come to the realisation that his lack of a Cambridge or Oxford degree or independent means would make it virtually impossible for him to aspire to a full-time career as a botanist, he does seem to have longed to have the opportunity to travel and to collect plants[37], with satisfaction and recognition coming his way through being the bearer of new species for eminent others to describe.

He went on: 'It is at present my intention to return shortly to the Australasian Colonies, and as heretofore I shall collect all that I see'.[38] His offer of plants may have been his way of putting pressure on Sir William to do the determinations, and of whetting Sir William's interest in him as a potential collector. Thus, he had wasted no time in informing Hooker that he had opportunities to collect 'which few persons can hope for'.[39]

Like Ronald Gunn in Tasmania and William Colenso in New Zealand[40], Oldfield was conscious that he had access to plants that Hooker might want and which would be difficult for the taxonomist to obtain otherwise. His willingness to collect for Hooker was eagerly reiterated in a later letter: 'When I leave England I shall certainly send to you anything which I consider interesting, as by those methods my labours may be of some avail to science'.[41] There is no way of knowing whether Oldfield said this in response to a request from Sir William to collect for him, nor is there any suggestion in this correspondence, implied or otherwise, that Oldfield was seeking remuneration. It is apparent that his primary motivation was to serve science and in that regard, he was following in the path of various gentlemen in the colonies who, as Endersby has said, 'botanized "con amore"'.[42] Nevertheless, Oldfield wanted his plants identified correctly and he appears to have been willing to offer specimens in return.

A further parcel of cryptogamic specimens was sent to Hooker sometime after 15 April 1850 and before 22 May 1850. In the covering letter Oldfield commented that a few of these plants 'had greatly puzzled him' although he 'did not think [them] new'.[43]

He believed that the scientific worth of his collections depended upon his ability to correctly identify the plants and to recognise species that were probably new to science:

> You will easily see that my collection is not very scientific, but while abroad I could never procure the necessary works to prosecute a proper research, and therefore scarcely know what to take or reject. I hope that the experience I have gained and my

[34] AO to WJH, 15 April 1850, English Letters, 1850, I–Z, fol. 154, RBGKA.
[35] Jim Endersby, 2008, p. 85.
[36] AO to WJH, 16 March 1866, English Letters, 1859–1900, vol. 97, NAT–OXL, fol. 289 (296), RBGKA.
[37] Chapter 10.
[38] AO to WJH, 15 April 1850, English Leters, 1850, I–Z, fol. 154, RBGKA.
[39] ibid.
[40] Jim Endersby, 2008, p. 89.
[41] AO to WJH, undated, Australian Letters, vol. 73, fol. 256–258, RBGKA. The content of the letters written from 14 Dorset Square, of which this was one, indicates that this undated communication was written after the letter of 15 April 1850 and before 22 May 1850.
[42] Jim Endersby, 2008, p. 85.
[43] AO to WJH. Hooker, undated, Australian Letters, 1850, vol. 73, I–Z, fol. 260 (491), RBGKA.

> increased knowledge of plants will render my next visit to Australia more successful in this branch, than my former. Perseverance, I think has not been wanting...The next time I visit Kew I shall not fail to see Mr Smith for I do not like to let slip any opportunity of acquiring knowledge which will avail me in my future wanderings.[44]

A veiled hint, perhaps, for him to be rewarded with taxonomic literature in return for specimens of interest. In this excerpt, there is the implication that he was responding to a suggestion by Sir William that he seek Smith's advice regarding what was required in the way of herbarium specimens and their presentation.

Also in this letter, Oldfield wrote of his correspondent's goodness in 'answering his question' (probably a reference to his earlier enquiry about a plant he found on the banks of the Murrumbidgee River and which he took to be a *Lythrum*).[45] This, he said, had emboldened him 'to put a few more [questions]' although he was 'aware that they will appear nugatory' to Sir William. Did this rather self-effacing tone simply reflect the deference that society demanded should be paid to someone of such elevated social rank or was it born of Oldfield's great respect for Hooker's high standing in the botanical sciences? Whatever the underlying reason he was not deterred in his quest for knowledge. An indication, perhaps, of his underlying confidence that Hooker would respond, probably deriving from an understanding that he (Hooker) was largely dependent upon colonial collectors for the plant material upon which he was building his scientific reputation. His confidence may have been strengthened as a result of Hooker questioning him about the occurrence of a particular genus[46], to which he replied:

> I have met with Dawsonia (D. polytrichoides I believe) but only on one occasion, when I procured one specimen only. I have looked over my plants but fear I have left it with other named plants in V.D.L.[47]

Whatever Hooker thought of Oldfield's queries, clearly the collector's interests extended beyond the cryptogams to include flowering plants. Thus, in addition to questions relating to a few fungi, Oldfield requested the specific names of the 'Morton [sic] Bay Cedar', which he thought to be 'Cedrola [specific name illegible] of Brown'[48]; 'the 'Loranthus [one of the mistletoes] on the Casuarinas'; and 'the broad leafed grass tree of Mount Wellington called "Richea dracophylla" but [he did not] think rightly [identified]'.[49] He asked whether Sir William knew of any species of *Hydrocotyle* in New Holland. His final question enquired whether 'your [sic] "Flora of Van Diemen's Land"' in the *Journal of Botany* was continued. Joseph Hooker authored this work.[50]

Apparently in response to an enquiry from Sir William, Oldfield responded that he was acquainted with Mr Ronald Gunn 'by name only' and 'chiefly known to me from his connexion with you'. He went on to say: 'In fact few of the inhabitants of any place where I have sojourned know that such a person as myself exists my pursuits not permitting me to join in much society'.51 This might support a view that in Tasmania he was engaged in an outdoors occupation, such as survey work, which required him to be constantly on the move. However, if his 'vocation' required him to spend time in one locality, as would have been the case with tutoring, he may have had no time for socialising if all his

---

[44] ibid.

[45] ibid.

[46] That Hooker queried Augustus on this genus is evident from Oldfield's response.

[47] AO to WJH, undated. Australian Letters, vol. 73, 1850, I–Z, fol. 260 (491), RBGKA.

[48] Stephen Hopper has suggested that this could be hoop pine (*Araucaria cunninghamii*), a coniferous cedar and regarded as one of the most useful trees found at Moreton Bay, personal communication, email attachment,17 December 2017, HOA.

[49] Although *Richea dracophylla* occurs in mountainous regions in Tasmania, Mary Cameron (ed), 1996, *A Guide to Flowers & Plants of Tasmania*, Launceston Field Naturalists Club, revised edition, New Holland Publishers (Australia), pp. 26, 40, the species Oldfield thought was different was probably *Richea sprengelioides* which he did collect on Mount Wellington, HAODb.

[50] AO to WJH, undated. Australian Letters, vol. 73, I–Z, 1850, fol. 260 (491), RBGKA. This paper was actually Joseph Hooker's 'Florae Tasmaniae Spicilegium, or contributions towards the Flora of Van Diemen's Land', *The London Journal of Botany*, vol. 6, 1847, pp. 106–125, 265–286, 461–478. Papers by JDH on aspects of the flora of Tasmania appeared in *The London Journal of Botany*, vols. 3 (1844), 4 (1846) & 5 (1846).

[51] AO to WJH, undated, Australian Letters, vol. 73, 1850, I–Z, fol. 260 (491), RBGKA.

spare time was spent wandering in the bush pursuing his natural history interests. The question arises, however, as to whether he avoided the company of others to escape the ignominy of not being accepted into the upper echelons of colonial society, circles in which Ronald Gunn moved with ease.

By 22 May, Oldfield was becoming uneasy: He wrote again to Sir William:

> I shall no doubt appear very importunate but my anxiety with regard to the naming of my plants will perhaps serve as a palliative. Should you not be already overstocked, I should like to send you the rest of my plants, for to have them named, I would willingly defer my departure for a few weeks...[When] you have done as much as you can with my plants (for I fear that some are so imperfect that it will be impossible for you identify them) if you will let me know I will call for them.[52]

It is not known whether Hooker agreed to identify the remaining specimens.

Oldfield reiterated his earlier offer to collect plants in Australia and corrected his earlier response to Sir William's previous query:

> I have just procured the characters of Dawsonia and I now find that I was in error to say that I had met with D. polytrichoides. I should think that the plant I so named is a species of polytrichum.[53]

Oldfield had one more question to ask which, in view of later taxonomic revisions, shows that he was painstaking with his plant identifications, considering the habits of plants he collected: 'Does the Aster microphyllus (class?) of La Billardière belong to Eurybia or (judging from its habit) to a distinct species?'[54] He stated further:

> I have marked No. 7 of those last sent as an Acrotriche, but could not determine the specific name, it seemed nearest to A. divaricata from which it differed greatly in its habit. Besides I am not aware that this last-named plant is found in V.D.L.[55]

In a subsequent undated note, Sir William must have been left in no doubt that Oldfield was much concerned that the plant identifications he sought would not be completed before he set sail for Tasmania:

> As my time for leaving England is very near at hand, I am obliged, though very unwilling to trouble you, to again address you. I only defer taking my passage to know whether you will be able to name a very few plants more, as I think I shall now have very little difficulty with the rest, though I much fear that more important engagements will scarcely afford you time to identify those you already have...The shipping season is so far advanced that I am doubly anxious on the subject, and if you have the goodness to leave a note or word with the Porter at the Museum [of Economic Botany], you would much oblige me. I shall be at the Gardens next Tuesday, should it be fine.[56]

Clearly, too, Oldfield was keener to have his plants identified than his various questions answered:

> With respect to the questions I asked, I fancy I have found answers to them: so, unless perfectly convenient, you need take no notice of them: yet permit me to say, that I treasure [? much] every remark of yours, and in that account, I should prefer having your answers.[57]

---

[52] AO to WJH, 22 May 1850, Australian Letters, vol. 73, fol. 259 (490), RBGKA.
[53] ibid.
[54] At the time of writing this biographical account, *Aster microphyllus* was considered not to belong to either the genus *Aster* or *Eurybia*. In 1867 George Bentham placed it in synonymy with *Olearia lepidophylla* but in 1916 J.H. Maiden and E. Betche retained it in the same genus but as a different distinct species, *Olearia microphylla* (APNI).
[55] AO to WJH, 22 May 1850, Australian Letters, vol. 73, fol. 259 (491), RBGKA.
[56] AO to WJH, undated, Australian Letters, vol. 73. fols. 260 (491), RBGKA.
[57] ibid.

Oldfield offered Sir William 'a pot of *Isolepis* [sp.]' from the summit of Mount Wellington which was 'just ready to flower'; a further inducement perhaps.

It seems from this letter that while he was waiting for Sir William's determinations, Oldfield was at his parents' home identifying his own specimens, possibly facilitated by newly acquired taxonomic works. There is no mention in his correspondence that he was given access to Sir William's library and herbarium at West Park. Had Oldfield been granted such a privilege, it is hard to imagine that he would not have referred to it in one of his letters. Thus, he seems never to have lost an opportunity to express his gratitude for any assistance afforded him by Sir William. Furthermore, had he had such access it would not have been necessary for him to ask Sir William whether the 'Flora of Van Diemen's Land' contributions to the *Journal of Botany* had been continued.

It is apparent that Oldfield submitted portions of his cryptogamic collection for identification on separate occasions and that he was happy to give Sir William any that he wanted.[58] What is not clear, however, is how many of the specimens Oldfield brought to London were sent to Sir William for determination and were retained by him. It is likely, however, that many of the specimens referred to in Joseph Hooker's *Flora Tasmaniae* (1860) were among those brought to England in 1850. This is suggested based on what we know of Oldfield's movements after 1850.

Shortly before 15 July 1850, Hooker returned Oldfield's cryptogams duly named. Oldfield was delighted and wrote thanking Sir William 'for the great pains [he had] taken' with his plants and for the gift of the copy of his *Musci exotici*[59], which Oldfield said he valued 'doubly for its intrinsic value, and as coming from its justly celebrated author'.[60] This work, that Oldfield said had already enabled him to identify three plants, was devoted to new foreign mosses. While there is no evidence of negotiations between Sir William and Oldfield over rewards in return for specimens, it seems obvious that Sir William's underlying motive in identifying Oldfield's plants and gifting this taxonomic work to him was to encourage him to continue collecting, classifying and sending specimens to Kew Gardens.

Oldfield offered to 'gladly undertake' whatever he could for Sir William wherever 'he [Oldfield] may go in future' and requested, due to the 'erratic nature of [his] movements' that Sir William address any correspondence care of Edwin at Wentworth Cottages, Argyle St, Hobarton. He also signalled his intention to 'firstly go to S. Australia, as I have a great desire to see the interior of New Holland, the time [for his departure from London] about the end of this month'.[61]

Oldfield also promised to 'take care to distribute the "Directions" [Sir William's *A Manual of Scientific Enquiry: Prepared for the Use of Her Majesty's Navy and Adapted for Travellers in General*[62]] where they [were] likely to do most good'.[63] As discussed previously, this document could not be located. A set of Hooker's instructions published in 1856 contained no directions as to the need to include on a label the date of collection or details of plant form or habitat.[64] Although Hooker was keen on producing colonial floras, his main concern was the correct classification and illustration of plants to facilitate the needs of nurserymen, curatorial staff of botanical gardens, including Kew Gardens, and others interested in the economic uses of plants.

---

[58] AO to WJH, 15 April 1850, English Letters, 1850, I–Z, fol. 154, RBGKA.

[59] WJH 1818–1820, *Musci exotici: containing figures and descriptions of new or little known foreign mosses and other cryptogamic subjects*, 2 vols., London.

[60] AO to WJH, 15 July 1850, Australian Letters, vol. 73, fols 258 (489), RBGKA.

[61] ibid.

[62] These instructions were originally published in the 1849 *A Manual of Scientific Enquiry: Prepared for the Use of Her Majesty's Navy and Adapted for Travellers in General* which was probably the document given to AO but a copy could not be found.

[63] AO to WJH, letter, 15 July 1850. AO to WJH, 15 July 1850, Australian Letters, vol. 73, fols 258 (489), RBGKA.

[64] See Chapter 2.

Fig. 8.3 Wood engraving of the interior of the Museum of Economic Botany, Royal Botanic Gardens, Kew, 1855

In 1863, after Oldfield returned to England for good, Bentham in his *Flora Australiensis* articulated the need for

> details of 'the time, place, and situation in which [a plant] was gathered; of the stature, habit and other particular relating to any tree, or herb of which the specimen is only a portion; of the kind of root it has; of the colour of the flower; or of any other particulars which the specimen itself cannot supply, or which may be lost in the process of drying' the plant.[65]

There is no evidence that Sir William and Oldfield had any face-to-face contact over the period he collected in Australia, although such was their intention. On the day before Oldfield's departure for Australia, Sir William arranged a meeting with him in the Museum of Economic Botany. In a letter written to Dr Joseph Hooker on 16 March 1866, Oldfield recounted what happened on that occasion:

> We were both punctual to the time, and one of the assistants pointed out Sir William to me, but as the meeting was for the discussion of certain matters relative to the disposal of my future collections, I had not the courage to address him, and on his part he did not know me from the other visitors to the Museum. So, I left England without having entered into arrangements which might have benefited me considerably.[66]

[65] GB, 1863, vol. 1, pp. xxxii.
[66] AO to JDH, 16 March 1866, English Letters, vol. 97, 1859–1900, NAT–OXL, fol.289 (296), RBGKA.

Oldfield explained that his courage flagged on this occasion because of 'my very great reluctance to intrude myself or my affairs on others'. While in this this letter, written 16 years later, there is some indication that it may have been his intention at their meeting to discuss remuneration for any plants he sent to Sir William in future, it was never specifically stated.[67] Oldfield's reticence on this occasion was to have an important bearing on his relationship with Kew Gardens later in his life.

Oldfield's South Australian collecting

Map 5 p. 132 – South Australia

Did he fulfil his desire to go to South Australia and venture to the mainland interior on his return voyage? A few specimen records have been found that confirm that he did collect there but they are all undated. They may, therefore, have been obtained during a single visit to the Colony or on more than one stopover during other journeys. Apart from three gatherings near Adelaide,[68] he obtained specimens at Kangaroo Island, Port Lincoln and Marble Range.[69] This suggests that he might have accompanied a supply ship calling at those places on its way to Coffin Bay, where a Mr Hawson was breeding Timor ponies for the Eyre Peninsula.[70] Oldfield may have made landfall there and walked about five kilometres inland to another of his collecting places, Marble Range.[71] The Old Baroota (Pastoral) Station, Mambray Creek, on the east coast of Spencers Gulf[72] from whence he may have ventured inland about 27 kilometres and gathered at least one specimen (*Acrotriche patula*) on Mt Remarkable.[73] There is also a Mt Remarkable close to Bremer Bay in Western Australia where he collected in 1861–1862, but according to *FloraBase,* the species occurs much further along the coast to the east of Esperance.

Was Oldfield's first acquaintance with Dr Ferdinand Mueller in Adelaide?

If Oldfield did visit Adelaide on his return from England he may on that occasion have made his first contact with Dr Ferdinand Mueller, a pharmacist and botanist, who was in the Colony at the time and who later became important in the development of Oldfield's career as a collector. Unfortunately, we cannot establish when and where they first met and started corresponding. None of Mueller's correspondence with Oldfield has been found.[74]

Oldfield might have come to know of Mueller's presence in South Australia through advertisements that he had placed in the *South Australian Register* and the *South Australian Gazette and Mining Journal* during September 1848, in which he appealed to the public for plant specimens at a price 'fixed at the lowest indemnifying rate' as it was his intention to prepare a natural history, 'with specimens of the different plants mostly indigenous to South Australia'.[75] Mueller does not appear to have commenced corresponding with Sir William and Joseph Hooker up to the time Oldfield returned from England so it is doubtful that he would have learnt of Mueller's presence in Adelaide from them.[76]

---

[67] ibid.
[68] HAODb.
[69] ibid. AO also stated in a letter to WJH that he had found *Gymnogramma leptophylla* in South Australia but did not say where, AO to WJH, 5 December 1863, English Letters, vol. 42, 1862–1865, L–Z, fol. 198, RBGKA.
[70] Coffin Bay Brumby Preservation Society Inc. undated, 'South Australia's wild Brumbys', viewed 12 November 2012 at www.coffinbaybrumby.org.au.
[71] HAODb.
[72] Department for Environment and Heritage2006, *Mount Remarkable National Park Management Plan*, Adelaide, p. 18, viewed 15 November 2012 at www.environment.sa.gov.au
[73] K000349098, *Acrotriche patula*, RBGK.
[74] R.W. Home *et al.* (eds), 1998, vol. 1, p. 551.
[75] ibid., pp. 103, 568.
[76] ibid., p. 44.

For health reasons, Müller (later Mueller) travelled from Schleswig-Holstein to South Australia in 1847 with his two sisters, Clara and Bertha. Müller's first choice was Madiera but why he changed his mind is not entirely clear.[77] The decision to migrate to South Australia has been reported as having been on the advice of fellow botanist, Dr Johann A.Ludwig Preiss[78], of Herzberg, Hanover who had just returned from Western Australia. There is, however, no evidence to confirm that Preiss and Müller ever met but Müller later claimed that it was after seeing some of Preiss' specimens collected in Western Australia that changed his mind.[79]

The twenty-two-year-old Müller arrived in Adelaide as a qualified pharmacist who in the previous year had received the degree of Doctor of Philosophy for a thesis on the Common Shepherd's Purse (*Capsella bursa-pastoris*).[80] It had not been the Müllers' intention to settle permanently in the Colony, but an outbreak of hostilities in Schleswig-Holstein induced them to remain. South Australia was the obvious choice. There were Prussians in the Colony, including thriving settlements of Prussians at Klemzig and Hahndorf. Müller immediately obtained employment as a chemist with Messrs Büttner & Heuzenroeder, German Chemists, Rundle Street, Adelaide and very soon afterwards he Anglicised his family name to Mueller.[81] He stayed with that firm until 24 December 1848, when he and a friend, F.E.W. Kirchauff, acquired from the South Australian Government a 20-acre block of land in the Bugle Ranges, between Strathalbyn and Mount Barker. He built a cottage in which he and Clara lived but farming proved uncongenial and after some months he returned to Adelaide to his former employment. During these years Müller spent a great deal of his leisure time exploring his adopted homeland and studying its flora as far north as the Flinders Range, around Lake Torrens, and to the eastward 'he penetrated the Murray scrubs'.[82]

If Oldfield and Müller did not meet in Adelaide, perhaps they became acquainted when they were both in Victoria between 1852–1853.

A 'Mr Oldfield' arrived in Sydney from Port Phillip on 19 December 1850 aboard the schooner *Phoebe*[83] and from what we understand of his intention to leave England about the end of July, this may well have been Augustus. Although there was a schooner by that name that travelled between England and Australia in the 1840s there is no record in *Lloyd's Register of Ships* of such a vessel making a voyage to Australia in 1850 or 1851[84] so it may have been a coastal ship from Port Phillip. No information has emerged as to why he might have been in Sydney on that occasion or when he returned to Hobart, where he was in August 1851.

---

77 Sara Maroske, 2005, Science by correspondence: Ferdinand Mueller and botany in nineteenth-century Australia, unpublished Dr. Phil. Thesis, Dept. History & Philosophy of Science: University of Melbourne, introduction, pp. 12-13.

78 Preiss's very large collection of about 200,000 plant specimens was described by Johann Georg Christian Lehmann and several European collaborators and published in Hamburg in parts in 1844–47 under the title *Plantae Preissianae Sive Enumeratio Plantarum Quas in Australasia Occidentali et Meridionale-Occidentali Annis 1838–1841 Collegit Ludovicus Preiss, Ph. Dr,* J. H. Calaby 'Preiss, Johann August Ludwig (1811–1882)', in ADBO, viewed 28 May 2007 at www.adb.online.anu.edu.au.

79 Sara Maroske, 2005, introduction, pp. 12-13.

80 *Capsella bursa-pastoris* is one of the most important drug plants of the family *Cruciferae* (M. Grieve, 'Shepherd's Purse' in *A Modern Herbal* which is still considered by herbalists to be one of the best specifics for stopping haemorrhages of all kinds, viewed 18 October 2012 at www.botanical.com. According to Stephen Hopper, 'Cruciferae is one of a small number of unconventional family names promulgated by Kew, finally replaced by the APG in modern times with Apiaceae, in line with the common and consistent practice of using the type genus as the base of the family name, personal communication to MHH, email attachment, 19 December 2017, HOA.

81 R.W. Home *et al.* (eds), 1998, vol. 1, p. 9.

82 Charles Daley, 1924, 'Baron Sir Ferdinand von Mueller, KCMG, MD, FRS, botanist, explorer, and geographer', *The Victorian Historical Magazine,* vol. 10, no. 1, May1924, pp. 23–25 and Edward Kynaston 1924, *A Man on the Edge: A Life of Baron Sir Ferdinand von Mueller*, Allen Lane, London, pp. 41–70.

83 *Maitland Mercury & Hunter River General Advertiser*, 25 December 1850, p. 2, DAN, NLA.

84 Lloyd's Register, *Lloyd's Register of Ships online 1849–1851,* viewed 24 March 2012 at www.lr.org.

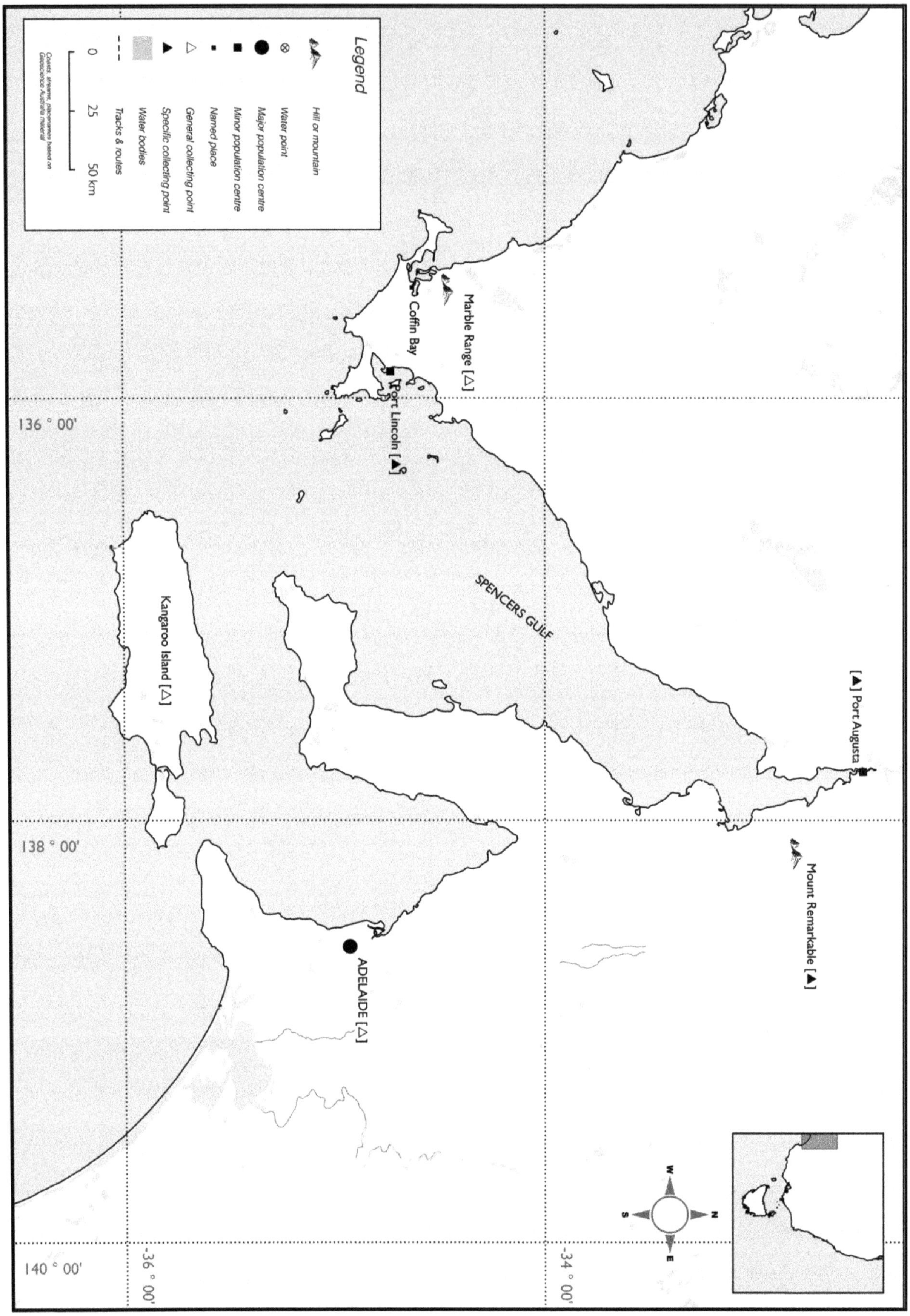

Map 5: South Australia

Fig. 9.0: *Calothamnus oldfieldii*

Chapter 9

Tasmania and Victoria 1851–1854

When Oldfield departed for England, rising wool prices were returning prosperity to the Colony[1] and the feelings of the Vandemonians about independence and self-government had progressively strengthened during his absence. Settlers were refusing to be taxed, advised or directed, and 'yet they clung to the advantages they enjoyed from the subsidy of the British Treasury, their easy sales to the commissariat market and their use of convict labour'. Nor would they pay rates to the elected Municipal Commission (Hobart Town), formed in 1847 to improve street lighting, paving and cleaning, in the belief that it was a ploy to secure tax for the up-keep of the convict gaols. Settlers' opposition to official rule had led to a polarisation of social groups and the hardening of class relationships. This was against the interests of emancipists who were convinced that it would lead to the permanent discrimination of those who had arrived in the Colony as convicts. In 1849 the Tasmanian Union was formed 'to assert emancipist rights in opposition'. While, on the one hand, the settlers held most of the advantages of wealth and influence, the emancipists, on the other hand, had only numbers and the help of Lieutenant-Governor Sir William Thomas Denison. By 1850, and through the meetings of the Masonic Lodges and Independent Order of Oddfellows, the emancipists proclaimed their loyalty to the Crown and Denison. The result overall was two sides espousing the same cause, each claiming the true inheritance of British, Christian, liberal and monarchist ideals but with the settlers asserting that they were the rightful inheritors and that the emancipists were interlopers who should leave or remain silent.[2]

The anti-transportation movement was also gaining considerable momentum. In 1847 free workmen had joined the Free Labour Union and struck over wages and competition for work. Along with the formation of the Tasmanian Union this was the beginning of class militancy in the Colony. Bad publicity in Britain had promoted the idea that 'the colony of Van Diemen's Land, being too evil for any emigrant who had the interest of his family in mind, was fit only for the remnants of society'. The hurt of knowing that their homeland was thought of in these terms in the outside world 'lay so deep within the consciousness of colonists that it was difficult for Denison or his superiors to understand their passion'.[3]

There is nothing in Oldfield's writings that would provide any indication as to where he stood in these debates. However, his family and educational background and his subsequent involvement in the Kingston Lodge of the Independent Order of Oddfellows[4] is indicative of someone who, while loyal to the British monarchy, would have been pro-self-government with some sympathy for the aspirations of those law-abiding emancipists who wished to remain in the Colony. Perhaps, however, he was too much taken up with earning his living and botanising in his spare time to be diverted by colonial

[1] Bruce Felmingham, 2005, p. 421.
[2] Peter Bolger, 1973, pp. 43–53
[3] ibid.
[4] See Chapter 10.

politics, energised in his scientific pursuits by what he had learnt in England and the positive communications he had had with William Hooker.

From Oldfield's own perspective one development that he may have welcomed, if he was able to afford the fees, was the establishment in 1849 of the Tasmanian Public Library in Hobart Town. It was funded by a government grant for books and the yearly subscriptions of members. The library was, according to John Levett,

> unashamedly middle class with upper class pretensions, sometimes courting viceregal patronage, and run by the trustees, (who were ever "gentlemen") largely for their own'. As it had been established and maintained by its committee, the library was 'inward-looking, classbound, and isolationist.[5]

Town and country members were encouraged to borrow books, with country members given double the loan period of their city counterparts as a concession to the difficulties of transport.[6] By 1850 an Act of Parliament legally established it.[7] Included in the library were books on natural history and philosophy.[8] Access to the Royal Society of Tasmania's collection of scientific works was restricted to members.[9]

When Augustus returned he found that his brother Edwin's circumstances had changed quite markedly.

Edwin's difficulties with the Colony's changing education system

Since 1839, the government's public education system was the focus of considerable controversy, which emanated from many quarters, and Edwin became caught up in the imbroglio. Franklin had established a Board of Education without ecclesiastical representation following the abandonment of a Board of Education he had formed in 1838, which had placed all schools under the control of the Church of England. Under his 1839 scheme, education was delivered through Free Day Schools run on the principles of the non-sectarian British and Foreign School Society (BFSS) system, excepting that 'the entire Sacred Scriptures' were to form part of the daily course of reading without 'commentary or exposition'.[10] These schools were subsidised out of the public purse at rates which depended on the number of students. Masters' salaries were held against these subsidies and, in addition, they were entitled to claim from parents one month in advance, weekly payments. These fees varied according to the number of children in the family and the location of the school.[11]

The Board recruited six Masters and Mistresses from the BFSS's Normal School in London and one master and mistress capable of taking charge of a Normal School to be established in Hobart Town because local teaching was judged to be 'illiterate and inefficient'.[12] Edwin was not one of these imports.[13] This plan for education was a turning-point in the history of education in Tasmania because it was an attempt at systematic education supported financially by the state. The method of paying teachers meant that it would be difficult to attract well-educated teachers because their incomes were unpredictable due to parents failing to pay fees.[14]

---

[5] John Levett, undated, cited in Heather Gaunt, 2010, Identity and nation in the Australian public library: the development of local and national collections 1850s–1940s, PhD thesis, University of Tasmania, Hobart, p. 53.

[6] fn. 87, ibid.

[7] Tony Marshall, 'State Library of Tasmania', in Alison Alexander (ed), 2005, pp. 345–346.

[8] *Wood's Tasmanian Almanack for 1857*, Walch and Sons, Hobart, p. 149.

[9] John Levett, 2010, p. 53.

[10] Derek Phillips 1985, *Making More Adequate Provision: State Education in Tasmania 1839–1985*, Government Printer for the Education Department of Tasmania, Hobart, pp. 17–19. In this work, there is a copy of the inaugural school regulations of Franklin's Board of Education of 1839. These cover the financial, disciplinary, secular and religious curricular components of the scheme.

[11] ibid., p. 18

[12] Lloyd Robson, 1983, vol. 1, pp. 343, 344.

[13] J.S. Bicheno to W.T. Denison, 13 December 1849, CSO 24/88/2652, TAHO.

[14] E.D. Oldfield to W.T. Denison, petition, 19 April 1849, Education Board Correspondence, CSO 26/3148, TAHO.

The new system had offended the Anglican clergy and their flock, believing that education was useless without religion and, thereupon, they withdrew their school supervisory services. Those others who were in favour of the BFSS form of education considered it was impracticable to establish separate, denominational schools in a small religiously diverse, and geographically dispersed, population such as then existed in Tasmania. It was also considered to be unjust and impolitic for the government to provide schools on principles so limited as to exclude from them the children of every denomination but one, namely the Church of England. The Clergy continued their attacks on the system throughout the remainder of Franklin's tenure and intensified them under Lieutenant-Governor John Eardley-Wilmot.[15] He was replaced by Lieutenant-Governor Sir William Denison in 1846.

Denison was instructed in 1847 by the new Secretary of State for the Colonies, William Ewart Gladstone, to change the education system. Questions and expositions by teachers on the Holy Bible were to be confined henceforward strictly to the meaning of scriptural texts according to the rule of the BFSS. Church schools were to be supported by the money of the State on the principle of the plan adopted in New South Wales.[16] This meant a return to funding of denominational schools from the public purse at the rate of a penny a day for each pupil in daily attendance with the remaining costs being borne from parents' fees or from other sources.[17]

Under the new arrangements Campbell Street School became an Anglican denominational institution with an Anglican minister in charge.[18] On behalf of the parents and guardians at the school Edwin petitioned Denison requesting that the non-denominational system of the BFSS continue.[19] Edwin responded to the changes by taking steps to establish his own private day school. On 16 December 1848, he placed a 'wanted to rent' advertisement in *The Courier* for '[a] house, or a good-sized room in a central part of the town [Hobart Town]', which would be suitable for such a purpose.[20] A Reverend Palmer became the Master of the Campbell Street School on 10 January 1849 and took over the school house.[21]

Edwin must have changed his mind about starting his own school at this time because Palmer engaged him to conduct the Campbell Street School's boys' branch at a reduced status on the understanding that it be run 'strictly in accordance with the British and Foreign School Society's system', that is, students could be of any religious denomination.[22] This was approved by Denison on 10 January 1849.[23]

On 19 April 1849, Edwin petitioned Denison saying that unless he received 'favourable consideration' in so far as making good his loss of income under the new arrangements it would 'be impossible for him to continue much longer the conduct of the School'. He explained that he had for a period of two years and upwards managed

> to the satisfaction of all parties, the Campbell Street School while under the inspection of the late Board of Education, and did, at the request of several persons, consent to continue to carry it on under the new Educational arrangements.

He went on to explain that he had found it necessary, during the previous quarter, 'to make up for the almost inevitable diminution in his Salary by increasing the payments made to him by the parents of the pupils' but he had found it impossible to do so. This was because some of the parents were so poor from want of employment and from other causes, 'that he was compelled to admit their children

---

[15] Lloyd Robson, 1983, vol. 1, pp. 343–345.
[16] ibid., pp. 395–398.
[17] Derek Phillips, 1985, p.21.
[18] *Wood's Royal Southern Kalendar, Tasmanian Register, and General Australasian & East Indian Official Directory*, J.W.H. Walsh, Hobart Town, 1850, p. 122, TAHO.
[19] E.D. Oldfield to W.T. Denison, Petition of parents and guardians of children attending the Campbell Street School, undated. CSO 24/88/2652, Education Board Correspondence, TAHO.
[20] Advertisement, *The Courier*, Hobart Town, 16 December 1848, p. 3, DAN, NLA.
[21] C. Bradbury to J.E. Bicheno, Colonial Secretary, 5 January 1849, CSO 24/88/2652, Education Board Correspondence, TAHO.
[22] ibid.
[23] Note unsigned indicating approval 'by the Lt. Gov.', 10 January 1849, ibid.

gratuitously'. Now that he was required to pay rent on his residence he thought it probable that his yearly income, after payment of rent, would amount to something less than £68, an amount that was 'positively inadequate to meet [his] necessary personal expenses'.[24] His petition was unsuccessful[25] because it would create a precedent by opening up the possibility of claims for compensation from all other Board of Education teachers. Bradbury did, however, note that '[h]ad the Board continued indeed it is probable Mr Oldfield would have been relieved, for his school was not satisfactory to them'.[26] Clearly, Edwin was not aware of this at the time he penned his petition. No report has been found that would provide the reasons for the Board's discontent with his services.

Shortly after the rejection of his personal petition[27], Edwin resigned.[28] He opened his own Commercial Academy for 'young gentlemen' on 25 March 1850.[29] This was located on the corner of Argyle and Brisbane Streets in Hobart Town, but which corner is not known.

Edwin's pedagogical approach was an enlightened one in line with the demands of many upper- and middle-class parents in the Colony. Thus, his 'plan of Instruction' aimed 'to instil a sound knowledge of the subjects studied, rather than to cause them to be mechanically repeated' which, as he said, 'was too commonly the case'. Furthermore, every opportunity would be taken 'of engrafting upon the minds of his Pupils, sentiments of sound moral principle'. The course of instruction included: 'English, Writing, Arithmetic, Book-keeping, Algebra, Geometry, Mensuration—with the other branches of the Mathematics, Geography, History and the Use of the Globes'. Latin, French and Drawing were charged as extras over and above the normal fee of four guineas for children under eight years of age, and above that age, six guineas.[30] The curriculum and his stated approach to education is indicative of someone who adhered to the philosophies of John Locke and other theorists of the enlightenment who held similar views.[31]

Edwin's marriage—1851

The historical record has so far yielded no insight into what Oldfield was doing after his return from England in December 1850 until 28 August 1851, when he was one of the witnesses of the marriage of Edwin to Miss Wilhelmina Susanna Sara Henrica Swifte (called Susanna), at the residence of her brother Allan William Godwin Swifte of Campbell Street, Hobarton. Apart from Allan and Augustus, the other witnesses were Susanna's brother, Bernam Henry Nassau Swifte, and Bernam's wife, Mary (née Bransgrove).[32] Bernam and Mary had been appointed master and mistress of the Back River school in 1847 and Allan, unmarried at the time of the wedding, was Clerk at Southport in the Huon Police Department.[33] The marriage ceremony was conducted by Mr William Day according to the rites and ceremonies of the Congregational Union of Van Diemen's Land.[34]

---

[24] E. Oldfield to W.T. Denison, 19 April 1849.

[25] J.E. Bicheno to E. Oldfield, 28 May 1849, CSO 26/3148, Education Board Correspondence, TAHO.

[26] C. Bradbury to J.E. Bicheno, 9 May 1849, CSO 26/3148, Education Board Correspondence, TAHO.

[27] J.E. Bicheno to E.D. Oldfield, 28 May 1849.

[28] E.D. Oldfield's resignation from the Campbell St School, CSO 26/3373, Education Board Correspondence, TAHO.

[29] Advertisement, *The Courier*, Hobart Town, 16 March 1850, p. 1, DAN, NLA.

[30] E.D. Oldfield, *Commercial Academy*, prospectus, pamphlet, undated, PQ 373.245 Old, W.L Crowther Library, TAHO. This document appears to have been published sometime after his Academy opened for business. Thus, Edwin's opening sentence read: 'In calling attention to his school, and in soliciting a continuance of the support which he has hitherto received…'

[31] See Chapter 1.

[32] Record of marriage of E.D. Oldfield and Wilhelmina Susanna Swifte, 28 August 1851. Extract from the Register Book of Marriages for the State of Tasmania and its Dependencies, 1851, Marriages in the District of Hobart, nos. 609 & 161, dated 25 November 1994, Registrar General Tasmania. In official and family records Edwin's wife's name is variously spelt Susanna, Susanah, Susannah, Susan. The spelling adopted here is that which was included as part of her full name on the extract of their marriage record. However, her signature on the register of marriages is given as 'Susan'.

[33] Elaine Taylor, Oldfield & Swifte family genealogist (personal communication, 29 May 2003, HOA). Taylor did not provide her sources regarding Bernam and Allan's appointments. Bernam was appointed 'to the Deputy Registrar of Births, Deaths, and Marriages for the district of New Norfolk' c. October 1851, *Colonial Times* (Hobart), 3 October 1851, p. 3, DAN, NLA.

[34] The official record of Edwin and Susanna's marriage refers to the ceremony being conducted according to the rites and ceremonies of 'Independents'. The term 'Independent' predominated in the colonies, but on the formation of the Congregational Union of England and

The couple's choice of a Congregational wedding ceremony suggests that Edwin had become disenchanted with the Established Church. Susanna's religious status when she arrived in Tasmania is not known but she may have been a Catholic; she had an Irish Catholic ancestral background.[35] If so, her apparent willingness to be married according to the rituals of the Congregational Union is perhaps indicative of dissension from the Catholic faith either on her part or that of her parents. Adherence to a dissenting sect like the Congregational Union tends to confirm that they were 'genteel' and 'respectable' and adhered to a strict code of manners in public and private domains, a strong work ethic and a Christian way of life.[36] This is also implied in Edwin's advertisement promoting his school to the parents of Hobart. If Perkin is correct, for Edwin, as a professional, and for Susanna, the daughter of a professional, their ideal society would have been a functional one in which trained and qualified expertise was the determinant and justification of status and power in society with employee selection by merit at all levels.[37] It is not difficult to imagine that with his family, educational and professional background Augustus held similar values.

Susanna was born in Namur, Belgium, on 12 October 1827, the daughter of barrister, Edmund Lewis Lenthal Swifte and his third wife, Susanna (née Robson).[38] During the Prince Regent's struggle to divorce Princess Caroline he engaged the services of Edmund as junior counsel. For his services to the Crown, Edmund was appointed, in 1814, to the sinecure of Keeper of the Regalia, which afforded him a spacious residence in the Tower of London.[39] The childhood home of Susanna, Bernam and Allan was the Jewel House. Two years after Susanna's mother died in 1834, Edmund re-married. The family was still in residence in the Tower on 30 October 1841, the night of the worst fire in its long history.[40] Edmund had four wives. Apart from Susanna they were Mary (née Daly), M.W. (née Atkins), and Isabella (née Ross). His first three wives were still married to him at the time of their deaths. According to Susanna, Edmund had 30 children.[41] So far, however, archival records have revealed only 28.[42] The birth places of these children indicate that Edmund spent most of his life in England with short periods in Belgium, France, Italy and New South Wales.[43]

On 'reaching womanhood', Susanna came out to Tasmania to one of her brothers, either Bernam or Allan, who had migrated to the Colony sometime earlier. It was there that she met Edwin. Marriage to Susanna and the establishment of his Commercial Academy no doubt improved Edwin's social class status within the Colony and by association, perhaps Augustus gained socially.[44]

In late October 1852, about two months after Edwin and Susannah's wedding, the first elections of the new partly-elected Legislative Council were conducted. Voting eligibility began with freehold ownership £100 and plural voting applied for multiple property owners. Voters had to be males over 21 years. As far as can be determined, neither Edwin nor Augustus met all the criteria for enfranchisement at that time.

Wales in 1832, and that of the Congregational Union of Van Diemen's Land in 1836, the latter name was made official, Patricia Ratcliff, 'Congregationalism', in Alison Alexander (ed), 2005, p. 84.

35 David James O'Donoghue, 'Swifte, Theophilus (1746–1815)', *Dictionary of National Biography, 1885–1900*, viewed 29 October 2012 at www.wikisource.org.

36 Leonore Davidoff & Catherine Hall, 1987, pp. 73, 76–106 & 397–402. See also Chapter 1.

37 Harold Perkin, 1972, pp. 221–230 & 258, 259.

38 Record for Susanna Swifte, Family Search, Internet Genealogy Service, Church of Jesus Christ of Latter Day Saints, viewed 20 April 2007 at www.familysearch.org.

39 Anon., 'The Tower of London: Interesting reminiscences', *The Mercury* (Hobart), 20 February 1917, p. 3, DAN, NLA. Exactly when and under what circumstances Swifte was engaged by the Prince Regent as junior counsel has not been ascertained. He may have played a role in the 'Delicate Investigation' of 1806 but since Swifte's appointment was not effected until 1814 it is possible that his sinecure was in recognition of services subsequent to the 'Delicate Investigation' undertaken in 1806 (Wikipedia, 'Caroline of Brunswick', *Wikipedia,* viewed 17 March 2014 at www.wikipedia.org.). Edmund Lenthal Swifte published a short account of his family's experience of a ghost in the Tower 'Ghost in the Tower', *Notes and Queries*, vol. 10, 2 September 1860, p. 192.

40 Anon., *The Mercury* (Hobart, Tas.), 20 February 1917.

41 ibid.

42 Noel Henderson, Henderson Family Tree, viewed 6 June 2011 at www.trees.ancestry.netscape.com.

43 ibid.

44 Anon. *The Mercury* (Hobart), 20 February 1917, p. 3.

Family developments back in England 1851–1856

The years 1851 to 1856 were marked by changes to the Oldfield family's circumstances in England. By the night of the *1851 England Census* (30 March 1851) Frederick, Mary, Marian, Frederica and Vincent had moved to Chelsea and were living at 1 Somerset Villas. Although Frederick's occupation is noted in his *1851 England Census* record as 'wine merchant' there is no evidence that he established another business of that nature after he left Dorset Street. Frederica was recorded as being unmarried and a teacher of French, German and music.[45] However, sometime during the three months prior to *Census* night, she married portrait and landscape painter Amos Hold[46] in Marylebone.[47] Intriguingly, the couple were married again in Wakefield, West Riding, Yorkshire, sometime between July and September of the following year.[48] Juliana was a resident governess at the Manor House in Ashby-de-la-Zouch, Leicestershire. Her pupils were the three teenage daughters of the head of household, the widowed Mrs M. E. Mammett, and the four much younger children of her widowed brother, J. V. Hall, who also resided in the house.[49]

By 1853 Frederick Snr's career had taken a different turn. Listed in the *Watkins Commercial & General London Directory and Court Guide for 1*853 as a commission and reversionary[50] agent, he was managing a business operating out of premises at 29 Bloomsbury Square, London.[51] According to Roy Porter, between 1841 and 1851 alone, the flood of immigrants to London accelerated the expansion of London's urban development. With numerous areas 'competing for capital and clients'; property development was cut-throat. Development continued to rely on the 'tried and tested building-lease system' whereby the landlord got his estate developed without tying up capital, and the property remained his own—an essential arrangement because London estates were generally entailed or held in trust; they could not be sold. For his part, the speculative builder acquired a prime site. Master carpenters, plasterers and plumbers pooled resources to reduce costs; a speculative builder needed just sufficient capital or credit to buy materials. The building-lease system shared risks and got things moving.[52] Obviously this was an opportune time for business men such as Frederick to become reversionary agents and cash in on the expanding but nevertheless complex leasehold business. Even though guidelines dating back to the beginning of the century were available[53], the preparation of leasehold documents and determining the value of the purchase and renewal of leases 'for Terms of Years certain and for lives' as well as determining the value of the reversion of estates after any such leases must have been challenging for Frederick who was not professionally experienced in those procedures.

---

[45] Record for Frederica Oldfield, *1851 England Census*, Class: HO107; Piece: 1473; Folio: 10; Page: 13; GSU roll: 87796, viewed 20 June 2005 at www.ancestry.com.

[46] Certified copy of the entry of death for Frederica Juliana Hold, Registration District: Doncaster; Sub-district: Doncaster in the County of York, DYC993484, given at the General Register Office, England.

[47] Records for Frederica Oldfield and Amos Hold, *England and Wales Free BMD Marriage Index*, March Quarter, 1851, vol. 1, p. 265, viewed 19 October 2001 at www.ancestry.netscape.com.

[48] Records for Frederica Oldfield and Amos Hold, ibid., September Quarter, 1852, vol. 9c, p. 42, viewed 19 October 2005 at www.ancestry.netscape.com.

[49] Record for J.M. Oldfield, *1851 England Census*.

[50] Reversion related to the lease of any lands or tenements (usually in consideration of rent, or other annual recompense) made for life, for years, or at will, or to annuities, sums payable in respect of a particular year or to a series of equal annual sums for a given period or until the death of the recipient, 'Reversion law', viewed 23 October 2005 at www.wikipedia.org.

[51] F. W. Watkins, *Watkins Commercial & General London Directory and Court Guide for 1853*, 2nd edn, London, Longman, Brown, Green, and Longmans, pp. 404, 680, London Directories from the Guildhall 1677–1900, Group 2, 1800–1855 (Excluding Post Office Directories), microfilm, vol. 205, NLA.

[52] Roy Porter, 2000, p. 128.

[53] Francis Baily, 1812, *Tables for the Purchasing and Renewing of Leases, for Terms of Years Certain and for Lives with Rules for Determining the Value of the Reversion of Estates After Any Such Leases; and for the Solution of Other Useful Problems: Adapted for General Use.*, 3rd edn., Royal Exchange, London.

Was Oldfield in Victoria during 1852

Map 3 p. 94 – Sydney to Melbourne

The historical record has remained silent on Augustus' movements after Edwin and Susanna's wedding until 1 March 1853 (see below). While he may have joined the staff of Edwin's school there is the possibility that he was working in Victoria.

Oldfield collected plants in Victoria and, if the few records of specimens we have located are any guide, he did not do any concentrated collecting anywhere in that Colony. However, a different picture may emerge once the records of all his Australian collections have been entered into Kew Gardens and the National Museum, London, herbarium electronic catalogues. He sent a packet of about 350 specimens of cryptogamic species on or about 1 March 1853 to William Hooker.[54] Amongst these may have been plants gathered in Victoria in 1852–53 but there is no indication in his letter that this was the case.

Apart from two single collections, one at Barwon River and the other at the Grampians[55], the specimen data we have found show that most of the places where Oldfield gathered plants in Victoria were either *en route* to, at, or near where, gold had been discovered c.1852. These localities were 'Mount Juliet'; 'Boolah' (location not found but given by Oldfield as being on the Victoria River, a stream about 15 kilometres to the west of the Omeo goldfield)[56]; 'Australian Alps'; 'Mount Barkers Range, Gippsland' (probably Barker Ridge, Gippsland near the road to Omeo)[57]; 'Mountain Hut Range [Peak]' (about 64 kilometres just ENE of Omeo); 'Budgee Budgee'[58] (about 56 kilometres northwards from Sale); 'Dandenong' and 'Bendigo'.[59] The geographical spread of these localities and apparent sparseness of the gatherings at each of them suggests that he had little time to botanise, and was limited in the amount of plant preservation equipment he could carry in his luggage, possibly because he was too busy working at some other occupation and was constantly on the move.

This could well have been the case if he had been a peripatetic digger. However, it seems more likely that he was a member of Surveyor George Smythe's party undertaking surveys of the Tambo River and roads to Omeo and Buchan in Gippsland, possibly having been engaged by Smythe when he was in Melbourne in late September 1851.[60] Oldfield's collections have lent some, albeit weak, support for such a hypothesis. Unfortunately, it cannot be confirmed as there is no record of him having travelled to Melbourne soon after Edwin and Susanna's wedding and details of the members of Smythe's survey party do not appear to have survived.

Oldfield was in Hobart toward the end of August 1851 and therefore could not have been with Smythe at the Buchan River, where he and his party were held up for six weeks during August and September of that year before abandoning a planned coastal survey to the east of the Snowy River due to flooding rendering it impassable.[61] Following his return to Melbourne, Smythe obtained permission to carry out the road and Tambo River surveys which he completed prior to 24 January 1852.[62]

---

[54] AO to WJH, 1 March 1853, Australian Letters, 1851–1858, vol. 74, fol. 149, 151, (152), RBGKA.

[55] HAODb. How Oldfield got to the Grampians is not known. There were two tracks marked on

[56] No Boolah has been found near the Victoria River but there is a Boola on the coast near Inverloch. Since the Victoria River and Boola are so far apart geographically it is likely Boolah was the name of a place during the gold rushes, which was near the Victoria River but the name has not been retained. The Victoria River is one of the headwaters of the Mitta Mitta River.

[57] 'Mt Barkers Range, Gippsland' has not been found on any gazetteer. Anita Davids, Geographic Names, Office of Surveyor-General Victoria, kindly conducted a search of the Register of Geographic Names and was unable to find a reference to Mt Barkers Range (personal communication, 10 May 2007, HOA). We believe it refers to what is now called Barker Ridge, Gippsland.

[58] There is a Budgee Budgee in New South Wales, about halfway between Lithgow and Dubbo, and another in Victoria about 40 kilometres north of Sale. There is no evidence that Oldfield was ever anywhere near the New South Wales Budgee Budgee.

[59] See HAODb for details of specimen records from all these localities.

[60] G.D. Smythe, Melbourne to Surveyor General, Melbourne, 23 September 1851, fol. 51/430, VPRS: 97/P/001; Unit: 000002, No. 3, Smythe, G. 1840–1853; Type: AIF; Locn: V/AK/162/03/02, Public Record Office of Victoria, PROV.

[61] ibid.

[62] G.D. Smythe to Surveyor General, Invoice for survey of Gippsland by contract, 24 January 1852, ibid.

CROSSING TONGIO GAP, THE OMEO TRACK.

Fig. 9.1 Crossing Tongio Gap, the Omeo Track, Victoria, 1880

It seems feasible to suggest that the most efficient way of undertaking this work and the subsequent coastal survey would have been for Smythe and his party to travel first to the source of the Tambo River in the Bowen Mountains in the southern reaches of the Australian Alps, then surveying along the river in a southerly direction, before turning eastward to survey the road to Buchan and beyond. The shortest route to Omeo was the very rough track that was the original route to the goldfields in the Yarra Ranges and Central Gippsland.[63] It passed Mount Juliet and Barker Ridge, two of Oldfield's collecting localities, ran through Jamieson and Mansfield to Omeo. From thence the party would have travelled through the bush to the source of the Tambo River. Mountain Hut Peak, another of Oldfield's collecting places, is a prominence about 32 kilometres to the east of the headwaters of the Tambo River in the Bowen Mountains which may have been used by Smythe as a trig point in the mapping of the river.

Smythe submitted to the Surveyor General his invoice for the Gippsland road survey on 24 January 1852.[64] He then went on to do the coastal survey east of the Snowy River to Cape Howe, which was finished some time before 4 February 1853. It was on that date he forwarded from the May Day Hills survey camp his field notebook and coastal plan to the Surveyor General.[65]

---

[63] James Flett, 1970, *The History of Gold Discovery in Victoria,* 2nd edn, Hawthorn Press, Melbourne, p. 102.
[64] G.D. Smythe to Surveyor General, 24 January 1852.
[65] G.D. Smythe, Camp Survey Paddock, forwarding plan and fieldbook of coast survey from Bernan [?] Inlet to Cape Howe, 4 February 1853, fol. 53.38, VPRS: 97/P/001; Unit: 000002, No. 3, Smythe, G. 1840–1853; Type: AIF; Locn: V/AK/162/03/02, PROV.

Fig. 9.2 25 kilometres from Ovens on the Ovens/Mount Beauty Road, Australian Alps

If Oldfield did assist with the coastal survey it must have been completed well before October 1852 when he was in Tasmania. His plant records suggest that the journey to Melbourne was undertaken via the track that approximated the road from today's Bruthen on the Tambo passing through Bairnsdale, Sale and Woodside. He gathered specimens at Budgee Budgee and Dandenong on the way[66], and probably returned to Hobart Town from Melbourne. Unfortunately, no record has been found of him travelling to Tasmania in 1852.

Augustus was probably the 'A. Oldfield' who travelled steerage on the *Yarra Yarra* from Launceston to Melbourne arriving on 20 October 1852.[67]

Previously we noted that we located a few records of specimens gathered by Oldfield to the north of Hobart along the Main Road, namely at Pontville and Jericho[68], which might have been obtained while he was in the Coal River area. However, perhaps that was not the case. They may have been gathered during coach stops on his way from Hobart Town to Launceston[69] to commence his voyage to Melbourne in October 1852 (Map 2 p. 84). Oldfield also gathered at least two plants from George Town at the mouth of the Tamar River[70] but he left no details as to when and why he was there. Perhaps on his way to Melbourne on this journey he made landfall there while the ship was going through exit formalities with the Customs, or he caught up with the vessel at George Town instead of

[66] J. Jones, 1850.
[67] Archives Office of Tasmania, Index of Departures 1817–1867, ref: POL2201.2, p. 205, viewed 15 April 2008 at www.archives.tas.gov.au.
[68] See Chapter 7.
[69] G. Hawley Stancombe, 1969, following p. 64.
[70] HAODb

boarding at Launceston. There is a record of a specimen of the benthic marine alga *Plocamium angustatum* with the locality 'Bass Strait'.[71] If he obtained it while he was waiting in George Town, he may have had time to do the approximately eight kilometres walk to the north coast and discovered the alga washed up on the beach. Of course, it could have been obtained anywhere along the north coast of Tasmania or even on the south coast of Victoria. The mystery might never be solved.

Oldfield visited the Ovens Diggings, which were located at the May Day Hills where Beechworth now stands; exactly when he was there he did not say although it was probably sometime after he landed in Melbourne on 20 October 1852. No records of plants collected by him at the Ovens have been found. He referred to his visit to the Ovens in a description of his observations of the behaviour of cattle which he drew upon to illustrate a point he was making in a discussion of phylogenetic differences between animal species in their response to danger. Thus, he recounted:

> When I first arrived at the Ovens diggings, there was a stockyard just filled with cattle for the butcher, and as long as they lasted, one was shot every day. After 3 or 4 had been thus killed, it became very difficult to get a fair shot at any of the remainder, for directly anyone approached them, all would turn their heads inwards, so as to be protected by the body of another; for seemingly they had learned to appreciate the cause of the death of their companions, and that the head was the part aimed at. When it came to the last bullock, the poor animal kept constantly on the move whenever a man approached the stockyard, but at length – being quite spent – it shared the fate of its fellows. In estimating the bearing of the foregoing fact, we must recollect that it is the nature of horned animals to face danger.[72]

We can only speculate as to why Oldfield was repeatedly at the stockyard to witness the daily kill or why and for how long he was in the district. His account would suggest, depending on the size of the stockyard and the number of beasts that could be accommodated there, that he was at the Diggings for quite a few days, possibly a week or more. Given his socio-cultural background, it is unlikely that he was one of the slaughtermen. His fascination with animal behaviour no doubt drew him to the stockyard day after day. This activity is not indicative of a person who was there to make his fortune from gold but more of someone either with time on his hands or attending the stockyard before he commenced work each day.

These events must have taken place at a time when the population was large enough to warrant a butcher killing a beast a day. After the first rush at the Ovens, the discovery of rich gold at Eaglehawk, Bendigo, in March 1852 left the Ovens almost deserted with a population reduced to about 60, but a few months later the population increased as a result of the gradual take-up of gold licenses at the rate of twelve per month, and further discoveries. By October the population had risen to 1,500[73] which was about the time Oldfield was probably there. By February 1853 the population was much higher because of the second rush at the Ovens beginning about 28 January 1853 and following the discovery of gold in the upper Yackandandah Creek at Stanley:[74]

Along all its innumerable gullies and creek beds 10,000 men were at work. All along the slopes and gullies there was the creaking of windlasses, the thudding of picks, the rattle of spades. Sluice-boxes thumped and rumbled, voices called, dogs barked, neglected children wailed, fire crackled, women argued, men fought.[75]

Since there were probably other types of meat (kangaroo, mutton, pork) available in the community at any time during its life and possibly more than one butcher, attempting to make any claims as

[71] MEL 1006082, *Plocamium angustatum*, Bass Strait, MELISR database, 18 November 2005, NHV, RBGM.
[72] AO, c.1861a, fn. 42, p. 622.
[73] James Flett, 1970, p. 64.
[74] ibid., p. 65.
[75] Edward Kynaston, 1981, p. 90

Fig. 9.3 The Ford Ovens River

to the approximate date Oldfield was at the Ovens based on his description of the slaughtering regime is not possible.

Oldfield may have re-joined Smythe's survey party at the diggings very late in October 1852. Before 3 May 1853 Smythe had undertaken surveys in the Ovens River district, including surveying and preparing plans of the proposed May Day Hills townsite[76], which he later re-named Beechworth.[77] Smythe was still at May Day Hills on 19 November 1853.[78] Oldfield, however, was back in Tasmania by 1 March 1853.

There were two routes Oldfield could have travelled between the Ovens and Melbourne. On each of these tracks there was the occasional inn where, according to contemporary advertisements, gentlemen and diggers might obtain good accommodation, with every attention to cleanliness and comfort, substantial meals, a well-stocked store, 'capital feed' for their cattle, stabling for their horses, a first-rate glass of grog, and the strictest attention to their wants all at a reasonable price.[79] Since there is evidence that Oldfield gathered plants at Bendigo, perhaps he travelled from Melbourne to Bendigo on his way to or from the Ovens and from Bendigo along the track to the Campaspe River, across the plains and then along the banks of the Broken River to Benalla and on to the Ovens Crossing Place (now Wangaratta). The other track on which he may have travelled either going to or coming from the Ovens was the most direct. It ran from Melbourne via Kilmore, Seymore, Benalla and the Ovens Crossing Place but we have not found any specimens from those settlements.

[76] G.D. Smythe, survey camp, May Day Hills, to Robert Hoddle, Surveyor General, Melbourne 3 May 1853, fol. 53/246, VPRS: 97/P/001; Unit: 000002, No. 3, Smythe, G. 1840–1853; Type: AIF; Locn: V/AK/162/03/02, PROV.

[77] Beechworth Visitor Information Centre, 'Beechworth's History', viewed 5 March 2013 at www.beechworthonline.com.au.

[78] G.D. Smythe to Robert Hoddle, Surveyor General from May Day Hill 19 November 1853, fol. 23/53, VPRS: 97/P/001; Unit: 000002, No. 3, Smythe, G. 1840–1853; Type: AIF; Locn: V/AK/162/03/02, PROV.

[79] Classified Advertisements, *The Argus*, 17 September 1853, pp. 3, 4, DAN, NLA.

Did Oldfield first meet Ferdinand Mueller in 1852 or 1853?

Mueller left Adelaide for Melbourne in August 1852 with the intention of establishing his own pharmacy in Victoria but this plan was soon abandoned. An introduction from the South Australian politician Francis Dutton to Victoria's Lieutenant-Governor, Charles Joseph La Trobe, resulted in Mueller's appointment to the new position of Government Botanist at a salary of £400 per annum.[80]

If Oldfield had not become acquainted with Mueller earlier in Adelaide, perhaps he sought him out c. October 1852 when they were both in Melbourne. However, there is also the possibility that they first met either at the May Day Hills or at the Ovens Crossing Place very soon after Mueller arrived there during the second week of February 1853.

Within a week of being offered the appointment on 26 January 1853, Mueller accepted and headed for the Australian Alps in the Great Dividing Range. He was accompanied by John Dallachy, who since 1849 had been Overseer of the Botanic Gardens, Melbourne and in 1852 had been promoted Superintendent of the Botanic Gardens, Melbourne. Dallachy had worked with William Hooker for some years at Kew Gardens and later managed a coffee plantation in Ceylon before taking up his overseer appointment in 1849.[81] Mueller and Dallachy set off from Melbourne on horseback. Leading pack-horses, they took the track across the mud-holed flats of Riley's swamp and headed for Kilmore.[82] In a letter to William Hooker (3 February 1853) from Darebin Creek, to the south of Kilmore, Mueller wrote:

> My first camp in a journey to the Australian alps, which to visit I prefered [sic] amongst the remoter localities for my first longer journees [sic], with a view of ascertaining as far as the lateness of the season and my somewhat limited means permit, the connexion likely existing between the vegetation of the only alps of this continent (Mount William of the Grampians being merely alpestrious) with those of V.D.L., N.Z., the Himalaya, Europe &c &c, a [sic] object that will offer, I hope, remarkable additions to Phytogeography.[83]

Travelling to the north-east, the gradually rising hills of the Great Dividing Range were on their right, and to their left, the edge of the vast, lightly treed plain that stretches up into Queensland. They most likely passed through Seymour and the Ovens Crossing Place (now Wangaratta) before exploring the Futter's Range, approximately 16 kilometres to the north and west of the 'Crossing'. This range, noted Mueller, 'rears, like some other granitic mountains in its vicinity, a host of very peculiar plants'.[84] They then explored the May Day Hills (now Beechworth), in the vicinity of the Ovens Diggings.[85] Mueller and Dallachy camped at or near the Diggings.[86] While they were there they may well have sought out Smythe for advice on routes through the Alps.

Regardless of when the meeting first occurred, Oldfield had started sending Mueller plants before then. Oldfield was one among many men and women across the country who over the years Mueller was Government Botanist enrolled as volunteer and paid plant collectors.[87]

---

[80] R.W. Home *et al.* (eds), 1998, vol. 1, p. 23.
[81] Alan Gross, 1972, 'Dallachy, John (1808–1871)', ADBO, viewed 15 May 2007 at www.adb.anu.edu.au.
[82] Edward Kynaston, 1981, p. 86.
[83] FM to WJH, 3 February 1853, Directors' Correspondence, vol. 74, Australia letters 1851–58, fol. 135, RBGKA in R.W. Home *et al.* (eds), 1998, vol. 1, p. 387.
[84] FM to John Foster, Colonial Secretary, 5 September 1853, C53/8928. VPRS 1189, Inward Registered Correspondence, VA856, Colonial Secretary's Office, Public Record Office, Victoria quoted in R.W. Home *et al.*, (eds), 1998, vol. 1, p. 155.
[85] Charles Daley, 1924, p. 27.
[86] FM to John Foster, 5 September 1853.
[87] See for example: Sara Maroske, 2014, 'A taste for botanic science: Ferdinand Mueller's Female Collectors and the History of Australian Botany', *Muelleria*, vol. 32, 2014, pp. 72–91, and Philip A. Clarke, 2008, *Aboriginal Plant Collectors: Botanists and Australian Aboriginal People in the Nineteenth Century*, Rosenberg Publishing, Kenthurst, New South Wales, pp. 108–111.

Hobart Town early 1853

No record of Oldfield's return passage across Bass Strait to Hobart Town has been found. However, he wrote to William Hooker from 10 Brisbane Street in that city on 1 March 1853. At that time Oldfield was recuperating from an injury to 'one of his eyes' which he sustained 'some months' previously and which, he said, had prevented him from using his microscope.[88] He left no explanation as to how, when and where this accident occurred. If he was with Smythe surveying at the Ovens River it could have happened sometime towards the end of 1852 or very early in January 1853 and there is the possibility he caused it, carelessly looking at the sun using a theodolite without a dark glass fitted.[89] The injury was possibly the reason why he was back in Hobart prior to the termination of Smythe's Ovens River contract. Perhaps it was for reasons connected with his injured eye that he was free to observe the daily slaughtering of the cattle. On the other hand, the slaughter yard may have been close to the survey camp from which he could observe the cattle.

Oldfield apologised in the letter for not having thanked Sir William for his kindness in getting his plants named:[90]

> but as I had a small collection of Cryptogamic plants for your inspection, I thought it better to wait until I could send it, and I take the opportunity of the departure of the Australasia to send the plants and to thank you and your son, Dr Hooker, for the Papers relating to the Botany of this island [Tasmania]. They have been of great assistance to me, having enabled me to clear off all my unnamed plants with the exception of some 20 or 30 that a friend of mine sent some time since to Dr. [Johan Karl] Müller [Mueller], who is at present in Port Phillip engaged in investigating the Botany of those regions.[91]

The letter does not say when he sent these plants to Mueller and there is no way of knowing whether he had learnt of Mueller's activities in Victoria from the man himself or through information that had reached him via a newspaper report or by some other means. If Oldfield was in Victoria from toward the end of 1852 until early 1853 and had met Mueller during that time, he may have sent the plants after he returned to Hobart.

The cryptogamic collection that Oldfield sent to Hooker on this occasion comprised about 350 specimens—'perhaps' 30 Lichens, 2 seaweeds, 8 or 10 ferns and the same of Lycopodiaceae,[92] and about 100 each of Mosses and Jungermannias'.[93] Oldfield entrusted carriage of the plants to a Mr Brent, Surgeon on the *Australasia*, 'in order that Sir William may have no trouble about the parcel'. Of the collection Oldfield commented

> among much that is worthless I think you will find a few that will repay you for your trouble of looking over, and as they are mostly in fruit they may be of use to your Cryptogamic friends.[94]

He assured Sir William that as soon as his eye recovered he would be able to do the determinations himself and indicated he did not expect Hooker to trouble himself about such plants 'as I can find

[88] AO to WJH, 1 March 1853, Australian Letters, vol. 74, 1851–1858, fol. 189, RBGKA.
[89] B. Ralph Chou, 1997, 'Eye safety during solar eclipses', adapted from NASA RP 1383 Total Solar Eclipse of 1999 August 11, April 1997, p. 19, viewed 6 May 2007 at www.sunearth.gsfc.nasa.gov.
[90] Presumably this reference is to the plants that were named by William when Oldfield was in England.
[91] AO to WJH, 1 March 1853, Australian Letters, vol. 74, 1851–1858, fol. 189, RBGKA.
[92] The *Lycopodiaceae* (class *Lycopodiopsida*, order *Lycopodiales)* is a family of primitive vascular plants, including clubmosses, Wikipedia, 'Lycopodiopsida', viewed 3 December 2014 at www.wikipedia.org.
[93] *Jungermannia is a genus of the class Hepaticae (liverworts) within the order Bryophyta, Encyclopaedia Britannica Online, 'Anthocerotidae', 'Bryophyta', 'Hepaticae', 20 May 2007 at www.britannica.com.*
[94] AO to WJH, 1 March 1853, Australian Letters, vol. 74, 1851–1858, fol. 189, RBGKA.

in [Karl] Müller[95], Syn. Musci[96] or Lindenberg, Syn. Hep.[97] my only present authorities in those departments'.[98]

Oldfield informed Sir William that should he, Oldfield, 'be in health next Spring' he would undertake an investigation of the 'South-west portion of this Island' with the assistance of 'two persons well versed in the Flora of Tasmania'; he did not give their names. He also reported that he had procured a 'Wards Case' with the intention of filling it with plants, 'first getting them well established in pots in their localities, and then in the Case. I have some choice ones already, but not enough to fill the Case'. He asked Sir William to provide him with a list of 'desiderata that I might procure them as occasion offered, and as I have use of a Greenhouse, I could easily establish them'.[99]

Oldfield concluded the letter on a note of self-deprecation and concern:

> I beg of you not judge me by either my old collection, or the present and to give up corresponding with me, but to wait until I have an opportunity of sending what I consider a good collection.[100]

The point Oldfield made in his letter to Sir William that he first established plants in pots 'in their localities' conjures up images of potted plants dotted around the countryside near where they were collected and probably hidden from the view of passers-by. They must have been in places that were relatively close to Hobart so that they could be tended and retrieved easily. Unfortunately, Oldfield did not say where the greenhouse was located or under what circumstances he had the use of it.

Oldfield in western Tasmania, 1853–54

Map 6 p. 149 – Western Tasmania

No information has emerged as to what he was doing to earn a living in 1853 after his return from Victoria. There may have been a few employment options available to him but his specimen collection data suggest one possibility.

We have located records of single specimens from Lake St Clair, approximately 108 kilometres north-west of Hobart, Frenchmans Cap, a mountain about 123 kilometres north-west of Hobart, and at the base of Quamby Bluff.[101] Oldfield also obtained a sample of quartz in clay slate 'near the sea level' from Point Hibbs, 40 kilometres south of Macquarie Heads on the west coast.[102] These were unsettled wilderness areas in the 1850s. It is unlikely, therefore, given the very few specimen records that have come to our notice and the geographical spread of their places of collection that Oldfield was there solely for botanising.

He may have been a member of surveyor James Sprent's party which was carrying out observations as part of the Trigonometrical Survey of Tasmania (1832–1855). It was in January 1853 that Sprent reported his failure at being unable to reach the summit of Frenchmans Cap.[103] Sprent had a gang of up to 11 men including convicts but their names are apparently not on record. The problems of access, maintenance of supplies and the conditions of weather and terrain 'were unbelievably difficult, particularly on the West and South West Coast.' His reports described the party being wet through day after day, tents leaking like muslin and many days passing without a single sighting being possible.[104]

---

[95] Although Dr Carolo (Karl) Müller, the eminent German botanist, undertook research and description of Australian bryophytes during the 19th century he never visited any of the Australian colonies.

[96] Carolo [Karl] Müller 1849–51 *Synopsis muscorum frondosorum omnium hucusque cognitorum,* A. Foerstner, Berolini [Berlin].

[97] C. M. Gottsche, J.B.G. Lindenberg, & Nees von Esenbeck 1844–1847, C.G *Synopsis Hepaticarum*, Hamburg.

[98] AO to WJH, 1 March 1853, Australian Letters, 1851–1858, vol. 74, 1851–1858, fol. 189, RBGKA.

[99] ibid.

[100] ibid.

[101] HAODb.

[102] Anon., 'Royal Society of Tasmania', report of March 1858 meeting, *The Courier* (Hobart Town), 24 March 1858, p. 2, DAN, NLA.

[103] James Sprent to Major [Hugh Calveley] Cotton, Deputy Surveyor General, 20 January 1853, microfilm, 1853, LSD1/1/81/ (29), TAHO.

[104] Larkin, Fred, 2001, 'Survey cairns, James Sprent and the Trigonometrical Survey of Tasmania 1833–1855', *Wildcare Incorporated,* Wildcare News Library, viewed 12 July 2010 at www.wildcaretas.org.au. This article appeared in *Wildtimes* No.11, 11 September 2001.

When Oldfield gathered *Richea sprengelioides* on Frenchmans Cap[105] it was almost certainly flowering.[106] If he was with Sprent, the party was probably attempting to climb the mountain sometime during December 1852 and prior to 20 January 1853[107], which falls within *R. sprengelioides*' flowering time.[108] The single specimen, *Trachymene humilis* subsp. *humilis*[109], obtained at Lake St Clair might have been gathered on their travels to or from Mount Olympus, which is just to the west of the lake. James Sprent and party had 'done Mt Olympus' by 21 February 1853.[110] This species' flowering time is from December to January.[111]

The specimen of *Australina pusilla* subsp. *pusilla* from Quamby Bluff[112] also may have been obtained during the survey. Quamby Bluff is 23 kilometres from Deloraine. Although Sprent said he was going from Circular Head to Deloraine on 6 April 1854[113], it has not been confirmed that he incorporated the Bluff in the survey. Nevertheless, he most likely did since it was the highest prominence in the Great Western Tiers. No botanical specimens have been found that were collected by Oldfield on the west coast but the quartz in clay samples may have been obtained during the above Survey. Sprent noted in a letter to Cotton dated 28 August 1854 that he intended to go to Port Davey by sea 'taking several months provisions' from whence he intended to proceed overland to Macquarie Harbour. He mentioned Point Hibbs as a place from which observations would be made.[114] If Oldfield was with the party at Point Hibbs he must have left it before Sprent wrote to Cotton from Macquarie Harbour on 15 January 1855 because Oldfield was residing in Brisbane Street on 18 December 1854 (see below).

There is a single specimen record of an *Ozothamnus selaginoides* collected at Table Mountain[115] but whether it was gathered could not be determined. The prominence is about ten kilometres northwest of Jericho and Oldfield could have visited it from there or he may have gone there with Sprent's survey team.

Hobart 1854

Despite Oldfield's self-deprecation over his 1853 consignment to William Hooker, that gentleman was pleased with it. On 9 April 1854, he wrote to Ferdinand Mueller:

> I have just received a most rich set of Cryptogamae from a Mr Oldfield in Van Diemen's Land. He is not neglectful of other plants but has a most extraordinary affection for the Acotyledons:[116] & we are very thankful for any one who will be at the pains to collect them: so I give him all the encouragement I can, & there are several charming things among them.

In the same letter, Sir William told Mueller that 'Dr [Joseph] Hooker is very busy commencing on the last of his 3 works the results of His Antarctic Voye [sic] the Flora Tasmanicae [*Tasmaniae*]'.[117]

In June 1854, 'A. Oldfield' of 'Brisbane Street' applied for 'any government position' [in Tasmania]. Unfortunately, the application documentation has not survived although the register record

[105] *Richea sprengelioides*, FM, 1858–59, *Fragmenta*, vol. 1, p. 38.
[106] The whereabouts of this specimen is not known to us. It might be among the specimens at Kew Gardens, details of which have not yet been digitised.
[107] James Sprent to Major [Hugh Calveley] Cotton, Deputy Surveyor General, 20 January 1853.
[108] Mary Cameron (ed), 1996, p. 26.
[109] MEL 643967, *Trachymene humilis* subsp. *humilis*, Lake St Clair, MELISR database, 18 November 2005, NHV, RBGM.
[110] James Sprent to Major [Hugh Calveley] Cotton, Deputy Surveyor General, 20 January 1853.
[111] Mary Cameron (ed), 1996, p. 26.
[112] JDH, 1860, *Flora Tasmaniae*, vol. 1, p. 371.
[113] James Sprent to Major [Hugh Calveley] Cotton, Deputy Surveyor General, 6 April 1854, LSD1/1/81/ (37), TAHO.
[114] James Sprent to Major [Hugh Calveley] Cotton, Deputy Surveyor General, 28 August 1854, LSD1/1/81/ (39), TAHO.
[115] HAOdb.
[116] 'Acotyledons' is a synonym of 'cryptogams' (*Cryptogamia*).
[117] WJH to FM, 9 April 1854, RB MSS M2, RBGML, quoted in R.W. Home *et al.* (eds), 1998, vol. 1, p. 177.

carried the note 'for promotion'.[118] This suggests that the applicant had previously been in the employ of the government but, if that was the case and as indicated previously, Augustus was never listed as being part of the civil service establishment.[119] While it cannot be ruled out that Alfred was the person seeking employment, we think it was more likely to have been Augustus since at the time he was living at the residential address given in the application.[120] If he was with Sprent, this application was probably made during a respite in the capital city.

It seems that Augustus had become friendly with Susanna's brother, Bernam. A series of advertisements promoting Bernam Swifte's school in Pontville appeared in *The Courier* (Hobart Town) between 18 December 1854 and 8 January 1855 inviting interested parents to obtain the prospectus and references from 'Mr Oldfield, No. 10 Brisbane Street'.[121] In order to provide this service for Bernam, Augustus must have been in Hobart continually during that period.

[118] Application for any government situation, A. Oldfield, Brisbane St, Hobart, June 1854, CSO 80/2, no. 52, TAHO and Candidates for employment under the government, A. Oldfield, Brisbane St, Hobart, June 1854, Colonial Secretary's Correspondence, CSO 80/1, no. 94, p. 50, TAHO.

[119] See Chapter 6.

[120] AO to WJH, 1 March 1853, Australian Letters, 1851–1858, vol. 74, 1851–1858, fol. 189, RBGKA.

[121] See for example, *The Colonial Times* (Hobart), 9, 12, 16, 19, 20, 21, 26, 28 and 30 December 1854 and 3, 6, 9, 10, 11 and 13 January 1855, DAN, NLA.

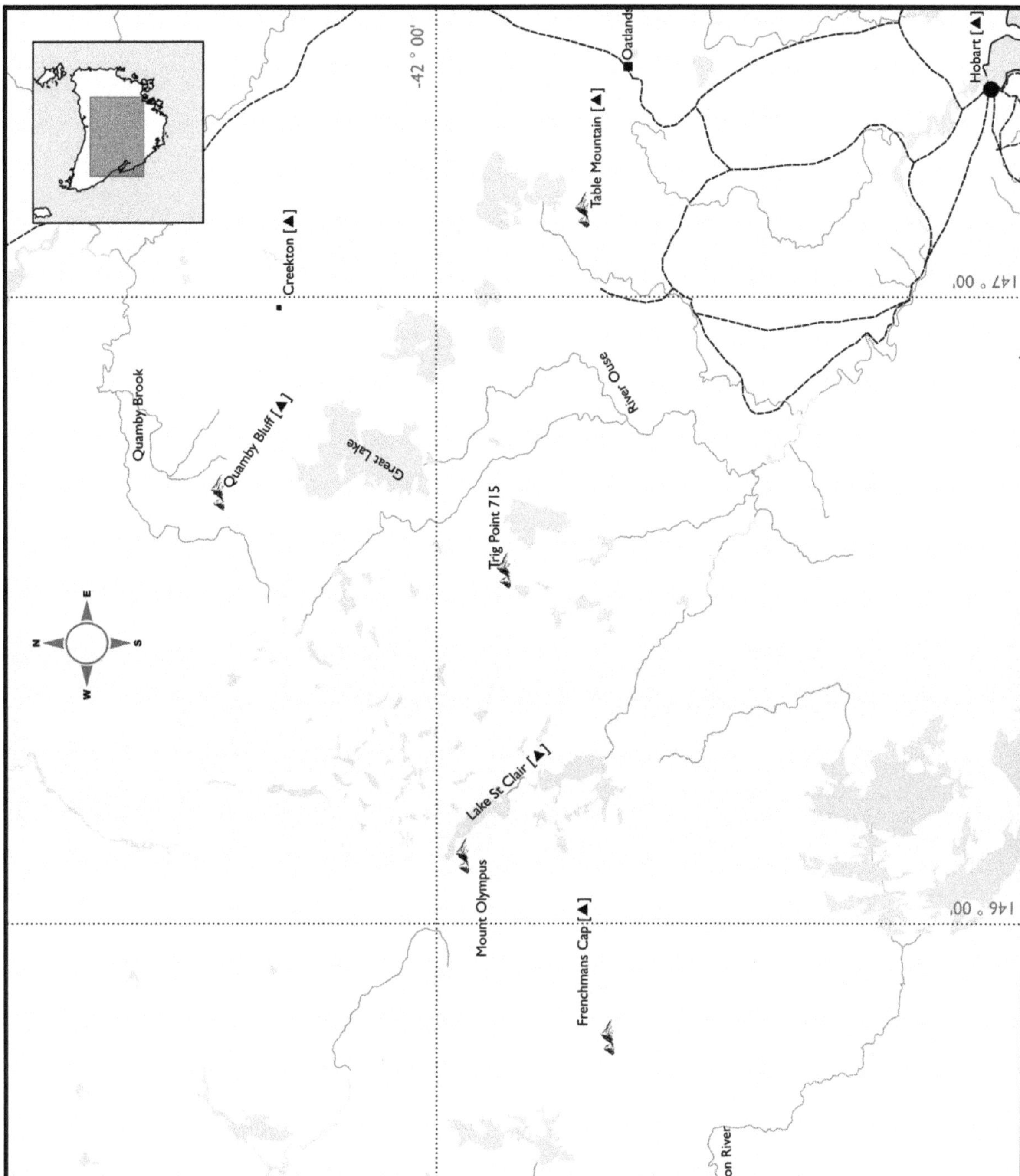

Map 6: Western Tasmania

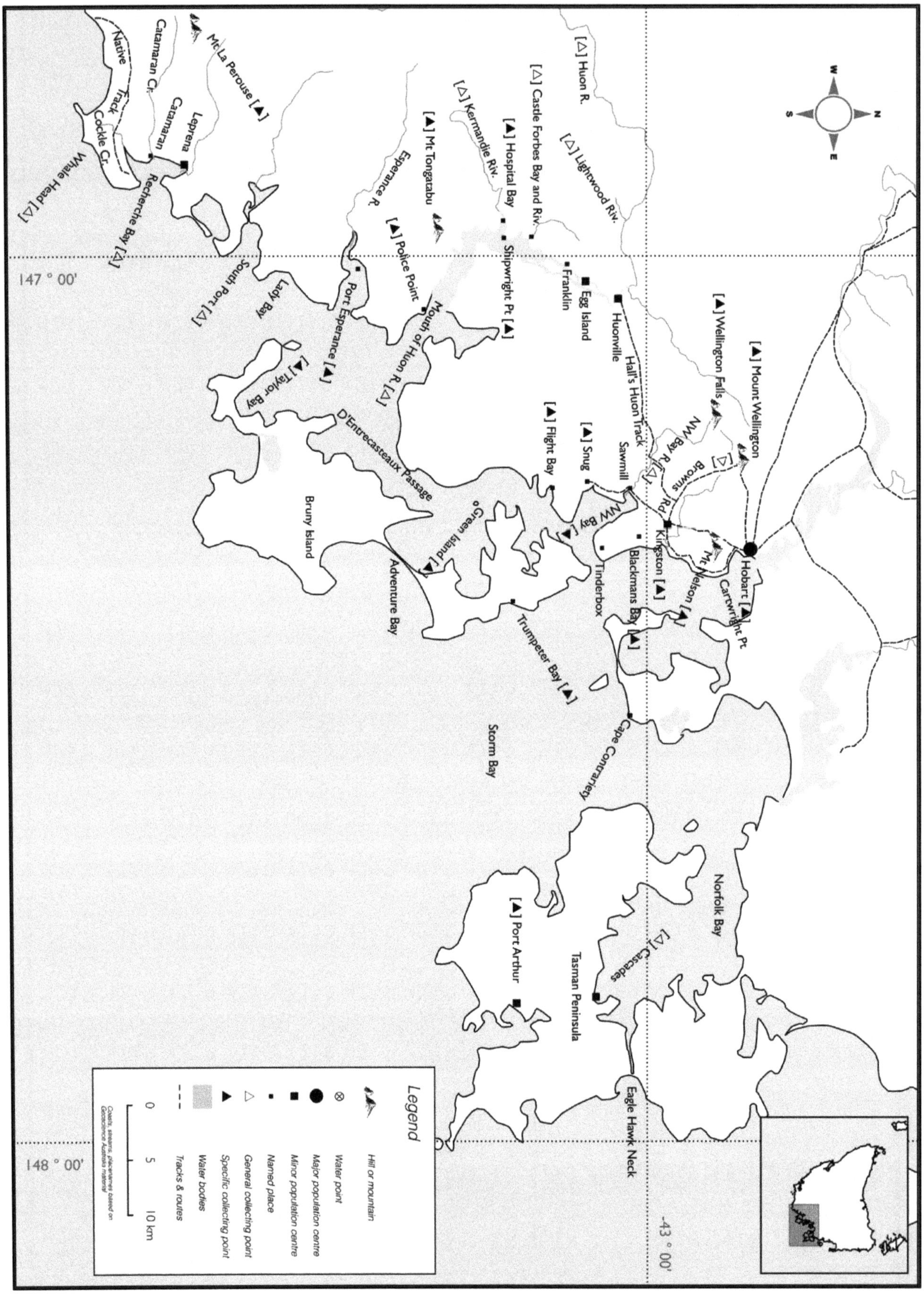

Map 7: Southern Tasmania

Fig. 10.0: Actinobole oldfieldianum

Chapter 10

South-east Tasmania 1855–1856

Oldfield was appointed to the government position of Schoolmaster of the second class at the newly established Browns River (Kingston) one-teacher public school. It was administered by the Board of Education (Tasmania), which funded it together with parents' fees.[1] He took up the post at the beginning of the 1855 school year.[2] It was an ideal location from which he could explore the flora of south-east Tasmania.

Kingston was a tiny settlement at the mouth of Browns River. The stream has its headwaters on Mount Wellington and flows eastward into the Derwent River estuary. At its mouth, there is fertile alluvial soil and reasonable anchorage for small boats. During the 1850s the main industries in the district were timber felling, fishing, fruit production and dairying. The settlement was about 15 kilometres south of Hobart Town and surrounded by intact natural forest. Since 1835, it had been linked to Hobart Town by a road that was constructed over the previous five years by a Mr Proctor.[3] The route zig-zagged over Mount Nelson into the Lea valley to follow approximately the course of Vincents Rivulet to Browns River, where a ford permitted access to the township.[4] By improving the area's links with the Colony's capital and major market, the opening of this road helped to increase the population and improve the local economy.

Soon after Proctors road was completed Lieutenant-Governor Denison, having refused to provide any financial assistance to Mr Proctor, declared the road a public thoroughfare.[5] In 1841, following submissions from landowners in the districts between Browns River and Hobart, construction commenced on the government-sponsored road from Hobart Town to Browns River via Cartwright Point (Taroona) using convict labour. Building this road necessitated the prior construction of a convict probation station to house the workforce, which at its peak in 1844 numbered some 300 men.[6] The then Deputy Adjutant General of Forces in Australasia, Colonel Godfrey Mundy, visited the area during the summer of 1850–1851. He noted that Browns River 'affords an object for equestrianism' and that the road from Hobart Town which

> is constantly nourished by convict labour, follows the outline of the bay, sometimes running along the beach, at others creeping round the steep face or sweeping round the level back of some headland, diving through a hill or striding over a gully;

[1] James Wood (compiler), *Tasmanian Almanack for 1855*, Walch & Sons, Hobart, p. 91. The Education Board provided for two classes of schools. Those of the first class were established in Hobart Town and Launceston, and in other localities where an attendance of eighty children was likely. Second class schools were established where it was estimated that the attendance of students would be fewer than eighty but more than twenty (ibid.). Julie Gardam stated that a Board of Education school was opened in Browns River in 1857 but it failed. Unfortunately, she did not indicate her source, Julie Gardam c.1988, *Browns River, a History of Kingston and Blackmans Bay*, Rotary Club of Kingston, Kingston, p. 16.

[2] The official record of Oldfield's appointment could not be located but the evidence comes from a letter from WHH to WJH, 31 March 1855, cited in Sophie C. Ducker (ed), 1988, p.210, and from the address on letters AO sent to WJH, 20 August and 26 December 1855, Australian Letters 1851–1858, vol. 74, 1851–1858, fols. 190, 191, RBGKA.

[3] Kingborough Council, 'Early settlement: the Kingborough Municipality', viewed 29 November 2012 at www.kingborough.tas.gov.au.

[4] Julie Gardam, 1984, *The History of Kingston*, [? J. Gardam, Kingston], p.7, TAHO.

[5] Kingborough Council, 'Early settlement: the Kingborough Municipality', viewed 29 November 2012 at www.kingborough.tas.gov.au.

[6] Julie Gardam, c.1988, pp. 13–14.

Fig. 10.1 Brown's River, near Kingston

> for a slice borrowed from the superfluity of a mountain and bestowed upon the hungry maw of a ravine is trifling work when half a dozen hundred hands can be thrown upon it by a word from the Governor.[7]

With the housing of convict labourers in the district during the first half of the 1840s and the influx of settlers, including emancipists, following the opening of the road in 1845[8] and the commencement of a coaching service, initiated by a Mr Joseph Fisher, the district began to grow markedly in productivity.[9] While the area was an important source of food for the people of Hobart Town, few farmers became extremely wealthy like the pastoralists on the northern sandplains.

On 27 January 1851, Denison proclaimed the settlement at the mouth of Browns River a townsite and named it Kingston, although it was more commonly referred to by its earlier name. When Oldfield took up residence, Kingston was described as 'an irregular and straggling country township of scattered cottages, some little better than a hut, and some well-built brick homes'. Fisher's Hotel was praised for its service.[10]

Mundy wrote positively about some of the agricultural and garden lands along the way. They were, he said,

> [a]s fine as I ever saw, the colour of the mould being precisely that of black rappee.[11] There were such fine crops of potatoes and onions in the alluvial hollows, and such fat sheep on the hillsides...The Brown's River potato is as well known in Australia as it is in Tasmania. Among many pretty and sterling-looking country homesteads looking over the bay, one was pointed out to me, somewhat superior to the rest, as the property of an emancipated prisoner.[12]

---

[7] G.C. Mundy, 1986, p. 35.
[8] Julie Gardam, c.1988, p. 15.
[9] Kingborough Council, 'Early Settlement: Kingston & Kingborough Area', viewed 6 June 2007 at www.kingborough.tas.gov.au.
[10] H. Butler Stoney, 1856, pp.84–85.
[11] A coarse kind of snuff.
[12] G.C. Mundy, 1852, facsimile 1986, p. 35.

What is known of the history of schooling in the Browns River area to the point where Oldfield took up his appointment reveals shaky beginnings. The first formalised education in the district commenced as a day school in 1833 in the home of Mr and Mrs Russell, who not only taught the children but also provided post office services. The school, subsidised by the government, was administered by the Church of England via a board composed solely of parishioners. When the government established the Board of Education in 1839 parents were expected to pay weekly fees to make up the whole of the teacher's salary. This imposition, along with local dissatisfaction, both with the Russells and with the inconvenience of their premises, resulted in a drop in attendance, and the school was discontinued in 1841. The Board of Education contended that attendance was too small to justify the expense of maintaining the school. Immediately following the closure, the families unsuccessfully requested the establishment of a new government-subsidised school and promised an enrolment of 36. [13]

In 1844 the local Anglican Church, St Clements, filled the gap and established a school funded by the church and parents' fees on its glebe allowance, in the apex of the present junction of Summerleas Road and the Channel Highway.[14] Mr H.J. Allen, formerly a merchant clerk, was appointed Schoolmaster and Postmaster, assisted in both departments by his wife. By 1847 the school had 36 pupils and was again being subsidised by the government, although still administered by the Anglican Church. In 1849 the Anglican Inspector of Parochial Schools, Reverend Davenport, found the children's scholastic performance satisfactory in all areas except the catechism. The school was then conducted in the Allens' home on the site of the present Kingston Hotel.[15]

Thomas Arnold, second son of Reverend Thomas Arnold, the former headmaster of Rugby School, was appointed Inspector of Schools in Tasmania in 1850. By 1853 he had advocated successfully for the return to the non-denominational system of education on lines similar to the system operating in New South Wales.[16] Further, despite Denison's repudiation of government responsibility for contributing to the cost of education, on Arnold's recommendation Denison established a Board of Inspection composed of a representative of each of the three major denominations and Arnold. The Board condemned the current reliance on parents to fund teachers' salaries. Denison backed down and government funding to education was increased substantially, including the purchase of textbooks for distribution to monitors-in-training and the 4,000 school children in Tasmania. These publications embraced the curriculum and, for older scholars, Arnold added poetry, prose and a textbook on agriculture, which included agricultural geology and chemistry. Arnold disapproved of gratuitous distribution of textbooks to all schools and proposed that they be sold to teachers to on-sell to their students but whether the Board of Education, established in 1851, adopted his suggestion could not be determined. Teaching in public schools was in 1855 still carried out under the monitorial system and factual learning was largely by rote.[17]

In 1850 the Browns River School continued to be administered by the Church of England.[18] By 1854 Mr Joseph Keen had replaced Mr Allen as Postmaster[19] but when this occurred is not known or who, if anyone, was the Schoolmaster.[20]

---

[13] Julie Gardam, 1984, p. 6, TAHO.
[14] ibid.
[15] ibid. and Julie Gardam, c.1988, pp. 15, 16.
[16] In 1848 in New South Wales a dual system of church and national schools, each under separate boards, was established. Four separate administrative units represented each of the four main denominations. The amount of money appropriated to each denomination was fixed by Parliament according to the proportion of the population belonging to that denomination, as determined by the last census statistics. The public-school system came under the Board of National Education, Penrith City Council 2004, 'Education in NSW in the 19th Century', viewed 9 December 2012 **at** www.penrithcity.nsw.gov.au.
[17] Lloyd Robson, 1983, vol. 1, pp. 460–62, and Derek Phillips, 1985, pp. 23–46.
[18] James Wood (Compiler), 1850b, *Wood's Tasmanian Almanack for 1850*, Walch and Sons, Hobart, p. 122, TAHO.
[19] James Wood (Compiler), 1854 *Wood's Tasmanian Almanack for 1854*, Walch and Sons, Hobart, p. 95, TAHO.
[20] No lists of public or ecclesiastic schools for that year have been found.

Based on brother Alfred's pay and conditions when he was appointed Schoolmaster at Kingston in 1860, Augustus' annual salary was about £50 plus anticipated parents' school fee receipts of £15 8s, totalling £68 8s. No house or pupil teacher allowances were paid by the government.[21] Given his very meagre income it is doubtful that Augustus would have agreed to purchase text books with the expectation that parents would buy them, particularly as in the Kingston district they had demonstrated their reluctance to pay school fees. By July 1856 Augustus was paying house rental of at least £10 per year (see below) but whether this was the case in 1855 could not be determined. Nevertheless, wherever he was living he appears to have had enough space to store, sort and prepare his specimens.

The Browns River public school was housed in a weatherboard building on the site of the present-day St Clements carpark.[22] The hours of attendance at the school were from 9 am to 12 noon and 2 pm to 4 pm weekdays, and 9 am to 12 noon on Saturdays. The curriculum Augustus was required to teach comprised reading, spelling, writing, arithmetic, English grammar, geography, history and singing.[23] Oldfield left no account of his experience as the Schoolmaster at Kingston.

Visit of Professor William H. Harvey, 1855

William Henry Harvey, Professor of Botany of the Royal Dublin Society, Keeper of the University Herbarium and eminent Irish algologist, collected algae and other plants in Australia between 1854 and 1856. During this period, he visited various coastal and other locations in the Swan River colony, Victoria, Tasmania and New South Wales. For ten years or more he and Joseph Hooker collaborated on the taxonomy of 'antipodean' algae. Between 1845 and 1855 they jointly published papers[24] on the algae of Antarctica, New Zealand and Tasmania and the section on algae in Joseph Hooker's *Flora Antarctica.*[25] Harvey also authored many other papers on the algae of New Zealand and Tasmania[26] and the piece on algae in Joseph Hooker's *Flora Novae-Zelandiae*.[27] During the early 1840s Harvey mastered the art of illustration and lithography, and, according to R.L. Praeger, Harvey became 'one of the most exquisite delineators of plants'.[28] Along the way, Harvey and Joseph Hooker developed a lifelong friendship and their collaboration on the taxonomy of Tasmanian and New Zealand algae was to continue for well over a decade.

During his visit to Launceston, Harvey spent time in the company of Ronald Gunn, who, while retaining a strong interest in natural history, was living in Launceston and in 1855 was elected to the Launceston seat of the Legislative Council. Harvey enjoyed a few days with another of Joseph Hooker's correspondents, the architect and landowner William Archer, botanising on his property 'Cheshunt', near Deloraine.[29]

Harvey was in Hobart Town in 1855 from 26 February until 1 March and from 15–21 March. The intervening period was spent collecting on the Tasman Peninsula to which he travelled by steamer.[30] He later wrote to William Hooker that while he was in the city he 'made a point to call on your

---

[21] Southern Board of Education Report for 1860, Tasmania, Report to the Legislative Council, Education Department of Tasmania Archives, EDTA. The introduction of an Apprenticed Monitors Scheme was announced early in 1851. Monitors had to be at least 13 years of age and training was for four years with annual salary commencing at £5 in the first year and rising annually by £2 10s to £12 10s in year four *Hobart Town Gazette*, 18 February 1851, p. 144, DAN, NLA.

[22] Julie Gardam, c.1988, p. 16.

[23] James Wood (Compiler), 1855, *Tasmanian Almanack for 1855*, Walch and Sons, Hobart, p. 91, TAHO.

[24] JDH and WHH, 1845, 'Algae antarcticae' and 'Algae novae-zelandiae', *London Journal of Botany*, vol. 4, pp. 249–276, 293–298 and 521–551; 1847a, 'Algae tasmanicae', *London Journal of Botany*, vol. 6, pp. 397–417.

[25] JDH and WHH, 1847b, 'Algae' in *The Botany of the Antarctic Voyage of H M Discovery Ships Erebus and Terror in the Years 1839–1843*, vol. 1 *Flora Antarctica,* Pt 2, Reeve Brothers, London, pp. 454–519, 165–193.

[26] WHH, 1849, 'Algae of Tasmania', *London Journal of Botany*, vol. 3, pp. 428–454, and 1846, 'Algae of Tasmania', *Tasmanian Journal Natural Science*, vol. 2, pp. 377–384, and vol. 3, pp. 153–159.

[27] WHH 'Algae' in JDH, 1855, pp. 211–266, 107–121.

[28] R.L. Praeger quoted in Sophie C. Ducker (ed), 1988, p. 4.

[29] WHH to Sisterella [Harvey's niece Mary Christy Harvey], 12 February 1855, cited in Sophie C. Ducker (ed), 1988, pp. 185–195.

[30] WHH to Hannah [Todhunter née Harvey], 9 February 1855; WHH to Sisterella, 12 February 1855, WHH to WJH, 24 February 1855 in ibid., pp. 181–185, 185–195 & 195–198 respectively.

correspondent Mr Oldfield' but found 'that he now resides at the Huon[31], where he superintends a school. I saw his brother, also a schoolmaster, and have since had a letter from himself [presumably Augustus]'. Harvey elaborated to Hooker on Oldfield's situation:

> His brother [Edwin] told me that Augustus has no taste for his present occupation, and a strong desire to be a natural history collector and traveller, for which he seems well fitted; that he has a competent knowledge of mathematics and practical astronomy, sufficient to enable him to map his course correctly, and that he is full of zeal for science of all kinds. I mention this to you, as you may possibly have it in your power to recommend him for some collector's appointment, should you be called on by Government for one or any of the Australian exploring expeditions, failing Drummond or otherwise.[32]

This would suggest that Harvey and Augustus did not meet, although we cannot be certain of this from what Oldfield wrote about the visit to Hooker a few months later:

> I am much obliged to you for inducing Dr Harvey to call on me, but at the same time much regret that I had nothing to show him, for I have not paid the least attention to the collection of Algae: as it is, I am the sole gainer by the introduction, as Dr Harvey kindly named all my little collection, and thus gave me a *point-d'appui* for future operations.[33]

We do not believe they came together. Thus, it is unlikely that Harvey would have mentioned receiving a letter from Oldfield but failed to say that they had met. Either Augustus, having heard that Harvey would return to Hobart, brought his algal specimens to the city while Harvey was on the Tasman Peninsula so that Edwin could show them to Harvey on his return, or Edwin was storing Augustus' collection and, in his absence, gave Harvey access to the material. If the meeting did not take place, Augustus' disappointment must have been intense.

Edwin's observation about Augustus' ambition to become a traveller and a natural history collector and his dislike of teaching is indicative of his brother's awareness of the frustration Augustus was experiencing at the time. Harvey's appeal to William Hooker on Augustus' behalf was no doubt prompted by Edwin's comments. It was just one of several occasions when, motivated by filial concern and awareness of Augustus' reluctance to actively advance his own interests, Edwin interceded for his brother.

Clearly, then, Augustus must have found it hugely tiresome listening by the hour to children reciting facts from question-and-answer texts and who, in turn, were no doubt bored with the experience. His antipathy may have been exacerbated if he shared Edwin's enlightened approach to education. It is not difficult to imagine that, during rote learning sessions, Augustus was drawn to the window and, gazing out over the heavily wooded hills and valleys surrounding Kingston, became lost in reveries of wandering along river banks, in the bush or on mountain tops searching for plants that might be of interest to 'men of science'.

If he found teaching tedious his evenings were more interesting. According to Edwin, Augustus 'devoted a portion of every day to systematic study'.[34]

---

[31] This is incorrect. Oldfield superintended a school at Kingston (Browns River).
[32] WHH to WJH, 31 March 1855, in Sophie C. Ducker (ed), 1988, p. 210.
[33] AO to WJH, 20 August 1855, Australian Letters, vol. 74, 1851–1858, fol. 190, RBGKA.
[34] Anon, [probably Edwin Oldfield], 16 July 1887.

A change of approach to collecting

Further evidence of Oldfield's frustration over the restraints being imposed on his collecting is manifest in his letter to Sir William of 20 August 1855:

> At the risk of appearing ungrateful for your many favours to me I have hitherto refrained from sending any more specimens, for, from my calling I have so few facilities for collecting that I was fearful that any I could send would be of any service to you. Even when the Xmas holidays allow me to make an excursion of two or three weeks, as I am obliged to go alone and carry provisions for that period, you may imagine that I cannot do very much. Besides in my attempts to penetrate towards the South-west [Tasmania], I was driven back by the Bushfires the year before last, and last year circumstances prevented me from making the attempt. However, it is a journey on which I have set my mind, and this year, if possible I will at least do something towards it.[35]

Despite his concerns about sending plants that would not be of use to Hooker, Oldfield nevertheless expanded the purpose of his collecting away from the focus of gathering only specimens of species that might be new to science:

> On the reading of Dr [Joseph] Hooker's "Introductory Essay to the Flora of New Zealand"[36] (for which and the vol. of the 'Icones Plantarum'[37] I return you many thanks) has considerably modified my views with respect to sending specimens, for on reading the Essay in question I consider that I might throw some light on the distribution of Tasmanian plants, and, though my collection may contain few or no new ones, yet there may be varieties or states which may throw light on old ones. For this end I have begun to cull from my herbal; 1st all plants of which I know not the place either with regard to order, genus or species. 2nd such plants as I have found here which none of my authorities refer to as indigenous to this Island. 3rd some which I think doubtfully named. The collection will be pretty extensive for I have done nothing towards naming the Graminaceae [grass family] and allied orders, and very little among the Compositae [now Asteraceae].[38]

Although Augustus extended his collecting focus, nevertheless discovering species new to science remained an important interest. As Edwin said after Augustus' death: 'The discovery of a new plant was to him a greater prize than that of gold'.[39]

Oldfield reported to Sir William that he had added very few Cryptogamic plants to his collection since he sent the parcel with Mr Brent. Adding to his other frustrations was not being able to obtain 'authorities' when he needed them:

> I regret that though I sent to England, through three different channels for [Carolo] Müller's[40] 'Syn. Musci'[41] yet as they all failed I have done very little towards naming them [that is, the specimens he sent with Brent]. Those that I have identified are some

---

[35] AO to WJH, 20 August 1855, Australian Letters, vol. 74, 1851-1858, fol. 190, RBGKA.

[36] JDH, 1853, *Introductory Essay to the Flora of New Zealand,* pamphlet, Lovell Reeve, London. This essay was reprinted as the introduction to JDH 1864, *The Handbook of the New Zealand Flora: a Systematic Description of the Native Plants of New Zealand and the Chatham, Kermadec's, Lord Auckland's, Campbell's, and Macquarrie's* [sic] *Islands*, Lovell Reeve, London.

[37] To what publication of *Icones Plantarum* (plant illustrations) he is referring cannot be ascertained. Perhaps it was WJH's 1854, *A Century of Ferns: Being Figures with Brief Descriptions of One Hundred New or Rare,* or his *Imperfectly Known Species of Ferns, from Various Parts of the World; a Selection from the Author's 'Icones Plantarum'*, W Pamplin, London.

[38] AO to WJH, 20 August 1855, Australian Letters, vol. 74, 1851–1854, fol. 191, RBGKA. According to Stephen Hopper, Compositae is '[a]nother of Kew's idiosyncratic family names, today recognised as Asteraceae', personal communication, email attachment, 19 December 2017, HOA.

[39] Anon, [probably Edwin Oldfield], 13 January 1887.

[40] Carolo Müller was a German bryologist.

[41] Carolo Müller, 1849–1851.

Fig. 10.2 Flight Bay (now Surges' Bay), Huon

British Ferns, and a few from the 'Musci Exotici'.[42] I must therefore wait the forthcoming of the 'Flora Tasmanica [*Tasmaniae*]'.[43]

He seems not to have had any contact with Gunn, Archer or other members of the Royal Society of Van Diemens Land who might have had relevant works that he could access. Even if he could have afforded the annual subscription, the Public Library in Hobart probably did not have most of the books he needed. Thus, according to Harvey who spent time in the Library: 'All the periodicals are taken in & a fair collection of new books, but the suite is by no means as extensive as would be desirable'.[44]

Oldfield ended the letter of 20 August 1855 by stating that he feared the consignment of specimens he proposed sending to Sir William would not reach him before the middle of January 1856 but hoped that it would be in time for it to be of service to Dr Hooker in preparing his *Flora Tasmaniae*.[45] No record of specimens freighted prior to Christmas 1855 has been found.

Expeditions south of Hobart to the Huon River and Bruny Island

Map 7 p. 150 – Southern Tasmania

Oldfield botanised in the dry, sclerophyll eucalypt forests on the gentle lower slopes of the coastal districts embracing Kingston, Browns River, Browns Road, which ran towards Mount Wellington from Kingston, to the south at Blackmans Bay, on the North West Bay track, at North West Bay, The Snug and further south at Flight Bay (now Surges' Bay).[46]

The North West Bay track branched off from the Browns River Road near Kingston and cut across country to the settlement (a sawmill and now Margate) toward the top of the west coast of

[42] This probably refers to WJH, 1818–1820.
[43] AO to WJH, 20 August 1855, Australian Letters, vol. 74, 1851–1854, fol. 191, RBGKA.
[44] WHH to Sisterella, 26 April, 1855, cited in Sophie C. Ducker (ed), 1988, p. 218
[45] AO to WJH, 20 August 1855, Australian Letters, vol. 74, 1851–1854, fol. 191, RBGKA.
[46] HAODb.

North West Bay, a distance of approximately 5 kilometres.[47] He could easily have walked there from Kingston in about an hour by travelling to the North West Bay settlement and the Snug along the same track traversed by Stoney during his 1851 tour of the Colony. However, Oldfield may have needed to ford the North West Bay River since the 'excellent wooden bridge' by which Stoney had crossed the stream was destroyed by the flood of 1854.[48]

Stoney was impressed when he first sighted the North West Bay from D'Entrecasteaux Channel:

> Leaving Kelly's Point at Bruni Island [sic.] and Tinderbox Bay on the mainland, you pass North-west Bay, a fine open one, with a pretty sandy beach; the shore is dotted here and there with some snug abodes of the settlers on the mainland, and a field now and then well cultivated.[49]

He was not so enthusiastic when later in the year he went overland to the Bay:

> This is a large arm of the sea, but, being shut in on all sides by gloomy and unpicturesque hills, it is not prepossessing. Approaching North-West Bay River, the soil sensibly improves in character, and the coarse, hungry sands we have passed give place to a rich red loam of great fertility. The farms here are small and strictly agricultural, the breeding of stock not being attended to...Falling into the head of North-West Bay is a fine little stream...Its earlier waters are collected from the first flat encountered on the summit of Mount Wellington.[50]

From The Snug, Oldfield may have travelled further south along the coast to Flight Bay in the D'Entrecasteaux Channel, although this could have been a stopping place on a boat trip to the Huon River or further to the south. When Stoney visited Flight Bay there was a sawmill operated by Mr Arthur and during that period the district was populous enough to support a public school.[51]

The most widespread forests in the area are dominated by White Peppermint (*Eucalyptus pulchella*) and Stringy Bark (*E. obliqua*) and are typical of fertile dry forest soils. Although no Oldfield specimen records of those species from this area have come to our attention, he did collect eucalypt species that dominated other forests within the area—Black Peppermint (*E. amygdalina*), White Gum (*E. viminalis*) and Silver Peppermint (*E. tenuiramis*)[52]—which subsequently have been cleared preferentially for housing and farmland. The understoreys of the dry forests vary from grasslands, sedgelands through to heathland and shrublands. The heathy understorey of Black Peppermint may have yielded the specimens of Showy Bossiaea (*Bossiaea cinerea*), Swamp Selaginella (*Selaginella uliginosa*) and Fan Sedge (*Lepidosperma inops*)[53] collected by Oldfield. However, he is more likely to have gathered them from heathland vegetation which is now restricted to a few small remnant patches, an example of which is reserved in the Peter Murrell Reserve at Kingston. This, too, may have been the location of the two of his collections of Western Peppermint (*Eucalyptus nitida*)[54], a species often associated with Button Grass (*Gymnoschoenus sphaerocephalus*) sedgeland, which in the Kingston area is now only found within a small part of the Peter Murrell Reserve. The absence of more recent records, suggests that Western Peppermint is no longer present in the catchments of the North West Bay River or Browns River.[55]

---

[47] John Tallis, 1851, *Van Diemen's Island*, Tasmania, map drawn and engraved by J. Rapkin, reprint, Map Supplies, Hobart, 1991.
[48] H. Butler Stoney, 1856, p. 86.
[49] ibid., p. 81.
[50] ibid., pp. 85–86.
[51] ibid., pp. 83, 285.
[52] Black Peppermint (*Eucalyptus amygdalina*), Browns River;White Gum (*E. viminalis*), Browns River; Silver Peppermint (*E.tenuiramis*), North West Bay& Blackmans Bay (HAODb).
[53] OXF unnumbered, *Bossiaea cinerea*; Browns River, (Dr Serena Marner, Fielding-Druce Herbarium, personal communication to MHH, email 27 January, 2006, HOA); MEL 2146201, MEL 2146201 *Selaginella uliginosa*, Browns River and MEL 2202506, *Lepidosperma inops* Browns River (MELISR database, 5 November 2005, NHV, RBGM); BRI 303614, *Lepidosperma inops*, Browns River, (HERBRECS Job Number-0506_115, Queensland Herbarium, Paul Robins personal communication to MHH, 28 July 2006, email, HOA).
[54] MEL 230746 & MEL 230747, *Eucalyptus nitida*, Browns River (MELISR database download, 5 November 2005, NHV, RBGM).
[55] Jayne Balmer, personal communication to MHH, email, October 2008.

Most of Oldfield's plants from Browns River, surrounding gullies and stony places, and further south to North West Bay (some from 'brackish swamps' and 'marsh rocks') probably were obtained while he was at the Browns River school or later when he returned to the town after his first trip to Western Australia.[56] His specimen records from these areas in HAODb are predominantly cryptogams, mostly mosses (Bryophyta) and liverworts (Hepaticae).[57]

He visited Trumpeter Bay on North Bruny Island and Taylors Bay, South Bruny Island. There are also records of specimens bearing the location 'Green Island, Taylors Bay, S. Bruni [sic] Island'. [58] However, the only island by that name in the vicinity of Bruny Island occurs within the D'Entrecasteaux Channel in Great Bay, about five miles (eight kilometres) due west of Smoothys Point on North Bruny, but Great Taylor and Little Taylor's Bay are both on the western coastline of South Bruny Island.[59]

Bruny Island is approximately 50 kilometres in length, and in all its inlets there 'is good anchorage'. At the time, there were several farms and many sawyers and splitters scattered over the area. The settlers, numbering between one and two hundred, 'suffered much for the want of good water'.[60]

Oldfield may have travelled there by ferry, as Stoney had done previously, leaving Tinderbox on the mainland, landing on the island at Kelly's Point, the northernmost tip of Bruny and from thence traversed southwards on foot or horseback. However, the specific locations of his specimens from the island suggest that he accompanied a trading ketch or small sailing craft and collected close to coastal settlements visited by the vessel along the way.[61] Current records give no hint as to the reason for the visit.[62] There may have been settlers on Green Island as it was leased for the first time in May 1868 by Robert Officer, Morton Allport and two others for acclimatisation purposes.[63]

Other collecting areas were various bays, promontories, streams and hills along and close to the Huon River from 'a rivulet by Lightwood Bottom', which is the lower part of Lightwood Creek, to the north-east of Huonville, and thence south-east along the Huon to where it emptied into the D'Entrecasteaux Channel.[64]

How many visits Oldfield made to the Huon is not known. He was there on 2 February 1853 when he collected a specimen of the moss *Zygodon hookeri* in a 'gully near the Huon'.[65] He was in the vicinity of Franklin during May 1855, while he was teaching at Kingston, when he contributed £1 to Tasmania's Patriotic War Fund.[66] The people of Tasmania were 'fanatically loyal to the Crown and Britain and outdid themselves during the Crimean War, donating huge sums of money to the imperial cause and emotionally reacting to the course of events.'[67] For someone whose income may have been under £70 per annum, Oldfield's donation to the Fund was significant and suggests that he retained strong ties to his homeland. Local social pressure could have been the driving force to donate.

To reach the Huon River from Kingston he probably travelled on one or more occasions along the 'Huon [bridle] Track', which apparently later became Halls Huon Track. This ran through what appears to have been a patchwork of dry eucalypt forest and woodland similar to the vegetation types

---

[56] See Chapter 14.

[57] Of the records of specimens collected in the Browns River, Blackmans Bay and North West Bay districts (77% publ., 33% herb.) cryptogams were overall the most numerous {*Bryophyta* (36% publ., 30% herb.); *Hepaticae* (51% publ., 18% herb.); Algae (3% herb.); Lichens (1% publ.); with the remainder comprising *Angiospermeae* (8% publ., 44% herb.), HAODb.

[58] HAODb.

[59] See for example, Mapping, Department of Environment and Planning (Tasmania), Hobart, Tasmania 1:100,000 Topographic Map, Edition 6, 1992 and John Tallis 1851.

[60] H. Butler Stoney, 1856, pp. 81.

[61] Audrey Hudspeth, 'Shipping and ports', in Alison Alexander (ed), 2005, p. 332.

[62] Jayne Balmer to MHH, October 2008.

[63] Green Island, Diagram from Actual Survey, Land Titles Office, attachment to memorandum, 7 April 1978, ref. M5/1, Department of Primary Industries, Parks, Water and Environment, Hobart, Tasmania,

[64] HAODb.

[65] MEL 1016372, *Zygodon hookeri*, 'gully near the Huon', 2 February 1853 (MELISR, 5 November 2005, NHV, RBGM).

[66] 'Patriotic War Fund', *The Courier* (Hobart Town) 5 May 1855, p. 3, DAN, NLA.

[67] Lloyd Robson, (updated by Michael Roe) 1997, *A Short History of Tasmania*, Oxford University Press, Melbourne, p. 25.

which occurred on the coastal slopes around Browns River and North West Bay.[68] Although some land along the river flats and rolling hills on either side of the estuary had been cleared for agriculture, the remaining native vegetation within the swamp land comprises mainly saline sedgeland and rushland.[69] The track joined the Browns Road from Hobart near Browns River, ran to the west of North West Bay and terminated a short distance to the south of the modern town of Huonville at the Ironstone Creek. Here he may have remained on the same side of the river, crossed it by ferry or taken another vessel to his destination.[70] This track was opened in February 1830 to enable parties in search of land 'the power of penetrating with facility into the heart of an unexplored country in the south-western [sic] portion of the Colony'.[71] By 1850, however, settlers were petitioning for a better road and a more direct route. The 28-mile track (from Hobart Town), they said,

> is nothing more than a single horse can travel in, is, for a distance of ten miles, the most rugged and difficult imaginable, fatiguing in the highest degree to a pedestrian and can only be travelled on horseback at a walking pace...A traveller [from Hobart Town] cannot reach the Huon on foot or by horse...under from 6 to 7 hours.[72]

A new road, however, was not commenced until 1857 and was not completed until c.1870.[73]

With one and a half days' respite from teaching each week of the school term it is feasible that Oldfield occasionally undertook the four to five hours walk along the track to the Huon River on a Saturday afternoon, did some collecting and returned to Brown's River the next afternoon. On such visits, it is unlikely that he had the time to penetrate along the Huon for more than five kilometres to the north or south of the end of the track. The longer Easter or Christmas vacations would have allowed him to collect further along the river.

While Oldfield may have preferred pedestrian excursions to the river, he probably travelled to many of his more inaccessible collecting places along the Huon further to the south east of Ironstone Creek on a trading vessel or on the little steamship *Culloden* which plied twice a week to the Huon from Hobart Town, remaining one night on the river.[74] Stoney described the journey after leaving Garden Island near the mouth of the river:

> The shores on both sides of the Huon are precipitous for a considerable way up, having very high land in the background covered with large timber. About five miles distant, you turn into a deep bay leading to Port Cygnet...Leaving Port Cygnet Bay, you again enter the river, and passing on the opposite side of Police Point, you proceed by Desolation Bay, where there are some clearings...In Flight's Bay are the sawmills of Mr Arthur, and in Hospital Bay those of Mr Hill. Above the Egg Island[s] the river forms into two channels, leaving a long swamp in between; which destroys the navigation in a great measure. Still the steamer can go up to Franklin Settlement...The highest navigable point, at present, is Victoria Settlement [now Ranelagh just north of Huonville],[75] where a town has lately commenced and a church built.

---

[68] S. Harris & A. Kitchener, 2005, Vegetation Map of Tasmania, opposite inside front cover. See also K.J. Williams & B.M. Potts, 1996, 'The natural distribution of *Eucalyptus* species in Tasmania', *Tasforests,* vol. 8, pp.39-149.

[69] (S. Harris & A. Kitchener 2005, pp. 389–390 & 410, 411)

[70] Sources quoted by Lyn Newitt stated that the track ended at Ironstone Creek, 1988, *Convicts and Carriageways: Tasmania Road Development Until 1880*, Department of Main Roads, Tasmania, Hobart, p. 224, Richie Woolley & Wayne Smith give Mosquito Point as the termination of the track, 2004, *A History of the Huon and Far South,* vol. 1, *Before the Orchards Grew*, Huon Valley Council, Tasmania, p. 55. This last is a short distance to the north of the Creek. See also J. Arrowsmith, 1842, *Van Diemen's Land*, map, copied from the original manuscript surveys in the Colonial Office and in the Van Diemen's Land's office, J. Arrowsmith, London, Allport Library, TAHO.

[71] *Hobart Town Gazette*, 13 February 1830, cited in Lyn Newitt 1988, p. 223.

[72] ibid., 1988, pp. 224–225.

[73] ibid., pp. 226–230.

[74] H. Butler Stoney, 1856, p. 79

[75] Anon., 'Huonville, Tasmania', viewed 19 November 2012 at www.wikipedia.org.

Fig. 10.3 Hall's Road to the Huon, 1858

> One mile above are the Falls; and the river is deep and navigable for small craft. There are a great many settlements up this river, and the timber, being of excellent quality, affords occupation to hundreds. For the most part the land is of a fine description.[76]

Apart from his gatherings along and close to the Huonat Castle Forbes Bay, Hospital Bay Shipwrights Point, Kermandie Rivulet and Police Point, Oldfield at times ventured further inland about five kilometres to the south of east of Police Point, to the densely vegetated wet eucalypt forest on Mount Tongatabu.[77] As far as can be ascertained this forest was dominated, as it is today, by Stringy Bark with its understorey of Myrtle Beech (*Nothofagus cunninghamii*), Sassafras (*Atherosperma moschatum*), Leatherwood (*Eucryphia lucida*) and Celery-top Pine (*Phyllocladus aspleniifolius*), Manfern (*Dicksonia antarctica*) and under which grew many ground ferns and other species.[78]

---

[76] H. Butler Stoney, 1856, pp. 82–84.

[77] HSODb.

[78] S. Harris & A. Kitchener 2005, pp. 372–373. See also K.J. Williams & B.M. Potts, 'The natural distribution of *Eucalyptus* species in Tasmania', 1996, pp. 82–84.

Fig. 10.4 Valley of Castle Forbes Bay Rivulet, Huon

Fig. 10.5 Kermandie Rivulet, Huon

Of the records of specimens that were gathered along the Huon River and its environs, cryptogams, mostly liverworts and mosses, are by far the most numerous of the published records while angiosperms dominate the herbarium records.[79] This may indicate that many of Oldfield's cryptogamic specimens from this area, which were transferred to the Natural History Museum from Kew Gardens, had not been digitised when we ceased searching online databases for his specimen records.

Oldfield's specimens for Joseph Hooker's proposed *Flora Tasmaniae*

For four months after writing to William Hooker in August 1855, Oldfield seems to have spent much of his spare time preparing specimens ready for shipping to London. On 26 December 1855, while still at Kingston, he wrote again to Sir William announcing that his long-promised contribution to Joseph's *Flora of Tasmaniae* 'is fairly on its road' to London, and trusted that it would reach him 'in good order and in time'. He offered it

> as the contribution of a poor student who chiefly laments his poverty as preventing him making as he would fain do in the cause of science. That the collection is imperfect I am fully aware, but I hope that it will prove not altogether unworthy of your acceptance. Since I packed the box I found two errors into which I have fallen, viz. in regard to the Velleia paradoxa R. Br. and the Orites diversifolia Id. How I came to make the former mistake I cannot imagine; but the latter plant is in such a different state from that hitherto known to me that I can easily account for the latter error. It is [indecipherable word] Cryptogamia than I thought but having no authority to judge by, I cannot speak as to its value in that department.[80]

Oldfield's frustration is palpable in this letter and once again he is self-effacing as to the quality of his collection. While his comments about the incorrect naming of the plants suggest an ongoing reflection upon his taxonomic decisions, the shock that he seems to have experienced upon discovering his mistakes perhaps reveals an underlying confidence in his classifications. If, however, the reference to his poverty was a veiled attempt to induce Sir William to offer some financial recompense for the collection it appears to have been at variance with the collector's usual practice. Thus, a few years later in 1861 when he did ask Joseph Hooker to help fund his last Western Australian expedition, Oldfield said, 'I am forced to depart from my general rule and seek, by the sale of specimens, some return for the great expense I shall be under in this undertaking'.[81]

In his letter of December 1855 Oldfield provided William with what he clearly thought were noteworthy features of the collection he was sending to London:

> You will find some Alga [sic] which were not in my collection when Dr Harvey so kindly arranged it. Of Musci you will have a good sprinkling not included in my former lot, and of Jungermanni[ae] the same. I think I have been most successful in this last named Family (and its neighbour the Marchantiace[ae])[82] for not only have I been so fortunate as (in many instances) to get them in fruit, but also to add materially to my collection. The Anthoceros punctatus[83] the two Fimbriaria and a species of Fossombronia I regard as my chief acquisitions in these Families. Among Ferns I regret to have nothing to send unless the Hymenophyllum. In my herbarium I have (with

[79] N=65 publ., 25 herb.: Mosses (20% publ., 15% herb.); Liverworts (48% publ., 4% herb.); Algae (20% herb.); Angiosperms (32% publ., 60% herb.), HAODb.

[80] AO to WJH, 26 December 1855, Australian Letters, 1851–1858, vol. 74, fol. 191, RBGKA.

[81] AO to JDH, 25 August 1861, Supplementary Foreign Letters 1865–1900, vol. 218, fol. 238, RBGKA.

[82] A family belonging to the *Hepaticae* (liverworts).

[83] This is a species of horned liverwort belonging to one of the genera of creeping annual or perennial plants of the class *Anthocerotopsida*. In some classification systems hornworts have been grouped as horned liverworts in the subclass *Anthocerotidae* (class *Hepaticae*), *Encyclopaedia Britannica Online*, viewed 25 July 2007 at www.britannica.com.au.

> great doubt) marked it as a variety of the H. Tunbridgensis. Among the Endogen[ae][84] I fear you will find little worthy of your attention, unless one or two Orchids should prove so. Should the Iridaceous[85] plant I have named Newmania rivularis be new I should like it to stand as it is, or in some manner in commemoration of Mr Newman.[86]

That gentleman was possibly Francis William Newman, Curator of the Royal Society of Tasmania Gardens (1847–1854).[87] It is not known why Oldfield wished to honour Newman in this way.

Once again Oldfield numbered most of the specimens he sent to Hooker with similar numbers on duplicates in his own herbarium, hoping that he would favour him with identifications because he was so far in arrears that he would 'never be able to arrange' his specimens properly. He went on: 'I have added a very few from my New Holland collections which through Dr Müller [Mueller] is better arranged than my Tasmanian and N. Zealand'.[88]

After all the trouble Oldfield took to establish the plants to send to London in the Wardian case he had disappointing news for Sir William:

> [H]aving filled it as you directed with the [indecipherable] and our rarest [? species] the person in whose care I left it in Hobart Town was insolvent during my absence and the case was sold: what became of it after that I cannot say.[89]

By this time Oldfield's personal herbarium must have been quite extensive but at no time did he specify his reasons for accumulating it. He was, however, by no means the only naturalist in Tasmania at the time to amass a private collection. Apart from Ronald Gunn, his fellow members of the Royal Society of Tasmania, William Archer and Joseph Milligan were also keen collectors of plants indigenous to the Colony and each had his own herbarium.[90] Their interest in botany seems to have been grounded in a fascination with the flora of Tasmania although, as Endersby has suggested, perhaps social ambitions were high on their agendas. Unlike Oldfield at this time, they were all correspondents of Joseph Hooker and sent many plant specimens to him. They were also contributors to *The Tasmanian Journal of Natural Science, Agriculture, Statistics, & etc.* and later to *The Papers and Proceedings of the Royal Society of van Diemen's Land*, mostly on botanical topics. Oldfield may have harboured ambitions of using his herbarium as a reference collection to assist him in preparing taxonomic papers but this has not become apparent in his writings. Perhaps, on the other hand, his aim was simply to accumulate a comprehensive reference collection so that he could better serve William Hooker and Mueller as a collector. A properly named herbarium would enable him to be selective in accordance with their needs. Whatever the explanation, it is clear that his passion for collecting was not diminishing with the passage of time.

The letter of 26 December 1855 to Hooker was written when Oldfield was on his Christmas vacation. In it he reported his intention of taking advantage of the break from teaching to go 'to South Port and thence to make a journey towards the S.W. as far as possible'. He went on:

> My time is very short (only 10 days) so that I cannot expect to do a great deal, but as I shall have C[harles] Stuart (who has collected for Mr Gunn and Dr Müller) with me, whatever I may send is likely to be valuable—You shall receive the results of this

---

[84] *Endogenae* is a synonym for Monocotyledons (Horace Mann, 'Notes on Endogenae [Class II: Monocotyledonous or Endogenous Plants], presumably for flora of Hawaii', Horace Mann Papers, Library of the Gray Herbarium, viewed 25 June 2006 at www.huh.harvard.edu.

[85] The *Iridaceae* are perennial herbs from rhizomes, bulbs or corms comprising about 80 genera and 1,500 species. Some of the more familiar members of this family are the genera *Freesia*, *Gladiolus* and *Iris*.

[86] AO to WJH, letter, 26 December 1855. 26 December 1855, Australian Letters, 1851–1858, vol. 74, fol. 191, RBGKA.

[87] Alex S. George, 2009, p.502.

[88] AO to WJH, 26 December 1855, Australian Letters, 1851–1858, vol. 74, fol. 191, RBGKA.

[89] ibid.

[90] G.T. Stilwell, 1969, 'Archer, William (1820–1874)', W.G. Hoddinott, 1967, 'Milligan, Joseph (1807–1884)' & T.E. Burns & J.R. Skemp, 1966, 'Gunn, Ronald Campbell (1808–1881)', ADBO, viewed 26 June 2007 at www.adb.online.anu.edu.au.

> journey as soon as possible and should these promise any great [indecipherable] to our Flora I will give a week or so more to the tour.[91]

A single specimen from Mount La Perouse in HAODb bears the date '1856'[92] which confirms that he carried out his intention.

Charles Stuart—Oldfield's co-collector in the Southport-Recherche Bay area 1855–1856 and 1857

Map 7 p. 150 – Southern Tasmania

Nothing is known of Charles Stuart's early background except that he was born in England. From 1842 Stuart collected plants for pecuniary gain at various times in Tasmania, chiefly for Ronald Gunn, and in South Australia in 1847. During that year, shortly after Ferdinand Mueller arrived in Adelaide, he and Stuart became acquainted. From that time Mueller periodically enlisted his services as a collector. Stuart had the reputation of being 'a man of good family and education, a neat and legible writer in correct English prose, and a careful and accurate observer of plants'. While he may not have been a nurseryman by training, he seems to have followed that occupation in Adelaide and Tasmania where there were plenty of job opportunities but the wages were low.[93]

Charles Daley has provided the most detailed account of Stuart's collecting activities in Tasmania, based on twelve letters written to Mueller between 22 May 1848 and April 1852. The original letters give no indication that Stuart knew Oldfield before or during that period. In 1848 Stuart went to Tasmania to find employment and carry on with his botanical activities. He had correspondents in South Africa and the Swan River colony who provided him with seeds and plants, possibly in exchange for material he sent to them.

The frustrating limitations which poverty imposed on Stuart in exercising his passion for botany were similar in many respects to those experienced by Oldfield. Both felt keenly that their straitened circumstances did not allow them the liberty to collect when and where they desired. As Stuart lamented to Mueller: 'What a pity it is, my dear sir, that our purses will not keep pace with our wishes in this interesting pursuit'.[94]

From 1848 until 1852 Mueller provided Stuart with practical help in sorting out difficulties in some of his business dealings in South Australia. On a few occasions Mueller financially assisted him, including lending him the fare for his passage from Adelaide to Launceston. Mueller also sent Stuart a gift of a magnifying glass which he had requested Mueller to purchase for him with the little money due to him from the sale of plants. On 20 July 1851 Stuart wrote despondently to Mueller:

> I have had nothing but misfortunes of late, which have quite unmanned me. I was for a long time ill, which plunged me into poverty, and when able to work I could get no employment worth speaking of. Add to which I, of course, was in debt.[95]

Mueller was clearly concerned, and in the following month provided further financial assistance, for which he refused repayment, and an offer of employment in South Australia. Stuart, replying from Tasmania, did not accept the offer because 'I have something in view here, which, if I succeed in, will place me in a situation where I shall have much liberty to enjoy my favourite pursuit.' In the meantime, he said he had no opportunity to read or study at night, and no one to speak to on 'the subject in which I take greatest pleasure'.[96] He was always extremely grateful for Mueller's help, promising to

[91] AO to WJH, 26 December 1855, Australian Letters, vol. 74, 1851–1858. fol. 191, RBGKA.
[92] MEL 10081, *Luzula australasica*, Locn: Mt La Perouse, 1856, AO, MELISR database, 8 March 2007, NHV, RBGM.
[93] C. Daley, 1935, p. 106. These letters were 'entrusted' to Daley by Mueller's niece, Mrs Sinclair, of Croydon, New South Wales (ibid., pp. 106–110; 132–137; 154–157). The original letters are now kept at RBGML, file RB MSS M36.
[94] C. Stuart to F. Mueller, letter, 16 April 1850, C. Daley, 1935, p. 136.
[95] ibid., p. 137.
[96] ibid., p. 155.

repay all the monies he received and to 'let slip no opportunity of obtaining [plants] for you and for no one else, as I shall not do it for profit, but the pleasure of having your correspondence'.[97] Between 1848 and 1852, the plant material he sent to Mueller from Tasmania seems to have been primarily, but by no means exclusively, cryptogams.[98]

Throughout these letters it is apparent that Stuart had the greatest admiration and respect for Mueller, feelings that he did not hold for Ronald Gunn for whom he had collected in the early 1840s:

> Respecting the specimens left with Mr Gunn he has in a most unhandsome manner made away with them, and then made a paltry excuse, saying that he thought I should not need them...but after the number of specimens I have given him, and information he has got from me, I should not have thought him so bad, but the fact is he is jealous of any person's knowledge of plants lest it should discover his defects.[99]

Stuart's low opinion of Gunn's behaviour and botanical knowledge is somewhat at odds with Gunn's reputation as a respected and helpful local botanist. Joseph Hooker, for example, had nothing but the highest praise:

> There are few Tasmanian plants that Mr Gunn has not seen alive, noted their habits in a living state, and collected large suites of specimens with singular tact and judgement. These have all been transmitted to England in perfect preservation, and are accompanied with notes that display important characters in the physiognomy of plants, such as few experienced botanists possess.[100]

If Stuart's derisive remarks about Gunn's botanical knowledge had any truth to them, Gunn's behaviour may have arisen from a conviction that Stuart was of inferior rank, therefore of no account and not worthy of the kind of honourable treatment afforded to Hooker. Furthermore, it would be interesting to know whether Gunn gave Stuart due recognition for the specimens he collected or whether he, Gunn, passed them on to the Hookers as a product of his own collecting. Whatever the situation, Stuart's admiration for Mueller seems to have been due to some large extent to the respectful way in which he treated his collector's botanical contributions and his readiness to help him in times of need.

In return for the plants he sent to Mueller, Stuart, like Oldfield, said he was satisfied with just receiving Mueller's determinations. Stuart also adopted a similar system to Oldfield's numbering of the specimens he sent while retaining duplicates bearing the same numbers, so that he could set his own collection in order or in case further information about the specimens was required.[101]

Apart from poverty and all that that entailed, and the dishonesty of a few customers, Stuart's letters have revealed some of the same disappointments and frustrations which Oldfield experienced due to geographical isolation, irregular and uncertain postal services, the length of time it took to receive replies to their letters from overseas correspondents and the lack of ready access to taxonomic 'authorities'.[102] Other difficulties experienced by Stuart probably also were faced by Oldfield at times—being unable, because of work commitments, to collect when plants were in a state that facilitated identification, severe weather conditions causing plants to be damaged to an extent that they were not worth collecting, flooded, unfordable streams preventing long excursions, and difficulties in finding accommodation with 'conveniences for preserving specimens'.[103] Shortly before 8 January 1849, a collection, comprising mostly cryptogams, which Stuart had carefully gathered, prepared and packed, was on its way to Mueller aboard the *Henry* when the consignment 'was lost with the vessel'.[104] There is no evidence that any of Oldfield's collections suffered a similar fate.

---

[97] ibid., p. 107.
[98] Stuart, Charles, MELISR electronic database download, 3 September 2007, NHV, RBGM.
[99] ibid., p. 132.
[100] JDH 1860, p. cxxv.
[101] C. Daley, 1935, pp. 108–110.
[102] ibid., pp. 133 & 136.
[103] ibid., pp. 107–108.
[104] ibid., p. 108.

A cloud hung over Stuart's reputation but the historical record appears to be silent as to why. George Bentham had received information about Stuart's background that caused Bentham to suggest to Mueller in 1862, a year before the publication of the first volume of *Flora Australiensis,* that 'precedents of [Stuart's] are not favorable to the frequent & full quotation of [his name]'. Mueller, who had known Stuart for more than ten years before he learnt 'the unfavorable episodes of his history', judged him fairly but was unforgiving:

> Altho' [he has] acted always *honorably & even generously* to me, more so than I could have expected from the monetary support which [he] received...Dr Milligan can tell you something about Stuart.... Had I known [his] former history when I *first* entered into communication with [him] [his] name...would less prominently have appeared in my writing. To get out of the dilemma I abbreviated [his] name However we must be good & charitable to our fellow beings & rather elevate th[a]n depress them. [He] has acted energetically & disinterestedly for me on many occasions. But [he is] aged [60 years] & it is very doubtful whether [he] will ever see in the distant parts in which [he lives], a copy of your work & should they feel overlooked, [he] can only blame [himself].[105]

According to Home *et al.*, despite Mueller's expectation that he would not see a copy of the *Flora Australiensis*, he was on his distribution list for the first volume and Bentham did not follow Mueller's advice but gave Stuart's name in full throughout the seven volumes of the work.[106]

Botanical activity southward from the Huon River prior to Oldfield's visits 1845–1856

Map 7 p. 150 – Southern Tasmania

Oldfield made two expeditions southward on the east coast from the Huon River between 1855 and 1857. Others had explored the flora of the area since 1845.

Dr Joseph Milligan, who at the time was Secretary of the Royal Society of Van Diemens Land, collected at Recherche Bay in 1848, and in other areas along the east and west coasts of Tasmania. He was an acquaintance of James Backhouse in the 1830s, and later a member of Gunn's collecting circle (1840s and 1850s). He presented a collection of flowering plants to the Tasmanian Museum in 1852, before leaving Tasmania in 1860. Many of his collections were incorporated in Gunn's herbarium, and thus are now found in major herbaria in Britain, France, USA and Australia.[107]

Charles Stuart had also botanised to the south of Huon prior to his joint expeditions with Oldfield. The dates on the labels of Stuart's specimens in the Nation Herbarium of Victoria show that he had been in Southport and Recherche Bay, and their environs, for varying periods every year from 1850 until 1856 except for 1854.[108] In February 1853, and during February and April 1856, he collected at the base of Mount La Perouse and in the neighbourhood of the mountain but no evidence has emerged that he ascended further up the mountain on any of these occasions. According to J. H. Maiden:

> This most meritorious botanist [Stuart] collected sea-weeds largely at Southport for Harvey (Phycologia Australica). Harvey figures Areschougia Stuartii, Harv., named in his honour.[109]

The majority of Stuart's Tasmanian specimens appear to be in the National Herbarium of Victoria. There are some in Kew Gardens and other herbaria in Britain, Russia, Austria and USA.[110]

---

[105] FM to GB, 24 September 1862, Kew Gardens correspondence, Australia, Mueller, 1858–70, fols 72–79, RBGKA cited in R.W. Home *et al.* (eds), 2002, vol. 2, p. 162. In the same paragraph Mueller also responded to a similar query from Bentham regarding George Maxwell, another of Oldfield's co-collectors. The editorial changes in square brackets were made so that the quotation referred to Stuart only.

[106] ibid., fn. 18, p. 162.

[107] A.E. Orchard, 1999, 'Milligan, Joseph (1807–1884)', *ADBO,* viewed 16 March 2010 at www.anbg.gov.au.

[108] MELISR, 3 September 2007, NHV, RBGM.

[109] J.H. Maiden, 1909a, pp. 28–29.

It seems, from Stuart's National Herbarium of Victoria specimen records, that when Oldfield first joined him, Stuart had been in the Southport/Recherche Bay area since November 1855.[111]

Southport and Recherche Bay expedition December 1855 to January 1856

Map 7 p. 150 – Southern Tasmania

Oldfield's time constraint of 'ten days' for the 1855–1856 trip with Stuart appears to be at odds with his earlier comment that during the Christmas vacations he was able to undertake journeys of two to three weeks. Some of the private school advertisements in local papers show that their school year commencement dates varied between 7 January and 22 January[112] but in that regard government schools were probably not flexible. If he left for the south soon after Christmas Eve 1855 when the school vacation commenced and it was to open three weeks later, he would have needed to be back in Kingston by around 13 January 1856. Whatever his reason for planning such a short trip, it was obviously not sufficiently compelling to prevent him from contemplating staying away longer if he thought it was warranted.

Oldfield had told William Hooker that, even when his Christmas holidays allowed him to make an excursion of two or three weeks, he could not 'do very much' since he was 'obliged to carry provisions for that period'[113] but why he did not say. Short supply of horses or lack of funds to cover hiring fees may have been contributing factors. However, on this expedition difficulties associated with transporting a horse by sea, the rough, heavily forested terrain, the steep, rocky mountain slopes and lack of suitable feed could have been significant reasons.

There was no road through the dense, wet eucalypt forest and woodland that connected the settlements at the Huon with the southern settlements, most, if not all of which at the time were located along the coast.[114] Given his time constraints and the difficulty of overland travel he almost certainly undertook the journey by sea. It is not known where on the coast he joined Stuart. There are six of Stuart's specimens in the NHV which were collected between 1 and 7 January 1856 at Southport so it may have been there that they first made contact.[115]

By this time on-shore whaling had ceased in the district and the convict probation station at Southport, established in 1841, was closed in mid-1848 as part of the process leading to the dismantling of the probation system in 1853.[116] The Van Diemens Land Company's coal mine at Recherche Bay on the Catamaran River had been abandoned in the late 1840s, mainly due to a decline in the quality of the coal.[117] Farming had been tried but with little success.[118] Nevertheless, from 1851 to 1853 the populations at Recherche Bay and Southport grew from 31 and 51, to 500 and 250 respectively due to the booming timber industry.[119] Since the early 1850s there had been strong demand for timber, brought about by the Victorian goldfields' insatiable appetite for construction and by the local ship-building industry. Timber leases had become available and in 1854 the area's first mechanised steam-powered sawmill was established at the Waterhole, near Ramsgate. Other mill towns south of the Huon started to emerge.

---

[110] Alex S. George, 2009, pp. 565–566
[111] MELISR database, 3 September 2007, NHV, RBGM.
[112] See for example, *Hobarton Mercury*, 5 January 1856, p. 4, and 7 January 1856, p. 4, DAN, NLA.
[113] AO to WJH, 20 August 1855, Australian Letters, 1851–1858, vol. 74, fol. 190, RBGKA.
[114] Bruce Poulson, 'The far south', in Alison Alexander (ed), 2005a, p. 129.
[115] MEL 2149857a, 1 January 1856; MEL 54540, 2 January 1856; MEL 54473A, 2 January 1856; MEL 2262121B, 4 January 1856; MEL 221493A, 4 January 1856; MEL 2092641A, 7 January 1856, Southport, MELISR database, 3 September 2007, NHV, RBGM.
[116] Richie Woolley and Wayne Smith, 2004, p. 95.
[117] ibid., pp. 163–168.
[118] ibid., pp. 58.
[119] ibid., pp. 140–142.

Fig. 10.6 Southport Narrows, Tasmania

By 1853, Ramsgate [now Cockle Creek] and Southport were flourishing with shops and other services catering for residents' needs.[120] There was a post office and a constabulary at Southport, which also served the Recherche Bay area. In 1853, four hundred men and women were rumoured to be living in 'rustic squalor' around Recherche Bay. Two public houses had been licensed and erected which

> soon earned a special notoriety with the regional constabulary at Southport. One local resident bemoaned the insults, obscene language, the exposing of persons day and night, with Saturday and Sunday all alike, and the landlord backing men to fight.[121]

The timber industry was still booming in 1856.

While unruly behaviour was characteristic of many people in Recherche Bay, there were compensations for Oldfield and Stuart. Food and other supplies could be obtained locally and possibly even reasonably comfortable accommodation in one of the bark huts or hotels in the area. However, they may have chosen the cheaper option of camping under canvas close to where they intended to collect.

The flora of the poorly-drained, peaty soils around Southport supported low, wet heathland vegetation with scattered emergent Swamp Peppermint (*Eucalyptus rodwayi*) or Swamp Gum (*E. ovata*) trees as well as extensive Button Grass (Tussock Sedge, *Gymnoschoenus sphaerocephalus*) moorland on poorly-drained flats and gentle slopes, in which there is a dense cover of Cyperaceae and Restionaceae [122] We found records of two species[123], Dwarf Wire-lily (Laxmannia orientalis) and Tiny Flannel-flower (*Actinotus bellidioides*), which indicate that he probably explored the wet heathland and buttongrass moorlands in this area.[124]

---

[120] ibid., pp. 142.

[121] Parry Kostoglou, 1993, 'Historic timber-getting between Cockle Creek and Lune River, Block 1', *Archaeology of the Tasmanian Timber Industry*, Report No. 4, mimeo., Tasmanian Forest Research Council and Forestry Commission, Tasmania, p. 10.

[122] Jayne Balmer to MHH, October 2008, and S. Harris & A. Kitchener, 2005, pp.128, 129 & 113, 114.

[123] For example, MEL 51875, *Laxmannia orientalis*, Southport; MEL 294455, *Actinotus bellidioides*, Southport, MELISR database, 18 November 2005, NHV, RBGM.

[124] Jayne Balmer to MHH, October 2008

Neither Oldfield nor Stuart left behind a journal of their expedition. Apart from one Oldfield specimen from Mount La Perouse dated 1856[125], the absence of collection dates on records of specimens gathered at and near Port Esperance, Southport and Recherche Bay, precludes any indication of which ones he obtained on his first and second visits.[126] Furthermore, there are no dated joint collections for the 1855–1856 expedition to confirm that they worked together. However, a few of Oldfield's Southport and Recherche Bay specimens are cited in the main body of Joseph Hooker's *Flora Tasmaniae*[127], whereas his Mount La Perouse specimens were referred to only in Additions and Corrections to Volume 1 of that publication. This suggests that Hooker had received some Southport and Recherche Bay material prior to those from Oldfield's later expedition.[128] The earlier gatherings were probably in the case of natural history specimens that Oldfield sent to London aboard the *Antipodes* on 26 January 1856.[129] Even allowing for Oldfield's return to Kingston prior to 13 January 1856, it would be surprising that the few specimen records of his that we have located from the coastal areas south of the Huon[130] are anywhere near the actual number he obtained.

At NHV there are 158 specimens collected by Stuart during January 1856; they were all gathered near Southport and Recherche Bay, most being from Southport.[131] This might reflect a situation where they focused most of their energies on that area but it is possible that many of Stuart's January 1856 gatherings from there were made after Oldfield returned to Kingston.

The beauty and diversity of the flora around Recherche Bay and the way it contrasted with the buttongrass moorland and wet heathland of the Southport district may have impressed Oldfield. The vegetation is mainly Stringy Bark or Western Peppermint wet forest with an understorey of Dogwood (*Pomaderris apetala*) and scrub species such as the Tea-tree Manuka (*Leptospermum scoparium*) and *Acacia* spp. Also occurring are patches of pure rainforest with differing floral associations; Myrtle Beech (*Nothofagus cunninghamii*) and Sassafras (*Atherosperma moschatum*) rainforest, ferns representing the greatest species diversity of vascular plants, Myrtle Beech–Celery-top Pine (*Phyllocladus aspleniifolius*) short rainforest with a tangled understorey of many species; and old-growth Stringy Bark or Western Peppermint forest over a mature rainforest understory. Oldfield probably obtained the crimson-flowered shrub Climbing Heath (*Prionotes cerinthoides*)[132] in one of these rainforest patches. Scrub communities, dominated by various species of *Leptospermum* and *Acacia* are common. On the poorly drained flats there are extensive wet heath and sedgelands, which very likely yielded his specimen of the Pink Swamp Heath (*Sprengelia incarnata*)[133] with the mat herb Matted Brooklime (*Gratiola nana*)[134] gathered on the wetlands along one of the creeks or the coast.[135]

Although we have not located any of Oldfield's plant collections from Lady Bay, a small settlement just to the north of Southport, he was there, but when and why he did not say. Perhaps his ship called there on one of his trips to collect with Stuart but since it was within easy walking distance of Southport he and Stuart may have gone there together. If they did, we have found no evidence to support such a suggestion. Lady Bay had previously been an onshore whaling station but by 1855 timber had become its main industry. Oldfield referred to his visit in a discussion of theories relating to the shaping of rock formations, which, he said, placed too much emphasis on glacial action without taking into consideration the possible action of wind and water and of other factors:

---

[125] MEL 10081, *Luzula australasica*, Mt La Perouse, AO, MELISR database, 8 March 2007, NHV, RBGM.
[126] HAODb.
[127] See for example, *Sematophyllum homomallum*, Southport; *Todea barbara*, Recherche Bay, JDH 1860, vol. 2, pp. 213 and 153 respectively.
[128] See for example, *Dracophyllum minimum* and *Euphrasia hookeri*, Mount La Perouse, Additions, corrections etc. (ibid., J.D. vol. 1, pp. 367 & 369 respectively).
[129] 'Shipping News', *The Courier* (Hobart), 26 January 1856, p. 2, DAN, NLA.
[130] HAODb.
[131] ibid.
[132] MEL 2187259, *Prionotes cerinthoides*, Recherche Bay, MELISR, 5 November 2005, NHV, RBGM.
[133] MEL 285409, *Sprengelia incarnata*, Recherche Bay, ibid.
[134] MEL 38740, *Gratiola nana*, plains behind Recherche Bay, ibid.
[135] Jayne Balmer to MHH, October 2008.

> It is not uncommon to meet with localities in the totally unsettled districts of Australia, that have all the appearance of having been formerly under cultivation, large areas being accurately divided into lands, and in some cases there exist what appear to have been diagonal furrows for the purpose of drainage. These 'ploughed lands'—as they are called by the colonials—are generally situated on hillsides, so that most probably, meteorological conditions have been efficient in their formation. In two particular instances however, such considerations are inadequate to explain the phenomena, for, in the first place, the localities were perfectly level, and in the second case, prior to my observation of them, they were occupied by a dense vegetation many centuries old, a fact proclaimed by the size of the plants there growing, as compared with the magnitude attained by the same forms in other seemingly undisturbed localities...One of the localities referred to near Lady Bay, South Huon, Tasmania had quite the appearance of having once been used as a garden, there being indications as of walks, beds, and even of lines of drainage. Indeed, so complete was the illusion, that had I not myself assisted in the burning the rank vegetation with which the place was covered, I should certainly have considered such indications as proof positive that that locality had been recently used as a garden.[136]

Oldfield and Stuart's evenings were probably taken up with pressing and labelling plants but there would have been ample opportunity during this visit for the two collectors to become better acquainted and to share botanical knowledge. At last Stuart had someone with whom he could speak on the subject in which he took 'greatest pleasure', and no doubt Oldfield was an eager participant in any discussions that offered the possibility of deepening his understanding of Tasmanian flora. It is not difficult to imagine that from time to time they commiserated at length over the limitations on their botanical activity imposed by poverty, unreliable postal services and environmental factors.

Oldfield apparently returned to Kingston in time to open the school at the start of the first term on or around 13 January 1856. Stuart, however, remained in the Southport/Recherche Bay area until at least April of that year.[137]

Oldfield's participation in the life of the Kingston community

While Oldfield was in Kingston during 1856, he involved himself in local affairs in positions for which his educational background no doubt fitted him well and which reflected the community's trust. He was Secretary of the group that organised the Kingston Regatta held on Friday, 1st February 1856. This event comprised six races for the following categories of craft: dinghies pulled by two youths under 16 years of age, and a steerer; unmarried native youths pulling in four-oared whale boats not exceeding 25 feet in length; five-oared whale boats; sailing boats under three tons; and barges carrying leeboards or centreboards, working on the Derwent. There was also a duck hunt competition. Entry fees were £1 except for the first race which was 10s and the first prizes ranged from £4 to £12.[138] Reflecting on the importance of sailing in 19th century Tasmania, regattas became popular events in the recreational calendars of many coastal metropolitan and regional centres, with the first held at Hobart in 1831.[139]

Oldfield was also Secretary of the Loyal Tasmanian Youth's Lodge, a sub-group of the Kingston Lodge of the Independent Order of Oddfellows (IOOF), from at least January to March 1856. IOOF was founded in Baltimore, Maryland, in 1819 and spread to many countries in the Americas, Europe,

---

[136] AO, c.1865a, fn. 53, p. 792.
[137] MELISR, dated 3 September 2007, NHV, RBGM.
[138] *The Courier* (Hobart), 30 January 1856, p. 3, DAN, NLA.
[139] Wendy Rimon, 2005, 'Regattas' in Alison Alexander (ed), 2005a, p. 305.

and Australasia where it still operates today. It is an altruistic secret fraternity derived from similar English Oddfellows service organisations. It came into being during the 1700s, when it was deemed odd to find people organised to give aid to those in need without recognition, and pursuing projects for the benefits of all human kind. Members contributed some of their hard-earned wages to a common fund, which they could use for unfortunate times such as sickness, losing a job and even death.[140] IOOF began in Australia in Sydney in 1836 and by the late 1850s there were nine lodges in Tasmania, mainly in and around Hobart.[141] When Oldfield relinquished this office is not known but involvement in the organisation indicates a concern for his own welfare and that of others and, furthermore, that he was held in sufficiently high esteem by his fellow members to have been elected Hon. Secretary.

A requisition, dated 14 July 1856, was published in the *Hobart Town Courier* of 21 July 1856 from 80 Electors of the District of Kingborough (of which Kingston was a part), offering Richard Cleburne, MLC, their sincere thanks for his past exertions as their representative in the present Legislative Council and promising their support if he would consent to become a Candidate to represent them in the House of Assembly under the New Constitution. Oldfield was one of the signatories to this requisition.[142] Such requisitions were required under the *1832 Reform Act* (England) as evidence that only candidates of the highest character and respectability offered themselves to the Electors. It appears that under that Act

> the most respectable Persons composing the communities to which the right of voting is now extended [should form committees of individuals entertaining similar views and opinions regarding the Policy of the Country; and should make all the necessary arrangements for inviting such persons to become their [sic] Representatives as may appear most acceptable to the Constituency.[143]

Cleburne was accepted as a candidate but was not elected.[144]

From the foregoing, Oldfield was entitled to vote as he was a householder paying rent of at least £10 per annum; furthermore, that he ranked among the most respectable people in the District of Kingborough.

Browns River school closed

The Browns River school closed in 1856 because the government grant, the source of Oldfield's salary, was discontinued. The exact date is not known but his subsequent movements indicate that it was probably before the end of June. This closure came about because Mr H.C. Cotton, the Inspector of Schools, Southern Board of Education, 'on his first inspection' found that only six pupils were attending.[145] According to Julie Gardam, the average attendance in 1856 was 'a healthy 21' but she provided no evidence to substantiate her claim.[146] The cost of tuition to parents may have been an important contributing factor; despite the impression left by Mundy of the district's apparent prosperity[147], poverty may have been the experience of many small farmers and loggers. Furthermore, if the Inspector called during harvest or some other time when farmers needed the assistance of their children in the fields, school attendance would have fallen.

---

140 Announcements, 'The second anniversary dinner of the Loyal Tasmanian Youth's Lodge, IOOF, Kingston', signed by Oldfield as Secretary, *The Courier* (Hobart), 26 January 1856, p. 1, 3 March 1856, p. 1, 6 March 1856, p. 1 and *Hobarton Mercury*, 7 March 1856, p. 1., DAN, NLA.
141 Geoffrey Blainey, 1991, *Odd Fellows: A History of IOOF Australia*, Allen & Unwin, Sydney, p. 1 & p. 20.
142 This requisition was reprinted in *The Courier* (Hobart), 29 July 1856, DAN, NLA.
143 J.I. Brash (ed), 1974, *Papers on Scottish Electoral Politics 1832–1854*, Scottish History Society, Edinburgh, p. 1.
144 *Hobarton Mercury*, 9 January 1857, DAN, NLA.
145 Reports of the Inspector of Schools, *Hobart Town Gazette*, no. 1202, 1857, p. 656, DAN, NLA. The date of closure of the school was not noted in this report. However, the Browns River school was included in the list of Tasmanian public schools in the *Tasmanian Almanack for 1855*, p. 91, but not in the list of schools in the *Tasmanian Almanack* for 1856.
146 Julie Gardam, c.1988, p. 16.
147 G.C. Mundy, 1852, facsimile 1986, p. 35.

Oldfield left no comment as to how he reacted to this turn of events, nor have we located any record which suggested that he was in any way responsible for the poor attendance. Any concerns that Oldfield might have had about his loss of employment and income must have been short lived because he soon took off to Western Australia. Since it is not known exactly when the school was closed, there remains the unanswered question of how long he remained in Tasmania and what he did in the interim. Nevertheless, his signature on the above published requisition suggests that he did not leave Tasmania until 14 July 1856 or shortly after. This journey seems to have been at his own expense.

The Loop, Murchison River, Western Australia

Part 3

Consolidating a sound reputation

1856

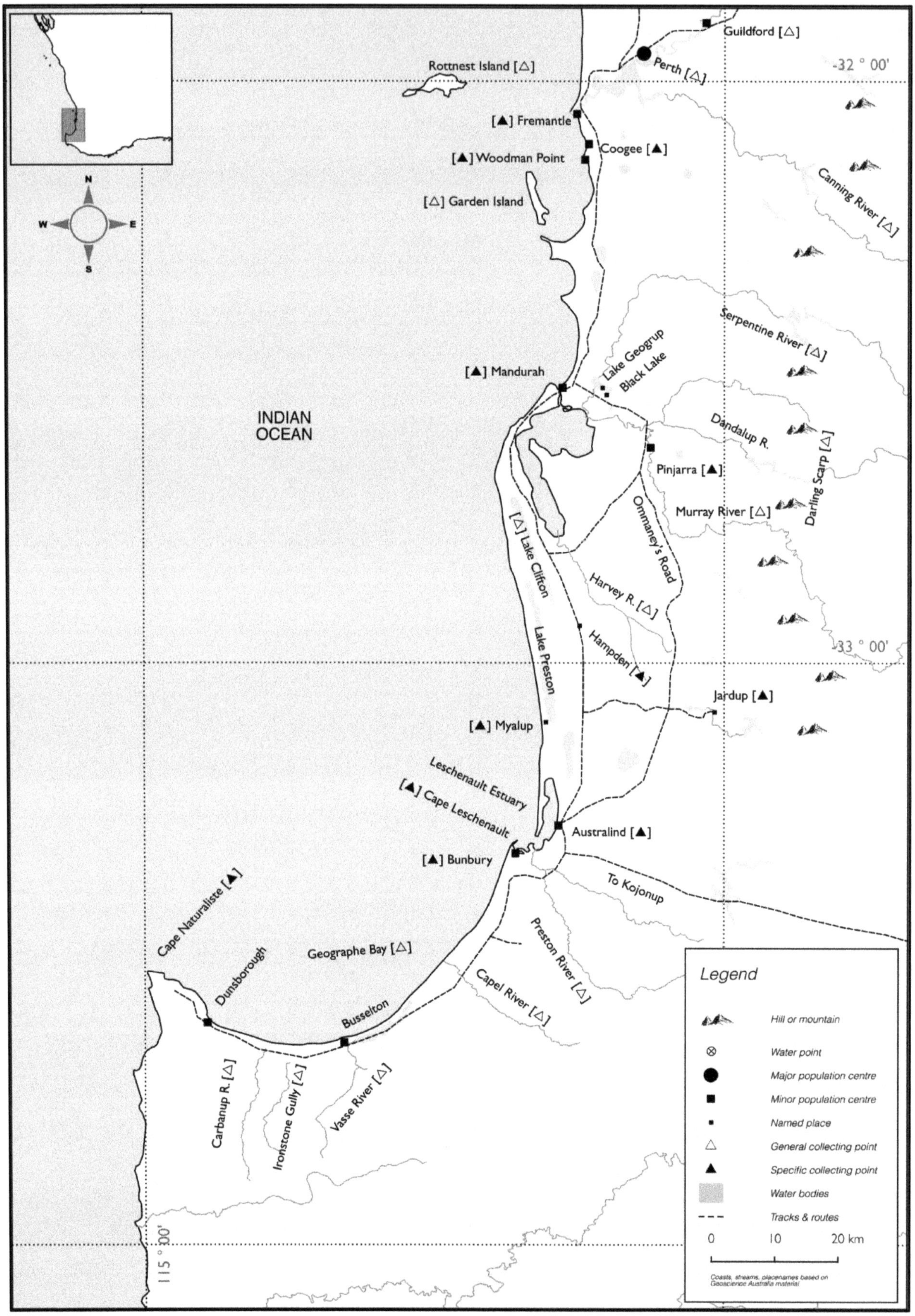

Map 8: Western Australia, Cape Naturaliste to Perth

Fig. 11.0: Isolepis oldfieldii

Chapter 11

Western Australia 1856

Cape Naturaliste to Swan River

Oldfield left scant details of his first visit to Western Australia in 1856 but there is no doubt that he was there at that time. He collected a specimen of *Eucalyptus caligata* [ms] at the Murchison River, which is dated September 1856.[1] Mueller, in the first volume of his *Fragmenta* (1858–1859), comprising ten fascicles separately published between March 1858 and December 1859[2], referred to specimens of 49 Western Australian species new to science that were collected by Oldfield at various localities from Cape Naturaliste[3] to north of the Murchison River, embracing the Swan Coastal Plain and the Geraldton Sandplain.[4] Oldfield must have gathered these during 1856. Thus, Mueller had not received any plants from Oldfield's 1858–1859 expedition to Western Australia by 15 June 1859[5] and in fascicle six of the *Fragmenta*, published or sent to press on 1 June 1859[6], there is the original description of *Vitis angustissima* based on a specimen collected by Oldfield at the Murchison River.[7] The last fascicle was published or sent to the printers on 2 December 1859 when Oldfield was still on his way back to Melbourne from his 1858–1859 Murchison River expedition.[8] A search of newspaper reports of freight imported into Hobsons Bay for the period July to December of that year revealed no evidence that any cases were shipped to Mueller from Western Australia during that time. It is, therefore, likely that all Oldfield's 1858–59 Western Australian collections accompanied him back to Melbourne as part of his luggage. Furthermore, he referred to the death, in 1857, of Carrotty, the Murchison River Aboriginal, to whom he attributed 'a large portion' of the *Watchandie* vocabulary listed at the end of his paper on Australian Aborigines.[9]

Oldfield's specimen records, particularly those referred to in Mueller's sixth fascicle, are the only indication that he went overland from Geographe Bay to the Murchison River in 1856, with a brief visit to Cape Naturaliste. What he hoped to achieve and to what extent, if any, Mueller influenced his movements is not known. The diversity and number of species in the Colony, as revealed in earlier Western Australian botanical publications, must have been a huge attraction, especially in the areas north of Champion Bay, which up to that the time had received little attention from botanists.[10] Since Mueller seems to have been an important recipient of specimens from this expedition, perhaps he had some involvement in planning the itinerary. On the other hand, Oldfield, through his own reading,

[1] NSW 154194, *Eucalyptus caligata* [ms], Murchison River, September 1856, AO, NHNSW. This species does not appear in either *FloraBase*, or in the *Australian Plant Name Index*. A Google search found it included as a yet undescribed species 'in ms' on K.D. Hill, 2004, 'The Eucalypts***Icons of the Australian Bush. Eucalink A Web Guide to the Eucalypts', viewed 8 May 2009 at www.plantnet.rbgsyd.nsw.gov.au.

[2] R.W. Home *et al.* (eds), 1998, vol. 1, pp. 577–580.

[3] One was from Cape Naturaliste, *Warburtonia potentillina* which is a synonym of *Hibbertia grossulariifolia*, APNI. The other, *Patersonia umbrosa* var. *xanthina*, was found '[i]n a marshy place near Geographe Bay', F. M., 1859–1859, *Fragmenta*, vol. 1, pp. 214 & 230. The flowering time of this species, according to *FloraBase*, is 'August to December' which, although flowering times of species vary, might lend some support to a claim that he was in the area in August.

[4] FM 1858–59, vol. 1.

[5] FM to WJH, 15 June 1859, Directors' letters, vol. 75, Australian and Pacific letters 1859–65, fol. no. 114, RBGKA cited in R.W. Home *et al.* (eds*)*, 1998, vol. 1, p. 465.

[6] ibid., p. 579.

[7] FM, 1858–59, vol. 1, p. 141.

[8] 'Shipping intelligence'. *The Argus*, 14 December 1859, p. 4.

[9] AO 1865b, p. 247.

[10] See Chapter 12.

may have concluded that his chances of finding species new to science in Western Australia would be very high. As the expedition appears to have been undertaken at his own expense, he should have felt free to go where his inclination took him regardless of any advice Mueller might have given about where to collect. Nevertheless, he may have been driven to impress Mueller by following his counsel in the hope of being taken into his employ as a full-time collector.

Oldfield apparently left Tasmania for Western Australia on, or shortly after, 14 July 1856, but it is not known by what vessel or exactly when he travelled to the Colony. Undertaken during the winter, his voyage around the south coast may have been even rougher and more 'dolorous' than when the Congregational minister John Legge Poore made the journey in May 1861 aboard the steamship *Balclutha*. He encountered

> very tempestuous weather, decidedly of the pitch, toss and tumble kind, so that comfort by day, and repose by night were things desired, but hardly to be expected. The great Australian Bight may be described, like Loch Goyle, as 'a dark and stormy water'. The strongest and most persistent winds blow from the south-west, and cause a long swell, which is at all times experienced. During the winter months weather is generally unsettled and tempestuous; gales of wind with squalls, shifting rapidly from north and north-west to south and south-west are frequent.[11]

Oldfield's specimen records suggest that he made landfall in Western Australia first at Geographe Bay sometime in early August 1856, at the town of Busselton on the Vasse River.[12] There was no jetty and as the ship dropped anchor out in the bay, he probably caught sight of 'The Tub', a beacon which indicated the Port. It was a 30-foot (9 metres) spar, topped with a large barrel containing a lamp, located near the corner of Marine Terrace and the northern end of Queen Street.[13] Other ships may have been moored nearby loading or unloading cargo to or from lighters. Agricultural products and timber, since 1850, were exported[14], and probably also whale oil, to the eastern colonies and the Swan River, and supplies for the settlers were imported from Fremantle and other Australian ports. Also, in the Bay there may have been American whaling and sealing vessels whose crews were busy processing their harvest of blubber on-shore, taking in fresh supplies of water and bartering with farmers for such commodities as potatoes, butter, cheese, meat, vegetables and wood in exchange for oil, molasses, tobacco, spirits, slush soap and many other items.[15]

For reasons that will be revealed in subsequent chapters, all Oldfield's specimens gathered from Cape Naturaliste to Coogee, just to the south of Fremantle, appear to have been obtained in August 1856.

---

[11] J.L. Poore, letter to the Editor, *The Perth Gazette and Independent Journal of Politics and News*, 6 September 1861, pp. 3–4.
[12] HAODb.
[13] Busselton Historical Society, 2007, *Historic Busselton*, 5th edn., Busselton Historical Society, p. 21.
[14] *Inquirer and Commercial News*, 14 May 1856, pp. 1 & 2 and Busselton Jetty Environment and Conservation Association, 'Busselton Jetty Timeline', Busselton Jetty, viewed 30 January 2013 at www.d1054021.myweb.westnethosting.com.au.
[15] Theodora Sanders, 1975, *Bunbury: Some Early History*, Roebuck Society Publication no. 16, Canberra, p. 70.

European interest in the flora of the Swan Coastal Plain prior to 1856

Map 8 p. 176 – Western Australia, Cape Naturaliste to Perth

Lying within Muir and Bibbulmun Districts of the Bibbulmun Province of the Southwest Floristic Region (SWAFR),[16,17]the flora of the Swan Coastal Plain was by no means unknown at the time of Oldfield's visit, particularly from Cape Naturaliste north to the Swan River. This plain of Quaternary deposits is essentially low-lying and extends from Cape Naturaliste in the south to near Jurien Bay north of Perth.[18] South of Perth it forms a low-lying belt, about 25 to 30 kilometres wide, bordered to the west by the shores of the Indian Ocean and associated dunes and to the east by the Darling Scarp (often referred to as the Darling Range) and the Precambrian Yilgarn Block. It is transected by rivers flowing west from the Darling Plateau interspersed by wetlands, and includes large estuarine systems and a number of lakes cut off from the sea by barrier dunes.[19] In summer there is little fresh surface water but dense undergrowth covers much of the landscape all year. In winter, heavy rains over a few short months fill the low areas and the rivers spread out across the sand plain in a maze of swamps.[20] Mainly *Banksia*, including Slender Banksia (*B. attenuata*) and Firewood Banksia (*B. menziesii*), occur in low woodland on leached sands with *Melaleuca* swamps, marked by Saltwater Paperbark (*M. cuticularis*) and Swamp Paperbark (*M. rhaphiophylla*), Flooded Gum (*Eucalyptus rudis*) and species of the sedge family (Cyperaceae), where the land is ill-drained.[21] On less leached soils there is woodland of Tuart (*E. gomphocephala*), Jarrah (*E. marginata*) and Marri (*Corymbia calophylla*).[22]

Willem de Vlamingh, during his 1696–1697 expedition in search of the Dutch East India Company vessel, *Ridderschap van Holland*, was the first to have recorded gathering botanical specimens, including seeds, gum and wood samples, at, and near, the Swan River. He took these to Batavia (now Jakarta) and although they were returned to Holland no trace of them has been found.[23]

The French exploring expedition led by Baudin in *Géographe* and *Naturaliste* anchored toward the southern end of Geographe Bay on 30 May 1801 and remained there until 9 June 1801. During their stay, the botanist Jean Baptiste Louis Claude Leschenault de La Tour, head gardener Anselme Riédlé, the gardener's assistants Antoine Guichenot and Antoine Sautier and probably others collected inland and as far north as the Wonnerup Estuary. This was discovered by François-Antoine Boniface Heirisson, midshipman aboard the *Naturaliste*, and explored by Hamelin and the expedition's scientists who collected botanical and other natural history specimens.[24] After returning south to Cape Leeuwin, Baudin took the *Géographe* north of Shark Bay. In the meantime, the *Naturaliste* sailed

[16] Paul Gioia and Stephen D. Hopper, 2017, 'A new phytogeographic map for the Southwest Australian Floristic Region after an exceptional decade of collection and discovery', *Botanical Journal of the Linnean Society*, vol. 184, pp. 11-12 & Appendix S.3, Fig. S.3.10. This map was the result of an evaluation of the robustness of the Western Australian Herbarium Database in elucidating patterns of diversity and phytogeographic maps of the Southwest Australian Floristic Region (SWAFR) after an 11-year period of exceptional specimen acquisitions. We are grateful to Stephen Hopper for drawing our attention to this recent paper, personal communication to MHH, 1 January, 2017, HOA.

[17] The Muir District is bounded by the south and west coasts from Albany to Bunbury and inland for 50 kilometres from the south coast to about 100 kilometres from Augusta along the west coast. The boundaries of the Jarrah District are the northern border of the Muir District, the west coast from Busselton to Perth and inland to the east from Albany, to Kojonup, Williams, Northam and Mogumber, ibid., p. 12 & Fig. S3.7, Appendix S3.

[18] J.S. Beard, 1990, *Plant Life of Western Australia*, Kangaroo Press, Kenthurst, New South Wales, p.88.

[19] Miranda Mockrin, 2001, 'Swan Coastal Plain Scrub and Woodlands (AA1205)', *World Wildlife Fund Full Report*, viewed online 31 January 2009, www.worldwildlife.org.

[20] Anne Brearley, 2005, *Ernest Hodgkin's Swanland: Estuaries and Coastal Lagoons of South-western Australia,* University of Western Australia, Nedlands, p. 164.

[21] ibid., and J.S. Beard, 1990, p.87–88.

[22] ibid., p. 87.

[23] Charles Nelson, 1994, 'Nicolaas Witsen's letter of 1698 to Martin Lister about a Dutch expedition to the South Land (Western Australia): the original text and a review of its significance for the history of Australian natural history', *Archives of Natural History*, vol. 21, (2), pp. 154–160, source Alex S George, 1999, *William Dampier in New Holland: Australia's First Natural Historian*, Bloomings Books, Hawthorn, Victoria, p. 155.

[24] Jean Fornasiero, Peter Monteath and John West-Sooby, 2004, pp. 39–68, and Leslie Marchant, 1982, *France Australe: A Study of French Explorations and Attempts to Found a Penal Colony and Strategic Base in South Western Australia 1503–1826*, Artlook Books, Perth, pp. 123–150.

north and Hamelin explored Carnac and Garden Islands and the Swan River but no plants were collected.[25] Some material from the first collection, which had been sent back to Paris with Hamelin on the *Naturaliste* in 1802, was referred to in Labillardière's *Novae Hollandiae Plantarum Specimen* (1804).[26]

It was not until 1827 that there was further European interest in the flora of the Swan Coastal Plain. In that year Charles Fraser, Colonial Botanist in New South Wales, accompanied James Stirling on his survey of the Swan River and supported Stirling's recommendation that it was a suitable place for settlement.[27] Three years later Fraser published his account of the collections he made on this voyage, which embraced the banks of the Swan River, Buache (Garden) Island, Geographe Bay and Cape Naturaliste in *Botanical Miscellany*.[28] Two sets of Fraser's specimens were subsequently examined and written up by Robert Brown.[29] According to Stephen Hopper, Brown's account published in 1832[30] 'was much better informed than the field identifications proffered by Fraser'.[31]

Many of the Western Australian Proteaceae species were introduced to Kew through William Morrison, a gardener, who came to the Swan River Colony in 1829 with Stirling and the first settlers. Morrison collected seeds and plants in the Colony and forwarded them to England for sale.[32] Following the establishment of the Swan River Colony in 1829, botanical interest in the flora of the Swan Coastal Plain to the north and south of Perth intensified.

On 30 July 1829 James Drummond, Honorary Government Naturalist and Curator of the Government Gardens, and nine others, accompanied Captain Mark J. Currie's four-day expedition south from Mangles Bay to a point approximately in line with Cape Bouvard, returning to the Swan River by boat.[33] They were the first Europeans to enter the Peel Inlet. They named the Serpentine River and Drummond probably gathered plants.[34] He relinquished his government appointments in 1835 and embarked on a career as a commercial collector. He went to live on his Helena Valley grant but changed this in 1836 for 'Hawthornden', a location of 2900 acres (1174 hectares), in the Toodyay valley.[35]

Dr Alexander Collie accompanied, as naturalist, Lieutenant Preston on an expedition from Cockburn Sound to Geographe Bay (8–19 September 1829) and collected plants for the Royal Gardens Kew.[36] Surveyor General John Septimus Roe explored in the vicinity of Cape Naturaliste and Port Leschenault (Bunbury) during the 1830s and, as was his wont when on his various expeditions, gathered plant specimens along the way.[37] The Western Australian Herbarium has a small collection of Roe's plants but all are without a specific locality.[38] Following his appointment in 1833 as Colonial Surgeon at the Swan River, Dr Collie collected around the Swan River area specifically for the Linnean Society and for the Haslar Royal Navy Hospital, Gosport, England and other herbaria nominated by

---

[25] Stephen D. Hopper, 2004, 'South-western Australia, Cinderella of the World's Temperate Floristic Regions 2', *Curtis's Botanical Magazine*, vol. 21 (2), p. 135.

[26] ibid., p. 136.

[27] A.E. Orchard, 'Fraser, Charles (sometimes 'Frazer') (1788–1831)', ANHBiog, viewed 25 January 2013, www.anbg.gov.au

[28] C. Fraser, 1830, 'Remarks on the botany, &c of the banks of the Swan River, Isle of Buache, Baie Geographe, and Cape Naturaliste', *Botanical Miscellany*, vol. 1, pp. 221–236.

[29] Stephen D. Hopper, 2004, p. 141.

[30] R. Brown, 1832, 'General view of the botany of the vicinity of Swan River', *Journal of the Royal Geographical Society*, vol. 1, pp. 17–21.

[31] Stephen D. Hopper, 2004, p. 144.

[32] J.H. Maiden, 1909b, p. 21.

[33] Mark J. Currie, 3 August 1829, 'Report of an excursion on the Main', 29–30 July 1829, in Joanne Shoobert (ed), 2005, *Western Australian Exploration,* vol. 1 *December 1826–December 1835: The Letters, Reports & Journals of Exploration and Discovery in Western Australia*, Hesperian Press, Carlisle, Western Australia, pp. 76–79.

[34] Anne Brearley, 2005, p. 163.

[35] Rica Erickson, 1966, 'Drummond, James (c. 1787–1863)', ADBO, viewed 13 June 2010, www.adb.online.anu.edu.au. Erickson has, perhaps, provided the most detailed account of Drummond's life and botanical collecting activities in her, 1969, *The Drummonds of Hawthornden*, Lamb Paterson, Osborne Park, Western Australia.

[36] A.E. Orchard, 'Collie, Alexander (1793–1835)', ANHBiog, viewed 8 May 2008, www.anbg.gov.au.

[37] A.E. Orchard, 'Roe, John S. (1797–1878)', ANHBiog, viewed 8 May 2008. www.anbg.gov.au, and R.L. Heberle, 'John Septimus Roe 1797–1878 Surveyor Explorer and Botanical Collector, 1786-1867', *History of Orchid Collecting in South Western Australia, 1791–1971*, The Species Orchid Society of Western Australia Inc., viewed 8 May 2008, www.members.iinet.net.au

[38] *FloraBase*.

Sir William Burnett who was professional head of the Royal Naval Medical Department.[39] Collie's main collection is now in the Royal Botanic Gardens, Kew, with duplicates at the herbarium of the Missouri Botanical Garden, St Louis.[40] The Western Australian Herbarium has a small collection of Collie's plants, including some from the Swan River.[41]

When Captain James Mangles visited the Colony in 1831 at the invitation of his cousin Ellen, Lady Stirling, he made extensive collections of seeds and specimens of the flora in an area approximating today's Perth Metropolitan Region. He was assisted by James Drummond and other settlers around the Swan River district—George Fletcher Moore, T. Carroll, Captain Richard Goldsmith Meares, and a Mrs. [?] Bull. Dried specimens, particularly from Drummond, were forwarded to Dr John Lindley, Professor of Botany at the University College, London, and some found their way to European herbaria. [42]

After spending from 1830 to 1834 at the Swan River colony, military surgeon Dr William Milligan went to India where he wrote a report on the Swan River colony, including a description of the plants, soils, geology, etc., and a second paper recording the edible plants at the Swan River. Hopper has referred to his botanical contributions as significant.[43]

Baron Karl (sometimes Carl) von Hügel, an Austrian army officer, diplomat, botanist and explorer, visited the Swan River colony in 1833–1834. He made considerable plant collections, including seeds for his garden, from along the coast near Fremantle, the 'Darlington hills' (Darling Range) and at King George Sound. The publication by Stephan L Endlicher *et al.*, *Enumeratio plantarum quas in Novae Hollandiae ora austro-occidentali...* (1837)[44] contains a list of all von Hügel's Western Australian herbarium specimens. Other specimens of his from the Colony were described in Endlicher's *Stirpium australasicarum herbarii hugeliania decades tres*[45] (1838). Hügel's collections are held by Naturhistorisches Museum Wien, Vienna, with duplicates in some Australian, British and European herbaria.[46] The Western Australian Herbarium holds a few specimens (possibly duplicates) from the Swan River and Fremantle.[47]

Georgiana Molloy, the wife of Captain John Molloy, Resident Magistrate at Augusta, made extensive collections after 1839 from near her new home, 'Fairlawn' at the Vasse, which were sent to Captain James Mangles.[48]

The German naturalist Dr Johann August Ludwig Preiss visited the Swan River colony from 1838 to 1842, during which time he gathered between 3,000 and 4,000 terrestrial and marine specimens from an area extending from about 100 kilometres north of Perth, around the Swan, the Darling Range, along the York Road and various other more distant localities to the south, including Geographe Bay, and south-east of Perth.[49] This material (with a small amount by other collectors) forms the basis for one of the landmark works on Western Australian botany, Johann Georg Christian Lehmann's *Plantae Preissianae sive Enumeratio plantarum quas in Australasia occidentali et meridionali-occidentali annis 1838–1841* (1844–1848).[50]

---

[39] Gwen Chessell, 2008, *Alexander Collie: Colonial Surgeon, Naturalist and Explorer*, University of Western Australia Press, Nedlands, pp. 99, 173.
[40] A.E. Orchard, 'Collie, Alexander (1793–1835)', viewed 8 May 2008, www.anbg.gov.au.
[41] *FloraBase*.
[42] R.L. Heberle, viewed 7 May 2008, www.members.iinet.net.au.
[43] Stephen D. Hopper, 2004, p. 145.
[44] S.L. Endlicher, E. Fenzl, G. Bentham & H.W. Schott, *et al.*, 1837, *Enumeratio plantarum quas in Novae Hollandiae ora austro-occidentali...*, G. Beck, Wien.
[45] S.L. Endlicher, 1838, *Stirpium australasicarum herbarii hügeliania decades tres,* J.P. Sollinger, Wien.
[46] Sophie C. Ducker, 1990, 'Early Austrian influence on Australian botany', in P.S. Short (ed), 1990, pp. 300–301.
[47] *FloraBase*.
[48] Alexandra Hasluck, 1955, pp. 147–207.
[49] G.G. Smith, 1982, 'Pioneer seashore botanists in Western Australia', *Early Days, Journal of The Royal Western Australian Historical Society*, vol. 8 (6), pp. 86–89. For further details on Preiss's collecting localities see N.G. Marchant, 1990, 'The Western Australian collecting localities of J.A.L. Preiss', in P.S. Short (ed), 1990, pp. 131–135.
[50] Johann Georg Christian Lehmann & Ludwig Preiss 1844–47, *Plantae Preissianae sive Enumeratio plantarum quas in Australasia occidentali et meridionali-occidentali annis 1838–1841*, vols 1 & 2, Sumptibus Meissneri, Hamburg. Preiss's collections are housed in at

In 1837, 1838, 1842 and 1844, James Drummond made collections around the Swan, initially[51] for James Mangles and later mainly for John Lindley.[52] His 1838 collection featured largely in Lindley's *A sketch of the vegetation of the Swan River*[53] and provided most of the 283 new species described therein.[54] Sir William Hooker, with whom Drummond had corresponded during the late 1830s and early 1840s, and other subscribers were sent specimens from his 1842 collection.[55] During May and June 1842, Drummond travelled to the Vasse district via Clarence, Mandurah, Pinjarra, and Australind in the company of Dr Joseph Harris, newly appointed surgeon to the Western Australian Company at Leschenault. While he was at the Vasse, Drummond stayed with the Molloys at Fairlawn. Later, in October 1842, Drummond, accompanied by John Gould's collector John Gilbert, journeyed to Augusta and on their way south they deviated from Pinjarrah [Pinjarra] to climb Mount William. The plants obtained on these journeys were dispatched to England as part of his 2nd Collection.[56] The largest holdings of Drummond's collections are at Kew Gardens and NHV with lesser numbers in 23 other herbaria.[57] Although employed by Gould to collect birds and mammals, Gilbert also obtained plant specimens while he was in the south-west of Western Australia in 1840 and 1842, including from around the Swan River. His specimens are at the Natural History Museum, London, Kew Gardens and the Naturhistorisches Museum Wien in Vienna.[58]

Prior to visiting the eastern colonies from mid-April to 20 July 1854, William Harvey gathered algae, along with other botanical material and marine animals, in the vicinity of Perth and Fremantle, and on Carnac and Garden Islands, but chiefly on Rottnest Island. His total algal collection from these localities was extensive, largely due to practical assistance with dredging or beach collecting provided by Captain Edmund Y.W. Henderson, Comptroller of Convicts, George Clifton, Inspector of the Imperial Water Police and the Colonial Secretary, Colonel William A. Sanford. Sanford, who left Western Australia in 1855, also afforded Harvey the use of 'a pilot boat of 6 tons with 2 men and the loan of a dredge, rakes and ropes for fishing'.[59] The herbarium at Trinity College Dublin is the primary repository of his collection but many of his specimens are held in other herbaria in Ireland and in England, Scotland, South Africa and USA.[60]

George Clifton became interested in Harvey's fieldwork of collecting and pressing seaweeds during the fifteen weeks the botanist spent at Fremantle. For the next nine years, Clifton sent Harvey 'a vast number of pressings'. In recognition of his contributions to the collections of the Herbarium of Trinity College, Harvey dedicated his first volume of *Phycologia Australica*[61] to Clifton.[62]

---

least 25 herbaria, A.E. Orchard, 'Preiss, J.A. Ludwig (1811–1883)', ANHBiog, viewed 8 May 2008, www.anbg.gov.au. For further information on the location of Preiss's specimens refer to this site.

[51] Rica Erickson, 1969, p. 168.

[52] J.H. Maiden, 1908, 'Records of Australian botanists—(a) general, (b) New South Wales', *Journal of the Royal Society of New South Wales*, vol. 42, pp. 60–111. Lindley was Assistant Librarian to Joseph Banks, Assistant Secretary of the Royal Horticultural Society and the first Professor of Botany at the University College, London.

[53] John Lindley, 1839, 'A sketch of the vegetation of the Swan River', Appendix to the first twenty-three volumes of *Edwards's Botanical Register*, James Ridgway, London.

[54] B.R. Maslin & A.S. George, 2005, 'James Drummond's *Acacia* collections from south-west Western Australia', *Curtin University of Technology, Department of Environmental Biology, Bulletin* 27, pp. 34–39, and Rica Erickson, 1969, p. 28.

[55] ibid., p. 168.

[56] ibid., pp. 67–74.

[57] A.E. Orchard, 'Drummond, James (1786 - 1863)', ANHBiog, viewed 11 January 2013, www.anbg.gov.au.

[58] Rica Erickson, 1975, pp. 43, 45–47, Isabella Tree, 2003, pp. 128-129, and A.E. Orchard, 'Gilbert, John (~1810–1845)', ANHBiog, viewed 8 May 2008, www.anbg.gov.au.

[59] WHH to Mary Christy Harvey, 21 April 1854, cited in Sophie C. Ducker (ed), 1988, pp. 106–114. See also G.G. Smith, 1982, pp. 92–96.

[60] For details of the herbaria involved see A.S. George 2009, p. 409.

[61] WHH 1858–1863, *Phycologia Australica*, vols. 1–5, Lovell Reeve, London.

[62] G.G. Smith, 1982, pp. 97–99.

Oldfield botanises from Cape Naturaliste to Perth—early August 1856

Map 8 p. 176 – Western Australia, Cape Naturaliste to Perth

Soon after disembarking, Oldfield must have discovered Busselton, located within the Sussex District, to be a sleepy hamlet with a very small population[63] livened at times by the carousing of whalers, ticket-of-leavers and others. Normally by day the silence was broken only by children playing in the street, the grind of iron-tyred sulkies, buggies or carts and the clip-clop of horses passing up and down.[64] Although the presence of whaling ships seeking fresh supplies of food and water and processing their catches onshore often added to the gaiety of the local elite when they entertained the officers at luncheons, dinners and balls[65], public brawls and crime were an ever-present problem due largely to the influx of sailors from American whalers and ticket-of-leavers.[66] Smuggling was rife. In exchange for fresh food and water the settlers avoided customs fees on imported goods, a situation facilitated by the lack of manpower to efficiently administer the collection of dues.[67] Local settlers had been involved in on-shore whaling since the 1840s[68] and the activities of American whalers close to the coast gave rise to anger among the locals involved in the industry.[69]

Despite Busselton's importance as a port and the townsite having been surveyed and gazetted in 1835[70], Oldfield could not have escaped observing that this was not reflected in the settlement's infrastructure; the streets were undrained dirt tracks and, arriving in winter, he may have found many inundated and difficult to traverse.[71] The pastoralists, farmers and townsfolk alike complained about the flooding and poor state of the streets and of the roads from the settlement to Bunbury, Perth, the Blackwood and to Albany.[72] Also, the people of the town objected to the cattle grazing on vacant lots.[73] There were few houses and hardly any government buildings. With the advent of a convict population after 1850 and the retaliation of settlers to Aborigines, resentful at the loss of their land and food sources, spearing them and their stock, there had been a police presence in the district since 1851. At the time of Oldfield's visit work on a gaol was either underway or completed. Previously the local lock-up was a rented room beside the home of one of the town residents and the perpetrators of serious crimes were taken to the Bunbury gaol.[74] Scattered along Queen Street, which ran from the Vasse to the port, and Albert Street, which crossed Queen Street toward its river end, were a few retail outlets, three hotels and a post office.[75] Near the southern end of Queen Street on the northern side of the river was St Mary's Anglican Church[76] and further on was a bridge across the river leading into the Vasse Road, which ran to Bunbury.[77]

In those days, it was usual for visitors to a district to make themselves and their business known to the local Resident Magistrate or Constable with supporting letters of introduction. By now Oldfield's association with Mueller had, we believe, developed to a point where that gentleman probably armed the collector with letters of this nature; if so, none has been found. To undertake such a call on the part-time Resident Magistrate, Lieutenant Colonel John Molloy, he would have walked or ridden

---

[63] By 1856 the Sussex District had a population of 256, most of whom lived at the Vasse and Wonnerup, J. Carroll, c.1972, *The Development of Busselton 1832–1872*, Thesis, Claremont Teachers College, p. 256, JSBL.

[64] C. Elliott, 1992, 'Albert Street, Busselton as it was before 1920', *Busselton Historical Society Newsletter*, October, pp. unnumbered.

[65] Rodger Jennings, 1999, *Busselton: 'A Place to Remember' 1850–1914*, The Shire of Busselton, Busselton, p. 47.

[66] ibid., p. 52.

[67] ibid., p. 43.

[68] ibid., pp. 16–20.

[69] ibid., p. 17.

[70] Landgate undated, 'Busselton', *History of Country Town Names*, viewed 15 July 2010, www.landgate.wa.gov.au.

[71] Rodger Jennings, 1999, p. 38.

[72] ibid., pp. 38 & 48–50.

[73] ibid., p. 38.

[74] ibid., pp. 10 & 35.

[75] Merle Emery, 1998, 'Busselton's earliest shops', *Busselton Historical Society Newsletter*, August 1998, pp. 1–2.

[76] Busselton Historical Society, 1978, *Historic Busselton*, 3rd edition, Busselton Historical Society, Busselton, p. 4.

[77] Rodger Jennings, 1983, *Busselton '...outstation on the Vasse' 1830–1850*, The Shire of Busselton, Busselton, p. 228.

three kilometres south, from the Port to Molloy's home from which he administered the Sussex District, and where he provided, to a limited extent, a court house and a customs and storage house.

Apart from the Resident Magistrate and his clerical assistant, who also acted as the Clerk of Courts and Tidewaiter[78], there was a postmaster. Mail deliveries were undertaken once a week on horseback to and from the post office at Bunbury.[79] From there, mail destined for more distant places was carried overland by horsemen along the Coast Road to Fremantle and to places in between, and mail from Bunbury and Fremantle was carried on the return journey.[80] Overseas mail was off-loaded at King George Sound and carried to and from Bunbury via Kojonup.[81] There were no coach services to settlements between Bunbury and Fremantle or between Bunbury and Albany.[82] Government positions were usually held by educated and often prominent settlers to augment the incomes derived from their businesses. By 1856 there were no resident medical officers or church ministers but a school was functioning.[83]

Oldfield left no details of what he did about his accommodation at Busselton but he did collect at and near the Vasse River.[84] He provided no indication of how far he ventured inland from and along the stream. Given his time constraints he probably focused his attention on the lower Vasse within a short distance of the Busselton townsite. Based on specimen data we have obtained, he appears to have spent longer around the river than at any other location between Cape Naturaliste and Coogee to the south of Fremantle.[85]

In Oldfield's time uncleared areas along the lower Vasse River were fringed with Flooded Gum, Albizia (*Paraserianthes lophantha*), Swishbush (*Viminaria juncea*), Spreading Sword-Sedge (*Lepidosperma effusum*) and other riparian species.[86] Away from the river the most noticeable trees on the sandy soils, limestone, swamps, and winter wetland areas were Tuart (*Eucalyptus gomphocephala*) and WA Peppermint (*Agonis flexuosa*) under which abounded a range of shrubs and smaller plants belonging to many families including the Myrtaceae, Proteaceae, Papilionaceae, Thymelaeaceae, Restionaceae, Cyperaceae and Orchidaceae.[87] Species from most of these groups are represented among Oldfield's Vasse River collections.[88]

He may have spent time gathering specimens around the Vasse–Wonnerup Estuarine lagoon system but his specimen records do not give a clear indication of this. This wetland is extensive, shallow, and nutrient-enriched and located on a narrow flat plain. It is separated from the ocean by a narrow system of low dunes and comprises two former estuaries—the Vasse and Wonnerup lagoons. Water in the wetland is fresh in winter and becomes saline in summer. The lagoons are fringed by samphire (Chenopodiaceae) and rushes (Restionaceae), with some Melaleuca woodlands on higher ground. No samphire specimen records from this area were found and the specimens of rush species he obtained89 are not listed in the RAMSAR information sheet for the estuary.90

---

[78] According to the *Concise Oxford Dictionary of Current English*, a Tidewaiter is a customs officer who boards ship on arrival to enforce customs regulations.

[79] Rodger Jennings, 1999, pp. 20–22.

[80] Theodora Sanders, 1975 p. 97.

[81] Donald S. Garden, 1977, *Albany: A Panorama of the Sound from 1827*, Nelson, Melbourne, p. 95.

[82] Joan Higgs, c.1950, Public Transport by Coach, Pt. 1, Thesis, Claremont Teachers' College, p. 6.

[83] Rodger Jennings, 1999, p. 41 & 61.

[84] HAODb.

[85] ibid.

[86] Richard Clark, 2007, *'Revegetation of the bank of the Vasse River at the Isaac Street Reserve, Busselton'*, Geographe Community Landcare Nursery, September 2007, viewed 19 January 2013 www.geographelandcarenursery.mysouthwest.com.au.

[87] Russell Smith, 2012, A Flora and Vegetation Survey for the Proposed Busselton-Flinders Bay Rail Trail, ekologica, prepared for ngh environmental, Bunbury, viewed 7 July 2012, www.environment.gov.au.

[88] See HAODb for details.

[89] ibid.

[90] Department of Environment, Australian Government, 'Vasse-Wonnerup System', *RAMSAR Wetlands information sheets,* viewed 19 January 2013, www.environment.gov.au.

Fig. 11.1 Looking east across Geographe Bay, nr Meelup

During the summer months, wildlife, particularly native and migrant birds, congregate to feed when many other wetlands are dry and barren.[91] If Oldfield visited the lagoons he was probably there too early to witness the extent of the annual influx of thousands of resident and migrant water birds of a wide range of species—more than 80 species have been recorded in this wetland.[92]

Before Oldfield left Busselton to walk to Cape Naturaliste, roughly 40 kilometres and a return walk of about four days (allowing for collecting and difficult walking conditions along the promontory), he needed some provisions and information from locals about the route. Perhaps he sought the assistance of Colonel Molloy or the police constable. Apparently, Augustus would have found it to be a comparatively short walk.[93] However, he may have gone on horseback, having managed to purchase a riding hack and a pack horse in the Busselton district. Horses were in short supply and consequently expensive, costing about £40[94] with an additional outlay if he needed to obtain a saddle. Such a purchase would have been considerable for an out-of-work schoolmaster whose annual income had been under £70, but horses would certainly have made the going easier and the journey quicker.

The botanist almost certainly travelled south-west to Cape Naturaliste along the sandy track known in those days as 'the Dunsborough road'. This track ran close to the beach bordering Geographe Bay[95] where he collected a few plant specimens at Ironstone Creek, Ironstone Gully and Carbunup River.[96]

---

[91] Anne Brearley, 2005, p. 247.

[92] Anon, undated, Protecting Ramsar wetlands, Priority site 12: Vasse-Wonnerup System, Australian Ramsar Site No. 38, State: Western Australia, Business Plan 2012–2013, *Caring for our Country*, Department of Agriculture, Fisheries and Forestry and the Department of Sustainability, Environment, Water, Population and Communities, Australian Government, Canberra, p. 3, viewed 19 January 2013, www.nrm.gov.au.

[93] Anon., ? Edwin Oldfield, 13 July 1887, p. 5.

[94] W.J. de Burgh, 1986, *The Old North Road*, Western Australian Experience Series, University of Western Australia Press, Nedlands, p. 91.

[95] Rodger Jennings, 1999, Dunsborough Road on Portion of Wellington Map 25 (1850) with roads taken from Wellington Map 33, Appendix C.

[96] HAODb.

Oldfield may have already encountered around the Vasse River many of the species of the coastal scrub that lined the Dunsborough Road. This bushland is dominated by WA Peppermint, Slender Banksia (*Banksia attenuata*), tall Tuart trees and in the wetter areas, Swamp Paperbark and Moonah (*Melaleuca preissiana*). The open heath-like vegetation found inland behind the sand dunes favours drier areas where the Christmas Tree[97] (*Nuytsia floribunda*) is locally dominant, while a better water table allows the development of eucalypt shrubs. Scrub-heath species include WA Peppermint, Fire-wood Banksia, Slender Banksia with *Allocasuarina* spp.[98] On the leeward slopes of the coastal dunes, shrubs such as the Thick-leaved Fan Flower (*Scaevola crassifolia*), Blueberry Tree (*Myoporum insulare*), *Acacia* spp. and, if stable, on their windward slopes dunes, a low dense thicket of Coastal Daisybush (*Olearia axillaris*), Panjang (*Acacia lasiocarpa*) and Coastal Honey Myrtle (*Melaleuca acerosa*) with *Spinifex* spp. closer to high water mark. Records in HAODb suggest that Oldfield focused his attention on shrubs, herbs and grasses behind the coastal dunes along this tract of land.[99]

The track, which initially traversed several farming properties, apparently ran along the beach terminating at Castle Rock, just to the north of the centre of the original Dunsborough townsite, and on the eastern coast of the Cape Naturaliste promontory[100], giving access to the whaling station at Castle Rock. Here Oldfield gathered specimens of three shrub species[101], probably in the nearby Jarrah woodland (*Eucalyptus marginata*).[102]

There was no track from the Dunsborough Road to Cape Naturaliste.[103] So Oldfield had no option but to traverse the dense bush on the promontory to the Cape or, less likely, along the beach inter-spersed with rocky outcrops and cliffs.[104] Whichever way he went the going would not have been easy for either horse or man. However, if he could explore at least some of the ten vegetation systems on the promontory he may have found the floral diversity exciting. Apart from the aforementioned Jarrah woodland, the other major vegetation systems include woodland running along the ridge of the prom-ontory from Bunker Bay in the north to south of Yallingup, where WA Peppermint and Marri (*Corymbia calophylla*) are co-dominant, coastal heath along the west of the promontory and in the vicinity of the Cape, and a mixed association of Jarrah, Marri and *Banksia* spp. across part of the base of the promontory.[105] All the records of his specimens which we have found from this area, indicate that he collected on the species-rich Cape Naturaliste promontory[106] at or near the Cape.[107]

Oldfield encountered a group of Geographe Bay Aborigines. When discussing interbreeding be-tween Aborigines and Europeans in his paper on Australian Aborigines, he recounted what he had been told by them of the fate of Monsieur Vasse, a French sailor lost in 1801 from one of Nicolas Baudin's two ships, the *Naturaliste*:

> This approach to the European type of features I have also observed among the na-tives about Geographe Bay, where the unfortunate Dr. Vasse was so cruelly deserted by his inhuman superior. According to the accounts of these natives, this unhappy naturalist lived many years among them, conforming to all their habits, and at length

---

[97] Noongar people from Perth held *Nuytsia* in awe, regarding them as sacred *Kaanya* tree where the *kaa-nya* or *wee-urn*, souls of the newly dead were to be found. They avoided every part of these trees, the shade provided by them and animals that rested beneath them. Further north, however, Noongars held different beliefs, their roots were regarded as a delicacy and they used the flowers for body decoration, S. D. Hopper, 2010, p. 337.

[98] J.S. Beard, 1990, pp. 89–91.

[99] See HAODb for details.

[100] Rodger Jennings, 1999, Map showing Busselton land holdings, 1858, Appendix K.

[101] HAODb.

[102] P.A.S. Valentine & N.J. Enright, 1975, 'The Cape Naturaliste Area, WA – An Environmental Resource Evaluation', *Geowest, Working papers of the Department of Geography, University of Western Australia*, no. 6, November, p. 20.

[103] Rodger Jennings, 1999, Appendix K.

[104] P.A.S. Valentine and N.J. Enright, 1975, pp. 20– 21.

[105] ibid., pp. 18–25.

[106] ibid., p. 20.

[107] HAODb.

Fig. 11.2 Cape Naturaliste promontory, Western Australia

> dying a natural death—"self-died" said an old native in return to my question as to the mode of his death.[108]

Although Oldfield referred erroneously to the sailor as 'Dr Vasse' and as a 'naturalist', the account he obtained was otherwise like that of George Fletcher Moore, related in the *Perth Gazette* in 1838 of Vasse's life with a local Aboriginal tribe and the circumstances of his death.[109]

Time limitations imposed by the 18 days or so it probably took him to travel from Hobart to Geographe Bay after 14 July 1856, and completing his overland journey from Perth to the Murchison River sometime in September, at the very latest by the end of the month, suggest that he travelled to Fremantle from Busselton during the second week of August. Furthermore, his schedule must have made provision for preparing for the next stage of his expedition to the north in either Perth, Fremantle or Guildford.

He could have kept to such a timetable by walking to Fremantle, some 224 kilometres, if he was able to cover about 32 kilometres a day. To accomplish this, he would have needed to leave Busselton by no later than 15 August and to reach Fremantle by 22 or 23 August, leaving about a week to prepare for his trip to the north. If he was on horseback the journey from Bunbury along the coast road to Mandurah and from thence to Fremantle with an overnight bivouac took about 32 hours plus the day or so for the trip to Bunbury from Busselton.[110] However, regardless of whether he was on foot or on horseback, it would have taken him longer as it seems he did not travel straight up the Coast Road to Mandurah but went inland via Pinjarra. Both walking and riding posed the problem of transporting his luggage—which could have included specimen cases, collecting, camping and navigation equipment and personal gear—unless, of course, he had a packhorse, or, he may have arranged for the ship on which he had sailed to carry it to Fremantle. By taking this last course, he would have run the risk of the vessel not arriving in time for him to reach the Murchison in September. Another possibility

[108] AO, 1865a, p. 219.
[109] Rodger Jennings, 1983, pp. 3–4. Jennings quoted Moore's letter in full.
[110] N.K. Perry, 1956, The History and Present Potentialities of the Old Coast Road from Mandurah to Bunbury, unpublished typed manuscript, pp. 10 & 15, JSBL.

was that he obtained carriage for himself and his luggage from someone travelling in a horse-drawn cart, although for reasons that will become apparent this is doubtful.

Whatever his mode of travel, his apparently tight time schedule placed restrictions on how long he could spend collecting and preserving plants. His journey from Geographe Bay to Perth was undertaken when it would be expected that the spring flowering had commenced and any uncleared place where he stopped would have yielded specimens useful for determination purposes. These he could gather very quickly and store temporarily in his vasculum, in which case the specimens would have required pressing before they spoiled. Alternatively, he might have stored them immediately in a press. Regardless of what he did, it is probable that the papers needed to be replaced with dry sheets and the plants labelled before the journey's end.

The patterning of the number of plant specimens he collected at each locality during the journey suggests that he was on his own and free to gather the occasional specimen when he saw something that interested him. Very few specimens appear to have been obtained at each of his collecting places *en route*, apart from those gathered at the Harvey and Murray Rivers.[111] These might have been places where he bivouacked. If he was travelling with a packhorse, perhaps he experienced much the same frustration that James Drummond expressed in 1843 in a letter to William Hooker, relating to a parcel of seeds that he had dispatched to Kew Gardens:

> [I] only regret that the number of species is not so great as I could wish, owing to the impossibility of collecting largely during a journey made with loaded pack-horses, except near the places where we stop for the night.112

Oldfield travelled to Bunbury along the Vasse Road, approximately 45 kilometres from Busselton, where he probably commenced his journey, passing Prospect Villa and St Mary's Anglican Church on the north side of the Vasse River and almost immediately crossing the bridge over the river. The road then skirted numerous farms[113] and ran through shady forests of tall Tuarts with understoreys of WA Peppermint, Slender Banksia, Sheoak (*Allocasuarina fraseriana*) and sedges, and drier areas where the vegetation was similar to that growing along the Dunsborough Road.[114] The winter rains almost certainly would have caused the road to be wet and boggy making for unpleasant travelling.[115] It crossed the Capel River, which like all rivers to the south of Fremantle, is fringed with Flooded Gum. Oldfield collected at, and near, this stream.[116]

Remaining on the Vasse Road, he would have entered Bunbury from the south along what was later renamed Spencer Street and which led into Stirling Street running east and west. Off it, in a northerly direction, were Victoria and Wittenoom Streets and transecting them, Stephen Street. Stores, inns and dwellings lined these unpaved thoroughfares but because of periodic flooding of a nearby lagoon, duckboards were used to allow for foot traffic along Victoria Street throughout the year.[117] A short distance to the west of the town were sand dunes and beyond, the Indian Ocean and the port on Leschenault Inlet (now Koombana Bay). Timber, sandalwood, horses and surplus farm produce were shipped from Port Leschenault, located to the south of the inlet, and through which a variety of goods were imported.[118]

---

111 HAODb.
112 James Drummond to WJH, 25 September 1842, *London Journal of Botany*, vol. 3, 1842, p. 266.
113 G.E. Clarke, 1946, *Early History of Bunbury*, G.E. Clarke, Bunbury, p. 35.
114 J.S. Beard, 1990, pp. 92–93.
115 Rodger Jennings, 1983, p. 279.
116 HAODb.
117 G.E. Clarke, 1946, p. 18.
118 ibid., pp. 24–26.

Fig. 11.3 Bunbury [looking north to the Leschenault Estuary], 1857

Oldfield stayed long enough in the district to make collections at and near the township, the Leschenault Inlet, near and at Cape Leschenault, which may be what is now known as Point Casuarina, and on the Leschenault Peninsula.[119] He also collected on the Preston River[120], possibly around where it emptied into the southern end of the Leschenault Estuary. Such an expedition could comfortably be undertaken in one day. The White Mangrove (*Avicennia marina*) grows on the tidal mud of the Leschenault Inlet, bordered by the Sea Rush (*Juncus kraussii*) and backed by trees of the Saltwater Paperbark, other paperbarks and Swamp Sheoak (*Casuarina obesa*). They are found intermingled with samphires, herbs and sedges of the salt marsh and fringing communities.[121]

As he travelled through Bunbury's streets on his way to the port he could have found the town quite a lively place—convict road gangs busy at work, ticket-of-leavers going about their masters' business, a visiting crew off an American whaler providing noisy, ribald and drunken behaviour, and farmers on horseback or in horse-drawn carts accompanied by their wives and children in town to shop, off-loading produce for shipping, or conducting other business. Aborigines selling clothesline props, gilgies and mushrooms, a constable riding through with a recently apprehended prisoner in tow and children playing in the streets all added to the activity of the town. Government presence was felt by the recently constructed (1855) limestone and shingle building on the corner of Wittenoom and Stephen Streets, which served as courthouse, post office, bond store and police quarters.[122] Operating out of this building was the Colony's lean regional bureaucracy—a Resident Magistrate, Chief Constable, Deputy Controller of Aborigines, Postmaster, Overseer of the Convict Store, a magistrate's clerk, tide-waiter and native police officer.[123]

[119] HAODb. The only Cape Leschenault that we have found is to the north of Guilderton 15 kilometres to the west of the Old North Road. For reasons that will become apparent we do not believe he ever visited that locality. Although not found on any map or gazetteer, Oldfield's collecting locality 'Cape Leschenault' was probably what is now known as Point Casuarina on the western side of Leschenault Inlet (now Koombana Bay). Leschenault Peninsula runs along the western shores of the Leschenault Estuary.

[120] HAODb.

[121] Anne Brearley, 2005, p. 22.6

[122] Department of the Attorney General, Court and Tribunal Services 2009, 'Bunbury Court House', viewed 23 February 2013, at www.courts.dotag.wa.gov.au.

[123] Anthony J. Barker & Maxine Laurie, 1992, *Excellent Connections: A History of Bunbury, Western Australia 1836–1990*, City of Bunbury, p. 73, and G.E. Clarke, 1946, pp. 33–37.

To access the Coast Road to Fremantle, Oldfield passed through the town. He gathered a few plants at and near Bunbury.[124] From Bunbury, the Coast Road ran along the eastern shore of Leschenault Estuary to another of his collecting places, Australind.[125] It was here that the failed Western Australian Land Company had planned a city and port to service its agricultural land development, a venture that was influenced by the fashionable colonising theory of Edward Gibbon Wakefield.[126]

Proceeding up the road from Bunbury, collecting at, and near, William Crompton's property 'Myalup' just to the north of Myalup Swamp and to the east of the Coast Road[127], and at Lake Clifton[128], Oldfield passed through Tuart woodland with an occasional Marri on some red soils, and patchy Redheart Moit (*Eucalyptus decipiens*) on shallow limestone. The small trees making up the lower layer were WA Peppermint, Slender Banksia and Sheoak. The Grass Tree (*Xanthorrhoea preissii*) and the trunkless cycad Zamia (*Macrozamia riedlei*) were also present. Many shrub, herb and climbing species occurred in the under storey[129] of which Oldfield gathered a few.[130]

In volume 1 of the *Fragmenta* (1858–1859), Mueller commented that he had named the species *Eremophila clarkei* in honour of 'Guil. Clarke' of 'Hampden', in recognition of the generous assistance he had rendered to Oldfield during his 'journey'.[131] A specimen of *Villarsia albiflora* was collected jointly by Clarke and Oldfield at Hampden.[132] There is also a specimen of *Grevillea centristigma* from Mandurah[133], which was given to him by a 'Mr Clarke' but where and when this presentation took placc is not known.

The good Samaritan was William James Clarke[134], the son of Ephraim Clarke, a surveyor and builder, who had taken up farming at Australind between 1844 and 1850.[135] Ephraim had also become a landholder in the districts of Leschenault, Mandurah, Preston and Wellington during the 1850s.[136] In the 1840s and 1850s he held 10 and 20 acres freehold respectively, associated with pastoral leases. These were 'Hampden' toward the north end and east of Lake Preston, and 'Jardup' in the foothills of the Darling Range, approximately half a kilometre to the east of the current townsite of Harvey to the south of the Harvey/Quindanning Road.[137] In 1850 Ephraim erected a wayside Inn with accommodation for travelers at Hampden.[138] William appears to have worked on one or more of these properties.[139] Oldfield gathered plants at and near Hampden, Jardup, Harvey River (probably near the present day town of Harvey) and Pinjarra.[140]

From the southern end of Lake Clifton there were two routes to Mandurah. One, the most direct, ran north along the Coast Road, which involved a ferry ride across the Murray River at Mandurah and

---

124 HAODb.
125 ibid.
126 J.S. Battye, 1924, *Western Australia: A History from Its Discovery to the Inauguration of the Commonwealth*, Oxford: Clarendon Press, Chapter 6.
127 A.C. Staples, undated c.1948, *Historical Development of the Harvey District up to 1895*, map 'The Harvey and south to Bunbury', unpublished MA thesis, University of Western Australia, opposite p. 1 and p. 121, JSBL. Myalup is now a beachside settlement at the southern end of Lake Preston to the west of the Old Coast Road.
128 HAODb.
129 J.S. Beard, 1990, pp. 90, 92.
130 HAODb.
131 FM, 1858–1859, vol. 1, p. 208. Mueller's actual statement was in Latin: '*Nulli risis E. Freelingii comparandam speciem nominavimus in honorem el Guil. Clarke, Hampdenensis, itinera Oldfieldiana generose foventis*'.
132 FM, 1860–61, vol. 2, p. 21.
133 Specimen sighted by A.S. George at Kew Gardens, personal communication, 2003, HOA.
134 'Guil' is an abbreviated Latin form of 'William' Alex S. George, personal communication, 7 July 2014.
135 Rica Erickson (ed), 1979, *Dictionary of Western Australians 1829–1914*, vol. 3, *Free 1850–1868*, University of Western Australia Press, Nedlands, p. 138.
136 ibid., p. 136.
137 Lands & Surveys Department (WA), Public Plan, c.1860–1870, Wellington 4 (506327) & Wellington 5 (506328) Conservation no. 4929, SROWA and Place 03171, Jardup Homestead, Shire of Harvey, State Heritage Register (WA) record, viewed 15 December 2017 at www.inherit.stateheritage.wa.gov.au..
138 'Local Intelligence', *Inquirer*, 17 April 1850, p. 1, DAN, NLA.
139 Rica Erickson (ed), 1979, vol. 3, p. 136.
140 HAODb.

the other, to the north east, crossed a ford on the Harvey River and then continued onto Pinjarra.[141] Both of these crossings were dangerous, if not impossible, during times when the rivers were in flood; an occurrence that was not uncommon during the winter months (June to August).[142] Furthermore, the ferry punt at the Murray River crossing was not safe and in need of scrapping.[143]

If, as might well have been the case, the Murray and Harvey Rivers were impassable due to flooding when Oldfield travelled up the road, the assistance that Clarke rendered him may have been to guide him along a less perilous route to Mandurah via Pinjarra. If so, he probably went on a road linking Hampden to [H.M.] Ommanney's Road (now the South Western Highway) to the south of the Harvey River Bridge[144], which Clarke probably developed to facilitate travel between his properties Hampden and Jardup. Today, the road from Hampden to the town of Harvey traverses Jarrah-Marri woodland, farmland and winter swamps but to avoid periodic flooding, it is doubted that this approximated the track that was used by the Clarkes to travel from Hampden to Jardup since it crosses the Harvey River.[145] They may have travelled a few kilometres south on the Bunbury Road and used a track approximating today's Forestry Road which runs to Ommaney's Road (South-west Highway) at Harvey. Ommanney's road then ran north to Pinjarra from whence he could access one of the tracks that ran along the Murray River to Mandurah. This diversion added about 32 kilometres to his journey to Fremantle.

Although much of the land around the gazetted townsite of Pinjarra, along the Murray River and toward the Darling Range had been taken up for pastoral or farming purposes by the time Oldfield passed through, the town itself was very much in an embryonic stage of development. There was a police station manned by one constable, the deserted military barracks and St John's Anglican Church, consecrated in 1848, located near the bridge over the Murray. There had been a licensed inn to the east on the opposite side of the river run by William Edmund Oakley from 1830 to the early 1850s.[146] John McLarty's hotel was closed in 1856 pending the opening of another in 1857 at his 'Blythewood' property a few kilometres to the south of Pinjarra. There may have been a few residents living in cottages in the settlement. The Murray River was navigable and was used by farmers to transport produce on their way to Fremantle.[147]

Oldfield obtained plants at Pinjarra[148] before continuing his journey but he may not have sought accommodation there, preferring to bivouac. His journey to Mandurah necessitated crossing the recently-completed, convict-built bridge over the Murray River at Pinjarra,[149] and then following the road north for about one and a half kilometres, before crossing the bridge over Dandalup River. From there he took one of two tracks that led to Mandurah. Records of his specimens obtained along the Murray and Serpentine Rivers[150] provide no indication of which of the two he followed. One, the older track, followed the river, before crossing the ford at Serpentine River just to the south of Black Lake, and the other went more to the north-west, past 'Ravenswood' farm, crossing the Serpentine at the north end of Lake Goegrup before continuing into Mandurah.[151] If the country was in flood he proba

[141] Ronald Richards, 1978, *The Murray District of Western Australia: A History*, Shire of Murray, Pinjarra, 'Roads in the Murray District, 1870', map, p. 360. The road from Lake Clifton to Pinjarra was in use in 1838, (ibid., 'Main tracks in use by 1838', map, p. 154).

[142] C. Elrick, A. Travers, T. Bond & R. Kay, 2008, *Coastal Zone Climate Change Risk Assessment and Adaptation Action Plan: Issues Paper, Report*, prepared for the City of Mandurah, Coastal Zone Management Pty Ltd, Perth., p. 9, viewed 22 January 2013, www.mandurah.wa.gov.au. Rainfall data are not available for the 1850s.

[143] Ronald Richards, 1978, pp. 289–290.

[144] A.C. Staples, undated c.1948, 'The Harvey and south to Bunbury', (map opposite p. 1). Today, there is a road that runs south-east from Hampden to Harvey (National Topographic Map Series, Sheet SI 50–6, Bunbury, 1: 100,000, Edition 1, Commonwealth of Australia, 1981.

[145] Paced Survey of Road between the Wellington & Williams Districts as marked by Acting Surveyor F.T. Gregory, October 1847, Lands & Survey Records, WAS 1861, Cons 5000, Item 245, SROWA. See also Ronald Richards, 1978, p. 360.

[146] ibid., pp. 301–302. Richards makes no mention of there being any retail stores in Pinjarra during the 1850s.

[147] Anne Brearley, 2005, p. 164.

[148] HAODb.

[149] Ronald Richards, 1978, p. 284. This bridge was completed in 1855 and was much sturdier than its predecessors.

[150] HAODb.

[151] Ronald Richards, 1978, pp. 153–156 and p. 360.

*Fig. 11.**4** Near Harvey River*

bly took the more northerly route. If so, a specimen of *Banksia attenuata* with the locality 'Lake near the Serpentine'[152] was probably obtained at Lake Goegrup.[153] The Serpentine River, of which these two lakes are part, feeds into the Peel-Harvey estuary that also receives water from the Murray and the Harvey Rivers. The fringing or salt marsh plant community of this estuarine system is subjected to periods of flooding and drying. [154]

The collector visited Mandurah (previously Peel Town).[155] He may not have stayed more than an hour or so because, on the face of it, there was not much on offer for the traveller. It was a tiny settlement, apparently without a retail store, situated on the narrow channel leading into the Peel Inlet from the Indian Ocean and mostly concentrated on the north side of the channel. The population was predominantly the remnants of the group of British men and women that the failed venture of Thomas Peel, in partnership with Solomon Levey, had financed to migrate to Western Australia to become small landholders. In return the Western Australian government agreed to provide the partnership with a total of 250,000 acres from near Woodman Point south to the Murray and east as far as Pinjarra with the coast forming most of its western border.[156] The only buildings in the settlement were a few

[152] GB, 1870, vol. 5, p. 548.
[153] Ronald Richards, 1978, p. 153–156 and p. 360.
[154] Anne Brearley, 2005, p. 159.
[155] HAODb.
[156] Ronald Richards, 1978, p. 30.

cottages, including Thomas Peel's 'Mandurah House'[157], the little Anglican church he built but which he reclaimed and converted into stables[158], Resident Magistrate David Smythe Murray's residence[159], a police station and Christopher Armstrong's inn.[160] The ferry plied the channel carrying customers, stock and produce to a landing on the south side near Henry Hall's cottage and the old army barracks.[161]

From Mandurah, Oldfield took the most direct route to Fremantle along the Coast Road[162] as evidenced by his plant gatherings at Warnbro Sound, Woodman Point and Coogee.[163] The road seems to have run close to the coast, behind the fore dunes of the Quindalup Dune System, and through vegetation comprising WA Peppermint and thickets of Summer-scented Wattle (*Acacia rostellifera*), Rigid Wattle (*A. cochlearis*), Coastal Wattle (*A. cyclops*), Chenille Honey Myrtle (*Melaleuca huegelii*), Coastal Hop Bush (*Dodonaea aptera*) and other species. There may also have been Rottnest Island Pine (*Callitris preissii*) low forest.[164]

From the Mandurah Road, Oldfield entered Fremantle via South Terrace and continued into the town, where he might have had time to explore before leaving for Perth. If so, like one visitor and many other new arrivals to the Colony, he may have thought Fremantle a dismal spot— 'North and South, as far as the view extended, there was nothing but miserable sand, covered by stunted bushes'.[165]

Nevertheless, with a population of about two thousand it must have presented as a busy place.[166] If he explored the settlement for an hour or more on his way to the river, he could not have failed to recognise that it was showing all the signs of becoming a substantial town. A walk along High Street revealed shops, a 'few decent looking' stuccoed stone and brick houses of the merchant families, as well as one or two hotels 'of fair proportions'. He would have passed the Anglican Church at the north end of High Street and possibly ventured to Henderson Street where some terrace cottages for enlisted warders had been finished and occupied. Further on, in Queen Victoria Street, the handsome Victorian Tudor/Colonial Fremantle Boys' School may have caught his eye.[167] In the streets, pensioner guards and convicts must have been very evident. Elevated to the east and above the town in Terrace Street, was the completed first wing of the imposing Convict Establishment[168] with convicts busily erecting the remainder of the institution. Proceeding to the Swan River estuary to the north of the town, he passed the makeshift timber and unmortared stone dwellings, which made up most of the town and were clustered around the base of the hill near the sea. This area bore 'a somewhat untidy, unfinished look inseparable from half-completed streets and unpaved footpaths'. Minor stores and open fruit and fish stalls were scattered in all directions.[169]

Among the townsfolk he passed Oldfield would have recognised the signs of a sharply demarcated social class structure—black clothing was the uniform of respectability and social status for both sexes, whereas the working-class were dressed in lighter clothing more suited to the hot climate.[170]

---

157 ibid., pp. 249–250.
158 ibid., p. 293.
159 ibid., pp. 294–295.
160 ibid., pp. 293–294.
161 ibid., pp. 51 & 289.
162 ibid., 1978, p. 360.
163 HAODb.
164 J.S. Beard, 1990, p. 90.
165 R. Reece & R. Pascoe, 1983, *A Place of Consequence: A Pictorial History of Fremantle*, Fremantle Arts Centre Press, Fremantle, p. 23.
166 J.K. Hitchcock, 1929, *The History of Fremantle: The Front Gate of Australia 1829–1929*, Fremantle City Council, Fremantle, p. 34.
167 John K. Ewers, 1971, *The Western Gateway: A History of Fremantle*, University of Western Australia Press, Nedlands, pp. 30–31 & 50.
168 F.K. Crowley, 1960, *Australia's Western Third*, Heinemann, Melbourne, p. 22.
169 R Reece & R Pascoe, 1983, pp. 26–27. This description was given by a Mrs Millet in December 1863 but the town probably had much the same appearance in the late 1850s.
170 ibid., p. 29.

Evident in the streets may have been the malaise that Reverend Joseph Johnston spoke of to a friend in 1856:

> I don't think I was ever in a place of its size so given up to profligacy of every kind. The old colonists are inveterate drunkards, the new ones are mostly convicts and pensioner guards and their families. The pensioners are reprobate old soldiers, who have grown grey in iniquity.[171]

Nevertheless, the cultural life of the community was quite rich. The Fremantle Literary Institute and the New Swan Lodge MUIO of Oddfellows had been operating since 1851, and amateur theatricals were performed quite frequently.[172] On his way to the jetty, the Perth and Fremantle cricket clubs may have been battling it out on Fremantle Green, the recreational and social centre of the community near the river jetty.[173]

The question then arises as to how he journeyed to Perth. If he was on horseback with a packhorse he probably opted to ride from Fremantle to Preston Point and cross the river about a kilometre, on the Fremantle side, from Point Roe over to Minim Cove by means of a horse ferry. From there the most frequently used track ran along a series of limestone ridge caps before entering open sandplain woodland, weaving its way through the present localities of Mosman Park, Claremont and Shenton Park to the north-west corner of what is now King's Park (Mount Eliza) and then downhill to connect with St Georges Terrace in Perth.[174]

The other alternative was to go by paddle steamer or yacht, which he may have done if he was on foot.[175] If so, the journey took him along the deep, dark channel of the lower reaches of the Swan River, bordered by vegetated high limestone cliffs or steep banks, before passing into the expanses of Mosman Bay, Melville Water, Matilda Bay and through the relatively small opening of the Narrows, on the banks of which stood Shenton's Mill, before emerging into Perth Water. William Harvey was impressed with what he saw when, the year before, he took a boat trip to Fremantle with the Colonial Secretary, William Sanford:

> The banks [of the river] are not very high, but pretty wooded with bare rocky places at intervals, some of them remind me of the shores of Muscross [sic][176], the bare worn limestone in the water & the luxuriant trees that spring out of it.[177]

The ferry journey finished at one of the jetties that stood at the foot of Barrack and William Streets.[178] If he travelled by the steamer *Lady Stirling* he disembarked at the William Street jetty, possibly to be faced with a line-up of drays belonging to Perth trading firms, waiting to cart imported goods unloaded from overseas ships and transported from Fremantle. Near the jetty there may have been a stack of sandalwood, brought from New Norcia, Toodyay, Northam, York and Beverley, ready to be carried down the river to the port on its way to Asian ports. Away to the west could be seen the old stone jetty at the bottom of Mill Street where he likely saw a cargo of stone being unloaded, and to the east the three-storey Commissariat at the bottom of Pier Street.[179]

With a population of only about 2,500 free settlers and convicts, Perth, the Colony's commercial centre and seat of government, was still a small settlement.[180] The central city block was confined to four or five streets running west-east—Mount Street which connected with St Georges Terrace, and

---

[171] ibid., p. 27.
[172] J K Hitchcock, 1929, p. 37.
[173] R Reece & R. Pascoe, 1983, pp. 25–26.
[174] Ruth Marchant James, 2001, 'From byway to highway: the linking of port and capital', *Early Days*, vol. 12, pt. 1, 2001, p. 15.
[175] Ruth Marchant James, 1977, *Heritage of Pines: A History of Cottesloe*, The Town of Cottesloe Council, Cottesloe, pp. 10–11.
[176] Harvey is probably referring here to the headland of Muckross or Kilrymont, in Fife, Scotland.
[177] WHH to Sisterella, 21 April 1954, cited in Sophie C. Ducker (ed), 1998, p. 108.
[178] These jetties can be seen on a steel-engraving by William French c.1855, Perth, in West-Australien vom Berge Eliza aus, steel engraving, produced by A.H. Payne, Leipzig.
[179] J.E. Hammond, 'The Perth water front', *Early Days*, vol. 2, pt. 11, 1932, pp. 16–17.
[180] F.K. Crowley, 1960, p. 54. According to Crowley the *Census of 1854* showed that Perth had a population of 2,407.

Fig. 11.5 Perth in West-Australien vom Berge Elizac, c. 1856

Howick (now Hay), Goderich (now Murray) and Wellington Streets.[181] These were crossed by six or eight streets running from the river northwards—Mill Street at the western end of the settlement, then to the east, King William, William, Barrack, Pier, and Irwin Streets and Victoria Avenue. The causeway across the Swan, which had been built in the 1840s, provided access to the southern districts and was located just beyond the east end of St Georges Terrace.[182] From the ferry the city may have impressed the newcomer as one of scattered cottages with some substantial houses interspersed among a few churches, and commercial and public buildings. For Oldfield, in search of lodgings, accommodation could be obtained at one of the boarding houses or hotels in the city centre.[183]

Later, perhaps, during a leisurely stroll around the city along its sandy or rubble streets, he might have passed the Police Station, the small Court House and public gardens on Barrack Street toward its river end. If he was not lodging there, perhaps he tarried in front of the Pier Hotel and Boarding House near the foot of William Street, watching townsfolk busily buying fish at boats pulled up on the bank. Venturing further to the north and crossing St Georges Terrace, confronting him were the town's most imposing buildings on a central square; the Military Barracks and, beyond, the Georgian-style Anglican Church of St George.

A little further to the east, on the opposite side of the Terrace, was the crumbling, termite-infested, one-storey Government House.[184] To the west along the Terrace were substantial dwellings, business houses, and the recently completed Perth Boys' School[185]; but not all the surveyed lots were built upon. Further to the west were the rifle butts on Mount Eliza. Scattered along the streets to the

[181] These streets were intersected by streets running from the south to the north, see for example, C.T. Stannage, 1979, *A New History of Western Australia*, University of Western Australia Press, Nedlands for an early Plan of Perth (undated) and Plan of Perth 1871–72, respectively inside the front and back covers.
[182] F.K. Crowley, 1960, p. 54.
[183] Trevor Tuckfield, 1975, 'Early colonial taverns and hotel', *Early Days*, vol. 7, pt. 7, p. 102.
[184] C.T. Stannage, 1979, p. 63.
[185] Margaret Pitt Morison and John White, 'Builders and Buildings', in C.T. Stannage (ed), 1979, p. 527.

west and east of Barrack Street were hotels, shops, eating rooms, boarding houses, churches and schools. The Swan River Mechanics' Institute, on the corner of Pier and Howick Streets next to the Freemasons' Lodge, was recently opened with the intention of 'diffusing a love of order, sobriety, industry, and good conduct' among free artisans and clerks. To the east, beyond Barrack Street in Murray Street, stood the recently opened, three-storey Colonial Hospital, the Servants Home/Poor House and the Roman Catholic St John's Pro-Cathedral in Victoria Avenue. Clearly visible to the north from Howick Street was the new Perth gaol situated on an extension of Barrack Street. Nearby were new cottages, mostly made of brick and inhabited by the lower classes, which had sprung up since the lakes behind the town had been drained earlier in the decade.[186] Perhaps Oldfield was of the same mind as agreed as one visitor who observed in1859:

> There are a good number of Merchants' stores in the place, most of whom do retail business as well as wholesale, but none but three of them have the appearance of places of business, and these are only an approximation to Melbourne shops, the others are just the ordinary dwelling houses with a garden in front, and many of them do their business in what may be called the offices, as if business were only a secondary consideration with them, and yet some of them do a large trade for the place, but without any show or the least attempt to attract customers.[187]

After experiencing the bustle of Fremantle, Oldfield may have thought Perth a very quiet place. As one traveller observed in the 1860s: 'The town looked so still that we wondered how the inhabitants employed themselves. I seldom noticed a vehicle in the street...I once saw someone on horseback and he was a mail man'.[188] As was the case in Tasmania, Perth's economy was highly dependent upon imports of clothing, furniture, household and office equipment, consumables and other manufactured goods from England and the eastern Colonies.[189] Nevertheless, there were some commercial and manufacturing businesses in the city and market gardens, flour mills, dairies, and butcheries within and close to the city that provided fresh produce to the ever-increasing population.[190]

Perhaps Oldfield drank a pint or two of ale at one of the inns with the intent of seeking information about the routes to the Murchison River and any traveller he might accompany. It would not have taken him long to realise from the locals that the Colony's governing structure closely resembled that of Tasmania in that it was 'the prerogative of the Imperial officialdom, the landed gentry, and the leading town merchants'.[191] There may have been talk of the Perth Town Trust and the government being at loggerheads over funding. Lack of finance had prevented the Trust from carrying out much-needed public works and this had been exacerbated because the Trust did not have access to convict labour.[192]

The impact of convictism across Perth was evident; their labour (recently extended to the Town Trust), as prisoners in work gangs and as ticket-of-leavers, was helping to develop public works and businesses. However, it is unlikely that Oldfield was there long enough to learn that 'the economic benefits' were spread unevenly and could not prevent the recession which beset the Colony in the mid-1850s.[193] The influence on Perth society *per se* was probably less obvious to the visitor. Many ex-convicts, although not allowed to join masonic lodges, agricultural societies, horse racing clubs, or the Mechanics' Institute nor attend 'gentry-run dances', had settled down 'comfortably' as teachers,

---

[186] C.T. Stannage, 1979, pp. 135–140.
[187] Archibald Sands, 1859, 'Description of Perth contained in a letter written by Archibald Sands, Esq., to his son in Melbourne, in 1859', *Early Days*, vol. 2, pt. 13, 1933, pp. 49–50.
[188] J.P. Stowe, 1996, 'The Forlorn Hope', unpublished ms, 1865, JSLBL cited in Sister Mary Albertus Bain 1996, *A Life of Its Own: A social and economic history of the City of Geraldton and the Shire of Greenough, 1846–1988*, City of Geraldton, Geraldton, p. 46.
[189] C.T. Stannage, 1979, pp. 124–134.
[190] ibid.
[191] ibid., pp. 185–190.
[192] ibid., pp. 152–154.
[193] ibid., p. 93.

artisans and small farmers. However, the imprisoned convicts and Aborigines led miserable lives at the hands of their gaolers. Crime, including brutal attacks on members of the police force and drunkenness, were rife among the impoverished ex-convicts and ticket-of-leavers and the incidence of suicide within their ranks had become so common that two years later the Mechanics' Institute held a debate on the subject.[194] Apart from a few Aborigines who had found employment as domestic servants, most of the local tribes people lived on the periphery of Perth society, hunting and camping in traditional ways close to the city around the Swan River. Drunken brawls occurred at times among them. This was all to the habitual consternation of the settlers.[195]

We have located herbarium and published records of Oldfield's plants taken from the vicinities of Perth, Mount Eliza, the Swan and Canning Rivers, Guildford, Fremantle and Rottnest Island.[196] Since none of his specimens from these localities was dated or referred to in Mueller's *Fragmenta* (1858–1859), it cannot be determined which were gathered on this visit or later in 1858–1859 when he was again in the area. Before 1856 the intensity of collecting around the Swan River district had resulted in hundreds of published descriptions of species new to science. Nevertheless, some of Oldfield's specimens from the area, although few, became types.[197]

It is unlikely that he had more than a few hours available on this occasion for collecting around Fremantle and Perth or that he had time to visit Rottnest Island, a place of incarceration for Aboriginal offenders about 18 kilometres off the coast of Fremantle. Shopping for durable food items—tea, flour, sugar, biscuits—for the journey north, and scouting around Perth and Guildford for transport or a traveller to accompany him on his journey, must have consumed much of his time. Nevertheless, whenever he had an hour or more to spare he may not have resisted the temptation to gather plants on the 'sandy plains [behind the town] covered with *Banksia* spp. & other trees, & an infinite variety of small shrubs & plants'.[198] Mount Eliza, one of his collecting places, and its surrounding bush, which later became King's Park, was close enough to Perth for this purpose—even today it is quite rich in flora native to the area.

To Oldfield, the embryonic stage of development of rural and urban settlements in the south-west of Western Australia would have been similar in many ways to that of the districts in south-east Tasmania. The settlers had an immense struggle in establishing their communities and industries in the face of a harsh environment; coupled with antagonism between them and the Aborigines, and the isolation of settlements through lack of infrastructure, did nothing to facilitate communication between those who were but a few miles apart, let alone the other Australian Colonies, or Europe and the Americas. Perhaps, however, he sensed that settlers were slowly succeeding in establishing farms, businesses and community institutions and, with the introduction of convict labour in 1851, settlers' demands for better roads and other facilities were starting to bear fruit.[199] In many instances, however, it was a long-term and often frustrating endeavour. With this first experience of the Colony behind him, it is reasonable to suppose that he wondered what lay before him on his journey to the more distant and isolated Murchison River.

---

[194] ibid., pp. 93–101.
[195] ibid., pp. 120–121.
[196] HAODb.
[197] For example, type herbarium specimens in HAODb included 1 holotype; 8 syntypes, 1 isotype, 3 types.
[198] WHH to Sisterella, 21 April 1854, cited in Sophie C. Ducker (ed), 1988, p. 106.
[199] Geoffrey Bolton, 2008, *Land of Vision and Mirage: Western Australia Since 1826*, University of Western Australia Press, Nedlands, pp. 25–44.

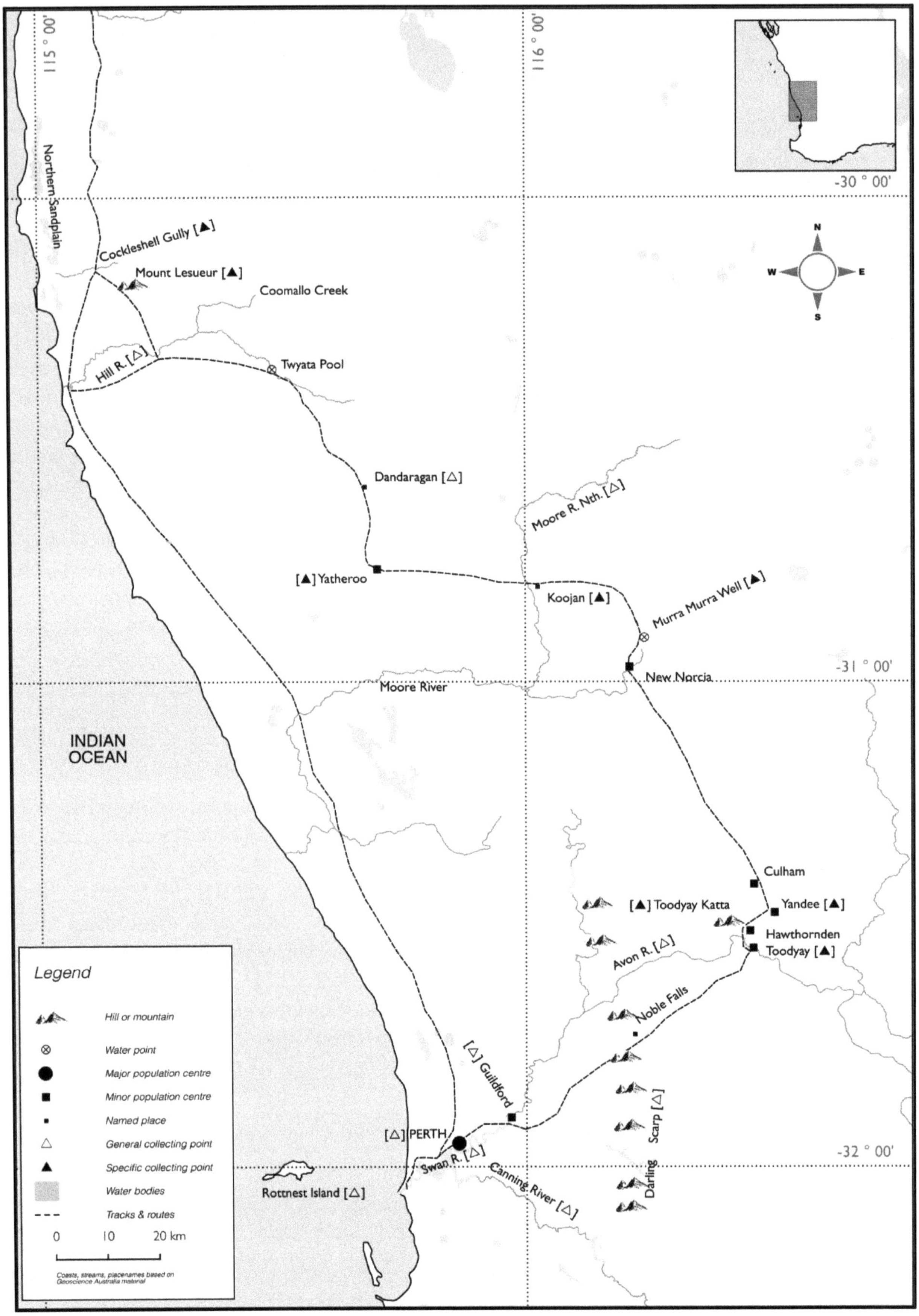

Map 9: Perth to Cockleshell Gully

Fig. 12.0 Leucopogon oldfieldii

Chapter 12

Western Australia 1856
Swan River to Geraldton

It appears that Oldfield travelled overland in 1856 to Geraldton, which is situated on Champion Bay (see below). At the time, roads from Perth to Champion Bay and the Murchison River were not much more than narrow sand tracks and there were no coach services. Oldfield's options for transport were limited, possibly restricted even further by time constraints. The mail was carried in stages to Port Gregory by mounted police constables stationed at Perth, Toodyay, Dandaragan, Irwin River and Geraldton and so mailmen with spring carts were not in the offing.[1] It is doubtful that he accompanied a carter with a horse or bullock dray delivering supplies to settlements between Perth and Greenough River or further north, or a drover with a spring cart or dray in attendance driving cattle or sheep to a property in the Victoria District; such journeys often took more than three weeks to reach Geraldton.[2]

It is highly unlikely that he went alone on foot on this, his first visit to such remote parts of the Colony. The distance to be covered was great, approximately 640 kilometres, and taking at least three weeks; too long to reach the Murchison River in September. Furthermore, a pedestrian journey and even going alone on horseback would be inadvisable since the area was still sparsely settled and he was not familiar with the watering points or the Aboriginal groups inhabiting the land between Perth and his destination. The only remaining options were to accompany a traveller either in his horse-drawn spring cart or on horseback with a pack horse—a journey of this nature to Geraldton took about eleven days.[3] It is reasonable to suppose that the botanist had his own horse and pack horse loaded with his food and camping and collecting equipment.

European interest in the flora to the north of Toodyay prior to 1856

Map 9 p. 198 – Perth to Cockleshell Gully

At the time of Oldfield's visit, the flora to the north of Toodyay had received very little attention from professional botanists, commercial collectors or settlers. Dr Johann August Ludwig Preiss, during his visit to the Colony between 1838 and 1842, collected at Toodyay and about 16 kilometres north of there at Samuel Phillips' property, 'Culham'. He does not appear to have gone further north although this might not have been the case since Neville Marchant reported one of Preiss's collecting places as 'Pitsacharding Spring' as an unknown location in the Victoria District.[4] James Drummond Snr, his son Johnston, another Victoria Plains pastoralist, and Captain W.W.J. Butcher, with a servant to help with the horses, undertook a collecting expedition in 1844, during which he claimed they had penetrated '75 miles from the Moore River nearly due north over a rich grassy country'. It is unlikely, however, that he and Johnston went any further to the north. In 1845, they made another journey to the north and east of Toodyay to Lake Walyormouring.[5]

---

[1] W.J. de Burgh, 1986, p. 93.
[2] ibid., p. 67.
[3] ibid., p. 63.
[4] N.G. Marchant, 1990, pp. 131–135.
[5] Rica Erickson, 1969, pp. 98–99.

There is an undated specimen collected by Augustus Gregory at the Murchison River in the National Herbarium of Victoria[6] but no evidence has been found that his brother Frank collected to the north of Perth. In 1836, Surveyor General John Septimus Roe led a party north from Perth for 161 kilometres. Three years later he led another group north to succour men from Captain George Grey's expedition.[7] He may have collected plants during these trips but there are no recorded specimens which bear any localities between Toodyay and the Murchison River in either the Kew online catalogue or the National Herbarium of Victoria catalogue.[8] Some of Roe's specimens were also sent to the Conservatoire et Jardin Botaniques de la Ville de Genève, Switzerland but according to its online herbarium catalogue, none were from between Perth and the Murchison River.[9]

George Maxwell, a commercial collector and later a colleague of Oldfield's, accompanied Lockier Burges in 1849 to the Geraldine when he (Burges) first took up his appointment as Mine Manager. Unfortunately,

> through the carelessness of one of the soldiers the grass [around the camp at the Geraldine] caught fire...Maxwell's tent caught fire and everything was consumed in a moment, his loss was very great as he had plants of over £50 which he prized very much.[10]

No records of specimens collected by Maxwell on this expedition have been found.

In October 1850 James Drummond Snr accompanied the first cattle and sheep drive to the new pastoral areas in the vicinities of the Irwin, Greenough and Bowes Rivers.[11] This expedition anticipated the publication in the *Perth Gazette* of new regulations for the lease and purchase of 'waste lands' of Western Australia'.[12] For ease of handling at watering points along the way, the stock was driven in two mobs and in each lot the cattle and sheep were driven separately. Driving the first mob were Lockier Burges and brothers Augustus and Henry Gregory. The personnel with the second mob were Thomas Brown and his younger son Kenneth, John S. Davis, James P. Walcott, Major Logue, a lad, Hamil [or Daniel] Hetherington, and three Aboriginal assistants, Courell, Judy and Arkle.[13] The plan was for the two droving parties to meet early in October at Dandaraga Spring, a new sheep station taken up by James Drummond Jnr.[14] During September, James Snr had accompanied his son James when investigating leases at the Hill River and he had remained in the region, making his base at Dandaraga Spring, from where he went on several short excursions to the east and west of Hill River.

Meanwhile, another party of horsemen who wished to examine land around the Upper Irwin planned to join Burges' party at Dandaraga Spring, and had gathered at Toodyay with the intention of joining the two droving parties. Led by surveyor Augustus Gregory, the group comprised Surveyor General Septimus Roe, Culham pastoralist Samuel Phillips, Narrier, an Aboriginal, and two of James Drummond Snr's sons, James Jr and John Nicol, as police escort and mediator with the Aborigines.[15] John was to take up his appointment as First Constable of a small force at Champion Bay. Drummond

[6] MEL 71759, *Brachychiton gregorii*, Murchison River, undated, A. Gregory (Catherine Gallagher, NHV RBGK to MHH, email 2 March 2012, HOA).

[7] Malcolm Uren, 1967, 'Roe, John Septimus 1797–1878', ADBO, viewed 3 March 2012 at www.adb.anu.edu.au.

[8] Both Roe and Gregory are included in the ANHBiog as having specimens in various herbaria, viewed 25 May 2009 at www.anbg.gov.au. Michael Henry O'Connor listed species reported by other explorers and visitors to mid-west coastal Western Australia (1697–1852) that are not referred to in the above online biography, Michael Henry O'Connor, 2001, pp. 81–82.

[9] Conservatoire et Jardin Botaniques de la Ville de Genève, Switzerland, online herbarium catalogue, viewed 3 March 2012 at www.ville-ge.ch.

[10] L.C. Burges, Journal 22 November 1849 to 17 February 1850, entry dated 5 December 1849, JSBL.

[11] Rica Erickson, 'Table of Plant Collections made by James Drummond', Appendix B, 1969, p.168.

[12] This drive anticipated the publication in the *Perth Gazette* (13 December 1850) of new regulations, given royal approval on 22 March 1850, for the occupation of what were known as the 'waste lands' of Western Australia, W.J. de Burgh 1986, p. 46.

[13] ibid., pp 44–58.

[14] Rica Erickson, 1969, p. 129.

[15] ibid., pp. 127–132. See also Sister Mary Albertus Bain, 1975, *Ancient Landmarks: A Social and Economic History of the Victoria District of Western Australia 1839–1894*, University of Western Australia Press,Nedlands, Western Australia, p. 53.

Snr joined them at Dandaraga Spring.[16] Their route to Champion Bay was chosen by Augustus and Henry Gregory and approximated the one that Augustus had forged when he led the 1848 settlers' expedition to the north of the Murchison River from Perth.[17]

James Drummond Snr spent the early part of 1851 accompanying John Nicol on many of his patrols around the Victoria District. During 1850 and early 1851 the botanist collected as far north as the Murchison River, around Champion Bay, and at various localities to the south.[18] On his way home from Champion Bay he stayed at Dandaragan (formerly Dandaraga) Station. According to Eliza Brown, the first European woman to travel overland to Champion Bay, she, her husband Thomas and their son Kenneth, camped at the station on their way north in June 1851 and spent the afternoon and evening with Drummond. He was residing in 'a leafy dwelling in every way suited to the wants of the inhabitant, with plenty of room for drying and arranging specimens'.[19]

Robert Austin, the government surveyor, led an expedition via Cow-cowing Lake, Mount Magnet, Mount Murchison & down the Murchison River to Geraldine lead mine, between 21 June to 25 November 1854. During this journey, a collection of approximately 50 plants & birds was obtained, most by Kenneth Brown. Austin later surveyed around Toodyay & Northam.[20]

Perth to Cockleshell Gully, Western Australia

Map 9 p. 198 – Perth to Cockleshell Gully

Specimen records indicate that Oldfield went to the Murchison River in 1856 and again in 1858 via Guildford, Toodyay, Dandaragan Station, Hill River and Geraldton before proceeding further north.

He must have soon become aware of changes in the floral association as he wended his way across the Swan Coastal Plain from Guildford to the Darling Scarp (Darling Range), and then on to Toodyay. The flora of the plain, familiar in many respects to that already described for south of Perth, features mainly *Banksia* low woodland on leached sands with patches of Grass trees (*Xanthorrhoea*), the occasional Christmas Tree (*Nuytsia floribunda*), and *Melaleuca* swamps where the soil is ill-drained. It also sustains woodland of Jarrah (*Eucalyptus marginata*) and Marri (*Corymbia calophylla*) on less leached soils and infrequent Wandoo (*E. wandoo*). Ascending the Darling Scarp from the plain, the forest becomes increasingly dominated by Marri-Wandoo woodlands. Occurring on the laterite and granite on the scarp and plateau are tall shrubs such as the Prickly Bitter-pea (*Daviesia horrida*) and Snail Hakea (*Hakea cristata*). Oldfield travelled through the area soon after the beginning of the Spring flowering. Although perhaps not at the height of its blooming, the bush was probably alive with colours of many different hues. It is not hard to imagine that as he jogged along he itched to be picking samples for his collection. The road, however, was in a bad state of disrepair and apparently there was no attempt to remove trees that had fallen across it and bridges often needed repair.[21]

Their first overnight stop may have been in the Darling Range, possibly at Noble Falls. The botanist gathered specimens near and in the Darling Range[22] but species distribution maps in *FloraBase* suggest that only about half of them were obtained along this road, while the remainder may have been gathered on his way from Albany in 1858.[23]

---

[16] A.R. Pashley, 2002, *A Colonial Pioneer, The Life & Times of John Nicol Drummond, Policeman, Magistrate, Pastoralist*, EDUCANT, Cloverdale, Western Australia, p. 43.
[17] W.J. de Burgh, 1986, p. 54, and Michael Henry O'Connor, 2001, p. 84.
[18] Rica Erickson, 1969, pp. 134 & 168. At the end of 1852, Drummond sent 14 sets of 225 specimens to Hooker and subscribers from Murchison River, Champion Bay and southwards to Dandaragan. This formed his 6th collection (ibid.).
[19] W.J. de Burgh, 1986, p. 56.
[20] Alex S. George, 2009, p.274.
[21] Sister Mary Albertus Bain, 1975, p.221.
[22] HAODb.
[23] See Chapter 15.

Fig. 12.1 Transitional Landform Zone, Darling Scarp

Still within the Jarrah District of the Bibbulmun Province of the Southwest Australian Floristic Region, [24] Oldfield proceeded in a north-easterly direction wending his way across the undulating plateau towards Toodyay. As he travelled along he would have noticed the woodland becoming lower and more open; Wandoo and Marri are replaced with York gum (*Eucalyptus loxophleba*) and brown mallet (*E. astringens*). Tall Jarrah stands on decrusted summits disappear as the soil becomes drier, Wandoo persists, Powderbark (*E. accedens*), and both Jam (*Acacia acuminata*) and Rock Sheoak (*Allocasuarina huegeliana*) emerge as small trees. This association persists, with Jam and York Gum becoming increasingly dominant further to the north until the track enters the coastal sandplains to the west of Koojan[25] a property on the north branch of the Moore River then owned by Walter Padbury.

They probably reached Toodyay late the next day and stayed there overnight. Oldfield gathered plant specimens at and near the township[26] but what he collected in 1856 could not be determined. This sojourn would have presented an opportunity to replenish supplies before setting out on the long trek to Champion Bay. North of Toodyay there were no settlements, apart from Champion Bay, where they could obtain food supplies and other needs. In any event, they could have supplemented their diets with game and perhaps fruits, seeds and yams from the bush. If there had been plenty of winter rain they may not have needed to carry much water. Thus, for their overnight bivouacs, they probably followed the usual practice of other travellers and headed for one of the riverine pools, springs or creeks that were located at various places along the track, with native grasses close by for their horses and where they could replenish their water supply.[27] In emergencies they could have obtained assistance from one of the few station homesteads along the way.

The 1850s were the heydays of old Toodyay, which was located on either side of the Avon River, a little over two kilometres downstream from the current townsite. The district, which had become prosperous after the depressed economic conditions of the 1840s, started to turn around and the lack of

[24] Paul Gioia & Stephen D. Hopper, 2017, pp. 11-12 & Appendix S3, Fig. S3.10.
[25] J.S. Beard, 1990, p. 79.
[26] HAODb.
[27] See for example, W.J. de Burgh, 1986, Chapters 1–6.

labour had been ameliorated by the introduction of convict labour in the early 1850s. The population of free settlers was increasing and included a range of tradesmen, merchants and storekeepers, publicans, clerks, civil servants, including a Resident Magistrate, Joseph S. Harris, a doctor, school master, over a hundred farmers, labourers and domestic servants, and the town, as the district's service centre, flourished as a result. The town boasted three inns, two schools, a chapel, a gaol and a convict hiring depot, which had been established in 1851 on the southern side of the river three kilometres from the town. In addition to ticket-of-leave men, the population was boosted by a detachment of pensioner guards stationed in recently-built, two-roomed brick cottages in the hiring depot grounds. The depot had become a centre for other public services, besides being a labour exchange, it included the fire brigade headquarters and the post office.[28]

Five kilometres north of Toodyay, the track north passed the pastoral property of James Drummond Snr and within a hundred metres or so of his residence, Hawthornden. In a letter to Sir William Hooker, Oldfield stated that the Common Maidenhair Fern, (*Adiantum aethiopicum)*,

> occupied a very limited area in W. Australia, being only found at the southern extremity of the Darling Range. Drummond did not know the plant, but I think that he must have found it in [sic] his trip to Port Augusta [?Augusta, Western Australia].[29]

This suggests that he and Drummond were personally acquainted but when they met is not known. It may have been in either 1856 or in 1858 on his way to the Murchison. It is difficult to imagine that Oldfield would not have taken the opportunity to call on Drummond at his home in 1856 since he was, at the time, the only professional botanist and one of only two commercial collectors to have worked as far north as the Murchison River. By now, he was an old man and no longer collecting but it seems he was 'still quite hearty'[30] and probably receptive to visitors.

From Hawthornden the track continued north, passing two more of Oldfield's collecting places, Toodyay Katta[31] and 'Yandee'[32], a property adjacent to Hawthornden, for about 10 kilometres to just beyond Samuel Phillips' 'Culham Station', where it veered to the north-west. It then traversed the Victoria Plains[33], past the Benedictine Monastery of New Norcia and onto the Murra Murra Rivulet (the same locality as Murra Murra Well on Donald Macpherson's pastoral property)[34], where some plants were taken.[35] On this leg of the journey Oldfield entered the Wongan District of the Transitional Rainfall Province, SWAFR.[36] The track then turned west, crossing the northern branch of the Moore River at 'Koojan' and into the Lesueur District of the Bibbulmun Province, SWAFR.[37] Oldfield gathered plants at and near Koojan[38] and at Moore River. This was the northern branch of the river because there is no evidence from other specimen records that he travelled up the lower section of the

[28] Rica Erickson, 1974, *Old Toodyay and Newcastle*, Toodyay Shire Council, Toodyay, pp. 148–174. Erickson provides population statistics for 1859 and district occupations for 1856 are inferred from those figures.

[29] AO to WJH, 5 December 1863, English Letters, vol. 42, 1862–1865, L–Z, fol. 198, RBGKA. No evidence of Drummond having collected in Port Augusta in South Australia has been found but he did visit Augusta in Western Australia (see Chapter 19).

[30] Percy U. Henn (ed), 1954, *Wollaston's Albany Journals (1848–1856) being volume 2 of the Journals and Diaries (1841–1856) of Revd. John Ramsden Wollaston*, Paterson Brokensha Pty Ltd., Perth, p. 219. Wollaston noted in his entry for 25 January 1856 that he had dined at Drummonds' the previous day and commented on Drummond Snr's state of health.

[31] HAODb.

[32] ibid.

[33] HAODb.

[34] Rica Erickson, 1971, *The Victoria Plains*, Lamb Paterson Pty Ltd., Osborne Park, Western Australia, p. 13.

[35] HAODb.

[36] Paul Gioia & S.D. Hopper, 2017, pp. 11–12 & Appendix S3. Fig. S3.10.

[37] ibid. The Lesueur District commences just above Perth and extends along the coast to Geraldton with its eastern boundaryfor the whole of its length about fifty kilometres inland, ibid.

[38] ibid. The locality on specimen labels is given as 'Culjong', 'Cujong' or 'Coujong' but for the following reasons we believe they are variant spellings of 'Koojan'. The ranges of species of other specimens bearing these localities, according to *FloraBase* distribution maps, do not extend above Geraldton and many not above Hill River (e.g. MEL 1534244 *Petrophile ericifolia,* MEL 2175961 *Lambertia multiflora*; 'Cujong'; MEL 672926 *Isopogon sphaerocephalus*; MEL 276696 *Aotus procumbens,* 'Near Coujong' MEL 2175262 *Grevillea pilulifera.*; and MEL 675064 *Hakea conchifolia*, 'Near Cujong, near the Hill River'). Also, one of Oldfield's specimen records, MEL 275746 *Verticordia patens* bears the locality 'Victoria Plains, near Cujong', MELISR database, 18 November 2005, NHV, RBGM. There is a record of an *Isopogon divergens* with the locality 'Culjong, Murchison River', Bentham, *Flora Australiensis*, 1870, p. 349. According to

Fig. 12.2 Toodyay Katta

Old North Road, which crosses the Moore River near Guilderton, or to the north-east branch of the river to the west of the Toodyay to Koojan track.[39]

A few miles after leaving Koojan, the track enters the vast northern sandplains through which it continues westwards to Yatheroo and then northwards towards Champion Bay. These plains extend from near Jurien Bay to the bases of the Cararang and Peron Peninsulas at Shark Bay.[40] From south to north and from the coast inland, the flora changes partly in response to decreasing rainfall. On first sight from a distance, this open country, covered in low shrubby heath (*kwongkan*)[41] with isolated small trees dotting the landscape away to the horizon, may have appeared dull and monotonous and possibly disappointing to a collector. Upon closer inspection, however, when he would have observed the extraordinary number of *kwongkan* species, which provide a spectacular flowering. The scrub-heath is under a metre in height, of which Sand Bottlebrush (*Beaufortia squarrosa*) and Round Fruit Banksia (*Banksia sphaerocarpa*) are generally the most prominent. The wattle *Acacia spathulifolia* dominates locally and there are occasional groves of small Illyarrie (*Eucalyptus erythrocorys*).[42] Although small, heath-like species persist across this plain, the Old North Road to the north of Hill River pass by swamps and occasional lakes, and through open flats and claypans with their park-like cover of *Eucalyptus, Casuarina* and *Acacia* species.[43]

The route passed Walter Padbury's pastoral lease 'Yathroo' (now Yatheroo)[44], where Oldfield gathered a few plants[45], and then ran north for about 19 kilometres to James Drummond's pastoral

FloraBase, this species occurs from south-east of Perth in a continuous belt to the north-west to above the Murchison River which appears to include 'Koojan', therefore where it was collected is unknown.

[39] *FloraBase* species' ranges of the 'Moore River' specimens embrace the Moore River at Guilderton, at 'Koojan' on the north branch the river and on its east branch about 10 kilometres west of the track near present day Bindoon (HAODb).

[40] J.S. Beard, 1990, p. 98.

[41] 'Kwongan' is now spelt 'kwongkan', see S.D. Hopper, 2014, 'Sandplain and Kwongkan: historical spellings, meanings, synonyms, geography and definition', in H. Lambers (ed.), 2014, *Plant Life on the Sandplains in "southwest Australia, a Global Biodiversity Hotspot*, pp. 23-33.

[42] W.J. de Burgh, 1986, p. 56.

[43] ibid., p. 60.

[44] Rica Erickson (ed), 1979, vol. 3, p. 653.

[45] HAODb. This locality is sometimes 'Yathroo' on specimen labels.

property Dandaraga Spring[46], which had a police station[47] and water for travellers. Once more Oldfield added to his collection.[48] From there the route veered north-west to Twyata Pool on Hill River and crossed Coomallo Creek, a tributary of Hill River, then proceeded south-west to Munbinea Crossing.[49] Oldfield obtained a specimen of a Sand Bottlebrush (*Beaufortia squarrosa*) '10 miles S of the Hill River' [50], which could have been taken anywhere to south of the river but most likely on or near the track south of Twyata Pool. Thus, there are no other records of specimens in HAODb bearing localities to the south of the track from Koojan where it enters the Old North Road on the coast. Oldfield gathered plants at and near Hill River.[51] A collecting place, 'Cabeenyong'[52], could not be located but he referred to *Ca-been-yong* as being a place near Hill River. The name signifies a locality 'where the *ca-been*, a kind of hawk, is common in summer'.[53] At Munbinea Crossing the track branched, one proceeding north to join Old North Road just below Cockleshell Gully, and the other continuing to the west, terminating at Hill River Crossing.[54]

A sample of Marri was obtained by Oldfield at Mount Lesueur[55], a prominence which is about five kilometres south of east of Cockleshell Gully. This indicates that at least on one of his journeys to the Murchison River, he crossed Hill River at the Munbinea Crossing where the track branched and ran north, and passed within about 3.3 kilometres to the west of Mount Lesueur and joined Old North Road a few kilometres south of Cockleshell Gully, where he obtained more plants.[56] They may have bivouacked at the Gully which, according to Eliza Brown, was 'an astonishing ravine'.[57]

Cockleshell Gully to Geraldton

Map 10 p. 207 – Cockleshell Gully to Geraldton

Oldfield made collections at other places along Old North Road[58] as it wended its way northwards across the northern sandplain to the Greenough River; Three Springs, near Three Springs, Lake Logue, Arrowsmith River, north of Arrowsmith River, near and on the banks of the Irwin River, and south of the Greenough Flats[59], which comprise two parallel alluvial plains (the Front and Back Flats) on the lower Greenough River. The area is approximately 10 kilometres wide, commencing 10 kilometres to the south of Geraldton and extending south-east along the Indian Ocean coastline for about 30 kilometres.[60]

Some, or all of these, localities may have been resting and watering places for Oldfield and his companion.[61] Eliza Brown considered Lake Logue 'a very refreshing piece of water abounding with ducks'.[62]

---

[46] Margaret McConnell, 1993, John McGuire & Carrick Moore, *Plateau, Plain and Coast: A History of Dandaragan,* Shire of Dandaragan, p. 20.
[47] W.J. de Burgh, 1986, p. 89–90.
[48] HAODb.
[49] No records of specimens collected from Munbinea Crossing have been located.
[50] HAODb.
[51] ibid.
[52] ibid.
[53] AO, 1865b, p. 291.
[54] W.J. de Burgh, 1986, map opposite title page.
[55] F. Mueller, 1860–61, vol. 2, p. 35.
[56] HAODb.
[57] W.J. de Burgh, 1986, p. 56.
[58] ibid., map opposite title page.
[59] HAODb.
[60] Michael Henry O'Connor, 2001, p. 47.
[61] W.J. de Burgh, 1986, pp. 53–56.
[62] ibid., p. 56.

Fig. 12.3 Sandplain looking south from laterite ridge, west of Koojan

Fig. 12.4 Cockleshell Gully

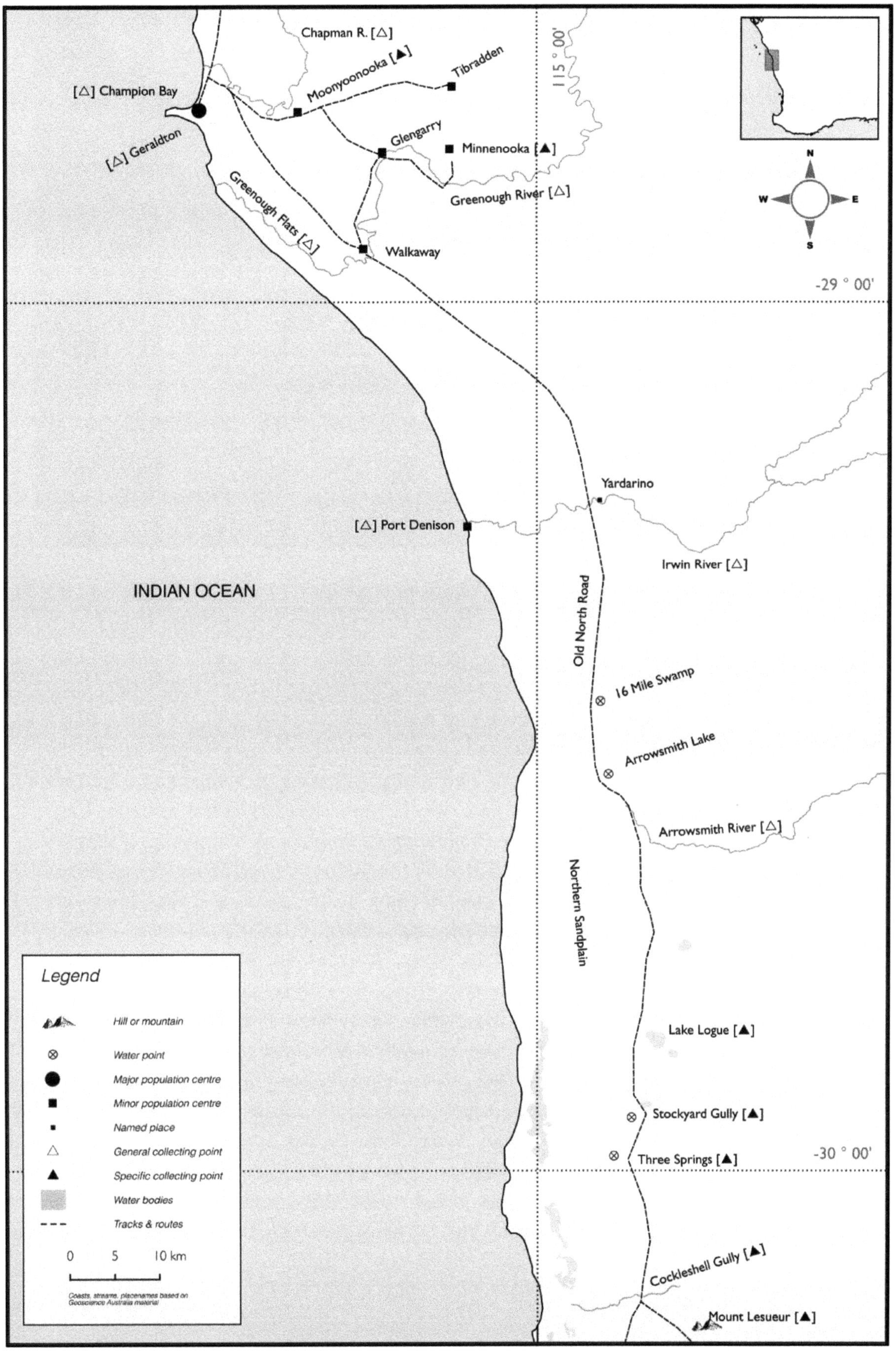

Map 10: Cockleshell Gully to Geraldton

Fig. 12.5 Lake Logue

In his unpublished manuscript Oldfield argued that Aboriginal origins and the peopling of the Australian continent by the ancestors of the *Alfoura* was by successive migrations of people of Malay origin, principally via New Guinea.[63] He speculated that:

> [T]hese hordes in their subsequent migrations seized on the facilities offered for subsistence by the physical conditions of the country, following, to some extent, the lines marking the distribution of plants...From all the observations hitherto made in Western Australia, it appears that, taking community of customs and similarity of dialects as indicative of identity of origin, the tide of emigration, starting from various points on the western coast, has followed lines trending nearly south-east; thus occupying strips of country possessing the same physical characters as the one from which it started. Thus the tribe located near Lake Longe [Logue] may be traced thence to the Australian Bight along a strip of country of nearly uniform nature, characterised as being the only known habitat of the Eucalyptus pleurocarpa. In like manner, the Eucalyptus gomphocephala marks the range of another tribe: and probably had attention been paid to this subject, many parallel cases might have been discovered.[64]

It is reasonable to assume that in his statement: 'From all the observations hitherto made in Western Australia...' he was referring to his own observations. As far as we are aware, he must have speculated on the range of Talyerak (*Eucalyptus pleurocarpa*) based on what Aboriginal groups told him. These may have included members of the *Amangu* tribe in the vicinity of Lake Logue and tribal groups along the south coast between Albany and Esperance, namely the *Minang*, *Koreng*, *Wadjari* and *Njunga*[65], who occupied territory of at least one or more of the places he visited along the south

---

[63] MHH considered AO's use of *Alfoura* to refer to the Aborigines of his day was intentional (MHH, 2005, p. iii).

[64] ibid. p. 3.

[65] Norman B. Tindale, 1974., *Aboriginal Tribes of Australia: Their Terrain, Environmental controls, distribution, limits, and proper names with an Appendix on Tasmanian Tribes by Rhys Jones*, Australian National University Press, Canberra, online reproduction of map of tribal boundaries published in this work, viewed 10 November 2010 at www.archives.samuseum.sa.gov.au.

Fig. 12.6 Original Old North Road Track, south of Irwin River

coast and where, according to *FloraBase*, *E. pleurocarpa* occurs.[66] He collected specimens of this species at Lake Logue and at the Kalgan River.[67] The distribution of *E. pleurocarpa*, as indicated by Western Australian Herbarium specimen records in *FloraBase*, is confined to either end of what Oldfield considered to be its range.[68] According to Alex George, *E. pleurocarpa* never occurred between Eneabba and the south coast.[69] However, according to Stephen Hopper, Oldfield's insight on Talyerak is a long overlooked but highly significant contribution on how Noongars of the south-coast perceive their territorial limits.[70]

The naturalist and his companion may have found the going slow and difficult if the flats and claypans were still wet and boggy after the winter rains.[71] Northwards along the track the taller species begin to dominate; Acorn Banksia (*Banksia prionotes*), Sandplain Woody Pear (*Xylomelum angustifolium*), Christmas Tree and Illyarrie, as well as the Slender (*Banksia attenuata*) and Firewood Banksia (*B. menziesii*). North of Arrowsmith River, Illyarries occur on the coastal limestone but less conspicuously. They are frequently embedded in thickets of the Summer-scented Wattle (*Acacia rostellifera*), White-stemmed Wattle (*A. xanthina*) and Dune Sheoak (*Allocasuarina lehmanniana*).[72] *Macrozamia fraseri*, a zamia, is an outstanding species along the Old North Road track beyond Arrowsmith Lake

66 See Chapter 20. These groups are part of the wider non-circumcising Noongar who occupy the south-west of Western Australia (see Alison Lullfitz et al., 2017, p. 202 for a map showing the limits of the Noongar territory.

67 MEL 1610974 and MEL 1610975, *Eucalyptus pleurocarpa*, Kalgan River; MEL 706642, *E. pleurocarpa*, Lake Logue (including 'Lake Lough' and 'Lake Louge' on specimen labels), MELISR database, 18 November 2005, NHV, RBGM, and NSW 340773, E. pleurocarpa, (download of Oldfield's collections at NHNSW. Karen Wilson, RBGS to MHH, email attachment, 23 October 2006, HOA).

68 This species has only been obtained on the west coast from north-east of Badgingarra to 7 kilometres south of Eneabba, a town about 7 kilometres south-east of Lake Logue, and along the south coast in a relatively narrow band on the Marine Plain from just west of Albany to Esperance, with one doubtful collection at Eucla on the Western Australian/South Australian border (refer to specimen record data, *FloraBase*).

69 Alex S. George, personal communication, 7 June 2014, HOA.

70 Alison Lullfitz, et al., 2017, p. 212. We are grateful to Stephen Hopper for drawing our attention to this recent finding, personal communication, email attachment, 22 December 2017, HOA.

71 W.J. de Burgh, 1986, p. 60.

72 J.S. Beard, 1990, p. 100.

and Sixteen Mile Swamp, with their over-storeys of picturesque paperbarks (*Melaleuca* sp.) and Casuarina (*Casuarina* sp.).[73]

The road crossed the Irwin River at 'Yardarino' and continued in a northerly direction, passing through the southern third of the Greenough back flats and very close to 'Allanooka' homestead before veering west to the ford at Old Walkaway.[74] A specimen of *Chorizema racemosum* has the locality 'J.P.W.'s residence, Mononoko'.[75] This can only refer to James Perry Walcott's residence at 'Minnenooka' about five kilometres north of Greenough River.

There is, we believe, ample evidence that Oldfield and James Walcott's brother Pemberton Walcott travelled north together from 'Minnenooka' during the Spring of 1856. In volume 1 of his *Fragmenta* (1858–1859) and in some later volumes, Mueller referred to specimens that were jointly collected by 'Oldfield and P. Walcott'[76] and 'Oldfield & Walcott'[77] from the Greenough River, and Drummond's Hibiscus (*Hibiscus drummondii*), which flowers from August to October, was taken from the vicinity of Greenough River.[78] Among other joint collections obtained to the north of Minnenooka were specimens destined to become the lectotype and isolectotype of *Logania spermacocea,* flowering August to December, from 'Moonyoonooka', Champion Bay.[79]

'Moonyoonooka' was Thomas Brown's pastoral lease which abutted his 'Glengarry' pastoral lease to the west and extended north-west to within about eight kilometres east of Geraldton on the road to 'Tibradden'.[80] Thomas Brown bred horses to trade with India and Mauritius.[81] There was no dwelling on Moonyoonooka, only a shearing shed just south of Tibradden Road.[82] Other jointly collected specimens taken on the way to Murchison River included *Myriocephalus oldfieldii*, flowering September to November[83], *Rhodanthe humboldtiana*, flowering July to November, from Champion Bay[84] and *Leptosema aphyllum,* flowering September to November, from 'White Peak', a pastoral property a few kilometres north of Geraldton.[85]

This gives rise to the question of when, and under what circumstances, Pemberton Walcott teamed up with Oldfield; neither of them left any details. Although no joint collections from south of Greenough River were located it seems likely that they travelled together from Perth. Thus, it is difficult to imagine why Oldfield would have deviated so far from the main track to Geraldton unless Pemberton invited him to accompany him to Minnenooka.

Walcott may have been in the Champion Bay district for family reasons or perhaps he was investigating the possibility of taking up a pastoral lease in the Victoria District; as yet he had not settled on farming in the lower south-west of the Colony. Initially, Thomas Davis, Major Logue and James Walcott stayed as working partners with Thomas Brown on 'Glengarry', a horse-breeding station on

---

[73] MHH personal observation, 1999.

[74] W.J. de Burgh, 1986, map opposite title page.

[75] MEL 629145, *Chorizema racemosum*, MELISR database, 18 November 2005, NHV, RBGM.

[76] See for example, *Burgesia homoclada* (now *Leptosema aphyllum*), White Peak and Murchison River, p. 222; *Breweria rosea* (now *Bonamia rosea*), at Murchison River and near Olenara, p. 233; *Pileanthus vernicosus*, Murchison River, p. 225, FM 1858/59 vol. 1.

[77] See for example, *Walcottia eriobotrya* (now *Lachnostachys eriobotrya*), Murchison River, p. 241; *Hibiscus elliottii* (now *Hibiscus drummondii*), Murchison River and near Greenough River, p. 220, ibid.

[78] ibid., p. 220.

[79] MEL 501242A & MEL 501242B, *Logania spermacocea,*'Mununoko, Champion Bay', and MEL 2193261, *Verreauxia reinwardtii*, 'Mununoko, Champion Bay', MELISR database, 18 November 2005, NHV, RBGM. The association of 'JPW's residence' with 'Mononoko' raises the question as to whether 'Mununoko' referred to the same place. In the NHV, RBGM herbarium database it is suggested that it refers to 'Moonyoonooka' and given that it is geographically much closer to Champion Bay than 'Minnenooka' we would support that proposal.

[80] Public plan, Victoria 1865, Cons. 5018 (507508), SROWA.

[81] Sister Mary Albertus Bain, 1975, p. 205.

[82] ibid., pp. 236, 271 & 344. The exact location of this shearing shed could not be determined. On one public plan, (Victoria, 1863, Cons. 5018, 506653, SROWA) there are Crown grants shown at the termination of each of the two tracks which ran from 'Glengarry' across Thomas Brown's leasehold property. We believe it is likely that the shed was on one of the locations at the end of the track closest to Geraldton since this would have been the shortest distance for transporting wool to the port.

[83] FM1862–1863, vol. 4, p. 157.

[84] ibid., p. 135.

[85] ibid., 1858–59, vol. 1, p. 222.

Fig. 12.7 Greenough River crossing, near Minnenooka

Greenough River about 20 kilometres south-east from Geraldton.[86] When Brown decided he would not live permanently at Glengarry but develop it for his son, Davis, James Walcott and others formed what was locally known as 'The Pastoral Company' and applied to take up two leases on their own account. On 19 March 1852, they were granted their first lease of 40,000 acres north-east of Glengarry. Davis established himself about 32 kilometres east of Geraldton at 'Tibradden', and James Walcott at Colilert Brook, which flows into the Greenough River east of Glengarry homestead. The Pastoral Company's headquarters were at Minnenooka.[87] During the 1850s others joined or left the Company. According to Bain, James Walcott moved to Minnenooka in 1857[88] but the timing of Oldfield's visit indicates that James was already there in August or September 1856. The Brown residence at Glengarry was on the north bank of Greenough River, about eight kilometres due west of Minnenooka.[89]

[86] Sister Mary Albertus Bain, 1975, p. 63.
[87] ibid., p. 175.
[88] ibid., pp. 176–177.
[89] Joint Operations Graphic, scale 1: 250,000, Series 1501, SH 50–1, Edition 1, 1983, Royal Australian Survey Corps, Department of Defence, Canberra, 1983.

Pemberton Walcott—Oldfield's co-collector north of Greenough River 1856

The eighth child of James and Johanna Walcott, Pemberton, was born in 1834, four years after his parents migrated to the Swan River colony from the English county of Shropshire. At the age of three he was sent to England into the care of his grandmother in Devon, and later educated at Blundell's School, Tiverton.[90]

He returned to Western Australia aged eighteen[91], c.1852, and joined his elder brothers Samuel and Robert on a 16,000-acre pastoral lease, which took in the flats along Moore River for 15 kilometres east of the 'Cowalla' boundary. He later took up land on Moore River near Samuel's property and not far from 'Mimegarra' where his eldest sister lived. Old North Road crossed the river only a few hundred metres from 'Cowalla' homestead and the Walcott residence was about eight kilometres east near the banks of the river.[92] Pemberton had little enthusiasm to follow Samuel, who during the early to mid-1850s had taken up a pastoral lease in the Irwin River valley, or to join the Pastoral Company with James on the Greenough River.[93]

Walcott's biographer, Joan Brockman, makes no mention of his 1856 visit to the Champion Bay and Murchison River districts with Oldfield, which resulted in at least 77 of his specimens being donated to the National Herbarium of Victoria.[94] Following this journey, in 1860 Walcott answered Augustus Gregory's call for volunteers to join an expedition to the north-west coast to assess the value of the land for pastoral purposes. No doubt Walcott's schooling in mathematics and science, which included natural history and astronomy, influenced Gregory's decision to appoint him as a volunteer for the collection of specimens of 'natural history and botany'.[95] However, his experience collecting with Oldfield may have contributed to his successful application but there is no confirmatory evidence.

Gregory, in his report of the expedition to the Governor of Western Australia, referred to the 'natural productions' discovered by the expedition: 'beds of the pearl oysters', sandalwood, indigenous fruits, tobacco and '[m]any beautiful flowers...which will be forwarded to some of the most eminent botanists, to be described and classified'. The expedition leader included a vocabulary of the language of the Aboriginal people at Nickol Bay assembled by Walcott. Also, in Nickol Bay, Gregory named two islands, Pemberton and Walcott, commemorating his naturalist and was high in his praise:

> Of the most valuable services rendered to the Expedition and to science by Mr P. Walcott I have already had occasion to refer, and I sincerely hope that his talents and zeal in the pursuits of Botany and Natural History may meet a more substantial reward than the thanks which are justly due to him.[96]

Species from this expedition, named in Walcott's honour, were originally described by Mueller and it appears likely that the entire botanical collection was sent to him.[97]

Mueller, in his paper on the plants collected by Walcott and Maitland Brown during the 1861 exploration of the region now called the Pilbara, referred to 'Mr Pemberton Walcott, who had already distinguished himself by forming a botanical collection at the Murchison River, towards Shark's Bay'

---

[90] Joan Brockman, 1989, 'Captain Pemberton Walcott', *Early Days*, vol. 10, part 1, 1989, pp. 13–17.

[91] ibid., p. 16.

[92] W.J. de Burgh, 1986, pp. 80–81.

[93] Joan Brockman, 1989, pp. 17.

[94] MELISR database, 05 December 2008, NHV, RBGM. There are several of Walcott's specimens in MEL without locality information.

[95] Joan Brockman,1989, pp. 17–18.

[96] F.T. Gregory, 1861, Report to the Governor of Western Australia on the North-West Coast Expedition of 1861, in Augustus Charles Gregory & Francis Thomas Gregory, 2002, *Journals of Australian Explorations by Augustus Charles Gregory and Francis Thomas Gregory*, Hesperian Press, Victoria Park, Western Australia, pp. 96–97.

[97] FM, 1863, 'A record of the plants collected by Mr Pemberton Walcott and Mr Maitland Brown, in the year 1861, during Mr F.T. Gregory's Exploring Expedition into North-West Australia in the year 1861', *Transactions of the Botanical Society*, vol. 7, 1863, pp. 479–500.

but there was no mention of Oldfield having been with him.[98] In an article on Western Australian botanists J.H. Maiden cited a species of the genus *Anthotroche* named in Walcott's honour the type of which was collected at the Murchison River, but he only referred to Walcott's 1861 Gregory expedition contribution.[99]

In 1862 Walcott became the second settler in the lower Warren River district when he took up land on One Mile Brook, a tributary of Big Brook, near the town of Pemberton, which, in 1916, was named in his honour. He did not continue farming and c.1873 moved to the metropolitan area, completed his master mariner's certificate and went on to have a successful career as Inspector of Pearl Fisheries, Sub-Collector of Revenue and an Officer of Customs.[100]

Following Gregory's expedition, Walcott kept up his botanical interests. The National Herbarium of Victoria holds specimens collected between 1863 and 1896 from various localities in the south-west, especially from Karridale and the Warren and Blackwood Rivers, and east of Albany, and from the north-west coast of the Colony, with most gathered on the Dampier Archipelago.[101] There are frequent references in the literature to him reporting on the diameter of the trunk of a 'gigantic' Karri, *Eucalyptus diversicolor*.[102]

Minnenooka to Geraldton with Pemberton Walcott

Map 10 p. 207 – Cockleshell Gully to Geraldton

The length of Oldfield's sojourn at the Minnenooka homestead, which was a bachelor establishment,[103] is not on record but he was probably there long enough to enjoy James's hospitality. Time enough perhaps to sort, re-paper and label his collections, take a refreshing bath, enjoy a meal or two of pigeon pie or a mutton, kangaroo or wild turkey roast with potatoes and other vegetables from the station garden[104] accompanied by fresh home-baked bread.

On their way from Minnenooka to Geraldton, Walcott and Oldfield jointly[105] and separately[106] collected plants at Moonyoonooka. They would have travelled there on the track that crossed Greenough River slightly south west of Minnenooka, continuing west, crossing the river again near the Glengarry homestead and just west of there, running north-west past Wizard Peak. They crossed both the Glengarry and Moonyoonooka leases to the east of the Greenough Back Flats and emerged on the Geraldton–Tibradden road just north of Moonyoonooka's shearing shed, about 11 kilometres east of Geraldton.[107]

The track was 1.6 kilometres or so west of the route taken by Captain John Lort Stokes and his 'large armed party' in December 1841, during the voyage of HMS *Beagle*, when they were returning to Champion Bay from Wizard Peak after having briefly explored the hill country between there and Mount Fairfax to the north. This return route took them across an 'undulating plain with patches of stunted woodland widely scattered' that was later to become part of Glengarry and Moonyoonooka leases. Just below the westward side of Wizard Peak the party discovered a 'native well' under a

[98] FM 1863, pp. 479-500.

[99] J.H. Maiden, 1909b, 'Records of Western Australian botanists and Australian botanists who have dealt with the Flora of Western Australia', *Journal of the Western Australian Natural History Society* for 1909, p. 26.

[100] Joan Brockman, 1989, pp. 19–24.

[101] MELISR database, 05 December 2008, NHV, RBGM.

[102] See for example, FM 1882b, *General Information Respecting the Present Condition of the Forests and Timber Trade Together with a Report on the Forest Resources of the Colony,* Government Printer, Perth, Western Australia, 1882, p. 6.

[103] James Walcott's marriage to Elizabeth Elliott took place on 5 April 1861, Rica Erickson (ed), 1979, p. 863.

[104] A meal of this nature was served up to a visiting survey party at Glengarry in the mid-1850s, Sister Mary Albertus Bain, 1975, p. 153.

[105] MEL 501241A & MEL 501241B, *Logania spermacocea* and MEL 2193261 *Verreauxia reinwardtii,* Walcott & Oldfield, MELISR database, 08 March 2007, NHV, RBGM.

[106] Oldfield: Moonyoonooka, Champion Bay (labels: 'Mununoko, Champion Bay', HAODb. Walcott: Moonyoonooka, Champion Bay labels: 'Monunooka, Champion Bay', 'Mounnoka', 'Munanoko, Champion Bay', 'Munonoke, Champion Bay', 'Mununoko, Champion Bay', 'Mununooka, Champion Bay', 'Near Moonyoonoka, Champion Bay', MELISR database, 05 December 2008, NHV, RBGM.

[107] Public Plans, Victoria 1863, Cons. 5018, Victoria District, (506653) and Victoria 1865, Cons. 5018 (507508), SROWA.

'drooping gum', encircled by 'deep holes from which the warran root [*Dioscorea hastifolia*] had been extracted'. Stokes provided an indication of the terrain and flora across the plain they travelled over, and which we believe was similar to that travelled by Oldfield and Walcott on their way to the Geraldton-Tibradden road.[108]

The bush in this area had never been cleared so it is reasonable to assume that the species noted by Stokes in 1841 were still there when Oldfield and Walcott passed through. Stokes reported that he had examined 'an extent of country little short of 40 miles, nearly the whole of which deserved the character of sterility.' He reported seeing 'no other woods or bushes' than the following:

> Narrow-leafed dwarf banksias
> Cabbage tree (*Nuytsia floribunda*)
> Small sheoak (Casuarina)
> Broom
> Wattle (or Acacia) of two kinds
> A species of Eucalyptus known by the appellation of drooping gum, of a dwarf and crooked description
> Blackboys (*Xanthorrhoea*), on south side of Wizard Hill only
> Tea-tree, of dwarf size, in two or three places.[109]

In some respects, the country which Stokes explored that summer would have presented a different picture to Oldfield and Walcott. According to the Acting Comptroller of Convicts, Lieutenant-General Henry Wray, who traversed the area in September or October 1856:

> I was very much surprised and delighted with the whole of the country from Mr Brown's place at Glengarry to the spring at the head of the flats—[i]n no part of the Colony have I seen so much good feed on so much good land. For miles the grass was up to the horses' girths and water abundant, and the land appeared to me of very good quality.
>
> The want of water in summer might I conceive be entirely obviated by sinking wells and raising the water by bullocks working in whims.[110]

Along the way Oldfield and Walcott may have encountered ticket-of-leavers and free men shepherding Brown's stock and one or more Aboriginal groups with whom communication was probably difficult.[111] According to Tindale, the *Amangu* occupied territory which embraced an area bordered by the Chapman River in the north, south to Hill River and to the east near Mullewa, Morawa, and Carnamah. The south-eastern boundary is not well defined and is somewhere north of Moora.[112] However, Michael O'Connor, who has reviewed the historical literature on Aboriginal group boundaries within the mid-west coastal region of Western Australia, found that the several studies which have attempted to delineate group boundaries are problematic. Thus, it is uncertain whether the determination was linguistic, ecological, or a mixture of both (for example, Norman Tindale).[113]

From Moonyoonooka to Geraldton Oldfield and Walcott passed into the lower limits of the Nanda District, Kalbarri Province, SWAFR. which continued in a north-westerly band along the coast to about Tamala Station and inland for about 50 kilometres and embraces the lower Murchison River. [114] As they proceeded into Geraldton they past through a succession of plant communities. Starting with low woodland of *Acacia* and *Banksia* species then with *Banksia* species and Christmas Tree (*Nuytsia floribunda*) on the upper slopes and hill crests, and an understorey primarily of Geraldton Wax

[108] J.L. Stokes, 'Expedition of the Beagle to Port Grey', *The Inquirer*, 22 December 1841, p. 2, DAN, NLA. This is a copy of a letter from Stokes to Surveyor General J.S. Roe.
[109] ibid.
[110] H. Wray to the Colonial Secretary, 15 October 1856, CSO, Inwards Correspondence, vol. 358, fols. 41–55, SROWA.
[111] Michael Henry O'Connor, 2001, pp. 215–216.
[112] Norman B. Tindale, 1974, p. 239.
[113] Michael Henry O'Connor, 2001, pp. 219–224.
[114] Paul Gioia & S.D. Hopper, 2017, pp. 11–12 & Appendix S3, Fig. S3.10.

(*Chamelaucium uncinatum*), followed by *Eucalyptus* savannah woodland with some sites containing River Gum (*E. camaldulensis*). There follows a narrow strip of low woodland of *Banksia Nuytsia* scrub with the Zamia (*Macrozamia riedlei*) also present. As the track descended into Geraldton from the Chapman River estuary it passed through coastal dune scrub, surrounding which were probably salt-tolerant species such as samphire (*Halosarcia* spp.) and the grasses *Sporobolus virginicus* and Couch (*Cynodon dactylon*).[115] No doubt there was a great deal to interest the naturalists as they wended their way to Champion Bay but they appear to have gathered very few plants. Two specimens were obtained at Chapman River[116] but when and where along the stream they were gathered is unclear.

Geraldton—September 1856

Map 10 p. 207 – Cockleshell Gully to Geraldton

There are records of Oldfield and Walcott's specimens that were given the locality 'Champion Bay' or 'C/Bay'[117] but, again, exactly where and when they were collected is unknown. As far as can be ascertained, Oldfield is the only one of the two collectors to have recorded a specimen, a *Patersonia juncea,* bearing the locality Geraldton[118], although there are some of his more distant collecting places associated with Champion Bay. For example, *Verticordia penicillaris*, 'Table Hill, Champion Bay' was gathered on the 1856 expedition.[119] The Bay only extends north from Point Moore peninsula to the mouth of the Chapman River, a direct distance of about 4.4 kilometres, which is further south than Table Hill.[120]

The need to obtain local knowledge about the territory to the north, and possibly to replenish supplies, would have rendered it necessary for the naturalists to visit Geraldton on their way to the Murchison River. If a strong south-westerly wind was blowing and churning up the sand it would have been most unpleasant travelling on the track, which ran south to Geraldton between dunes from the direction of the mouth of the Chapman River.[121] Furthermore, a sand patch was gradually forming across the track and closing the approach to the town.[122]

Almost surrounded by wind-battered, moving sand hills devoid of vegetation, Geraldton was a desolate place with unpaved streets and few buildings. Some of the original vegetation may have remained around the settlement—*Spinifex* grass and Coast Saltbush (*Atriplex isatidea*) forming dense thickets along the dunes bordering the beach. Further back, on the more stabilised dunes, there may have been 'climax thickets' of Summer-scented Wattle, White-stemmed Wattle, Dune Sheoak (*Allocasuarina lehmanniana*), Lilac Hibiscus (*Alyogyne huegelii*) and Tangling Melaleuca (*Melaleuca cardiophylla*). If there had been recent fire in these thickets, which is very common, a regression to a pioneer stage of low heath of Panjang (*Acacia lasiocarpa*), Coastal Honey Myrtle (*M. acerosa*) and

[115] This road runs through land north of Greenough Flats and is outside the parameters of Michael Henry O'Connor's 2001 study. However, his historical study of vegetation types was correlated with the land forms given in L. Gary Rogers, *Geraldton Region Land Resources Survey*, Land Resources Series No. 13, Natural Resources Assessment Group, Agriculture Western Australia, 1996, Geraldton Region Sheet 2 Land Resource Map No. 14/2. The plant communities along the Moonyoonooka–Geraldton Road have been inferred for the relevant landforms by applying O'Connor's historically derived plant communities for those land forms on Greenough Flats, Michael Henry O'Connor, 2001, pp. 183–184, 193–194, 196 & 198.

[116] MEL 1521403, *Scaevola globulifera* and MEL 236839, *Boronia coerulescens* subsp. *spicata-spinescens*, Chapman River, MELISR database, 18 November 2005, NHV, RBGM.

[117] AO: HAODb; Walcott: MELISR database, 05 December 2008, NHV, RBGM.

[118] *Patersonia juncea*, Geraldton, Coll. AO, D. Geerinck, 1974, 'Revision of the Australian Iridaceae', *Bulletin du Jardin Botanique National de Belgique*, vol. 44, 1974, pp. 29–60.

[119] FM 1858–1859, vol. 1, p. 226. See HAODb for further examples.

[120] An ordinance of 1860 was issued by the Governor giving the limits of the port at Champion Bay, Sister Mary Albertus Bain, 1975, p. 140. See also Director of Survey, Department of Defence, Joint Operations Graphic 1: 250000, Series 1501, SH50–1, Commonwealth of Australia, 1982.

[121] Sister Mary Albertus Bain, 1975, p. 19.

[122] H. Wray to Colonial Secretary, 15 October 1856.

Scaevola spp. may have occurred.[123] None of these species, however, is represented in the records we found of Oldfield's and Walcott's Champion Bay specimens.

Geraldton had been officially established in 1850.[124] The discovery of galena (lead sulphide) in the bed of the Murchison River in 1848 had led to the formation of the Geraldine Mining Company and directly to the creation of the townsites of Geraldton and, further to the north in 1853, Port Gregory and an associated townsite, Pakington, near Hutt River. This period coincided with opening the Victoria and Murchison districts to new settlers.[125] Soon after the arrival of convicts to Western Australia in 1851, the first ticket-of-leave men to work in the area were employed at the Geraldine lead mine on the Murchison River, about 50 kilometres north-east of Port Gregory.[126] Among those men were some with mining experience from Cornwall and the North Country in England. Two years later, a convict hiring depot was established within the newly gazetted town of Lynton overlooking the Hutt Lagoon, about 10 kilometres south-east of Port Gregory. Many pastoralists and other settlers availed themselves of its services.[127]

Laying out the townsite of Geraldton in 1849 was a challenge for Augustus Gregory. The site selected for the settlement was shifting sand hills devoid of vegetation, and this meant he had no other option but to plot streets between these dunes. This made for some departures from the favoured grid pattern.[128] Perhaps Oldfield and Walcott agreed with Joseph S. Harris, the Resident Magistrate at Toodyay, deputising for the first Victoria District Resident Magistrate William Burges who on holiday, when he expressed the view in 1856 that 'the present station is useful as a landing place but is sterile and unsuited for any other purpose'.[129]

Overlooked by Flagstaff Hill, most of the buildings in the settlement were situated along or near the beach front. Apart from the 99th Regiment's wooden barracks and Lieutenant Lempster R. Elliot's cottage 'on a bend in the bay' on what is today Marine Terrace, the settlement comprised a number of other buildings along the east side of the thoroughfare.[130] By 1856 these included a post office, 'a building divided into two stores belonging to the Government' (which probably served as the commissariat)[131], George Shenton's general store, and nearby, two sheds, one owned by William Burges for holding hides and wool from his pastoral property awaiting shipment, and the other belonged to the Geraldine Mining Company, which stored ore and equipment.[132] Two pensioner guards and their families, who had arrived in 1856 with the first convicts to the town, were 'living in a most uncomfortable manner. Their barrack...a wooden building divided into two by a blanket partition' was located on Gregory Street.[133] In the same year, according to Harris, there were also several small cottages, housing three pensioners and their families and a gaol.[134] The Colonial Secretary, Frederick P. Barlee, who was at the Bay in October 1856 while Oldfield and Walcott were at the Murchison River, reported:

> The Gaol…is without exception the very worst Public Building I ever saw, it is simply a long hut quite useless as a place of security as prisoners placed there have to be manacled hand and foot to prevent their escape. To say that it swarms with Fleas and

---

[123] J.S. Beard, 1990, p. 100.
[124] Sister Mary Albertus Bain, 1996, p. 45.
[125] Fiona Bush & Philip Griffiths, 1998, 'Geraldine Lead Mine', Documentary Evidence, Assessment of Cultural Heritage Significance, Register of Heritage Places, Permanent Entry, Heritage Council of Western Australia, p. 5, viewed 16 March 2009 at www.heritage.wa.gov.au.
[126] ibid.
[127] Sister Mary Albertus Bain, 1975, pp. 120–121.
[128] Sister Mary Albertus Bain, 1996, pp. 43–46.
[129] J. Harris to Colonial Secretary, 8 April 1856, CSO, Inwards Correspondence, vol. 368, SROWA cited in (ibid., p. 48).
[130] Sister Mary Albertus Bain, 1975, p. 38 and 1996, p. 45. According to Bain, surveyors' pegs were so covered with sand that a number of buildings were erected on streets (ibid.).
[131] H. Wray to Colonial Secretary, 15 October 1856.
[132] Sister Mary Albertus Bain, 1996, pp. 43– 59 and 1975, pp. 130–131.
[133] H. Wray to Colonial Secretary, 15 October 1856.
[134] J. Harris to Colonial Secretary, 8 April 1856.

> Vermin would be a poor description of what the place is, and it is really cruel and unchristian like to confine any human being for twenty-four hours in such a wretched place.[135]

At the time of Oldfield and Walcott's visit there was no public jetty but one was built the following year.[136]

There was also no school so those children whose parents could not afford private tuition spent much of their days playing in the dunes or on the beach. Living conditions were generally harsh. Cottages were sparsely furnished, often with packing cases serving as chairs and storage places, and were frequently overcrowded with the need for beds to be in the kitchen and for children to sleep on straw mattresses on the floor. Keeping such dwellings clean and tidy, according to Bain, 'must have been a nightmare...The constantly blowing wind with its burden of fine sand and/or topsoil would find its way into every nook and cranny'.[137] Difficulties with issuing stores from the Commissariat arose because local government had no control over Commissariat staff.[138] Although there was no hotel in the town, ticket-of-leavers, pensioners, other settlers and Aborigines occasionally obtained contraband liquor from visiting whalers and carters, or by stealing from the Commissariat store or settlers. The ensuing wassail often resulted in brawls that required police intervention.[139] Apart from these occasional disturbances and the bustle of visiting carters who brought ore from the mines or agricultural products to the port for shipment, or taking on goods imported by ships from the south for delivery around the district, it appears to have been a quiet place with few residents in the streets at any one time.

James Walcott may have encouraged Oldfield and Pemberton to make themselves known to William Burges, who had returned from leave in May[140], to obtain up-to-date news of disturbances among the settlers and tribal groups further north, and advise where to camp and access mail services. If they did succeed in catching up with him at the Bay they were lucky; by his own admission his 'principal residence' was his saddle.[141] As Victoria District Magistrate, Burges was required to travel over a wide area, from Irwin River in the south to Murchison River in the north.[142] His duties were not confined to court hearings. He spent considerable time in the field investigating crimes, mostly those alleged to have been committed by ticket-of-leavers and Aborigines, ensuring that mail services from the south were maintained, and dealing with all manner of other administrative issues.[143] He also held the positions of Sub-Collector of Customs at Port Gregory and Visiting Magistrate at the Lynton Convict Depot. From 1854 he was under instruction to reside at Port Gregory, much to the chagrin of settlers to the south and south-east who considered that he would be better placed at his property 'Knockbrack' on the Bowes River but a petition requesting this had no immediate effect. Knockbrack was well managed by Thomas Williams and his son, John; with the aid of ticket-of-leavers he ran a successful business.[144]

---

[135] Frederick P. Barlee to Governor Arthur Edward Kennedy, 6 October 1856, CSO, inwards correspondence. vol. 358, fol. 23, SROWA, cited in Sister Mary Albertus Bain, 1975, pp. 218, 219. Even so, the new lock-up had not been built by January 1858, William Burges to Colonial Secretary, 8 January 1858, CSO, inwards correspondence, Resident Magistrate—Champion Bay, vol. 413, fol. 13, SROWA.

[136] Geraldton Port Authority, 'Geraldton Port – history and handy facts', viewed 17 February 2013 at www.gpa.wa.gov.au.

[137] Sister Mary Albertus Bain, 1996, pp. 266–268.

[138] William Burges to Colonial Secretary, 9 March 1854, 13 March 1856 and 11 June 1855, CSO, Inwards Correspondence Resident Magistrate—Champion Bay, vol.309, fol. 93; vol. 367, fol. 42, vol. 339, fol. 48, SROWA.

[139] Scattered through the correspondence between William Burges and the Colonial Secretary are reports of drunken disturbances at Geraldton, CSO, Inward Correspondence, Resident Magistrate—Champion Bay, 1852–1859, vols. 256, 248, 309, 339, 367, 382, 413, 435, SROWA.

[140] William Burges to Colonial Secretary, 14 May 1856, CSO, Inwards Correspondence, vol. 367, fol. 65, SROWA.

[141] William Burges to Colonial Secretary, 12 June 1855, CSO, Inwards Correspondence, Resident Magistrate—Champion Bay, vol.339, fol. 49, SROWA cited in Sister Mary Albertus Bain, 1996, p. 183.

[142] Tanya Suba, *The Thematic History of the Town of Northampton*, Shire of Northampton, 1993, p. 42.

[143] See for example, William Burges to Colonial Secretary, 1856, CSO, Inwards Correspondence, Resident Magistrate—Champion Bay, vols. 367 & 368, SROWA.

[144] Sister Mary Albertus Bain, 1975, p. 183.

There was no public accommodation in Geraldton for travellers or their horses.[145] While they may have received a warm reception from townsfolk—visitors often helped to relieve the tedium of geographical isolation—meals and accommodation were probably not on offer. Ships from the south bringing goods to replenish the Commissariat and settlers' larders were not particularly regular and shortages were a frequent occurrence.[146] Settlers had to depend on wells in their own backyards or on rainwater tanks, which became depleted rapidly during the hot, dry summer months.[147] If householders managed to grow vegetables, it is unlikely that there was any surplus to their needs. Thus, even while Oldfield and Walcott were in Geraldton, indigenous plants (yams and fruits), and fauna (kangaroos and smaller marsupials, fish, crustaceans and birds), probably formed the mainstay of their diets. If they needed to purchase additional provisions—flour, tea, medicines—Shenton's store may have proved disappointing.[148] The naturalists probably needed to find a place to camp. Although fodder grasses could have been in short supply, Point Moore amongst the dunes and where there was a good supply of water would have been a suitable camping place.[149] There was, however, another place where they may have camped.

Dr Robert Thirkill Foley (1805–1868), one of Oldfield's contacts at Geraldton

Oldfield was to become acquainted with the medical practitioner Dr Robert Thirkill Foley during one of his visits to the Victoria District but when and where they first met is not known. Perhaps James Walcott had spoken of Foley's knowledge of the customs and language of the local Aborigines and suggested that the pair call on the doctor while they were in Geraldton. However, Pemberton may have known of Foley's interests before he arrived at Minnenooka. In 1851, Edward R. Parker, a settler at York, had published extracts of letters he had received from Foley in the *Perth Gazette*, giving details of what he had found out about the local Aboriginal groups and the terrain of the country surrounding Champion Bay.[150] Parker was high in his praise of the doctor's accomplishments:

> No gentleman in this country ever exerted himself in examining soils, proving the seasons, watching the climate, and in searching into the habits, customs, manners and languages of the aborigines of Western Australia, than my friend Dr Foley.[151]

Since 1853 Foley had resided about 5.3 kilometres north-east of the Geraldton settlement on a 17-acre allotment (Location 13) on the south bank of Chapman River.[152] This property, which is now part of the suburb of Strathalbyn, was close to the track that led into Geraldton from Mullewa Springs. Over the years it became 'an oasis of self-sufficiency', with a small vineyard, fruit trees, a substantial garden of 1.5 acres, a lime kiln, stables and plenty of fodder for horses. Two olive trees planted by Foley have survived to this day and are massive. Foley may have invited the naturalists to camp there.[153]

---

[145] In 1858 William Burges reported that 'a house of public accommodation with a Publican's Licence is much needed at the townsite of Geraldton', Burges to Colonial Secretary, 10 March 1858, Inward Correspondence, CSO, Resident Magistrate—Champion Bay, vol. 413, fol. 28, SROWA.

[146] See for example William Burges to Colonial Secretary, 2 September 1854, CSO, Inwards Correspondence, Resident Magistrate—Champion Bay, vol. 309, fol. 187, SROWA.

[147] Sister Mary Albertus Bain, 1996, p. 141.

[148] William Burges to Colonial Secretary, 7 August 1854, CSO, Inwards Correspondence, Resident Magistrate—Champion Bay, vol. 309, fol. 178, SROWA. In this communication Burges reported there were no stores in Geraldton from which 'rations' could be purchased. This may have changed by 1856 but no evidence has been found that by then Shenton was selling foodstuffs.

[149] Sister Mary Albertus Bain, 1996, p. 142.

[150] *Perth Gazette and Independent Journal of Politics and News*, 31 January 1851, p. 3, 14 February 1851, p. 3 & 21 February 1851, p.3, DAN NLA.

[151] ibid.

[152] Public Plan, Victoria, Cons. 3869, Victoria 076, 1853, SROWA.

[153] Geraldton Local Studies Department, 2001, *Geraldton: 150 Years, 150 Lives*, Geraldton Regional Library, Western Australia, p. 65.

Through Dellagah, a local Aboriginal guide who had 'attached himself to Foley', the doctor had learned, and reported on, the culture, including the vocabulary, of the local tribes.[154] In October 1850, Parker reported that Foley had informed him that 'he knew about 700 words of their language'.[155]

Foley arrived at Swan River colony in *The Protector* on 25 February 1830. He was surgeon to the 21st Regiment, and also in private practice in the Avon Valley. With Lieutenant Lempster R. Elliot and a detachment of the 99th Regiment he went to Champion Bay in late 1849 as surgeon, and at times took charge of the detachment in Geraldton during Elliot's brief absence. Following the 99th's return to Perth in 1853 he remained behind and practised privately in the district. From 1856 until the settlement was closed in 1857, Foley also served as medical attendant to the convict depot at Lynton.[156] Oldfield's initial meeting with the doctor, therefore, may not have been at his home on the Chapman River but at Geraldton or Lynton.

It seems reasonable to suggest that Foley's vocabulary of the 'Champion Bay tribe', which Oldfield later published and attributed to its compiler[157], was given to him on this his first visit to the district to assist him in his efforts to recruit Aboriginal assistance with his collecting. Although the *Watchandie* of the lower Murchison River and the *Amangu* people of Champion Bay appear to have different dialects of the *Nhanda* language[158] their territorial proximity probably meant they could understand each other. Foley's list includes useful everyday words: parts of the body and physical and mental states (e.g. hungry, afraid), some kinship terms, terrestrial bodies, meteorological phenomena, cardinal points, topography, common plants and animals, food gathering implements, and various kinds of action (e.g. to go, to swim). The historical record has not revealed the extent of Oldfield's relationship with Foley. However, at some stage they had at least one philosophical interchange. Oldfield referred to this when discussing 'the accidental blending of races':

> At Champion Bay, in Western Australia, I was much surprised to find in some of the *old* natives features nearly approaching the European type, although those parts had been settled but a very few years. I mentioned this fact to a medical gentleman who informed me that he had made the same observation, and could account for it in no other way but by supposing that a ship which had sailed from Calcutta to Swan River in the early days of the colony, and had never since been heard of, was lost in these parts, and that some of her people who had escaped had mingled with this tribe, a surmise strengthened by the traditions of the natives, who to this day call Perth (the capital of Western Australia) Ca-cut-ta, having probably mistaken the place of departure for that of the destination of the rescued people: added to this, they often asserted that in the event of the departure of the whites from among them, there were many of their females whom their laws would permit them to eat, they having white blood in their veins.[159]

Although Foley may not have been the only Medical Attendant in the district at the time of this discussion[160], his longstanding interest in the local Aborigines and Oldfield's known association with him points to him being the 'medical gentleman' to whom Oldfield referred.

---

154 Sister Mary Albertus Bain, 1975, pp. 179 & 238

155 Edward R. Parker to the Editor, *Perth Gazette and Independent Journal of Politics and News*, 31 January 1851, p. 3, DAN, NLA.

156 Geraldton Local Studies Department, 2001, p. 65.

157 AO 1865b, pp. 297–298. The vocabulary was attributed to 'Dr R.J. Foley'. He was baptised Robert Thirkill Foley, 5 February 1805, Huntley, Gloucestershire, England, Michael H. O'Connor, personal communication, 14 June 2005.

158 Michael Henry O'Connor, 2001, p.224.

159 AO 1865, p. 218.

160 There was a Medical Attendant, Mr F. Lonson, at Port Gregory who was appointed during 1856 but he resigned on 14 January 1857 (W. Burges to Colonial Secretary, 14 January 1857, Colonial Secretary's Office, Inward correspondence, Resident Magistrate–Champion Bay, vol. 387, fol. 185, microfilm, SROWA). Whether, like Foley, he was a qualified surgeon and took an interest in Aboriginal culture has not been ascertained. Foley resigned his official position as surgeon in Geraldton toward the end of 1859 (W. Burges to Colonial Secretary, 31 August 1859, CSO, Inward correspondence, Resident Magistrate—Champion Bay, vol. 435, fol. 122, microfilm, SROWA).

How long Oldfield and Walcott remained in Geraldton is not known. Their apparently tight travel schedule and the dreariness of the settlement probably meant they were there for no more than a day or two before they set out for the Murchison River.

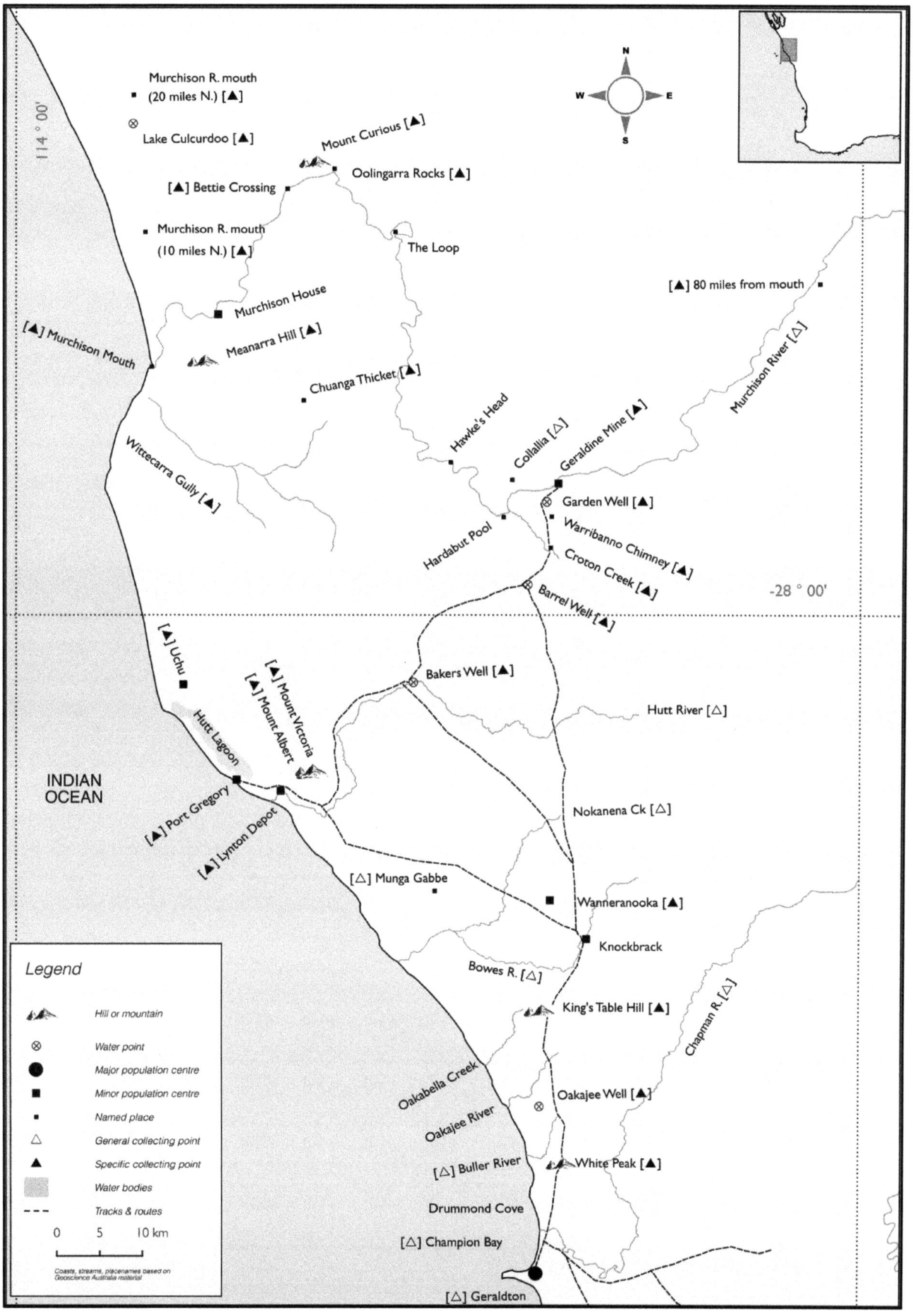

Map 11: Geraldton to the Murchison River

Fig. 13.0 Darwinia oldfieldii

Chapter 13

Western Australia 1856

Geraldton to the Murchison River

The road north from Geraldton commenced on the same track that Oldfield and Walcott most likely used to enter Geraldton.[1] They may have met up with a convict road party of 20 men working to improve the ford across the Chapman River.[2] Nearby, the sandy track, which posed a great challenge for heavily laden bullock-drays, became firmer and, according to Wray, from that point onwards to the Geraldine Mine was 'not bad'.[3] This road left the coastal dunes just north of the river and entered a series of low limestone hills covered with red, black or yellow sandy soil[4] extending to Hutt River and running parallel to the coast.[5] The terrain supported a more open shrub land dominated by Summer-scented Wattle (*Acacia rostellifera*) and Acorn Banksia (*Banksia prionotes*) associated with several other *Acacia* and *Banksia* species.[6]

A little further on, the road passed the track into the police station at Smugglers Cove (now Drummond Cove), where they may have hoped to meet up with John Nicol Drummond. Prior to coming to Champion Bay, he had been Inspector of Police at York and was apparently respected by Aboriginal people with whom he had dealings because of his excellent marksmanship with the rifle, his ability to live off the land, communicate with them in their own languages and ameliorate conflicts between them and settlers.[7] This reputation had by no means gone unnoticed by settlers living between Perth and Champion Bay[8], so the chances of Oldfield and Walcott, prior to reaching Geraldton, being aware of Drummond working at the Cove were high. It is difficult, therefore, to imagine them passing Smugglers Cove without calling in at the police station to seek more information about the tribal groups and terrain to the north. Where to find water, suitable grass for their hacks, edible bush plants, and what they needed to do to assist them in their relationships with the local Aboriginal groups must have been matters of great interest to them.

In 1856 Drummond held several offices: Sub-collector of Customs, Magistrate of the Colony, Commissioner of the Peace, Guardian of Natives and Deputy Superintendent of Police for the Northern District.[9] The police station was comprised of quarters for a second mounted policeman and an Aboriginal tracker, stables and a gaol for Aboriginal offenders. Under the superintendency of the Resident Magistrate, William Burges, Drummond had charge of one European and three Aboriginal constables and five horses. In a private capacity, he held a pastoral lease of 6,000 acres, which extended from Oakajee River in the north, south past Buller River and east to include White Peak, after which Drummond's property was named. The police station reserve at Smugglers Cove and the nearby

[1] Public plan, Victoria, 1863, Cons. 5018, (Tally no. 506653) and public plan, Victoria District, Sheet 2, Cons. 5017, (Tally no. 507475), SROWA.
[2] William Burges to Colonial Secretary, 15 October 1856, CSO, Inwards Correspondence, Resident Magistrate—Champion Bay, vol. 367, fol. 96, SROWA.
[3] H. Wray to the Colonial Secretary, 15 October 1856.
[4] L G., Rogers, and National Landcare Program (Australia), 1996, *Geraldton region land resources survey*. Report 13,Soil-Landscape Map of the Geraldton Region, Department of Agriculture and Food, Western Australia, Perth, p. 20.
[5] Public plan, Victoria, 1863, Cons. 5018 (Tally no. 506654), SROWA.
[6] J.S. Beard, 1990, p. 100.
[7] A.R. Pashley, 2002, p. viii.
[8] Rica Erickson, 1969, pp. 38–41.
[9] A.R. Pashley, 2002, pp. 57–61.

land owned by Drummond were within the boundaries of his pastoral lease.[10] He lived with his wife, Mary (née Shaw), at 'Redcliffe', his ten-acre freehold property at the mouth of Buller River, about four kilometres north of Drummond Cove.[11]

The geographic area that Drummond policed coincided with Burges's administrative district. This meant police patrols frequently took Drummond and his police constables away for lengthy periods. If Walcott and Oldfield did call at the station there may not have been anybody to receive them. In that event, they may have decided to try to see him at his home. Disappointment could have been offset if they found the reputedly 'tall, graceful and very beautiful' Mary[12] hospitable and of assistance. Botany would have excited mutual interests. Previously, she had sent Drummond Snr seeds of local plants, some of which he distributed to 'all garden lovers in the colony as well as to India and other countries'.[13] At one time or another Oldfield gathered plants at Buller River, possibly where the track north crossed that stream or perhaps near Redcliffe.[14]

A little further north the road veers slightly north-east[15] into undulating country of granites, granulite and flat-topped mesas. Red loamy soils support a scrub of acacias and hakeas with scattered trees of York Gum (*Eucalyptus loxophleba*) and River Gum (*E. camaldulensis). Allocasuarina campestris* thickets occur on gravelly soils. On the laterite, there is scrub heath of Champion Bay Poison (*Gastrolobium oxylobioides*), *A. campestris*, *Dryandra fraseri* var. *ashbyi* (=*Banksia fraseri* var. *ashbyi*), Spreading Coneflower (*Isopogon divergens*) and others. Communities on the sandy areas of the mesas include scrub heath of *Banksia* (including *Dryandra*), *Acacia*, *Dryandra* and *Allocasuarina* species. The side slopes of the mesas have thickets of *Melaleuca megacephala* and *Hakea pycnoneura* on the stony areas.[16]

It was here that that the road passed into Drummond's 'White Peak' station. Oldfield collected specimens which bear the label 'White Peak' and 'near White Peak'.[17] One of these was collected jointly with Walcott on the prominence.[18] They probably called in at the copper mine that had been operating for about a year within the pastoral lease. This mine was on a freehold block excised from the lease.[19] The White Peak Copper Mining Company was floated by Drummond and George Shenton in 1855.[20]

Oldfield collected specimens at or near Oakajee Well[21], six kilometres to the north of White Peak station, but on which of his visits is unclear. The pair may have rested there and replenished their water supplies.[22] However, a gathering of carters could have rendered the place so unpleasant that they did not dally for longer than necessary. Writing about three years later, William Burges did not have a good word to say about them:

> [T]he carters are too idle and lazy to assist themselves, they all keep Natives to look after their horses or bullocks, and they lounge their time away under the trees, I never saw in my life so good for nothing a set of men...I saw a lot of them the other day at Oakagee [sic] well, they complained of want of water but they would not even clean out the well, and wasted so much water as would have supplied their teams [for the want of] their caulking the old trough at the well.[23]

---

[10] ibid., p. 76.
[11] Sister Mary Albertus Bain, 1975, p. 100.
[12] ibid., p. 99.
[13] Rica Erickson, 1969, p. 154.
[14] HAODb.
[15] Cons. 5018 (Tally no. 506653), Victoria, 1863.
[16] L. Gary Rogers, 1996, p. 22.
[17] HAODb.
[18] *Leptosema aphyllum*, White Peak, P. Walcott & Oldfield, FM 1858–1859 vol. 1, p. 222.
[19] A.R. Pashley, 2002, p. 79.
[20] Tanya Suba, 1993, p.33.
[21] HAODb. No Walcott plants from this locality have been found.
[22] H. Wray noted in 1856 that 'There is good water always at the Oakagee Spring', H. Wray to Colonial Secretary, 15 October 1856.
[23] William Burges to Colonial Secretary, 12 January 1859, CSO, Inwards Correspondence, vol. 435, fols 56 & 57, SROWA.

Fig. 13.1 White Peak from the west

Approximately 11 kilometres further on was another of Oldfield's collecting places[24], the mesa Table Hill (sometimes referred to as Kings Table Hill).[25] No record has been found of a plant collected by Walcott at or near this locality.

An additional 11 kilometres north of Table Hill, the track reached the Bowes River and William Burges' sheep station, 'Knockbrack', which at the time was 'covered with good feed' and where 'in summer time [it had] very good water for cattle and sheep'.[26] Burges' residence was about a kilometre east on the banks of the Bowes River. It is very likely that Oldfield and Walcott called in there, especially if they had been unsuccessful in meeting the Magistrate in Geraldton.

Burges, who never married, was fond of entertaining and if he was at home when they arrived they would have been made welcome, particularly as one of the visitors was the brother of James Walcott with whom Burges had had a long-time association. Burges' establishment was impressive. His residence was a substantial stone dwelling reflecting his status as the local Resident Magistrate and successful pastoralist. It had large lofty rooms, including a well-appointed dining room, with double doors opening out onto a wide verandah, which served as a passageway to give access to each room. Running at right angles was a detached building comprising a serving-room, kitchen and pantry. Beyond was the dairy, a tool-shed and servants' quarters forming another side of a large quadrangle. Further to the north were the men's quarters, lofty stables, blacksmith's shop, a barn and saddlery room.[27] Perhaps the visitors were served a delicious meal of mutton in joints, pies and curries that had been so much enjoyed by the Lieutenant-Governor, Charles Fitzgerald, and his party when they visited Knockbrack in 1852.[28]

---

[24] *Verticordia penicillaris,* Table Hill, Champion Bay, Oldfield, FM 1858–1859, vol. 1, p. 226 & GB, 1866, vol. 3, 26. The flowering time, which accounts for all specimens recorded as flowering in the Western Australian Herbarium's *FloraBase,* is September to October, again well before the time Oldfield made his return journey from the Murchison River.

[25] Cons. 5018 (Tally no. 506653), Victoria, 1863.

[26] H. Wray to Colonial Secretary, 15 October 1856.

[27] Sister Mary Albertus Bain, 1975, pp. 180–181.

[28] ibid., p. 151.

There are records of Oldfield's specimens being found at and near Bowes River[29] and more of them than from anywhere else between there and Geraldton, but none were referred to in Mueller's *Fragmenta* (1858–1859). There are no Walcott specimens from this locality in the National Herbarium of Victoria. If the number of Oldfield's specimens does reflect his collecting activity at Bowes River in 1856, the pair may have tarried there for a day or so, possibly while Walcott was scouting for land for himself or his brother James.[30]

Three miles north was the Wannerenooka Copper Mine[31], another of Oldfield's collecting places.[32] It was one of several lead and copper mines in the area, around which had developed a thriving community of miners and their families. Oldfield left no indication of whether they availed themselves of the services of 'Dr' Joseph Lucas Horrocks' store and post office in the settlement. Horrocks was a conditionally pardoned convict who, following his arrival in the Colony in 1852, had gained medical knowledge through his work in the infirmary of Fremantle's convict establishment. Armed with this experience he was appointed Medical Superintendent at the Lynton Convict Depot in 1853. He resigned that position in 1855 and established business interests around Wannerenooka and 'attended to all the medical needs of the district'.[33]

There are also specimens which bear the locality 'Yanginooka [Yanganooka]' and 'Yanginoo Range'.[34] Yanganooka may have referred to the copper mine south-west of Yanganooka Hill, a prominence about five kilometres north of Wannerenooka and within one of James Drummond Jnr's pastoral leases.[35] 'Yanginoo Range' may have been a misspelling of 'Yanganooka Hill'. There is also a Yanganooka Well roughly 11 kilometres north of the northern end of Hutt Lagoon.[36] However, on Oldfield's labels 'Yanginooka' and 'Yanginoo Range' are often associated with other places quite distantly separated, geographically, which throws into doubt whether they were collected at the copper mine or the nearby prominence. Thus, Oldfield variously associated 'Yanginooka' and 'Yanginoo Range' with north and south of, and at, Murchison River, south Port Gregory, south Champion Bay and '? near Fremantle'. The distribution of each species given in *FloraBase* is of no assistance in identifying where, specifically, the specimens might have been collected. In his *Watchandie* vocabulary Oldfield refers to '*Yan-gin-no-ka*' as '[a] grassy valley'.[37] 'Yanginooka' when associated with a particular-named geographical feature such as the Murchison River might, therefore, have been a descriptive term for the locality rather than the actual name of it. However, this would not satisfactorily explain 'Yanginoo Range' which, according to Oldfield's labels referred to a range of hills that stretched from the Murchison River to south Champion Bay. As far as can be determined, the only gazetted range of hills between these two places and near the track that Oldfield and Walcott followed, is what is now known as the Menai Range which runs south from Lynton for a few kilometres.

The localities of specimens mentioned in the *Fragmenta* (1858–1859) and those of Walcott's in the National Herbarium of Victoria, and the flowering times of some of these, indicate that the pair went to the Geraldine via Port Gregory (see below) by following the track from Geraldton, which veered to the north-west just south of Wannerenooka Mine, past two of Oldfield's collecting localities, Nokanena and, about a kilometre further on, Munga Gabbe.[38] It then crossed Hutt River and from thence ran to Port Gregory[39], roughly following today's Northampton to Port Gregory road.

---

[29] HAODb. No record has been found of a specimen collected by Walcott at the Bowes River.
[30] Public plan, Victoria 18, Northampton. Cons. 5018 (506666), SROWA.
[31] Public plan, Victoria, 1863, Cons. 5018 (506653), SROWA
[32] HAODb.
[33] Tanya Suba, 1993, p. 54.
[34] ibid.
[35] Public plan, 1870, Victoria, Victoria District, Sheet 1, Cons. 5018, (507810), SROWA.
[36] Joint Operations Graphic, 1: 250 000, Geraldton Australia, Series 1501, Sheet SH50–1, edn 1, Royal Australian Survey Corps, 1983.
[37] AO, 1865b, p. 293.
[38] HAODb.
[39] Public plan, Victoria, 1870, Cons. 5018, (507810). SROWA

For about 14 kilometres after leaving Wannerenooka Mine the road passed through terrain and vegetation similar to the flat-topped mesas and undulating country that occur to the north of Drummonds Cove. It then traverses, for nearly 25 kilometres, a rich scrub heath coastal association of Summer-scented Wattle, Sandplain Cypress (*Callitris arenaria*), Common Woolly Bush (*Adenanthos cygnorum*), Christmas Tree (*Nuytsia floribunda*), Sandplain Woody Pear (*Xylomelum angustifolium*) and various species of *Banksia* and *Grevillea*. Hills and mesas support thickets dominated by *Melaleuca megacephala* and *Hakea pycnoneura* with some *Allocasuarina campestris* and Broom Bush (*M. uncinata*). Valley slopes on sandstone are covered with thicket and scrub heath, with Hutt River lined with River Gum, Swamp Sheoak (*Casuarina obesa*) and Swamp Paperbark.[40]

Before reaching Port Gregory, the road passed Lynton Convict Hiring Depot.[41] It was 'in a narrow valley, rock strewn, infertile, often waterless and in the summer quite airless'.[42] The original intention was that the occupants of the Depot would provide a founding population for the town and construct a road between Port Gregory and the Geraldine Lead Mine. However, the site was unsuitable for such a settlement, which had been recognised in official circles, and by the time the naturalists arrived, there were moves afoot to relocate the Depot and its inhabitants.[43]

Captain Henry Ayshford Sanford, who oversaw the Lynton Convict Hiring Depot from 1853 to 1854, accumulated pastoral leases totalling over 150,000 acres near Port Gregory and the lower Murchison River between 1853 and 1856.[44] One of Sanford's leaseholds (Murchison Location No. N236, 20,000 acres) was sold to Robert Mace Habgood in 1858. It was later to become 'Murchison House Station'.[45] Sanford also held a mining lease (Henrietta Mine) on Murchison Location 2 close to, and west of, the Geraldine Mine.[46] He purchased this after he had accompanied Frank T. Gregory on his exploration of the upper reaches of the Murchison River in 1857.[47] Sanford is reported to have leased 48,000 acres surrounding Lynton townsite and purchased Victoria Location No. 10, of 150 acres, on which his house was built.[48] He was apparently quite a successful farmer. He had grown wheat, barley and oats, ran a flock of sheep and had over 100 horses.[49] After he resigned his position as Superintendent of the Convict Hiring Depot in 1854 he engaged in whaling out of Port Gregory until 1856 and employed Charles von Bibra as his farm manager.[50] Sanford also had a carting contract from the Geraldine.[51] After Sanford returned to Fremantle in 1856, von Bibra rented his house at Lynton where he and his wife resided and operated a public house.[52]

Oldfield and Walcott may have stopped at the public house for refreshments and to catch up on the latest news about the track to the Murchison River and other matters that may have been of concern to them: Were the Aborigines friendly? Was there a local tribesman who could act as interpreter? Could they obtain fresh food at the Geraldine or anywhere else along the way?

Of those specimens from near Wannerenooka only one, *Halgania bebrana*, is referred to in the *Fragmenta* (1858–1859).[53] This plant was given to Oldfield by 'Domina de Bebra' and bears the

---

[40] L. Gary Rogers, 1996, Geraldton Region Sheet 1, Land Resource Map no. 14/1.
[41] ibid.
[42] Sister Mary Albertus Bain, 1996, p. 122.
[43] H. Wray to the Colonial Secretary, 15 October 1856.
[44] Applications for Depasturing Leases, Lands & Surveys RED, nos. 698, 700,701, 702, SROWA.
[45] Sister Mary Albertus Bain, 1996, p. 103. Bain referred here to Location 2 as being a Crown grant on what later became known as 'House Station'. However, this was pastoral lease N236 Victoria & North 2A, (Tally no. 506638), 1860, SROWA. Lot 2 was the Henrietta Mine.
[46] A.C. Gregory, Eye Sketch of Part of the Murchison River and Chain Survey of Locn 1, 1850, Cons. 3869, Victoria 0001, SROWA. Although the date of this plan is 1850, it seems that Sanford's mining lease was recorded on it sometime after Frank T. Gregory returned to Perth following his 1857 expedition.
[47] Sister Mary Albertus Bain, 1996, p. 103. Bain referred to this as a depasturing lease but it was a mining lease. Victoria & North 2A, Cons. 5018, (Tally no. 506638) 1860, SROWA.
[48] Tanya Suba, 1993, p. 76.
[49] ibid.
[50] Sister Mary Albertus Bain, 1996, p. 22.
[51] Tanya Suba, 1993, p. 76.
[52] William Burges to Colonial Secretary, 30 September 1856, CSO, Inwards Correspondence, Resident Magistrate—Champion Bay, vol. 367, fol. 93, SROWA.
[53] FM 1858–1859, vol. 1, p. 209.

Fig. 13.2 Captain Sanford's house, Lynton, 2001

locality 'In Western Australia, near Wannerenooka between Champion Bay and Port Gregory'. The donor was almost certainly Mrs Jane Charlotte von Bibra (née Gill)[54], wife of Charles von Bibra of Lynton. Mrs von Bibra is reported to have given the specimen to Oldfield when he visited Murchison House Station in 1859.[55] However, the species was described in volume 1 of Fragmenta which was published before Mueller received Oldfield's 1858–1859 Murchison collection.[56] Furthermore, the flowering time of *H. bebrana* is August to November which would suggest that Oldfield obtained the plant from Mrs von Bibra on his way to the Murchison.[57] This, of course, would depend upon whether a fresh specimen was presented to the collector. If it was a dried, pressed cutting it might not have been collected earlier. In any event, the flowering times of three other plant specimens referred to in Mueller's Fragmenta (1858–1859), all collected by Oldfield between Champion Bay and Port Gregory or Bakers Well, tend to confirm that the route they took to the Murchison was via Port Gregory.[58]

Lynton was established on sandy soils overlaying Tamala limestone that supported a vegetation of thickets and scrub heath, with the nearby Hutt River lined with River Gum, Swamp Sheoak and Swamp Paperbark.[59] The settlement comprised a cluster of seven well-built limestone buildings, including living quarters for pensioners and other officials, lock-up and associated enclosed yard, depot, commissariat, bakehouse and blacksmith's building.[60] Close by, to the north, was von Bibra's residence, a two-storey building that overlooked Hutt Lagoon.

[54] We are grateful to Alison Vaughan, Collections Information Officer, National Herbarium of Victoria, for providing Mrs von Bibra's given names and her birth and death dates, email to MHH, 25 April 2009, HOA.

[55] Sara Maroske and Alison Vaughan, 2014, 'Ferdinand Mueller's female plant collectors: biographical register', *Muelleria*, vol. 32, p. 103.

[56] See Chapters 17 & 18.

[57] The flowering time considered all the specimens of this species in the Western Australian Herbarium's *FloraBase* that were stated to be flowering at the time of collection, viewed 9 March 2009.

[58] *Anthobolus foveolatus*, between Port Gregory and Bowe River, flowering June to October; *Ptilotus helichrysoides*, near Baker Well, flowering July to November; and *Maireana planifolia*, between Geraldine and Port Gregory, flowering time May to September (FM, *Fragmenta,* vol. 1, 1858–1859, pp. 212, 237 & 213 and *FloraBase*. The flowering times considered all the specimens of this species in the Western Australian Herbarium's *FloraBase*, viewed 10 March 2009, that were stated to be flowering at the time of collection.

[59] L. Gary Rogers, 1996, 'Soil-Landscape Map of the Geraldton Region, Sheet 2' and p. 21.

[60] Heritage Council of Western Australia, Lynton Convict Hiring Depot (1853–1856), Register of Heritage Places: Interim Entry, Data Base No. 1915, p. 4, viewed 11 March 2009 at www.register.heritage.wa.gov.au.

Sanford employed a gardener, probably a ticket-of-leaver, to grow vegetables for the Depot on one of his leases which joined Bakers Well reserve. Although this was successful only during the winter months when there was a reliable source of water, there may have been produce available when the naturalists passed through in spring.[61]

If the travellers got into conversation with any of the local officials or ticket-of-leavers, they may have been shocked to learn of the conditions under which the inhabitants were living. In addition to isolation and a harsh climate, there was a shortage of fuel, the well water at Lynton townsite had become brackish, and during the summer months there were no vegetables. Sanitary arrangements were minimal earth closets. The inevitable outcome was that many of the inhabitants suffered from several diseases, including scurvy. Drunkenness, no doubt the result of disappointment and boredom, was quite common among the ticket-of-leavers and pensioner guards. Alcohol was readily available in the district. Apart from von Bibra's public house, Francis Pearson, who erected and worked the smelter at the Geraldine Mine, had operated a wayside inn at Lynton since 1853 and, in the whaling season, sold contraband liquor.[62]

Both Walcott and Oldfield obtained specimens at Hutt River in 1856[63] and Oldfield also made other collections near the river[64] but when he did so isn't known. Most of his plants from this area bear the locality 'South Hutt River' but there is no stream with that name on any gazetteer or map so it is assumed that he is referring to the southern portion of the river, which, according to a comment on one of his 1856 plant labels, extended to near Baker Well.[65] The river flows into the sea south of Port Gregory.

We have not located any of Walcott's specimen records from or near Port Gregory but there are many of Oldfield's.[66] Only one of these, a Tangling Melaleuca (*Melaleuca cardiophylla*) was referred to in the *Fragmenta* (1858–1859).[67] The flowering time of this species does not provide a clue as to when it was obtained.[68] In any event it would be surprising if, having reached Lynton, the pair did not explore Hutt River Lagoon, a very large and intriguing expanse of water north of, but not connected to, Hutt River. In the Lagoon, there occurs naturally a single-celled micro-alga, *Dunaliella salina*, which, when it becomes trapped in salt crystals, renders the water pink.[69] The Lagoon, rich in salt deposits—15 centimetres deep and 99 per cent pure—was mined using convict labour.[70] It does seem, however, that in 1856 he collected at least one specimen, *Anthocercis intricata*, at Culla, located somewhere between Port Gregory and the Murchison River, but its precise location is unknown. *Culla* in Oldfield's *Watchandie* vocabulary refers to 'a spring under a hill'.[71]

He gathered plants at Uchu, near Port Gregory, and Wanmeeneenu, near Uchu.[72] His *Watchandie* vocabulary refers to '*U-chu*. A swamp near the sea'.[73] According to Mrs Robin Simkin, a local resident, Uchu is a place just beyond the north end of the Hutt Lagoon.[74] This may have been where, in 1839, George Grey and his party, during their walk south from the mouth of the Murchison River,

[61] Sister Mary Albertus Bain, 1996, p. 21.
[62] American whalers frequently arrived from the USA with their vessels packed with contraband in May when the humpback whales swarmed into the Great Australian Bight from the Antarctic and from there followed them along the coast to the north where they bred in warmer waters, ibid., p. 135.
[63] *Siebera cirrosa*, Hutt River, A. Oldfield, *Ptilotus helichrysoides;* Near South Hutt River, e.g. near Baker Well, A. Oldfield (FM. 1858–1859, vol. 1, pp. 231 & 237) and MEL 1585181A; *Waitzia acuminata* var. *acuminata*, South Hutt River, Walcott (MELISR database, 05 December 2008, NHV, RBGM); MEL 275381A, *Verticordia picta* 'South Hutt River', P. Walcott & A.F. Oldfield (MELISR database, 18 November 2005, NHV, RBGM).
[64] HAODb.
[65] *Ptilotus helichrysoides*, near South Hutt River, e.g. near Baker Well, AO, FM 1858–1859, vol. 1: p. 237.
[66] HAODb.
[67] *Melaleuca cardiophylla,* Port Gregory, AO, FM 1858–1859, vol. 1, p. 225.
[68] This species flowers from August to February based on specimens in *FloraBase* that are stated to be flowering.
[69] Greening Australia & Alcoa, 'Hutt River: River in profile', *Rivers of Western Australia*, viewed 12 March 2009 at www.rowa.org.au.
[70] Tanya Suba, 1993, p. 81.
[71] FM, 1858–1859, vol. 1, p. 212.
[72] HAODb.
[73] AO, 1865, p. 293.
[74] Mrs Robin Simkin, personal communication, May 2011.

came upon native wells north of Hutt Lagoon and evidence of extensive digging in land overgrown with Warrine (the yam, *Dioscorea hastifolia* called by the *Watchandie Ad-ju-co*[75]).[76] We doubt that Oldfield visited this place in 1856 on his way to the Murchison River since it is quite a few miles north of the track they took to the Geraldine.

From the Lagoon, the track continued westward to Port Gregory and the adjacent township of Pakington, gazetted in 1854[77], passing through sand dunes with vegetation similar to the coast at Geraldton.[78] Pakington port had been officially approved by Governor Fitzgerald about a year after his visit in 1852, but no port facilities had been established at the time of Oldfield and Walcott's visit. Boats anchored in the harbour were attached to a permanent buoy from the stern.[79] The narrow harbour, with three navigable passages (Leander, Gold Digger and Hero), and enclosed by a reef, was unsafe for vessels exceeding 200 tons to enter under sail, and for a vessel of any size 'to bring to, if she enters with a fair wind of any force'.[80] Scattered through William Burges' correspondence with the Colonial Secretary are occasional reports of ships becoming wrecked on the reef; one of these was the schooner *Sara* a few months prior to Oldfield and Walcott's visit.[81] The 'best entrance' according to Wray, was

> the Leander passage, and any vessel entering with the wind from the South West must luff up close under the reef, and unless she is being handy would be very likely to be set down on the point which runs out from the land side of the Port.[82]

By 1856 the only inhabitants of Pakington were two conditionally-pardoned men in the service of von Bibra as boatmen.[83] At times, however, Port Gregory was busy with carters unloading lead, salt or wool and hides from their drays on the beach[84] and loading the cargo into tenders for uptake by a vessel, or with receiving cargo for delivery to settlers, the Geraldine Mine and the Lynton Convict Hiring Depot.

Pakington was located behind the sand dunes of Hillock Point, opposite Gold Digger Passage[85] about a kilometre north of Port Gregory. Nearby, Sanford and his partner Joshua Harwood operated a whaling station. Harwood's crew was apparently living in Sanford's storehouse, built on Lot 1 of the Pakington townsite, at the time Oldfield and Walcott were in the area.[86] In addition to these local whalers, others from America, France and eastern Australian Colonies visited Port Gregory during winter and spring to fill their water casks, barter liquor and other goods for fresh produce and to boil down blubber on the beach in their huge iron trypots.[87] As they came over the dunes from Lynton, the naturalists may have found the stench of boiling blubber overwhelming[88] but perhaps it signalled the possibility of purchasing some desired contraband.

---

[75] AO, 1865, p. 294.

[76] George Grey, *Expeditions in Western Australia 1837–1838, Journals of Two Expeditions of Discovery in the North-west and Western Australia, during the years 1837, 38 and 39, under the authority of Her Majesty's Government*, vol. 2, pp. 8–22 and *Map and Chart of the West Coast of Australia; from Swan River to Shark Bay from the Surveys of Capts Grey, Wickham and King*, vol. 1, compiled by John Arrowsmith, T. & W. Boone, London, 1841.

[77] Jennifer Rodrigues & Ross Anderson, 2006, *Pakington Whaling Station*, Report no. 214, Department of Maritime Archaeology, Western Australian Museum, p. 9, viewed 15 March 2009 at www.museum.wa.gov.au.

[78] cf. L. Gary Rogers, 1996, 'Soil-Landscape Map of the Geraldton Region', Sheets 1 & 2..

[79] H. Wray to the Colonial Secretary, 15 October 1856.

[80] ibid.

[81] William Burges to Colonial Secretary, 21 July 1856, CSO, Inwards Correspondence, Resident Magistrate—Champion Bay, vol. 367, fol. 85, SROWA.

[82] H. Wray to the Colonial Secretary, 15 October 1856.

[83] William Burges to Colonial Secretary, 13 March 1856, CSO, Inwards Correspondence, Resident Magistrate—Champion Bay, vol. 367, fol. 42, SROWA.

[84] H. Wray to the Colonial Secretary, 15 October 1856.

[85] There are apparently no historical references which pinpoint the location of Harwood's processing area or try works, although contemporary sources suggest the station was opposite Gold Digger Passage, Jennifer Rodrigues and Ross Anderson, 2006, p. 2.

[86] ibid.

[87] Sister Mary Albertus Bain, 1975, pp. 62–63, 137 & 140.

[88] ibid., p. 140.

Fig. 13.3 Sand dunes behind Port Gregory

Fig. 13.4 Ad-ju-co (Dioscorea hastifolia), The Loop, Kalbarri National Park

In front of the naturalists was a 70-kilometre trek across country from Lynton to the Geraldine Mine.[89] As yet there was only a rough track to the Mine, which was one of Wray's tasks when he was in the District in 1856 to recommend a route. While he cautioned about the difficulty of provisioning a construction party, his opinion was that

> a good line of road might be struck out running up the valley of the Hutt as far as Baker's [sic] Well avoiding the sandplains between Lynton and that place...After passing Baker's Well a line across the sand plains might be run in such a way as to pass close to fair road making material nearly all the way, to about 9 miles distance from the mine—when the line enters on a loamy clay, requiring little but drainage to stand a very considerable amount of traffic.[90]

No evidence has been found that this road was constructed.

The route Wray proposed appears to have followed closely that of the then-existing track that Oldfield and Walcott travelled.[91] After leaving Lynton the track ran north-west along the northern side of Hutt River, across Sanford's lease and continued in a north-westerly direction through terrain of sandy and gravelly rises and occasional swamps and soaks.[92] It passed south of Mount Albert and Mount Victoria where, at some stage, Oldfield gathered plants[93], probably from among thickets of *Melaleuca megacephala* and *Hakea pycnoneura*. The track then entered sand plain with leached grey sands over gravel, gravelly sands, and pale yellow and deeper leached grey sands supporting a rich scrub heath association of Summer-scented Wattle, Sandplain Cypress (*Callitris arenaria*), Christmas Tree, Sandplain Woody Pear and *Banksia* and *Grevillea* species.[94] On the way, Oldfield made some collections at and near Baker Well[95], including at least one in 1856.[96] The track continued through similar terrain running north-east before crossing into the predominantly deep yellow sand plain[97] and passing Barrel Well situated on a small tributary of the Murchison River about 14 kilometres from the Geraldine. This well was an important source of drinking water for the Mine's inhabitants, taking 'two men and a team of four bullocks, constantly employed, to draw water from the...well' and cart it to the mine each day.[98] Records have been located of Oldfield's specimens gathered south of, near to and at the well but when he did this is unknown.[99]

What remains of the scrub heath on the deep yellow sand on this section of the track, which included what was to become the Kalbarri National Park, is rich in component species (see Map 1 p. 25). These feature the Sceptre Banksia (*Banksia sceptrum*), Acorn Banksia, Sandplain Woody Pear, Sandplain Cypress and the White Plume Grevillea (*Grevillea leucopteris*), with occasional strips of the Oldfield's Mallee (*Eucalyptus oldfieldii*), Malallie (*E. eudesmioides*) or *E. arachnaea* subsp. *arachnaea*.[100] As Oldfield and Walcott passed across this country they must have encountered species that had become familiar to them further south.

---

89 Fiona Bush and Philip Griffiths, 1998, 'Geraldine Lead Mine', Documentary Evidence, Assessment of Cultural Heritage Significance, Register of Heritage Places, Permanent Entry, Heritage Council of Western Australia, p. 5, viewed 16 March 2009 at www.heritage.wa.gov.au.

90 H. Wray to the Colonial Secretary, 15 October 1856.

91 Victoria, 1863, Cons 5018, (Tally no. 506653), SROWA.

92 L. Gary Rogers, 1996, 'Soil-Landscape Map of the Geraldton Region', Sheet 2.

93 HAODb.

94 L. Gary Rogers, 1996, 'Soil-Landscape Map of the Geraldton Region', Sheet 2, & p. 21.

95 HAODb.

96 A *Ptilotus helichrysoides* from 'near Baker's Well' (FM 1858–1859, vol. 1, p. 237). Mueller gave the location as '*In locis glareosis collium prope flumen South Hutt, e.g. prope Baker's Well*' [In gravel, hill near South Hutt, e.g. near Baker's Well].

97 L. Gary Rogers, 1996, 'Soil-Landscape Map of the Geraldton Region', Sheet 2.

98 Sister Mary Albertus Bain, 1975, p. 149.

99 HAODb.

100 J.S. Beard, 1990, p.105–113. Beard (p.113) depicts these Eucalypts, including *Eucalyptus redunca*. This is more likely *E. redunca* var. *melanophloi*, since renamed *Eucalyptus arachnaea* subsp. *arachnaea* based on the isotype (at MEL) collected by Oldfield at Cheeanga thicket, Murchison River, M.I.H. Brooker, and S.D. Hopper, 1991, 'A taxonomic revision of *Eucalpytus wandoo*, *E. redunca*, and allied species (E. series *Levispermae* Maiden – Myrtaceae) in Western Australia, *Nuytsia*, vol. 8 (1), pp. 1-189. We are grateful to S. D. Hopper for drawing our attention to this taxonomic revision, attachment to email, 23 December 2017.

Fig. 13.5 300 metres south of Barrel Well

About 10 kilometres south of Geraldine Mine the naturalists entered country of gently undulating to rolling rises, and low hills with many rocky ridges on hillcrests with long gentle slopes and red sand alluvial terraces of local rivers. The main plant community of this country is scrub of Jam (*Acacia acuminata*) with Kurara (*A. tetragonophylla*), *Dodonaea* sp. and Needle Tree (*Hakea preissii*) on scattered patches of York Gum woodland and Rock Sheoak on rocky soils, which continued until the track terminated at Geraldine Mine. There Swamp Sheoak and River Gum line the banks of Murchison River.[101]

A few hundred metres off the track, 'at the Gum Forest, about three miles W.S.' from the mine

> the Company...[had] erected a roasting and two large smelting furnaces, with large chimney 105 feet above the works [Warribanno Smelter], together with good dwelling houses for workmen, &c. Nearly all the lead risen from the mine for the last three years [1853–1856] has been smelted here, it being the most convenient spot for wood, and in the high road leading from the mine to Port Gregory.[102]

About a kilometre further on towards the Mine, the track passed Garden Well[103], where an elderly expiree, William Sandford, grew vegetables which he supplied to the miners all year round. The site was 'a camping ground for carters and others seeking work at the mine'.[104] One of Oldfield's specimens with the locality 'Garden Well, Collallia, Murchison River'[105] confirms that he was there on one of his visits to the Murchison. He also gathered other plants at or near 'Collallia'.[106] Neither 'Garden Well' nor 'Collallia' has been located on any gazetteer. In his list of *Watchandie* place names Oldfield

[101] L. Gary Rogers, 1996, Geraldton Region Sheet 1 & p. 20.
[102] Fiona Bush and Philip Griffiths, 1998, viewed 16 March 2009 at www.heritage.wa.gov.au, p. 6. This quotation is from a Geraldine Mine Prospectus of 1856 but no further details provided. Today, the smelting works ruins are referred to as the 'Warribanno Lead Smelter Complex'.
[103] H. Wray to the Colonial Secretary, 15 October 1856, CSO, Inwards Correspondence, vol. 358, fols. 48, SROWA.
[104] Sister Mary Albertus Bain, *A Life of Its Own*... 1996, p. 28.
[105] MEL605452, *Salsola tragus* subsp. *tragus*, Garden Well, Collallia, Murchison River, A. Oldfield, MELISR database, 18 November 2005, NHV, RBGM.
[106] HAODb.

refers to '*Col-lail-li-a*' as 'a grassy flat with flooded gums'. Mr Stan Gratte, Wonthella, Geraldton[107], who has had a longtime interest in the history of the Murchison River district, informed us that the pastoral lease (Location N257) of W.C. Glass was known as Collallia. On an earlier plan, the southern boundary of this lease was the Murchison River and included Geraldine Mine.[108] William Nancarrow's lease immediately east of Glass' station is indicated as being '[g]um forest'[109], which probably referred to River Gum (*Eucalyptus camulduensis* subsp. *arida*) and was very likely to have been the species that Oldfield called 'flooded gums'.[110]

Perhaps Oldfield and Walcott camped at Garden Well. There was 'good dry feed in the neighbourhood of the mine'[111] and the convenience and luxury of being able to obtain fresh vegetables and a close-by source of 'good' water might have been compelling reasons to base themselves there. Perhaps, however, the Mine Captain, Martin Hosken[112], allowed them to camp at the mine where their collections and equipment could be safely stored, either at the Manager's residence or in the office while they were away exploring the Murchison River.

Oldfield collected at and near Geraldine Mine[113] but only two of his specimens are confirmed as being obtained in 1856. One of Walcott's specimens was also from the Mine.[114] There is a published record of *Eremophila maculata* [*E. maculata* subsp. *brevifolia*][115] attributed to Oldfield from '300 miles above the Geraldine'.[116] There is no other evidence to support a claim that he or one or more members of the *Watchandie* group ventured that far above the Mine. Nor is it likely that any tribal groups from that far north would have heard of Oldfield's botanical interest or indeed, been welcome in *Watchandie* territory or would have described the locality in terms of being '300 miles' to the north of the mine. Then there is the question of whether the cutting would be in a condition suitable to be a herbarium specimen after such a long pedestrian journey. It is probably an error on either Oldfield or Bentham's part. Unfortunately, we have not been able to track down any herbarium specimen of *E. maculata* to check the label.

The Geraldine Lead Mine shaft was in the bed of the Murchison River and 'the lode dips down at a very sharp angle'. According to Wray,

> it has hitherto been worked along the surface, and when the river comes down in the winter, the mine fills with water—and is for weeks together not workable...such was the case at the time of our visit.

Shafts had been sunk for copper but 'were not carried out sufficiently to afford anything but mere surface indications'.[117] Oldfield and Walcott reached the mine shortly after Wray's visit so the shaft may still have been flooded and the miners otherwise engaged, if at all.

With the sinking of mine shafts in 1849, some, or all, of the structures built on the site prior to 1856 included an engine house, a knapping floor and a mine office on the north bank and the Mine Captain's house on the hill beyond. On the south bank were huts, among which was probably the long

---

[107] Personal communication, 4 April 2005.

[108] Public plan, Victoria & North 1, Cons 5018, (Tally no. 506637), c.1858, and Victoria and North 2a, Cons 5018, (Tally no. 506638), 1860, SROWA.

[109] Public plan, Victoria & North 1, Cons 5018, (Tally no. 506637), circa 1858, SROWA.

[110] Today 'Flooded Gum' refers to *Eucalyptus rudis* which occurs along rivers on the Swan Coastal Plain and other places in the south-west of the State. There are records in *FloraBase* from just to the north of Moora where identifications have not been confirmed as either River Gum (*E. camuldulensis*) or a subspecies of *E. rudis* but almost certainly none of these last occur near the Murchison River. Stephen Hopper agrees that *E. rudis* does not occur in the lower Murchison River area. There is, he states, 'a broad zone intergradation between *E. rudis* subsp. *rudis* and *E. camulduensis* subsp. *arida* extending from Northam to the lower Irwin River, north of which is pure *E. camulduensis* subsp. *arida*', personal communication to MHH, email attachment, 23 December 2017.

[111] H. Wray to the Colonial Secretary, 15 October 1856.

[112] Tanya Suba, 1993, p. 30.

[113] HAODb.

[114] *Pityrodia verbascina* Benth and *Dodonaea larraeoides,* AO (FM 1858–1859, vol. 1, pp. 219 & 233) and *Brachychiton gregorii*, MEL 71762A, Walcott (MELISR database, 05 December 2008, NHV, RBGM).

[115] *Florabase*. We are grateful to Stephen Hopper for drawing our attention to this subspecies which, according to *Florabase*, is the only *E. maculata* to occur 300 miles, personal communication to MHH, email attachment, 23 December 2017.

[116] GB, 1870, vol. 5, p. 29.

[117] H. Wray to the Colonial Secretary, 15 October 1856.

Fig. 13.6 Murchison River at the Geraldine Mine

shed which provided living quarters for the ticket-of-leavers.[118] When in full production, the establishment must have been a very busy place—men down the shaft extracting ore, others on the knapping floor pounding the ore and loading the crushed material onto drays, with carters urging their bullock or horse teams with their heavy burdens to cross the river and move down and up the sandy, river worn banks to the track to the smelter on the other side.

Assistance from the *Watchandie*, the Aborigines of the lower Murchison River

Map 1 p. 25 and Map 11 p. 221

Oldfield spent some considerable time during his 1858–1859 visit moving around the Murchison River area in the company of members of the *Watchandie* tribe[119] and there is also evidence that he and Walcott spent time with them while they were in the area in 1856.

Walcott was adept with the flute.[120] The following passage in which Oldfield recounts the reaction on the part of some *Watchandies* to what appears to have been a deliberate attempt by Walcott to produce dissonant sounds:

> When the Watchandies first heard the notes of a flutina, the effect on them was remarkable. This instrument was first used to scare them as they slept round the fire. At the first horrid, discordant blast they all started from their sleep, and on a repetition of the unearthly sound one and all decamped into the bush, declaring, on the morrow, that they had been roused from their sleep by the screechings of many in-gnas [malevolent spirits of the dead].[121] Afterwards, on hearing it played, they stopped their ears, begging the performer to desist as the noise made them sick. On being asked

[118] Fiona Bush and Philip Griffiths, 1998, p. 4.
[119] See Chapter 2.
[120] Joan Brockman, 1989, p. 20.
[121] In the *Watchandie* belief system *in-gnas* infest 'the whole land... and their places of abode are thickets, caves, springs', AO, 1865b, p. 229.

> what caused the sound, they at once pronounced it to be a "Tindokamya" [Native Cat's horn][122], and that the sounds were produced by wounding that animal, an operation effected by pressing down the keys, the intensity of the sound emitted being proportionate to the amount of injury inflicted.[123]

Oldfield did not say where and when this incident took place but his account suggests that he and Walcott spent some time with the *Watchandie* as they moved around the country. Why, then, might the pair have been motivated to frighten the group as they slept? They may simply have been amusing themselves at the expense of the *Watchandie.* If that was the case they were amply rewarded. On the other hand, perhaps it was one of several occasions when Oldfield deliberately evoked an emotional response with the object of finding out more about their belief system. If this was the case, it could be argued that in this instance he had some success. It seems the *Watchandie* bore the pair no deep grudge since they still communicated with the naturalists next day, when the performance was repeated and the listeners were pleading for the playing to cease and explained why they disliked the sound so intensely. Even though they begged the performer to stop the noise, from what Oldfield has intimated, the *Watchandie* were not convinced that it was not '*Tindokamya*'.

It has been claimed elsewhere that Oldfield's investigative approach to understand the *Watchandie* was not always ethical by today's standards of anthropological practice. When faced with this kind of provocative behaviour, possibly arising from a combination of Eurocentrism and a naïve belief that he was being true to scientific method, one wonders why the *Watchandie* continued to have anything to do with him, but they did. Their tolerance may have had more to do with the material rewards he offered than with their liking of him. On the other hand, perhaps Oldfield appealed to their sense of humour to dissolve any anger that he caused or they may have forgiven him his ignorance. He left no details.[124]

In his account of a revenge killing of a member of the *Watchandie* tribe in 1857 near Port Gregory, Oldfield provides a clear indication that during his first visit he spent awhile in the company of at least one member of the *Watchandie* group, Carrotty, whose trust he had gained and with whom he was able to communicate on a more than basic level:

> The Watchandie tribe, then located near U-chu, were visited by a man of the War-ra-na-noo-ra tribe, who remained with them some days, joining in all their hunting expeditions and amusements. One day, however, one of the Watchandies (to whom the nickname of Carrotty had been given by the Europeans) declined going with his fellows to the chace [sic], whereupon the stranger, pleading fatigue, said he would stay behind also, to keep Carrotty company. The after proceedings were witnessed by a child, who for some cause had not accompanied the women who had gone forth to dig yams. About noon the girl observed that both Carrotty and the stranger were apparently fast asleep; but, after a time, she observed the latter crawling stealthily towards the Watchandie, and, arrived near him, stand up and hurl his spear into the heart of the sleeping man. She ran off, in the twofold intention of avoiding a like fate and of apprising the men of what had transpired, but ere they could reach the spot, the murderer was secure amongst his own tribe. Poor Carrotty, whose death was unavenged up to the time of my leaving the Murchison, was one of the most intelligent natives yet known, and to him is due a large portion of the vocabulary of the Watchandie tribe hereafter given.[125]

---

122 In AO's *Watchandie* mammal vocabulary *Tin-do-kat* refers to the Native Cat, *Dasyurus geoffroii*, but with no explanation of the meaning of the suffix *'-kayama'*. It might refer to 'horn' i.e. being the animal's howl of pain, ibid., p. 294.

123 ibid., pp. 226–227.

124 MHH 2005, p. xvi.

125 AO, 1865b, p. 247.

Although Oldfield does not give a specific date for this occurrence, the detail he provides of the circumstances of the murder suggest that he heard it when he returned to the Murchison in 1858 from a member of Carrotty's tribal group who remembered the treachery clearly. The nature and extent of the *Watchandie* vocabulary and grammar in Oldfield's paper 'On the Aborigines of Australia' indicates that not only was Oldfield quite an accomplished linguist but that he, and possibly Walcott, must have been in Carrotty's company for more than a day or two. Perhaps he acted as their guide and interpreter while they travelled through *Watchandie* territory. Oldfield clearly respected Carrotty and must have spent many hours with him questioning him and recording the phonology, syntax and vocabulary of his language.[126]

The apparent equanimity of the *Watchandie* in their dealings with Oldfield and Walcott is a measure of their ability to relate to them in a friendly and non-threatening way, despite the discordant flute playing. This is surprising in view of the difficulties experienced by the indigenous inhabitants in those early days of colonisation of the Murchison River district. It had only been seven years since the area began to be settled by Europeans and it was not until the early 1850s that pastoral activity got underway. As the pastoralists took over Aboriginal lands across the Victoria District, the tribal people experienced increasing difficulty to follow their traditional way of life. This drove many to live on the fringes of settlements and into the employ of the colonists, under rules of social control about which the Indigenous people had no or minimal understanding. Although there were those who were friendly, and even loyal, and of assistance to the settlers, relationships between those who preferred their nomadic way of life and the colonists were fraught with conflict. Stealing pastoralists' stock and stores and the aggressive behaviour of some tribesmen aroused fear and anger in the settlers. The Aboriginal people, for their part, bitterly resented the loss of large game animals and the damage caused by the settlers' sheep and cattle to their water sources, yam beds and other important food plants in fertile areas.[127] Conflicts, at times resulting in fatalities inflicted by one side on the other, were common. William Burges and the constabulary were kept busy trying to keep the peace in a situation where Aboriginal weapons were no match for those of the colonists and where, even though the Magistrate and the Chief Constable tried to administer justice fairly, colonial law in the main was relentlessly enforced in favour of the settlers, to the detriment of the Indigenous people.[128]

Oldfield and Walcott along the Lower Murchison River

Map 11 p. 221 – Geraldton to the Murchison River

As far as can be ascertained, Oldfield and Walcott followed the course of the Murchison from the Geraldine or 'Collallia' down to its mouth, close to where Kalbarri is located. According to Bain:

> From the earliest days of exploration, a bush track had been created along the Murchison River which flowed beyond the Mine in a south-westerly direction. At a given point, parties left the river bed and headed for Noondamurra Pool, which was described by an old Murchison identity, Charlie Groessler, as receiving its water from ground water seepage, so was not prone to salt, and was a permanent, deep water oasis.[129]

---

[126] ibid., pp. 289–297.

[127] Sister Mary Albertus Bain, 1975, pp. 76–88.

[128] The correspondence between William Burges and the Colonial Secretary provides evidence of the lengths to which Burges and Drummond went to keep the peace and be fair in their dealings with the Aborigines, Resident Magistrate: Champion Bay, Inward Correspondence, Resident Magistrate—Champion Bay, 1852–1859, CSO vols. 256, 248, 309, 339, 367, 382, 413, 435, SROWA.

[129] Sister Mary Albertus Bain, 1996, p. 94. While the Murchison River runs south-west from the Geraldine, it only does so for about 6 ½ kilometres, it then runs south for about 6 kilometres and then north-west for over 48 kilometres, after which it turns south-west for more than 32 kilometres before reaching its mouth (Joint Operations Graphic, Geraldton Australia, 1: 250 000, Series 1501, Sheet SH50-1, edn 1, Royal Australian Survey Corps, 1983.

No track along the river has been found or marked on any early or modern public map. In former days one of the water bodies near the Murchison River may have been called 'Noondamurra Pool' but we have not found it on any early map of the area or on a recent gazetteer.[130] Groessler may have been mistaken and was referring to the Hardabut Pool, which is the only pool south-west of the mine.[131] Since few colonists had visited the area by 1856, it is likely that the track along the river was originally forged by the *Watchandie* over many hundreds of years. If so, we have not found any literature records to confirm this although in other parts of the Colony, settlers' tracks often trailed Aboriginal paths, and often following rivers.[132] Of all the localities on the records of Oldfield's specimens in HAODb, those associated with the Murchison River are by far the most numerous. Unfortunately, most them do not give a specific indication of where along or near the river they were obtained.[133] This is understandable since very few places in the district had been named by Europeans at that time and Oldfield relied to some extent on Aboriginal names, many of which have not found their way into gazetteers. There are herbarium records of specimens that bear the localities 'G'nia' and 'Yanarie' associated with the Murchison River (for example, 'Yanarie, north of the Murchison River').[134] These are probably *Watchandie* descriptor terms. Thus 'G'nia' could be a misspelling of *Y'nia* referring to a sandy bushy hill and 'Yanarie' to *Yan-a-rie* a dry grassy flat.[135]

The river flows across about 3.5° longitude and several markedly different floral associations, with the ranges of some species extending across two or more degrees of longitude. The lack of locality specificity reduces to some extent the scientific value of many of Oldfield and Walcott's specimens and of course, our understanding of where they went in the area. Nevertheless, it is possible to give some indication of this from some of their specimens referred to in Mueller's *Fragmenta* (1858–1859).

Whatever side of the river the track ran towards the mouth, the terrain it traversed from the Geraldine to about three kilometres past Croton Creek, near Hardabut Pool, gently slopes to the sandy but intermittently rocky shore and river bed, where the path may have run a little above the flood line. For the next two kilometres past the mine, the naturalists passed Sanford's Henrietta Mine and two other mining leases in the name of C.R. Prinsep (Murchison Locations No. 5 and 10).[136] After Croton Creek the river flows into gorge country where the stream has cut deeply into the Tumblagooda Sandstone. The gorge intersects the sand plain and is rugged with spectacular scenery consisting of precipitous cliffs, steep rocky slopes and gentle sand-covered benches; possibly making it necessary for the track to pass along the plateau that borders the tops of the cliffs. The sandplain north and south of the stream is essentially the same as elsewhere although there are discernible differences in the flora, varying in response to the decreasing rainfall from south to north and from west to east.[137]

Aboriginal paths from the main track may well have provided the naturalists with access to permanent riverine pools with fringing vegetation of River Gum and Swamp Sheoak where they might have bivouacked and swum. It would not be until they reached the upper tidal limit of the estuary that fish, such as Black Bream (*Acanthopagrus butcheri*), became an important component of their diet.[138]

---

[130] A Noondamarra Pool is marked on an early public plan as being on the Greenough River, Cons 5018, Victoria (Tally no. 506664), SROWA. There is a Noodamurra Pool on the Greenough River, Joint Operations Graphic, Geraldton Australia, 1: 250 000, Series 1501, Sheet SH50-1, 1983, which is probably the same place.

[131] ibid.

[132] Martin Gibbs, 1995, *A Report on the Aboriginal Ethnohistory of the Warren-Blackwood Region, Southwest Australia*, imprint, p. 15, JSBL.

[133] HAODb.

[134] ibid.

[135] AO 1865b, p. 293. *Yan-a-rie* is not to be confused with *Yan-er-rie* which in Oldfield's *Watchandie* vocabulary means dung of Wild Dog ibid., p 295.

[136] Public plan, Victoria 0001, 1850, Cons. 3869, SROWA.

[137] J.S. Beard, 1990, pp. 104–105.

[138] Anne Brearley, 2005, p. 267.

Further up the river, tiny freshwater fish species occur in freshwater pools.[139] In common with Aboriginal groups and settlers, Oldfield and Walcott would have relied on potable water arising from springs in the Tumblagooda Sandstone back from the river.[140] Although the water from these springs did find its way into the river[141], except for when the river was in flood, its lower reaches were saline as Oldfield attested:

> The Western Australian rivers, Irwin, Hutt, Murchison, Gascoign [Gascoyne], Glenelg, etc., which drain immense areas of level country, are liable to inundations in the fore part of summer, which proves that the interior parts of New Holland are subject at certain epochs to very heavy spring rains, the floods caused by these being a long time ere they reach the settled parts near the coast. In 1857 the effects of the summer flood of the Murchison were felt at Col-lail-ya seventeen days before they were perceptible at Oo-lin-yur-ra, a direct distance of forty miles only. The same inundation brought from the interior a species of fish hitherto unknown in the lower parts of the Murchison, proving that towards the sources of that river there must exist permanent pools of fresh water; for as the floods subsided and the pools in the lower part resumed their saline character, these fishes all died. The greatest floods have but little effect in changing the waters of these pools, the fresh water from the interior apparently rolling over the saline without much disturbing it, the only perceptible difference in the quality of the water being due to the filling up of the pools by the fresh water of the flood, as the pools are emptied by evaporation they resume their former saline condition. As to the time elapsing between these inundations, the same law seems to obtain as that ruling periods of drought in other parts of the continent; they seem to occur every five, seven or nine years, while the periods of the greatest floods probably embrace centuries.[142]

The dominant vegetation of the rocky places comprises such trees as Jam, *Jacksonia cupulifera* and the Rock Sheoak (*Allocasuarina huegeliana*). Jam trees, with scattered River Gum, occupy lower flats in the gorges, and the river itself is lined by River Gum and Swamp Sheoak, with occasional Lesser Bottlebrush (*Callistemon phoeniceus*) in the river bed.[143]

On their way to or from the river mouth, Oldfield and Walcott deviated toward Mount Curious, a conical-shaped prominence rising from the sandplain approximately three and a half kilometres to the north-west of the northerly point of the lower Murchison, to the west of the Geraldine Mine, and about four kilometres west of the current eastern boundary of Murchison House Station. The hill is visible across the *kwongkan* sandplain for many miles to the east and south. Mount Curious and, about three kilometres to the south-west, '*Yoaddalinu*' (Mount Yooadda Darlinoo) are significant sites in *Nanda* mythology, but Oldfield gives no indication that he was aware of their significance to the *Watchandies*.[144] There is no evidence that Oldfield and Walcott climbed Mount Curious but nearby on the dry, sandy plain Walcott obtained an *Anthotroche walcottii*.[145] Some or all of Oldfield's specimens collected north of and near Mount Curious may have been gathered in 1856.[146]

---

[139] David L. Morgan and Howard S. Gill, 2004, 'Fish fauna in inland waters of the Pilbara (Indian Ocean) Drainage Division of Western Australia—evidence for three sub-provinces', *Zootaxa*, 636, pp. 1–43.
[140] Anne Brearley, 2005, p. 264.
[141] ibid.
[142] AO, 1865b, pp. 233–234.
[143] J.S. Beard, 1990, p. 109.
[144] Gavin Jackson, 1996, *The Report of an Aboriginal Heritage Study of the Nanda Area Central, Kalbarri Region, Western Australia*: Prepared for the Nanda Aboriginal Corporation, Barrel Well Community, Anthropos Australis, Perth, p. 44.
[145] MEL 41166A, *Anthotroche walcottii*, P. Walcott, MELISR database, 05 December 2008. The original description gives the locality as "*In planitiebus arenosis ad montem Curious*" [On sandy plains at Mount Curious], FM 1858–1859, vol. 1, p. 123.
[146] HAODb.

Fig. 13.7 Hawke's Head, Murchison River

Bonamia rosea was collected jointly by Oldfield and Walcott at Oolinyurra; a place not found on any map or in any gazetteer. Oldfield placed *Oo-lin-yur-ra* at 64 kilometres 'direct distance' downstream from *Col-lail-ya* [W.C. Glass's property 'Collallia'] and noted:

> [T]he word *Oo-lin-yur-ra* (a place on the Murchison) is compounded of the word *Oo-lin-yu,* which signifies a species of swamp tea tree [the Wiry Honey Myrtle] (*Melaleuca noematophylla*, [sic], F. Muell.[147]) and the particle *wa* transformed into *ra* for the sake of euphony, that locality abounding with that shrub.[148]

Assuming that 'direct distance' is a straight-line measure, 64 kilometres from 'Collallia' would place 'Oolinyurrah'[sic] slightly north-west of Bettie Pool, a body of water in the Murchison River. However, tracing the river on a map[149] with a 'map wheel' puts Oolinyurra near the river south of Mount Curious, which is a few miles north-east of Bettie Pool.

Our plant data records suggest that Oldfield might have based himself at Oolinyurra for a few days and collected from there in different directions north, south east and west.[150] Walcott may have been otherwise engaged during the time that Oldfield travelled further towards the coast as there do not seem to be any Walcott National Herbarium of Victoria specimens for that tract of territory. If Walcott was not with Oldfield what might he have been doing? It is on record that his brother James applied for pastoral leases totalling 26,000 acres in the lower Murchison River area in 1858[151], so it is feasible that Pemberton was scouting around the area for suitable land for a pastoral lease for James and possibly for himself.

---

[147] According to *FloraBase* the correct spelling is *Melaleuca nematophylla*.
[148] AO, 1865b, p. 291.
[149] Royal Australian Survey Corps, Ajana Australia, Scale 1: 125,000, Series 1501, Sheet SG 50–13, edn 1, Director of Survey, Department of Defence, Canberra, 1982.
[150] HAODb.
[151] J.C. Walcott, Application to occupy crown land, 1858, Lands and Surveys Red Book no. 703, p. 11, SROWA.

Fig. 13.8 Mount Curious from the west

Oldfield obtained specimens at and near Yalthoo Springs[152], 24 kilometres downstream from Mount Curious, and approximately one and a half kilometres inland north of the river on Murchison House Station. Other specimens bear locality names which are similar ('Yatthoo', 'Yarrhoo', 'Yadskoo', Yadthoo and 'Yattoo')[153], which may have been variants of Yatthoo. The most frequently used name is 'Yatthoo' which is *Watchandie* for a large grassy plain[154], and the term may have been employed as a descriptor in the absence of a recognised place name. Two of Oldfield's plants obtained in 1856, which are labelled as being obtained at 'Yattoo', has one label stated to be 'in the vicinity of the Murchison River'. These were a specimen of Pixie Bush (*Eremophila oldfieldii*) and one of the Slender-flowered Eremophila (*E. graciliflora*).[155]

Once the naturalists, if indeed Walcott was still with Oldfield, resumed the path along the river they were probably close to Murchison House Ford about 18 kilometres upstream from the river mouth. By now the pair would have noticed that the river was tidal. From there to its mouth, the estuary is a sequence of shallow pools and elongate bars.[156] They may have shared explorer George Grey's very favourable impression of the estuary when he visited the area in 1839, after he and his twelve companions were shipwrecked near Red Bluff. According to Brearley, Grey's description, 'still captures the essence of the [riverine] estuary':

> [April 1] ...On reaching the valley...we found a small stream, and following this to the northward for about a mile, came out upon on of the most romantic and picturesque-looking estuaries I had yet seen; its shores abounded with springs, and were bordered by native paths, whilst the drooping foliage of several large sorts of Casuarinas, the number of wild swans on its placid bosom, and the natives fishing in the distance, unconscious of our presence, imparted to the whole scene a quiet and a charm which

[152] HAODb.
[153] ibid.
[154] AO, 1865b, p. 293.
[155] FM 1858–1859, vol. 1, p. 208.
[156] Anne Brearley, 2005, p. 264.

> was deeply felt by those who had now for so many days been either tossed about by the winds and waves, or had long been wandering over barren and inhospitable shores...
>
> We followed the shores of the estuary to the northward and eastward until we saw a point, where it appeared to separate into two branches...We...traced the shores of the estuary to its mouth, which turned out to be the opening we saw in the morning: this mouth is completely sheltered by a line of breakers and reefs, which, although they present a most formidable appearance from the sea, can be doubled by keeping pretty close along the shore in approaching the mouth of the river. Owing to this reef there are no breakers in the bar; but its mouth is very narrow, and so shoal that I doubt if a boat could be got in at any other time than high water: some of the sailors with me, however, thought otherwise; but there is, at all events, convenient landing at this point under the shelter of the reef...[157]

Oldfield collected at and near the river mouth in 1856.[158] As previously noted, there are records in HAODb of other specimens (all undated) taken by him from the river mouth, to its north and south and along the estuary.[159]

Expedition to Lake Culcurdoo

Map 11 p. 221 – Geraldton to the Murchison River

It is not possible to provide a firm indication of how long Oldfield and Walcott took to reach the mouth of the river from the Geraldine. Searching for new species, sometimes a few miles from the river, coping with rough terrain, pressing and organising specimens which possibly had to be done in daylight, hunting for game for themselves and pasture for their horses and finding suitable places close to fresh water to bivouac were clearly all time-consuming activities. Under these circumstances the journey down river could have taken them three weeks, arriving at the mouth towards the end of October or beginning of November.

From the river mouth, the pair went to Lake Culcurdoo (Oldfield label variants: 'Culcurda', 'Coolcurda', 'Coalcurda'), where they gathered at least one plant each.[160] The lake is 32–40 kilometres slightly north-east of the river mouth, about a day's journey on horseback. There are other Oldfield specimens (all undated) from that locality and two others from '10 miles N. of Culcurdoo' that may have been obtained on that visit[161], but in later chapters it becomes apparent that the latter were taken during his second visit. Their expedition to Lake Culcurdoo was possibly undertaken sometime during November. By that time the going may have been uncomfortable. It was late Spring and the days were probably extremely hot, the temperature at times rising to 40°C or more, and apart form small groves of River Gum at wet depressions in the land such as at Lake Culcurdoo itself,[162] there was virtually no dense shade to provide relief from the blazing sun. Undertaking even a relatively short journey north of the river at that time of the year necessitated reliable water sources along the way. The *Watchandie* and *Malgana* people of Shark Bay visited each other periodically for ceremonies and other purposes,

[157] George Grey, 1841, vol. 2, pp. 3–5, part of the above quotation in Anne Brearley, 2005, pp. 262–263.

[158] *Senna pleurocarpa* var. *pleurocarpa* and *Corynotheca micrantha* var. *micrantha*, mouth of the Murchison River; *Eremophila clarkei* and *Pityrodia oldfieldii*, near mouth of the Murchison River, FM, 1858–1859, vol. 1, pp. 223, p. 215, 208 & 234 respectively.

[159] See fn. 131.

[160] *Verticordia pholidophylla*, Coalcurda [Culcurdoo] north of Murchison River, A. Oldfield, FM 1858–1859, vol. 1, p. 227, and MEL 2214395, *Burchardia congesta*, [Lake] Culcurdoo, P. Walcott, MELISR database, 05 December 2008, NHV, RBGM.

[161] HAODb.

[162] S. D. Hopper, personal communication to MHH, email attachment, 23 December 2017.

Fig. 13.9 Sandplain, south of Lake Culcurdoo

and the naturalists would have followed a coastal path used by those groups along which there were springs, soaks and then-functioning gnammas, some containing up to 50 litres of water.[163]

Approximately 11 kilometres north of the river mouth, the coastal cliffs become extremely high and steep in a long unbroken line. These cliffs are named after the Dutch ship, *Zuytdorp*, which was wrecked there in 1712. The cliffs' slopes are a stony scree with a vegetation of scattered succulents. A windswept belt of limestone and sand extends behind the cliffs containing wind- and salt-tolerant low heath thickets of small Southern Diplolaena (*Diplolaena dampieri*), Umbrella Bush (*Acacia ligulata*), Coastal Daisybush (*Olearia axillaris*), Thick-leaved Fan Flower (*Scaevola crassifolia*) and Yellow Tailflower (*Anthocercis littorea*) with small ground plants of *Atriplex*, *Sclerolaena*, *Frankenia* and *Carpobrotus* species.[164] East of this belt lies undulating sandplain of *kwongkan* with patches of dense scrub, the dominant species being Tangling Melaleuca and clusters of Mallee (*Eucalyptus sp.*).[165] According to Stephen Hopper, 'there is also the occasional presence of small stunted groves of Illy-arrie (*Eucalyptus erythrocorys*) that occur, for example, on the State Barrier fenceline close to the coast above the Zuytdorp cliffs'.[166]

---

[163] Modern day maps show the locations of springs to the north of the Murchison River near the coast, see for example, Australian Surveying & Land Group 2000, Ajana, 1: 250,00, SG50–13, Western Australia, Edition 2, Map 98/047, Commonwealth of Australia).
Phillip Playford, 1996, describes water availability near the Zuytdorp wreck site to the north of Lake Culcurdoo, p. 76.
[164] J.S. Beard, 1990, p. 100 & 105.
[165] Phillip Playford, 1996, *Carpet of Silver: The Wreck of the Zuytdorp*, University of Western Australia Press, Nedlands, p.76.
[166] Personal communication to MHH, email attachment, 23 December 2017.

Did Oldfield and Walcott venture to the upper reaches of the Murchison River?

Map 11 p. 221 – Geraldton to the Murchison River

Two undated Walcott plant records are puzzling and raise the possibility that either he alone, or in the company of Oldfield or others, also ventured far to the east of the Geraldine to the 'Upper Murchison River 360 to 370 miles inland' (*Eremophila maculata* subsp. *brevifolia*), and to the 'Sources of the Murchison River' (*Leptochloa digitata*).[167] Neither of these is apparently a sandplain species, only occurring much further east on the Mulga (*Acacia aneura*) low woodland plains.[168] It was beyond our resources to undertake a comprehensive search of *FloraBase* to ascertain the distributions of all the Murchison River taxa represented in our Oldfield database, for the purpose of finding out whether any came from the upper reaches of the river but are not found on the sandplains along its lower reaches. Nevertheless, a more limited search was conducted to assess whether we had records in our database of any of the species in Beard's list of the more important taxa occurring in the inland Mulga low woodland. There are some of Oldfield's specimens of species whose ranges extended from the sandplains of the lower reaches of the Murchison to its upper reaches and beyond[169], but none whose ranges are limited to the Mulga woodland.

Although the possibility cannot be discounted at this stage, we believe it is highly unlikely that either, or both, of them went that far east in 1856. In the first place, there would have been insufficient time considering the journey to the west of the Geraldine and north to Lake Culcurdoo and for Oldfield to have completed the journey back to Tasmania and be on Mount La Perouse at the end of February 1857. Secondly, the only evidence we have that Oldfield ventured much further upstream past the Geraldine is in an 1863 letter to Sir William Hooker in which he said: 'With regard to the Isoetes[170], the most northerly point at which I detected it was on the Murchison, about 80 miles [126 kilometres] above its mouth'.[171] When he went there is not known but it is most likely when he was with the *Watchandie* in 1858–1859. Eighty miles upstream is about where the Coolcalalaya Station homestead is located and approximately 20 kilometres to the west of what has been officially recognised as the eastern border of the *Nanda* territory. Thirdly, Oldfield and Walcott ventured into what was then uncharted territory, where water sources and pasture availability were unknown.[172] By the end of February 1857, when Oldfield was no longer in Western Australia, the farthest any known surveyor/explorer had penetrated to the east along the Murchison River was in 1854 when Assistant Surveyor Robert Austin reached and named Mount Murchison, about 288 kilometres from the river's mouth, before turning north-west in his abortive attempt to reach the mouth of the Gascoyne River.[173] It was not until March 1857, when Francis Gregory, who had been surveying locations along the lower reaches of the Murchison, decided to extend the exploration of the country beyond the point that had been reached by Austin. His aim was to assess the extent of pasture following the heavy rains in 1857, which caused the river to flood, but he turned back after reaching Mount Murchison.[174]

Gregory's field book shows that he was undertaking surveys around Geraldton and the Murchison River from November 1856 to June 1857.[175] It is possible that Oldfield and Walcott met up with the

[167] MEL 83292A, *Eremophila maculata* subsp. *brevifolia*, and MEL 1717099, *Leptochloa digitata*, MELISR database, 05 December 2008, NHV, RBGM.

[168] *FloraBase*.

[169] J.S. Beard, 1990, pp. 188–189.

[170] *Isoetes* is a genus of plants commonly known as the quillworts. They are mostly aquatic or semi-aquatic in clear ponds and slow-moving streams, 'Isoetes', viewed 19 January 2011 at www.wikipedia.org.

[171] AO to WJH, 5 December 1863, English Letters, vol. 42, 1862–1865, L–Z, fol. 198, RBGKA.

[172] F.T. Gregory, 1857, 'The Murchison River 1857' in Augustus Charles Gregory & Francis Thomas Gregory, 2002, p. 34.

[173] Robert Austin, 1855, Accompanying Map to *Journal of Assistant-Surveyor R. Austin: Commanding an Expedition Sent by the Government to Explore the Interior of Western Australia, Forth and East of the Settled Districts for Extensive Tracts of Fertile Land Available for Pastoral and Agricultural Purposes...*, Government Printer, Perth.

[174] F.T. Gregory, 1857, pp.34–36.

[175] F.T. Gregory, Field Book 9, Cons 3401, SROWA.

Fig. 13.10 Fringing vegetation, Lake Culcurdoo

surveyor somewhere north of Geraldton but it could not be confirmed. Walcott did not accompany Gregory to Mount Murchison[176], so perhaps at a much later date he obtained the specimens himself or received them from someone else.

Return to Tasmania

The historical record has revealed no details of Oldfield's return journey to Tasmania, nor is it known what Walcott did immediately after the Lake Culcurdoo expedition. 'By a small monetary contribution' from Mueller, Oldfield was 'induced to make his trip to Mount Laperouse [sic]'.[177] He was there at the end of February 1857.[178] In the next chapter once again he teamed up with Charles Stuart. For whatever reasons this occurred, there is little doubt that mutual support in the face of danger and trauma would have been an important consideration. If it was a prior arrangement the timing of Oldfield's arrival at their meeting place could have been crucial to the success of the expedition.

There is only one instance of what appears to have been a direct voyage from Hobart to Fremantle on record for 1856–1857: this was in February 1856[179] and well before Oldfield came to Western Australia. The voyage took 27 days but, because of the direction of the prevailing winds, it probably would have taken less time to travel from Fremantle to Hobart. Since there do not appear to have been any ships travelling from Fremantle to Hobart between November 1856 and January 1857, he probably had to change ships at some ports along the way. This could have taken two weeks or longer than the direct voyage, depending on how many changes of vessel he needed, and how long he had to wait

---

176 F.T. Gregory, 1857 pp.34–36.
177 FM to GB, 1 May 1866, Kew Gardens correspondence, Australia, Mueller, 1858–70, fols 196–201, RBGKA. A copy of this letter was kindly provided by Jason Benjamin, University of Melbourne Archives, with the approval of Dr Sara Maroske, from The Correspondence of Ferdinand von Mueller Project database 17 July 2006.
178 See Chapter 14.
179 Fremantle Shipping Register 1833–January 1881, Acc. 1076/1, microfilm, SROWA.

for the departure of a ship to carry him further towards his destination.[180] In order to be in Recherche Bay before 28 February 1857, the voyage back to Hobart, with time allowed for preparations for the forthcoming field trip to the south-east, may well have required him to sail from Port Gregory or Champion Bay very early in December 1856.

No evidence has been found that Oldfield freighted any plant material to Hobart or to Melbourne while he was in Western Australia in 1856.[181] His specimens, therefore, probably accompanied him as personal luggage. Given the uncertain shipping timetables and unable to book a berth well in advance, it must have been a stressful journey if he had agreed upon a firm date to meet Stuart.

---

[180] Dr Graham Henderson, Marine Archaeologist, Perth, Western Australia, personal communication, 21 March 2013.

[181] Apart from this case, other cases consigned to 'A. Oldfield' were imported at Hobsons Bay during 1857. None of these contained Western Australian specimens because the *[Samuel] Appleton,* and the *Coleroon* travelled from London via Cape Horn and the *Jacob Badger* from London via Ceylon (*The Argus*, 14 March 1857, p. 5, 19 November 1857, p. 4, October 1856, p. 4, DAN, NLA) which did not call at Perth (Fremantle Shipping Register December 1833–January 1881, Acc. 1076/1, microfilm, SROWA). Whether these were for Augustus, Alfred or someone else is not known.

King George Sound, Western Australia

Part 4

Collecting for Dr Ferdinand Mueller

1857–1862

Fig. 14.0 Leionema oldfieldii

Chapter 14

Tasmania and Victoria 1857–1858

On 12 November 1856, a young lad, Henry Chamberlayne, was murdered near Kingston, Tasmania. This boy was a student of 'Mr Oldfield's school at Kingston'. The teacher testified in the Supreme Court on 5 February 1857 that Henry had not attended school on the day his body was found.[1] As Augustus was in Western Australia at the time of the homicide, the schoolmaster was probably his brother, Alfred, but this could not be confirmed. In May 1860, Alfred was appointed schoolmaster to the re-opened government-funded school at 'Brown's River'.[2] A school at Brown's River was established 'experimentally' in 1857 and failed, 'not more than six pupils attending'.[3] The government grant was discontinued after the first visit by the Inspector of Schools but it is uncertain how long the school remained open. Although Augustus was in Tasmania at the time of the trial, he was apparently in the south-east of the Colony (see below).

On his return to Tasmania from Western Australia he resided once more at Kingston.[4] Perhaps he moved into the school house with Alfred who, like Augustus, had not married.

Collecting on Mount La Perouse with Charles Stuart

Map 7 p. 150 – Southern Tasmania

Since overland travel to the south-east coast was still slow and difficult, it is almost certain that once again Oldfield journeyed by ship from Hobart or Kingston to meet his colleague. Stuart's specimen records in the National Herbarium of Victoria indicate that he had remained near Southport and Recherche Bay after Oldfield left in January 1856, and had collected, with varying intensity, in every month that year except May and August. Stuart was collecting in Southport during January 1857 and it may well have been there where the pair met again.[5]

In February 1857 Stuart obtained a specimen at Whale Head. There is also an undated Oldfield plant record from the same locality[6] which suggests they went there together before their ascent of Mount La Perouse towards the end of February (see below). Whale Head is a promontory on the coast about six kilometres tending south-west of the mouth of Cockle Creek, from whence they probably set out on their trek. Perhaps for about two kilometres they followed the Aboriginal track that ran along the south side of Cockle Creek and then left the path and headed due south across country to Whale Head, over terrain which for the most part was quite hilly and heavily forested.

---

[1] *Colonial Times* (Hobart), 7 February 1857, pp. 2–3, DAN, NLA.

[2] Southern Board of Education Fourth Report, Report for 1860, *Notes and Proceedings of the Legislative Council*, paper no. 34, Appendix E, Nominal Return of Teachers in the Employ of the Southern Board of Education on 31 December 1860, TAHO.

[3] Report of the Inspector of Schools, 1857, *Hobart Town Gazette*, no. 1202, p. 656, TAHO.

[4] This will become apparent later in this chapter.

[5] MELISR database, 3 September 2007, NHV, RBGM.

[6] MEL 2187381A, *Pentachondra involucrata*, February 1857, Browns Lookout near Whale Head, C. Stuart, MELISR database, 3 September 2007, NHV, RBGM, and *Pentachondra involucrata*, Whale Head, AO, FM 1867–1868, vol. 7, p. 57.

Fig. 14.1 South Coast Track, Tasmania

Labillardière, Delahaye and possibly Ventenat had traversed the Aboriginal track in January 1793 on their expedition to the south coast during D'Entrecasteaux's second visit to Recherche Bay.[7]

Oldfield and Stuart set out on their expedition inland to Mount La Perouse toward the end of February 1857.[8] It is not known what route they took or where they started from; perhaps there was no need for them to seek local advice in that regard as it appears Oldfield had been at the mountain in 1856[9], and Stuart earlier that year on 19 and 20 April.[10] There is no evidence that either collector had climbed the mountain during these visits. Contemporary 19th century maps[11] show no pathways to the mountain, although there must have been timber tracks through the forested areas behind Recherche Bay where pit-sawyers and shingle splitters had been operating for many years. A modern contour map[12] suggests that the easiest and most direct route would have been from the vicinity of what was later the settlement of Leprena, just south of the mouth of D'Entrecasteaux River on the west bank of Pigsties Bay, a northerly extension of Recherche Bay. The direct distance to the prominence from Leprena is approximately 12.5 kilometres. There is very minor difference between the most direct

[7] Edward Duyker, 2003, p.141. Duyker considers it likely that Labillardière and his party travelled the course of the present South Coast Track which provides bushwalkers with access to South Cape Bay and the eastern limits of the Western Tasmania Wilderness Area.
[8] One of Stuart's specimens at the National Herbarium of Victoria, MEL 1603977A, *Eucalyptus amygdalina*, was collected at the base of Mount La Perouse and dated 27 February 1857, MELISR database, dated 3 September 2007, NHV, RBGM.
[9] See Chapter 10.
[10] ibid.
[11] See, for example, John Tallis, 1851.
[12] Department of Environment and Land Management, Tasmania, 1994, 'Tasmania 1: 1000,000 Topographic Map, South Cape, 8210', Edition 1.

Fig. 14.2 Cockle Creek

distances to the summit of the mountain from either Catamaran, at 13.5 kilometres, or Cockle Creek, at 14.5 kilometres. There are no major rivers to cross and only gently rising contours; terrain immediately to the west of each of those settlements for about 4 or 5 kilometres, is relatively flat with patches of heavy forest interspersed with light bushland. Beyond this, the steep slopes would make travel much slower and more tiring than from the more northerly route from the Bay. Today, there is a track to Mount La Perouse that commences at the termination of Lune Road to south of Lune River.[13]

It was during the French expedition's second visit that Labillardière attempted to reach Mount La Perouse, which forms part of the great chain of mountains stretching from north-west to south-east of Tasmania. Probably starting out from the expedition's Cockle Creek watering place, the party travelled west through open country. At the first foothills, they struck dense rainforest and turned north. After a 'very laborious walk' through forest the party reached and ascended a mountain, thought by them to be Mount Leillateah, a prominence to the south-east of Mount La Perouse and in the same mountain range.[14] From there they travelled in a north-westerly direction, but it became apparent that the dense rainforest presented an almost impenetrable barrier, which would take much longer to traverse than their provisions would allow and they abandoned their attempt to reach the mountain.[15]

Oldfield and Stuart did not record anything about the challenges they faced climbing up the mountain or during their ascent. Today, the area around the mountain is wilderness and in much the same condition as it was in the mid-nineteenth century. Although we do not know the route the botanists took, Peter Grant, from Parks and Wildlife Service Tasmania, has kindly provided a description of his experience of climbing Mount La Perouse. Since the landscape has apparently changed little since the 1850s, this will give the reader an indication of what the collectors experienced as they made their way to the summit:

---

[13] Department of Environment and Land Management, Tasmania, 1997, Tasmania 1: 1000,000 Topographic Map, Huon, Edition 6.

[14] Mount Leillateah, Tasmania, Bonzle Digital Map, accessed 15 July 2017 at www.bonzle.com. Mount Leillateah could not be located in the Geoscience Australia Place Names gazetteer.

[15] ibid., pp. 144–145.

> Through a quirk in geology, the long ascent of Mount La Perouse begins atop one of Australia's longest descents. Exit Cave worms its way 23 km into the Ordovician limestone beneath these Southern Ranges. Lesser holes pepper the forest and unwary walkers in this wild area have disappeared without trace. But such chilling thoughts are soon distracted by the steepness of the uphill walk. It is all climb, through still, damp, and overwhelmingly green forest. The hue is not just on the mossy forest floor or in the leaves but at every level, from roots to trunks to branches to fungi. There are fluorescent greens, sombre olives, glaucous blue-greens, lively gold-tinged greens and deep-sea greens. The very air has a greenish tinge.
>
> But thoughts of colour are unlikely to distract you for long from the relentless climb. There is nothing to do but hoist and haul yourself through, around and sometimes over the trees. After an hour, perhaps two, of this sweat-raising effort you may sense a change in the air, and in the scale and type of trees. As you break out of forest and into a more open tea-tree and paperbark scrub, there are glimpses of the distant coast. The scent of decay-tinged damp is replaced by a sharper, cleaner air that somehow smells blue. After the shy light of the forest the sun here can be blindingly bright. If the air is also calm – and sunny, calm conditions are not the norm here – you may find yourself sharing the air with myriad flying insects: beetles, dragon flies, butterflies and a host of Diptera. Most won't trouble the walker, though March flies may show an interest in any exposed skin. As will leeches.
>
> Although the scenery has changed, the incline of the track has not. And the more open going is countered by sections of deep mud that stand between here and the blandly named Hill 1. By the time you have trudged up to this bald, wind-battered eminence, you will have gained nearly 800m in altitude as well as a good thirst. While you take your break, the other bare hills of Moonlight Ridge are still ahead. By blunt logic these are named Hills 2, 3 and 4. But they are dwarfed by the majestic mountains that start to dominate the scene above: The Hippo, La Perouse, Cockscomb [Ridge].
>
> Geological dramas from many eras have helped create the landscape around you. Even in (geologically) recent times – the last couple of million years – glaciers carving off Mount La Perouse and nearby peaks have dozed out deep valleys, gouged cirques, deposited moraines and left resistant rock standing above the fray. In the last several millennia, although the glaciers have gone, wind, water and ice have continued to chip away at the landscape. As you traverse Moonlight Ridge and close on Mount La Perouse itself, you climb through sedimentary rocks scraped almost clean by glaciers. First are the often softer, Permian siltstones, then the more recent Triassic sandstones. Down valley to your left Arndell Falls plunges from a cliff at the edge of a later dolerite intrusion. Although it's tempting to leave the track here and head straight for the growing bulk of Mount La Perouse, the more oblique approach via the Pigsty Ponds is far easier. Just as you start to climb from these moraine-dammed lakes onto Maxwell Ridge, a cairned route to the left leads you up the long-humped slope to the remarkably bare and flat-topped "peak" of Mount La Perouse.
>
> The sandstone here was laid flat on a sandy plain, and despite all the ructions of the last 200 million years, has remained remarkably flat to this day…[16]

Although Stuart is on record as having previously collected a plant on Mount La Perouse, together the two collectors appear to have been the first botanists to have reached the summit and gathered

[16] Peter Grant, Parks and Wildlife Service, Tasmania, unpublished ms, 2008 (with kind permission of the author).

Fig. 14.3 Swallows Nest Lake, Mt Laperouse

plants for scientific purposes. By 27 or 28 February 1857 they were on Mount La Perouse above the foothills, a climb from the plains of approximately 600 metres, and possibly higher.[17] They were on the summit, a further 557 metres, on 1 March 1857.[18] This date was given on most of Stuart's Mount La Perouse specimen labels, which suggests that they were there for only one day before moving on to another mountain to the west, where Oldfield collected at least one plant.[19] This peak was possibly Knife Mountain which lies west of Mount La Perouse at a direct distance of approximately 2.6 km from its summit.[20]

Of the 72 herbarium, and 58 published, records we have located of Oldfield's specimens taken from or near the mountain[21], the majority were angiosperms.[22] An additional angiosperm is recorded as having come from the 'Southern Alps', which probably refers to the mountain range that includes Mount La Perouse.[23] If records in the Oldfield database properly reflect the patterning of his collection from this locality, it appears to be at variance with his collections from other parts of Tasmania where cryptogams were much more frequently represented. The reason is not known. Perhaps it was the result of Oldfield selecting only those species that were unfamiliar to him or he was acting on Dr Mueller's instructions. On the other hand, it may have corresponded with the relative occurrence of cryptogams, angiosperms and gymnosperms on the prominence.

Oldfield did leave behind a record of his experience a-top the alpine area he visited on that occasion – the feelings it evoked and something of the plants he observed:

---

[17] MEL 1603977A, *Eucalyptus amygdalina*, base of Mount La Perouse, 28 February 1857, Charles Stuart, MELISR database, 3 September 2007, NHV, RBGM. However, the jointly collected holotype of *Leionema oldfieldii*, MEL 4822A, on 27 February 1857, is given the locality 'Mount La Perouse', Oldfield types, MELISR database, 23 December 2011, NHV, RBGM.
[18] For example, MEL 594990A, *Erigeron pappocromus*, Summit of Mount La Perouse, 1 March 1857, Charles Stuart, MELISR database, 3 September 2007, NHV, RBGM.
[19] MEL 73595, *Cheesmania radicata*, Mountain westward of Mt La Perouse, AO, MELISR database, 8 March 2007, NHV, RBGM.
[20] Department of Environment and Land Management, Hobart, 1994, Topographic Map, edn. 1, Sheet 8210.
[21] HAODb. 'Mount La Perouse' is given variously as 'Mount Laperouse', 'Mt de la Perouse' and 'Mount Lapeyrouse' on plant labels, ibid.
[22] Publ. 100%., 72% herb., HAODb.
[23] MEL 2227548, *Geum talbotianum*, Southern Alps of Tasmania, AO, MELISR database, 3 March 2007, NHV, RBGM.

> In the southern parts of Tasmania is a mountainous district abounding in tarns, for the most part perhaps, the craters of extinct volcanoes. The silence of the desert is horrible, but the stillness that broods over these mountain-lakes is truly awful. The human voice at its ordinary pitch, could be heard distinctly and intelligibly at a distance of three or four miles. Not a bird, not even an insect, broke the dead silence that reigned supreme in that lonely, yet lovely spot; not even a tiny beetle skimmed over the unruffled waters of those lakes, for the wind that whistled or moaned among the lofty pinnacles of the overhanging mountains, had no power to waken their waters from their apathetic slumbers, by endowing them with that semblance of life, motion, with its attendant phenomena of sight and sound, so deeply buried were they among the sheltering hills. Among all the gorgeous vegetation that surrounded these tarns, not a single one of the aquatic—or even hydrophilous—plants so common in similar localities in other parts of the island, was to be found. Flags, Sedges, and even the almost universal Juncus communis (common rush), all were absent, their places being filled by Mosses, Hepaticae, Geum, Diplopeltis[24], Dichopetalum [= *Dichsciadium ranunculaceum*], several species of Senecio, these being intermixed with large clumps of Astelia (A. alpina and A. stylosa [= *Milligania stylosa*]), Gahnia and Oreobolus.[25]

Oldfield's deep emotional connection with his natural environment is apparent in this passage and was evidently born out of a passionate interest in the world around him. His expressed desire to serve science, in any way he could, may have derived from this inquisitiveness, fuelled by a perception that by becoming the discoverer of species new to science he could achieve an elevated status within the scientific community and possibly a fulltime career as a botanist. We know from William Harvey that he had ambitions to be a collector, but there is no evidence that he saw this activity leading to social advancement. It is perhaps puzzling that he does not appear to have offered his services as a collector to any of the wealthy local naturalists. Perhaps despite his poverty, he was unwilling to become the servant of those he considered to be his equal socially and who, like himself, were dependent upon overseas botanists for plant determinations and confirmation of species new to science. He may well have come to the conviction that he would be better served in his quest for knowledge and recognition by going, as he did, directly to the most eminent authorities in Australia and Britain.

Back to Kingston

It is not known when Oldfield returned to Kingston but it was probably soon after his expedition with Stuart to Mount La Perouse. The dating on Stuart's records in the National Herbarium of Victoria show that he did not collect in the Southport–Recherche Bay area again until 2 October 1857.[26] On his arrival back in Kingston Augustus was probably greeted with the news that a rumour was circulating criticising Edwin for having too many pupils in his school. According to Edwin, the assertion had been set afloat by an 'Interested Party' and which had 'in a great measure, injured him: for it [had] come to his attention that several persons would have, otherwise, placed their children under his Tuition'.[27] In March his rebuttal appeared in the press. While no details of his system of teaching were presented, he said that it was

---

[24] This genus has not been recorded in Tasmania, Department of Primary Industries, Parks, Water and Environment, *Natural Values Atlas*, viewed www.naturalvaluesatlas.tas.gov.au, 21 September 2017.

[25] Oldfield provided this description in a discussion challenging Charles Darwin's theory accounting for the dispersion of certain aquatic and marsh plants, AO, 1865a, pp. 795–797.

[26] MELISR database, dated 3 September 2007, NHV, RBGM.

[27] Public notice, *The Hobart Town Mercury*, 13 March 1857, p. 1, DAN, NLA.

> better adapted to a large than a small number…[and] inspection of his books would, at once, bear him out in his assertion, and evidence, for several years, a support as steady as it has been unsolicited.[28]

He called upon his friends not to allow to 'go uncontradicted any unfounded statements respecting him or his establishment'.[29] The damage seems to have been short-lived; Edwin's school thrived for many years to come.

Augustus sent a selection of his Mount La Perouse specimens to Mueller who, as has been indicated previously, received them in time for some to be referred to in his *Fragmenta* (1858–1859).[30] One of these, 'the little Epacrideous plant', *Dracophyllum minimum* F. Muell., 'the only one Mr Oldfield noticed', was the subject of subsequent taxonomic discussion between Mueller and William Hooker. [31]

The collector was apparently at pains to expeditiously prepare his Mount La Perouse specimens for freighting to London so they were with Joseph Hooker in time for him to refer to some of them in his *Flora Tasmaniae*, a work that was published in 1860.[32] Since Oldfield seems to have endeavoured to identify the plants before sending them to Hooker and Mueller, the task of getting his collection ready for shipment could well have taken a month or more after the expedition to the south-east.

On 8 August 1857, *The Hobart Town Mercury* carried an advertisement issued by the Audit Office requesting among others, 'A. Oldfield' to make application at the Colonial Treasury 'for payment of their respective accounts for which warrants have been passed since the 1st inst.'[33] No information has emerged as to the nature of the accounts or whether Alfred or Augustus was the creditor.

Once again Oldfield involved himself, if somewhat marginally, in Kingston's community affairs, but no evidence of him having obtained full-time employment in Tasmania has been found. He was appointed to be the Deputy Returning Officer for taking a poll on 13 May 1857 at Oyster Cove in the Huon District[34], for which service he no doubt received a small remuneration.

Apparently discontent had been mounting for some years among certain Anglican parishioners in Kingston over the behaviour of their pastor, Reverend E. Freeman. Things came to a head on 3 September 1857 when, at a 'Synodical Meeting' held at the Kingston Parsonage, Freeman 'openly accused the mover of a certain resolution' of 'an attempt to burke the meeting'. The outcome was the publication in the *Hobart Town Mercury* on 2 October 1857, of 'The Kingston Memorial' addressed to Dr Francis Russell Nixon, Lord Bishop of Tasmania, stating their grievances. The complainants stated that Freeman

> has long been so involved in secular pursuits as to preclude his devoting that time and attention to their spiritual wants which they consider they have a right to look for from a disciple of the Cross and a faithful minister of their holy and beloved Church.[35]

They called upon the Bishop to 'remove the Rev. E. Freeman from this parish, and to nominate as his successor some discreet and faithful minister of God's Word'.[36] Augustus was a signatory to this document.

On 12 October 1857 Freeman published a petition in the *Hobart Town Mercury* that was signed by over 100 parishioners from Kingston, North West Bay and Bruny Island. Among these were some who claimed they had signed the previous Memorial 'by mistake' and regretted their error. While

---

[28] ibid.

[29] One Emanuel Levy, a former student who had gained a Newcastle Scholarship at the High School, publicly expressed his gratitude to Edwin, 'Original correspondence', *The Hobart Town Daily Mercury*, 2 August 1858, p. 3, DAN, NLA.

[30] FM, 1858–1859, vol. 1, pp. 3, 15, 38–39.

[31] FM to WJH, 12 May 1858, Directors' letters, vol. 74, Australia letters 1851–1858, fol. 178, RBGKA, from the Ferdinand von Mueller Correspondence database, copy kindly provided by Dr Monika Wells, 7 June 2001, HOA..

[32] JDH, 1860, Additions and Corrections, vol. 1, pp. 363–367, 369 & 371.

[33] *The Hobart Town Mercury*, 12 August 1857, p. 3, DAN, NLA.

[34] *The Hobart Town Mercury*, 13 May 1857, p. 2, DAN, NLA.

[35] 'The Kingston Memorial', *The Courier* (Hobart), 2 October 1857, p. 3, DAN, NLA.

[36] ibid.

Oldfield made no such an admission, he was among the signatories indicating that they were perfectly satisfied with him as their 'Spiritual Pastor' and asserted that the previous Memorial was framed and signed by individuals, part of whom either did not understand its meaning, or who cannot be considered of the most Christian spirit.[37]

Just where Oldfield really stood in this imbroglio and why he backed down from supporting the original complaint against Freeman we might never know, but it is difficult to imagine that he 'did not understand its meaning'. He is listed as 'schoolmaster' in the second petition. If he was privately employed in that capacity at the time he may have been under threat of dismissal if he did not change his position.

The Bishop established an investigation by the Archdeacon and other clergy to enquire into the charges. Their report and the Bishop's conclusion were tabled at the Anglican Synod in Hobart on 20 July 1858[38], but his decision could not be found in local newspapers. The Rev. E. Freeman was in Franklin during December 1858 touting support for one of the candidates in the forthcoming parliamentary elections[39] but, unlike prior to the complaint going before Synod, there appears to be no evidence from subsequent announcements in *The Courier* that he was still undertaking parish duties in Kingston.

Augustus was one of 106 'inhabitants' of Kingston, North West Bay and its environs who signed and published in the *Hobart Town Mercury* of 11 November 1857, an 'Address' to Constable Henry Beresford, who was leaving the district, congratulating him 'on the result of six years' residence in [that] district, during which time by his 'exertions' he was the 'means of saving many lives', and on his untiring zeal in the pursuit of criminals.[40]

Augustus must have needed paid employment after so long in the field but the historical record is silent in this regard. Since he appears to have remained in Kingston for most of 1857 he probably was employed privately as a tutor or as a teacher in a local community school.

Oldfield's donations to the Royal Society of Tasmania

Sometime after his return from Mount La Perouse, Oldfield donated fragments of quartz collected by him on that prominence to the Royal Society of Tasmania. The newspaper account of the meeting (January 1858) at which these presentations were reported, referred to Oldfield as being 'a most zealous and successful botanical collector and student, particularly in the department of cryptogamic botany' who had 'promised to contribute to the Society's herbarium specimens of anything new or rare which may fall in his way'.[41] Oldfield donated a copy of Joseph Hooker's 'Introductory Essay to the Flora of New Zealand' to the Society, which was received at its meeting in February 1858.[42]

At the Society's March meeting it was noted that Oldfield had made a further presentation to the Society of 12 dried plants from Mount La Perouse[43] and some rock fragments from Point Hibbs on the west coast, the channel of the D'Entrecasteaux Rivulet (eight or nine kilometres from the head of Recherche Bay), from plains about 300–450 metres above sea level inland from Recherche Bay toward Mount La Perouse and from the summit of the mountain.[44] This collection is no longer in Tasmania[45] and its whereabouts could not be determined.

---

[37] *The Hobart Town Mercury*, 12 October 1857, p. 2, DAN, NLA.
[38] *The Courier* (Hobart), 22 July 1858, p.3, DAN, NLA.
[39] 'The Franklin election', *The Courier* (Hobart), 22 December 1858, p. 2, DAN, NLA.
[40] *Hobart Town Mercury,* 11 November 1857, p. 4, DAN, NLA.
[41] *The Courier* (Hobart), 18 January 1858, p. 2, DAN, NLA.
[42] *The Courier* (Hobart), 15 February 1858, p. 3, DAN, NLA.
[43] See HAODb.
[44] *The Courier* (Hobart), 24 March 1858, p. 2, DAN, NLA.
[45] Alex Buchanan, Director, Tasmanian Herbarium (Tasmanian Museum and Art Gallery), University of Tasmania, personal communication to MHH, email, 12 December 2008, HOA.

These donations suggest that Oldfield had been accepted as a member of the Society. His brother Edwin, in a letter to Joseph Hooker after Augustus's death, stated that he was a 'late member' of the 'Royal Society'.[46] Augustus was never a member of, or a contributor to, the Royal Society in London.[47] Edwin was probably referring to the Royal Society of Tasmania but Augustus is not listed in surviving membership records.[48]

In Mueller's domain 1857

When James Drummond declined Augustus Gregory's invitation to accompany him on his North Australian Exploring Expedition, Ferdinand Mueller was invited to replace him. He accepted and was granted unpaid leave from his position as Government Botanist. The party left Sydney in July 1855. A few months after Mueller's return to Melbourne in May 1857, Mueller was appointed Director of the Botanic Gardens, Melbourne while still retaining his post as Government Botanist, and Dallachy assumed the title of curator.[49] For Mueller, this increased responsibility came without additional remuneration. He set about energetically developing the garden by overseeing an extensive series of capital works and increasing its plant diversity by exchanging seeds and plants with gardens in many other parts of the world. Nevertheless, he saw the primary role of the Garden as a support to his scientific office. In 1857–1858 he established an aviary for song birds at the southern end of the lagoon and a menagerie on the south-western side of the lagoon to the east of the Director's office.[50] He accepted the appointment of Director of the Zoological Garden on 7 July 1858, again without any increase in his salary.[51]

In 1857 Mueller planned to travel to London to gain experience at Kew Gardens that he considered indispensable for him to undertake his long-dreamt task of writing a flora of Australia.[52] In a letter to William Hooker, dated 15 September 1857, Mueller stated his intention not to postpone the publication of his essay on tropical *Eucalypti* until he arrived in England, 'altho I may have failed in recognizing a few old species of that intricate genus'. In this letter, there is indication that Oldfield had become one of his valued correspondents:

> Several of my botanical friends in Australia are enquiring eagerly about its issue [i.e. Mueller's essay on tropical *Eucalypti*], all approve of the cortical system, which I adopted for the first time in arranging the species of tropical Australia, and they wish to extend its application over those species, which surround them. Mr [? Augustus] Gregory in Western Australia, Mr Oldfield in Tasmania, two or three gentlemen in New South Wales, and as many in remote parts of Victoria and South Australia, all intend to subject these obscure trees to a renewed scrutiny upon different principles[.][53]

The 'cortical system' was a system of arranging and identifying eucalypts incorporating information on their bark. Usually the distinguishing characters of eucalypts came from the flowers and fruits that were often inaccessable, and using the bark to get a rough sense of where the trees fitted into the classification system was practical in the field - hence its value to collectors like Oldfield.[54]

---

46 E.D. Oldfield to JDH, 27 July 1887, Herbarium Presentations to 1900, vol. 2, fol. 627, RBGKA.
47 Rupert Baker, Library Manager, Royal Society of London, personal communication, email 9 May 2011, HOA.
48 See Chapter 7.
49 Allan Gross, 1972, ADBO, viewed 5 October 2017 at www.adb.anu.edu.au.
50 R.T.M. Pescott, 1982, *The Royal Botanic Gardens Melbourne: A History from 1845 to 1970*, Oxford University Press, Melbourne, Oxford University Press, Melbourne, p. 21.pp 52 & 89.
51 R.W. Home *et al.* (eds), 1998, vol. 1, p. 33.
52 Edward Kynaston, 1981, p. 115.
53 FM to WJH, 15 September 1857, Directors' Letters, Australia Letters, vol. 74, 1851–58, fol. no. 169, RBGKA, cited in R.W. Home *et al.* (eds), 1998, vol. 1, p. 322.
54 Sara Maroske, personal communication, email to MHH, 2 October 2017, HOA.

DR. MUELLER [FROM A PHOTOGRAPH BY FRITH].

Fig. 14.4 Ferdinand Mueller, 1865

Unfortunately, any observations Oldfield might have made are no longer available; his correspondence with Mueller was destroyed.

Difficulties with getting his botanical works published in England caused Mueller, commencing in March 1858, to begin producing his *Fragmenta* as a series of fascicles.[55] This work provided the vehicle for him to put to press in Australia Latin descriptions of new species found by himself or members of his ever-increasing network of collectors. Mueller was actively seeking to develop a comprehensive continent-wide herbarium collection in Melbourne and there can be little doubt that this endeavour was motivated in part by his ambition to produce an Australian flora and, in part, to increase the number of his published new species' descriptions, and species' range extensions, in his *Fragmenta*. For the flora, he would need access to specimens of species already described and, if possible, specimens of species that were new to science. Mueller was a prolific correspondent with many written exchanges with Victorian government officials, overseas scientists and collectors in the various Australian colonies. In January 1858, Mueller told William Hooker that his annual correspondence amounted to at least 2,000 letters.[56]

In his monthly report for February 1858 to the Chief Secretary, William Haines, on the 'proceedings of the Government Botanist and of the work performed at the Botanic Garden' Mueller reported that he had received correspondence on 'scientific points' from 'Mr Oldfield of Kingston, Van Die-

[55] R.W. Home *et al.* (eds), 1998, vol. 1, p.577.
[56] ibid., p. 34.

men's Land' and others, including Charles Darwin.[57] The Government Botanist reported in his monthly report to the Chief Secretary for March 1858 that he had received from Oldfield a Brush Wallaby (?*Macropus rufogriseus*) for the menagerie and that he had written to Oldfield and others on matters 'calculated to promote the progress of this department'. Again, Mueller gave no indication of the topics covered in these letters.[58]

On 4 February 1858 and 27 April 1858, four and two cases respectively, addressed to 'A. Oldfield', were offloaded at Hobsons Bay from the *Child of the Regiment* and the *Thames* respectively. Both these ships were from London and since there is no evidence that the vessels had called at Launceston or Hobart before coming to Hobsons Bay it is assumed that they came direct from England, possibly to be freighted on to Tasmania on another ship. It is not known to which brother they were directed.[59] Whatever the contents of the cases it seems, in total, they held a significant amount of material. The relatively impecunious state of both brothers would suggest that the family in England was sending much needed supplies.

Writing to William Hooker on 14 June 1858, Mueller said that he had given more notes on Tasmanian plants to Dr Hooker, some of which were from Oldfield. Mueller also told his correspondent that Oldfield was 'just here [at present in my cottage]'.[60] Home *et al.* stated that their editorial addition (that is, the words in square brackets) was 'obscured by binding'.[61] In a letter to George Bentham, dated 5 February 1866, Mueller said of Oldfield that he had 'lived fifteen month free of expense in [his] house and shared [his] hospitality such as it was…' and '[a]fter staying as a guest under my roof for more than a year, I engaged him as collector for the Department in W. Australia'. [62] In an earlier letter to Bentham Mueller said:

> Aug. Oldfield, a thorough Gentleman, was engaged as collector for my department for a journey to the Murchison River, on which he was out for more than a year. A second time he collected supported by my department in the [most] south west parts of Australia.[63]

This meant that his extended stay with Mueller probably commenced about two months after his return from Mount La Perouse in early March 1857, following his stint as Returning Officer on 13 May 1857, and ended about 15 months later, on 15 July 1858 (see below) when he sailed to King George Sound and was in Western Australia for more than a year. However, there are newspaper records which show that Oldfield returned to Tasmania, possibly for about a month, during October and November 1857. His various donations to the Royal Society of Tasmania may have been made during this visit since they were all acknowledged during the first quarter of 1858.

Neither Mueller nor Oldfield left details explaining Oldfield's stay with Mueller. Perhaps he was there to sort and properly mount his Mount La Perouse and first Western Australian collections. If he

---

[57] FM to William Haines, 1 March 1858, E58/1877, unit 745, VPRS 1189 inward registered correspondence, VA475, Chief Secretary's Department, Public Record Office of Victoria, quoted in R.W. Home *et al.* (eds), 1998, vol. 1, p. 362.

[58] F. Mueller to John O'Shanassy, Monthly Report on the labours performed at the Botanic Gardens and on the proceedings of the Gov. Botanist during March 1858, 1 April 1858, F58/2731, unit 745, VPRS 1189 inward registered correspondence, VA475 Chief Secretary's Department, Public Record Office, Victoria, excerpt kindly provided by Dr Monika Wells, The Correspondence of Ferdinand Mueller Project, personal communication to MHH, 7 June 2001, HOA. In his annual report to Parliament, tabled 3 November 1858, Mueller named AO among contributors to the herbarium, the botanic garden and to the menagerie in the Melbourne Botanic Garden. A list of the live birds and mammals presented to the menagerie during the year is included in Mueller's report but they are not linked to their donors, Annual Report of the Government Botanist and Director of the Botanic Garden, Victoria – Parliamentary Papers – Votes and Proceedings of the Legislative Assembly 1858–9, no. 17, vol. 2, pp. 1–27, Government Printer: Melbourne, quoted in R.W. Home *et al.* (eds), 1998, vol. 1, pp. 419, 424 & 426.

[59] 'Shipping intelligence', *The Argus*, 4 February 1861, p. 4, and 27 April 1858, p. 4, DAN, NLA.

[60] FM to WJH, 14 June 1858, Director's Letters, Australia Letters, vol. 74, 1851–1858, fol. 181, RBGKA cited in R.W. Home *et al.* (eds), 1998, vol. 1 p. 387.

[61] ibid., fn. 15.

[62] FM to GB, 5 February 1866, Kew Correspondence, Australia, Mueller, 1858-70, fols. 196-201, RBGKA cited in R.W. Home *et al.* (eds), vol 2, 2002, p. 337.

[63] FM to GB, 24 September 1862, Kew correspondence, Australia, Mueller, 1858–70, fols. 72–79, RBGKA, quoted in R.W. Home *et al.* (eds), 2002, vol. 2, p. 162.

Fig. 14.5 Director's House Royal Botanic Gardens Melbourne

was undertaking this work, Oldfield would have dealt with his specimens in Mueller's office-residence.[64]

Since c.1854, Mueller's home and office were located in one building in the Botanic Gardens Melbourne. It was a stone, single-storeyed construction with separate quarters for a male servant. Shortly after Mueller took up residence the house was extended.[65] This building was the one in which Oldfield resided during 1857–1858. Mueller moved to his own cottage at 28 Arnold Street, South Yarra during 1863 when work on a second storey of the Director's residence was in progress. This building was the home of subsequent Directors of the Botanic Gardens from 1873 until 1992.[66] While the interior was re-modelled several times, the outer structure has been faithfully maintained to this day: 'It has a delightful Georgian-influenced facade with a charming neogothic porch.'[67]

Notwithstanding Oldfield's presence in Mueller's house for such a lengthy period, the pair were not on intimate terms. Mueller said he 'scarcely [knew] anything of his family relations':

> I discussed never anything with him except some points of characters & distribution of plants. But as his eyesight was so impaired he was seldom positive in any information. The bustle in my department indeed left me little time to converse with him & we parted comparatively as strangers.[68]

In the same letter Mueller told Bentham that he had given Oldfield 'a present in cash of between £20–30 privately'.[69] Since it was mentioned in the context of this letter it seems that the money was given while he was under Mueller's roof but this could not be confirmed.

There is little doubt that the breadth of Mueller's official responsibilities, his dedicated attention to plant taxonomy, and to communicating to the public information on aspects of economic botany and

[64] Helen Cohn, Library Manager, RBGML to MHH, email, 17 May 2010, copy in HOA.
[65] R.T.M. Pescott, 1982, p. 39.
[66] Helen Cohn to MHH, 17 May 2010.
[67] R.T.M. Pescott, 1982, p. 39.
[68] FM to GB, 5 February 1866.
[69] ibid.

zoology, as evidenced by his prolific correspondence and many journal and newspaper articles, left little time for conversation with his lodger.

Oldfield's own passionate interest in botany may have meant that he did not want to waste the precious time he had alone with Mueller talking about his personal affairs when he could be discussing matters of mutual scientific interest. However, more probable was his apparent reluctance to bring himself and his affairs under the notice of others.[70]

The collector did tell Mueller that he had to give up his 'teachership' in Tasmania because of ill health[71] but the nature of the condition is not known. It is doubtful that he hid the fact he was the schoolmaster when the Brown's River School was closed in 1856, as it was apparently because of too few students and not his poor performance. Mueller was aware that Augustus's brother, Edwin was still [1866] a teacher in Hobart, but was 'personally not known' to him (Mueller). By 1857 Edwin had been running his Commercial Academy on the corner of Argyle and Brisbane Streets for seven years.

Augustus' poor eyesight must have been a source of deep concern to him. No details have emerged in his writings as to the nature of his ophthalmic disease or what steps he took to have it treated. However, about ten years later Mueller referred to it as 'his [Oldfield's] glaucomatous affection'.[72]

Glaucoma is a group of eye diseases causing optic nerve damage which may lead to partial or total blindness. In glaucoma, increased eye pressure usually plays a role in damaging the delicate nerve fibres of the optic nerve. Not all types of glaucoma are characterized by eye pressures; in some cases, the optic nerve may suffer damage with resulting visual field loss even though normal eye pressures are maintained.[73] Therapies, however, were largely in the rudimentary stages of development and the causes of the disease were not understood.[74] Whatever the condition, the accident to his eye that Oldfield suffered in the early 1850s may have been a contributing factor. Perhaps glaucoma was not the ophthalmic condition he suffered. The harsh physical conditions under which he labored, particularly at the Murchison River, might have given rise to other diseases causing loss of sight. Repeated periods of dehydration may have resulted in the formation of early onset cataracts, which were slowly and increasingly clouding the lenses of his eyes. Another likely disease was trachoma causing him bouts of untreated bacterial infection from bush fly attacks and poor sanitation leading to trichiasis. This is characterised by scarring of the cornea and conjunctiva, causing the eyelid to turn inwards, making eyelashes scratch the cornea which eventually becomes opaque, causing poor vision and eventually irreversible blindness.[75]

Mueller's generosity to his collector was reminiscent of the financial and other support he had afforded Charles Stuart a few years earlier. In Oldfield's case, it was no doubt out of kindness as well as gratitude for the valuable material he had obtained in Western Australia, particularly as Mueller recognised that government and his personal monetary payments hardly covered his expenses.[76] The Government Botanist was grateful:

> Oldfields [sic] searches for plants in W. Australia were very creditable to him & I think I have done him full justice in my writings. The 2 or 3 plants bearing our joint authority [*Eremophila clarkei* and *Halgania bebrana*] were named by myself. I liked

[70] See, for example, AO to JBH, 2 February 1874, English Letters, vol. 97, NAL–OXL, 1859–1900, fol. 296 (303), RBGKA.

[71] FM to GB, 5 February 1866, Kew Correspondence, Australia, Mueller, 1858-70, fols. 196-201, quoted in R.W. Home *et al.* (eds), 2002, vol. 2, pp. 337-8.

[72] FM to JDH, 12 March 1868, Australia correspondence,, Mueller, 1858–70, fols 307–8, RBGKA, copy of extract provided by Dr Monika Wellsfrom the database of The Correspondence of Ferdinand von Mueller Project, personal communication to MHH, 7 June 2001, HOA.

[73] *University of Michigan Kellogg Eye Centre*, undated, 'Glaucoma', viewed 24 March 2010 at www.kellogg.umich.edu

[74] Success in MICROphth 2004, 'Glaucoma' in *A Historical Tour of Ophthalmology*, viewed 24 March 2010 at www.mrcophth.com.

[75] Sightsavers, undated, 'Causes of blindness', viewed 20 March 2011 at www.sightsavers.org.

[76] FM to John O'Shanassy, 10 August 1859, L59/8146, unit 747, VPRS 1189 inward registered correspondence, VA475 Chief Secretary's Department, PROV, quoted in R.W. Home *et al.* (eds), 1998, vol. 1, p. 481.

> to give him every encouragement in my power and did not spare my private means for the purpose.[77]

It is unlikely that Mueller's gift of cash and free accommodation came without the pair agreeing on tasks that Oldfield would perform in return. Members of his family were adamant that the collector would not under any circumstances accept charity.[78] Given Mueller's eagerness to publish descriptions of new species, he would have been well served in that regard by having his collector sort, mount and label his Western Australian plants for storage in the herbarium ready for subsequent scientific examination. Otherwise he may not have been able to complete the task quickly because of his limited personnel resources. The collector must have been pleased to have had the opportunity to prepare his specimens ready for Mueller's consideration, although as Mueller said: '[H]is eyesight was so impaired he was seldom positive in any information'. By inference he was having great difficulty using the microscope but Oldfield left no comments concerning the problems he was experiencing.

Throughout the volumes of Mueller's *Fragmenta*, species descriptions include references to the forms and the habitats of many of the plants from which Oldfield took specimens. This information was apparently drawn from separate notes the collector 'made at the time of their gathering'.[79] Despite every endeavour we have been unable to locate them. Many of his specimen labels in the National Herbarium of Victoria bear brief habitat descriptions in English[80], which in Mueller's *Fragmenta* were Latinised. Part of Oldfield's work may have been to extract this information from his notes and to include it on the specimen labels.

Oldfield probably enjoyed living within the confines of the Botanic Gardens. Located on the south side of the Yarra River, about a kilometre east of the city, the site consisted of 'undulating land running down to the river' and 'was well described at the time as "a small beautiful valley"'. Not only would he have been able to take pleasant and interesting walks through the garden areas, developed by John Arthur and John Dallachy since 1846, but also those more recently planted and landscaped under Mueller's direction.[81]

By now the Botanic Gardens, Melbourne, had been under Mueller's directorship for approximately three years. His concept of such a garden was that it should be 'an establishment for the diffusion of knowledge', for acclimatization of foreign plants, for propagating native plants, and 'as a healthy locality for recreation':

> As an universal rule, it is primarily the aim of such an institution to bring together with its available means the greatest possible number of select plants from all the different parts of the globe; to arrange them in their impressive living forms for systematic, geographic, medical, technical or economic information, and to render them accessible for original observations and careful records.[82]

Mueller had taken control of gardens that comprised plantings of many indigenous and exotic species, expanses of lawn, and a well-developed system of walkways, some of which encircled the large lagoon, the northern bank of which ran close to the Yarra River. It was one of the 'few pleasant public areas in [the] young city and had become a focal point for the inhabitants' to gather for recreation, picnics, public celebrations and charity functions.[83]

By the time Oldfield stayed in the Gardens, the new Director had extended the area under cultivation by one and a half hectares to just over seven hectares. A comprehensive system of drainage was

[77] FM to GB, 5 February 1866.
[78] See for example Juliana Oldfield to JDH, 31 May 1875, vol. 97, English Letters, 1859–1900, NAT–XL, fol. 297 (304), RBGKA.
[79] AO to JDH, 16 March 1866, English Letters, vol.97, 1859–1900, NAT–OXL, fol. 289 (296), RBGKA.
[80] MELISR database, 08 March 2007, NHV, RBGM.
[81] R.T.M. Pescott, 1982, p. 10 & chapters 2 & 3.
[82] FM, *The Objects of a Botanic Garden in Relation to Industries*, lecture delivered at the Industrial and Technological Museum, Melbourne, Mason, Firth, McCutcheon, Melbourne, 23 November 1871, quoted in R.T.M. Pescott, 1982, p. 42.
[83] ibid., pp. 25–27.

installed and many walks and gardens were lined with brick or stone gutters.[84] Surrounding Mueller's house to the east and west was a 0.8 hectare 'system garden', in which 1700 indigenous and exotic plants had been established, representing more than 100 of the principal families. To the north was a collection of fruit trees and vines, and to the south an arboretum of oaks. Work on the pinetum, on the barren area facing the western portion of the lagoon, was well under way. Two hundred and twenty-six trees had been planted, many rare and 'all equally calculated to beautify the area'.[85] Mueller also had formulated a design for the large reserve between the Gardens proper and Brice's Bridge, and he extended the lines of Blue Gums (*Eucalyptus globulus*) along the banks of the river. Mueller also had charge of the Zoological Society of Victoria's menagerie located across the river in Richmond Park.[86]

During 1858 the Gardens became more accessible to visitors with the provision of several new entrance gates and the erection of an iron footbridge across the Yarra River near Melbourne Cricket Ground that had been established in 1853. This crossing linked directly to the Melbourne and Suburban Railway Company's Botanic Gardens station, close to where the line from Melbourne branched at Swan Street (Richmond), one branch going to Hawthorn and the other south to Chapel Street (Windsor).[87]

Stimulated by the advent of the Victorian gold rushes, Melbourne's infrastructure had expanded in response to the huge increase in population from approximately 30,000 in 1850 to 165,000 in 1858.[88] Suburbs had been established and the city itself had become more identifiable as a capital. It was claimed that 'by now the city had reached its final form'. Most of the land had been sold and the vocational enclaves settled. The eastern end of Collins Street had attracted members of the medical profession, while along eastern and central Collins Street insurance companies, banks and building societies were clustered. The Mechanics' Institute was also in that street. In Bourke Street, there were theatres & music halls while the inner section of Little Collins Street was the domain of the legal profession. Shops of all kinds and hotels made up most of the remaining commercial outlets.[89] Scattered around the city were some prominent landmarks, for example, some places of worship were prominent landmarks[90]: on the corner of Collins and William Streets was the bluestone and sandstone Anglican Cathedral Church of St James[91], and on a block of land bordered by Grey Street West and Gisborne, Albert and Lansdowne Streets, the new Roman Catholic St Patrick's Cathedral.[92] Scattered around the city were other substantial and notable buildings providing services to the community.

Although Oldfield may have found the walk from the Botanic Gardens along the banks of the Yarra River into the city pleasant enough—the stream's pollution by sewage and factory discharges had only just begun[93]—exploring the city's streets could have been very disagreeable indeed:

> The familiar spectacle arose of private wealth amid public squalor—the great mansion standing in its glorious private garden, blazing light provided by its own gas supply, while outside were streets knee-deep in mud, undrained, badly lit, stinking with uncollected horse manure and household sewage.[94]

While most of the central streets had been made passable for general street traffic and footpaths constructed, there were still dangerous places. During the winter months flooding was a common occurrence and was severe enough at times to result in drownings. Melbourne became the city of mud,

---

[84] ibid., p. 43–45.
[85] ibid., p. 45.
[86] ibid., map at p. 43.
[87] ibid., p. 51.
[88] Michael Cannon, *Life in the Cities: Australia in the Victorian Age: 3*, Thomas Nelson, Melbourne, 1975, p. 14.
[89] Records & Archives Branch, City of Melbourne, 'History of a city', 1997, pp. 20–21, viewed 7 January 2014 at www.melbourne.vic.gov.au.
[90] ibid.
[91] St James Old Cathedral, 2011, 'History', viewed 7 January 2014, at www.sjoc.org.au.
[92] Catholic Archdiocese of Melbourne, undated, 'A Story of St Patrick's Cathedral', viewed 12 February 2010 at www.cam.org.au.
[93] Michael Cannon, 1975, vol. 3, p. 28.
[94] ibid., p. 24.

which 'caked the boots of every pedestrian who braved its winter streets, splattered the clothes of every newcomer, [and] was trailed in slushy footprints through every shop and dwelling in town'.[95] 'When summer storms came the business part of the city was overwhelmed with blinding dust, and for long intervals you could not even see across the street'. The Council's contribution was to employ a man to hose down the roads until powdered manure ran liquid in the gutters.[96] Added to this was the ever-present rotting household garbage and the uncontrolled nuisance of householders' pigs, dogs, goats and fowls wandering the streets, feeding on waste organic matter and in doing so, spreading rubbish around the streets.[97]

People were dying from diseases directly caused by the Melbourne City Council's inattention to sanitary matters.[98] This neglect of public health by the Council, which had been incorporated in 1842, was, as in other Australian colonial capitals, mainly due to the unwillingness of rate payers to 'pay higher rates or accept civic intervention for the common good' and the government's refusal to provide financial support for the day-to-day running of the city. The class-based Council elective system gave multiple votes to wealthy ratepayers, thereby ensuring that no man could stand for election unless he possessed £1,000 in property.[99] This meant that 'industry and commerce came first and the physical welfare of urban residents a long way behind'.[100]

Despite the unsavoury city environment and the poor condition of the roadways, the streets of the city's central business district were 'bustling with "an American, go-ahead spirit"'[101], which Oldfield might have thought exciting. Pedestrians and horse-drawn vehicles abounded. Middle- and working-class shoppers thronged to the Western Market, corner of Collins and William Streets, for their fruit and vegetables and the many shops, commercial houses and street vendors catered for their other purchases or business needs.[102] However, presumably on fine dry afternoons between three and five o'clock, 'the city's more exclusive "Block" in central Collins street was "ablaze with its crowds of colonial fashionables and celebrities...passing the hour in an easy, careless, lounging, gossiping manner"'.[103] The social distinctions in the city's road transport were imported direct from England, although for every horse owner in that country there were ten in Australia. '[P]ractically every house had to have its own stables, trough and hay storage shed, no matter how simple'. Also plying the streets were many delivery drays and 'jingles' of various types providing carriage for those who needed it—these were two-wheeled covered drays in which one or two passengers sat with the driver, and another two or three sat on a bench facing the rear.[104]

Another trip to Western Australia planned

Despite the considerable British and European interest in Western Australian plants up to this time, with substantial collections of its indigenous species housed in herbaria in Britain and Europe, the flora of the vast expanse of the ecologically-diverse western Colony remained largely unexplored.[105] Apparently Mueller had come to realise that financially supporting Oldfield would amply reward him to go once more to Western Australia to collect for him. Oldfield's reports and the richness of his

[95] Andrew Brown-May, 1998, *Melbourne Street Life: The Itinerary of Our Days*, Australian Scholarly/Arcadia and Museum Victoria, Kew, Victoria, pp. 32 & 42–43.
[96] David Kennedy Jnr., 1876, *Kennedy's Colonial Travel*, Edinburgh, quoted in Michael Cannon, 1975, vol. 3, p. 50.
[97] Michael Cannon, ibid., p. 28.
[98] ibid., p. 24.
[99] ibid., p. 22.
[100] ibid., p. 26.
[101] ibid., p. 31.
[102] ibid., pp. 32–38.
[103] ibid., pp. 31–32.
[104] ibid., pp. 52–55.
[105] Roger Underwood, 2011, *A Botanical Journey: The Story of the Western Australian Herbarium*, Department of Environment and Conservation, Perth, pp. 1 & 31–68.

1856–1857 collections probably enhanced Mueller's belief that the naturalist would add many specimens to the Victorian government's herbarium and would 'thereby obtain new material for the *Fragmenta'*.[106]

By this stage in their relationship Mueller must have fully appreciated Oldfield's passion for botany and discovering species new to science, his collecting skills, powers of observation, bushmanship and his ability to involve Aborigines in his work, all of which were necessary for such an undertaking. The historical record, however, has not revealed Mueller's detailed instructions to Oldfield.

A passenger signified only by 'Oldfield' travelled to King George Sound aboard the Royal Mail steam ship *Emeu*, departing from Hobsons Bay for the Suez on 15 July 1858[107] with the homeward (Britain) mails.[108] This person was almost certainly Augustus since the date of his departure accords with Mueller's statement in his monthly report for December 1859 to the Chief Secretary, William Nicholson, that Oldfield was in Western Australia for eighteen months and had arrived back in Melbourne 'during the month'.[109]

Mueller's great disappointment

Following the British Government's refusal to finance a Flora of Australia, William Hooker's successful advocacy gained the financial backing of the colonial governments of New South Wales, Queensland, Victoria and South Australia.[110] Mueller's dream of writing the Australian Flora was shattered about four months after Oldfield left for Western Australia. George Bentham wrote to Mueller from London, on 23 August 1858, informing him that he, Bentham, had agreed to write the official Flora of Australia.[111] Apparently this was the first Mueller had heard of the decision and he was extremely upset. It was perhaps even more infuriating to have received the news from the very person chosen to author the work. On the same day that Bentham wrote to the Government Botanist, Joseph Hooker sent a letter to Mueller in which he gave strong indication that he was of the opinion Mueller was not methodical enough in his writings:

> [I]t is extremely difficult already to refer to your published descriptions & I am sure I omit referring to some that I would not give the go bye to – but life is short & books are long & the indifference of men of science to the convenience of posterity & contemporaries in publication is leading to considerable confusion.[112]

This is a puzzling state of affairs since it is quite clear from Bentham's letter that the appointment had already been made; he must have thought Sir William had already informed Mueller of the decision.[113] At the time of writing, however, Bentham was on a tour in the west of England and possibly out of touch with what the Hookers were doing. Joseph Hooker's letter may have been written with the intention of letting Mueller know what he thought of his taxonomic writings as a warning precursor prior to breaking the news about the appointment.

---

[106] FM to WJH, 15 June 1859, Director's Letters, Australian and Pacific Letters, vol. 75, 1859–65, fol. 114, RBGKA, cited in R.W. Home *et al.* (eds), 1998, vol. 1, p. 465.

[107] 'Shipping intelligence', *The Argus*, 16 July 1858, p. 4. Date of arrival not known.

[108] The *Emeu* did not call at Adelaide on the way to King George Sound. Homeward mails from South Australia were dispatched to Melbourne per the screw steamer *Havilah* on 10 July 1858 for loading on the *Emeu*, 'Shipping news', *The South Australian Advertiser*, 17 July 1858, p. 2.

[109] F. Mueller to William Nicholson, 5 January 1860, 060/223, unit 748, VPRS 1189/P, Inward Registered Correspondence, VA 475 Chief Secretary's Department, Public Record Office, Victoria (copy kindly provided by Dr Monika Wells to MHH, from the database of The Correspondence of Ferdinand von Mueller Project 7 June 2001, HOA).

[110] Ray Desmond, 1995, pp. 216–217.

[111] GB to FM, 23 August 1858, RB MSS M4, Library, Royal Botanic Gardens, Melbourne, quoted in R.W. Home *et al.* (eds), 1998, vol. 1, p. 405–406. In this letter Bentham stated that the Flora was to be funded by the British Government but this apparently was not the case, Ray Desmond, 1995, pp. 216–217.

[112] JDH to FM, 23 August 1858, RB MSS M3, Library, Royal Botanic Gardens, Melbourne, cited in R.W. Home *et al.* (eds), 1998, vol. 1, p. pp. 403–05.

[113] ibid., fn. 94, p. 36.

Fig. 14.6 George Bentham, c. 1855

In any event, no official notification to Mueller of Bentham's appointment has been found. It was publicly stated that Bentham was chosen over Mueller because Joseph Hooker believed it would not be possible to successfully undertake a work of such magnitude in Melbourne when the major earlier Australian collections were housed in England.[114] By deciding not to go to England to examine earlier collections, Mueller had apparently ruled himself out of contention.[115] However, by the end of the 1850s 'the Kew botanists also saw him as too different from them in his views on taxonomy tobe a suitable author'.[116]

George Bentham—the Hookers' choice to author *Flora Australiensis*

Bentham, like William Hooker, was a self-taught botanist. His interest in botany was kindled while living in France c.1814, when he successfully identified a plant by using his mother's copy of de Candolle's *Flore française*. For some years he managed his father's estate near Montpellier and during that time began his botanical studies. In 1825, he went to London where he met several important botanists including Robert Brown. In Glasgow, he met William Hooker and George William Arnott, with whom he later visited the Pyrenees. Bentham's *Catalogue des Plantes Indigènes des Pyrénées et du Bas-Languedoc* was published in 1826. After the French estate was sold the family returned to England. For a few years he served as secretary to his uncle, Jeremy Bentham, and then entered Lincoln's Inn. In 1826, he was elected a fellow of the Linnean Society. From 1829–1840 he was secretary of the Royal Horticultural Society. During that period, in 1831, he was called to the Bar but never practised. In 1833, upon his inheritance of wealth following the deaths of his father and uncle, he concentrated on botany. Bentham's first major botanical work, *Labiatarum Genera et Species*, appeared between 1832 and 1836. After he donated his herbarium of more than 100,000 specimens to

[114] ibid., p. 37.
[115] ibid.
[116] Sara Maroske, personal communication, email to MHH, 2 October, 2017, HOA.

Kew Gardens in 1854, William Hooker invited him to establish permanent quarters there. Apart from numerous contributions to botanical literature, Bentham was particularly associated with the preparation of Kew's colonial floras: *Flora Hongkongensis* (1861) and *Flora Australiensis*, 7 vols (1863–1878).[117]

[117] N.T. Burbidge, 1969, 'Bentham, George (1800–1884)' in ADBO, viewed 9 February 2011 at www.adb.online.anu.edu.au.

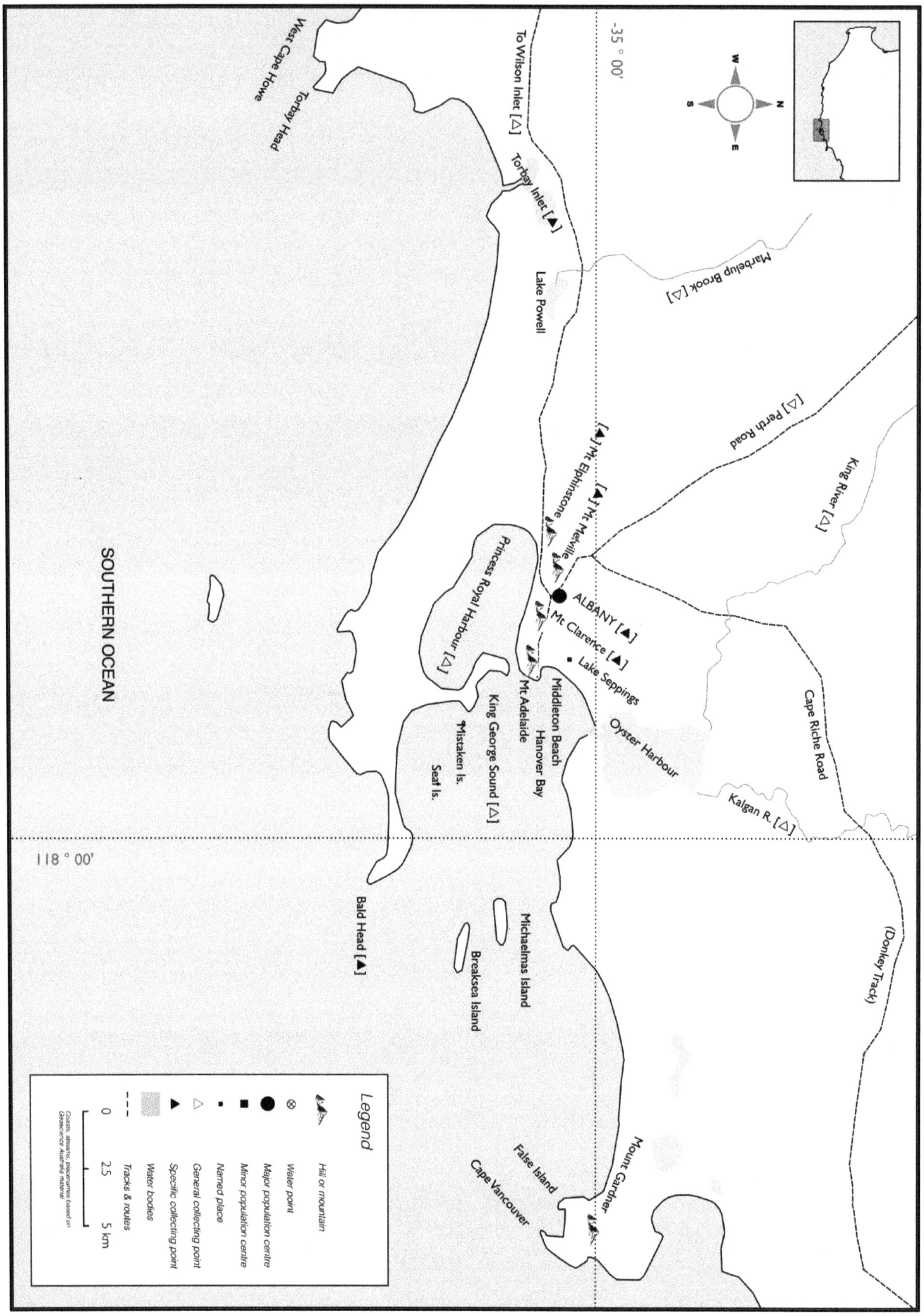

Map 12: Western Australia, Albany

Fig. 15.0 Hakea oldfieldii

Chapter 15

Western Australia 1858–1859

King George Sound to the Swan River

Give or take a day or two, the journey from Hobsons Bay to Albany took nine days[1], arriving in King George Sound about 25 July 1858. The steamship *Emeu,* like other vessels, entered the Sound between Bald Head and Breaksea Island, bringing into view the new lighthouses on Breaksea Island and Point King[2] (See Map 12 p. 268). Oldfield probably stood on the deck gazing with interest at the surrounding landscape: low hills of granite of the mainland and islands sloping into the sea and capped with coastal heath and stunted bushes of *Taxandria, Banksia, Melaleuca* species and species of many other genera. If he remained on deck as the *Emeu* was piloted through the narrow channel between Point Possession and Point King, into Princess Royal Harbour, there is little doubt that he noted the Harbour's southern shore of coastal dunes and narrow beach.

Albany July/August 1858

Map 12 p. 268 – Western Australia, Albany

The town of Albany lies within the Muir District, Bibbulmun Province, South Western Australia Floristic Region (SWAFR).[3] It includes the low coastal granite hills and peninsulas from Point Hillier near Denmark and east to Bremer Bay, the coastal Mounts Manypeaks and Gardner and, inland, the Porongurup Range and Mounts Barker, Lindesay and Frankland. It is a significant, although scattered, eco-zone, which contains habitats and occurrences of many vegetation communities that are endemic, localised or restricted.[4] According to Beard, the modern coastline is controlled in the south, from Cape Riche to Point Hood at the western end of Doubtful Island Bay, by a chain of large granite bosses, some of which rise to over 150 metres above sea-level.[5]

From his ship Oldfield probably observed, as Archdeacon John Ramsden Wollaston had noted ten years previously:

> The site of the town is beautifully chosen, between two hills [Mount Melville and Mount Clarence] sloping down to the white sands of the harbour, of which it commands a full view...The white stone houses are picturesquely scattered about...[6]

Moored in the Bay about 40 metres offshore was the hulk of the *Larkins*, which served as storage for the P&O Coal Depot that was situated on the foreshore on the fringe of the settlement, east of York Street.[7] Prominent landmarks clearly visible from the ship were the Convict Depot at the foot of

[1] J.L. Poore, letter to the Editor, *The Perth Gazette and Independent Journal of Politics and News*, 6 September 1861, p. 4, DAN, NLA.
[2] Donald S. Garden, 1978, *Southern Haven: A History of the Port of Albany, Western Australia*, Albany Port Authority, Albany, pp. 8–9 and 15–19.
[3] Paul Gioia and S.D. Hopper, 2017, pp. 11–12 and Appendix S3, Fig. S3.7.
[4] Sarah Barrett, Sarah Comer, Nathan McQuoid, Meghan Porter, Cameron Tiller and Deon Utber, 2009, 'South coast ecodistricts and their relationship to fire', in *Identification and Conservation of Fire Sensitive Ecosystems and Species of the South Coast Natural Resource Management Region*, South Coast Resource Management, Department of Environment and Conservation (now Department of Parks and Wildlife), Perth, pp. 11-16., p. 13 and Fig. 1, p. 16.
[5] J.S. Beard, 1990, pp. 139 and 146.
[6] Percy U. Henn (ed), 1954, p. 40.
[7] Donald S. Garden, 1977, p. 142.

PRINCESS ROYAL HARBOUR, KING GEORGE'S SOUND, WESTERN AUSTRALIA.

Fig. 15.1 Princess Royal Harbour, King George's Sound, 1857

Parade Street and, overlooking the foreshore, the mansion on Stirling Terrace built in the 1850s by trader, whaler, pastoralist and sandalwood cutter, George Cheyne.[8]

There being no deep-water jetty, the ship anchored some distance out in Princess Royal Harbour; Oldfield and his fellow passengers, their luggage and any cargo destined for Albany and the hinterland were lightered ashore[9], almost certainly landing them on the jetty at the foot of Osnaburg Street (now Bridges Street).[10]

The naturalist's initial experience of Albany might have been rather depressing. While winter days can be sunny and welcoming, Albany in July and August is one of the wettest and coldest places in Western Australia.[11] If it was overcast and raining when he disembarked, he may have wished himself back onboard the *Emeu* whilst he trudged through the town's unsealed, sodden sandy streets looking for accommodation.

In other ways, too, he may have found the general ambience of the settlement disappointing. As one observer remarked:

> The little town is surrounded by scrub, or is rather actually built upon it, unreclaimed land intermingled with the houses and gardens. A closer inspection modified previous conclusions as to the neatness, revealing indeed among its poorer houses a general un-tidiness invisible from the deck of the steamer.[12]

The townsite had been laid out in 1851 by the Assistant Surveyor, Philip Chauncy. His plan shows that, with two exceptions, all the lots along the foreshore and Stirling Terrace from Parade Street to Osnaburg Street had been taken up, and the same was true for the southern half of York Street.[13] Oldfield may have thought that the town was too quiet and relatively backward; there were no public conveyances to save walking and the town offered little of interest. According to Donald Garden, 'visitors found it difficult to empathise with the somnolence which they sensed hanging over

---

[8] ibid., p. 166.
[9] Donald S. Garden, 1978, pp. 8–9 and 15–19.
[10] Donald S. Garden, 1977, p. 79.
[11] Bureau of Meteorology, Australian Government, 'Albany weather – monthly climate statistics', viewed 4 June 2009 at www.bom.gov.au.
[12] *Perth Gazette*, 20 May 1863, CSO, vol. 256, fol. 107, SROWA, cited in Donald S. Garden, 1977, p. 146.
[13] Philip Chauncy, 1851 plan, Cons 3868/001, SROWA, reproduced in John Dowson, 2008, *Old Albany: Photographs 1850–1950*, National Trust of Western Australia, Perth, pp. 20–21.

the town like a deadly pall'. Apparently, they did not appreciate the peace and quiet of the community 'which many of its residents found to be a major attribute'.[14]

The naturalist may have headed for one of the three or four hotels for an ale or two and to glean information about accommodation options in the town from the other customers, among whom were ticket-of-leavers, whalers, sealers and local laborers. No doubt he soon discovered that minimal lodgings and little entertainment were on offer in all the hotels except for the Chusan (later the London Hotel).[15] Situated on the north side of Stirling Terrace, this large prefabricated, two-storey wooden building had been brought from Singapore by John McDonnell in 1855, to cater for passengers on Peninsula and Orient Company (P&O) steam ships stopping at the port to coal-up.[16] Where Oldfield resided whilst in Albany is not known, although perhaps he had much the same experience as William Harvey who, in Albany in 1854, was 'very comfortably housed and boarded for 7/- p[er] day'.[17]

While the Convict Hiring Depot at the foot of Parade Street may have aroused his curiosity; a browse through St John's Anglican Church, a few shops, and a look at other buildings along the lower end of York Street, Stirling Terrace, then the waterfront, where most of the settlement was concentrated, would not have taken Oldfield long.[18] He would have soon found that, except for Stirling Terrace which overlooked the waterfront, and York Street, which ran up quite a steep hill from the shore at right angles to the Terrace, and a few other bits of track along the waterfront, Albany's streets had not yet been cleared[19] and there were only a few cottages and larger dwellings above the Terrace to the east and west of York Street.[20] If an 1864 photograph of Albany is any guide, he was familiar with brick and stone one- and two-storey Georgian-style houses and cottages. Most of them were without verandahs and lean-to kitchens that were to become a common feature of Australian workers' homes. Dotted among these buildings were a few bungalows with deep-pitched roofs and wide front verandahs.[21] At the foot of Mount Clarence, behind his mansion, George Cheyne had recently established the first flour mill in Albany.[22]

Perhaps Oldfield joined his shipboard companions watching some of the local Aborigines who were around the town 'in great numbers' in 1858, amusing onlookers with their spear and boomerang throwing in return for small payment of a 'kikpence' [sixpence].[23] They were smeared with 'red earth and oil' which gave them 'a most horribly rancid smell' government regulations required them to be 'decently' clad. An indication of what this meant can be seen in the series of posed group photographs of King George Sound Aborigines taken by visiting photographer Gustav Adolf Riemer in 1877, at what is thought to have been the Albany recreation ground. Many of the men and women are depicted wearing long kangaroo cloaks and some of the men were unclad apart from pieces of cloth or kangaroo hide suspended from belts around their waists and covering their genitals. A few women are dressed in long voluminous wraps of cloth, which may have been purchased from a local store or given to them by members of the community.[24] Oldfield possibly tarried after the other spectators left and made efforts to engage the assistance of the performers in his botanising activities. A comment by Joseph Hooker in a letter to Charles Darwin suggests that Oldfield was successful in that regard at the

---

[14] Donald S. Garden, 1977, p. 147.
[15] ibid., pp. 122–123 and 146–147.
[16] ibid., p. 123.
[17] WHH to Hannah Todhunter, 13 April 1854, cited in Sophie C. Ducker (ed), 1988, p. 80.
[18] John Dowson, 2008, pp. 32–33.
[19] Donald S. Garden, 1977, p. 129. Surveyor Philip Chauncy had marked out the town of Albany in 1851, John Dowson 2008, pp. 20–21.
[20] ibid., pp. 38–39. Dowson states the two photographs were taken c.1864, apparently the earliest that show dwellings on the rises to the north of Stirling Terrace. We used Wollaston's account of cottages dotted around, Percy U. Henn (ed), 1954, p. 40.
[21] John Dowson, 2008, pp. 32 & 38–39.
[22] Rica Erickson (ed), 1979, vol. 3, p. 128.
[23] *Inquirer and Commercial News*, 10 November 1858, p.4, DAN,NLA. Dowson (2008, p. 35) refers to 'likpence' but this appears to be in error.
[24] John Dowson, 2008, pp. 50–53.

Sound. Thus, Hooker noted that Oldfield had 'lived amongst & observed natives shrewdly from K.G.S. to Sharks bay, & W. up to Wide bay [not located]'.[25]

Another place where Oldfield and his fellow passengers might have sojourned awhile was the school for very young Aboriginal children near Middleton Beach. This was popular with visitors in the 1850s and 1860s and was run by Anne Camfield, wife of the local Resident Magistrate, Henry Camfield. Its aim was to remove the children from their parents so they could be trained in the attitudes and behaviour of British civilisation. It was run along the lines of British industrial schools and offered a basic training in 'useful skills'.[26] Most of the settlers' children attended the government school, housed in the old commissariat store on the waterfront.[27]

There was no community newspaper at the time but, as he moved around Albany mingling with its residents, including possibly a courtesy call on Camfield, he would have observed that a hierarchical class system, like that found in other larger Australian districts, was alive and well within the community. Wealthy pastoralists from the hinterland, some of whom had also established highly successful trade and commercial businesses in the town centre, were recognised as the elite of the region. The second rank were financially less successful, or newly arrived: mainly professional men, the smaller business men and government bureaucrats. Below them, 'lowly' office and shop workers and finally the labouring classes.[28] The complement of bureaucrats serving the government's interests at the port, and the settlement and surrounding district, reflected that of other seaboard towns in the Colony. Although some Aboriginals were incorporated into the economic life of the European community in lowly occupations such as laborers and domestic servants and, with a few exceptions, were not included in Albany's communal social life at any level with most living in poverty on the fringes of European settlements.[29]

Oldfield probably learnt that the town was recovering from the devastating effects that the closure of the coal depot had had on its economy when the P&O steam shipping service came to a halt in 1855, after the British Government called upon shipping companies for the charter of vessels to carry troops and war materials to support its Crimean War effort. The consequent loss of employment and business and the cessation of the Convict Hiring Depot in 1855, following the withdrawal of the convicts, left the community reeling and without a regular mail service to the Colony. However, when Oldfield arrived the coal and convict hiring depots were back in full swing. P&O had resumed its shipping service soon after the war ended in 1856 and shortly afterwards the convicts returned.[30]

The social life of the community was very restricted. Well-educated men and women of the middle-class missed the music and books that were part of their lives they led back in their home country. The wealthier families held balls and private sporting and indoor entertainments[31] but probably recreational outlets for the lower classes were confined to hotel gatherings, impromptu games, fishing and picnics.

[25] JDH to C.R. Darwin, 21 December 1862, letter 3856, Darwin Correspondence Project, viewed 30 April 2013 at www.darwinproject.ac.uk.
[26] Donald S. Garden, 1977, p. 148.
[27] ibid., p. 177.
[28] ibid., pp. 166–172.
[29] ibid., pp. 80-86. 239.
[30] ibid., pp. 121–126.
[31] ibid., p.176.

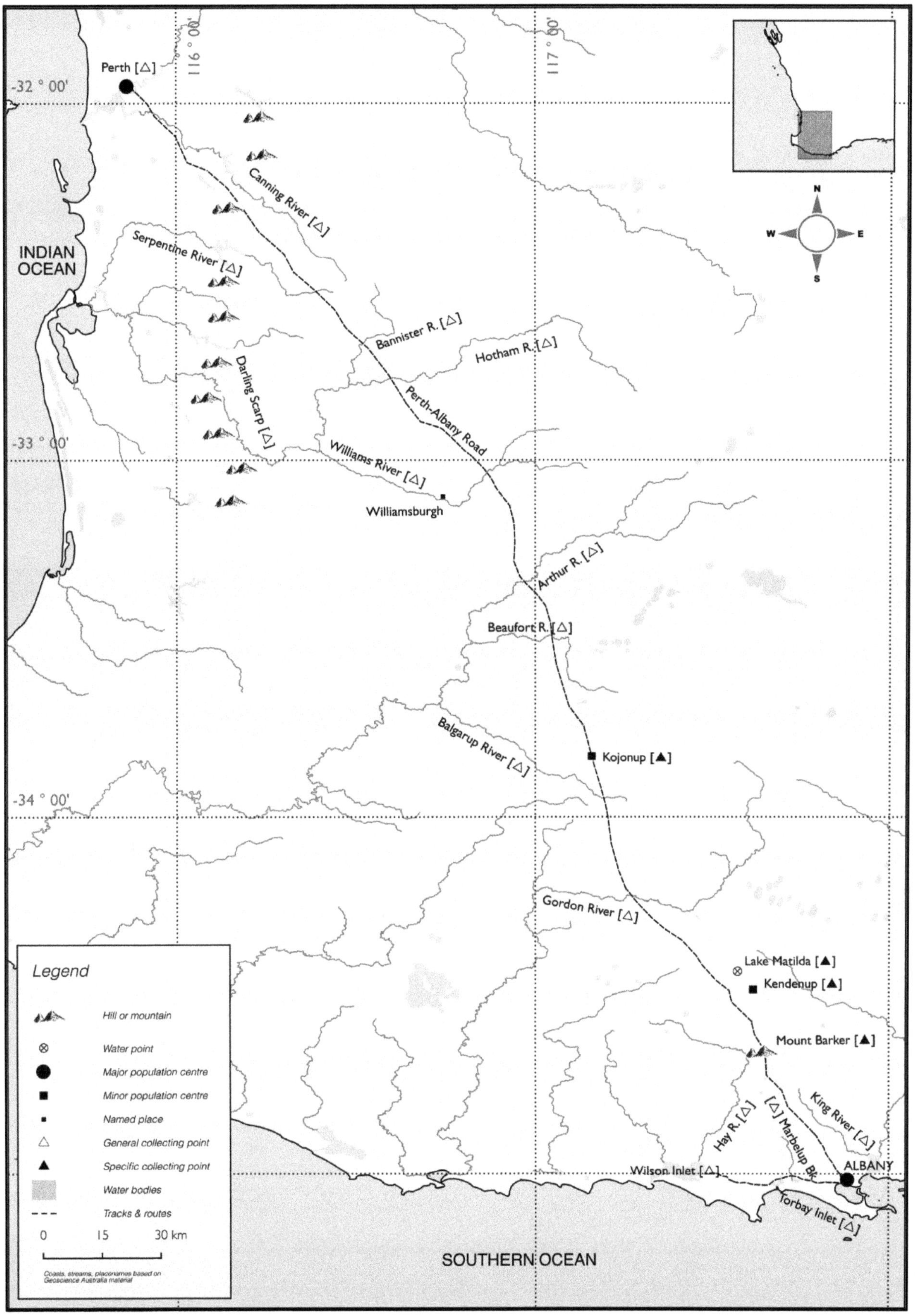

Map 13: Albany to Perth

European interest in the flora of King George Sound and environs, along the coast to the west and Albany to Perth Road prior to 1858

Map 13 p. 273 – Geraldton to the Murchison River

Oldfield followed a long line of botanical collectors who had worked near, and to the north, east and west of, King George Sound and many species had already been described by the time he arrived.

On 28 September 1791, a British naval expedition under the command of George Vancouver, discovered King George Sound on the voyage out from England. While at the Sound in September and October 1791, Archibald Menzies, a surgeon-naturalist with the expedition, collected plants that were later incorporated into Joseph Banks' herbarium.[32]

Ten years later, on 6 December 1801, Matthew Flinders in HMS *Investigator* first sighted the coast of New Holland at Cape Leeuwin and anchored in King George Sound two days later. On board were the naturalist Robert Brown, artists Ferdinand Lucas Bauer and William Westall and gardener, Peter Good. They collected extensively at King George Sound, Princess Royal Harbour, King River, and Torbay Inlet from 9 December 1801 to 5 January 1802.[33] From the time of his return to England until his death in 1858, Brown published many descriptions of Western Australian species that were new to science, including those species he brought back from Western Australia; in volume 1 *Prodromus Florae Novae Hollandiae* (1810)[34] (an uncompleted Australian flora) and *Supplementum Primum Prodromus Florae Novae Hollandiae* (1830)[35], also W.J. Aiton's *Hortus Kewensis* (1811)[36] and other publications such as the *Transactions of the Linnean Society of London* and the *Botanical Magazine*. Brown's manuscript writings on Western Australian plants are in his *Memoranda Botanica*.[37] While the first set of Robert Brown's Australian collection, which probably included Ferdinand Bauer's specimens, was housed in the Natural History Museum, London, duplicate sets were distributed in the second half of the nineteenth century. The second set went to Kew Gardens, and the third to the Royal Botanic Garden Edinburgh; small sets were distributed to twenty other herbaria throughout the world.[38]

On their return journey to England from Port Jackson, the expedition, under the command of Nicolas Thomas Baudin, anchored in King George Sound. The *Casuarina*, which had replaced *Naturaliste* and was under the command of Sub-Lieutenant Louis de Freycinet, arrived on 13 February 1803 and *Géographe*, under the command of Baudin, two days later. Many plant specimens were gathered by the expedition's botanists, Leschenault and Perón, while surveys were undertaken of the Sound, Oyster Harbour, Kalgan River, Flinders Peninsula and Mount Gardner. After they left King George Sound they may have carried out botanical collecting near West Cape Howe.[39] According to Hewson, the expedition was extremely rewarding in the volume and quality of the botanical and zoological

[32] Helen Hewson, 1999, p. 30.
[33] T.G. Vallance *et al.*, 2001, pp. 91–108.
[34] Robert Brown, 1810, vol. 1.
[35] Robert Brown, 1830, *Supplementum Primum Prodromi Florae Novae Hollandiae: Exhibens Proteaceas Novas quas in Australia Legerunt DD. Baxter, Caley, Cunningham, Fraser et Sieber; Quarum e Siccis Exemplaribus Characteres Elaboravit Robertus Brown*, Richard Taylor, London.
[36] W.J. Aiton, 1811, *Hortus Kewensis; or a Catalogue of the Plants Cultivated in the Royal Botanic Garden at Kew*, 5 vols., Longman, Hurst, Rees, Orme & Brown, London.
[37] R.M. Barker & W.R. Barker, 1990, pp. 46–52. See also Janette Gathe, Mollie Edgecombe and Bevan Carter, 2001, *Robert Brown: Botanist with Matthew Flinders* (Bevan Carter, Bassendean, Western Australia, pp. 29–55) for illustrations of plants originally described by Brown. See also Greg Keighery and Neil Gibson, 2005. 'The Flinders Expedition in Western Australia: Robert Brown, the plants and their influence on W.A. botany', in Juliet Wege, Alex George, Jan Gathe, Kris Lemson and Kath Napier (eds), 2005, *Matthew Flinders and His Scientific Gentlemen*, Western Australian Museum, Perth, p. 105.
[38] T.G. Vallance, *et al.* pp. 14–15.
[39] Jean Fornasiero *et al.*, 2004, pp. 246 –265.

collections. The herbarium specimens were passed to Labillardière but the result was only a minor contribution to science.[40]

Allan Cunningham, on instructions from Joseph Banks, accompanied Captain Phillip Parker King's four expeditions to survey and chart various parts of the eastern, northern, north-west and south coasts of Australia. King George Sound was visited during the first expedition, in the *Mermaid*, and the fourth, in the *Bathurst*. During the first visit, they anchored in Frenchman Bay on 20 January 1818, and while they were there Cunningham collected on Seal Island and Flinders Peninsula. The next day the ship moved to just inside the entrance to Oyster Harbour, where it remained until the end of their stay on 1 February 1818. From his diary descriptions, it seems that Cunningham collected near Mount Martin, along Middleton Beach, at the foot of, and on, Mount Adelaide, along the eastern, north-eastern and western shores of Oyster Harbour, Green Island, Lake Seppings and along the western bank of Kalgan River for a little over seven kilometres above its mouth. During the fourth expedition, on 24 December 1821, King returned to the former anchorage just inside Oyster Harbour, where Cunningham collected over much the same ground as he had done previously. On 3 January 1822, the anchorage was moved to Frenchman Bay and from there he collected on Mistaken Island and again on Seal Island and Flinders Peninsula, this time to Bald and Limestone Heads.[41] Cunningham made significant botanical collections during his voyages with King. Apart from their intrinsic historical value, Cunningham's specimens are important because many were selected to describe species new to science and they were also extensively cited by George Bentham in *Flora Australiensis* (1863–1878).[42]

From 1823–1825 and again during 1828–1829, William Baxter, an English gardener and of whom very little is known, collected seeds and plants for nurserymen and private individuals along the southern coast in Western Australia, including at King George Sound. On his second trip Charles Fraser, Colonial Botanist in New South Wales, arranged for Baxter to supply material to the Botanic Garden Sydney as well as to English employers. On Baxter's return to Sydney he tried to cheat Fraser out of his due portion of the collections, but Fraser called in the Collector of Customs and forced the honouring of the agreement. William Hooker, Robert Brown and the English botanist and horticulturalist, Robert Sweet, were among the recipients of Baxter's collections.[43]

Explorations which embraced the area around King George Sound, Princess Royal Harbour, Oyster Harbour and Kalgan (then French) River, were continued in 1826 and 1827 by Major Lockyer when he was sent to establish a settlement at the Sound. Others followed with explorations to the north, north-west and west of the Sound and although many of them commented in their journals on the flora of the lands they collected, few collected botanical specimens.

In 1829, Dr Thomas Braidwood Wilson, medical practitioner and settler, together with Thomas Fellowes, Private William Gough, Mokaré (an Aboriginal guide), and Thomas Wood, explored to the north-west of Albany (then Fredericks Town) for about 112 kilometres and then to the west and south-west, discovering and naming Lake Katherine, Kent and Denmark Rivers and Mounts Lindesay and Hallowell, returning to Albany along the coast via West Cape Howe.[44] Wilson was a friend of Allan Cunningham and brought seeds and roots of many Western Australian plants to him at the Botanic Garden Sydney.[45] John Kent apparently also collected plant specimens and seed during this expedition

---

[40] Helen Hewson, 1999, p. 66. According to Hewson, Labillardière is believed to have used some of the material in the compilation of his *Novae Hollandiae Plantarum Specimen*, 2 vols, Paris, 1804–6.

[41] Suzanne Curry *et al.*, 2002, pp. 10–12 & 73–74.

[42] ibid., p. 1.

[43] Helen Hewson, 1999, p.101. According to R.M. & W.R. Barker, 1990, pp. 55–56, there are 'numerous references to Mr Henchman's [an English importer of New Holland plants] collector, Mr Baxter, throughout Robert Sweet's *Flora Australasica*, 1827–1828, Ridgway, London. Baxter's collections are housed in the MNHL and Kew Gardens with a few duplicates in other herbaria.

[44] T.B. Wilson to Captain Collet Barker, 15 December 1829, cited in Joanne Shoobert (ed), 2005, vol. 1, pp. 103–110.

[45] Gwendoline Wilson, 'Wilson, Thomas Braidwood (1792–1843)', ADBO, viewed 7 May 2008 at www.adb.online.anu.edu.au.

but their whereabouts is unknown; Alex George queried whether Kent was the collector of specimens from King George Sound that have been attributed to Phocion-Aristide-Paulin Verdier.[46]

Captain Collet Barker, commanding officer of the military settlement at King George Sound, and accompanied by John Kent, Private William Gough, convicts Thomas Lovegrove and James McKone, and Mokaré, explored to the north-west of Albany in February 1830. They followed Sleeman River to its mouth in Wilson Inlet and then proceeded further west of the Hay River mouth, and towards what is now the town of Denmark, returning to Fredericks Town (Albany) via Mount Warrumbup (now Weedon Hill),[47] Mount Lindesay and Hay River.[48] Barker made a short journey to Michaelmas Island at the entrance of King George Sound on 7 and 8 May 1830, and an excursion to West Cape Howe on 3 and 4 June 1830. According to A.S. George, Barker tried to preserve plants for P. Lindesay at Raffles Bay, Northern Territory but there is no indication of success. He collected specimens including mosses and seaweeds at King George Sound but the whereabouts of his specimens are unknown except for eight pieces of moss attached to the page of his journal for 15 August 1830.[49]

Surveyor General John Septimus Roe surveyed the south coast in December 1831, whilst in the company of His Excellency the Governor Sir James Stirling, Lieutenant William Preston and four others. He went north to the Porongurup Range and then explored Mount Pwackenback (Mount Barker), and then a short distance to the north-east before turning north-west and then south-south-west to Hay and Sleeman Rivers, Wilson Inlet, and east to West Cape Howe and Torbay before returning to Fredericks Town [Albany] along the coast.[50] As noted in Chapter 11, a specific collection locality is not available for his specimens in the Western Australian Herbarium.

Whilst he was Resident Magistrate in Albany (1831–1832), Alexander Collie collected plants near King George Sound. A small collection of his specimens from the area have been located in the Western Australian Herbarium but further details of the places they were obtained are not available.[51]

In February 1835, in the company of Rivett Henry Bland, George M. Cheyne and five others, Roe explored in the vicinity of King and Kalgan Rivers, visiting Moorilup, near Kendenup, on the Kalgan River. The group then travelled south-west to Mount Barker and from there south-west along Hay River and, although not specifically mentioned in his journal, Sleeman River as well.[52]

Later that year, from 19 October to 21 November, in the company of Sir James Stirling and ten others, Roe led an overland expedition from the Swan River to King George Sound via Canning River, Kelmscott, the Darling Range, Bannister River, Williams River and then east-south-east, crossing the Arthur River and proceeding to the west end of the Stirling Range. They then headed west-south-west to Lake Katherine, William Bay, Wilson Inlet and Torbay.[53] He probably obtained plant specimens on these journeys and sent them to Kew Gardens[54] but this has not been determined.

Baron Charles von Hügel gathered plants at King George Sound during his visit to the Colony in 1833–1834.[55] These collections were described by Stephan L. Endlicher *et al.* in *Enumeratio Planta-*

---

[46] A.S. George, 2009, pp. 438, 583.

[47] S.D. Hopper, personal communication to MHH, email attachment, 23 December, 2017. According to Hopper this can be worked out from D.J. Mulvaney & Neville Green, 1992, *Commandant of Solitude: the Journals of Captain Collet Barker 1828-1831*, Melbourne University Press at the Miegunyar Press, Carlton, Victoria, p. 258.

[48] Collet Barker, 'Journal of exploration of the Wilson Inlet area [February 1830]' and 'Log of an expedition west of King George's Sound, February 1830', cited in Joanne Shoobert (ed), 2005, vol. 1, pp. 140–152.

[49] A.S. George, 2009, p. 280.

[50] J.S. Roe, 'Journal of expedition to the northward and westward of King Georges Sound in December 1831', cited in Joanne Shoobert (ed), 2005, vol. 1, pp. 274–289.

[51] *FloraBase*

[52] J.S. Roe, 'Journal of an expedition to Moorilup, the Hay and the Sleeman Rivers, & etc. February 1835', cited in Joanne Shoobert (ed), 2005, vol. 1, pp. 410–418.

[53] J.S. Roe, 'Journal of an expedition from Swan River overland to King George's Sound, 19 October to 21 November 1835', cited in ibid., pp. 456–497.

[54] A.E. Orchard, 'Roe, John S. (1797–1878)', ANHBiog, viewed 8 May 2008 at www.anbg.gov.au, and R.L. Heberle, undated, 'John Septimus Roe 1797–1878 surveyor, explorer and botanical collector, 1786–1867', viewed 8 May 2008 at www.members.iinet.net.au.

[55] S.L. Endlicher *et al.*, 1837, and Endlicher 1838, contain lists of all von Hügel's Western Australian herbarium specimens, Sophie C. Ducker, 'Early Austrian influence on Australian botany', in P.S. Short (ed), 1990, pp. 300–301.

rum Novae Hollandiae (1837) and *Stirpium Australasicarum Herbarii Hügeliana Decades Tres* (1838).

George Maxwell lived in Albany from 1838 until his death in 1879. He enjoyed a reputation as a highly skilled bushman with a deep understanding of the area east and west of Albany and his services were often called upon by those seeking suitable grazing land and mining opportunities. For varying periods, he was a pastoralist, a sandalwood cutter, a mailman, and, towards the end of his life, he had a retail business on Stirling Terrace in Albany selling seeds, plants, shells, minerals and other natural history specimens.[56] From 1838 to 1858, and during the following two decades, he gathered plants and seeds around King George Sound, east and west along the south coast and north along the Albany Road as far as Hotham River.[57] Unfortunately, many of Maxwell's specimens in the National Herbarium of Victoria are undated and bear either no locality or only general regional details (for example, 'S.W. Australia'[58]) so it is not possible to provide a comprehensive indication of where he collected. Available records show that Maxwell obtained a few plants along the coast west of King George Sound at Torbay, Hay and Franklin Rivers, Wilson Inlet and around Nornalup Inlet.[59] Nornalup Inlet specimens may have been obtained in 1845 when the cutter *Roseanna,* en route from Fremantle to Adelaide, called at Nornalup Inlet where Maxwell, along with fellow passengers G.W. Landor, J. Leake and C. Crutchley disembarked, with the intention of examining the area for its suitability for pastoral purposes.[60] Maxwell's specimen records in the National Herbarium of Victoria include some from various places north of Albany on the road to Perth, including nine kilometres and 40 kilometres north of Albany, Mount Barker, Lake Matilda, Gordon River and Gordon Plains, Balgarup River and Kojonup.[61] These may have been obtained while he was plying the mail between Perth and Albany from 1852 to 1854.[62] Substantial numbers of his specimens collected prior to the mid-1850s are located in the Natural History Museum, London, and at Kew Gardens. He then became one of Mueller's collectors and, as a result, the National Herbarium of Victoria held his main collection from that time.[63] The main published sources citing Maxwell's specimens are George Bentham's *Flora Australiensis* and Mueller's *Fragmenta.*

In 1840 James Drummond travelled from his property, Hawthornden near Toodyay, to King George Sound via York, 'the Williams' (Williamsburg or Williams River) and south along the Albany Road to the Sound, collecting at various localities along the way. At Albany, he met up with Ludwig Preiss and together they botanised around the Sound. Drummond apparently returned to Hawthornden along the same route.[64] Preiss had also made collections at various places on his way to the Sound from Perth, including Beaufort River, Kojonup, Balgarup, south of Kojonup, Gordon River, Lake Matilda, Kendenup and Mount Barker.[65]

The ship carrying John Gould's collector John Gilbert, returning to Sydney from his first expedition to Western Australia (1839–1840), stayed long enough in King George Sound to allow Gilbert to make a trip inland. He collected around the Sound and along the Albany Road as far as Kojonup.

[56] W.G. Henderson, 2011, 'George Maxwell: bushman and naturalist', *Early Days*, vol. 13, pt. 5, p. 642.
[57] For example, MELISR database, 02 April 2009, NHV, RBGM. There is a map showing Maxwell's collecting places in W.G. Henderson, 2011, p. 630.
[58] For example, MELISR database, 02 April 2009, NHV, RBGM. MEL holds 3,306 Maxwell specimens.
[59] ibid. and an undated list of Maxwell's specimens referred to in the first four volumes of George Bentham's *Flora Australiensis* kindly made available to W.G. Henderson by Dr Neville Marchant, then Curator of the Western Australian Herbarium.
[60] Lee & Geoff Fernie, 1989, *In Praise of a National Park: The Origins and History of the Walpole-Nornalup National Park*, Lee and Geoff Fernie, Walpole, Western Australia, p. 12.
[61] MELISR database, 02 April 2009, NHV, RBGM.
[62] W.G. Henderson, 2011, pp. 636–637.
[63] ibid., p.626.
[64] Rica Erickson, 1975, pp. 55–57.
[65] N.G. Marchant, 1990, pp. 131–135.

Although his primary focus was on birds and mammals, he apparently obtained botanical specimens during this journey.[66]

Drummond returned to King George Sound in mid-November 1843 with his son Johnston and an assistant, travelling along the Albany to Perth Road. On the way south, they examined the country about Beaufort and Gordon Rivers and the Porongurup Range, and attempted to explore Stirling Range, but watercourses were dry and they were forced 'to return to water' before proceeding to King George Sound, Cape Riche and thence back to Albany. Before leaving for Toodyay, they 'dug some sods from the swamps of Albany in which were growing the world-famous [Albany] Pitcher [P]lant *Cephalotus follicularis*'.[67] Around March 1847, Drummond collected once more at the Sound.[68]

William Harvey arrived in King George Sound from Ceylon [Sri Lanka] on 7 January 1854.[69] During January and February 1854, he gathered algae from Princess Royal Harbour and King George Sound and, although he said that the Sound was 'by no means a favourable locality for Algae', during the first three weeks he was in the district he collected and dried about 2500 specimens representing about 70 species. He also gathered terrestrial plants in and around Albany but bemoaned the fact that his visit was in high summer when most of these species were not in flower.[70] The Herbarium of Trinity College, Dublin, where Harvey was curator from 1844–1866, holds most of his collections.[71]

Collecting near Albany and west to Wilson Inlet

Map 12 p. 268 and Map 13 p. 273

Apart from his 1858 visit to King George Sound, Oldfield was there again in 1861–1862 and once again the collections he made on each occasion cannot be separated because of the almost total absence of collection dates. It is likely, however, that most of his specimens from this area were obtained in 1858 since he was considerably longer in the Colony on that expedition than in 1861–1862.

Mueller almost certainly encouraged Oldfield to seek out George Maxwell on his arrival in Albany on the assumption that the naturalist would be able to assist the visitor with local contacts and practical information relating to bushmanship and Aborigines in that part of the country.[72] It seems likely, therefore, that the pair met soon after Oldfield made landfall in Albany, as Maxwell was in Albany and its environs during July and the first two weeks of August 1858.[73]

Oldfield apparently collected very little around Albany townsite[74] instead focusing his attention on the surrounding district where he appears to have penetrated territory visited by earlier collectors. Most of their specimens were in herbaria in Europe so it seems reasonable to suggest that Mueller, anxious to build his own herbarium, had instructed Oldfield to go over the same ground as his predecessors.

The area surrounding Albany lies within the Muir District, Bibbulmun Province, SWAFR.[75] It comprises Jarrah-Marri forest on duricrusted plateau and loam soils in the west, grading to Marri (*Corymbia calophylla*) and Wandoo (*Eucalyptus wandoo*) woodlands on drier laterite-free soils to the

---

[66] E.M. Whittell, 1949, 'The visits of John Gilbert, naturalist to the Swan River Colony, *Early Days,* vol. 4, pt. 1, p. 23 and Alex S. George, 2009, 'Gilbert, John', p. 386.

[67] Rica Erickson, 1975, pp. 90–91.

[68] ibid., p. 168. This journey will be referred to in Chapter 20, in which Oldfield's 1862 journey to the Great Australian Bight is discussed.

[69] WHH to Jane Gray, 2 February 1854, cited in Sophie C. Ducker (ed), 1988, pp. 77–80.

[70] WHH Harvey to W. Hooker, 29 January 1854, Mrs Asa Gray, 2 February 1854, Hannah Todhunter, 22 February 1854, cited in Sophie C. Ducker (ed),1988, pp. 72–83.

[71] Sophie C. Ducker, 1972, 'Harvey, William Henry (1811–1866)', ADBO, viewed 1 May 2013 at www.adb.anu.edu.au.

[72] W.G. Henderson, 2011, p. 639.

[73] There are 60 specimens in MEL collected by Maxwell, from July to 17 August 1858 at King George Sound and environs, within about a 12.8 kilometres radius of Albany. He was well away from Albany on 18 August 1858 to the north of Mount Barker at Milyunup Lake, MEL 2087008, and then east at Wilyunegup [? Wilyungulup, a crossing on the Blackwood River], see, for example, MEL 4468, MELISR database, 02 April 2009, NHV, RBGM.

[74] HAODb.

[75] Paul Gioia and S.D. Hopper, 2017, pp. 11–12 and Appendix S3, Fig. S3.7.

east. There are extensive areas of swamp vegetation in the south-east dominated by Swamp Paperbark (*Melaleuca rhaphiophylla*), Saltwater Paperbark (*M. cuticularis*) and Mauw[76] (*E. occidentalis*). On the more favourable valley soils Swan River Blackbutt (*E. patens*) joins the forest dominants. On stream banks, the Flooded Gum *(E. rudis)* is characteristic and in the upper reaches of valleys on the plateau, Bullich (*E. megacarpa*) forms woodland. Noticeable smaller trees may include Bull Banksia (*Banksia grandis*), Sheoak, Christmas Tree/*Moojar*[77] (*Nuytsia floribunda*), Snottygobble (*Persoonia longifolia*), Spreading Snottygobble (*P. elliptica*) and Woody Pear (*Xylomelum occidentale*). Zamia (*Macrozamia riedlei*), Grass Tree (*Xanthorrhoea preissii*) and Kingia (*Kingia australis*) are also evident in many areas. Beneath the small trees is a plethora of shrub and smaller species representing many genera, including *Hakea, Acacia, Banksia, Hibbertia, Hovea, Petrophile, Pimelea, Verticordia, Conostylis, Lepidosperma, Clematis, Hardenbergia and Kennedia.*[78]

During at least one of his visits to King George Sound, Oldfield climbed Mount Clarence.[79] In early photographs taken in the 1850s[80] this prominence appears in the distance and it seems to have been covered in low heath-like species due to recent Aboriginal and/or settler burning.[81] Thus, in 1801 Robert Brown noted: '[t]he wood on the side of the hill of a considerable height. The most lofty trees resinifera [= Jarrah, *Eucalyptus marginata*][82] and Casuarina'.[83] Today the vegetation is dominated by low woodland comprising Marri, Jarrah and Yate (*E. cornuta*), WA Peppermint (*Agonis flexuosa*), Bull Banksia and Showy Dryandra (*Banksia formosa*), beneath which grow *Hakea, Taxandria, Melaleuca, Adenanthos, Banksia, Grevillea* and many other species.[84] The distant view of the hill in the photographs may give the wrong impression of the height of the vegetation since Oldfield obtained at least two species, Jarrah (*E. marginata*)[85] and the Yate, (*E. cornuta*)[86], that can be tall enough to be conspicuous at a distance. However, most likely his specimens were from stunted wind- affected plants on shallow soils over granite.[87]

He also ascended and gathered plants on Mount Melville[88], which today has a somewhat different flora from that on Mount Clarence although there are many species that occur on both. Jarrah and Marri are the dominant species on Mount Melville in deeper soils off the granite outcrops. Noticeable in the understorey, among others, are bushes of Winged Wattle (*Acacia alata*), Tassel Flower (*Leucopogon verticillatus*), and the Dwarf Sheoak (*Allocasuarina humilis*). Below these are smaller plants including *Astroloma, Sphenotoma, Olax* and *Borya* species.[89] The beautiful viscid plant painted by Ferdinand Bauer, *Anthocercis viscosa* subsp. *viscosa,* according to Hopper, represents a striking shrub emergent from herbfields and dwarf shrubs on Albany's granite outcrops.[90]

---

[76] Referred to as Flat-topped Yate in *Florabase,* this tree is known as *Mauw* not Yate by south coast and Great Southern Noongars (for example, C.G. von Brandenstein, 1988, *Nyungar Anew: Phonology, text Samples and Etymological and Historical 1500-word vocabulary of an artifially recreated Aboriginal Language in the South-west of Australia*, Department of Linguistics, Research School of Pacific Studies, Australian National University, Canberra), According to Hopper, 'Yate is strictly *E. cornuta*. Much confusion has been generated by white people perpetuating the word "yate" when talking/writing about *Eucalyptus occidentalis*. S.D. Hopper, personal communication to MHH, email attachment, 23 December 2017.

[77] According to S.D. Hopper, *Moojar* is the name by which Noongar people south of Perth refer to *Nuytsia floribunda*, personal communication, to MHH, 24 December 2017, HOA. See S.D. Hopper, 2010.

[78] J.S. Beard, 1990, pp. 77–78.

[79] HAODb.

[80] For example, John Dowson, 2008, p. 38.

[81] S. D. Hopper provided the explanation for the heath-like appearance of the flora on Mount Clarence in early photographs, personal communication to MHH, email attachment, 23 December 2017, HOA.

[82] Correct name provided by S.D. Hopper, ibid.

[83] T.G. Vallance *et al.*, 2001, p. 94. '*Eucalyptus resinifera*' is in error. It is an east coast species (ibid.) but they did not suggest a species name.

[84] This information on the flora of Mount Clarence was kindly provided by S.D. Hopper, personal communication to MHH, email attachment 23 December 2017 and Ruth Moir from the records of the Albany Herbarium, a branch of the Western Australian Herbarium, personal communication to MHH, 2 July 2009, HOA.

[85] MEL 703855, *Eucalyptus marginata* subsp. *marginata,* MELISR database, 08 March 2007, NHV, RBGM.

[86] MEL 58308, *Eucalyptus cornuta,* (ibid.).

[87] S.D. Hopper, personal communication to MHH, email attachment, 23 December2 017.

[88] HAODb.

[89] Ruth Moir to MHH, personal communication, 2 July 2009.

[90] Personal communication to MHH, email attachment, 23 December 2017.

Fig. 15.2 Western face, Mt Clarence, Albany

Oldfield obtained many specimens at and near King George Sound and along the adjoining coast and districts[91] but most of their locality descriptions do not permit a more precise indication of where he gathered them. He did, however, specify on plant labels the names of a few places; 'Princess Royal Harbour'[92] and to the west, 'Mount Elphinstone'.[93]

Some of his herbarium specimens bear the locality 'Manjinup', 'Manjanup' or 'Manjimup', all of which have been tentatively recorded in the National Herbarium of Victoria as being from 'Manjimup'.[94] As far as we are aware Oldfield was never in that district. 'Manjimup' is thought to be derived from a place, called by the Aborigines '*manjin-up*', for the place where *manjin*, a bulrush (*Typha orientalis*) with an edible plant root grew.[95] However, based on taxon distribution maps of the species in *FloraBase* and the history of the use of 'Manjimup' by European settlers, it appears more likely that the specimens were collected at a place (precise locality unknown) where this species also grew, either near King George Sound or to the west at Cape Naturaliste, Geographe Bay. [96]

Oldfield visited Bald Head.[97] At a distance of a little over ten kilometres south west in a straight line from Albany, it provided him with a spectacular example of the bold granitic headlands of the south coast which alternate with shallow curving sandy bays backed by sand dunes on low country.

---

91 HAODb.

92 ibid.

93 ibid.

94 MELISR database, 08 March 2007, NHV, RBGM.

95 S. D. Hopper, Personal communication to MHH, email attachment, 23 December 2017. According to Hopper, it is well known that *manjim* or *yaindyap* are alternative Noongar names for bulrush.

96 Manjimup, a town site about 294 kilometres south of Perth on the South West Highway, was not named until 1910 and was renamed Manjimup in 1915, nor do we believe it was J. Mottram's property, 'Manjimup House' thus named in the 1860s. Manjimup is thought to have derived from the Aboriginal word *manjin[up]*, for the place where a plant (species unknown) with an edible root grows, Geoffrey Higham, 2005. *Gone West: Explorer's Guide to Places from Western Australia's Past*, Geoproject Solutions Pty Ltd,, Winthrop, Western Australia, p. 56. *FloraBase* taxon specimen record maps show that three of the species with a Manjinup location— *Taxandria spathulata*, MEL 979; *Sphaerolobium grandiflorum*, MEL291015; *Hibbertia depressa*, MEL 2231265, MELISR database, 08 March 2007, NHV, RBGM—are south coastal species, including King George Sound. Although the distribution of the other two species—*Acacia cyclops*, MEL 2073473, and *Taxandria linearifolia*, MEL 3364, (ibid.)—extend further to the north along the coast and the latter inland from the south coast, they, nevertheless, occur along the south coast around Albany.

97 HAODb.

Sand also mantles the granite headlands. Most of the sand has become stabilised and vegetated but some is still mobile, as on Flinders Peninsula. Windswept coastal heath covers the consolidated sand, on which in Spring and Summer, the pink flowers of the *Pimelea ferruginea* and the yellow blooms of the dwarf form of the Bull Banksia provide brilliant colour against the grey foliage of the Coastal Daisybush (*Olearia axillaris*) fronted by the very deep blue Southern Ocean.[98] If, as he probably did, Oldfield went overland from Albany to Bald Head collecting *en route*, he traversed part of today's Torndirrup National Park, crossing through vegetation dominated by WA Peppermint before emerging onto the coastal heath.

Although the naturalist collected specimens from along King River[99], he apparently made more substantial gatherings from the lower Kalgan River, commencing at the river mouth on Oyster Harbour.[100] There are records of specimens with the locality 'Upper Kalgan River', which were probably obtained during his 1861–1862 expedition to the Colony.[101] Given the time constraints associated with this expedition, he most likely collected along the King and lower Kalgan Rivers in 1858. This area is still within the Muir District of the Bibbulmun Province, SWAFR.[102]

After passing the shipwrights' yards[103], located at the lower ends of both the King and Kalgan Rivers, Oldfield probably followed a native track wending slightly upslope from the rivers on dry soil in woodlands of Jarrah, Marri, WA Peppermint and Albany Blackbutt (*Eucalyptus staeri*)[104] Fringing vegetation along the estuarine portion of Kalgan River is dominated by Saltwater Paperbark trees surrounded by Coast Saw-sedge (*Gahnia trifida*) and shore rushes. Further on, after the water becomes fresh there is a much greater variety of species with fringing forest that includes Swamp Paperbark, Marri, Jarrah, *Acacia* sp. and WA Peppermint. The understorey includes species of *Adenanthos*, *Hovea*, *Lysinema*, *Leucopogon*, *Acacia* species and many others.[105]

Expedition to Wilson Inlet

On one of Oldfield's visits to Western Australia Oldfield undertook an expedition west of Albany as far as Wilson Inlet. There is some evidence, not confirmed, that he and George Maxwell went there in 1858. Thus, a comment by Maxwell on the label of his '1858' specimen of *Eucalyptus megacarpa* which he collected 'on the margin of Wilsons' (i.e. Wilson Inlet) suggests that he and Oldfield may have viewed the tree together at the Inlet:

> Mr Oldfield says that it [*E. megacarpa*] approaches upon the Van Diemen Black Gum. I believe it to be the splendid Blue Gum [ie. Karri, *Eucalyptus diversicolor*] on the Franklands River or Deep River about 40 miles west. Specimens in flower I will send when obtainable.[106]

Furthermore, Maxwell obtained *Eutaxia virgata* 'near Marbalup' (Marbelup Brook) in 'September 1858'.[107] On the basis of these records, and given what we understand of Oldfield's movements for the remainder of his time in Western Australia during 1858–1859, that he took a short trip to Wilson Inlet sometime in September with Maxwell seems feasible.

[98] J.S. Beard, 1990, pp. 69–70 and 75.
[99] HAODb.
[100] ibid.
[101] See Chapter 18.
[102] Paul Gioia and S.D. Hopper, 2017, pp. 11–12 and Appendix S3, Fig. S3.7.
[103] Gordon de L. Marshall, 1994, Memories of maritime Albany: Les Douglas et al., Report—Department of Maritime Archaeology, Western Australian Maritime Museum, No. 53, p. 51.
[104] S.D. Hopper, personal communication to MHH, email attachment, 23 December 2017.
[105] *Wikipedia,* 'Kalgan River', viewed 2 September 2009 at www.wikipedia.org
[106] MEL 1609798, *Eucalyptus megacarpa,* dated 1858, MELISR database, 02 April 2009, NHV, RBGM.
[107] MEL 2048985, *Eutaxia virgata*, September 1858, ibid.

If the collectors were together on this trip, it is likely that they were travelling in Maxwell's spring cart. While it is possible that he made the journey by boat, we believe his specimen records suggest he went overland along the approximately 51 kilometres from Albany west to Wilson Inlet sand track that F.T. Gregory surveyed in 1855.[108] It followed closely today's Lower Denmark Road from Albany and crossed Marbelup Brook. According to Gregory's plan, the track ran through 'rugged ironstone hills densely timbered with stunted mahogany and sheoak forest, banksia, xanthorrh[o]ea and tall scrub'.[109] It skirted wetlands such as Lake Powell (not shown on Gregory's map) and the Torbay Inlet into which Marbelup Brook drains.

Following Gregory's track, their route to Wilson Inlet would have taken them further into the Muir District, Bibbulmun Province, SWAFR.[110] It is an area of high rainfall, varying width and comprised of dissected, undulating country of small relief, hard-setting, loamy soils, alternating with leached sandy soils. Magnificent tall forests of Karri (*Eucalyptus diversicolor*) occur on deep loams while Jarrah-Marri forest is on the leached sands. Extensive paperbark (*Melaleuca rhaphiophylla*) and sedge swamps occupy the valleys. In places, Karri is found with the giant Yellow Tingle (*E. guilfoylei*), mainly inland from Broke Inlet, and Red Tingle (*E. jacksonii*) in the Walpole area. Some other tree species have a restricted habitat, such as the Red Flowering Gum (*E. ficifolia*) which is confined to the south coast on dune sand although an outlier is known from the Stirling Range.[111] Conspicuous under the top canopy (generally about 70 metres) are Karri Sheoak (*Allocasuarina decussata*), Bullich (*E. megacarpa*) and smaller trees such as WA Peppermint, Bull Banksia (*Banksia grandis*), Granite Banksia (*B. verticillata*) and Snottygobble. Among the taller plants are shrubs such as Karri Wattle (*Acacia pentadenia*), Water Bush (*Bossiaea aquifolium*), Chorilaena (*Chorilaena quercifolia*), *Pimelea clavata*, Zamia and Grass tree. There are many smaller species representing such genera as *Acacia, Bossiaea, Chorizema, Crowea, Hakea, Hibbertia, Leucopogon, Anigozanthos, Dampiera, Lepidosperma, Patersonia, Scaevola, Clematis,* and *Hardenbergia.*[112]

Marbelup Brook, where Oldfield gathered some plants[113], is lined with freshwater species such Swamp Paperbark (*Melaleuca rhaphiophylla*), Jointed Twigrush (*Baumea articulata*), Yate (*E. cornuta*) and WA Peppermint.[114] The stream drains into the Torbay Inlet, the head of which is less than a kilometre south of the road and another of Oldfield's specimen localities.[115] The Inlet lagoon is shallow and surrounded by a margin of dense saltwater tolerant species, the Saltwater Paperbark (*Melaleuca cuticularis*) and the Sea Rush (*Juncus kraussii*). Behind this an occasional Slender Banksia (*Banksia attenuata*) is present in the flats.[116] The showy Shining Fanflower (*Scaevola nitida*) is also present near the shores of the Inlet. According to Gregory, west of Marbelup Brook the road passed through an 'open level swampy valley producing rushes' and then across 'level sandy country thinly covered with Banksia and Xanthorrh[o]ea'. To the north-west of the road was sandy country covered with 'mahogany & red gum [Marri] forest and further to the west behind the shore of the Inlet were 'rocky hills covered with thick forest of mahogany & red gum'.[117]

---

108 Wilson's Inlet-Road to Albany and adjacent country compiled from surveys and sketches by F.T. Gregory, 1855, Record series 237 Survey Plans–Roll Plans, 1 January 1841–1 January 1930, Cons. 3846, Item no. 082, SROWA, and F.T. Gregory, Field Notebook 13, item, Gref/13, Item 253, Cons. 3401, Lands & Survey Department Records, SROWA. It should be noted that although the above survey plan (in two parts) was photocopied for Bill Henderson at the SROWA, the first half of it showing the road from Princess Royal Harbour to just west of Marbelup Brook could not be located at the SROWA in 2013.

109 ibid.

110 Paul Gioia and S.D. Hopper, 2017, pp. 11–12 and Appendix S3, Fig. S3.7.

111 S.D. Hopper, personal communication to MHH, email attachment, 23 December 2017.

112 ibid., pp. 59–87.

113 HAODb.

114 Anne Brearley, 2005, p. 427.

115 HAODb.

116 Anne Brearley, 2005, p. 426.

117 Wilson's Inlet-Road to Albany and adjacent country compiled from surveys and sketches by F.T. Gregory, 1855.

Fig. 15.3 Mouth of the Hay River, Wilson Inlet

Gregory's road terminated at a small farming settlement just east of Nornalup Point, on the south-eastern shore of Wilson Inlet. It was possibly a base for pastoral activities.[118] There do not appear to have been any other official roads leading to or from the settlement. Oldfield went to the north and west around the Inlet, possibly following the shore or a local track, and crossed and collected at Hay River which flows into the Inlet north-west of Nornalup Point.[119]

From the specimen records in our Oldfield database he appears to have obtained more specimens near Wilson Inlet than at any other place along the road[120]; the database shows no evidence of his collecting along the shores of the Inlet, which is fringed by Sea Rush and Saltwater Paperbark, or of the community behind them on flatter ground and dominated by Creeping Brookweed (*Samolus repens*), sedge (*Ficinia nodosa*) and Bare Twigrush (*Baumea juncea).* His plants from around the area, including Hay River, all seem to be forest, heath and woodland species, among which were two specimens from Wilson Inlet that were to become syntypes of Karri.[121] Exactly where he gathered these specimens is not known but he may not have gone far from the settlement. Today, remnant patches of Karri woodland occur close to the western and north-western shores of Wilson Inlet and there are stands of the species at various places near the northern shore to the east, for example, at Karri Point near Hay River.[122]

Absent from Gregory's description of the vegetation is mention of Karri. Its appearance and height are so different from Jarrah and Marri (*Corymbia calophylla*) that it would seem he never came across it near the Inlet or on rises to the north of the road. Nor does he comment on the vegetation beyond Denmark River, which enters the Inlet on the north-western shore. This may mean that he never went beyond there.

---

[118] ibid., both references.
[119] HAODb.
[120] ibid.
[121] MEL 1608410, MEL 1608411, *Eucalyptus diversicolor*, MELISR database, 08 March 2007, NHV, RBGM.
[122] Ruth Moir and Libby Sandiford, Albany Herbarium, personal communication to MHH, 30 November 2009, HOA.

Allowing for time spent exploring the Inlets and collecting, Oldfield and Maxwell's journey to and from Albany could have taken about six days. The beauty of the changing vegetation and terrain would have made for pleasant travelling if the weather had been fine but most uncomfortable if it rained heavily, a not uncommon occurrence along that part of the south coast.

Albany to Williams River

Map 13 p. 273 – Albany to Perth

From Albany, Oldfield went to Perth. Such a journey could have taken eleven days or more depending on his mode of travel and how much time he spent collecting along the way. How he travelled is not known. One option was to accompany Robert Toovey and his police escort[123] carrying the mail to Perth and places on the way. Another was to walk, either with a packhorse or push cart; making for a very slow and tiring journey. The third, and perhaps the most likely, is that he purchased a hack and a packhorse in Albany and travelled on his own. The number of plants he gathered at some places along the road suggest that at times he stopped and botanised as he travelled along and stayed at some places longer than Toovey would have been willing to tarry. We assume that the last option was his mode of travel.

Although most of the plant material Oldfield obtained at King George Sound and nearby areas was deposited eventually in Mueller's herbarium[124], Mueller does not appear to have received any specimens from this expedition by June 1859.[125] It is unlikely that Oldfield took his collections from Albany with him on his journey to Perth. He may have left them in Maxwell's care, thinking that he would return to Albany after his Murchison River expedition; Maxwell was waiting for him on 8 March 1859.[126] On the other hand, perhaps he forwarded them to Perth by ship to avoid carrying a bulky wooden or leather case, or, like James Drummond[127], he may have feared the plants would suffer damage during the rough overland journey.

There is no specific mention in William Harvey's letters of how he got his specimens from the south coast to Perth. He did, however, say that his belongings took up most of the space in George Maxwell's spring cart[128] which suggests he took his collection with him. Oldfield's collecting places along the Albany-Perth Road appear to have coincided with most of Harvey's camping places when he travelled to Perth with the mailman George Maxwell in April 1854.[129] Harvey considered Maxwell to be 'a very intelligent man and a bit of a botanist—that is, he knows a good many plants...and we are very good friends'.[130]

As revealed by his specimen locality records, Oldfield went from King George Sound to the Colony's capital, via the Albany-Perth Road, the route closely following today's Albany Highway. The exact date on which Oldfield set out for Perth is not known. He is on record as having been at the Murchison River by at least 15 December 1858.[131] To be there by then, he needed to leave Albany during the third or fourth week of October at the latest. This allowed the eleven days for the trip to

---

[123] Rica Erickson (ed), 1979, vol. 3, p. 839.
[124] MELISR database, 08 March 2007, NHV, RBGM.
[125] FM to WJH, letter, 15 June 1859, Directors' letters, vol. 75, Australian and Pacific letters, 1859–65, fol. no. 114, RBGKA, cited in R.W. Home *et al.* (eds), 1998, vol. 1, p. 465.
[126] George Maxwell to FM, 8 March 1859, K59/2683, unit 746, VPRS 1189 inward registered correspondence, VA475 Chief Secretary's Department, Public Record office, Victoria, cited in R.W. Home *et al.* (eds), 1998, vol. 1, pp. 454-455.
[127] Rica Erickson, 1969, p. 91.
[128] WHH to Hannah Todhunter, letter, 13 April 1854, cited in Sophie C. Ducker (ed), 1988, p.93.
[129] ibid., pp.93–105.
[130] WHH to Reuben Harvey, 13 March 1854, cited in Sophie C. Ducker (ed), 1988, p. 91.
[131] MEL 2078490, *Acacia puncticulata,* 15 December 1858, Murchison River, MELISR database, 08 March 2007, NHV, RBGM.

Perth[132], a week in Perth preparing for the next leg of his expedition to the north and the approximately four-week pedestrian journey, which he is reputed to have undertaken to the Murchison.[133]

One outcome of the establishment of the coaling depot at Albany was the upgrade of the Albany-Perth road to facilitate conveyance of mail to the capital and places in between. At the same time, a shorter route was developed, which ran north from Kojonup to Perth via Kelmscott. The previous longer routes from Albany to the city had gone via Kojonup and Bunbury and the other via Kojonup, Williamsburg and York. The new road, built using convict labour, by-passed Williamsburg. It was completed by mid-1853[134] but for the most part it was still a dirt track.[135]

The extension of the road to Perth via Kelmscott brought a spate of pastoralists into the southern region in the 1850s.[136] Probably as a consequence of this there may have been a fair amount of traffic along the road at the time of Oldfield's visit apart from mail carriers—Aboriginal groups moving in tune with their customary traditions, kangaroo hunters, sandalwood cutters, clergymen, pensioner guards, ticket-of-leavers, shepherds and pastoralists pursuing their respective vocations, laborers in quest of employment, and wagons and carts bringing supplies to settlers or exporting wool, kangaroo hides and sandalwood to the ports of Albany and Bunbury.

The naturalist left no account of his journey in April 1854 to Perth but one of William Harvey's letters to his sister, Hannah Todhunter,[137] in which he described his journey to Perth with George Maxwell, affords some insight into what Oldfield may have experienced four years later, particularly as his collecting localities are all mentioned as Harvey and Maxwell's stopping places. Nevertheless, travelling in Spring, Oldfield's experience of the bush along the road must have been somewhat different floristically from that of Harvey, who covered that territory during Autumn when few plants were in bloom and environmental conditions much drier. Climate statistics for the area are not available for the 1850s but recent official historical records would suggest that it was probably fine with temperatures mild most days making for pleasant travelling, although he may have endured some wet days and nights, some of them very cold indeed.[138] With the days warming up, bush flies could have become increasingly annoying as he advanced towards Perth[139] possibly necessitating delaying the preparation of his evening meal until after the sun went down. As he moved up the road he probably supplemented his diet with the meat of kangaroos, wallabies and possums and by that time, he may have been aware of which plants were nutritious and edible. In some of the rivers there were freshwater crayfish, Marron (*Cherax sp.*) to be caught and savored.[140]

After leaving Albany, Oldfield travelled through Jarrah-Marri Forest to just north of Mount Barker. About ten kilometres north of Albany he entered the Marine Plain Ecodistrict. Along the Albany-Perth Road this plain extends north about 65 kilometres nearly to the Gordon River and comprises two interconnected tracts, one to the south, the other directly above to the north. To the west, the tracts stretch out for about 50 and 75 kilometres respectively. To the east the plain runs along the coast and becomes relatively narrow varying in width over its length from 65 kilometres at its widest point (north of Albany) to about 12 kilometres at its narrowest, where it terminates 36 kilometres south-east of Ravensthorpe. This ecodistrict has coordinated drainage systems that have formed linear alluvial

132 WHH to Hannah Todhunter, 13 April 1854, cited in Sophie C. Ducker (ed), 1988, p. 93.
133 Anon. ? Edwin Oldfield, The Argus, 13 July 1887, p. 5.
134 Donald S. Garden, 1977, p. 116.
135 WHH to Hannah Todhunter, 13 April 1854, cited in Sophie C. Ducker (ed), 1988, p. 96.
136 John Bird, 1979, *The Last Coach*, John Bird, Perth, p. 9.
137 WHH to Hannah Todhunter, 13 April 1854, cited in Sophie C. Ducker (ed), 1988, p. 93.
138 Australian Bureau of Meteorology, 'Climate WA', viewed 22 October 2009 at www.bom.gov.au.
139 Jim Heath, 1989, *The Fly in Your Eye*, Viacorp, pages unnumbered, viewed 22 October 2009 at www.viacorp.com. This book was originally published by J&E Publishing, Perth.
140 See, for example, H.G. Cowin, c.1971, *The Williams*, Williams Shire Council, p. 13.

and breakaway colluvial soils which, in addition to several duplex soils, support a range of vegetation systems.[141]

The surface of the plain is marked by numerous small depressions, which fill with water and become swamps in winter and several rivers, including Pallinup and its tributary Corackerup Creek, Bremer River, and Gairdner River to the east. These streams have cut deep and steep-sided trenches into the plain. This coastal plain is broken by various ranges and hills rising from its surface, including Stirling Range in the west and Barren Ranges further east.[142] Beard refers to the plain as being 'as flat and monotonous' but Hopper disputes this:

> An extraordinary statement in light of data presented in Gioia and Hopper (2017). Especially the Pallinup system between the Stirling and Fitzgerald River National Parks is as species rich as either of its more famous upland neighbours. Topography is much more varied than Beard's 'flat and monotonous' describes. There are many more granite outcrops and small hills, each with locally endemic plants, than is acknowledged it seems by Beard.143

Harvey's description indicates that from Albany, Oldfield passed through 'a succession of wooded low hills & sandy scrubby flats', past a large lagoon and then on to the 'nearest water' at the '25th milestone' (366 kilometres from Perth), where he may have camped on the first night.144 Following a night under canvas or snuggly rugged-up before the fire, as Harvey had done, an early breakfast and pack-up, Oldfield's next stop may have been Mount Barker, a mere eight kilometres further on. Mount Barker is within the Muir District, Bibbulmun Province, SWAFR145 just west of the narrow passage of marine plain separating the two tracts mentioned above.146 Harvey described the area near the prominence as

> a chain of higher hills covered with a thin forest of gum-trees, with open places and grazing ground for sheep here & there—Some of the trees were nearly 4 feet in diameter for nearly 30 feet or so, to the 1st branch—These appeared to me large, then—being fresh from the K. G. Sound forests, but I have seen many bigger since.[147]

It seems from the number of Oldfield's specimen records from Mount Barker that we have located, it was one of Oldfield's more important collecting localities in the Great Southern.[148] According to Stephen Hopper, this is understandable because it is so rich in plant species that it is 'now seen as a fantastic floristic hotspot'.[149] Since Oldfield was in the vicinity again in 1861–62 some or all his collections from this locality may not have been collected on his way to Perth in 1858.

Oldfield collected at Captain John Hassell's pastoral station, 'Kendenup'[150], but whether he was there during this journey or in 1862 on his return to Albany could not be determined. This property is about 1.6 kilometres east of Albany Road, 27 kilometres further north from Mount Barker, close to a road which ran east to the Stirling Range.

Hassell had taken up Kendenup in 1840 but by 1858, when Oldfield passed through, it was being managed by his son, John Frederick. Since 1840, apart from three years in England in the 1860s, Captain Hassell had lived mostly at Albany, where he had a business importing merchandise and

---

[141] Sarah Barrett *et al.*, 2009, p. 14 & Fig. 1, South Coast Ecodistricts.
[142] J.S. Beard, 1990, pp. 139–144.
[143] Personal communication to MHH, email attachment, 23 December 2017, HOA.
[144] Oldfield gathered MEL 14367 *Chaetanthus aristatus*, '25 miles from Albany' which may have been on this road but it cannot be confirmed, MELISR database, 08 March 2007, NHV, RBGM.
[145] Paul Gioia and S.D. Hopper, 2017, pp. 11–12 and Appendix S3, Fig. S3.7.
[146] Sarah Barrett *et al.*, 2009, p. 14 & Fig. 1, South Coast Ecodistricts.
[147] WHH to Hannah Todhunter, 13 April 1854.
[148] HAODb.
[149] Personal communication to MHH. email attachment, 23 December 2017.
[150] HAODb. 'Kinderup' on labels but this place could not be located on any gazetteer and as far as can be ascertained it refers to Kendenup. This is supported to some extent by the locality on two Oldfield specimen labels, MEL 1525946, MEL 1525941, that refers to 'Kinderup, K G Sound Road', MELISR database, 8 March 2007, NHV, RBGM.

Fig. 15.4 Summit, Mount Barker

station supplies, managing the finances of his properties and arranging wool shipments to England.[151] In accordance with successive business partnership arrangements between 1840 and 1850, he had pursued a policy of building up his flocks and exploring for good grazing land with potable water. By 1850 he owned, or leased, 63,000 acres (25,495 hectares) mainly at Kendenup, Hay River, Kojonup and Jerramungup.[152]

Maxwell and Harvey went to Kendenup on their way to Perth in 1854, not to deliver mail but to 'beg the loan of a horse', because the one they had refused to draw their cart further than about 43 kilometres from Albany.[153] Harvey walked the section of the road on the marine plain between Mount Barker and Kendenup and described the vegetation along the way:

> The road now leading thro. more hilly forest land with narrower & deeper valleys between—The trees were chiefly *Red Gum* [Marri, *Corymbia calophylla*], a large-growing & well-limbed tree, not unlike an Oak in figure, & the *White-Gum* [*Wornt*,[154] *Eucalyptus wandoo*], a light Ash-like tree with a white trunk & blueish [sic.] green leaves—This was the commonest Tree & continued very common for 100 miles—there were open spots of grassy land; few plants in flower & most places recently burned, the ground bare & black & the small shrubs reduced to cinders or dead sticks.

He was generously treated by the overseer to dinner and a bed for the night in the hut in which he had eaten the evening meal.[155]

---

[151] Merle Bignell, 1971, *First the Spring: A History of the Shire of Kojonup*, University of Western Australia Press, Nedlands, p. 33 and C.W. Hassell, 1972, 'Hassell, John Frederick Tasman (1839–1919)', ADBO, viewed 10 August 2009 at www.adb.online.anu.edu.au.

[152] Cleve W. Hassell, 1972, *The Hassells of Albany*, unpublished ms., Albany History Collection, Albany Public Library, pp. 16–23. See also Donald S. Garden, 1977, pp. 71–74.

[153] WHH to Hannah Todhunter, 13 April 1854, cited in Sophie C. Ducker (ed), 1988, p. 94.

[154] The local Noongar name is *Wornt*. *Wandoo* is a Noongar name for the species from New Norcia, S. D. Hopper, personal communication to MHH, email attachment, 23 December 2017, HAODb.

[155] WHH to Hannah Todhunter, 13 April 1854, cited in Sophie C. Ducker (ed), 1988, pp. 95–96.

Fig. 15.5 Lake Matilda

Travelling during late Spring, it is unlikely that Oldfield passed recently burnt areas although if there had been a fire within the last two years or so, many plants may have been in new leaf and some of them in flower.

After leaving the turn-off to Kendenup on the Albany Road, Oldfield passed through country where Harvey noted the 'White Gums' (*Wornt*/Wandoo) were more numerous than before and 'much infested with a scarlet flowered Loranthus'[156], probably the Stalked Mistletoe (*Amyema miquelii*) which flowers from February to December.[157] After ten kilometres, Oldfield stopped at another of his collecting spots[158], Lake Matilda, where Harvey reported seeing for the first time Zamias,[159] a cycad that would have been familiar to Oldfield since they occur commonly among the Jarrah forests around Albany and to the west.

If Oldfield did not stay at Kendenup the previous night, he may have bivouacked at the lake where there was plenty of freshwater.[160] It has since become mildly brackish and, even though the remnant bush surrounding about half of the lake is degraded, it is a beautiful spot and a haven for water birds of various kinds. Today, the most noticeable species fringing the lake are Saltwater Paperbark and Beaded Samphire (*Sarcocornia quinqueflora*) and present in the bush behind the shore are Mauw, Swamp Banksia, Manna Wattle (*Acacia microbotrya*), *Rhodanthe citrina*, Honey Bush (*Hakea lissocarpha*), Blind Grass (*Stypandra glauca*) and other species.[161] Harvey reported flocks of Red-tailed Black Cockatoos (*Calyptorhynchus banksii naso*) 'screaming from tree to tree' and with the night came 'shriller screams from Mosquitoes and...[a] great many Kangaroo-rats...coming hopping

[156] ibid., p. 96

[157] According to *FloraBase, Amyema miquelii* appears to be the only hemiparasitic species with red flowers on eucalypts occurring between Kendenup and Kojonup.

[158] HAODb.

[159] WHH to Hannah Todhunter, 13 April 1854, cited in Sophie C. Ducker (ed), 1988, p.97.

[160] ibid.

[161] Ruth Moir and Libby Sandiford, Albany Herbarium, personal communication to MHH, 30 November 2009.

close to us'.[162] Oldfield may have had similar experiences; if so, pressing and labelling his plants might have been a task he deferred until the morrow.

From the lake, Oldfield went on to Gordon River, which crosses the road 27 kilometres further on. Harvey noted that the road

> lay through undulating forest chiefly of white Gums [Wornt], the ground and under shrubbery being mostly burnt.[163] Among the unburnt shrubs, one was conspicuous & very refreshing to the eye by the bright yellow-green of its leafless branches…It is either Leptomeria or Exocarpus.

After three hours walking Harvey came across 'a succession of open thinly scrubbed plains [the Gordon River Plains], separated by swamps overgrown with large bushes & small trees' where

> [t]he most common trees were what they here call 'tea-tree' that is a Melaleuca Cuticularis [Saltwater Paperbark], which grows in every swamp forming a thickly branched, crooked, flat-topped small-leaved tree with a profusion of small white flowers near the ends of every branch (when in blossom) & bark several inches thick very soft & white, & feeling of fine thin plates like tea-paper—A small Cypress like Callitris—Lambertia inermis—a variety with pale primrose-coloured flowers & many wild Cherry (Exocarpus) trees…[with] twiggy branches. In one of the Swamps...Banksia Meisneri—a very charming & rather rare species of small size, a foot or two high, with heathlike leaves & heads of flowers no bigger than the top of [Harvey's] thumb...The soil of the plains was miserably thin, a little sand & ironstone gravel on the Ironstone Conglomerate—It fed, however, a great variety of rigid leaved small shrubs as the fernlike ground Banksias & Dryandras—thorny & round leaved Hakeas—small Myrtaceae, & scrubby Eucalypti, with here & there a scattered tree...[164]

Now within in the Jarrah District, Bibbulmun Province, SWAFR,[165] Oldfield collected on Gordon River Plains[166], which, according to Harvey, would have continued until he 'entered the hills among which the Gordon flows, & where the usual forest of Gum trees, white [*Wornt*/Wandoo] and red [Marri] &c again closed me in'.[167] The Marri-Wandoo woodland along the Albany-Perth Road lies east of the Darling Range and tends to be lower and more open than it is on the western side of the Range. Along the way, Jam and Rock Sheoak appear as small trees, sometimes locally abundant. The Grass Tree can be particularly common in these woodlands and in wet clayey depressions may be found stands of Mauw (*Eucalyptus occidentalis*) with York Gum (*E. loxophleba* subsp. *loxophleba*) usually replacing Marri and Wornt/Wandoo[168] in valley bottoms.[169]

When Harvey arrived at the Gordon River he was treated hospitably by Mr Augustus Gregory,[170] 'who with a road-party [comprising convicts and ticket-of leavers][171] is here encamped putting up a bridge for Govt.'. He dined on kangaroo steaks, prepared by one of Gregory's charges, and slept in

---

[162] WHH to Hannah Todhunter, 13 April 1854, cited in Sophie C. Ducker (ed), 1988, p. 97.

[163] The ground and under shrubbery being mostly burnt suggests the local Noongar practice of burning to encourage the growth of fresh grass to attract macropods such as the Western Grey Kangaroo, *Macropus fuliginosus*. S. D. Hopper, personal communication to MHH, 23 December 2017.

[164] WHH to Hannah Todhunter, 13 April 1854, cited in Sophie C. Ducker (ed), 1988, pp. 97–98.

[165] Paul Gioia and S.D. Hopper, 2017, pp. 11–12 and Appendix S3, Fig. S3.7.

[166] HAODb.

[167] WHH to Hannah Todhunter, 13 April 1854, cited in Sophie C. Ducker (ed), 1988, p. 97-98..

[168] Somewhere to the north of Kojonup the northern word 'Wandoo' is applied to *Eucalyptus wandoo*, S. D. Hopper, personal communication to MHH, email attachment, 23 December 2017, HAODb.

[169] J.S. Beard, 1990, pp. 84–87. See also M.I.H. Brooker and D.A. Kleinig, 1990, *Field Guide to the Eucalypts: South-western and Southern Australia*, Inkata Press, Melbourne, for distributions of these species of *Eucalyptus*.

[170] Harvey refers only to a 'Mr Gregory'. It was Augustus Gregory who was involved in the building of the King George Sound-Perth Road, Merle Bignell, 1971, p. 72.

[171] ibid. According to Bignell, Gregory oversaw the ticket-of-leave men whose task was to clear the line, and Vincent, the gaoler, in charge of the colonial prisoners who erected the bridges along the route. If this was the case, Gregory and his party must have been assisting the convicts when Harvey came upon them.

Gregory's tent.[172] Oldfield also may have camped near the banks of the Gordon River, perhaps arriving early enough to collect plants and to shoot, skin and butcher a kangaroo ready for dinner. In the Henderson Oldfield database, his gatherings from Gordon River Plains and from the vicinity of the river itself were more numerous than at any of his other collecting places along the Albany-Perth Road.[173] According to Stephen Hopper:

> This again reflects how rich the flora of these plains really is. In Oldfield's time it was poorly collected, and still is, because people believed it was species-poor compared with the Stirling Range and coastal regions. Oldfield knew enough about the SW flora to recognise a place full of new species.[174]

In 1861 (October) Oldfield travelled along the Albany-Perth Road to Mount Barker and apparently followed a track from there westwards to Lake Muir, and from there east and north possibly following a track to the confluence of the Gordon and Frankland Rivers.[175] The absence of dates of collection on all specimen records relating to localities from Albany to Mount Barker along the Albany-Perth Road does not permit the identification of other plants gathered on each occasion. Except for his collections from the Serpentine and Canning Rivers, all his other specimens from along that road could only have been collected in 1858.

Travelling further north from Gordon River through 'wooded undulating country as before'[176] with plentiful occurrence along the track of plants (*Gastrolobium* spp.) that were poisonous to sheep[177], Oldfield's next gatherings appear to have been from near and at Balgarup[178], a deserted pastoral property on the Balgarup River, which crossed the road 42 kilometres north of Gordon River.[179] He might well have bivouacked there overnight near the ruins of the old homestead, bypassing a waterhole one and a half kilometres off the road, about 22 kilometres north of Gordon River where Harvey and Maxwell camped.[180]

From 'Balgarup' it was only 18 kilometres further up the road to Kojonup, which had been first settled in 1837 as a military outstation.[181] On entering this remote little hamlet, centred on the Kojonup Spring, Oldfield would have been confronted immediately by the most prominent building, a small stone military barracks, housing some Pensioner Guards and their families. A little further away to the north there were fields of ripening wheat or corn on Pensioner Guards' ten-acre grants. On at least three of these were two-room, stone cottages roofed with 'strong-bound timber' and 'thatched to be perfectly water tight'.[182] Harvey referred to the 'the town of Kogonup [Kojonup]' as 'a sort of Malines, (I mean Mechline)[183], of the colony, a place everyone talks about as a centre from which roads branch'[184]; it was here that roads took off to Bunbury, York and Perth.

Oldfield obtained some plants at and near Kojonup.[185] Here, he entered the Narrogin District, Bibbulmun Province, SWAFR through which the road appears to have run until a little to the north of Williams.[186] Since at that time, there was no hotel at Kojonup, he may have camped, or, as Harvey had

---

172 WHH to Hannah Todhunter, 13 April 1854, cited in Sophie C. Ducker (ed), 1988, pp. 98–99.
173 HAODb.
174 Personal communication to MHH. email attachment, 23 December 2017, HAODb.
175 see Chapter 19.
176 WHH to Hannah Todhunter, 13 April 1854, cited in Sophie C. Ducker (ed), 1988, p. 99. 4.
177 Merle Bignell, 1971, p. 37.
178 HAODb.
179 Merle Bignell, 1971, p. 37.
180 WHH to Hannah Todhunter, 13 April 1854, cited in Sophie C. Ducker (ed), 1988, p. 99..
181 Merle Bignell, 1971, p. 20.
182 This suggestion is based on a report by Captain [? Major] John Bruce, Officer-in-charge of Convict Stations, who visited the settlement in August 1854, CSO, 303, 9 September 1854, SROWA cited in Merle Bignell, 1971, pp. 66–67.
183 A reference to Mechelen (or Malines) in Belgium, a town which lies on the major urban and industrial axis Brussels–Antwerp, about 25 km from each city.
184 WHH to Hannah Todhunter, 13 April 1854, cited in Sophie C. Ducker (ed), 1988, p. 100.
185 HAODb.
186 Paul Gioia and S.D. Hopper, 2017, pp. 11–12 and Appendix S3, Fig. S3.7.

Fig. 15.6 Beaufort River flood plain

done, slept 'on a stretcher in a hut'.[187] Perhaps there were opportunities for him to chat to some of the folk in the settlement and to augment his supplies with 'corn and even flour from [Pensioners'] hand mills' and possibly some vegetables and milk from those who had surplus to their needs.[188]

From Kojonup, Oldfield could have found the 48 kilometre journey to Beaufort River difficult and tiring, particularly if it was a warm sunny day. Thus, for Harvey, the road was

> very straight—carried right ahead up the hills, down the hills, like an Ant-track, taking every obstacle in the face—& as it lay across a sandy plain and over sandy hills it was heavy walking & cruel riding—walked almost the whole way, & yet it was the worst day on the horses.[189]

Harvey and Maxwell interrupted their journey to have their midday meal 'beside a well of excellent water' where the botanist measured 'a cabbage tree (Nuytsia)... [*Moojar*/Christmas Tree (*N. floribunda*)] ten feet in circumference'[190]. According to Stephen Hopper:

> Many moojars are still to be seen on this stretch of Abany Highway, left standing in paddocks by Noongar people engaged to clear for farmlands. They dared not touch or destroy such a sacred tree.[191]

Oldfield may have stopped for lunch at the same spot. Along the way, perhaps he met Toovey waiting while the police escort inspected a bridge. The primary duties of the officers, W. Smith and Edward Barron, who were stationed at Beaufort River, were to escort the mails, each completing half of each journey to Perth. However, they had been instructed to pay 'every attention' to all the bridges on the Albany Road 'within their neighbourhood' by visiting all of them 'twice a week...to remove all dried grass or other combustible matter'[192], to save those wooden structures from being destroyed by

[187] WHH to Hannah Todhunter, 13 April 1854, cited in Sophie C. Ducker (ed), 1988, p. 100.
[188] Merle Bignell, 1971, pp. 67.
[189] WHH to Hannah Todhunter, 13 April 1854. cited in Sophie C. Ducker (ed), 1988, p. 100.
[190] ibid.
[191] S. D. Hopper, personal communication to MHH, 23 December 2017. See S.D. Hopper, 2010
[192] A.R. Pashley, 2000, *Policing Our State: A History of Police Stations and Police Officers in Western Australia 1829–1945*, EDUCANT, Cloverdale, Western Australia, p. 75.

fire. Assuming the day's travel had been as grueling as Harvey and Maxwell's, Oldfield, like them, may have thankfully bivouacked beside Beaufort River, where he collected a few plants.[193]

North of the river Oldfield crossed Beaufort Plains. Perhaps he halted for water at a spring near the road ten kilometres north of the river, but was possibly as disappointed as Harvey and Maxwell had been when they found '8 or 10 dead Kangaroos in all stages of decay in the well'. The spring 'issue[d] from under an overhanging rock and the K[angaroo]s coming round to drink, tumble[d] over' into the well below.[194] From the plains, Oldfield 'ascended a grassy range of hills with ironstone on the summit and then descended to the Arthur River'[195], a distance of about 27 kilometres from Beaufort River. Although not running when Harvey and Maxwell visited, the stream contained 'excellent water' in a [riverine] water-hole a mile long'.[196] There were pastoral leases in the district but there was no police station or town settlement near the river.[197]

In this vicinity Oldfield gathered a few plants.[198] If he was on his own, he may have stayed there overnight and found it a pleasant place to bathe, catch up on field notes, press and label his new specimens and change the papers of those already in his presses.

Williams River to Perth

Map 13 p. 273 – Albany to Perth

About 35 kilometres north of Arthur River, Oldfield crossed Williams River where there was no town settlement. However, legend has it that as early as 1851 there existed a structure 'purporting to be a wayside inn and situated a little south of the present hotel', but no documentary evidence has been found confirming its existence[199] nor did Harvey mention it. There may have been 'several public buildings'[200] situated on a government reserve, that had been declared for public utilities in 1855, when Oldfield passed through but there was no police presence.[201] Had he stayed at Arthur River the night before, Williams River was probably another of his camping spots. Oldfield did obtain some plants at and near the stream.[202] The surrounding country, according to Harvey, 'was chiefly sandy'.[203]

A little to the north of Williams River the road entered the Jarrah District, Bibbulmun Province, a little to the north of Williams.[204] Oldfield travelled to Bannister River through 'rather an open forest chiefly of white gum with an under-storey of grass trees of all dimension, & many branched' where he may have encountered, as Harvey had done, 'an abundance of Kangaroos standing to look & then hopping away, young & old in troops thro. the woods'.[205]

On the way, Oldfield traversed and gathered specimens at and near Hotham River.[206] Perhaps he remained there overnight, given that it was about 29 kilometres further on to Bannister River, where he again made a few collections.[207]

From Bannister River, it was 16 kilometres to Serpentine River. Specimens labelled as coming from, at, or near Serpentine may have come from there or, as indicated earlier, from where one or

---

[193] HAODb.
[194] WHH to Hannah Todhunter, 13 April 1854, cited in Sophie C. Ducker (ed), 1988, p. 101.
[195] W. Crossman, 19 January 1853, Letterbooks, 8 March 1852–3 July 1855, fols. 5, 6, 7 8, microform, JSLB, cited in Merle Bignell, 1971, p. 62.
[196] WHH to Hannah Todhunter, 13 April 1854, cited in Sophie C. Ducker (ed), 1988, p. 101.
[197] Anon., 'Arthur River, Western Australia history', viewed 6 May 2013 at www.wikipedia.org.
[198] HAODb.
[199] H.G. Cowin, c.1971, p. 4.
[200] ibid., p. 4.
[201] Landgate (Western Australian Land Information Authority), 'Williams', *History of country town names,* viewed 22 September 2009 at www.landgate.wa.gov.au.
[202] HAODb.
[203] WHH to Hannah Todhunter, 13 April 1854, cited in Sophie C. Ducker (ed), 1988, p. 101.
[204] Paul Gioia and S.D. Hopper, 2017, pp. 11–12 and Appendix S3, Fig. S3.7.
[205] WHH to Hannah Todhunter, 13 April 1854, cited in Sophie C. Ducker (ed), 1988, p. 102.
[206] HAODb.
[207] ibid.

Fig. 15.7 Darling plateau, off Albany Highway

other of the two Pinjarra-Mandurah roads crosses the lower reaches of the stream[208] Oldfield, like Harvey, could have found the next 42 kilometres after Bannister River 'a weary journey...across hills and vallies [sic] of conglomerate ironstone, part of the Darling range' and continued for another 19 kilometres to the pool of excellent water at the '41st milestone' and a welcome place to stay for the night.[209]

Given that Oldfield also crossed the Darling Scarp and plateau on his way to Toodyay, it is not known where on the 'Darling Range' his specimens, assigned that locality, were from, but it is reasonable to suggest that some of them were gathered during this trip along the Albany Road. As he progressed up the road Oldfield may have noticed that the Marri-Wandoo woodland extended for about 250 kilometres to the north of Kendenup, further than Harvey had indicated. At about that point on the Albany-Perth Road Wandoo starts to become less frequent and the Jarrah-Marri association increasingly dominates as the road traverses the slopes of the Darling plateau.

From the 41st milestone the road continued 'to traverse the Darling Range of the same rough and hard hills as before', but, as Harvey had experienced, Oldfield may have discovered that 'the flora was more interesting'. In bloom, 'abundant and handsome' may have been *Adenanthos barbiger*

> with much the look of a scarlet Salvia...In several paces large patches of Kingia (Rush-trees) [*Kingia australis*] supplanted Grass trees [*Xanthorrhoea sp.*] but the lat-

[208] see Chapter 11.
[209] WHH to Hannah Todhunter, 13 April 1854, in Sophie C. Ducker (ed.). 1988, p. 102.

ter mostly prevailed. The trees were chiefly the Mahogany Gum tree, or Yarra [Jarrah (*Eucalyptus marginata*).][210]

The road gradually began to descend until it 'finally left the hills thro. a valley running westward for some miles'. In the lower part of this valley Harvey came across 'one of the rarest Australian sights, a small perennial stream [Neerigen Brook] that runs the whole year round'. He mentions reaching 'a small cottage owned by a laboring man who has a little garden, & a room where travellers may eat their dinners'.[211] The proprietors were John and Isabella Boult and the name of the property was 'Paradise' (Canning Location 34) which was some half a kilometre up the hill from the junction of the Albany and Bunbury Roads. They sold up soon after Harvey and Maxwell called in. The 789 hectares, comprising Canning Location 31, was leased by Thomas Saw from Henry Vincent in 1856 for six years. In the same year Saw built a larger farmhouse on the site on this property at the junction of the Albany and Bunbury Roads, which was called 'Narrogin' after the Neerigen Brook. In partnership with his brother-in-law, Henry Gibb, he applied for a publicans' license on 30 November 1856 and set aside part of his house for the sale of liquor. This became known as Narrogin Inn.[212] Since then there has been a hotel of the same name on that site. It is difficult to imagine that Oldfield by-passed the opportunity for refreshments and company.

Near the Inn, Albany Road entered the Swan coastal plain and passed through the district of Kelmscott where a townsite had been gazetted in 1830. However, at the time Oldfield passed by, the town lots were mostly unoccupied and there were only scattered farms and pastoral leases along and near the Canning River.[213] The bridge over the river, which was built in 1853 or 1854, was located about eight kilometres from the Narrogin Inn and about 11 kilometres 'as the crow flies' upstream from where the river enters the Swan River estuary. As noted previously[214], Oldfield collected along the Canning River but exactly where could not be determined. When he visited Perth in 1856, and again in 1858, he may have taken plants from along the river near where it entered the Swan River estuary, and probably also near the Canning River bridge (an upgraded version of which is still located on the Albany Highway within the City of Gosnells), on his way to Perth from Albany. From the bridge, it is only 18 kilometres to Perth.

Harvey and Maxwell stayed overnight by Canning River; apparently on the southern side; Oldfield may have done likewise and his experience of the terrain from the bridge into Perth was probably similar in many respects to that of Harvey who gave the following account:

> Crossing the Canning by a new bridge we entered an extensive grass plain thinly sprinkled with large trees & partly settled along the banks of the river, where there is a considerable extent of cultivated land—Daylight coming on showed us the low flattish range of hills we had yesterday finished passing & which now terminated the Eastern view—also the plains on which we had travelled & which now reminded me of a desolated Irish nobleman's Park—a sort of Castle Rackrent Domain[215] where some fine trees were still standing in the neglected fields. Our road was a mere cart-track thro. the clay or sand & varied at the will of the driver—Here & there was a broken fence & a ½ ruined house...The road soon entered the deep sands which surrounded the town of Perth and thro. which we had to pull for 4 or five miles—Grass-trees & Kingias & Zamias of large size abounded & a low forest of large Banksias gradually thickened as we approached the City, till from the top of a hill about 2 miles

---

[210] ibid.

[211] ibid, p. 102.

[212] Jennie and Bevan Carter, 2011, *Settlement to City: A History of the Armadale Districts and its People*, City of Armadale, pp. 41–42 & 53–54.

[213] ibid., pp. 27–70.

[214] See Chapter 11.

[215] *Castle Rackrent*, a novel by Maria Edgeworth (1800) which, according to Sophie Ducker, gives a remarkable picture of a profligate landlord in Ireland in the 18th century, Sophie Ducker, 1988, fn. 16, Notes, Chapter 2, 'The Journey to the Swan River Colony', p. 375.

> out, we caught the first view of the Swan shining thro. a glade in the woods, no unwelcome sight after our 11 days of bush travelling—We crossed the river by a long causeway & bridges [the Causeway], & a gently rising hill led us up the principal street [St George's Terrace].[216]

Arriving in Perth on about 18 August 1858, and progressing toward the centre of town from the Causeway along the still-unsealed, and largely unlevelled, Adelaide and St Georges Terraces[217], Oldfield would have noticed that it was still a small settlement with little vehicular traffic. Since his last visit, there were a few more cottages, substantial residences and commercial buildings lining the route. The Anglican 'Deanery', completed in 1859, was possibly under construction on the corner of Pier Street and St Georges Terrace. On the right-hand side of the Terrace, approximately half way between Government House and Pensioner Barracks at the western end of the Terrace, was the newly opened Perth Church of England Collegiate School, known today as the 'Cloisters' and the Colony's first secondary school for boys. This two-storey brick Gothic building with Tudor embellishments had been established during the year by the first Anglican Bishop of Perth, Mathew Blagden Hale, funded with a grant from the Society for Promoting Christian Knowledge and the Bishop's own donation.[218] Wandering around the settlement later on it must have come to his attention that there were more commercial and residential buildings along the other main streets of the capital, and cottages on the drained land north of the town's centre. However, due to instructions from the Colonial Office in London, the number of government buildings was about the same as it had been two years before.[219] Any of the August issues of the *Perth Gazette* would have informed him that Perth and Fremantle stores were flush with a vast array of goods that had recently arrived in the *Kenilworth* from London; paper, saddles, medicines, clothes and many other items that might have been useful for his journey north.[220]

---

[216] WHH to Hannah Todhunter, 13 April 1854. cited in Sophie C. Ducker (ed), 1988, pp. 103–204.
[217] C.T. Stannage, 1979, p. 155.
[218] Heritage Council of Western Australia, 'The Cloisters', Place 02119, Assessment Documentation, State Heritage Register, viewed 28 October 2009 at www.inherit.stateheritage.wa.gov.au.
[219] C.T. Stannage, 1979, pp. 140–141.
[220] *Perth Gazette*, issues of 6, 13, 20 & 27 August 1858, DAN, NLA.

Fig. 16.0 Myriocephalus oldfieldii

Chapter 16

Western Australia 1858–1859

Perth to Murchison River

The duration of Oldfield's stay in Perth, where he lodged, and how he occupied his time while he was there, are again matters for speculation. He must have needed a few days to recuperate from his journey from Albany, prepare for his expedition to the Murchison River, sort and pack his specimens and arrange for their storage until his return. It appears that he did not send any to Mueller at this stage.[1] He may have done some collecting around Perth and Fremantle during his stay but as indicated earlier, this cannot be determined.[2]

Edwin reported that Augustus had walked from 'Perth to North-western Australia, where for a year he mixed freely with the natives, acquiring much knowledge of their language, and habits'.[3] However, in another obituary he stated that he had undertaken the journey 'with a saddle and pack-horse'.[4] These two statements are probably not necessarily incompatible since travelling with a pack-horse would have meant proceeding at walking pace. By referring to the length of time he was in the north-west, Edwin could only have been alluding to Augustus' 1858–1859 journey to the Murchison River. Undertaking this journey alone is indicative of someone who was physically strong, supremely confident of his bushmanship and of his ability to quickly establish friendly relationships with Aboriginal tribal groups, shepherds and settlers encountered along the way. The knowledge and contacts gained during his previous trek to the Murchison River no doubt bolstered his confidence.

Once again, there is the question of what luggage he took. A journey of a year must have necessitated a supply of clothes, medical goods and certain foods, as well as camping and collecting equipment including insect-proof cases, possibly made of metal, in which to store his specimens safely.

Since Mueller initially, and later the Victorian Government, bore the expense of the expedition, Mueller must have wanted Oldfield to collect across all seasons to maximise the number of specimens of species collected whilst in flower.

At the time of Oldfield's 1858–1859 visit to Western Australia there does not appear to have been any collecting for scientific purposes north of Perth to the Murchison River since he and Walcott were there in 1856–1857.

Perth to Geraldton, October–November 1858

Map 9 p. 198 and Map 10 p. 207

All the records in the Oldfield database of specimens collected between Perth and where the track to 'Minnenooka' left the Old North Road, indicate that he traversed the same road from Perth on his way to that point in 1856 and 1858. Although on both his journeys to the Murchison River he apparently

[1] FM to WJH, 15 June 1859. Directors' Letters, vol. 75, Australian and Pacific Letters 1859–1865, fol. 14, RBGKA cited in Home *et al.*,1998, p. 465
[2] See Chapter 11.
[3] Anon. ? Edwin Oldfield, The Argus, 13 July 1887, p. 5.
[4] Anon. ? Edwin Oldfield, *Bairnsdale Advertiser and Tambo and Omeo Chronicle*, 16 July 1887, p. 3, DAN, NLA.

16.1 Hill River at Old North Road crossing

came and went along the same route, he collected very few specimens[5]. This may have been more a reflection of the limitations imposed on his botanising activities by travelling under difficult conditions than on his interest in the flora.

If, as appears to have been the case, he set out from Perth in late October or early November, the weather conditions were no doubt much hotter and drier than when he went north in the spring of 1856. On his previous journey he would have become familiar with the 35 watering places between Doghole Pool, about 10 kilometres south of Yatheroo, and Walkaway.[6] Although the spacing of wells and other sources of freshwater varied, on most days he was within reasonably easy reach of at least one. Nevertheless, it would have been imperative for him to carry two or three days' water supply for himself and his horses as there was always the possibility of unforeseen circumstances—dry freshwater pools and wells, or illness preventing him from travelling for a day or so.

Since 1856, agricultural pursuits in the Toodyay district were becoming increasingly important. The area of land planted to cereal crops and hay, and the numbers of horses, pigs and goats were increasing significantly.[7] This was not the case across the Victoria Plains to Hill River, where pastoral endeavours still dominated and there had been no further granting of freehold locations since 1856.[8] Therefore, except within the Toodyay district, there had been few, if any, additions to the number of isolated and scattered homesteads and shepherds' huts along the way to Geraldton.

Oldfield did report having been at Stockyard Gully, north of Hill River on the Old North Road, in 1858–1859 when discussing the colonists' lack of interest in ascertaining the full range of plant species toxic to stock. Two species of *Gastrolobium* had been identified as toxic but he believed there

---

[5] see Chapter 12.

[6] Martin Gibbs, 1994, *Historic Watering Holes Survey: The Old North Road—Wanneroo to Walkaway Coastal Stock Route Western Australia*, National Trust of Australia (WA), Perth, mimeo, pp. 17–18. May Connor revised and expanded the original report, March 1988.

[7] Rica Erickson, 1974, p. 163.

[8] Rica Erickson, 1971, pp. 22–23.

were probably others and possibly 'rusts and brands'[9] as well as 'mechanical' causes. He also described one instance of the latter which had occurred at Stockyard Gully,

> a locality that had been used as a halting place by drovers for the last 10 or 12 years, no loss imputed to poison, having been sustained there during that period. Just before I arrived in that vicinity, a settler had lost 25 oxen there in one night, and as this was the first recorded instance of the kind I was naturally anxious to ascertain what plants had been the cause of so serious a loss, but after a careful search extending over two days, I failed to discover, not only any of the common poisons of the colony, but even a plant of a suspicious nature, so that, in this case at least, the cause of death must have been purely mechanical, a fact of which I was satisfied by an inspection of the contents of stomachs of some of the deceased cattle, hard woody matter predominating in them.[10]

The lengths to which Oldfield went on this occasion demonstrates how great was his desire to understand phenomena that puzzled him.

Travelling at that time of the year, we can only imagine how he coped with the constant attention of hundreds of bush flies and the intensifying heat during the day as he moved northwards. Earlier explorers' complaints abound of flies being 'at you all day, crawling into your eyes, up your nostrils and down your throat'.[11] Perhaps for the sake of comfort he wore over his hat a net that covered his face and neck to ward off the flies, and he may have sucked on a pebble to stimulate the release of saliva, thereby ameliorating a dry mouth due to thirst; this practice having been learnt from Aborigines. He probably arose before daylight, had breakfast, packed up and got underway by 5 am or earlier, trekked until midday when the sun's rays were at their most intense, rested in the shade of a tree for a few hours, and moved on until after sunset when the flies had abated, allowing him to light a fire and prepare and eat his evening meal in peace. Trees that provided deep shade, however, were few and far between on the sand plain west and north of Koojan, making the journey across those areas even more difficult. There can be little doubt that he doused himself with cooling water at wells, bathed in the various rivers he crossed on the way, welcomed the shade of the trees along their banks and cursed loudly when wells were dry. Where there was no shade he may have constructed a temporary shelter by throwing a light tarpaulin over tall shrubs. No matter how physically strong he was, there must have been times when fatigue, aching limbs, burning feet or a thumping headache from excessive sun exposure forced him to take a break and wish that he was at his journey's end.

After crossing the south end of Greenough Flats, the Old North Road terminated at the Old Walkaway (previously Wagawa) ford. Since it is unlikely that he had reason to go to Minnenooka, as James Walcott had moved to the Murchison River in 1858[12], Oldfield probably then travelled across the Back Flats along what had become the main road from the south to Geraldton. Three records of plants with the locality 'Greenough Flats' have been found.[13]

The exact location of the road across Back Flats, which in 1858 was just a sandy track, has not been ascertained but it is thought to have been closely aligned to today's Kennedy Road, which crosses near the western border of the alluvial section of Back Flats.[14] The track apparently approximated the route laid down by George Grey in 1839, when he and his party walked south from Gantheaume Bay, and it was traversed again in 1846, by Lieutenant Benjamin Franklin Helpman on his way south to examine the recently discovered coal seam at Irwin River.[15] Following this road,

---

[9] 'Brand' refers to a kind a kind of blight on leaves etc. with burnt look, H.W. Fowler and F.G. Fowler (eds), 1970, p. 143.
[10] AO, 1865a, fn. 46, p. 667.
[11] Jim Heath, c.1989, pp. unnumbered.
[12] Sister Mary Albertus Bain, 1975, p. 177
[13] MEL 68085, *Alyogyne huegelii* var. *huegelii* & MEL 68084, *Hibiscus drummondii*, MELISR database, 08 March 2007, NHV, RBGM, and *Jasminum calcareum*, GB 1869, vol. 4, p.297.
[14] We have been unsuccessful in locating survey plans showing the track.
[15] Dr Michael O'Connor, personal communication to MHH, 27 November 2008. See also Sister Mary Albertus Bain,1975, pp. 16–19.

Fig. 16.2 Eucalyptus erythrocorys on Old North Road track

Oldfield started at the Old Walkaway ford and went in a north-easterly direction through the alluvial break in the Tamala limestone before turning north-west and going in that direction across Back Flats.

Pastoral leases over the Back and Front Flats had been granted since December 1850, when the new regulations came into effect for the occupation of 'waste lands' of Western Australia.[16] Much of the Back Flats to the south, was held by Hammersley & Co.'s Cattle Company, and at the northern end by Thomas Brown.[17] Ticket-of-leavers and free men shepherding Brown's stock may also have crossed his path. By the time Oldfield passed through in 1858 the area under pastoral use had retracted to the Back Flats, the Front Flats having been set aside for agriculture.[18] Recognising the rich alluvial soil on the Front Flats, the Governor, Sir Arthur Kennedy, had instructed surveyor Frank Gregory to lay out roads and lots along the river banks for tillage purposes in 1857. In the same year, Hammersley & Co purchased a series of freehold blocks on the Front Flats before the land sale held later that year exercising pre-emptive rights of purchase. The Pensioner Guards, who had languished at Lynton since the closure of the Depot, were offered land in a series of allotments 15 kilometres south of Champion Bay (Geraldton). However it was not until January 1858 that the last residents of the Depot were shipped out[19]

According to historical records, crossing the alluvial plain and low river terraces of the Back Flats, Oldfield passed through uncleared savannah-woodland of York Gum and *Acacia* species such as Summer-scented Wattle (*Acacia rostellifera*), and occasional stands of River Gum (*Eucalyptus camaldulensis*) over an open ground grass layer of *Themeda triandra* and tussocks of *Austrostipa* sp. Along the way he may have encountered Paper Daisies (*Rhodanthe* spp.) in flower, and extensive

[16] W.J. de Burgh, 1986, p. 46. According to de Burgh, these regulations provided for the granting, right of renewal and compensation for improvements of Tillage Leases within pastoral leases and could be let in certain areas up to 320 acres at a minimum annual rental of two shillings an acre and Pastoral Leases, for grazing only, at ten shillings per 1,000 acres.

[17] The Cattle Company had been formed when Lockier Burges was granted grazing rights over 60,000 acres on Irwin River, Sister Mary Albertus Bain, 1996, p. 86.

[18] Michael Henry O'Connor, 2001, p. 274.

[19] Martin Gibbs, 2007, 'Lynton: convicts, landscape and colonisation strategies in midwest Western Australia', *Australasian Historical Archaeology*, vol. 25, p. 62. Also, Dr Michael O'Connor, personal communication to MHH, 27 November 2008, and Sister Mary Albertus Bain, 1975, pp. 224–226.

diggings for the yam Warrine (*Dioscorea hastifolia*) and other species.[20] Perhaps at times he diverged from the track after it entered the Back Flats and climbed the non-alluvial Tamala limestone ridge to the west, that runs parallel to and separates Back and Front Flats.[21] Along this ridge before he reached the north end of the Back Flats, there was low woodland of *Banksia* and *Acacia* species with Christmas Trees (*Nuytsia floribunda*) on the upper slopes and hill crests, and a mainly 'scrubby' undergrowth perhaps including Geraldton Wax (*Chamelaucium uncinatum*) and Warrine. The Zamia was also present.[22]

After leaving the alluvial plain, he may have crossed through thickets of *Acacia* spp., Tangling Melaleuca (*Melaleuca cardiophylla*) and Stinkwood (*Jacksonia sternbergiana*) and areas of Warrine to the west, or to the east, savannah woodland with pockets of River Gum and other *Eucalyptus* spp. until he was a few miles south from the Moonyoonooka to Geraldton track. Here he would have entered *Acacia-Banksia* low woodland[23] before reaching Chapman River estuary and the unconsolidated sand track leading south into Geraldton.

He appears to have contacted *Amangu* who inhabited this territory but he apparently experienced difficulties in communicating with members of some tribes in the Mid-West Region. Thus, he noted

> so great is the difference between the dialects even of contiguous tribes, that I have found my knowledge of the Watch-an-die dialect totally useless, as a means of communication with a tribe inhabiting a locality only fifty miles from the country of the Watch-an-dies.[24]

However, it seems his knowledge of the language enabled him to understand the language of the *Amangu*. Thus, when referring to Aboriginal flood stories, Oldfield recounted the following that was probably told to him by *Amangu* people:

> A settler on the Irwin [possibly Lockier Burges of Irwin House], in Western Australia, has been repeatedly warned by the blacks of the insecurity of the spot on which he has built his house and premises, and although they are one hundred feet above the bed of the river, and at the very least fifty feet above the height attained by the greatest inundation caused by that river hitherto experienced by the whites; yet, from the appearance of the plains in that neighbourhood, there can be no doubt but that the country for many miles around is subject to floods, and someday the value of the caution will be properly estimated.[25]

Thus, it has been suggested that the language he recorded in 1856-57 with the assistance of the *Watchandie* tribesman, Carrotty, was a mixture of a central dialect of the *Nanda* language and *Amangu*, the southern *Nanda* dialect.[26]

If Oldfield visited Geraldton on his way to the Murchison, and he very likely did to replenish his stores and possibly to catch up with Dr Foley and William Burges, he would soon have been aware of at least one important advance in Geraldton's infrastructure since his previous visit. The first town jetty, extending 244 metres due north from Gregory Street, had been completed in 1857.[27] Otherwise the settlement had not changed much; it was still the same desolate place with unpaved streets[28], few

---

[20] Michael Henry O'Connor, 2001, pp. 193–195.
[21] L. Gary Rogers, 1996, p. 48, and Geraldton Region Sheet 2, Land Resource Map no. 14/2.
[22] Michael Henry O'Connor, 2001, pp. 193–195.
[23] ibid., pp. 179–200, and Fig. 6.1, p. 157.
[24] AO, 1865b, p. 231.
[25] AO, 1865b, pp. 233.
[26] Juliette Blevins, c.2001, *Nhanda: An Aboriginal Language of Western Australia*, University of Hawai'i Press, Honolulu, p. 3.
[27] Sister Mary Albertus Bain, 1996, p. 55.
[28] By 1877, only two of the cross streets in Geraldton were made of limestone, and that for only five chains in each street. The terrace along the foreshore had a ribbon road made for 4 kilometres and the road leading out of Geraldton was not smooth enough for pleasant driving in wagonettes, buggies or spring traps, J.E. Hammond, 1958, *Western pioneers: The Battle Well Fought*, Hesperian Press, Victoria Park, W.A, p. 93.

buildings and no lock-up, and it continued to be without licensed accommodation for travellers.[29] However, he may have been pleased to explore John Bryant's new general store and find an increased range of goods available in the town. Bryant had established the shop in anticipation of trade from the settlers on Greenough flats and was successful in obtaining a gallon licence.[30]

It may not have been long before it came to Oldfield's attention that over the previous ten or twelve months there had been an escalation in conflicts in the Victoria District between the settlers and local tribesmen, mostly arising from the slaughter of stock by Aborigines and the sexual predation of tribal women by Europeans. Among these incidents, a freeman, Stephen Morgan, had been convicted of assaulting a tribal man who tried to prevent Morgan 'from seducing his woman'[31], and approximately 80 of Thomas Brown's sheep had been slaughtered by Aborigines [? *Amangu*] as they came from a corroboree.[32] Oldfield was probably particularly interested in events of this nature that occurred in the district to where he was heading. Following the alleged spearing in July 1858 by Aborigines [? *Watchandie*] of 'some of Mr. [von] Bibra's and Mr James Walcott's cattle' on the Murchison River[33], the police party sent to arrest 'the native cattle rustlers returned after a fortnight empty handed'. During the search, however, two Aborigines were shot by police who claimed self-defence.[34]

About the time Oldfield arrived at Geraldton, William Burges responded to a request from Sir George Grey, by then Governor of the Cape Colony, South Africa, for seeds of several species but the 'only ones in the list he could send were the square capsuled Eucalyptus'.[35] No other record of plant or seed collecting in the district has been found for the period after Oldfield left the Murchison in 1857 and his return in 1858.

If Oldfield perceived the possibility that regardless of his previous friendly relationships with the *Watchandie* these conflicts had caused them to be hostile to all Europeans, it may have been with some trepidation that he set out for the Murchison River. By now summer had set in; day after day of clear blue skies and relentless burning sun made access to frequent, reliable water supplies a necessity.

Geraldton to the Murchison River—1858

Map 11 p. 221 – Geraldton to the Murchison River

Since 1856 there had been no changes to the route to the Wannerenooka mine, although it seems that in April 1857, a convict road party was macadamising the road near the White Peak mine. How much of the road from Geraldton to White Peak had been improved in this way could not be ascertained but William Burges was moved to comment that the ground near White Peak was so hard 'it wasn't worth the expenditure'.[36]

When ordered to relocate to Albany, John Drummond refused to give up his mining and pastoral interests in the Champion Bay district and resigned his position as Deputy Superintendent of Police for the Northern District in May 1857.[37] Under these circumstances it would have been unlikely that he could provide Oldfield with up-to-date information on the state of relationships between the Aborigines and the settlers to the north beyond Wannerenooka. Nevertheless, Oldfield might have called in at

[29] William Burges to Colonial Secretary, 10 March 1858, CSO, inwards correspondence, vol. 413, fol. 28, microfilm, SROWA.
[30] Sister Mary Albertus Bain, 1975, p. 301.
[31] William Burges to Colonial Secretary, 29 January 1858, CSO, inwards correspondence, vol. 413, fol. 20, microfilm, SROWA.
[32] William Burges to Colonial Secretary, 3 January 1858, CSO, inwards correspondence, vol. 413, fol. 1, SROWA.
[33] William Burges to Colonial Secretary, 10 March 1858, and Charles von Bibra to William Burges, 14 July 1858, CSO, inwards correspondence, vol. 413, fol. 56, microfilm, SROWA.
[34] William Burges, Resident Magistrate to Colonial Secretary, 1 September 1858, CSO, inwards correspondence, vol. 413, fol. 78, microfilm, SROWA.
[35] William Burges to Colonial Secretary, 2 December 1858, CSO, inwards correspondence, vol. 413, fol. 119, microfilm, SROWA. This description did not permit an identification of the species since there are numerous square-fruited *Eucalyptus* species occurring in the Mid-West Region, for example, Fluted Horn Mallee, *E. stowardii* and *E. gittinsii*.
[36] William Burges to Colonial Secretary, 10 April 1857, CSO, inwards correspondence, vol. 387, fol. 202, microfilm, SROWA.
[37] A.R. Pashley, 2002, p. 60.

Redcliffe in the hope that Drummond did have some news and to catch up on Mary's botanical activities before proceeding up the road to Oakajee Well.

The well, however, could have been dry due to want of maintenance.[38] In this event perhaps Oldfield called in at the new Police Station on Table Hill to replenish his water supply from the well dug by Constable (?Samuel)[39] Farmer who was stationed there, but it may have been dry. In January 1859, Burges reported that Farmer had complained that the well should be reserved for the exclusive use of the police as the carters were draining it to such an extent he 'would not have enough for the summer'. The well was subsequently closed to the public.[40]

It is not known whether Oldfield called at William Burges' property, Knockbrack, on his way through[41], nor for certain the route he took from Bowes River to the Geraldine Mine. From previous experience he was aware of the wells along the roads running from the mine to the Geraldine via Lynton, Port Gregory and Barrel Well. It would be expected, therefore, that instead of taking the more direct track from Wannerenooka to Barrel Well, he travelled the same route to the Geraldine that he and Walcott had traversed in 1856, particularly as there appear to be no records of specimens in the Oldfield database from along the Wannerenooka to Barrel Well road. This suggests that he went via Lynton, Port Gregory and Barrel Well. Furthermore, as he was aware that at times *Watchandie* people were near Port Gregory, he may have decided to go that way with the object of prevailing upon them to let him accompany them as they moved around the country. No details have emerged as to how Oldfield coped with the journey from Wannerenooka to the Geraldine or where he joined the *Watchandie*. He apparently reached the Murchison River sometime during December. It remains a mystery as to where he camped on arrival but it was probably either at Garden Well or at the Geraldine Mine.

[38] William Burges, Resident Magistrate to Colonial Secretary, 12 December 1859, CSO, inwards correspondence, vol. 435, fols. 56 & 57, microfilm, SROWA.
[39] Rica Erickson, 1979, vol. 3, p. 263.
[40] William Burges to Colonial Secretary, 12 December 1859.
[41] See Chapter 10.

Fig. 17.0 Eucalyptus oldfieldii

Chapter 17

Western Australia 1858–1859

At Murchison River

Captain Sanford sailed to England in 1858, leaving von Bibra in charge of his Port Gregory district interests. Prior to his voyage Sanford sold the lease on Henrietta Mine on Murchison River to Habgood in 1858, and shortly afterwards, von Bibra leased the farm and Sanford's cottage to Andrew Brand, established a small mercantile business for whalers and miners at the Port[1] and became the manager of Habgood's station ('Murchison House') on the lower Murchison River. Included in the stock were several English thoroughbred horses. Subsequently, von Bibra sub-leased the property until Habgood's death and his own retirement in 1875.[2]

According to Albertus Bain, when James Walcott moved to Murchison River he joined von Bibra's cattle station (Murchison House)[3] but in August 1858 Burges referred to 'Mr Walcott's Station "Yackanew"'.[4] Three depasturing leases in the Murchison River district, totalling 26,000 acres, were approved in Walcott's name in the last half of 1858[5] but where they were located could not be determined.[6] Oldfield collected at and near Yakinu.[7] However, most Yakinu-related specimens were stated to have been from Port Gregory that is nowhere near James Walcott's property 'Yackanew' which as far as we can ascertain, was somewhere near the Murchison River. 'Yakinu' probably has nothing to do with Walcott's 'Yackanew' but is simply a habitat descriptor. Thus, Oldfield's *Watchandie* place name vocabulary includes '*Yăk-in-u'* the meaning of which is given as a spring with fire-trees[8] [*Nuytsia floribunda*].[9] Oldfield spent a considerable amount of time over the next twelve months with the *Watchandies* moving with them around their territory, collecting plants and observing their way of life. He was not, however, without the company of Europeans at Murchison River during that period. In the summer [January and/or February] of 1859 he was residing with 'gentlemen' but did not reveal who they were or the location of the property.[10] Given his prior contact with James Walcott and Charles von Bibra, it is probable he was with them on von Bibra's Murchison House Station.[11] In the vicinity there were a few other pastoral stations but it seems they were not partnership leaseholds—A.J. Ogilvie's property (Murchison location N1051) was adjacent to Murchison House to the east and

[1] Sister Mary Albertus Bain, 1996, p. 23.
[2] ibid., p. 103.
[3] Sister Mary Albertus Bain, 1975, p. 177. Bain does not provide a source reference in support of her statement regarding Walcott joining von Bibra.
[4] William Burges, Resident Magistrate to Colonial Secretary, CSO, inwards correspondence, 30 August 1858, vol. 413, fol. 79, microfilm, SROWA.
[5] Applications for Leases 1858, 1080 6,000 acres, 1081 10,000 acres, 1131 10,000 acres, Cons 5,000, Lands & Surveys RED 703, SROWA.
[6] All the digitised plans in SROWA relevant to the north of Geraldton for the 1850s and 1860s were searched.
[7] HAODb
[8] AO 1865, p. 293
[9] Oldfield referred to this species as fire tree, AO, 1865b, p. 281.
[10] AO, 1865b, p. 280.
[11] See Chapter 16.

Fig. 17.1 Southern border, Murchison House Station.

Francis Du Boulay's station was immediately to the south on the other side of the river.[12] Francis and his brother Julius were in partnership with Alex Menzies and James Walcott at 'Minnenooka' on the Greenough River 1857–1858.[13]

In supposing that Oldfield was with von Bibra and Walcott, the conditions under which he lived must have been primitive. Von Bibra's original stone and mud-brick cottage was built by him with convict labour in 1858, and convict-built shearers' quarters were constructed in 1860.[14] It is possible that, if the cottage was complete by the time Oldfield arrived, he was accommodated in it. However, it is a very small building and it may have been necessary for him to provide his own shelter nearby in a tent or, more likely, a V-hut. These structures

> were actually a 'V' reversed made by two strong poles with a fork at one end placed in the ground from 12 to 16 feet apart. The length of the poles depended on the height required for the dwelling; a long ridge pole was then placed in the forks. Other poles were placed in a sloping position against the ridge pole, with one end nailed or tied to it with rope yarn and the other end buried in the ground. Rough wattles as battens were then tied or nailed length-wise across these poles or rafters and the whole covering was thatched over with rushes or grass tops from the blackboy bush. The back of the hut was then fitted up with rushes and portion of the front also, but leaving a gap to serve as door and window, and in winter to admit the warmth of the fire blazing outside.[15]

Under these circumstances it is likely his plant storage cases, and other equipment not required while he was away collecting, were safely stored in the cottage.

---

[12] Victoria and North 1860, Cons 5018, Vict. 001, (Tally no. 506638), SROWA. Although this map is dated 1860 the leases appear to have been operating for at least three years.

[13] Sister Mary Albertus Bain, 1975, p. 177.

[14] Murchison House, undated, 'History', viewed 20 June 2013 at www.murchisonhousestation.com.au.

[15] Sister Mary Albertus Bain, 1975, p. 65. Her description was based on one provided by Lieutenant H.W. Bunbury 'writing back to England in the 1830s'—no details of the source reference were provided.

Fig. 17.2 C. von Bibra's cottage, Murchison House Station

He used a hack for at least part of the time he was in the area but it is not known whether it was his or borrowed locally.[16] Von Bibra, for example, bred horses, probably for the Indian Army and for local settlers.[17] On 8 March 1859 George Maxwell, writing from Albany to Mueller, reported:

> I am still anxiously looking out for Mr Oldfields [sic] arrival. I hope he [is] safe and reaping a rich harvest; let me know when you heard from him last, should he not arrive within a month from this date I will proceed in a boat myself to the Westward.[18]

Obviously, Maxwell was under the impression that Oldfield's stay was for a period of six months only, and it appears from this communication that there was an arrangement for Oldfield to return to Albany in March and for them to travel together to the west along the south coast. Oldfield, however, did not return to Albany after his expedition to the Murchison River district unless, of course, his ship stopped there on its way to Melbourne.

Mueller approved Oldfield extending his stay at Murchison River for a further six months, from 1 January to 30 June 1859[19], and was anxiously awaiting the collector's return in the latter half of June. Thus, in a letter to William Hooker dated 15th of that month:

> I have hope to receive an interesting supply of plants from Mr Oldfield, who went on my private expense to the Murchison, West Austr, from whence I expect him back every day. I shall thereby obtain new material for the Fragmenta, and as soon as I have Dr Hookers [sic] 500 identifications of Indo-Australian species, I shall learn

[16] AO, 1865b, p. 281.
[17] See Chapter 16.
[18] George Maxwell to FM, 8 March 1859.
[19] FM to John O'Shanassy, 10 August 1859, L59/8146, unit 747, VPRS 1189 inward registered correspondence, VA475 Chief Secretary's Department, Public Record Office, Victoria cited in R.W. Home *et al.*, (eds), 1998, vol. 1, p. 481.

> what is new of the rest of the Gregorian[20] collections & would then add the new species also for the Fragmenta, by which an important volume of Australian plants would be finished.[21]

As far as can be ascertained, there is no evidence that Oldfield sent Mueller any specimens from this expedition prior to his return to Melbourne later in the year.

It was not until 10 August 1859 that Mueller requested Chief Secretary, John O'Shanassy to permit him to

> defray out of the wages fund placed under my [Mueller's] control for the current quarter the sum of £78, being payment to Mr Oldfield, our botanical collector in Western Australia from January 1st til June 30th, 1859 at the rate of 10/- for every work day, including both remuneration for his services and travelling expenses.
>
> The reason for my delaying to apply for this sum arose from the circumstance that Mr Oldfield was meditating of entering into the services of the survey-department of Western Australia at the beginning of the year and it is only now...that he intimated the acceptance of my offer.[22]

Mueller went on to plead his case:

> I may add that the payment of 10/- a day will barely defray the outlay of the journey and that his collections from a botanically new field will be as well of great interest to Science as of value to our herbarium, whilst the seeds will be most acceptable to us for our garden and for interchange.[23]

This expenditure was approved by the Under Secretary, J. Moore, on 16 August 1859.[24] Although it cannot be confirmed, the money was probably sent to Oldfield while he was still in Western Australia, otherwise it is difficult to imagine how he coped financially since he did not return to Victoria until December 1859. While he could have lived very frugally at Murchison River, camping and largely relying on the local flora and fauna for his food, after a year on the six-month's expenses income from Mueller's private purse, an injection of funds in August or September must have caused Oldfield to remain in the field after 30 June 1859. It is not known whether he did this with Mueller's agreement.

Mueller's reference to Oldfield 'meditating of entering into the services of the survey-department of Western Australia' lends some support for prior speculation that earlier on he was employed on surveys in Victoria and Tasmania. It would seem he was offered a position but by whom and what it involved remain a mystery.[25]

In that same letter, Mueller argued the case for the 'Australian Flora' to be produced jointly by Bentham and himself:

> Probably you will say, my dear Sir William, that it is difficult to carry a work under joint authorship out at such a distance, where in the case of discrepancy of opinion the differences cannot be settled verbally. But I think it might be very well arranged, that each authors [sic] writing[s] are distinguished by his initials, so that he may be responsible for it alone. If under such arrangements the "plants of Australia" can be published, I need not say that I shall feel proud to work with one of the most learned & amiable Botanists of the age.[26]

---

[20] This is probably a reference to the collections Walcott made during the North Australian Exploring Expedition (1855–57) with Augustus Gregory.

[21] F. Mueller to W. Hooker, letter 15 June 1859, Directors' letters, vol. 75, Australian and Pacific letters 1859–65, fol. 114, RBGKA, cited in R.W. Home *et al.* (eds), 1998, vol. 1, p. 465.

[22] FM to John O'Shanassy, 10 August 1859.

[23] ibid.

[24] ibid.

[25] Relevant records of the Lands and Survey Department at the SROWA were searched unsuccessfully.

[26] F. Mueller to John O'Shanassy, 10 August 1859.

Oldfield at the Murchison River 1858-59

Map 11 p. 221 – Geraldton to the Murchison River

With a few exceptions, the impossibility of identifying which of Oldfield's specimens obtained near and between Wannerenooka, Port Gregory and the Murchison River were collected on his 1856 and 1858/1859 expeditions was discussed in Chapter 13. There are 23 specimens from his expedition to the north of Perth that are dated as collected in 1859; all but one, and possibly another, were from near Munga Gabbe and further to the north. Of these, 10 bear only the date '1859' and were obtained on Murchison River: Three Springs, Murchison River (location not found, it may be Three Springs on the Old North Road)[27]; 1.5 miles south of the River; and at 10 miles to the north of Lake Culcurdoo **and** near Munga Gabbe.[28] The remainder have a more precise date of collection. He was three kilometres south of Murchison River on 4 February, along Murchison River in March, at Port Gregory in April, at 'Oolinyurrah' a little to the north-west of Bettie Pool on Murchison River on 23 April, at Yanarie near Murchison River (location not found but it may refer to the habitat in which the plant was found since '*Yan-a-rie'* is *Watchandie* for 'a dry grassy flat')[29] on 14 May 1859 and on the Murchison in June.[30] It will become apparent in the following pages that he recorded his whereabouts at some other times during the year.

Moving around the Murchison River area with the *Watchandie*

Oldfield left no details of how long he spent travelling around the Murchison River area with the *Watchandie* or exactly where he went. While he was with them he participated in their everyday life to the extent that his collecting activities allowed. His contribution to our knowledge of the socio-cultural life of the *Watchandie* has already been referred to in Chapter 2 and is discussed in much greater detail elsewhere.[31] In this chapter we try to describe, using his own words, something of his relationships with the *Watchandie*. We do not try to provide a comprehensive account of what he discovered about their culture and society; there is far too much information in his paper. With few exceptions, Oldfield did not give the date and location of the events he described. Therefore, there may be incidents referred to in this chapter which occurred in 1856.

His paper *On the Aborigines of Australia* is testimony that he not only took careful note of what he saw but also attempted to attain understanding through questioning, taking part in the Aborigines' activities, and at times deliberately arousing their reactions to situations he purposefully instigated. He admitted that at times he had difficulty in comprehending what he was told by tribesmen:

> The natives are very apt to get...excited when striving to explain anything, more especially when endeavouring to make an European understand the point of a story, and they generally contrive to be unintelligible from the very great pains they take to be otherwise.[32]

---

[27] MEL 1611568, labelled *Eucalyptus eudesmioides* subsp. *eudesmioides,* Locn: MELISR database download, 8 March 2007, NHV, RBGM, but according to *FloraBase* currently there are no subspecies. The range of this species is from the north of Shark Bay to at least as far south as Wongan Hills and Mogumber so it is not possible to know for certain whether it came from the Murchison River area or further south at Three Springs on Old North Road. However, since it seems unlikely that he returned overland to Perth it was possibly from the Murchison River area.

[28] HAODb.

[29] AO, 1865b, p. 293.

[30] HAODb.

[31] MHH 2005, pp. iii–lviii.

[32] AO, 1865b, p. 286.

Fig. 17.3 Sandplain, south of Murchison River

Fig. 17. 4 Murchison River estuary

Nevertheless, although there are significant gaps in his understanding of the *Watchandie's* socio-religious and kinship systems, there is much accord between what he has described of their lifeways and those of traditional Aboriginal groups recorded from other parts of Australia. These were discussed in chapter 2.

His paper was based on field notes.[33] The detail with which he recorded his observations suggests that he wrote the notes very soon after the events took place, possibly while he was with the horde or shortly afterwards. This work would have been an added burden to the imperative task of pressing and labelling his botanical specimens at the end of each day to ensure that they remained in a suitable condition for scientific investigation. He must have been frustrated at times when high winds disturbed his papers, plants and seeds as he worked under the limited illumination of a lamp or the camp fire, and when the inquisitiveness and joking remarks of his *Watchandie* companions interrupted his concentration. Sadly, he left no record of his field methods and of the difficulties under which he worked.

He did not comment on how he coped with the nomadic life of the *Watchandie.* What equipment, for example, did he have with him when he accompanied them as they went hunting or changed their place of abode, travelling in 'Indian file', 'the males preceding, along well beaten tracks, leading to their place of destination' while the women acted as 'beasts of burden, carrying all the moveables belonging to themselves and lords [?]'.[34] On such occasions

> [t]he woman always carries a bag or net, which contains the goods and chattels of her husband, his arms, of course, excepted. I have examined many of these boo-tas, and in all cases their contents were nearly identical. A few packages of pigments, some flints, a piece of ty-a-lo [the gum of the Grass tree[35] *Xanthorrhoea drummondii*], sedges and kangaroo tail, sinews for sewing, with opossum teeth for needles, some scallop shells for drinking from, and a few other articles indispensable to the existence of the savage, constituted the whole of the riches of these outcasts of nature. This bag, a few sticks for digging up roots, with perhaps a child (in a bag also), make up the load that the female is obliged to carry in all the journeyings of this essentially nomadic race.[36]

Bush-hardened, Oldfield's needs for comfort, safety and collecting were probably confined to a blanket and groundsheet for sleeping, a knife, a few utensils for cooking and eating, a container of water, tomahawk, rifle, his vasculum, and a back or shoulder bag for carrying his notebook, plant press, seed papers, labels, magnifying lens, secateurs, a few items of clothing, tea, sugar and flour and other small items he considered indispensable. He might have carried all his own belongings without having to prevail upon women to assist him, particularly if he took few or no comestibles and followed the same diet as the *Watchandies,* gathering fruits and other plants as he went along and playing his part by shooting game and sharing it with the group. While camped in one place for a time, Oldfield may have been able to trust those members of the tribe remaining behind to look after the few belongings he did not need while he was out collecting.

In referring to particular instances of assistance rendered to him by individual Watchandies, Oldfield only named two of them, namely the aforementioned Carrotty, and a youth, 'Choe-do-da'. Nevertheless in his vocabulary he did place on record the names of twelve male and seven female Watchandies, one of which was Choe-do-do. It is perhaps reasonable to suggest that some, if not all, of the other names on the list were those of people comprising the horde who welcomed him into their midst and aided him in his botanical and ethnological endeavours. The males were, Ning-gar-ra,

---

33 AO, in a letter to JBH, stated that he had 'worked up his notes' on Australian Aborigines, 4 November 1863, English Letters 1859–1900, vol. 97, NAT–OXL, p. 278, RBGKA. These notes have not been found.

34 AO, 1865b, p. 289.

35 ibid., p. 268

36 ibid., pp. 267–268.

G'nak-na, G'nany-urdie, Waar-dam-ar-za, G'nan-jo, It-ter-ic, G'nong-o, Guth-thic, Œ-do, Wam-æ-tic, Chæ-do-do, Woon-dang-a. The females, Yat-pan-yu, In-die In-die, E-li-a, Y'nowl-lo, Per-rog, Mang-al-q'nar-ra and Mung-a.

Oldfield was there when the camp awoke in the morning and saw that 'the first care of the [*Watchandie* male] was to collect the fragments of the previous night's meal, which consisted chiefly of bones, that he pounded between two stones

> chewing or sucking out the marrow or other nutriment there may be in them, and then throwing the bone portions which remain to the dogs, their only share of the game they may have captured.

Having finished 'this very unsatisfactory repast', the tribesman took 'his arms and sallie[d] forth in quest of whatever may be in season, animals, birds, fish, etc.'[37] Oldfield observed that on expeditions of this nature the *Watchandie* man, was

> content with a spear [*Wee-ra*], wamra [*U-ra-ga* or throwing apparatus], and dowak [or *Chuna*, a throwing stick, about two feet long, made of heavy wood and having a flint fixed at one end], the rest of his arms and valuables being left at the head-quarters of the tribe under the especial protection of his Ba-been (friend or chosen companion), and on all occasions when obliged to lay aside his arms for ever so short a time, it is to the care of this individual that he consigns them.[38]

The naturalist, on occasions, accompanied the men when they hunted. As he walked along with them he noticed that

> every faculty is enlisted in the cause of providing for the day's subsistence; sight, hearing, and even the sense of smell, are all exerted to the utmost stretch in this important duty. Every track left by an animal is attentively studied; if it be fresh they follow it, if old, they wistfully scan the country in the direction of the track; but seeing there is no moving thing they go on, for the day is too short and their necessities too pressing to allow them to hunt on speculation only.[39]

In the Murchison River area where, according to Oldfield, the kangaroo and emu were scarce, their capture was 'generally effected by stratagem':

> Sometimes, during the dry season, when the greater part of the watering places are dried up, the natives lie in wait at springs frequented by such game, and spear them as they come to drink. At other times they stalk the kangaroo, emu [*Dromaius novaehollandiae*], and wild turkey [Australian Bustard, *Ardeotis australis*], carrying a thick screen of boughs and approaching very stealthily until within spearing distance, then dropping the screen they do their best to secure the game. The turkey is a very difficult bird to capture, being very wary, and its habit of squatting in the high grass with its long neck so bent as to have a great resemblance to a crooked stick, that even the piercing eye of the black man is often deceived, increases this difficulty; for it frequently happens that the bird is aware of the presence of its enemy before he has discovered the proximity of such a much-sought-after delicacy.[40]

Unless very successful at the outset in their search for game, the men generally travelled from 32 to 48 kilometres in the course of the day, 'never stopping for an instant, [except] to examine the whereabouts of animals'.[41] This raises the question as to how Oldfield managed to botanise and keep up with the *Watchandies* as they hunted. Perhaps his collecting was opportunistically undertaken when

---

[37] AO, 1865b, p. 270.
[38] ibid., p. 266. It should be noted that the *Watchandie* terms for the implements mentioned in this quotation, as stated in Oldfield's *Watchandie* vocabulary have been included by MHH in square brackets, ibid., p. 295.
[39] ibid., p. 270.
[40] ibid., p. 272.
[41] ibid., p. 271.

they paused to make an assessment as to the location of a mammal, reptile or bird or when they returned to camp each evening. He may at times have only stayed with them for part of the day before remaining behind to collect at his leisure.

He noted the only method that he was aware of hunting [for small animals] which involved the 'simultaneous action of men, women and children':

> Taking advantage of the direction of the wind, they fence in a portion of the leeward side of the thicket chosen for their operations. In this fence, which is neatly made of brush-wood, they here and there leave openings at which the men station themselves, armed with their spears and dowaks. All being thus prepared, the women and children set fire to the windward side of the thicket at many points simultaneously, shouting all the while to drive back the game which might otherwise escape by passing through those portions of the thicket which the fire has not yet reached. By their endeavours the whole of the windward side is soon one continuous line of fire, which being carried forward by the wind, the heat and smoke compel all the animals within their range to run to leeward, where, being arrested by the fence, they seek egress at the openings left for the purpose, and are secured by the men stationed there.[42]

He observed and described the ways which the *Watchandie* manufactured and used their tools, weapons and clothes, harvested water and their methods of fishing.[43] He also referred to the operation of a system of barter with other tribes, whereby 'they merely exchange the natural productions of one country for those of another':

> The imports of the Watchandies consist of a sort of net for fishing, shell drinking vessels, and a much-esteemed kind of flint from the North, boomerangs, shields, red ochre and wooden bowls from the South, and a kind of pipe-clay (used by women to denote mourning), from the East. In return for these articles, they supply the North men with ty-a-lo [gum of the grass tree, *Xanthorrhoea* sp.][44]...to the South men they give the beautiful rose-coloured tufts of a species of cockatoo...and flints; while to the Eastern men they send flints only.[45]

While the men were employed in hunting kangaroos and emus or fishing, or were scouring the thickets in search of eggs, the women were

> performing their allotted task of collecting whatever vegetable productions may be in season; of course, while thus occupied, they must frequently capture small animals or find birds' nests with eggs or young in them.

Oldfield described some of the plants the *Watchandie* women collected and how they prepared them for consumption. The most important being *ad-ju-co* (*Dioscorea hastifolia*, called *warrine* further south) and *u-ra* (*Typha* sp.). Other less important plants included

> the tubers of several orchids, the corms of two or three species of Hameodorum [sic.] [*Haemodorum simulans*, *H. discolor* and *H. brevisepalum*],[46] which impart their orange-red hue to the teeth and lips of those eating them; the seeds (wit-to) of various species of acacia; (This wit-to [*Acacia cyclops*][47] is very pretty, the dark black coat of

[42] ibid., p. 273.
[43] ibid., pp. 261–267.
[44] Oldfield noted here that the exchange of *ty-a-lo* with the men to the north seems to indicate that the Murchison is the northern limit of *Xanthorrhoea* in the south-west of Western Australia, ibid., p. 269.
[45] ibid., p. 269.
[46] According to S. D. Hopper, *Haemodorum simulans* and *H. discolor* are the most common *Haemodorum* species in the *Watchandie* territory and all three species after fire are in abundance, personal communication to MHH, email attachment, 23 December 2017.
[47] This is the only species that answers this seed description, ibid.

> the seed forming a fine contrast with the bright scarlet of the arillus.)[48] and, finally, the drupes of certain species of astroloma and other epicridaceous plants.[49]

In the evenings, when the *Watchandie* men returned with the spoils of their hunt, which variously included kangaroo, wild dog, reptiles, or eggs of the *ignoro-oo* (Malleefowl, *Leipoa ocellata*) and the women with the yams, seeds and fruits from their day's foraging, Oldfield watched as the food was divided according to custom 'among those...who have been concerned in its capture' and each having received his share 'cheerfully walks off to the fire to cook it':

> When feeding, the female takes her station behind and a little to the left of her lord, and her face turned nearly in the contrary direction to his. There she sits, eagerly watching from the corner of her eyes every mouthful that he takes, but never asking for anything. Now and again the man throws a morsel over shoulder, which piece of bone or gristle ... is adroitly caught and greedily devoured by the woman.[50]

If women obtained any small animals or birds' eggs during the day, 'they never appeared at the evening meal'. Oldfield inferred from this that any game the women obtained were 'eaten as soon as they [were] secured, otherwise it would be difficult to imagine how they keep up their condition':[51]

> Many times during a meal, the man responsible for killing a kangaroo or emu was pressed to narrate over again the circumstances under which his good fortune occurred, having to enter into all the minutiae of the case, the exact locality, relative position of himself and the game, and other matters vastly interesting to the huntsman. During the evening compliments [were] continually showered upon him, and these he receive[d] with great modesty.[52]

Although apparently not with the intention of supplementing the *Watchandies*' diet, he and one or more of his European companions did so on at least one occasion when

> [d]uring the summer of 1858-9 the Murchison was visited by great numbers of kites, the native country of these birds being Shark's [sic] Bay. As other birds were scarce, we shot many of these kites merely for the sake of practice, the Watchandies eagerly devouring them as fast as they were killed.

One day, continued Oldfield, this feeding frenzy was temporarily terminated when

> a man and woman, natives of Shark's [sic] Bay, came to the Murchison, and the woman, immediately recognising the birds as coming from her country, assured us that the natives there never kill them, and that there they are so tame that they will perch on the shoulders of the women and eat from their hands. On seeing one shot she wept bitterly, and not even the offer of the bird could assuage her grief; for she absolutely refused to eat it. No more of these kites were shot while she remained amongst us.[53]

For this woman the kite may have been her totem. Since the 1930s, there has been wide recognition among anthropologists that traditionally the major focus of Aboriginal religion has been totemic.[54] Aboriginal totemism is

> a view of nature and life, of the universe and man which colours and influences the Aborigines' social groups and mythologies, inspires their rituals and links them to the past. It unites them with nature's activities and species in a bond of mutual life-giving, and imparts confidence in the vicissitudes of life...It is a relationship between a per-

[48] An aril (or *arillus*) is an accessory seed-covering in certain plants, H.W. Fowler and F.G. Fowler (eds), 1970, p. 61.
[49] AO, 1865b, p. 278.
[50] ibid., p. 61.
[51] ibid., p. 277.
[52] ibid., pp. 271–272.
[53] ibid., p. 225.
[54] C.H. Berndt & R.M. Berndt, 1964, p. 189.

son or group of persons on the one hand and a natural species or several species, that is, a part or parts, of nature on the other.[55]

Oldfield learnt that cannibalism was practised occasionally, for the most part when, during droughts, the *Watchandie* suffered severely from hunger as a result of a severe shortage of game and 'vegetable substances fit for food'. He did not actually witness any such occasions but described in some detail what he had been told by tribesmen of a few instances of this nature, the victims variously including one of their own and Europeans.[56]

Now and then, during the winter when the day's hunting had been eminently successful, and 'they were gorged with their evening repast', the tribesmen diverted themselves with song and story round the fire; 'but it too frequently happen[ed], that instead of such amusement, they fain to sleep away the calls of appetite'. Oldfield stayed to listen and watch proceedings. On occasions, the *Watchandie* regaled each other with 'the most extraordinary tales' of their bravery. They told, for example, of the malign spirits of the dead, the *In-gnas,* they had seen during the day's hunt 'giving minute descriptions of them, specifying even three white hairs in the tips of their tails, and in all respects painting them in the blackest of colours'. The narrator may have told 'how he fought for hours with the spirit, until hugging it close (which mollifies its supernatural powers), he killed, and as a climax of wonder, cooked and ate it'. The audience received such tales 'in perfect faith, merely now and then giving forth a click with the tongue and teeth, a sound indicative of wonder, for they are too polite to utter a word of dissent'.[57] There were other stories Oldfield heard and recorded: 'The Story of the Stout Blackfellow'; 'The Story of the Moon and her Daughter'; and 'The Story of the Native Cat'.[58] He learnt that some favoured stories and songs had come from afar, being passed on from tribe to tribe, to the point, noted Oldfield, that the *Watchandie* had little understanding of their origin and meaning.[59]

The activities of Europeans were sometimes incorporated into their songs and dances. The *Watchandies* composed this song in recognition of Oldfield's considerable interest in the genus *Eucalyptus*:

"*Toota babeen, Yandre babeen,*
Collaille babeen, Warrigeera."

which means, according to Oldfield's *Watchandie* vocabulary:

Spearwood [probably *Eucalyptus arachnaea* subsp. *arachnaea*] friend,
Yandre[60] [probably *Eucalyptus fruitcosa*][61] friend,

River Gum [*E. camaldulensis*] friend,
Sandplain Woody Pear [*Xylomelum angustifolium*] friend.[62]

Stephen Hopper suspects that the *Watchandie's toota* (spear-wood)[63] is *E. arachnaea* subsp. *arachnaea* since it well known as a source of spearwood elsewhere in *Noongar* country.[64]The 'continual struggle for existence', stated the naturalist

[55] A.P. Elkin, 1954, *The Australian Aborigines: How to Understand Them*, Angus and Robertson, Sydney, p. 133.
[56] AO, 1865b, pp. 285–289.
[57] ibid., p. 241.
[58] ibid., pp. 259–261.
[59] ibid., p. 258.
[60] *Yandre* is not mentioned in Oldfield's *Watchandie* plant vocabulary but *Yandee* refers to *Eucalyptus fruticetorum* which name, according to APNI, is not current, being a synonym of the South Australian species, ***Eucalyptus odorata.***
[61] S. D. Hopper, personal communication to MHH, email attachment, 23 December 2017.
[62] AO, 1865b, p. 258. Oldfield noted that this song could be easily translated with the aid of the vocabulary at the end of his paper. 'Yandre' does not appear in the list so it is assumed this was a printing error and that it should have been '*Yandee*'.
[63] A. Oldfield, 1865b, p. 258.
[64] S.D. Hopper, personal communication to MHH, email attachment, 23 December 2017.

> left but little opportunity for the *Watchandies* to relax. It was only during the summer months, when food was plentiful, 'that anything like systematic amusement' such as throwing the boomerang was indulged in. Dancing was an integral part of these activities.[65]

Occasionally, at the time of the full moon, the *Watchandies* held what Oldfield referred to as 'grand corrobory [corrobboree]'. These seem to have been for amusement and were secular in nature and also at times emulated the behaviour of Europeans. He described one such event held at a place called *Y'ni-a* (*Watchandie* for 'a sandy, bushy hill'[66]) by the Murchison River, near to where the collector was staying and which he apparently witnessed. The participants of this lively performance, created by an old tribesman, were males of the horde. The dance imitated the actions of a party of whalers capturing a whale at Port Gregory.[67]

Oldfield noticed that '[f]idelity to their lords [was] not one of the prominent virtues' of the women. While 'the elders, gorged and listless after a hearty supper, [were] enjoying the songs and stories', he often heard the subdued whistle of assignation from some neighbouring thicket, and the women, making some excuse, 'strut[ted] off to their younger lovers'. This caused him to comment: 'I should rather think that the men wink at this little irregularity of conduct; for, as they were once young themselves they must know the signification of that whistle'.[68]

Oldfield described initiation practices, betrothal of girls at birth, defloration of virgins prior to marriage, marriage, mother-in-law avoidance, punishments and killings for infringements of marital and other laws and the differential treatment of children, men and women after they died. Although he went into some detail about the division of labour between men and women, referring to women as their husbands' 'slaves', he had little to say about *Watchandie* children and nothing about the respective roles of husbands and wives in the childrearing process.

The collector became very much aware that the *Watchandies* explained their cosmogony, the advent of catastrophic events, such as fire, flood, violent storms, disease, death and spiritual life after death in accordance with a system of beliefs in a constellation of supernatural beings and that these beliefs influenced much of their behaviour.

He seems to have been told some elements of an important creation myth; the full narrative may have been kept from him because it belonged to the sacred and secret life of the group. These beliefs embraced *Ca-di-ja*, the abiding place of two great divinities, *Nam-ba-jan-die*, who sprang from the earth having had no mother, and *Bad-ja-ban-die* of whom the former was probably the chief, for, said Oldfield, in speaking of them he is always mentioned first. In some of the versions the naturalist was told there appears a son of *Nam-ba-jan-die*, who, under the name of *Tar-lo Tond-da* performed many wonderful feats. Oldfield was unable to find out where *Ca-di-ja* was thought to be located. The place was, however, represented to him as idyllic, attainable only by the good, namely those members of the tribe who had been canonically buried (that is, according to traditional law). If a female went to *Ca-di-ja*, Oldfield wrote, 'it was only as a portion of the man's goods and chattels', though he had 'never heard of her presence there'. Apparently, no supplications were made to *Nam-ba-jan-die* and *Bad-ja-ban-die* nor any attempts to propitiate them, 'since they were of such an exalted nature, that the affairs of this world were beneath their consideration; they [took] no heed of the black man, either for good or evil'.[69]

---

[65] ibid., p. 256.
[66] ibid., p. 293. This place has not been found on any map or gazetteer.
[67] ibid., p. 257. '*Y'ni-a*' has not been found on any gazetteer, map or Western Australian Lands & Survey Department plans. In Oldfield's *Watchandie* vocabulary '*Y'ni-a*' refers to *A sandy, bushy hill* (p. 293).
[68] ibid., pp. 251–252.
[69] ibid., pp. 228–229.

The naturalist was witness to the *Watchandie's* dread of the evil spirit, *War-roo-goo-ra*. This spirit, represented with long horns and a tail, dwelled in the nethermost regions called *U-ta*; where they thought this was located Oldfield never discovered. Pouring sand from the hand evoked *War-roo-goo-ra*, and in revenge for being called from *U-ta* he stopped up with shells the ears and anus of the person using this invocation, thereby causing deafness and constipation, 'a fact,' said Oldfield,

> no doubt often demonstrated to the satisfaction of the blackman by cunning soothsayers producing in triumph such articles which have been extracted by sorcery from their bodies, where they have been the cause of these maladies.

When its anger was aroused, *War-roo-goo-ra* 'instigated all grand calamities which befal [sic] mankind'. Excessive and continuous unpropitious weather, thunder and lightning, floods and tempests and individual misfortunes such as deafness and constipation were among the manifestations of its wrath. During at least one violent thunderstorm Oldfield watched as the *Watchandie's* fear of *War-roo-goo-ra* overpowered their dread of the subordinate *In-gnas*, as they [*Watchandie*] sought

> shelter in the haunted caves to screen themselves from his observation. There in silent terror, they prostrate[d] themselves with their faces to the ground, waiting until the spirit, having expended his fury, [retired] to U-ta without having discovered the place of their concealment.[70]

Oldfield was once severely reprimanded by a party of *Watchandies* for speaking during a thunderstorm, 'for they declared that if I were not silent *War-roo-goo-ra* would find out our retreat, and do us all manner of mischief'.[71]

He observed that the *Watchandies* shunned as much as possible graves, thickets, most caves, and many springs because in their estimation these places were infested by malign *In-gnas* 'of diverse kinds' since one and all were apparently 'striving to do all imaginable mischief to the poor black fellow'. 'Often,' noted Oldfield,

> when travelling with natives I have taken advantage of caves to sleep in, but could never induce them to follow my example, their fear of ghosts leading them to prefer bivouacing [sic] round the fire, naked as they were, rather than seek such shelter against the cold night dews of those climates.[72]

He was told that *In-gnas* were the souls of 'departed black men who, from some cause, have not received the rites of sepulture [sic]', such as those who were slain in fight and their bodies left to rot in the sun, or to be devoured by wild dogs, while the spirits of all men not of their own tribe 'are enrolled in this ghostly army and in consequence are constrained to wander around the place of their death'. These spirits, with the exception of those haunting springs, were represented as being of human form, but having long tails and long upright ears:

> Many of the watering places were avoided on account of the *In-gnas* infesting them; but these spirits, although included under the same general name, are different in their nature and origin from the ordinary In-gnas, which never seem to frequent watery localities...Some of these aquatic spirits have the form of serpents, others of alligators, the latter perhaps a traditionary relic of that northern country from which the natives formerly migrated. Every deep, muddy pool is infested by one of these monsters, whose powers for mischief seem peculiarly active by night; for if by any chance the natives are obliged to pass by such places after night-fall, they carry bundles of blazing bark...and shout most loudly, in order to intimidate the spirits residing therein.

70 ibid., pp. 229–230.
71 ibid., p. 230.
72 ibid., p. 239.

> Some such pools they will not enter for any consideration, even in broad daylight; but such localities are always destitute of aquatic fowl.[73]

The naturalist also discovered that the *Watchandies* believed that disease and death, ill-success in hunting, loss of personal property, in fact, most of the misfortunes that can befall a man were due to the power that hostile tribes possessed over the spirits and demons that infested every corner of the land. Furthermore, that the *Watchandies* regarded the *Angaardi* tribe to the north-east of them 'as the greatest proficients in this art of enchantment', and consequently held those natives in the greatest dread. The name given by the *Watchandies* to this 'power of enchantment' was *boollia*, and Oldfield described 'various methods of eliciting this essence from the bodies of the different individuals claiming the privilege of its possession'.[74]

'There was a pool...near the Murchison, about which such terrible stories were current, that it was shunned even by Europeans'. '[One day] [i]t so happened', recounted Oldfield,

> that a bird, of a kind quite unknown to whites and blacks, was shot and fell into this pool, and to secure so valuable a specimen of Natural History, the most tempting offers were made to the latter to induce them to seek for it. In vain were flour, pipes, tobacco, sugar, clothing, and other coveted objects promised to them: they one and all refused the proffered bribes, alleging that whoever should enter the pool would sink down miles in the mud, and eventually be devoured by a monstrous snake, and thus become an In-gna. Determined not to be thwarted, two of us resolved to make the trial by ourselves, and, in spite of the warnings of the blacks, we entered the Stygian pool, and to our surprise found, that far from being unfathomable as the natives had asserted, after sinking in the mud a foot or so, we came to a firm bottom, and thus were able to wade over the whole of it. The blacks, who at first had looked on in speechless terror at what they considered our foolhardiness, seeing that no harm befel [sic] us, by degrees ventured into the water, and together we searched the whole place narrowly; but, I regret to say, without success. From this time forth that pool was entirely disenchanted, and I often saw the natives, after this occurrence, cross it by night without any torches or shouting. We used frequently to jeer the blacks on this subject, but they persisted in the assertion that it was formerly in the state they had described, but that we had made Boollia to destroy the monster and raise the bottom.[75]

According to Oldfield, when a *Watchandie* felt unwell he at once applied to that individual in whose *boollia* he had the greatest faith, and this person, in consideration of a certain reward, undertook the cure. Those who were 'the greatest in this art of charming away diseases' often produced a stone or a shell, which they asserted had been 'drawn by their enchantment from the sick man's body', where it has been the cause of his illness:

> I have seen this trick performed several times without being able to detect the imposture, though it was to that end that I witnessed it, having taken especial care that no such objects were at hand, while both patient and operator were in a state of perfect nudity.[76]

That payment for this service was expected and made, Oldfield was confident

> for having given to a native a silk pocket-handkerchief as a reward for some botanical specimens he had procured for me, seeing him without that article, I asked what he had done with it, when he told me that he had given it to another native for having made Boollia to cure him of the tooth-ache.

[73] ibid., p. 238.
[74] ibid., p. 235.
[75] ibid., pp. 238–239.
[76] ibid., p.236.

Further confirmation came when, after Oldfield had eased the same man's headache by 'performing such manipulations' as he had seen others make use of, his patient rewarded him with yams and appeared much gratified to see the healer eat them.[77] In the normal course of events *Watchandie* men adorned themselves in various ways:

> Sometimes necklaces, made of the hair of the rock kangaroo, are worn by the men, passed several times round the neck, and laid in fantastic loops on the breast. Sometimes a few feathers placed in the hair serve as an ornament, and crowns and garlands made of flowers or rushes often serve the same purpose. The women never wear any kind of adornment.[78]

Oldfield observed that in preparation for ceremonies, including funerals, the *Watchandies* went to greater trouble to 'decorate' their bodies.

Oldfield described what he referred to as the *Watchandie's* 'grand semi-religious Festival of *Caaro*', a fertility ceremony that was performed in spring and in which it seems outsiders were not involved. In connection with this event he made a statement which inexplicably contradicts his references to the sexual activities of the *Watchandies* during the normal course of events:[79]

> Like the beasts of the field, the savage has but one time for copulation in the year, a season marked out by nature, and determined by the abundance of food and the comparative ease with which it is to be procured, as well as by the genial warmth of that season.

He did not say where the event took place. He did, however, refer to 'my informant' in the context of describing this event.[80] While there is no way of knowing for certain, the detailed description which he has provided suggests he witnessed the ceremony and that the informant remained with him throughout the proceedings. In any event, it would seem that the participants accepted his interest in their affairs and trusted him not to make light of, or interfere with, what to them were important life-sustaining activities.

The preparations commenced at the time of the first new moon after the yams were ripe. The *Watchandies* laid a stock of all kinds of food 'sufficient to subsist them during the continuance of the festival'. Then

> [o]n the eve of the feast the women and children retire from the company of the men, shouting as they go, "Ow-ee, Ow-ee," and henceforth, until the conclusion of the ceremony, the men are not permitted to look on a female, but sometimes, when their store of food prove insufficient, this law is a little infringed.

The men then rubbed their bodies with a mixture of charcoal, ashes, and wallaby-fat; after which, having dug a large pit in the ground, they retired to rest, 'not, however, before they are gorged with the good things provided for the occasion'. Early next morning they re-assembled and decorated themselves with a mixture of ochre and emu-fat, 'dressing their hair with fine shavings and wearing garlands of *My-a-lie* [*Eucalyptus eudesmioides*] and *A-rum-ba* [*Acacia* sp.]'. The whole day was spent adorning their bodies and feasting frequently, but towards evening the real ceremony began:

> The men danced around the pit they had dug, shouting, singing, and some few whistling '(this they never do in their common corrobories [sic]), and thus they continue all night long, each in turn snatching a few moments for rest and gormandising'.
>
> Every figure of their dances, every gesture, the burden of all their songs, is calculated to inflame their passions. The pit is so dug and decorated with bushes as to represent

[77] ibid.
[78] ibid., 1865, p. 268.
[79] ibid., pp. 250–251.
[80] ibid., p. 231.

> the private parts of a female: as they dance they carry the spear before them to simulate priapus: every gesture is obscene.

At the conclusion of the ceremony '*Aumanno-maddijubat-wabayadia*' (copulation)[81] occurred. The participants [then] placed sticks in the ground to mark the scene of the ceremony which, henceforth became a tabooed place, 'and any looking on it, inadvertently or not, will infallibly sicken and die'. For some time after the feast the men who had participated wore shavings in their hair to distinguish them as *Caa-ro* men.[82]

In an apparent effort to elicit some understanding of how strongly a particular taboo was held, Oldfield deliberately broke one in the presence of a tribesman. When taking a short journey in company with *Choe-do-da*, Oldfield was anxious to visit the grave of a Shark Bay native, named *Bau-bing-a* who, having come to those parts for the 'purpose of merry-making', had recently died. His death was imputed 'to the enchantments of a much dreaded *Angaardi* man, named *Mubino*, and had not as yet been avenged and was not entitled to a 'canonical burial':

> In vain the youth used all his powers of persuasion to deter me from the rash act; in vain he pictured all the horrors I should view, and the evils I should incur by such reckless conduct. I was resolved to go, if only to show him the groundlessness of his fears. Finding me thus determined, as he had not the slightest intention to follow me to what he considered certain death, pointing out the direction in which the grave lay, and at the same time indicating a certain rocky point as our place of rendezvous, he departed, not, however, before he had made a final appeal to my fears; but finding that fruitless, he asked me to observe in what condition was the grave, and also whether the wild dogs had broken into it.[83]

Oldfield found the grave, one of the crudest he had 'ever yet seen in those parts: a hole had been scratched in the hard, gravelly soil, the body thrust into it, and stones and logs piled over it'. The grave was intact, except at the eastern corner, where the dingos had been scratching, but the ground was too hard to allow them to reach the body. When he re-joined *Choe-do-do*, Oldfield terrified him with an account of all the *In-gnas* he had seen; and, finally, as a climax to his horror, 'I turned towards the grave, and shouted as loudly as I could, "*Baubinga*! *Coo*!" [*Baubinga*! Hullo!]'. Whereupon the youth ran away and Oldfield did not see him again for several days:

> At our next meeting he bitterly reproached me for having uttered *that* name. "What name?" I asked. "The *Chockie*-[name for a group of the *Malgana* of Shark Bay] man's," he replied. "What *Chockie*-man's?" I demanded. "The dead man's," he answered. "What was his name?" I again asked; but "I don't know," "I forget," "He had none," and so on, were the only answers I could get from him when I pressed him on this point; nor could I induce him by any means to utter the awful sound of a dead man's name, for by so doing he would have placed himself in the power of the malign spirits.[84]

And so Oldfield invoked a reaction that revealed to him how strongly these taboos were held.

Oldfield questioned 'the wonder' with which '[t]he expertness of the natives in tracking any living thing in the bush having observed that they are 'often at fault'. Apparently in an attempt to demonstrate that they are not infallible in that regard he put one *Watchandie* man to the test with near disastrous results:

> One day I observed a native come in from the bush bearing a large bundle of yams in his hand. These he secreted under a tree, and then went towards the river to quench

[81] The full meaning of '*Aumanno-maddijubat-wabayadia*' cannot be translated using Oldfield's *Watchandie* vocabulary but he does give '*Wab-a-yad-i-a*' as the verb 'to copulate', ibid., p. 296.
[82] ibid., pp. 230–231.
[83] ibid., pp. 237-238.
[84] ibid.

> his thirst. During his absence, I removed the yams to another place, and went to the fire, over which I jumped, having taken the precaution of effacing all my tracks by means of a bough which I burned. When the man to whom the yams belonged returned, he was not long before he discovered his loss, and the swept place leading to the fire fixed his suspicion on a black standing there, but who was ignorant of what I had done. Without saying a word, he went to the place where he had left his spear, and returning with it, hurled it at the supposed thief, who, however, contrived to evade the weapon. Thereupon a wordy war ensued, to which I put an end by producing the yams.[85]

Oldfield did not relate what happened after he returned the yams but if the owner did not react with anger he must have exercised great restraint. Apparently, Oldfield thought this incident satisfactorily demonstrated the point he was making. However, to this reader, at least, it simply demonstrates that the naturalist was quite expert in hiding his tracks and because of this it was an inappropriate test for assessing the skill of the *Watchandie* man in locating animals, including humans, who made no such attempt.

With the promise of rewards, the *Watchandies* were enrolled in the business of obtaining plants. They answered his questions regarding their names for various species of plants and animals and where they were collected. He did not always receive a satisfactory answer:

> A native once brought me in some specimens of a species of eucalyptus, and being desirous of ascertaining the habit of the plant, I asked, "A tall tree?" to which his ready answer was in the affirmative. Not feeling quite satisfied, I again demanded "A low bush?" to which "Yes" was also the response. I then told the man to indicate with his hand the height attained by the plant in question; but as I had made two opposite assertions, neither of which his sense of politeness would allow him to contradict, he first held his hands a few inches from the ground, saying, "Very small", and then spreading them towards the zenith, said, "Very large", thus leaving it an open question; nor was it until I visited the locality inhabited by this species, that I ascertained its exact height.[86]

Apparently, some members of the horde were much more helpful than others and he cited the instance of a deaf and dumb boy of about nine years of age who

> evinced considerable powers of understanding. To test these powers, I have often shown him a bunch of flowers, making signs to him to fetch me similar ones, but not the same; an errand which he always successfully accomplished. In all respects, he proved of greater service to me than others in full possession of their faculties.[87]

The following account suggests that the nature of the inducements offered may have influenced to some extent the amount of collecting the *Watchandies* undertook:

> One day, as a reward for some specimens brought in by a party of natives, a mess of melons and sugar boiled together was prepared for their delectation and set before them in a large shallow tin dish. Squatting round this, each began to help himself from that portion of the vessel immediately before him, leaving by far the largest space to an old man of the company, none trenching on that portion of the dish in front of another. Towards the end of the feast, there was still a large quantity remaining in the space before the old man, yet none meddled with it, although they could not refrain from evincing their partiality for the savoury mess by rubbing their part of the dish with their fingers and then licking them. At length, even this solace was denied them,

[85] ibid., pp. 254–255.
[86] ibid., pp. 255–256.
[87] ibid., p. 224.

Fig. 17.5 From Meanarra Hill looking north-west towards the Murchison River estuary

> for the old man, who was certainly the weakest of the party and the least dextrous in the use of weapons, took the vessel in his hands and licked it quite clean. All professed themselves highly gratified with their feast, and proffered to bring in an infinity of botanical specimens on conditions of receiving another such.[88]

Oldfield described in some detail a collecting expedition that he took during the summer of 1859 to a 'a celebrated waterless thicket [*Chu-ang-a*][89] about twenty-five miles from the Murchison' and close to a

> neighbouring thicket called Whih-ta-ra-ra [perhaps more properly Whit-to-ka-ra from its abounding in that species of acacia, the seeds of which are called Whit-to by the Watch-an-dies], where there are numerous springs.[90]

Chu-ang-a, now referred to as Junga is near the Junga Dam, close to Kalbarri/Ajana Road. [91] An Ordinance Map[92] shows an extensive patch of medium woodland to the south and south-west of Junga Dam to where the northern branch of Wittecarra Gully (stream)[93] rises. This woodland probably includes the *Chu-ang-a* and the *Whit-to-ka-ra* thickets mentioned by Oldfield. With the exception of a small cleared area around the dam, it is surrounded by low scrub. Within two kilometres of Junga Dam, *Eucalyptus* species collected by Stephen Hopper and others included *E. loxophleba* subsp. *supralaevis*, *E. dolichocera*, *E. arachnaea subsp. arachnaea*, *E. rigidula*, Dongara Mallee, *E. obtusiflora* and Northern Sandplain Mallee, *E. gittinsii*. As it is not known exactly where within the area the *Chu-ang-a* thicket is located its location on Map 11 (see p. 221) is given as the same as for Junga Dam. However, the authors visited the area in 2010 and about 500 metres west of the Dam found a

[88] ibid., p. 284.
[89] ibid., pp. 280–283.
[90] ibid., p. 238. In his *Watchandie* vocabulary Oldfield gives 'Whit-ta-ka-ra' as 'a thicket with springs', ibid., p. 293.
[91] We are grateful to Dr Philip Playford for pointing out that this place is now referred to as Junga, near Junga Dam, personal communication to MHH, 4 June 2005.
[92] Joint Operations Graphic, Royal Australian Survey Corps, December 1983, Ajana, Series 1501, Sheet S.H. 50.13, 1: 250,000, for the location of Junga Dam.
[93] Wittecarra Gully is referred to as a stream in GAPNA.

stand of very straight and relatively slender trees, surrounded by dense thicket dominated by *Melaleuca* spp., which we thought that at some stage may have been suitable for spears. A photograph of them has been seen by Stephen Hopper who said:

> Yes they are probably *E. arachnaea.* I'd need to see buds and fruits to be certain. Moreover, the distinctive tree form Ian Brooker [and] I named as subspecies *arrecta.* The photo indicates a range extension further north than I was aware. The mallee *E arachnaea ssp arachnaea* is much more common in the Kalbarri NP area. Perhaps the thicket at Junga (Cheanga) was especially favoured by the Watchandie because it had ssp *arrecta* with taller straighter stems than the mallee ssp *arachnaea*?[94]

There are three specimen records which bear the locality 'Near ? Chuanga' and 'South of Chuanga' respectively.[95] One of the gentlemen with whom Oldfield was then residing volunteered to accompany him and persuaded a number of *Watchandies* to go with them

> some for the purpose of carrying a small keg of water, for none was to be found between the river and the thicket Chu-ang-a, others to act as guides into it, for there is but one place at which horses can enter, and the rest merely taking advantage of our journey thither to procure a supply of sticks fit for spears, for the production of which this place is celebrated.[96]

Whence this expedition commenced and who the gentleman was that accompanied Oldfield are matters for speculation. Since von Bibra and Walcott were still residing together at Murchison House Station in the summer (January or February) of 1859 they both may have been there at the time of the expedition to *Chu-anga-a* thicket and it was one of them who went with Oldfield, starting out from Murchison House. The question arises as to why von Bibra and Walcott were still residing together in the summer (January or February) of 1859. They may have had some joint business arrangement or perhaps it was safer for Walcott to live on an already established station until he had a residence and staff of his own.

Oldfield's account of the expedition continues:

> The Watchandies took this opportunity of going, under our protection, to a locality otherwise but little visited by them for fear of the Anguardies [*Angaardis]*, who occasionally frequent the same place and for the same purpose, i.e. to procure spears; and it is probable that these latter visit Chu-ang-a quite as seldom as the Watchandies do, for fear of coming in collision with *them.* Besides this, the thicket is in very bad repute on account of the number and malignity of the spiritual inhabitants imputed to it; so that, as a safeguard from these, our presence was further acceptable.[97]
>
> Starting in the late afternoon, no doubt to avoid the heat, it took the party some seven hours ride over a very heavy sand plain to reach the verge of the very dense thicket. By following their guides, they 'penetrated with comparative ease into the mystic interior of the thicket and camped on a small grassy plain encircled with dense scrub'. The following day when the sun was high, the party began to 'experience thirst and depression of spirits caused by the stifling heat of the place':
>
> Not a leaf was stirred in the slightest degree by a breath of wind, and the whole region seemed totally devoid of animal life, even the troublesome insects, common in other parts, were absent.[98]

---

[94] S.D. Hopper, personal communication to MHH, email 27 January 2018, HOA.
[95] HAODb.
[96] ibid., p. 238. In his *Watchandie* vocabulary Oldfield gives '*Whit-ta-ka-ra*' as 'a thicket with springs', ibid., p. 293.
[97] ibid., p. 280.
[98] ibid., p. 281.

Fig. 17.6 Black-stemmed Mallee, Chu-anga-a thicket, nr Junga Dam

Nevertheless, they managed to get through the day and the next night comfortably with the supply of water they had brought with them. Next day, however, their circumstances worsened. In Oldfield's words:

> As is usual in these latitudes, and at that season of the year, the following day was perfectly cloudless and without the slightest breeze, and as the sun rose higher and higher so did our sufferings from thirst and lassitude increase. Our horses, of course, suffered equally with ourselves, and their weakened condition from want of water and food (for they were so thirsty they refused to eat the burnt-up grass) rendered it useless for us to attempt to cross the heavy sand plain between us and the Murchison before the cool of the evening.

According to Stephen Hopper: '...this [description] is one of the unique highlights of Oldfield's ethnography. Nowhere else in my reading of literature on Noongars and related people, nor in discussions of oral history with elders, have I come across a description of how a water soak in sandplain is excavated.' At Hopper's suggestion Oldfield's full description is quoted in full for this reason.[99]

When the *Watchandies* 'one and all' insisted that no water was to be found nearer than one day one man told them that 'he knew of an old native well' some few miles off. Leaving the horses behind the party set off to find it:

---

[99] Personal communication to MHH, email attachment, 23 December 2017, HOA.

As anything was preferable to the state of listless inactivity into which we had fallen, and the trip, at worst, would serve to while away the time which seemed dreadfully long, we all resolved to go upon speculation, for the black fellow was very doubtful whether we should find water there. In due time we arrived at the well, or, rather, where the well had been; for there existed but very slight evidences that the natives at some distant period had dug there for water, but it is doubtful whether they had found any. The aspect of the place was certainly promising – the native broom (*Jacksonia scoparia*) and the fire tree (*Nuytsia floribunda*), two very good indications of the neighbourhood of water, grew there in profusion. As it was necessary to dig in order to discover whether water was to be had there or not, the natives set to work in the following manner: One stood in the centre of what had once been the well, three others grouped themselves round him, and the rest, formed into a circle, surrounded the other four. The man in the middle, stooping down, began to scoop up the sand with his hands, throwing it backwards between his legs towards those immediately near him, at each effort changing his place, so that he continually circulated round the intended excavation, the direction in which he moved being from east to west. Those outside of him acted in like manner but did not circulate so quickly, and those in the outer circle did the same, but changed their position very slowly. In this way, the faces of all being towards the centre of the well, they continued to labour until they had excavated a funnel-shaped pit twelve or fourteen feet in diameter at the top and nine or ten feet in depth. They then desisted from their work, and procuring a long stick, which they jagged at one end, they thrust it down several feet in the centre of the hole; but as they found on withdrawing it that the small quantity of sand brought up among the jags was as dry as that on the surface, they declared it useless to make any further efforts, for the water lay too deep to be attainable by digging.

They then held a long consultation as to what we should next do: some wereof opinion that we had better return to the thicket where we had left the horses,while others proposed that we should try our fortune at a place some four miles distant, of which the only one of the party acquainted with the locality, gave greater hopes than of the last place tried. On the matter being referred to our arbitration, we gave our decision in favour of the party inclined to go forward, which being acquiesced in by the rest, we again started, When we arrived at our destination, we found it nearly of the same character as the well we had just left, excepting that the native broom was absent, and the signs of former digging were more manifest. The black fellows acted in precisely the same manner as they had before. When the stick had been thrust in and withdrawn, it was found to be slightly damp, whereupon they redoubled their efforts and soon came upon sand sufficiently moist to cohere when pressed tightly in the hand. It was now no longer necessary to carry the large pit any farther down, as a much smaller hole dug in its bottom would serve our purpose. But now a difficulty arose in the minds of the natives, and one not at all appreciated by us. With *them* the question now was, in which corner of the larger pit the smaller one should be dug, so as to be sure of finding water while to *us*, not so influenced by imagination as they, it seemed perfectly immaterial, for water was evidently to be had by sinking anywhere within the above space, for the whole of it was equally moist. But the black man never trusts to chance, he must have a reason, good or bad, to guide him in every action; and, consequently, they at once proceeded to discuss this knotty point in regular form. The first proposition was to dig it on the western side of the pit; for the sea being in that direction it was probable that water would be found towards that quarter; but this plausible proposal was immediately scouted, and its author ridiculed on the ground that though water would be found in the direction suggested, yet, coming from the sea, it would be salt, and therefore unfit for drinking.

The next proposition was to the effect that it should be in the eastern part of the pit, for the reason that the Angaardies dwelling in that direction have plenty of boollia, and can make it rain whenever they please, consequently, they are never short of this element. This proposal seemed to decide the matter, for they were on the point of digging the required hole in the eastern part of the pit, when an old man expressed a fear lest these much dreaded Angaardies should turn sulky if their rights were thus infringed, and, in revenge, use their terrible powers of enchantment against the Watchandies, and thereupon this idea was at once abandoned.

One sage proposed the north-west, for all the rain came from that quarter, and this suggestion would have been adopted, had not another proposed the south, contending that the whites coming from that quarter must have found plenty of water on their journeys, consequently that desideratum was to be found in the specified direction. This compliment to ourselves carried the day; the hole was dug, the water, not of the purest description, was found, we all took our fill, and some hours after, we joyfully quitted Chu-ang-a.

When the natives wish to procure water from great depths, as from fissures of rocks, they attach a bundle of grass to a string of sufficient length, and lowering the bundle into the water, let it remain there until saturated, then, drawing it up very gently, they shake the drops from it into a large wooden bowl, repeating the operation until they have procured sufficient for a draught.[100]

Apart from accompanying the *Watchandie*s as they moved around their territory or enticing them to accompany him and others on special expeditions for the purpose of collecting, Oldfield apparently undertook quite lengthy trips on his own. To these ends, at times he questioned members of the horde about the terrain, including watering points, that he planned to traverse. This he did in 1858–1859 when he was 'desirous of making a journey to Shark's [sic] Bay'.[101] Since he was already familiar with the route between Murchison River and Lake Culcurdoo, the information he sought from the *Watchandies* probably related more to the route north of the lake.

From the foregoing, it can be deduced that Oldfield was quite relaxed with the *Watchandies* and, on the whole, they with him. This is evidenced by the generous way they responded to his questions about the flora and their way of life, allowed him to accompany them as they moved through their territory, put up with his occasional trickery and mockery and chided him when he did not behave the way they believed he should. They may have perceived him as being somewhat different from most, if not all, other Europeans with whom they had come into contact during those very early days of the colonisation of the Murchison River. Perhaps they recognised that his interest in their way of life was genuine and that what they told him and what he observed would be taken seriously. Nevertheless, the rewards he offered, the gun that he carried[102] and the 'boollia', which they thought he possessed when he demonstrated he had no fear of malign spirits and could cure a headache, may have caused them to be careful not to upset him.

It is quite common for a non-Aboriginal person who lives and works with tribal people for any period to be assigned a place within their kinship system. This facilitates the development of deeper relationships between the outsider and members of the group by imposing rights and obligations on

[100] AO, 1865b, pp. 281–283.

[101] ibid., p. 219.

[102] Oldfield noted that among Tasmanian Aborigines 'A very favourite locality for the development of this power [boollia] is that from which [they] imagined that the soldiers (by enchantment of course) drew the fire, noise, and smoke which they put into their muskets, from seeing them put their hands behind them to procure cartridges, ibid., p 235.

both parties and defining who he or she could marry.[103] Occasionally an outsider will submit to initiation. That Oldfield appears to have been unaware of the terms defining relationships within the group suggests that he was either deliberately kept outside the kinship system or the practice of assigning outsiders a place within it had not commenced in the Murchison River area.

With the promise of rewards, members of the horde with whom he was travelling obtained plants for him. Nevertheless, perhaps the greatest service the *Watchandies* rendered relating to his collecting activities, was by allowing him to accompany them as they moved around their territory; to places where he might have been loath to venture without protection from visiting *Angaardis* and *Choe-kies* and knowledge of water sources. This enabled him to extend the geographical range of his collecting and provided additional opportunities to enhance his understanding of the plants and their habitats from the Aborigines' perspective.

There are recorded instances from various parts of Australia of Aborigines joining colonists' exploring expeditions and assisting naturalists with the collection of plants[104] and of Europeans becoming members of Aboriginal tribes[105] during the 19th century. At the time, however, Oldfield appears to have been unique among botanical collectors in that he moved around a tribal territory with a horde for the specific purpose of furthering his botanical endeavours. If, as is often the case when an outsider becomes involved in the life of a tribe, the naturalist was accorded a place within their social structure (the details of which are not known), he gave no indication that this had occurred.[106] His writings suggest that he did not carry the burden of the fear and stereotype that the vast majority of European settlers apparently held of the indigenous people of the Colony. As noted in chapter 2, for them to tolerate his presence for such a lengthy period must be an indication of someone who had learnt to relate to them in a friendly, respectful and accommodating way. Although this speaks well of his interpersonal skills and possibly linguistic facility, it is clear that there were times when he tested their patience and good humour.

Murchison River to Shark Bay

Map 14 p. 329 – Murchison River to Shark Bay, Western Australia

Local legend has it that Oldfield was the first European to reach Shark Bay via an overland route from Murchison River.[107] There is evidence that he did make the journey but whether he was the first European to do so is a matter for speculation.

There is a record of a single, undated specimen from Shark Bay in the Henderson Oldfield database.[108] Other specimens are recorded as being gathered between Murchison River and the Bay in 1859—one from Lake Culcurdoo[109], and one from 16 kilometres away from Murchison River.[110] There are also four other undated records of specimens from various unspecified places between Murchison River and Shark Bay, which might have been obtained on his way to, or from, the Bay in 1859 but could have been gathered in 1856.[111] Perhaps a stronger piece of evidence comes from Oldfield's letter to Sir William Hooker in which he stated:

[103] W.E.H. Stanner, 1965, in Max Charesworth *et al.*, 1984, p. 145.
[104] R.M. Barker & W.R. Barker, 1990, in P.S. Short, 1990, pp. 37–85, and Philip A. Clarke, 2008, Chapters 1, 4– 9.
[105] See for example, R. Brough Smythe, 1878, *The Aborigines of Victoria*, vol. 1, John Ferres, Govt. Printer, Melbourne, p. 25.
[106] MHH, 2005, p. xxi.
[107] Mr Calum Carruth, Murchison House Station, personal communication to MHH, 5 May 2009.
[108] MEL 82380B, *Eremophila glabra* subsp. *albicans*, Shark Bay MELISR database download, 8 March 2007, NHV, RBGM.
[109] PERTH 1623613, *Verticordia pholidophylla,* 1859, Lake Culcurdoo, *FloraBase*.
[110] MEL 1611566, *Eucalyptus eudesmioides* subsp. *eudesmioides*, Locn: 10 miles north of Lake Culcurdoo, Date coll.: 1859 MELISR database download, 8 March 2007, NHV, RBGM.
[111] HAODb.

> Nephrodium unitum [=*Cyclosorus interruptus*][112] first makes its appearance in W. Australia in springy ground at the South Hutt, Port Gregory, and is abundant in similar localities hence as far north as Freycinet's Peninsula (Shark's [sic] Bay) to my knowledge.[113]

He also referred to the Prickly Sow Thistle (*Sonchus asper*) as being 'abundant in wet places all over the southern parts of Australia, and certainly extends as high as Shark Bay on its western shores'.[114] On 10 August 1859, in a letter to Chief Secretary, John O'Shanassy, Mueller stated that Oldfield had returned from 'the botanical exploration of the interior of Western Australia towards Sharks [sic] Bay'.[115] Presumably, since Oldfield did not arrive back in Melbourne until December 1859, Mueller was referring to his return to the Murchison River region from his Shark Bay expedition, but does not confirm that the collector reached Shark Bay. The dating of Mueller's letter to the Chief Secretary, considering the remoteness of the Murchison and mail irregularities, suggests that Oldfield undertook this expedition c. June 1859.

The 1859 specimens indicate that he travelled along the coastal route used by the *Watchandies* and the *Choe-kies* when they visited each other and by which track we believe he and Walcott went to Lake Culcurdoo. With apparently only a single plant specimen from the Bay, the possibility that it was brought from there by an Aborigine cannot be wholly discounted but we believe this to have been unlikely. Although *Watchandies* and *Choe-kies* visited each other from time to time[116] we doubt that the specimen would have remained in a condition suitable for preservation, and use for scientific purposes, after the journey of four or more days, covering approximately 154 kilometres.

Oldfield interacted periodically with *Choe-kies*. He reported the names of two members of that group without reference to their sex, namely, *Why-na, Bare-bing-a* [elsewhere *Baubinga*][117] He also obtained some information from them about their weapons, fishing techniques and the size of the roots of the yam, *Dioscorea hastifolia*, that grew in their territory, but he makes no mention of having accompanied them as they moved around their territory or that of the *Watchandies*. All his reported communications with the *Choe-kies* seem to have taken place when they came to the Murchison River for 'merrymaking' [religious ceremonies or recreational corroborees] with the *Watchandies*.[118] To that extent, therefore, it could be safely claimed Hooker's statement that Oldfield had 'lived amongst & observed' Shark Bay natives shrewdly'[119] is true, although whether he interacted with the *Choe-kies* while he was there could not be confirmed.

Was he, then, the first European to travel overland from the Murchison River to Shark Bay? Prior to 1859 numerous Europeans had visited the Bay, some of whom had botanised in the area. William Dampier described the vegetation and gathered some plant specimens on Dirk Hartog Island in 1699.[120] Naturalists with the Baudin expedition collected plants on Bernier Island during June–July 1801 and on the mainland in March 1803.[121] When the *Uranie* called at Shark Bay in 1818, one of the ship's surgeons and naturalists, together with the apothecary and botanist, Charles Gaudichaud-Beaupré, obtained many specimens over a period of two weeks. Unfortunately, all the specimens were lost when the *Uranie* was wrecked on the Falkland Islands on 15 February 1820.[122] Alan Cunningham

---

[112] ***Cyclosorus interruptus*** is a fern. It occurs further south than the Hutt River at Gingin Brook and Moore River, Alex S. George, personal communication to MHH, 7 June 2014.
[113] AO to WJH, 5 December 1863. English Letters, vol. 42, 1862–1865, L–Z, fol. 198. RBGKA.
[114] A. Oldfield, 1865a, p. 823.
[115] FM to John O'Shanassy, 10 August 1859.
[116] AO, 1865b, pp. 225 & 242.
[117] ibid., p. 293.
[118] ibid., p. 242.
[119] JDH to C.R. Darwin, 21 December 1862, Letter DCP-Lett-3856. Darwin Correspondence Project, viewed 22 January 2018 at www.darwinproject.ac.uk. The Western Australian 'Wide Bay' has not been located.
[120] Alex S. George, 1999, p. 27. In this work George describes all of Dampier's Australian specimens that have survived and which are now housed in the Fielding-Druce Herbarium at Oxford University, England.
[121] Jean Fornasiero *et al.*, 2004, pp. 55 & 258.
[122] Stephen D. Hopper, 2004, p. 139.

also collected on Dirk Hartog Island in January 1822, when he was with Lieutenant Phillip Parker King on his fourth hydrographic survey expedition (26 May 1821 to 25 April 1822).[123]

There had been overland exploring expeditions north of the Murchison River prior to 1859 but none had reached the lower shores of Shark Bay. Thus, neither the overland settlers' expedition led by Augustus Gregory in 1848[124] nor the exploring expedition led by Robert Austin in 1854[125] got to Shark Bay, although Austin went there by sea with Captain Elliot, with a detachment of the 99th Regiment, in 1850.[126] In May 1858, Frank Gregory and his party went overland to the northern end of Shark Bay and explored the mouth of the Gascoyne River but their route did not take them close to the southern end of the Bay.[127] Therefore, Oldfield may have been the first European to travel overland from the Murchison River to that part of the Bay.

Why then did his visit Shark Bay apparently result in so few specimens? It may have been that there were few plants in flower when he passed through the area, or that he was particularly selective. There is always the possibility that his specimens were lost or damaged or he was so intimidated by *Choe-kies* with whom he was not familiar that he returned quickly to *Watchandie* territory. On the other hand, the 'botanical exploration of Sharks [sic] Bay' referred to by Mueller suggests that the purpose of his trip was simply to assess the flora of the area pending a decision by Mueller to fund a lengthier collecting expedition. Mueller might have had to wait to augment his herbarium's collection from the area until he visited there in 1877.[128]

Return to Melbourne

Oldfield sailed for Adelaide from Perth in the brig *Champion* on 11 November 1859[129] but when he left Murchison River and how he returned to the Colony's capital is not known. He was not listed among the passengers of any of the vessels that sailed to Perth from Champion Bay between August and the first week of November 1859[130], which suggests that he travelled overland to the south; this cannot be confirmed from his specimen records. There is an undated collection record in the Oldfield database of a Blueberry Tree (*Myoporum insulare*) from Port Denison[131], which is about 69 kilometres south of Geraldton. While he may have deviated there from the Old North Road on his way to Geraldton, it could have been gathered when a ship on which he was travelling to Fremantle stopped at the port.

Oldfield's voyage to Adelaide took a lengthy 25 days. There is some evidence that the *Champion* called at King George Sound on the way where it may have spent some time taking on cargo.[132] Whether he and Maxwell met on this occasion is not known, nor are there any specimen records in the Oldfield database that would indicate that he collected around Albany during this stopover but he probably could not risk the temptation.

---

123 Suzanne Curry *et al.*, 2002, p. 71.

124 Augustus Charles Gregory, 1848, 'The settlers' expedition to the northward from Perth, under Mr Assistant-Surveyor A. C. Gregory, 1848', in Augustus Charles Gregory and Francis Thomas Gregory, 2002, pp. 13–30.

125 Peter J. Bridge, Kim Epton, Marion Hercock & Sheryl Milentis (eds), 2009, *'The Finest Goldfields in the World': The Austin Expedition to the Murchison in 1854*, Hesperian Press and the Western Australian Explorers' Diaries Project, Carlisle, Western Australia, p. 112.

126 ibid., p. 53.

127 Francis Thomas Gregory, 1858, 'Gascoyne River 1858', in *Augustus Charles Gregory and Francis Thomas Gregory*, 2002, pp. 37–51.

128 Alex S. George, 2009, pp. 251–259 & 495. In this work, George notes (p. 258) that Mueller explored 'N towards Shark Bay' and later (p. 495) that he went from 'Geraldton to Shark Bay'.

129 *The Perth Gazette and Independent Journal of Politics and News*, 18 November 1859, p. 2.

130 'Shipping intelligence', *The Perth Gazette and Independent Journal of Politics and News*, for the months of August, September and October 1859.

131 MEL 2260758, *Myoporum insulare*, Port Denison, MELISR database download, 8 March 2007, NHV, RBGM.

132 'Shipping News', *South Australian Advertiser*, 30 November 1859, p. 2. It was reported that the *Champion* from King George Sound was expected.

The *Champion* arrived in Adelaide on 6 December 1859[133] and Oldfield continued his journey on the screw steamer *Havilah,* which sailed for Hobsons Bay on 11 December[134], arriving there on 13 December.[135] At most he would have had three full days in Adelaide during which time he may have explored and collected in the surrounding bush, for some or all of the South Australian specimens referred to in Chapter 4 but this could not be verified.

[133] 'Shipping news', *South Australian Advertiser*, 7 December 1859, p. 2.
[134] ibid., 12 December 1859, p. 2.
[135] 'Shipping intelligence', *The Argus*, 14 December 1859, p. 4.

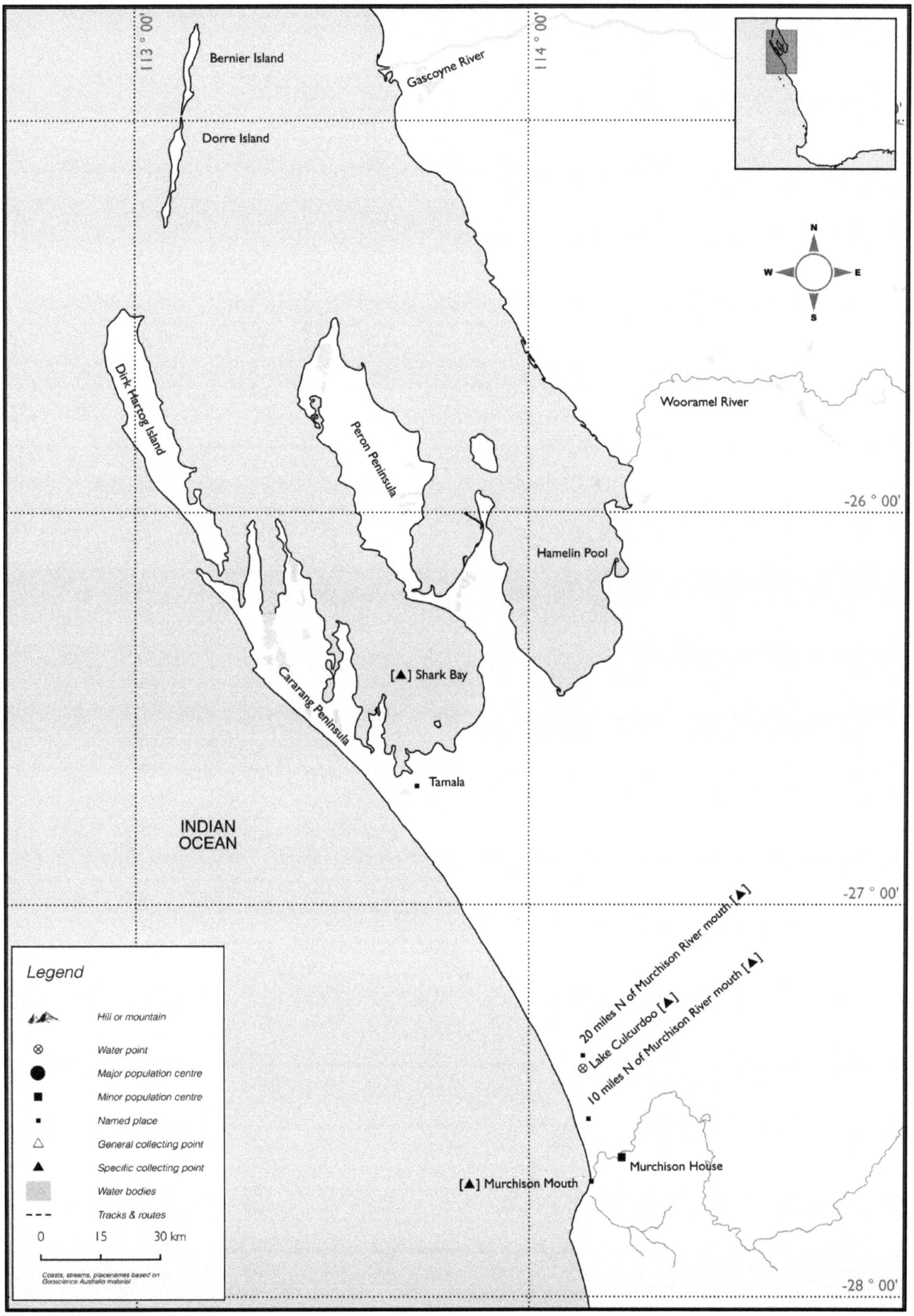

Map 14: Murchison River to Shark Bay

Fig. 18.0 Acacia oldfieldii

Chapter 18

Victoria and New South Wales 1859–1861

Although eager to examine Oldfield's plants from Western Australia, Mueller must still have been smarting from his failure to convince the Hookers, even with support from the Governor-General of the Australian Colonies, William Denison, and the Lieutenant-Governor of Victoria, Sir Henry Barkly (1815–1898), that he had the necessary skills, talents, industry and botanical resources to author the Australian Flora. Mueller fought long and hard[1] but by 15 November 1859 he had surrendered his claim for sole authorship and tried for second authorship when the Victorian Government voted £1,000 toward the cost of production. This offer was turned down.[2] Nevertheless, at Barkly's urging, he agreed to co-operate with Bentham by lending him specimens from his herbarium at the Botanic Gardens, Melbourne. Bentham later acknowledged that he could not have written *Flora Australiensis* without Mueller's help, and on the title page his 'assistance' for the communication of 'notes and description' was prominently recognised.[3] The two botanists, however, 'differed considerably in their approach to classification and engaged in vigorous debates during the preparation of the work'.[4]

Oldfield arrived back in Melbourne 11 days after the last fascicle of volume one of Mueller's *Fragmenta* (1858–59)[5] was either sent to the printers or made available to the public.[6] All ten fascicles had been published since 1 March 1858[7], three before Oldfield left in July 1858, with references to nine of his Tasmanian specimens appearing in the first two and one from the Murchison River, Western Australia in the third. How happy and excited he must have been on his return when he discovered that Mueller had cited 49 of his specimens from his 1856 Western Australian expedition, including some joint collections with Pemberton Walcott. Thirty-nine of these are types and two are syntypes of current species.[8] Oldfield's dream of finding species new to science had become a reality.

The Government Botanist was pleased with the material Oldfield obtained in 1858–1859. On 5 January 1860, Mueller wrote to the Chief Secretary (Victoria), William Nicholson:

> [W]e have received most valuable and interesting contributions towards [the new herbarium] from Mr Oldfield, who during the month returned from 18 months journey to the northern districts of Western Australia.[9]

Mueller had allowed Oldfield to retain a set of duplicates of the plants he had obtained in Western Australia in 1858–1859 but an embargo was placed on their distribution.[10] It seems, however, that

[1] Edward Kynaston, 1981, pp. 201–211.
[2] JDH to FM, 26 December 1859, RB MSS, RBGML, cited in R.W. Home *et al.*, 1998, vol. 1, pp. 506–509.
[3] GB, 1863–1878, vols. 1–7, 1967 reprint.
[4] R.W. Home *et al.*, 1998, vol. 1, p. 37.
[5] F. Mueller, 1858–1859, vol. 1, p. 208.
[6] R.W. Home *et al.*, 1998, vol. 1, p. 580.
[7] ibid., pp. 577–580.
[8] HAODb.
[9] FM to William Nicholson, extract of ms of 'The Annual Report of the Government Botanist and Director of the Botanical and Zoological Garden', 5 January 1860, 060/223, unit 748, VPRS 1189/P inward registered correspondence, VA475, Chief Secretary's Department, PROV, copy of extract from the database of The Correspondence of Ferdinand von Mueller Project, provided by Dr Monika Wells to MHH, 7 June 2001, HOA. In the published version of this report it was stated that Oldfield had been away for 17 months but according to our calculations it was very close to 18 months, Annual Report of the Government Botanist and Director of the Botanical and Zoological Garden, in Victoria – Parliamentary Papers – Votes and Proceedings of the Legislative Assembly 1859–60, vol. 4 (No. 37), 1–16, Government Printer: Melbourne.
[10] FM to GB, 5 February 1866, Kew correspondence, Australia, Mueller, 1858-70, fols. 196-201, RBGKA, cited in R.W. Home *et al.* (eds), vol. 2. 2002, p. 337.

Fig. 18.1 The Botanic Gardens [Melbourne] 1863

Mueller did not want to hold the remainder in his herbarium. The day after Oldfield returned to Victoria, the Government Botanist promised to send William Hooker a set of the specimens[11] but it is not known when they were despatched.

In April 1860, Mueller wrote to Joseph Hooker on some discoveries he had made from Oldfield's collections:

> The flora of West Austr is richer as you had the means of ascertaining. Mr Oldfield brought from thence Heterodendrum oleifolium, Nitraria Billardierii & other Eastern Australian forms & [presumably from George Maxwell][12] from E. Mt Barren & the Phillips River I have many identical Murray desert plants!'[13]

While Oldfield was in Melbourne he received the news via Mueller that William Hooker had sent him (Oldfield) a copy of Joseph's *Flora Tasmaniae* to Hobart. Regretfully it had never been received. Oldfield apparently requested Edwin to try and locate it. In response, he wrote to the Colonial Secretary, William Henty, on 9 March 1860 asking whether he had it in his power

> to give any information upon the subject, and beg to mention, as an excuse for my so doing, the facts of my brother having received, upon a previous occasion, a communication from Sir William through your office.[14]

The Colonial Secretary's Department replied within a week informing Edwin that no such publication had been received.[15]

[11] FM to WJH, 17 December 1859, Directors' letters, vol. 75, Australian and Pacific letters 1859–65, fol. 120, RBGKA, cited in R.W. Home *et al.* (eds), 1998, vol. 1, p. 505.

[12] As far as we have been able to ascertain, Oldfield did not venture that far east of Albany until 1861–62 but Maxwell had been along the south coast many times during the previous decade or so.

[13] FM to JDH, 20 April 1860, Kew Gardens correspondence, Australia, Mueller, 1858–70, un-numbered folio after f. 44, RBGKA, copy of extract of letter from the database of The Correspondence of Ferdinand von Mueller Projectprovided by Dr Monika Wells, letter to MHH 7 June 2001, HOA.

[14] E.D. Oldfield to Colonial Secretary, 9 March 1860, Colonial Secretary's Department (CSD), 1/1/92, no. 2454, TAHO.

Publication of Flora Tasmaniae—1860

Joseph Hooker, who had been appointed Assistant Director at Kew Gardens in 1855[16], had commenced periodically publishing fascicles of his *Flora Tasmaniae* in 1855, the entire work being made public in 1860 as *The Botany of The Antarctic Voyage of H.M. Discovery Ships Erebus and Terror, in the Years 1839–1843, under the command of Captain Sir James Clark Ross, Kt., RN., F.R.S. & LS., etc.,* Part 3, *Flora Tasmaniae*. This comprised Volume One: the Dicotyledons and Volume Two: Monocotyledons and Acotyledons. Mueller appears to have received the fascicles as they became available, although some went astray in 1858, causing Hooker to send his 'only remaining coloured copy'.[17] Mueller proffered some corrections[18], and 'a few remarks' on Leguminosae, 'perhaps partially useful for Dr Hookers [sic] Tasmanian Flora', and on various Tasmanian species, some of which were from Oldfield.[19] Joseph Hooker appears to have been grateful.[20]

Through his ongoing association with Mueller since 1856, Oldfield had almost certainly become aware of these fascicles and of the use of his specimens in the work. Thirty-six of his plant records appear in Volume One. The placement of all except two of these in the Additions and Corrections indicates, that the majority were late arrivals (16 of them from Mount La Perouse) that could not be incorporated in the main body of the text. In Volume Two, 287 of his specimens were cited, of which 91 per cent were cryptogams.[21] While this reflects Oldfield's particular interest in cryptogams, it may signify the relative neglect of this group by his predecessors, favouring instead the more obvious and perhaps attractive flowering plants and ferns so sought after by botanic gardens and nurseries in England.

Oldfield must have been gratified to find that so much of his material had been examined by Hooker and referred to in the work.

Oldfield returns to Hobart

Oldfield returned to Hobart on the *City of Hobart*, arriving 2 April 1860.[22] He was also there in July 1860 when he was among the many patients and friends who were signatories to a letter in *The Mercury* (Hobart) congratulating T.C. Smart, Surgeon of Macquarie Street, Hobart, 'on the result of the late action' brought against him.[23] It afforded the correspondents 'great satisfaction to offer [their] testimony to [his] unblemished honor [sic] and reputation as a Member of the Medical Profession'. Why Augustus was in Tasmania at that time is unknown.

---

[15] Colonial Secretary to E.D. Oldfield, 14 March 1860, ibid.

[16] Ray Desmond, 1995, p. 425.

[17] J. Hooker to F. Mueller, 20 December 1858, RB MSS, Royal Botanic Gardens Melbourne Library, cited in R.W. Home *et al.* (eds), 1998, vol. 1, pp. 433–444.

[18] F. Mueller to W. Hooker, 1 February 1857, Directors' letters, vol. 74, Australia letters 1851–58, fol. 157, RBGKA, and J. Hooker to F. Mueller, 20 December 1858, RB MSS, RBGML, cited in R W Home *et al.* (eds), 1998, vol. 1, pp. 263–265 & 433–444.

[19] F. Mueller to W. Hooker, 15 March 1858, and 14 June 1858, Directors' letters, vol. 74, Australia letters 1851–58, fols. 176 & 181, RBGKA, cited in ibid., pp. 373-74 & 387.

[20] See, for example, J. Hooker to F. Mueller, 3 June 1857, RB MSS, Library, RBGML, cited in R.W. Home *et al.* (eds), 1998, vol. 1, p. 295.

[21] JDH, 1860, vols. 1 & 2.

[22] Sue Towers & Margie Knight, 1993, Unassisted immigrants & coastal passengers to Hobart Tasmania 1829-1865 Index, Macbeth Genealogical Books, Hampton, Vic., Film 8, p. 86, MB2/39/25, microfiche, SLWA. There is a newspaper report that the City of Hobart arrived in Hobart on 2 April 1860. He is not listed among the passengers but he may have been among the 23 passengers travelling steerage, 'Shipping intelligence', The Hobart Town Daily Mercury, 3 April 1860, p. 2, DAN, NLA.

[23] A 'Mr Oldfield' arrived in Hobart on the steamer *Tasmania* on 25 March 1861 as a steerage passenger from Sydney, but since he was travelling from Sydney and not Melbourne it is more likely to have been another Oldfield, possibly Edwin or Alfred, *The Mercury* (Hobart), 11 July 1860, p. 3, DAN, NLA.

Fig. 18.2 Original National Herbarium of Victoria, [completed 1860]

Back in Melbourne

When Augustus returned to Melbourne is unclear, but since he was invited to join the Victorian Exploring Expedition, which was charged with investigating the country between Cooper's Creek and the Gulf of Carpentaria, he was probably there in July or early August. However, under the leadership of Robert O'Hara Burke, the expedition left Melbourne on 20 August 1860 without Augustus, because of his 'unwillingness to travel in company'. According to Edwin, 'With the exception of a month's exploration of Mount Perouse [sic] in Tasmania, in company with one person [Charles Stuart], all his journeys were performed alone'.[24] It seems Edwin was unaware of the time Augustus spent in Western Australia with Pemberton Walcott.

As a 'result of constant recommendations' on the part of Mueller, a herbarium was built on the Domain, a short distance from the western entrance to the Gardens. It was completed early in 1860 but it was necessary to allow the building to dry thoroughly before occupying it, and hence the transfer of specimens did not commence until late in 1860, and was completed in mid-1861.[25]

Some Melbourne colleagues

Although it seems that Oldfield and Mueller did not converse much, there is evidence that Oldfield had more than a passing acquaintance with some of Mueller's other collectors.

In his major work, *Critical Revision of the Genus Eucalyptus*, Joseph Maiden stated that (William) Lockhart Morton, pastoralist, explorer and inventor, 'was a friend of Oldfield'[26] but left no details regarding the circumstances of their relationship. Morton travelled extensively throughout Victoria, New South Wales and Queensland, examining the agricultural and pastoral potential of the areas he visited. He had migrated to Port Phillip, arriving in August 1841, and between 1845 and 1855 was involved in pastoral stations, firstly on the Morton Plains (near Birchip) and later, near Kyneton on the Plains of Thalia and the Salisbury Plains in 1853. He abandoned these properties in 1855 and moved to Melbourne. In 1859, he explored country north of Rockhampton, Queensland and in June

[24] Anon., probably Edwin Oldfield, 'An Australian Naturalist' [obituary], 16 July 1887, p. 3.
[25] R.T.M. Pescott, 1982, pp. 54-55.
[26] J.H. Maiden, 1909–1933, vol. 3, pt. CXLIII, p. 173.

and July 1861 made an extensive tour of the Wimmera and Mallee regions in Victoria.[27] From what we have ascertained of the timing of their respective movements around Victoria and New South Wales, the only time their paths may have crossed outside Melbourne was at, or near, Kyneton if, as has been suggested, Oldfield returned from the Ovens to Melbourne via Bendigo in 1852 or 1853.[28] Morton gave Mueller the plants he collected on his expeditions[29] and these included a specimen of *Eucalyptus ovata* from Twofold Bay, which was later sent to Maiden.[30] Oldfield also collected at Twofold Bay from 15 June 1861 (see below) but we do not know exactly how long he remained there, although he was in Hobart on 25 August 1861.[31] Since Morton was much further to the west in the Wimmera and Mallee region in June and July 1861, it is highly unlikely that they met at or near the Bay.[32] It seems possible that sometime, while Oldfield was in Melbourne between 1858 and 1861, he and Morton interacted, possibly at Mueller's residence. If so, the question arises as to whether they met for the first time then or simply renewed their friendship. Clearly both would have shared the common interests of botany, bushmanship and travel.

Among the specimens attributed to Oldfield at Kew Gardens and Oxford University, are six from the Hastings and MacLeay Rivers north of Sydney.[33] One of the specimens, a tufted perennial grass (*Paspalum scrobiculatum*) from Hastings River, was labelled as collected by Dr Hermann Beckler, the labels of the others were annotated 'Hb. Oldfield' but no collector was cited.[34] These may have also been collected by Beckler. Beckler worked for a time for Mueller as his herbarium assistant, and from September 1859 to mid-1860 as a collector around the Hastings, MacLeay, Richmond and Clarence Rivers in northern New South Wales. Then, on 20 August 1860, he left Melbourne as the doctor and botanist with the Burke and Wills Victorian Exploring Expedition.[35] Beckler was so dismayed by Burke's erratic behaviour, which led to the resignation of George Landells, that he too offered to resign. Although accepted, he agreed to remain with the men and supplies at Menindee. Burke, William John Wills, John King and Charles Gray reached the Gulf of Carpentaria, but Gray died on the way back to Cooper's Creek where Burke and Wills also died, leaving King the sole survivor.[36]

Oldfield no doubt became acquainted with Beckler at Mueller's residence when Beckler returned from northern New South Wales in mid-1860, and before he went on the expedition with Burke and Wills. It is unlikely that the two naturalists came into contact after that, since Beckler was still at Menindee on 13 November 1862[37] by which time Oldfield was in England.[38] After testifying at the Royal Commission of Inquiry into the sufferings and deaths of Burke and Wills, and the 'causes of the lamentable result of the expedition'[39], Beckler returned to Germany and spent the rest of his life as a village doctor.[40]

---

[27] J. Anne Hone, 1974, 'Morton, William Lockhart (1820–1898)', ADBO, viewed 13 May 2010 at www.adb.online.anu.edu.au, and J.H. Willis & Helen M. Cohn, (1993), 'Morton, William Lockhart (1820–1898)', *Flora of Victoria*, Vol. 1, Chap 5, 'Botanical Exploration of Victoria', sourced from ANBG, viewed 13 May 2010 at www.anbg.gov.au.

[28] See Chapter 9.

[29] J. Anne Hone, 1974.

[30] J.H. Maiden, 1909–1933, vol. 3, pt. CXLIII, digital p. 173.

[31] A. Oldfield to J. Hooker, 25 August 1861, Supplementary Foreign Letters 1865–1900, vol. 218, fol. 238, RBGKA.

[32] J.H. Willis & Helen M. Cohn, 1974.

[33] K000644226, *Xanthorrhoea macronema,* Hastings River, H. Beckler, OXF unnumbered, *Syzygium smithii, Leptospermum laevigatum,* Hastings River, Oxford University; K unnumbered, *Cordyline stricta* and K unnumbered, *Paspalum scrobiculatum*; K unnumbered, *Orites excelsa,* Macleay River, Kew Gardens, HAODb.

[34] Juliet Wege, Australian Botanical Liaison Officer, Kew Gardens, personal communication to MHH, email 30 November 2005. HOA.

[35] F. Mueller to G. Bentham, 14 September 1862, Australia, Mueller, 1858–1870, fols. 72–9, RBGKA, cited in R.W. Home *et al.* (eds), 1998, vol. 2, pp. 162–162, and 'Introduction' (ibid.), p. 12 and A.E. Orchard, 'Beckler, Hermann (1828–1914)', *Biography*, ANBG, viewed 13 May 2010 at www.anbg.gov.au

[36] Alan Moorehead, 1963, *Cooper's Creek*, Hamish Hamilton, London, pp. 81–99 and 113–123.

[37] ibid., p. 185.

[38] F. Mueller to W. Hooker, 24 May 1862, Directors' letters, vol. 75, Australian and Pacific letters 1859–1865, fol. 161, RBGKA, copy from the database of The Correspondence of Ferdinand von Mueller Project, copy provided by Dr Monika Wells, Department of History & Philosophy of Science, University of Melbourne to HH, 7 June 2001, HOA.

[39] Alan Moorehead, 1963, p. 160.

[40] A.E. Orchard, 1999, 'Beckler, Hermann (1828–1914)'.

While he was illustrating Mueller's *The Plants Indigenous to the Colony of Victoria* (1860–1865), Ludwig Becker, scientist, artist, botanical illustrator and lithographer, probably became acquainted with Oldfield in Mueller's residence prior to August 1860. Becker illustrated some early parts of Mueller's *Fragmenta.* He became a council member of the Victorian Society of Fine Arts in 1856, and in 1859, of the Philosophical Institute of Victoria, to which he contributed many scientific papers. Becker joined the Burke and Wills Exploring Expedition as artist, naturalist and geologist. Despite his efforts at Menindee to inform the Expedition's Committee in Melbourne that the exploration could be a failure, his warnings were ignored. Becker became lame, and although he had stated in his application when joining the expedition, that he was accustomed to bush life and its hardships, he could not withstand the exhaustive effects of scurvy and dysentery. He died and was buried at Bulloo, 13 kilometres south of Cooper Creek on 28 April 1861.[41]

There are also records of two specimens from the Gulf of Carpentaria noted as being in Oldfield's herbarium but collected by Diedrich Henne.[42] On the recommendation of Ferdinand Mueller, Henne, as botanical collector, accompanied the Queensland Government's Relief Expedition in search of Burke, Wills, King and Gray. Henne was in the party led by William Landsborough. At the time Henne joined the Expedition he was an assistant to Mueller in the herbarium at the Botanic Gardens, Melbourne. Landsborough and his party left Brisbane on 24 August 1861 in the SS *Firefly* for the Gulf of Carpentaria but she ran aground on the Great Barrier Reef. She was re-floated and was taken in tow to the Gulf by the *Victoria.* Henne spent much of his time in the Gulf at the base camp on Sweers Island. He left there on the *Victoria* on 12 February 1862 and arrived at Williamstown on 30 March 1862, which was about the time that Oldfield returned from his last visit to Western Australia. Henne brought back six cases of botanical specimens for Mueller's herbarium. Oldfield and Henne probably got to know each other during 1860–1861, while they were both working in Mueller's herbarium with the *Paspalum* given to Oldfield after he returned from Western Australia early in 1862.[43]

Changes in the family's circumstances 1859–1861

During the first half of 1859, Frederick, Mary and Marian Oldfield left 29 Bloomsbury Square, London and relocated to 1 Dunsford Villas, a stand-alone, two-storey, modest dwelling in a middle-class housing development on Merton Road, in the then semi-rural town of Wandsworth, London. The extension of railways and the establishment of stations sparked the construction of middle-class villas and respectable terraces in suburbs, which were 'delightful retreats for the busy man'—Wandsworth and further on Richmond, Wimbledon and Kew among those desirable locations.[44] No doubt the move was made in the hope that Frederick's health would benefit from the cleaner air of his new surroundings. He was suffering from advanced pulmonary consumption[45], almost certainly caused by a bacterial micro-organism, the tubercle bacillus or *Mycobacterium tuberculosis*, which was a leading cause of death in 19th century England.[46]

Augustus may have had letters from Mary, Edwin and Alfred while he was in Western Australia bearing news of his father's deteriorating condition and of the family's change of residence. On his return to Melbourne he probably received news of the frustration and anger that brother Vincent had

[41] Marjorie J. Tipping, 1969, 'Becker, Ludwig (c.1808–1861)', ADBO, viewed 13 May 2010 at www.adb.online.anu.edu.au, and A.S. George, 2009, 'Becker, Ludwig (1808–1861)', p.287.

[42] K000659084, *Drosera petiolaris* and K unnumbered, *Paspalum scrobiculatum*, Gulf of Carpentaria, D. Henne, Kew Gardens.

[43] Burke & Wills Web, 'Diedrich Henne (1825– 1913)', viewed 26 June 2010 at www.burkeandwills.net.au. A letter from J.H. Maiden (*The Sydney Morning 3Herald,* 7 February 1913, p.5) briefly outlining Henne's botanical career is quoted at this URL.

[44] Roy Porter, 1994, pp. 274-285

[45] Record for Frederick Oldfield, No. 209, 18 November 1860, Copy of An Entry of Death, Registration District: Wandsworth, 23 April 2009, DYC298203, General Register England. HOA.

[46] Mary Wilson Carpenter, 2009, *Health, Medicine & Society in Victorian England*, Greenwood Publishing Group, pp. 60–61, viewed 28 April 2010 at www.books.google.com.au.

caused his father after they vacated their home in Bloomsbury Square. Frederick was apparently too ill and geographically distant to directly manage the settling of his accounts and the disposal of the contents of the house. He left these tasks to Vincent who, in his father's view, did not carry them out conscientiously.

Vincent was the recipient of a series of letters from his father with quite specific instructions, commencing sometime in the first half of 1859 and continuing over succeeding months.[47] Vincent's tardiness in responding to these requests drew forth admonishments from Frederick, one of which was particularly vitriolic:

> Again, what is the reason my requests are not attended to?... Has Mr Gronow paid you? You have not condescended to let me know what any of the things fetched, what was sold and what was bought in. I consider your silence on many subjects not only unjustifiable, but as treating me as an imbecile, or a person unworthy of your esteem, not to say affection. What reason or motive you have to usurp such a position I cannot conceive.[48]

Frederick's concern is understandable since the catalogue of the private disposal sale at 29 Bloomsbury Square shows that the items on offer were of great value and indicative of him being well-to-do and of the large size of the house he had vacated. There is also evidence in the catalogue that whilst engaged as a reversionary agent he was still actively involved in the wine trade. Listed were 'upwards of 400 Dozens of Matchless Port, Sherry, and other Superior Wines, seldom to be met with'. Among the other items for sale were 55 items of furniture and floor coverings, 26 lots of 'plate and plated' articles, and sundry other items, a collection of Nicholas and Gaspar Poussin engravings, and 81 paintings stated to have been 'collected at a great expense by the present proprietor'. This implies that they were purchased and not obtained to satisfy gambling debts. Among the paintings were works by Titian (1), Rubens (3), Canaletto (4), Correggio (1), Breughel (1), Rembrandt (2), Gainsborough (1), Raffaelle (1) and van Eyk (1).[49] By today's standards they would have been worth a fortune but to what extent Frederick profited from their sale remains a mystery.

Available records do not reveal the reason for Vincent's behaviour regarding these matters. He had married Sarah Jane Friend on 25 February 1857, and at the time of the birth of their first child, Jonathan Howard on 25 January 1858, they were living at 27 Everett Street, Bloomsbury. Not long after this event, Frederica wrote lovingly to both Vincent and Jane [the name (or 'Janey') she was referred to by her husband and the rest of the family]:

> I was glad to learn from Marian that you are well and that baby is so much stronger. I hope, if it be for the good, that he will be spared to you...Janey must not think me unmindful of her because I have not written for really my time is so taken up & when I have not lessons I feel so tired and averse to writing—it makes me feel more poorly.[50]

Vincent may already have been conducting his business as a wine merchant at 22 Change [Exchange] Alley, London.[51] Although his home was close to Bloomsbury Square his shop and his family could have left him little time for other business activities. However, extrapolating from subsequent events, to be briefly referred to as they unfurl, he was a loving father but unreliable as a provider and as a source of support to his wife. It seems that early in their marriage Vincent was not conscientious

---

47 Frederick Oldfield to Vincent Oldfield, 11 July 1859, files 0127–31; fragment, no date, file 0306; 29 August 1859, file 0332; file 0332, 29 August 1859, PFA.

48 Frederick Oldfield to Vincent Oldfield, 14 September 1859, file 0314, PFA.

49 Catalogue of Frederick Oldfield's 'Private Sale of Superior Wines, a Valuable Collection of Paintings by the most Eminent Masters, A Few choice Prints, Excellent Modern Furniture, together with a Small Service of Plate, &c., &c., and Other Useful Articles, 29 Bloomsbury Square, WC1.', c. June 1859, PFA.

50 Frederica Oldfield to Vincent and Jane Oldfield, 5 April ?1858, Files 0309–0310, PFA.

51 *Post Office London Trades Directory for 1861*, English Census Directories Project (ECDP), G Kingsley & Nick Vine Hall, Melbourne, microform, NLA.

about paying his creditors.[52] Perhaps, then, it is not surprising that Vincent could not be depended upon to do his father's bidding.

Frederick died on 18 November 1860 at 1 Dunsford Villas in the presence of Juliana[53], and almost certainly Mary and Marian were nearby. He was buried six days later, on 24 November at the Brompton Cemetery, Earl's Court, following a church service (possibly Anglican) conducted by Reverend A. Badger.[54] Frederick's grave is marked by a broken marble column, which can signify the loss of the head of the household, a life cut short or ending before its prime.[55] Whoever ordered the headstone left no explanation as to what, in Frederick's case, its message was intended to convey. He certainly had been the head of the household but it would be difficult to assert, particularly in those days, that at 74 years of age his life had ended before its prime.

Only three of the siblings were living in London: Juliana[56] and Marian were at 1 Dunsford Villas and Vincent was either residing or operating a business from 5 Lambs Conduit Place (now Guildford Place) near the Foundling Hospital, Bloomsbury.[57] No details have emerged concerning Juliana's occupation at the time.

By 1860, Frederick Jnr and Mary Jane had moved from New York to San Francisco and were occupying their residence with their 21-year-old married daughter, Lavinia Gage, and her two-year-old son, Californian-born Frederick. The name, whereabouts and occupation of Lavinia's husband are not known. He may have enlisted in one of the armies fighting in the American Civil War (1861–1865). Frederick Henry Howard's vocation was given as 'Professor of Music' in the Census record but it appears that it was not a well-remunerated profession; his accumulated assets at the time were valued at US$200.[58]

During the first quarter of 1856, Frederica's husband, Amos Hold, died.[59] The couple were childless and after Amos' death Frederica remained in Yorkshire. She was still there in 1861, living with her mother-in-law, Mary Hold, and her sister-in-law, Hannah Hold, in the small village of Ardsley, a few miles from Barnsley, and, was earning her living as a 'Professor of Music'.[60] Depending upon whether the sad, but perhaps not unexpected, news of their father's death was brought to Australia by steam or sailing ship, the earliest it could have reached Augustus, Edwin and Alfred was about the middle of February 1861.[61] If Frederica was able to travel down from Yorkshire in time, the siblings would have been represented at their father's funeral by only half their number.

Signed by Frederick as 'F.H. Oldfield' on 11 July 1860 together with a codicil, dated 25 October 1860, his will made provision for Mary and all his children except Vincent. The historical record is silent on the reason for this omission. It may have been out of pique since Frederick's complaints about Vincent's inaction seem to have commenced by the 29 August 1859.[62] On the other hand, another scenario is that Vincent had already received his share to enable him to finance the establishment of his wine business. Apart from leaving sufficient household goods for the 'absolute use and

---

[52] For example, I. Geldard, Surgeon, to V.H. Oldfield, 1859, invoice for medical services rendered, File 0133, PFA, and H.E. Blaxall, Blaxall Family Grocery and Italian Warehouse at 33 Lamb's Conduit Street, London, to ?V.H. Oldfield, File 0318, 12 April 1861, PFA.

[53] Certified copy of an Entry of Death for Frederick Oldfield, no. DXZ901359, given at the General Register Office, 2 October 2001, Registration District: Wandsworth, in the Sub-district of Wandsworth in the County of Surrey, no. 228 for 1860, HOA.

[54] Copy of Register of Burials in the West of London and Westminster Cemetery, Earl's Court, Old Brompton, no 25489, 24 November 1860, provided by Laurence Epstein, Cemetery Supervisor, Brompton Cemetery, letter to Helen Henderson, 23 October 2002, HOA.

[55] The broken column was a common symbol used in gravestones from the latter part of the 19th century. These columns were not broken after they were created, but were deliberately carved to appear broken. Broken columns were often used as monuments or memorials, or as the main stone in a family plot, rather than as the headstone of an individual. Richard Stockton, College of New Jersey, undated, 'Art and Architecture of New Jersey', viewed 17 April 2010 at www.ettc.net.

[56] Certified copy of an Entry of Death for Frederick Oldfield, no. DXZ901359, 2 October 2001, no. 228 for 1860, HOA.

[57] Envelope addressed to 'Mr Oldfield, 5, Lambs Conduit Place, Foundling', post marked '7 Mr [March] 1860', PFA.

[58] Record for F.H. Oldfield, 1860, United States Federal Census, viewed 11 April 2010 at www.ancestry.com.

[59] Entry of Death for Amos Hold, 1856, Death, Burial, Cemetery & Obituaries, England, Wales Free BMD Death Index 1837–1915, viewed 10 April 2010 at www.ancestry.com.

[60] Record for Frederica Hold, *1861 England Census*, Household schedule no. 19, folio 28, p. 4, Town: Ardsley, County: Yorkshire, Registration District: Barnsley, Source Information: RG9/3443, viewed 9 November 2005 at www.ancestry.com.

[61] Robin Haines, 2003, p. 270.

[62] Frederic Oldfield to Vincent Oldfield, 29 August 1859, File 0322, PFA.

benefit' of his 'dear wife', he directed that the proceeds of the sale of other household furniture, wine, plate, paintings, prints, watches, clocks, jewellery and whatever [other items] he possessed at the time of his decease, his freehold property in Doncaster (or lease), various other properties and effects together with the cash at his on bank, debts due to him when collected, and the proceeds of his reversions as they fell due, be bequeathed in the following manner: In lieu of 'the share and shares in the general funds' left to Marian, Frederica, Frederick, Edwin and Juliana, Mary was to receive an annuity of £70 'from the day of [her husband's] decease during her natural life payable quarterly'. Augustus and Alfred each were also to receive life-long annuities of £35 and £25 respectively to commence immediately after Frederick's death 'thinking by this to put them in a better position, they having only to trust to the world'. The total value of Frederick's estate available to his beneficiaries was approximately £5,000.[63] Juliana, Frederick Jnr, should he be in England, and Mr Harry Fell, the Junior of Montpelier [indecipherable], Brompton, were appointed co-executors of the will and Juliana, the Treasurer of the general fund.[64]

Although Vincent was not named in the will, the lawyer handling the sale of one of Frederick's properties, 'the Victoria', insisted on receiving Vincent's full name, address and occupation along with similar details for all the other siblings.[65] No record has been found of this sale or whether Vincent gained financially from it. At the time, Frederica reported to Vincent that she had told the lawyer that 'you were engaged at Covent Garden Theatre – so he put you down so! I now remember Wm G. told me he thought it was the Royal [Royal Opera House]'.[66]

For those days Frederick's estate may have been quite substantial for a middle-class family but any attempt to calculate its worth in terms of today's pound sterling is too problematic to attempt.[67] Perhaps, however, it did not reflect the degree of wealth implied by the catalogue of the sale of Frederick's artworks, household goods and wine stocks conducted the previous year and of the other assets claimed by him in his will. Subsequent correspondence suggests that the amount received by Marian from her father's estate made for a very frugal lifestyle, requiring her to live with one or other of her siblings. Nevertheless, Frederick's sons and daughters were probably grateful to receive their legacies, particularly the brothers in Australia and America.

Edwin was the most prosperous and his school was thriving. Four years previously (1856) he had expanded his Commercial Academy to accommodate boarders and more day pupils by adding to his premises in Brisbane Street the adjoining house in Argyle Street.[68] He was also conducting evening tuition.[69] Nevertheless, his family was growing and no doubt placing an increasing drain on his and Susanna's financial resources, which the money from his father could have alleviated somewhat. By 1860, he and Susanna had four surviving children[70]; they were to have four more.[71]

[63] Board of Inland Revenue and predecessors: Estate Duty Office and predecessors: Registers of Legacy Duty, Succession Duty and Estate Duty, Frederick Oldfield, IR 26,1860, N–R, fols 1366, NAL. The estate was valued for probate at under £3,000 but this did not include Mary's annuity (if this was for a life expectancy of 20 years the total value would have been £1,400) nor does it tally with the total amount (£3,633) that the probate document indicates was available to the other beneficiaries.

[64] Will and testament of Frederick Howard Oldfield, 11 July 1860, and Codicil, 25 October 1860, copy obtained, 28 January 2002, from the Postal Searches and Copies Department, Probate Registry, Court Service, York, England, HOA.

[65] ? Frederica Hold to Vincent Oldfield, date indecipherable, letter fragment, Files 0225–0226, PFA. The contents of this letter indicate very clearly that Mary Oldfield was no longer living in Wandsworth and that it was written before probate was declared on Frederick's will. Mary went to live with Frederica in Doncaster soon after Augustus returned to England in early 1862 and probate was not finalised until September 1862, Board of Inland Revenue and predecessors: Estate Duty Office and predecessors: Registers of Legacy Duty, Succession Duty and Estate Duty, Frederick Oldfield, IR 26,1860, N–R, fols 1366, NAL.

[66] Frederica Oldfield to Vincent Oldfield, date indecipherable c.1861–1862, Files 0225–0226, PFA.

[67] George P. Landow and Jim Skipper, undated, 'Wages, the Cost of Living, Contemporary Equivalents to Victorian Money', Wages, the Cost of Living, Contemporary Equivalents to Victorian Money, viewed 17 April 2010 at www.victorianweb.org.

[68] Advertisement, *Hobarton Mercury*, 7 January 1856, p. 4, DAN, NLA.

[69] Advertisement, *Hobart Town Daily Mercury*, 3 May 1858, p. 1, DAN, NLA.

[70] Mary Howard, Edwin Lenthall, Edith Juliana Susan and Frederick Augustus, *Colonial Tasmanian Family Links database*, TAHO, viewed 15 July 2013 at www.archives.tas.gov.au.

[71] Grace Henrica, Bernam Howard and Violet Louisa, ibid. Their first-born, Douglas Allan, was born on 6 June 1852 and died in infancy, 16 August 1852 and their eighth child, Rose Adele, was born in 1868 and died the following year, ibid.

Edwin was making his presence felt through the press. In August 1859, he wrote to the Editor of the *Mercury* (Hobart) protesting that competition for the recently introduced National Schools Scholarships was open only to pupils of the National [government] Schools. This complaint was on the grounds of unfairness to pupils in private educational establishments and to those of poor parents. In the first place, he argued, it provided opportunities for many wealthy students in government schools to be candidates for scholarship that were denied to those from private schools. Secondly, he considered that competition between children of well-to-do parents in government schools and those of poor families put the latter at an unfair disadvantage. Thirdly, National Schools were designed for the needy and parents of students who could afford the fees should pay them, 'thereby relieving the government of a part of the expense incident to the maintenance of the School'.[72] This letter drew a long repost in *The Hobart Town Mercury* from 'C' who seems to have largely misunderstood Edwin's argument, wrongly believing that he was inferring that the National Schools were inferior to private schools and that they should not be open to all those who choose to attend.[73] Reading between the lines, Edwin's letter gives the impression that he, as a private provider of education, did not believe he was operating on a 'level playing field' insofar as the competition from government for well-to-do students was concerned, which would be alleviated by National Schools charging fees to those who could afford them.

About 17 months later, Edwin wrote to the Editor of *The Mercury* in response to a review of a recently published school text, *Ireland's Geography of Oceania,* which recommended the adoption of the work 'in unqualified terms'. He claimed the text was so full of factual errors as to 'destroy, in [his] mind at least, any general confidence in its statements, manifesting, as many of them do, an unpardonable amount of carelessness, if not ignorance'. He considered the publication, from which he cited over 30 errors

> not only useless, but a dangerous guide to the learner; as well as an intolerable bore to the Teacher, who must be ever upon his guard, lest ideas at variance with truth, and contrary to scientific arrangement, be imbibed by his Pupil.[74]

Alfred was appointed, on 1 May 1860, school master at the re-opened government-funded Browns River (Kingston) school. He was classified as a Teacher of the Third Class[75] with a salary of £50 per annum and an expected additional sum of £15 8s. derived from parents' fees. Special allowances provided to some teachers for accommodation and supervising pupil teachers were not made available to him.[76] Thirty-six pupils were enrolled for the first year. By the next year there were 71 pupils scattered through grades 1 to 5. An inspector's report was not, however, positive:

> [I]nstruction is elementary. The children are mostly dull and backward. Sixteen learn writing, 12 the first four rules of arithmetic, 4 the higher rules, 21 geography, 15 grammar and scriptural history, 13 sewing.[77]

His apparently poor showing as a teacher may have resulted in dismissal. He left Browns River sometime during 1861 and was replaced by Henry (?Hatherley) Tucker.[78] Alfred was probably in

---

[72] 'Original correspondence', *The Hobart Town Daily Mercury*, 20 August 1859, p. 2, DAN, NLA.

[73] 'Original correspondence', *The Hobart Town Daily Mercury*, 24 August 1859, p. 3, DAN, NLA.

[74] E.D. Oldfield, *The Mercury*, 18 February 1861, p. 3, DAN, NLA.

[75] Teachers of the Third Class were those employed in schools averaging an attendance of 20–35 scholars, Southern Board of Education, Report for the Year 1859, Board of Education Reports 1842–85, Tasmanian Department of Education, pp. 5–7, EDTA.

[76] Southern Board of Education Report for 1860, Tasmania, Report to the Legislative Council, Appendix E, Nominal Return of teachers in the Employ of the Southern Board of Education on 31st December 1860, pp. 20–21, EDTA. Julie Gardam referred to a Mr Harry Oldfield as being the first teacher, Julie Gardam, 1988, p. 16, but this does not accord with official records.

[77] Julie Gardam, 1984, pp.6–7.

[78] Fifth Report of the Southern Board of Education, Tasmania. Report to H.E. the Governor-in-Chief of the Colony of Tasmanian and its Dependencies, Appendix E, Nominal Return of Teachers in the Employ of the Southern Board of Education on 31st December 1861, pp. 16–17, EDTA.

Kingston when he learnt of his father's death but was aboard the *Derwent Hunter* bound for Sydney on 5 August 1861[79]; his reason for making the journey is not known.

Frederica, in the letter written to Vincent concerning the sale of 'the Victoria' (prior to probate being effected on Frederick's will in 1862), expressed concern that, due to the deeds having to be signed by all the siblings, there would be 'delay enough'. There was already a buyer and apparently 'Alfred's business' being already 'a year or 2 behind owing to the I R [?Inland Revenue]...& no funds to meet it' rendered the sale imperative so that Alfred's share of the proceeds could be used to offset the taxes he owed. [80] No record has been found of the nature and location of this business but the interest of the Inland Revenue indicates that it was in England.

If the news of Frederick's death reached Augustus in February or March 1861, he was still in Melbourne and unemployed. At the same time, Augustus and Alfred may have learnt the details of their respective legacies but, despite Frederick's direction that the annuities were to commence immediately after his death, the brothers would wait until about December 1862 for their first payments, which had been effected in England on 3 September 1862[81]; a long time to wait when experiencing straitened circumstances. The value of Alfred's legacy was about £10 per annum less than those of Augustus and his other siblings. This suggests the possibility that an angry reaction from Alfred was the root cause of the 'dreadful scene' that had 'shaken [Augustus'] nerves'.

Collecting wood samples for the 1862 International Exhibition, London

Map 3 p. 94 – Sydney to Melbourne

A 'Mr Oldfield' arrived in Hobart on the steamer *Tasmania* on 25 March 1861 as a steerage passenger from Sydney.[82] On 15 June 1861 an 'Oldfield' left Hobart for Twofold Bay travelling steerage, aboard the *Tasmania*.[83] This passenger was almost certainly Augustus because he stated that in 1861 he had been employed collecting specimens of Australian timbers for the 1862 London International Exhibition and referred to being thus engaged in the vicinity of Twofold Bay[84], a locality on the south coast of New South Wales approximately 48 kilometres north of the New South Wales–Victoria border. If so, what had he been doing in New South Wales earlier in the year?[85] Perhaps he was there in response to one of the advertisements placed in local newspapers by the Commissioners for the International Exhibition (New South Wales Branch), offering to purchase specimens of the Colony's indigenous woods[86] or an advertised teaching or survey position. Most likely, however, Mueller was contacted, by the Commissioners, for advice on a suitable collector and he recommended Oldfield. No official records have emerged, which would lend support to any of these explanations.[87]

Whatever the circumstances, the naturalist must have been pleased to be out in the bush again, being paid for what he like doing best and having the opportunity to contribute to Australia's displays at such a great exhibition. Mueller was at Twofold Bay in September of the previous year[88] so perhaps, impressed by the variety of the timber species occurring in the area, he influenced Oldfield to go there.

---

[79] 'Cleared out', 'Shipping', *The Mercury*, 6 August 1861, p. 2, DAN, NLA.
[80] Frederica Hold to Vincent Oldfield, date not decipherable, c.1861–1862, Files 0225–0226, PFA.
[81] Board of Inland Revenue and predecessors: Estate Duty Office and predecessors: Registers of Legacy Duty, Succession Duty and Estate Duty, Frederick Oldfield, IR 26,1860, N–R, fols 1366, NAL.
[82] Sue Towers & Margie Knight, 1993.
[83] 'Shipping: Port of Hobart Town', *The Mercury*, 6 June 1861, p. 2, DAN, NLA.
[84] AO, 1865b, p. 263.
[85] He could not have taken his passage from Melbourne since there is no evidence in *The Argus* shipping reports for March 1861 that the *Tasmania* called at Hobsons Bay.
[86] No such advertisement has been located but others offering to purchase manufactured goods and artworks have been found. See for example, *The Sydney Morning Herald*, 9 May 1861, p. 2, DAN, NLA.
[87] The 1862 London International Exhibition records held at the Public Records Office of New South Wales (The Rocks Branch) were-searched unsuccessfully for evidence of his employment by the Commission and of his wood samples having been sent to England.
[88] Alex S. George, 2009, p. 256

Thirty-six countries from Europe and Asia and other parts of the world, including each of the Australian Colonies, contributed to the 1862 International Exhibition, also known as the Great London Exposition, which was held in a building that occupied 26.1 acres (6.5 hectares) on the southern part of the Royal Horticultural Society Gardens, London. Initiated by the Royal Society of Arts, Manufactures and Trade, and organised by a Royal Commission from 1861, the erection of the huge iron and glass edifice, which featured two huge domes, was largely financed on a surety of £250,000 obtained by public donation. The exhibition featured over 28,000 exhibitors embracing economically useful natural products, manufactured goods, industrial equipment and the arts. It attracted 6.1 million visitors yielding receipts (£459,632) which were slightly above cost (£458,842), leaving a total profit of £790. [89]

Following the Exhibition 'a large collection' of specimens of Australian woods were donated to the Museum of Economic Botany, Kew Gardens. Of these, the 'Geography' of 63 are given in the Kew Economic Botany Collection[90] database as 'New South Wales' only. Furthermore, although, with some exceptions, species' names are recorded, none has the name of the collector included in its provenance.[91] A search of the specimen location maps associated with *PlantNET: New South Wales Flora Online* (SYD) and *Australian Plant Name Index* (CANB) found that the range of species of 21 of the samples included the Twofold Bay area.[92] Therefore, while it seems likely that Oldfield's specimens are among them, it is not possible to identify which are his.

Sixty-four samples of woods from 'New South Wales south' were sent from the 1862 Exhibition to Captain F. Fowke for testing their strengths and properties[93] at the South Kensington Museum (now the Natural History Museum) in London. The museum retained these. Fowke's publication[94] has been sighted and although the test samples may have included those collected by Oldfield, attempting to link the test species with species in our database was too problematic. Thus, individual species were, in the main, identified by their common names which vary over time and from place to place.

The samples in the Kew Economic Botany Collections, which are unequivocally from New South Wales, are of varying size. Those that are classified as being in the largest category, of which there are 44, are about 1.8 metres in length.[95] The task of obtaining such big pieces of timber must have required Oldfield to engage the assistance of another person. Thus, even if he used an axe to fell the tree, the excision of the sample would have involved the use of a cross-cut saw to ensure pieces of timber of regular size to facilitate their transport and display. Then there was the challenge of conveying them from the forests to the port of Eden for shipping to Sydney; for this purpose, a horse and spring cart were probably employed.

Whaling ships had operated in Twofold Bay since 1791 and shore whaling commenced there in 1828.[96] '[G]reat numbers of runaways [convicts]...[succeeded] in making their escape from the Colony' via the many vessels visiting there. Boyd Town, the original settlement in the Bay, was founded by Benjamin Boyd in 1843 to service his 14 pastoral properties on the Monaro plains, and as a base from

---

[89] Adrian Duncan, 'The 1862 London Exhibition', viewed 4 May 2010 at www.mcgee-flutes.com.

[90] For a brief overview of the history of the Kew Economic Botany Collection see Alex S. George, 2010, 'Australian type material in the Economic Botany Collection, Royal Botanic Garden Kew', *Muelleria*, vol. 28 (2), pp. 163–171.

[91] Donations from the 1862 exhibition of plant specimens from New South Wales Court, Museum Entry Book, 5 February 1863, p. 60, kindly provided by Dr Mark Nesbitt, Jodrell Laboratory, Kew Gardens (email to MHH, 28 April 2010, HOA). The Economic Botany Collection database records include 218 Australian woods received from the 1862 exhibition. These are arranged by genus and species but collector's name and only the Colony of origin are noted in the records.

[92] PlantNET, viewed 30 April 2010 at www.plantnet.rbgsyd.nsw.gov.au.

[93] 'The colonial woods in the International Exhibition', *The Sydney Morning Herald*, 2 November 1862, p. 3, DAN, NLA.

[94] F. Fowke, 1867, *Tables of the Results of a Series of Experiments on the Strength of British Colonial and Other Woods Exhibited at the International Exhibition, 1862; Made at the South Kensington Museum with His Report on Similar Experiments in 1855*, Science and Art Department of the Committee of Council on Education, Her Majesty's Stationery Office, London, 1867. See for example, Table 1, pp. 1–11, viewed 13 August 2014 at *www.google.com.au*.

[95] The Economic Botany Collection database records, RBGKA, printout kindly provided by Dr Mark Nesbitt, Jodrell Laboratory, Kew Gardens, email to MHH, 28 April 2010.

[96] H.P. Wellings, 1996, *Eden and Twofold Bay: Discovery, Early History and Points of Interest 1797–1965*, Eden Killer Whale Museum, Eden, New South Wales, p. 5.

Fig. 18.3 Twofold Bay, New South Wales 1874

which to conduct his shipping and whaling operations. Boyd erected a wool store, hotel, houses, stores, salting and boiling-down works, jetty, a whale spotting tower at the settlement, and a whaling station at East Boyd. When his banking interests failed and he lost a long and costly court case in 1846, he left the Colony for the Californian goldfields.[97] Today, all that is left of Boyd Town are the remains of his whaling station and spotting tower.

The town of Eden, situated across the Bay and north of Boyd Town, was surveyed in 1842 and became a port for the export of wool, beef, tallow, extract of wattle bark, neatsfoot oil, and other agricultural products from the hinterland to Sydney, Hobart, and Port Albert. By the time Oldfield visited, it was well established with a customs house, police station and lockup, post office and government school. The population of the town had grown considerably when it became the port for the Kiandra goldfields; gold was first discovered there in 1859 and was followed by a rush over the next twelve months.98

Apart from any wood samples Oldfield may have collected, we have located 25 herbarium and published records of his specimens from Eden and Twofold Bay.[99] These embrace 19 species, including 14 *Eucalyptus* species. Eleven were of tree species, which grow to heights of 30 metres or more, four less than 30 metres and four were shrubs or bushes. However, only three of these species are represented in the wood samples at the Museum of Economic Botany: Blackbutt (*Eucalyptus pilularis*), Woollybutt (*E. longifolia*) and Stringy Bark (*E. obliqua*). He may have collected some of the herbarium specimens to enable the positive identification of the wood samples he obtained; if so, it is puzzling that there are not more of the larger species represented among the Museum's (Kew Gardens) wood samples.

[97] G. P. Walsh, 1966, 'Benjamin Boyd (1801–1851)', ADBO, viewed 5 May 2010 at www.adb.anu.edu.au.
[98] H.P. Wellings, 1996, pp. 6, 15–18 & 22.
[99] HAODb.

Fig. 18.4 Nature reserve, south of Eden

The distribution of the species represented in our Oldfield database suggests that he confined his collecting to within a radius of under eight kilometres.[100] He gathered specimens of species that occurred, according to *PlantNET*, on the coast and adjacent ranges, including headland heath[101], warm rainforest gullies along the coastline[102], open grassy or sclerophyll woodland[103], and dry sclerophyll forest or woodland.[104]

Oldfield was apparently always on the lookout for evidence of Aboriginal uses for the plants in the areas he visited. Thus, for example, in the context of discussing 'the vast expenditure of time and labour [involved in] the construction' of a kind of Aboriginal 'shield about three feet in length and six inches broad, with a handle behind, the whole cut out of solid wood', Oldfield drew upon an observation he had made near Twofold Bay. There he discovered 'a tree of the kind called "stringy bark" (Eucalyptus obliquus L'Her) [now referred to as *E. obliqua*]' from which, at first sight, he thought he might cut a sample

[100] Craig Dickman, New South Wales Department of Environment and Wildlife Service, Eden, personal communicationto MHH, email, 19 May 2010, HOA.

[101] For example: Unnumbered OXF, *Banksia spinulosa* var. *spinulosa*, Twofold Bay, HAODb, MEL 2153026, *Melaleuca armillaris* subsp. *armillaris*, Twofold Bay, MELISR database download, 8 March 2007, NHV, RBGM.

[102] For example: MEL 2227346, *Elaeocarpus reticulatus*, Eden, MEL 1609136; *Eucalyptus cypellocarpa*, Twofold Bay; MEL 703540, *E. pilularis*, Twofold Bay; MEL55048 *Pomaderris aspera*, Twofold Bay; MEL 61014 *Syzygium smithii*, Twofold Bay, ibid.; NSW328644, *E. muelleriana*; NSW 336014, *E. sieberi*, Twofold Bay, (NHNSW); and *E. obliqua*, Twofold Bay, J.H. Maiden, 1909, vol. 1, p.t 2, p. 46.

[103] For example: *E. bridgesiana*, Twofold Bay, GB 1866, vol. 3, p. 243; *E. goniocalyx*, Twofold Bay, ibid., p. 229.

[104] For example: MEL 230966, *E. longifolia*, MELISR database download, 8 March 2007, NHV, RBGM.

> but on going to the opposite side of the tree a large excavation in its trunk at once convinced me to the contrary. The cavity in question was ten or twelve feet from the ground, was about five feet long, three feet broad, and about the same in depth, its length not being perpendicular to, but making an angle of about sixty degrees with the horizon, thus following the direction of the grain of the wood. In this cavity was something I could not make out, so piling up logs, etc., I climbed up to it and then discovered that the object which had puzzled me was the rude outline of a native shield. The workman [probably a *Yuin* tribesman][105] had completed his labour so far, that, by the removal of a few more inches of the solid wood, the implement could have been removed from the tree, and the work of finishing and ornamenting it performed at his leisure; but he had abandoned it at this point, probably on account of some imperfection discovered in it, or, quite as likely, he may have fallen a victim to the spear of an enemy ere he could complete the arm that might have saved his life. Some idea of the amount of labour expended on the fabrication of the shields of the New Hollanders may be formed from the consideration that, in digging out the one in question, after allowing liberally for the increase of the tree during the many years that must have elapsed since the native abandoned his work (the genus Eucalyptus grows rapidly only when young...), not less than thirty cubic feet of timber had to be removed, absolutely scraped away, the only tool used being a flint fixed on a stick. The excavation had been made so deep in order to cut the shield from the old wood, and that many years had elapsed I infer from the whole appearance of the affair, and also from the fact that it is many years since the blacks in those parts were under the necessity of working with such an imperfect implement as was evidently the case here, for the wood was absolutely scraped away.[106]

Did he not take the wood sample from the tree out of respect for the tribesman and his labours or was the excision of the wood for the shield cut too great to enable a suitable wood sample to be taken? The historical record does not relate.

Captain Fowke tested 12 specimens of wood from the Hunter River[107] and there are many species represented in the samples of wood in the Museum of Economic Botany, whose ranges, according to *PlantNET*, seem to include the Hunter River area. At some stage Oldfield collected near that river but he left no indication of when he was there. Apart from one specimen of the Grey Myrtle or Ironwood (*Backhousia myrtifolia*), the remaining nine are either shrubs or herbs.[108] Nevertheless, it cannot be ruled out that he obtained Hunter River wood samples tested by Fowke.

There is no record detailing the delivery arrangements for the samples of wood destined for the Exhibition. A chest of apparel from Twofold Bay addressed to 'Oldfield' was among the imports at the Port of Hobart on 5 August 1861[109], which suggests that he returned to Tasmania.

Another trip to Western Australia planned

During 1861 and 'entirely at [his] own risk', Oldfield planned a 'botanical trip to some parts of Western Australia not previously visited by any collector' to obtain plants and seeds. Mueller advanced £50 towards his expenses and agreed to buy specimens to that amount. Professor Frederick McCoy, Professor of Natural Science at the University of Melbourne, Government Palaeontologist and

[105] The *Yuin* are the traditional owners of land from the Shoalhaven to the New South Wales/Victorian border, Anon., undated, 'Saltwater people of the coast', *Koori Coast, Living Knowledge,* viewed 24 March 2014 at www.livingknowledge.anu.edu.au.

[106] AO, 1965b, pp. 263–64.

[107] F. Fowke, 1867, see for example, Table 1, pp. 1–11.

[108] HAODb.

[109] 'Shipping', *The Mercury*, 5 August 1861, p. 2, DAN, NLA.

Director of the (government) Museum and Applied Science, undertook to purchase £25 'worth of objects of Natural History generally' obtained on the expedition to add to the collections that were held at the University.[110] In an attempt to obtain further financial support, Oldfield wrote to Joseph Hooker from Hobart on 25 August 1861, asking whether he would purchase a set of any plants and seeds he collected since 'I am forced to depart from my general rule and seek, by the sale of specimens, some return for the great expense I shall be under in this undertaking'. He requested Hooker to write to him in Perth if he could assist him but to exactly where he should address the reply, the collector did not say. Possibly as an inducement to Hooker, Oldfield noted in the same letter:

> By the next vessel (about 'Xmas) I forward my own Herbarium to London, and should you think of supplementing the 'Flora of Tasmania' it will be of some service to you, especially in Cryptogamiae, and will be at your service.[111]

Apart from any ideas he may have entertained regarding encouraging Hooker to purchase specimens obtained in Western Australia, it seems that at that stage Oldfield was planning to return to England; otherwise why would he have parted with his precious herbarium?

A month later, 24 September 1861, Mueller wrote to Joseph Hooker in which he said: 'Mr Oldfield is gone to Cape Lewin [sic]', the junction of the south and west coasts in Western Australia.[112] Mueller's statement suggests that Oldfield returned to Melbourne before embarking on his journey west. Three cases of books and apparel addressed to 'Oldfield' were sent to Melbourne from Hobart on the *City of Hobart* on 4 September 1861.[113] Although we believe Alfred to have been in New South Wales at the time, he cannot be discounted as the addressee. It seems likely, however, that Augustus had either packed all his belongings for storage in Melbourne, in readiness for his departure to England immediately after his return from Western Australia, or the cases contained clothes and books he wished to take on the expedition, but these seem inappropriate to include in his luggage for such an expedition. Exactly when and on what vessel he went to Melbourne could not be determined.

Oldfield had timed his visit to Western Australia to enable him to take advantage over the next three months of a later flowering period in the colder climate of the extreme south-west and south coast of the Colony.

---

[110] AO to JDH, 25 August 1861, Supplementary Foreign Letters, 1865–1900, vol. 218, fol. 238, RBGKA.
[111] ibid.
[112] FM to JDH, 14 September 1861, Australia, Mueller, 1858–1870, unnumbered letter after fol. 49, RBGKA, copy from the database of The Correspondence of Ferdinand von Mueller Project provided by Dr Monika Wells University of Melbourne to email to MHH, 7 June 2001, HOA.
[113] 'Shipping. Port of Hobart Town', *The Mercury*, 5 September 1861, p. 2, DAN, NLA.

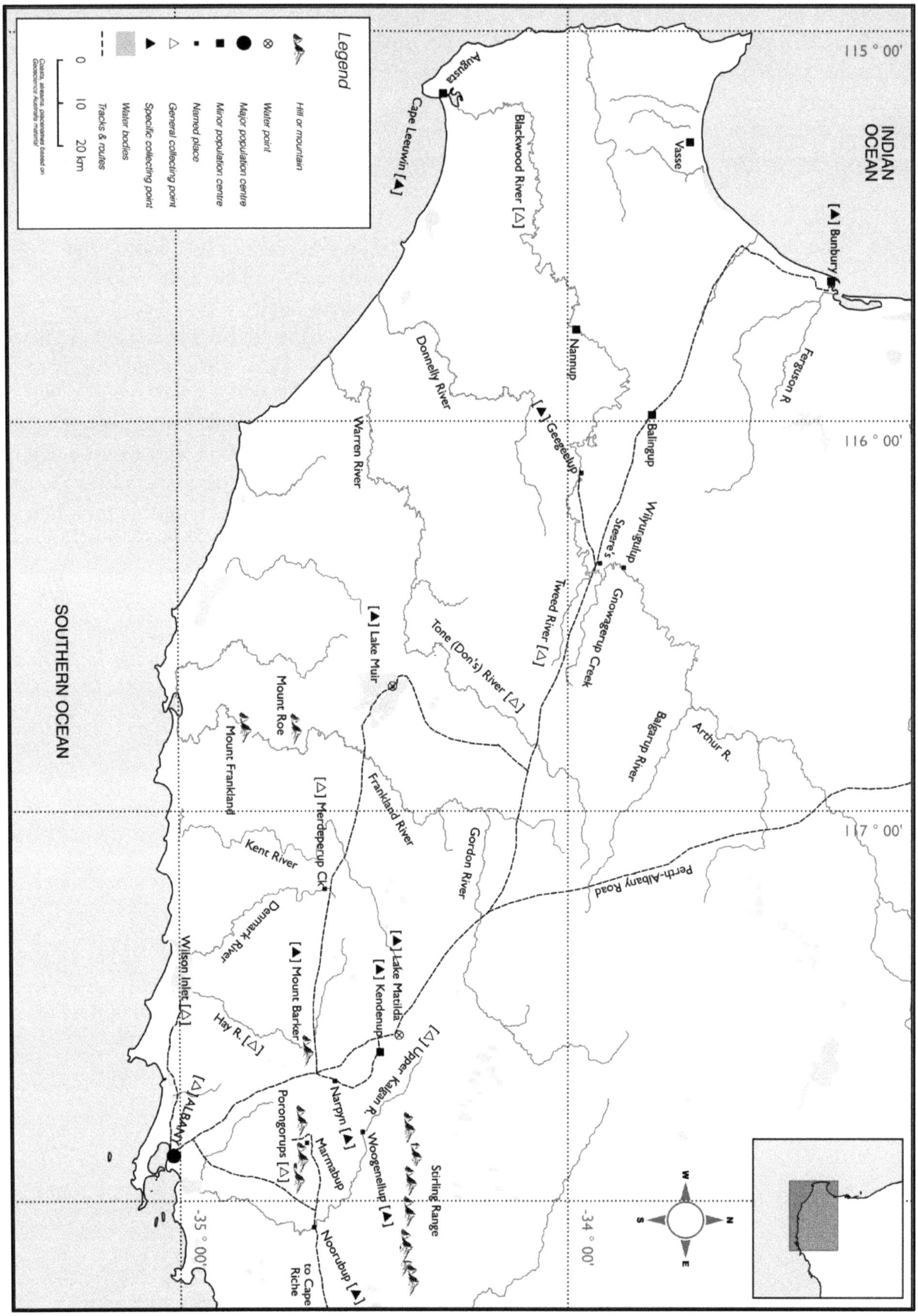

Map 15: Albany to Upper Kalgan and Cape Leeuwin

Chapter 19

Western Australia 1861–1862

King George Sound to Upper Kalgan and Cape Leeuwin

Fig. 19.0 Senecio oldfieldii

A passenger 'Oldfield' travelled steerage from Melbourne to Albany on the *Balclutha*1, which left Hobsons Bay with the 'homeward' mails on 23 September 1861.[2] There is little doubt that this was Augustus since the ship sailed the day before Mueller wrote to Joseph Hooker telling him that 'Mr Oldfield is gone to Cape Lewin [sic]'.[3] He probably disembarked during the first week of October 1861.[4] Journalists on board may have enlivened the journey. They travelled on mail boats to Albany to obtain the latest news from Western Australia and from vessels bringing newspapers from Europe. During the voyage to Adelaide they would condense the news into 10,000 words ready to transmit by telegraph to Melbourne and Sydney. At Adelaide, a race would ensue as mounted couriers rushed to the telegraph office in an endeavour to enable their respective newspapers to be the first to put ground-breaking news on the streets.[5]

From the deck, as the ship steamed into Hanover Bay, Albany, Oldfield may have noticed some changes since he was there in 1858. A few more cottages and larger houses dotted the rises east and west of York Street and there were some additional commercial buildings along Stirling Terrace, York Street and the water front.[6] Casting his eye along the foreshore he probably observed that the P&O Depot's floating infrastructure had expanded. In the early 1860s another hulk, *Kingfisher,* was acquired to supplement the company's coal storage space on the *Larkins.* Two iron lighters, imported in 1859 to carry coal between the ships and the storage places, could also be seen either moored out from the Depot or busy at work in the Bay.[7]

There was still no deepwater jetty or wharf and it must soon have become apparent to Oldfield that once more his luggage had to be lightered to the jetty at the foot of Osnaburg Street.[8] The roads leading up to Stirling Terrace from the jetty remained sandy tracks, but the walk along the Terrace would have been much easier. During 1861, the Terrace was the first to be properly laid and gravelled; the other roadways of the town were awaiting such treatment and were very difficult to traverse, particularly after rain.[9]

---

[1] 'Shipping Intelligence', *South Australian Adverttiser*, 15 September 1861, p. 2, DAN, NLA.

[2] 'Shipping Intelligence Hobson's Bay', *The Argus*, 24 September 1861, p. 4, DAN, NLA.

[3] FM to JDH, 24 September 1861, Australia, Mueller, 1858–1870, unnumbered letter after fol. 49, RBGKA. Copy of letter provided by Dr. Monika Wells, Ferdinand von Mueller Correspondence Project, personal communication to MHH, 7 June 2001, HOA.

[4] According to J.L. Poore, the journey in 1861 from Hobsons Bay to King George Sound aboard *Balclutha* took nine days, J.L. Poore, letter to the Editor, *The Perth Gazette and Independent Journal of Politics and News*, 6 September 1861, pp. 3–4, DAN, NLA.

[5] Donald S. Garden, 1977, p. 157.

[6] A pictorial image for 1861 to compare with those in 1858 was not found.

[7] Donald S. Garden, 1977, p. 142.

[8] Adam Wolfe, 1994, *The Albany Maritime Heritage Survey 1627–1994*, prepared for The Heritage Council of Western Australia, The Albany Port Authority and The Albany Maritime Heritage Association, mimeo., pp. 122–123, JSLB. In 1864 a new and larger jetty was completed just below Spencer Street, ibid., p. 99–100.

[9] Donald S. Garden, 1977, p. 130.

Fig. 19.1 Albany, King George Sound, 1863

Whilst looking for accommodation, Oldfield discovered that the town was not as quiet as it had been when he was there in 1858. The population had expanded somewhat because of the increased demand for labour by P&O at its coaling depot with the injection of more ticket-of-leave men. While the streets were busier, there was a downside. There was a severe and growing imbalance of males to females in the town, which had resulted in an escalation of sexually-related improprieties and criminal offences, and public drunkenness and fighting were common.[10]

In the coming days, he no doubt discovered that the P&O depot had a marked effect on another aspect of the town's social structure. According to Donald Garden, the depot was in some respects a separate and closed community, with its own structured society—the agent at the top, then, in descending order, officers, minor officials and labourers. The personnel were kept separate from the town to some extent by resentment from some of the older settlers. There was also a marked social distinction between the professional officers of the P&O establishment and the commercial men of the town; the former looked down upon the latter, a common occurrence regarding professionals' views of those 'in trade' in Victorian society.[11]

Since Oldfield's last visit, little had changed in the community's involvement in the arts and literature; these remained the private province of the elite and the educated middle-class. Local interest in the flora and fauna was barely on the socio-cultural landscape, possibly confined to George Maxwell and perhaps those others looking for suitable depasturing and farming land. No local newspaper was being published yet. Sport, however, had become increasingly popular on a purely impromptu basis whenever teams could be put together.[12] Oldfield may have stopped to watch a boat race in the Bay or a cricket match on the beach or on a piece of cleared, vacant land. Recently completed, the Roman Catholic community had its church and first resident priest, but, apart from the Anglicans, other

[10] ibid., p. 170.
[11] ibid., pp. 170–171.
[12] ibid., pp. 176 and 190.

religious groups were still awaiting their churches.[13] The government school remained in the old commissariat and two private schools catered for children whose parents had higher educational aspirations.[14]

The number of local *Minang* people that were gathered around the settlement was probably noticeably smaller than when he was there in 1858. In August 1860, following the visit of the P&O steamer *Salsette,* residents boarding the vessel contracted Scarlatina (Scarlet Fever). Once on shore the disease ran quickly through the community; scarcely any young European escaped and one female settler died. Every one of the 60 or so Aboriginals who lived near the town contracted the disease, and at least 29 died. Scarlatina, venereal disease, and continual clashes between settlers and the Aborigines devastated and depleted the Aboriginal population. They were fast becoming a group of fringe-dwellers, reduced to begging and reliance on government handouts.[15]

Soon after his arrival Oldfield would have set about locating George Maxwell. Maxwell's residence at the time is not known, but if he was in Albany, Oldfield may have found him having a pint or two in one of the hotels. Maxwell was fond of a drop. William Harvey observed, when on his return to Albany on 3 August 1854, he 'met Maxwell on horseback, half seas over'.[16] Oldfield's next mission may have been a courtesy call on the Resident Magistrate, Alexander Cockburn-Campbell, who had recently replaced Henry Camfield.[17]

It will be recalled that Mueller had agreed to pay Oldfield up to £50 for plants and seeds obtained on this expedition, and Professor Frederick McCoy, £25 for natural history specimens 'generally'[18], to offset his own financial outlay relating to the expedition. Mueller was keen to incorporate plants from Australia generally into the botanic garden so the collection of seeds was possibly high on Oldfield's agenda.

Mueller's letter to Joseph Hooker (14 September 1861) suggests that, in the first instance, Oldfield was to go to Cape Leeuwin[19], but by what route the Government Botanist did not say. Apart from Oldfield's statement to Joseph Hooker (25 August 1861) that he was going 'to some parts of Western Australia not previously visited by any collector'[20], neither he nor Mueller provided further details of the route he was to take to the Cape. During this visit to the Colony, Oldfield visited the Upper Kalgan River area, the far south-west of the Colony and eastward along the coast from Albany to near the west end of the Great Australian Bight. This, then, begs the question of whether and to what extent these areas had not been traversed by botanical collectors prior to 1861.

European interest in the flora of the Upper Kalgan River and the far south-west of Western Australia prior to 1861

Map 15 p. 346 – Albany to Upper Kalgan and Cape Leeuwin

From June 1828 to December 1835, the Upper Kalgan River was included in the explorations of about 14 expeditions.[21] Among the explorers were Surveyor General J.S. Roe, Alexander Collie and Captain Joseph Wakefield[22] who at times sent specimens to Kew Gardens. To date, none are mentioned in the online Kew Herbarium catalogue, nor do the labels on Roe's or Collie's specimens in the Western Australian Herbarium give any indication that they were gathered in this district. Surveyor and drafts-

[13] ibid., p. 177.
[14] ibid.
[15] ibid., pp. 159–160 & 177–179.
[16] WHH to W.A. Sanford, Colonial Secretary, 31 July 1854, cited in Sophie Ducker, 1988, *The Contented Botanist…*, p. 127.
[17] Donald S. Garden, 1977, p. 126–127.
[18] See Chapter 18.
[19] FM to JDH, 14 September 1861.
[20] AO to JDH, 25 August 1861, Supplementary Foreign Letters, vol. 218, 1865–1900, fol. 238, RBGKA.
[21] Joanne Shoobert, 2005, vol. 1, and Marion Hercock *et al.*, 2011, vol. 2.
[22] A.S. George, 2009, p. 586.

man Alfred Hillman explored the area; he had an interest in indigenous plants[23], but it is not known whether he sent any to institutions for scientific purposes. As far as can be ascertained

James Drummond Snr did not collect on the Upper Kalgan River when, accompanied by George Maxwell, he visited the Stirling Range in 1846. However, Maxwell may have obtained plants when Drummond sent him on an errand to 'Neurabup' (now Noorubup), on the Upper Kalgan River.[24] Many of Maxwell's specimens in the National Herbarium of Victoria were obtained along Kalgan River but most of them have no collection date; those from the Upper Kalgan which are dated were all obtained after Oldfield returned to England in 1862. It seems, therefore, that the area near the Upper Kalgan was little known floristically when Oldfield went there in 1861.

The Blackwood River is the longest river in Western Australia, rising at the confluence of the Arthur and Balgarup Rivers about 35 kilometres north-west of Kojonup, and flowing in a south-westerly direction through the towns of Bridgetown and Nannup until it discharges into the Southern Ocean via Hardy Inlet near Augusta.[25]

Captain John Molloy, Assistant Surveyor John Kellam and Thomas Turner were the first to venture up the Blackwood River from Hardy Inlet. This they did in September 1830 for about ten miles before returning to the estuary.[26] In September 1834 Thomas Turner and his brother George set out to trace Blackwood River to its source; they managed to reach only about fifty miles above the estuary, about ten miles south of Nannup.[27] As far as can be ascertained no further attempts were made to trace the river's course until May 1845, when Augustus Gregory made a brief excursion down the Blackwood River from what he referred to as 'the junction of Arthur and Beaufort' Rivers.[28] He followed the river for approximately 171 kilometres to about where the bridge over the river at Bridgetown is now located.[29] Gregory may have obtained plants but no records have been found.

In January 1842, Major Frederic Chidley Irwin, together with Messrs Singleton and Rivett Henry Bland were the first Europeans to travel overland from King George Sound to the Vasse via the Blackwood River. They struck the river near Southampton, about 12 kilometres due south of where Balingup is today, and, after a short excursion of about four kilometres to the Balingup Brook, they followed the Blackwood River down to Lewana, approximately 15 kilometres south-west of Balingup. From there they headed north-west through the Cundinup district to the Vasse.[30] None of these expeditioners appear to have collected plants.

There was, however, some dedicated collecting by Georgiana Molloy, the wife of Captain John Molloy, who made extensive collections near Augusta (1836–1839). This material was sent to James Mangles.[31] His specimens are at the Natural History Museum, London, Western Australian Herbarium and the National Herbarium of New South Wales.

In 1842–43 James Drummond and John Gilbert travelled down the inland track from the Vasse to Augusta with the intention of exploring the Blackwood River to the north, but they did not commence

[23] Alfred Hillman, File of Botanical Specimens—Wildflowers, Accn. no. 2003/44A–Z, The Royal Western Australian Historical Society Inc. None of these pressed specimens has provenance and they are, therefore, only of historical interest. Hillman worked as a surveyor and draftsman in Western Australia from 1836-1860, Rica Erickson, 1979, p. 395.

[24] Rica Erickson, 1969, pp.114–115.

[25] Anon., 'Blackwood River', *Wikipedia*, viewed 28 September 2013 at www.wikipedia.org

[26] J. Molloy, 1830, 'Report of Captain Molloy on neighbourhood of the Blackwood River', cited in Joanne Shoobert, 2005, vol. 1, pp. 172–173.

[27] Thomas Turner, 1834, 'Mr Thomas Turner's Journal of an Expedition to trace the Blackwood River to its source 21st to 29th September 1834', cited in ibid., pp. 368–374.

[28] The Hillman River joins the Arthur River and southwards the combined stream joins Beaufort River and the three become the Blackwood River so the exact location of Gregory's starting point is not certain (South West Western Australia, 1: 750,000 map, 4th edn, Hema Maps Pty Ltd, Eight Mile Plains, Queensland, 2007.

[29] Augustus Gregory, 1845, 'Gregory's excursion down the Blackwood River'. Extract from private letter of Ass. Surveyor Gregory to Surveyor General, giving account of an excursion down the Blackwood River, May 1845', cited in Marion Hercock, *et al*., 2011, vol. 2, p.351.

[30] Rivett Henry Bland, 1842, 'R.H. Bland's trip from Albany to the Vasse River, January 1842', cited in ibid., pp. 294–297.

[31] Alexandra Hasluck, 1955, pp. 147–207.

this journey.[32] They made an unsuccessful attempt to cross from Augusta to Albany[33], after which they returned to the Vasse. In January 1844 specimens from this expedition were sent to William Hooker.[34]

In early May 1845, a party led by one of the Gregory brothers (probably Augustus) followed Arthur River downstream from its junction with Beaufort River (near today's town of Duranillin). They then traced the upper and middle Blackwood River, and made a superficial examination of the country along the river, briefly noting the terrain and water sources. It seems that the expedition ended near where the town of Bridgetown was later established.[35]

Setting out from their camp at Balgarup on 31 March 1846, a party, again led by Augustus Gregory, started on an expedition to trace Gordon River to its source and follow up the stream which runs into the Irwin Inlet. They discovered that Gordon River flowed into Frankland River and concluded that they were one and the same stream; this conclusion was not officially adopted. They traced the Frankland to 'Nornalup', ascended Mounts Roe and Frankland but, owing to almost impenetrable thickets, they were obliged to return to Balgarup.[36] Gregory may have gathered plant specimens during this journey but none have been found.

On his return journey to Perth from his exploration to the Stirling Range in 1848[37], Surveyor General John Septimus Roe crossed Blackwood River, about 38.5 kilometres west of Kojonup, and then travelled another 27 kilometres to a branch of Collie River:

> At 12 miles from the Blackwood river the white gums [*Eucalyptus wandoo*] and mahogany [*E. marginata*] forest began to show some very good timber of the latter description and it increased in both quantity and quality as we proceeded northwestwards.

Then, for 12 to 16 kilometres, the party followed 'on a devious course to the N.N.W. though hilly country...well timbered'. They came to Ferguson River and continued for five and a half kilometres through steep and sharp hills. After this the country opened out across the Swan coastal plain and became level, making the rest of the journey to Bunbury relatively easy going.[38] As with his previous journeys Roe collected plant specimens and seeds along the way, which were sent to William Hooker at Kew Gardens.[39]

In 1858 George Maxwell obtained a few specimens at Wilyungulup, a crossing on Blackwood River a little north of Gnowangerup Creek.[40]

Apparently, professional botanical collectors had only paid attention to the river near Augusta, and there was very little interest in collecting plants by expeditions undertaking geographical explorations further upstream.

---

[32] See Topographical map drawn by Thomas Turner in 1851, reproduced in Anne Brearley, 2005, p. 310. The original map, County Sussex, Western Australia, 1851, 1953/ooD6, Art Gallery of Western Australia.

[33] Rica Erickson, 1969, pp. 83–88.

[34] ibid., p. 168.

[35] ? Augustus Gregory, 1845, in Marion Hercock *et al.*, 2011, vol. 2, p. 351.

[36] A.C. Gregory, 'Excursion to the eastward of Kojonup and down the Gordon River by Assistant Surveyor A.C. Gregory April 1846', *Exploration Diaries*, 1836–1846, vol. 3, unpublished transcripts, pp. 764–765, JSBL.

[37] A.E. Orchard, 1999, 'Roe, John S. (1797 – 1878)', ANHBiog, viewed 8 May 2008 at www.anbg.gov.au, and R.L. Heberle, undated, 'John Septimus Roe 1797–1878…', viewed 8 May 2008 at www.members.iinet.net.au.

[38] A.C. Frost, 1979, *Baylya-Balinga: A History of Balingup, W.A.*, Donnybrook-Balingup Shire Council, Western Australia, pp. 5–6.

[39] WJH, (ed), 1854a, 'Introductory note', J.S. Roe, 'Report of a journey of discovery into the interior of Western Australia, between 8th September 1848 and 3 February 1849', *Hooker's Journal of Botany and Kew Garden Miscellany*, vol. 6, p. 42.

[40] MELISR database, 02 April 2009, NHV, RBGM.

Albany to the Upper Kalgan River–October 1861

Map 15 p. 346 – Albany to Upper Kalgan and Cape Leeuwin

In late 1861, Oldfield made a short trip to the Upper Kalgan River. The only dated specimen from this journey shows that he visited Noorubup, Upper Kalgan River in October 1861.[41] Maxwell may have accompanied him on this excursion but in the absence of any evidence to support such a claim, it is assumed that he went on his own. Although he obtained specimens along the Albany-Perth Road and to the east towards, and along, the Upper Kalgan River (i.e. Narpyn, Kendenup, Woogenellup, along the Upper Kalgan River, Noorubup and at the Porongurup [Range]),[42] they provide no clear indication of the route he took to and from the river as there is only one specimen with a date of collection recorded (see above).

Given that he appears to have gone west to Cape Leeuwin via a track that commenced on the Albany-Perth Road from close to Mount Barker, he may have gone to the Upper Kalgan first, on horseback with a packhorse, travelling along a track that ran from Albany, crossing the King River and continuing in a northerly direction towards Noorubup, John Hassell's property on the Noorubup Creek,[43] a tributary of the Upper Kalgan River. At the time this was a sandalwood carriers' track that later became the Chester Pass Road.[44] Ethel Hassell wrote of her journey along this route to the Kendenup/Jerramungup road when, as a newly wed in 1878, she travelled with her husband to her new home at Jerramungup.[45] A 'Noorubup' homestead could not be located. It was probably a station outpost and appears to have been the most southerly of Oldfield's collecting places on the river.[46]

As he approached Upper Kalgan River he entered the Mallee Region, which was still within the Marine Plains Ecodistrict.[47] The Mallee Region is characterised on its western margins by Mauw (*Eucalyptus occidentalis*), Tall Sand Mallee (*E. eremophila*) with scrub heath and *Allocasuarina* thickets on residual plateau soils. The vegetation along Upper Kalgan River is very different from that on its lower reaches which we have suggested was traversed by Oldfield in 1858. The river's source is west of Stirling Range and rises north-east of Lake Matilda where the trees and understorey of the Jarrah-Marri forest to the west become lower and give way to the vegetation of the Mallee. Fringing the river are species such as Mauw, Flooded Gum (*E. rudis*) and varieties of *Banksia* and *Hakea* are found.[48]From Noorubup he could have collected to the west on the north side of the Porongurup Range from the north by taking the track that ran from Myabup to Cape Riche via Noorubup.[49]

Oldfield would have found the vegetation of the Porongurup Range vastly different from that of the Mallee region to the east and north. Twelve kilometres in length, the granite domes of the Range rise over the plain to about 670 metres at their highest point. Karri trees cover the upper slopes and grow exclusively on a deep red soil known as Karri loam. This mountain range is one of the 'islands' of Karri forest that receive the rainfall necessary for the species' survival because of the drying and warming of areas over much of its former range. As well as the Karri, 750 plant species occur in the Porongurups—many of the understorey flowers and shrubs are typical of the Karri forest but the greatest percentage grow within the Jarrah-Marri and other woodland which dominate the laterite soils

[41] MEL 96452, *Drosera leucoblasta*, October 1861, Noorubup, Upper Kalgan River, MELISR database, 18 November 2005 NHV, RBGM.
[42] HAODb.
[43] Plantagenet 08 - Plantagenet 15 [80 chain plan, Tally No. 506630], July 1876 (1876–1896)
[44] Carolyn Thomson, Graham Hall and Gordon Friend (eds), 1993, *Mountains of Mystery: A Natural History of the Stirling Range*, Department of Conservation and Land Management, Como, Western Australia, p. 37.
[45] Ethel Hassell, 1975, My Dusky Frends, C. W. Hassell, Perth, Western Australia, pp. 1–6.
[46] HAODb.
[47] Sarah Barrett *et al.*, 2009, p. 14 and Fig. 1, South Coast Ecodistricts.
[48] Waters & Rivers Commission, 1999, *Albany Waterways Resource Book: Waterway Life*, viewed 24 March 2014 at www.rivercare.southcoastwa.org.au.
[49] Plantagenet 08 - Plantagenet 15 [80 chain plan, Tally No. 506630], (1876–1896), SROWA.

Fig. 19.2: Near Chester Pass Road at Kamballup golf course

Fig. 19.3 Karri (Eucalyptus diversicolor), Porongorup Range

Fig. 19.4 Near Takalarup Crossing, Upper Kalgan River

of the lower slopes.[50] The granite outcrops have a very distinctive flora from the adjacent forests and woodland, including several rare species such as the glossy-leaved herbs, *Ornduffia calthifolia* and *O. marchantii*.[51]

It is proposed that he returned to Noorubup along the Myabup-Cape Riche track and then proceeded northward along the Upper Kalgan River collecting along the river to as far as Woogenellup, another of Hassell's properties.[52]

From Woogenellup, he could have travelled to Kendenup. No track is shown on public plans connecting the two properties but there probably was one as they were both Hassell's pastoral leases.[53] From Kendenup he may have travelled to Mount Barker via another of Hassell's pastoral leases, Narpyn (Narpund on some early public plans).[54] where he gathered at least two plants.[55]

Mount Barker to Cape Leeuwin–October to December 1861

Map 15 p. 346 – Albany to Upper Kalgan and Cape Leeuwin

Although none of Oldfield's plant records from Mount Barker to Cape Leeuwin bear a date of collection, Mueller's statement that Oldfield had 'gone to Cape Le[eu]win' confirms his intention to go there during the spring of 1861. Maxwell was at Mount Barker during October 1861[56], which could have placed him in the district immediately after Oldfield's tour to Upper Kalgan River. However, this is insufficient evidence to assert that they went together to the west. Maxwell's specimen records in the National Herbarium of Victoria show that he had visited some of the area traversed by Oldfield (namely, Tweed, Tone and Blackwood Rivers) but none bore a date of collection. Maxwell had reason to be in the area on other occasions prior to October 1861. For example, he applied for a depasturing lease in 1859 over land somewhere near the Tweed and Blackwood Rivers, which was granted for seven years.[57] In the absence of evidence to the contrary, it is assumed that Oldfield undertook the expedition on his own.

Instead of making the 46 kilometres journey back to Albany following his Upper Kalgan trip, Oldfield may have proceeded to the Old Bush Inn which William Cooper had built in 1860 beside the Albany Road, north of Mt Barker, and commenced his journey west from there. The inn had become a place for Royal Mail coaches to change horses and a welcome resting place for travellers. For local settlers, ticket-of-leave men and shepherds it was developing as a centre to build their community life.[58]

The localities on Oldfield's specimen records suggest that he travelled to the north of west from Mount Barker, possibly following an unsurveyed track that ran close to Merdeperup Creek, one of his collecting places[59], before it crossed Kent River and then proceeded almost due west to Lake Muir[60], another of his collecting localities.[61] Following the discovery of Lake Muir in 1852, by Thomas and John Muir[62], their brother Andrew established a farming and pastoral business in the area. Other

---

[50] Department of Conservation and Environment Western Australia (now Department of Parks and Wildlife), 2009, 'Porongurup National Park', information brochure, viewed 18 November 2013 at www.parks.dpaw.wa.gov.au.
[51] S. D. Hopper, personal communication to MHH, email attachment, 24 December 2017, HOA.
[52] Plantagenet 16 [80 chain plan, Tally No. 506635],(1883–), SROWA.
[53] ibid.
[54] South West [506065], AU WA S979-cons4900 S42/1 1883-09-24, (Accumulation) SROWA.
[55] HAOdb.
[56] MEL 2177793, *Melaleuca* sp., near Mount Barker, Swan Road (? Albany-Perth Road), October 1861, George Maxwell, MELISR database, 2 April 2009, NHV, RBGM.
[57] W.G. Henderson, 2011, vol. 13, pt. 5, p. 639.
[58] Heritage Council of Western Australia 1999, *Mount Barker – Settlement and Development of the Mount Barker District*, Heritage Trail pamphlet, H88/021, August 1999, viewed 26 June 2010 at www.plantagenet.wa.gov.au.
[59] HAODb. 'Merdeperup' is also the name of a locality.
[60] Today's Muir Highway follows a similar route.
[61] HAODb.
[62] Joanna Moore, 2010, 'Manjimup and the Muirs', *Landscope*, vol. 25, no. 4, p. 26.

Fig. 19.5 Vegetation fringing Lake Muir

brothers had properties to the west. A network of roads and bridges was developed to move their cattle and produce to Andrew Muir Snr's store in Albany.[63]

The country around Lake Muir is poorly drained and comprises lakes and swamps. The Lake supports a narrow zone of open-scrub, sedgeland and low shrubland at or near its margins. The dominant low shrubs are the samphires (*Sarcocornia quinqueflora* and *Halosarcia lepidosperma*); the wetland scrub is dominated by tall shrubs of Saltwater Paperbark (*Melaleuca halmaturorum*), and some Swamp Paperbark (*M. ericifolia*) and Mohan (*M. viminea*). Other wetland plants near the lake margins include Spreading Sword-Sedge (*Lepidosperma effusum*), Coast Saw-Sedge (*Gahnia trifida*), *Schoenus submicrostachyus*, Narrow-leafed Wilsonia (*Wilsonia backhousei*) and Flooded Gum. Surrounding this narrow zone is open Jarrah-Wandoo woodland.[64]

In about 1856, Andrew Muir Jnr built a two-roomed paperbark hut at 'Nagabup' on the edge of Lake Muir. In the same year, Thomas became the first European to settle further to the west when he established a farm at Deeside, 32 kilometres south-east of today's town of Manjimup.[65] Thomas became, in about 1872, one of Mueller's correspondents and sent him plant material for the National Herbarium of Victoria.[66] Mueller had been appointed a hereditary baron by the King of Württemberg in 1871 and from then on was known as Baron Sir Ferdinand von Mueller.[67] Thomas entertained von Mueller at Deeside and Lake Muir during his visit to the Colony in 1877.[68]

From Lake Muir, Oldfield probably proceeded north to the track that ran west from near where Gordon River crossed the Perth-Albany Road and continued to the Vasse-Bunbury Road emerging

---

[63] National Trust of Australia, 'Deeside Homestead Group', *National Trust Register of the Built Heritage of Western Australia,* viewed 27 September 2013 at www.traces.duit.uwa.edu.au.

[64] Department of Environment, Australian Government undated, 'Muir–Byenup System', *RAMSAR Wetlands information sheet*, viewed 24 March 2014, www.environment.gov.au, and Per Christensen, 1992, *The Karri Forest: Its Conservation Significance and Management*, Department of Conservation and Land Management, Como, Western Australia, Appendix 6, p. 181.

[65] Joanna Moore, 2010, p. 26.

[66] ibid.

[67] Deirdre Morris, 1974, 'Mueller, Baron Sir Ferdinand Jakob Heinrich von (1825–1896)', ADBO, viewed 16 June 2010 at www.adb.online.anu.edu.au.

[68] Joanna Moore, 2010, p. 26.

Fig. 19.6 East bank, Tone River, near Tonebridge

west of the confluence of the Gordon and Frankland Rivers. He moved further west to Tone River, a tributary of Warren River.[69] In doing this he traversed woodland of Wandoo in the valleys and lower slopes, with Mauw (*Eucalyptus occidentalis*) occurring in some of the damp areas and Flooded Gums common along watercourses, with Jarrah confined to lateritic hill tops merging with Wandoo on the upper slopes.[70] Perhaps the most feasible explanation for his journey to Lake Muir via Merdeperup Creek is that previous collectors had not been in the area. It does not appear to have been a shorter route than going along the track that terminated on the Vasse-Bunbury Road from where it commenced on the Albany-Perth Road.

He botanised much more intensively at Tone River than he had at Lake Muir, and this would suggest that he camped near the river. By now he was well into territory where there had been virtually no prior botanical collecting. This may account for his relatively large number of Tone River published specimen records in the Henderson Oldfield database. Of these 159 records, 51 (32 per cent) were sourced from taxonomic publications.[71]

From Tone River, the track ran north-west and within a few kilometres, was roughly parallel to Tweed River, one of the 41 tributaries of the Blackwood River. The predominant vegetation complex back from the river is Jarrah-Marri woodland interspersed with Wandoo and an undergrowth of sclerophyllous shrubs and herbaceous plants. The riparian flora mainly consists of Flooded Gum while Swamp Paperbark forms a dense understorey at the edge of the stream in some places.[72]

---

69 HAODb.

70 South West ARM Region Appraisal Team, 2005, *Boyup Brook-Upper Warren Area Catchment Appraisal*, Resource Management Technical Report 319, Department of Agriculture and Food, Perth, Western Australia, p. 16.

71 HAODb. Tone River is a tributary of the Warren River and does not go anywhere near the sea.

72 D. Wong & G.M. Blake, 2009, *Salinity Situation Statement: Tweed River & Gnowergerup Brook*, Water Resources Technical Series, Report no. WRT41, Department of Water, Western Australian Government, p. 4, viewed 5 June 2011, at www.water.wa.gov.au.

Fig. 19.7 5 kilometres north-west of Tonebridge

There were depasturing leases for most of the length of the Tweed and Oldfield may have stopped to chat with shepherds along the way. The track probably crossed the upper reaches of the river on one of John Hassell's blocks (lease 739), then traversed George Edward Egerton-Warburton (lease no. 318) and John Henry Monger's leases (leases 578 and 911) further west.[73] As well as his pastoral interests, Monger was a merchant with a head office in Perth and branches at Fremantle and York.[74] He held two leases, one with a southern border along the north side of Tweed River and the other an adjacent block that embraced the confluence of the Tweed and Blackwood Rivers in the south and extended on either side of the Blackwood north as far as the confluence of the Blackwood with Boyup Brook. Hassell had two other blocks on the south side of the Tweed, which was the entire northern border of one block and part of the second and which continued west along Blackwood River (leases 749 and 1184).[75]

Oldfield's collections along and near Tweed River do not appear to have been as extensive as those from Tone River, but this may have been due to the similarity of the understorey vegetation in both places. Nevertheless, of the 68 specimen records we have located 18 (27 per cent) that were sourced from taxonomic publications.[76] Here again, this possibly reflects minimal, or no, prior collecting in the area.

---

[73] Sussex & Blackwood River, undated (c. late 1850s), Cons. 4926, Item Sussex-01, (506223), SROWA.
[74] R.P. Wright, 1974, 'Monger, John Henry (1831–1892)', ADBO, viewed 5 June 2010 at www.adb.online.anu.edu.au.
[75] Sussex & Blackwood River, undated (c. late 1850s), Cons. 4926, Item Sussex-01, (506223), SROWA.
[76] HAODb.

In his unpublished manuscript, Oldfield, in a discussion of the factors inhibiting and facilitating the extension of species' ranges, referred to a plague of locusts in 1860, which descended upon a farm on the Tweed '150 miles [240 kilometres] from their original habitat' that he had probably heard about from a local resident.[77] In an associated footnote, Oldfield gave an account of the somewhat amusing and impractical response of one farmer to an impending plague of locusts, but we doubt he resided near the Tweed or Blackwood Rivers since the narrator made reference to women returning 'from the township'. There was no settlement of that nature in the vicinity at the time:

> I know of one settler living at some distance from the sea, who, seeing this plague steadily approaching his domain, made almost frantic efforts to arrest its progress. It so happened that a small stream flowed between his farm and the infected districts, and the first bright idea that struck him, was to tether fowls on the solitary bridge that spanned the brook, that they might devour the locusts as they assayed to cross the stream, for he seemingly overlooked the possibility of their using their wings for that purpose. In spite of this ingenious precaution, some few contrived to pass the stream, and at once guessing by what means, he issued strict orders to the female part of his household, to shake their crinolines before attempting to cross the bridge on their way from the township, for, as he justly remarked, those garments are admirably adapted to harbour vermin. Finding himself still foiled, he invented a mathematic-entomologic formula, by the aid of which he discovered the 'centres of origin' of the locusts, and forthwith proceeded to travel the 'Bush' accompanied by a bullock dray, on which was carried an immense cauldron. On arriving at a locality indicated by his formula, he placed his cauldron in position, and having heated a quantity of water liberally drenched the suspected locality, firmly persuaded that he thus destroyed ova [fertilised eggs][78] by millions. But all was in vain: either the locusts flew across the stream, or the fowls or the females were remiss, or the formula indicated imaginary localities, or the water was not hot enough, for at length the enemy invaded his estate, leaving his erstwhile existing garden and fields, a blackened wilderness.[79]

Near the junction of the Tweed and Blackwood Rivers Oldfield probably turned north-east along a track that ran close and parallel to Blackwood River through Monger's lease no. 911, forded the Gnowergerup Brook before almost immediately turning west and traversing the Blackwood.[80] He may have called in at James George Lee Steere's homestead (known locally as Steeres'), which lay close to the track near the western bank of the river and to its confluence with the Gnowergerup Brook. In 1860 Steere (later Lee Steere) entered into partnership with Monger in his grazing property (Lease nos. 469, 911 and 912)[81] along the Blackwood River. When Oldfield passed through, James and his wife, Catherine Anne, were in residence, having come from Bunbury earlier in the year. They named the property 'Jayes' after George's ancestral estate in Surrey. Their homestead, built in 1861 of bricks and shingles, appears to have been completed prior to the collector's arrival. It is unlikely that the Steeres could provide Oldfield with fresh food supplies. As recent settlers, they had little time to become, as they did in the years to follow, self-sufficient in vegetables, milk and probably fruit. The major focus of their activities in those early years was wool production.[82]

---

[77] AO, 1865a, fn. to p. 806.

[78] Ova are female egg cells before fertilisation. Since fertilisation in locusts takes place through copulation and the fertilised eggs Are laid in burrows,, Oldfield appears here to be referring to fertilised eggs, D.G. Mackean, 'The Desert Locust', *Biology Teaching Resources*, viewed 8 January 2017 at http://www.biology-resources.com.

[79] ibid.

[80] Sussex & Blackwood River, undated (c. late 1850s), Cons. 4926, Item Sussex-01, (Tally no. 506223), SROWA.

[81] ibid.

[82] A. Schorer, 1968, *History of the Upper Blackwood*, South West Printing and Publishing Co., Bunbury, p. 25.

Fig. 19.8 Blackwood River near Jayes homestead

There is no evidence that Oldfield continued along the track from Steeres' to Busselton[83] or that he went from Busselton along the road that ran south just west of there to Cape Leeuwin.[84] It was noted previously[85] that he collected at and near the Vasse River, but we have not been able to confirm whether his specimens from there were obtained during either, or both of, his 1856 and 1861 expeditions, but it seems likely that he only visited Cape Naturaliste and the Vasse in 1856. With one exception (see below) there is no way of knowing where along the stream he obtained his Blackwood River specimens. Therefore, the route taken to Cape Leeuwin once he reached Steeres' is largely a matter for speculation.

Apparently, the intent of Oldfield's 1861 collecting expedition was to go to 'some parts of Western Australia not previously visited by any collector'.[86] Up to that time, no professional botanical collectors had worked along the river from Steeres' to just short of Augusta, nor has any evidence been found that any settlers had collected and sent material to von Mueller or to European and British herbaria. Therefore, it would seem reasonable to suggest that that stretch of the stream was one of those 'parts' and was the primary object of his visit.

Oldfield obtained False Boronia (*Phyllanthus calycinus*) on the Blackwood River '90 miles [144 kilometres] from the sea'[87], which has been calculated to be near Geegeelup, then a tiny settlement that was to become the gazetted town of Bridgetown. He may have gone there by travelling along the track that ran south-west from Steeres' to Geegeelup. The route appears to have traversed Monger's lease (no. 911), unclaimed forest away from the river and then another of Monger's leases (no. 912) that was roughly bisected by the Blackwood with its western border close to Geegeelup.[88] Apart from

[83] Sussex, Cons. no. 4926, Item Sussex 3, Series 9789, (506200), SROWA.
[84] Thomas Turner in 1851.
[85] Chapter 11.
[86] AO to JDH, 25 August 1861, Supplementary Foreign Letters, vol. 218, 1865–1900, fol. 238, RBGK.
[87] *Phyllanthus calycinus*, Blackwood River, 90 miles from the sea (GB1973, vol. 6, p. 105.
[88] Blackwood River, undated (c. late 1850s), Cons. 4926, Item Sussex-01, (506223). The track is not shown on this public plan but is shown partially on Wellington and Nelson, Public Plan of Roads: Bunbury to Blackwood and with branch roads from Balingup to Jayes (Steere's), District Survey Officer, Bunbury, Survey Department of Western Australia 4 September 1860, Accession no. 3869, Item No. Roads 127, SROWA.

Monger, Walter Padbury and one or two others who held pastoral leases in the Geegeelup district[89], there were few resident settlers at that time. In 1857, John Blechynden settled on the river near Austin Ford (west of the bridge at the southern end of Bridgetown, now located east of the current bridge).[90] In the same year Edward Godfrey Hester was at 'Blackwood Park' to the north-west of the settlement.[91] Rightly or wrongly it is assumed that the collector followed the river from Geegeelup to its mouth on the Hardy Inlet.

Oldfield apparently obtained more specimens along the Blackwood than he had on any of the previous rivers encountered since he left Mount Barker.[92] Of the 227 Blackwood River plant records we have located, 87 (38 per cent) were referred to in taxonomic publications[93]; again, reflecting the small amount of previous collecting in the area. It may be expected that by following the river from Geegeelup to Hardy Inlet his gatherings would have been more numerous. However, it was late in the flowering season and many species would have been setting fruits or seeding. He may, therefore, have spent most of his efforts in collecting seed material for Mueller.

For most of the way below Geegeelup the Blackwood River was surrounded by dense vegetation of which, by 1861, little or no clearing had been undertaken. It is unlikely, therefore, that Oldfield would have been able to follow its course to Augusta if there had been no Aboriginal pathways close to the river, along which he could travel with minimal impediment all the way to his destination. Easy access to the river throughout his journey would have ensured a constant supply of fresh water with marron, gilgies, fish and water birds to supplement any food he carried until he reached the tidal channel at about today's Alexandra Bridge.

Many early settlers in the south-west, and elsewhere in the Colony, depended on Indigenous knowledge and use of Country and its resources. Drawing upon this knowledge, often with the assistance of Aboriginal guides, they followed Aboriginal paths and made use of their wells while looking for suitable grasslands and watering points for domestic and pastoral purposes.[94] Thomas Turner's account of his journey up the river above its navigable channel, indicates that for the most part his party travelled along the stream's left bank with some brief sorties into the country to the east.[95] The party found the vegetation and rocky terrain along and away from the river banks difficult and tiring to penetrate although, at times, Aboriginal firing of patches of the country made for relatively easy walking.[96] He reported that on one occasion his party followed a path away from the river to a 'swampy place'.[97]

The *Wardandi,* whose territory was encompassed by the coast from Bunbury to Cape Leeuwin and eastward by the Blackwood River to within a few kilometres south-west of Greenbushes and from there across to Bunbury[98], maintained a number of paths between Vasse and Wonnerup, Augusta and Nannup, and along the coast and the Blackwood River.[99] The Turners' apparent reluctance to venture far from the river during their travel upriver may have caused them to miss any pathways that followed it a little further inland. Turner supported the notion that the *Wardandi* did have access points along the river. While most of the stream, particularly in the lower reaches, was an impassable barrier to people without boats, an intersection of tracks occurred at the places where the river could be crossed (Hut Pool and the mouth near Augusta). They became focal points of traditional camps where

[89] Public Plan, Sussex & Blackwood River, undated (c. late 1850s), Cons. 4926, (506223), Item Sussex-01, SROWA.
[90] Just below Ford House, Eedes Terrace, Bridgetown.
[91] C. Gaines, c.1970, *Bridgetown: One Hundred Years of History*, mimeo., C. Gaines, JSBL, p. 3.
[92] HAODb.
[93] ibid.
[94] Sylvia J. Hallam 1979, *Fire and Hearth: A Study of Aboriginal Usage and European Usurpation in South-western Australia*, Australian Institution of Aboriginal Studies, Canberra, pp.66–71.
[95] Thomas Turner, 1834, pp. in Joanne Shoobert (ed), 2005, vol. 1, pp. 368–374.
[96] See for example ibid.
[97] ibid.
[98] Norman B. Tindale, 1974.
[99] Brad Goode & Colin Irvine 2006, *A Survey of Aboriginal Social Water Requirements for the Southern Blackwood Plateau and the Scott Coastal Plain, Southwest, Western Australia*, report prepared for the Department of Environment, Western Australia, p. 10.

ritual and other customary activity often took place, at times performed jointly by the *Wardandi* and their 'next door' neighbours, the *Pibelmen/Bibbulmun.*[100] The land of the *Pibelmen/Bibbulmun* was bounded on the west and north by the Blackwood from about Cape Leeuwin to the north-east as far as Bridgetown (Geegeelup) then directly south-east to near Tenterden and from there south-west to Broke Inlet and west along the south coast to near Cape Leeuwin.[101] No descriptions have been found of any paths running along and close to the river on the *Pibelmen/Bibbulmun* side of the river. However, there must have been one or more along the north-south axis of their territory, including close to the river, to satisfy their need for relatively easy travel around their country for relaying message sticks[102], and gaining ready access to the potable water and wealth of food that the river provided all year round.

If Oldfield did undertake the journey down the Blackwood from Geegeelup, he probably employed the services of a member of the *Pibelmen/Bibbulmun* tribe to guide him through their territory; a wise course of action if he was travelling alone. For Oldfield, used to moving around the country with Aborigines, such an undertaking would have been well within his comfort zone although, if his guide could not speak English, Oldfield's *Watchandie* language is unlikely to have been of much assistance.[103] Nevertheless, he was an experienced linguist so communication difficulties may not have been insurmountable.

Although since 1861 areas of land along the Blackwood River have been cleared for farming, from Bridgetown (Geegeelup) to the Hardy Inlet, the river still flows through large tracts of open forest composed of Jarrah, Jarrah-Marri mixtures with Swan River Blackbutt and Yate occasionally occurring in small patches by themselves. Common understorey species include Bull Banksia, Sheoak, Snottygobble (*Persoonia longifolia*) and Spreading Snottygobble (*P. elliptica*). On grey sands, Slender Banksia, and Woody Pear (*Xylomelum occidentale*), and on damper sites WA Peppermint may be present. Common shrubs include Zamia, Grass Tree, Graceful Grass Tree (*Xanthorrhoea gracilis*), *Acacia* spp., *Bossiaea* spp. and *Hakea* spp.[104] In Oldfield's era, however, near Augusta there was high open Karri forest. When James Drummond and John Gilbert visited Augusta during their 1842–43 collecting expedition, they found James Woodward Turner: 'trying to burn up some of the huge Karri trees he had cut down. The work was slow. Many of the trees were sixty feet and more to the first branch and about six to eight feet in diameter'.[105] Today, due to clearing for farming and timber felling, only a few small stands of Karri remain near Augusta.[106] Nannup has stands of Karri on moist high rainfall sites of deep red loams.[107]

By the time Oldfield journeyed down the Blackwood, the very heavy rain of the winter months would have lessened considerably and for ease of travel he may not have carried a tent, perhaps relying instead on a light canvas sheet for shelter or nothing at all. Although his journey may have been blessed with fine, clear days, the summer dry was still a month or two away and periodic rains[108] could have made travelling and bivouacking unpleasant, causing difficulties with writing his notes, and pressing and drying his plants. Nearly 30 years previously Thomas Turner and his party experienced discomfort due to rain:

---

[100] ibid., p. 15.
[101] Norman B. Tindale, 1974.
[102] J.E. Hammond, 1933, *Winjan's People: The Story of the South-West Australian Aborigines*, Imperial Printing Co., Perth, p. 17.
[103] According to J.E. Hammond, who during the last three decades or so of the 19th century spent many years in the company of Aborigines across most of the south west, the language of tribal people to the north of Geraldton was very different from those spoken by the group he referred to as the 'South-west Tribe' which inhabited the land from just north of Perth, south east to Albany and bounded by the south and west coasts, ibid., p. 16.
[104] Per Christensen, 1992, pp. 13–14.
[105] Rica Erickson, 1969, p. 85.
[106] Bradley Commins, Ranger, Department of Environment and Conservation, Augusta, personal communication to MHH, 19 August 2010.
[107] Blackwood Environment Society c.2000, pp. 83–84.
[108] 'Bridgetown comparison', Climate Statistics for Australian Locations, Bureau of Meteorology, Australian Government, viewed 14 August 2010 at www.bom.gov.au.

> Sunday night looking inclined for wet we pitched a tent with our largest blanket across a fallen tree, we were all more or less wet with our days march, - the weather was mizzling rain all night, which wetted all our bedding & took us a long time to dry on Monday morn^{g}, which prevented our starting so early as we could wish.[109]

As he moved along the river, Oldfield would have observed that the course of the stream was marked by a repeating pattern of pools and rapids, interspersed, where topography permitted, by floodplains and at times by islands dividing the stream and forming lagoons where the water flow is slowed; all providing habitats for a variety of water birds and aquatic animals.[110]

The combined flora of the lower storeys in these two forest types is very rich in species diversity. There are many monocotyledon, eudicotyledon[111] and cryptogam families, each represented by taxa varying in number.[112]

The Flooded Gum dominates the riparian flora of the river. Occurring on moister sites along the river, on loamy soils and along creek lines are Swan River Blackbutt, Swamp Banksia (*Banksia littoralis*), Swamp Paperbark and probably the Grey Honey Myrtle (*Melaleuca incana*); in the understorey, many rushes and sedges.[113] Although flowering was on the wane, Oldfield's plant specimens from along the river[114] suggest that the plethora of small shrub, herbaceous and creeper species of the upper bank and further into the forest were probably still producing an attractive display of colour.

While his gatherings include species from 15 families of monocotyledons, 35 of eudicotyledons, and 2 of ferns, no records of cryptogams have been found.[115] This is puzzling given his earlier special interest in this group. Perhaps those he came across were not in a suitable state for taxonomic purposes or Mueller had encouraged him to focus on the other three subclasses.

If he crossed the Blackwood at Hut Pool and headed into *Wardandi* territory along one or more of their paths that headed west, sooner or later he would have come to the settlers' Busselton–Augusta road. Following this road to the south eventually brought him close to West Bay on the Hardy Inlet, the northernmost point of the Augusta townsite. A narrow channel from West Bay empties to the east into a wide basin of water, which, in turn, flows to the south into a narrow channel[116], the western shore of which forms the eastern boundary of the townsite. This channel leads to the mouth of the Blackwood, its flow into the ocean usually hindered by the periodically shifting sandbars across the river mouth.[117]

On the other hand, if a member of the *Pibelmen/Bibbulmun* tribe was his companion it is likely that they remained in his territory on the eastern side of the river until they reached its mouth. Following that route would have meant crossing Scott River which empties into Hardy Inlet at the southern end of Molloy Island but the locations of fords across the stream have not been determined. The Google Earth image of the area would suggest they might have had to follow the stream to the east for a few kilometres before it was narrow and shallow enough to cross. From there they would have proceeded along the river to the west, skirting the western end of the chain of swamps, which had prevented James Drummond and John Gilbert from travelling overland from Augusta to Albany in 1842–1843[118], turned southwards along the estuary, past the western shores of Swan Lake and an

[109] Thomas Turner, 1834, cited in Joanne Shoobert (ed), 2005, vol. 1, pp. 369.
[110] Blackwood Environment Society, c. 2000, pp. 30,55, 69, 77–81.
[111] The taxononmic classification Dicotyledon has now been replaced by Eudicotyledon, which accounts for 75% of all known Angiosperm species. Monocotyledons comprise 22% of all known Angiosperm species. The remainder of the Angiosperms includes a taxonomic group of 5 orders with the ending "ales"and the Magnoliids, Ben G. Bareja, 2012, ' Monocots, Dicots and Eudicots distinguished', Crop Farming, viewed 8 January 2017 at http://www.cropsreview.com. S.D. Hopper kindly drew our attention to this change in the classification of Angiosperms, personal communication to MHH, 24 December, 2017, HOA.
[112] Per Christensen, 1992, pp. 17–23.
[113] J.S. Beard, 1990, pp. 77-80, and Blackwood Environment Society, c. 2000, pp. 87–89.
[114] HAODb.
[115] See ibid. for details of the records of species we have found with the locality at or near the Blackwood River.
[116] Thomas Turner, 'Augusta, 1830', watercolour on linen, JSBL.
[117] Anne Brearley, 2005, pp. 317–321.
[118] Rica Erickson, 1969, p. 85.

elongated, narrow stretch of dead water just to the south[119] before crossing the sandbar at the river mouth.

The shallow tidal flats along the inlet widen into the larger basin with more complex vegetation along the banks. Fringing vegetation extends down to the water's edge—along the banks, growing into the water, salt-tolerant rushes, including *Juncus kraussii* and *Baumea juncea,* and close to the Saltwater Paperbarks, while such sedges as *Lepidosperma tetraquetrum* and *L. effusum*, and the Tall Kangaroo Paw (*Anigozanthos flavidus*) are confined to the upper reaches. On higher ground, in damp, swampy places, there are paperbarks (*Melaleuca* spp.) or Samphire (*Salicornia* spp.), and in drier areas, Zamias, Kingia and WA Peppermints. Gradually these are replaced by less salt-tolerant trees, Swamp Paperbarks and *M. hamulosa* and as the ground level rises, Flooded Gums succeed them, and finally there is Jarrah-Marri low open forest.[120] None of Oldfield's specimens in our database are recorded as having been obtained at the inlet. However, one, *Juncus caespiticius*, labelled as being from the Blackwood River, could have been gathered on saline sand near the inlet.[121]

Before venturing on to Cape Leeuwin, he probably contacted one or more of the very few residents in Augusta. The settlement was near several properties along the western shores of Hardy Inlet where James Turner, the Bussell brothers and John Molloy had previously lived.[122] The historical record is not clear for the number of people living in Augusta in 1861, but by 1864, according to Surveyor (possibly Robert) Quin, there were only two individuals or families (surnames: Brennan and Longbottom) in the settlement.[123]

To reach the Cape from the settlement, he had to travel 10 kilometres to the south, skirt Flinders Bay, and then enter the promontory. Cape Leeuwin is a projecting tongue of granite marking the meeting point of the Indian and Southern Oceans. Slopes behind the Cape have wind-pruned vegetation in which the grey-leaved *Olearia axillaris* is conspicuous but the wattles *Acacia littorea* and *A. cyclops* are dominant and, nearer the sea, *Scaevola crassifolia* and *S.* [124] Woody species such as Rottnest Tea-tree (*Melaleuca lanceolata*) and Peppermint (*Agonis flexuosa*) may be stunted to less than a metre on exposed sites, but a little further inland they grow into tall shrubs or trees in gullies and sheltered locations.[125] The heath flora shares marked similarities with that between and behind granitic headlands to the east.[126] The promontory had been part of one of James Turner's pastoral leases on which he ran stock.[127] He cut tracks through the dense coastal heath along the windswept slopes of the promontory to the Cape.[128] Since it had been some years since Turner ran stock there (1847)[129], Oldfield may have found the tracks overgrown and difficult to traverse.

Five published and 23 herbarium records of Oldfield's specimens with the locality 'Point Irwin' have been found.[130] Today, two places in Western Australia bear this name, both west of Albany. One is on Hardy Inlet at Augusta and the other further east just south of Peaceful Bay.[131] The geographical ranges, as indicated by Western Australian Herbarium specimen distribution maps for the taxa in *FloraBase* do not settle at which Point Irwin Oldfield's specimens were collected, nor indeed, whether they were obtained to the west of Albany. Thus, according to *FloraBase*, only seven of the 17 species

119 Anne Brearley, 2005, p. 315.
120 ibid., pp. 325–326.
121 *FloraBase*.
122 Alexandra Hasluck, 1955, pp. 72, 123 & 139–140, and Thomas Turner, 1851, reproduced in Anne Brearley, 2005, p. 310.
123 ? Robert Quin cited in Gail J. Cresswell, 1990, *The Light of Leeuwin*, Augusta-Margaret River Shire History Group, Margaret River, p. 59. Cresswell only provided surnames. Peter Brennan (1817–1890) held land in Sussex District and Stephen Longbottom (1818-1902) was a storekeeper in Busselton and a pastoralist, but there is no mention of either of them living in Augusta, Rica Erickson, 1979, vol. 3, p. 67.
124 J.S. Beard, 1990, p. 75.
125 Friends of the Cape to Cape Track, 'Flora', viewed 15 August 2010 at www.capetocapetrack.com.au.
126 J.S. Beard, 1990, p. 75.
127 Sussex, Cons. no. 4926, Item Sussex 3, Series 9789, (506200), SROWA.
128 Graham Davis, Curator, Augusta Museum, personal communication to MHH, 23 August 2010.
129 Rica Erickson (ed), 1979, 'James Woodward Turner', vol. 3, p. 852.
130 HAODb.
131 see GAPNA.

Fig. 19.9 Cape Leeuwin promontory

represented in Oldfield's Point Irwin collection in our database occur near Point Irwin at Augusta and two at the more easterly Point Irwin, one of which is a *Eucalyptus albida*, which is the only specimen of this species in *FloraBase* from west of Albany.[132]

The ranges of all the species in the collection, including *E. albida*, extend east of Albany, many beyond Esperance. Furthermore, comparison with species lists derived from the Department of Environment and Conservation's website, NatureMap,[133] show even less prevalence of occurrence of the species in Oldfield's collection at the Point Irwin at Augusta with four more at the other Point Irwin. It would seem, therefore, that at the time Oldfield was collecting along the south coast, there was another Point Irwin to the east of Albany but such a locality has not been found on any archival map, nor is it known to Landgate's nomenclatural section.[134] Bentham referred to *Leucopogon gibbosus* as having been from 'Point Irwin, Doubtful Island Bay' and this might be where all the Point Irwin specimens were obtained,[135] with the exception of *E. albida* the specimen of which appears to bear an incorrect collection locality as it is a wheatbelt species that doesn't occur anywhere near the south coast at Doubtful Island Bay or Pt Irwin.[136] Oldfield visited Doubtful Island Bay on his expedition to the east of Albany in 1861–62.

Only one record of a specimen collected by Oldfield at Cape Leeuwin has been found. This was a Grass Tree (*Xanthorrhoea preissii*), specimens of which he also gathered at the Gordon and Blackwood Rivers.[137] Why then would he choose a species, which was common in many of the localities he had visited in the south-west of the Colony? Perhaps he suspected that those he took from the above localities were of a new species and therefore worth the effort of bringing specimens of such a bulky plant back to Melbourne. This is suggested because Mueller did use all these specimens to describe a

---

[132] *FloraBase*.
[133] Department of Environment and Conservation 2013, *NatureMap: Mapping Western Australian's Biodiversity*, viewed 8 October 2013 at www.naturemap.dec.wa.gov.au.
[134] Brian Goodchild, Nomenclature Section, Landgate personal communication MHH, 17 October 2010.
[135] GB, 1869, *Flora Australiensis*, vol. 4, p. 189.
[136] S. D. Hopper personal communication to MHH, email attachment, 24 December 2017, HOA.
[137] *Xanthorrhoea preissii* was collected by Oldfield at Blackwood River, Gordon River and Cape Leeuwin, FM, 1863–64, vol. 4, p. 110.

new species, *Xanthorrhoea pecoris.*[138] However, this name is not current and is a synonym of *Xanthorrhoea preissii.*[139]

It is not known how long Oldfield remained in the Augusta district or by what route he went back to Albany. The shortest, but probably the least comfortable, was to retrace his steps back to Geegeelup along the Blackwood and then return to the Sound via Steeres', Tweed and Tone Rivers and Mount Barker; a journey which would have brought him to Albany about 27 November 1861. This may have been a preferred option, particularly as he had plans to venture east of Oyster Harbour sometime in December 1861. If he was in the company of a *Pibelmen/Bibbulmun* tribesman they may have had an arrangement for both to return to Geegeelup. A longer route was to go north to Busselton and then east across to Steeres' via Balingup, and an even longer one was to take the overland route to Albany via Busselton, Bunbury and Kojonup. He could have journeyed back to Albany by sea, but the infrequency of ships visiting Augusta renders this unlikely and more so if the difficult, and dangerous, task of loading his horse or horses onto a vessel moored offshore had to be undertaken.

In terms of the number of specimens collected, it is not possible to come to conclude whether or not Oldfield did justice to the effort and expense of his expedition to Cape Leeuwin via the Blackwood River. The very few records of specimens we have located from Cape Leeuwin and the environs of Augusta suggest that Hardy Inlet and further south do not appear to have been a focus for the expedition. One explanation for this is that the flora was well known from the collections of James Drummond Snr and Georgiana Molloy. However, this may not have been the case as Mueller was not the recipient of their material. He was keen that the herbarium at the Botanic Gardens, Melbourne, embrace the flora across the nation so it is doubtful that he would have encouraged Oldfield not to collect there. Floristic conditions may have prevented him from gathering plant cuttings for pressing. Given the lateness of the flowering season he may have had little alternative but to concentrate on gathering seeds from as many species as possible around Augusta and along the Blackwood River but this cannot be confirmed. Furthermore, collecting, labelling and packaging 'objects of natural history' for Professor McCoy could have been time consuming, although no such material from this expedition has been found (see above). Heavy rain may have also slowed his journey down the river. It is possible that he cut short his visit to Augusta and the Cape Leeuwin promontory to get back to Albany in time to keep any arrangements he had made for his trip east of Oyster Harbour along the coast.

---

[138] ibid.
[139] *Xanthorrhoea pecoris*, APNI.

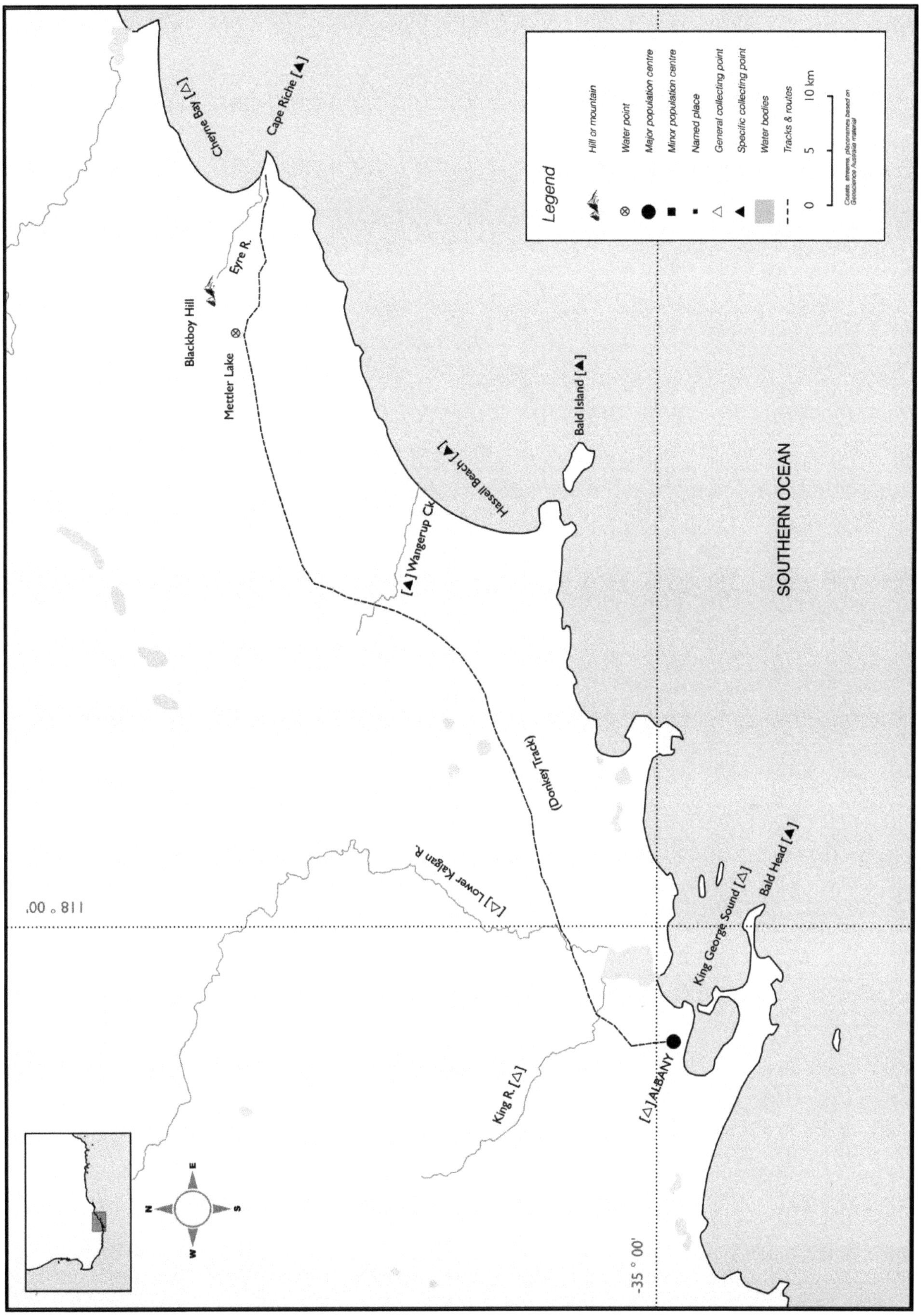

Map 16: Albany to Cape Riche

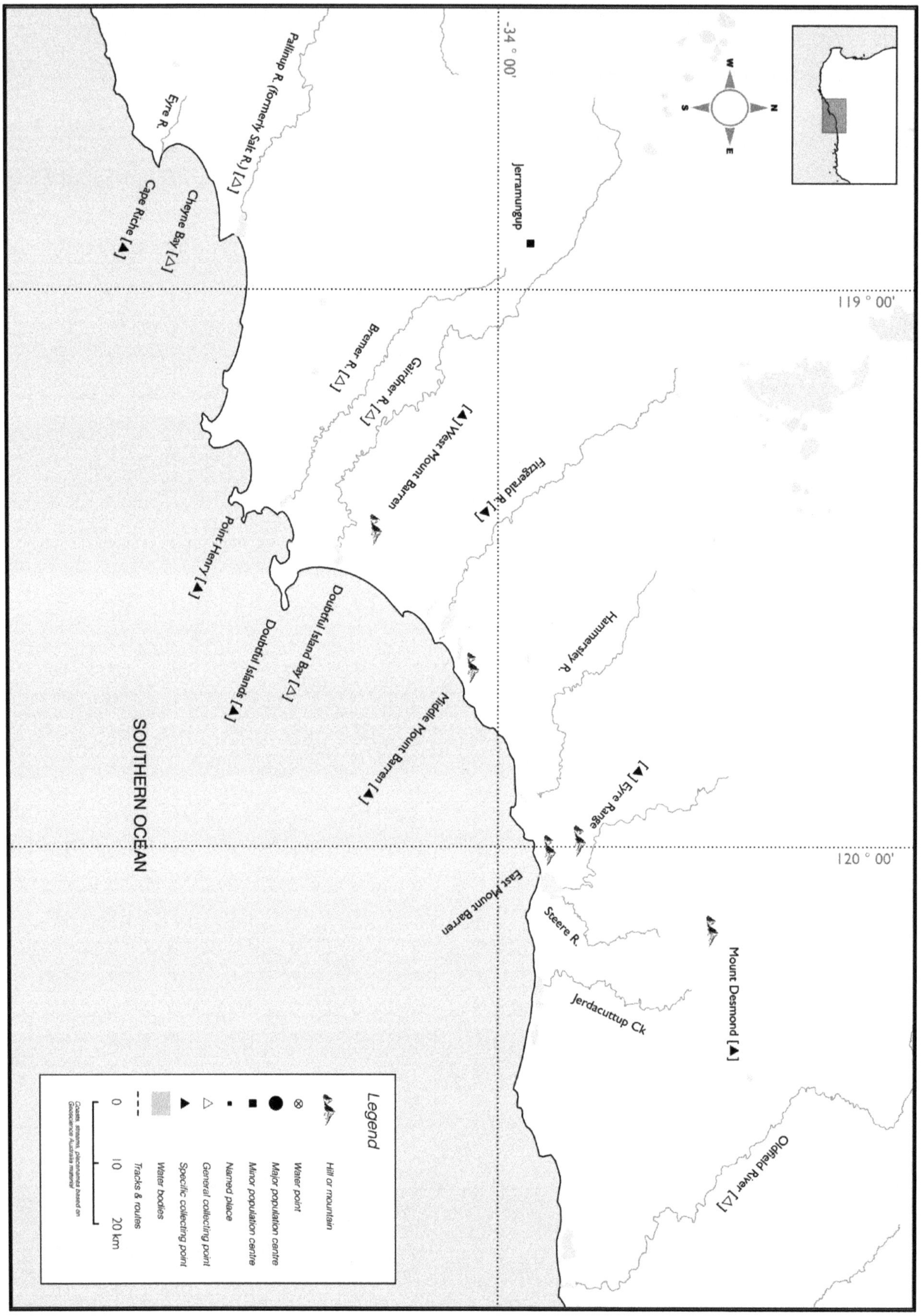

Map 17: Cape Riche to the Oldfield River

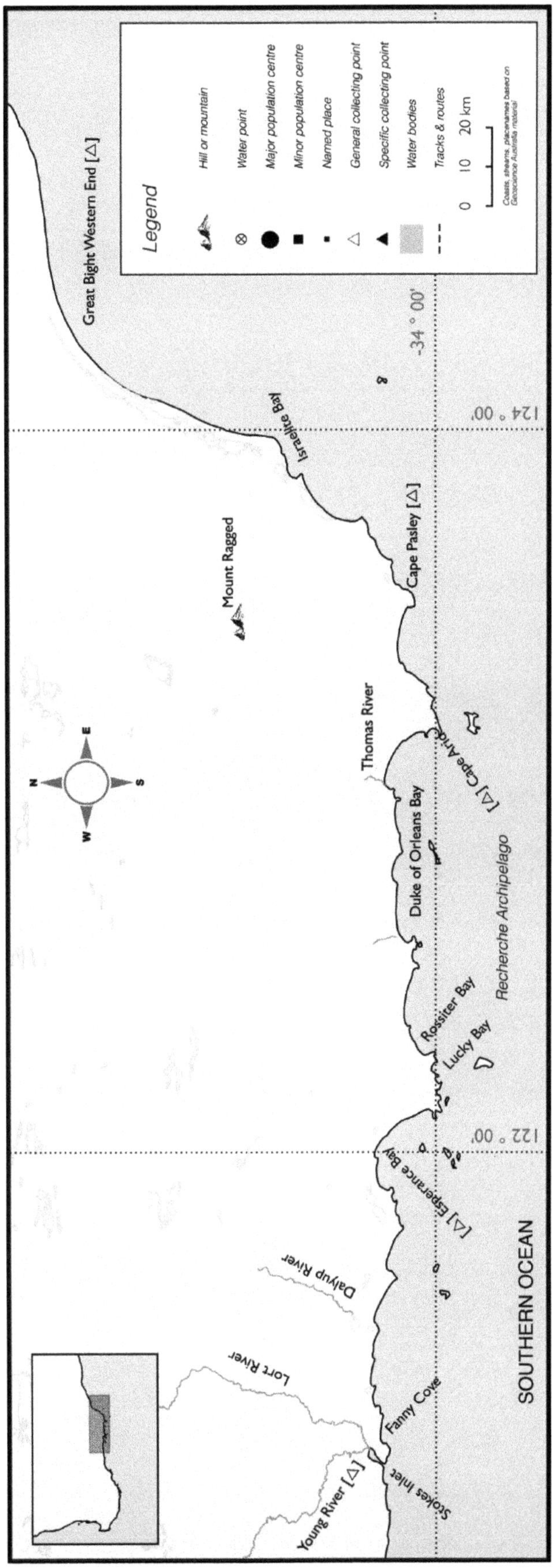

Map 18: Stokes Inlet to the Great Australian Bight

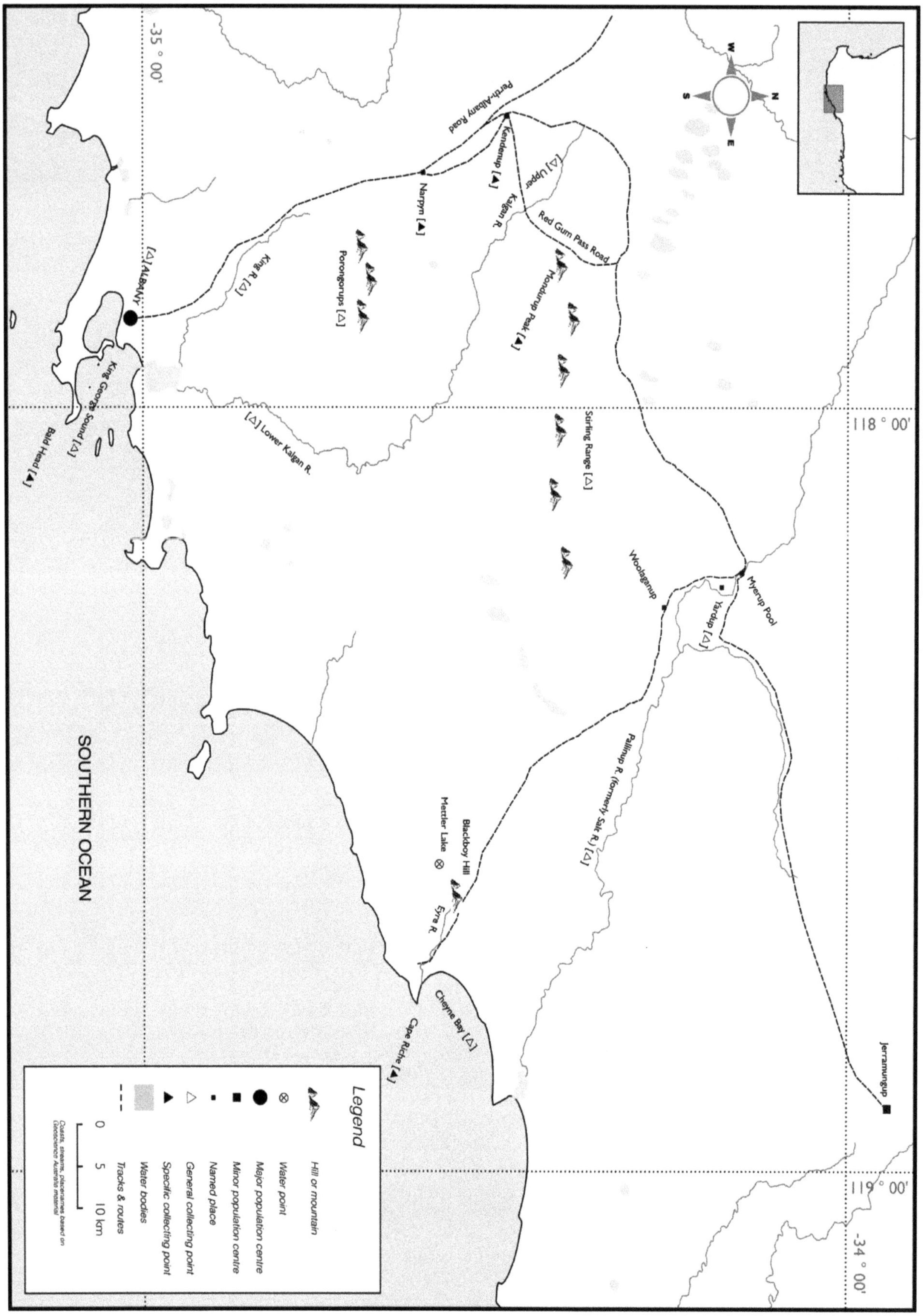

Map 19: Cape Riche back to Albany

Fig. 20.0 Calytrix oldfieldii

Chapter 20

Western Australia 1861–1862

Albany to the Great Australian Bight

Oldfield set out on an expedition to the east soon after he returned to Albany from Cape Leeuwin. Many specimens bear testimony to his undertaking such a journey, and that he did so at that time is borne out by his statement that he had botanised 'in the vicinity of Doubtful Island Bay in the early part of 1862'.[1] On the basis of some specimen records it would seem that Maxwell and Oldfield joined forces on that trip but only one joint collection has been found.[2] Five of Maxwell's specimen records in the National Herbarium of Victoria bear dates within the period December 1861 to February 1862. These show that he was at Stokes Inlet in December 1861[3] and in the Stirling Range[4] in January 1862. Both were places where Oldfield collected (see below). His records of plants gathered to the east of Oyster Harbour, although undated, indicate that they went to Cape Riche[5] and collected close to and at various places along the coast, possibly as far as Israelite Bay (see below), and returned to Albany via the Stirling Range. Maxwell's dated specimens would suggest that the pair set out together at the beginning of December 1861; we assume that they did.

Oldfield's obituary in the *Argus* stated that he had walked 'from Adelaide to King George's Sound'.[6] The copy the authors received was corrected in his brother Edwin's handwriting to 'from King George's Sound towards Adelaide'. There is no evidence that this was a pedestrian expedition or that he went beyond the western end of the Great Australian Bight.

European collectors east of Oyster Harbour to Cape Arid pre-1861

Map 16 p. 367, Map 17 p. 368 and Map 18 p. 369

Labillardière, Ventenat, Deschamps and Riche, the naturalists on the D'Entrecasteaux expedition, were the first to make collections on the south coast east of Oyster Harbour. This was in December 1792 on Observatory Island, in the Recherche Archipelago where the expedition was anchored, and during two visits to the mainland near Esperance Bay.[7] Labillardière used Western Australian specimens in the compilation of his *Novae Hollandiae Plantarum Specimen.*[8]

Between 9–16 January 1802, Matthew Flinders was east of King George Sound near the Recherche Archipelago, where Robert Brown, Ferdinand Bauer and Peter Good collected on the mainland at Lucky Bay and Frenchman Peak, and on the Archipelago on Middle and Goose Islands. From there the expedition sailed eastward to Port Jackson and plant collections were made at various locations along the coast.[9] Flinders returned to the south-west coast during his circumnavigation of the continent

[1] AO, 1865a, fn. to p. 484.
[2] *Acacia incurva*, Mount Desmond, George Maxwell, also Oldfield, GB, 1864, vol. 2, p. 322.
[3] MEL 66199, *Cheiranthera filifolia var. filifolia*, river entering Stokes Inlet (Young River), December 1861, coll. G. Maxwell (MELISR database, 02 April 2009, NHV, RBGM)
[4] See for example, MEL 2259432, *Stylidium repens*, 7 January 1862, Mongerup (now Mondurup Peak, Stirling Range), Maxwell, (ibid.)
[5] HAODb.
[6] Anon. ? Edwin Oldfield, '*The Argus*', 13 July 1887, p. 5.
[7] Leslie Marchant, 1982, pp. 93–99.
[8] Helen Hewson, 1999, p. 66.
[9] T.G. Vallance, *et al.*, 2001, pp. 109 – 117.

between July 1802 and July 1803; on this voyage, the naturalists revisited Middle and Goose Islands between 17–21 May 1803 and obtained specimens.[10]

On the return journey to France from Port Jackson in 1803, the expedition, under the command of Nicolas Thomas Baudin, undertook surveys along the south coast from Mount Gardner to Bald Island. On the south coast Leschenault[11] and Guichenot[12] gathered plants at King George Sound.

During Captain Phillip Parker King's first voyage of exploration (22 December 1817 to 29 July 1818) Allan Cunningham landed on Middle Island, Recherche Archipelago[13] but it seems that his plant collection may have been disappointing.[14]

From 1823–1825 and again in 1828–1829, the English gardener William Baxter obtained plants and seeds at Lucky Bay and Cape Arid east of Esperance Bay. His collections are housed at the Natural History Museum and Kew Gardens with a few duplicates in other herbaria.[15]

In November 1840, Ludwig Preiss travelled from King George Sound to Cape Riche[16] where, according to James Drummond Snr, 'Preiss in a visit to the farm of the worthy proprietor [George Cheyne] found many of his rarest plants'.[17] On this occasion Preiss's collecting places included Mount Many Peaks, Konkoberup (now Mount Melville) near Cape Riche, Cape Riche and Cheyne Beach.[18]

Drummond Snr headed once more for the south coast late in 1846. This time he was accompanied by George Maxwell, whom the botanist had engaged as an assistant and companion. They travelled from York along 'the King George Sound Road' until they 'got from it a distinct view of the Toolbrunup Hills [now Stirling Range]'. They then headed for Cape Riche via Mount Mongerup (now Mondurup Peak), toward the western end of the Range, and Mount Talyuberup 'to the south east', Myerup, a large pool in the Pallinup (Salt) River north of the eastern end of the Stirling Range, and Kojaneerup Spring to the south. From the Cape, they journeyed eastward along the coast for five days to Doubtful Island Bay and then on to West Mount Barren.[19] In May 1848, Drummond undertook yet another trip to the south, this time on his own, with the intention of travelling further east than he and Maxwell had gone previously. However, according to Erickson, 'the Barren Ranges and their hinterland were so rich in new species of plants that there was no need to go further'.[20]

In 1848, Surveyor General John Septimus Roe explored eastward to Russell Range.[21] During this journey he visited Mount Madden, Bremer Range, Peak Charles, Mount Ridley and Mount Ragged, returning to Cape Riche via the Fitzgerald and Pallinup Rivers.[22] On their way to Cape Riche the party came across one of Drummond's camp sites but made no contact with him.[23] From the Cape they proceeded to Myerup Pool, at the north-east end of Stirling Range, where they called on George Maxwell at his sandalwood camp nearby.[24] Maxwell conducted his sandalwood pulling business from 1847 to 1849.

During the 1840s, apart from his journey with Drummond in 1846, George Maxwell undertook journeys east of Oyster Harbour and made plant collections, which he sent to Kew Gardens and apparently to the Natural History Museum, London, before he became one of Mueller's collectors and

---

[10] ibid., pp. 405–406.
[11] Alex S. George, 2009, pp. 452–453.
[12] ibid., p. 398.
[13] Suzanne Curry, *et al.*, 2002, p. 10.
[14] Marsden Hordern, 1997, *King of the Australian Coast: The Work of Phillip Parker King in the Mermaid and Bathurst 1817–1822*, The Miegunyah Press at Melbourne University Press, Carlton South, p. 59.
[15] Alex S. George, 2009, p. 286.
[16] ibid., pp. 519–520. See also G.G. Smith, 1982, pp. 86–89.
[17] Rica Erickson, 1975, p. 116.
[18] N.G. Marchant, 1990, pp. 131–135.
[19] Rica Erickson, 1975, pp. 114 – 117 and 168.
[20] ibid., p. 120.
[21] R.L. Heberle, undated, 'John Septimus Roe 1797–1878…', viewed 8 May 2008 at www.members.iinet.net.au
[22] Alex S. George, 2009, p. 533.
[23] Rica Erickson, 1969, p. 121.
[24] J.S. Roe, 1854, vol. 7, 1855, p. 146.

correspondents sometime during the 1850s.[25] Unfortunately, there are scant details of the sequencing of his travels to the east since, as noted previously, many of his specimen records do not include date and locality of collection. *Eucalyptus lehmannii* was collected at Cape Riche in 1840.[26] Maxwell received a fee of £20 from John Hassell in 1847, for the discovery of land suitable for pastoral purposes along the Gairdner River to north-west of Bremer Bay, and for information about a suitable shipping point at Doubtful Island Bay.[27] He collected the Qualup Bell (*Pimelea physodes*) in 1848, which bears the locality 'East Mount Barren, the Gairdner River and Middle Mount Barren'.[28] It seems likely he visited all three places at that time but at which one he collected the specimen is a mystery.

In the following year, Maxwell discovered what he thought was coal in the vicinity of Fitzgerald River and carried samples back to the Resident Magistrate at Albany. On his return from Mount Ragged, Surveyor General Roe investigated and confirmed a coal find in the Fitzgerald River valley. On reaching Perth, Roe ordered Augustus Gregory to revisit the area and obtain samples of the coal. He and his brother, Henry, left Perth on 22 March 1849; Maxwell was engaged to lead the party to where he had found coal. He travelled on the *Champion* from Albany to Doubtful Island Bay where he met the brothers. Maxwell and Roe had found different deposits but on investigating the quality of each deposit tests proved them to be unsatisfactory for industrial or domestic purposes.[29]

At the invitation of George Cheyne, William Harvey left Albany for Cape Riche on 27 February 1854 accompanying two of Cheyne's carts. In one of the carts he had a seat when he chose to ride 'but as they travelled barely 3 miles an hour and it was more amusing to walk, [he] went mostly on foot' and gathered plants along the way. In letters to his sister Hannah Todhunter, his brother, Reuben, and William Hooker, William Harvey left quite a detailed account of his journey and of the flora that captured his interest along the way and near the Cape. Given the mileages he gave for distances between camping places and the time taken for the journey, the route taken to the Cape appears to have been along the original Cape Riche Road from Albany (referred to locally as the Donkey Track) that ran to Jerramungup crossing the lower Kalgan River just above Oyster Harbour.This track traversed closely the route taken by today's Albany to Jerramungup Road. Harvey was very disappointed with the small number of algal species along the coast; although he had 5,000 dried specimens there were 'only 120 species among them'.[30]

Maxwell harvested seeds from trees and shrubs for Sir George Grey, Governor of the Cape of Good Hope in 1857; some of these were from the Fitzgerald and Phillips Rivers.[31] During that year he collected at Stirling Range, Pallinup and Fitzgerald Rivers, West Mount Barren, Middle Mount Barren, and Yungunup [Pool] to the west of Bremer Bay[32] and in April 1861, at Bremer Bay and possibly Cape Riche in June of that year.[33] He made further collections at various places from the

[25] Alex S. George, 2009, 'p. 480.

[26] MEL 1609489, *Eucalyptus lehmannii*, Cape Riche, 1840, George Maxwell, MELISR database, 02 April 2009, NHV, RBGM.

[27] W.G. Henderson, 2011, vol. 13 (5), p. 634.

[28] MEL 102904, *Pimelea physodes*, East Mount Barren, the Gardner River and Middle Mount Barren, MELISR database, 02 April 2009, NHV, RBGM.

[29] The deposit Maxwell discovered was carbonaceous but was not coal. The coal Roe found burnt successfully but produced far too much white ash.

[30] W.H. Harvey to Hannah Todhunter, 9 March 1854; to William Hooker, 12 March 1854; to Reuben Harvey, 30 March 1854, cited in Sophie C. Ducker (ed), 1988, pp. 83–88, 88–90, 90–93. It should be noted that the distance between Albany and Kalgan River given by Harvey as 31 miles (50 kilometres); it is about 12 miles. He may have mistaken Wongerup Creek, about halfway to Cape Riche, for the Kalgan River.

[31] W.G. Henderson, 2011, pp. 637–638.

[32] G. Maxwell, 1858, MEL 2069631 *Comesperma spinosum*, Fitzgerald Ranges and near West Mount Barren; T.G. Maxwell, 1858, MEL 1058503 *Boronia scabra* subsp. *scabra*, Fitzgerald River and Salt River [= Pallinup River]; G. Maxwell, 1858, MEL 268974, *Gompholobium venustum*, Middle Mount Barren Ranges; G. Maxwell, June 1858, MEL 2170976, *Banksia baxteri*, Stirling Range to Salt River, MELISR database, 02 April 2009, NHV, RBGM.

[33] G. Maxwell, April 1860, MEL 80422, *Daviesia striata*, South West Bay (Bremer Bay). He may also have visited Cape Riche in June 1861. See G. Maxwell, June 1861, MEL 64439, *Grevillea fasciculata*, 'from[?]'Korrinup Range [not located] to Cape Riche', but since the locality is given as a range, place of collection cannot be determined, (ibid.).

eastern shores of Oyster Harbour to Cape Arid between 1862 and 1879.[34] Mueller wrote of Maxwell after his death in 1879:

> Encouraged by myself, he undertook several extensive journeys over then untrodden ground, eastward as far as the Great Bight, and thus found many new plants and enabled us to extend the known limits of the range of many rare species, as recorded in the *Flora Australiensis*.[35]

It can be concluded that prior to and after colonisation in 1826, there had been considerable European interest in the flora east of Albany, particularly along the coast. Extensive collections had been made by professional botanists and collectors before 1861 at various localities east of Oyster Harbour as far as Cape Arid, including some islands of the Recherche Archipelago. However, Maxwell appears to have been the only collector who sent Mueller specimens prior to 1861. Earlier collections were sent to Britain and Europe. It would be reasonable to suggest, therefore, that Mueller was keen to have material from this botanically diverse region and encouraged Oldfield and Maxwell to undertake the expedition to the east of Oyster Harbour.

Oldfield and Maxwell travel to Cape Riche

Map 16 p. 367 - Albany to Cape Riche

Oldfield collected at Wongerup Creek[36], which was traversed by the Donkey Track[37] at approximately 63 kilometres east of Albany. Based on this record it is assumed this was the route they took to Cape Riche. Once again, the mode of travel is a matter for speculation. While Oldfield and Maxwell may have gone to Cape Riche on horseback with one or two packhorses, carrying their collecting and other equipment, food, water, and other necessary supplies would have been much easier in Maxwell's spring cart.

Moving along the track they went in a north-easterly direction from Albany, crossing the King and Lower Kalgan Rivers and travelling through Jarrah-Marri forest until they were within a few miles of Wongerup Creek and about where they entered the Mallee heathlands of the Marine Plain Eco-system.[38] About 25 kilometres[39] beyond Wongerup Creek their route took them south along one of the tracks to Cape Riche, one of which came to be known as the Sandalwood Track.

To the east of Albany, the flora of the Marine Plain Eco-system is very rich in species, many being endemic to the area. Owing to the weak development of laterite a 'shallow' soil type is uncommon, but the 'moderate' and 'deep' types can be readily distinguished underlying respectively the widespread Mallee heath (*kwongkan*) and a *Banksia-Lambertia* association. The most conspicuous species of the Mallee heath are Talyerak (*Eucalyptus pleurocarpa*) and Lerp Mallee (*E. incrassata*), dominance passing from the former to the latter in a west to east direction. Patches of Christmas Tree/*Moojar* (*Nuytsia floribunda*) occur in depressions. In areas where the depth of sand in the topsoil is greater, the Mallees drop out and are replaced by Proteaceae and Christmas Tree. In the west

[34] ibid.

[35] F. Mueller, 1880, letter to the *Gardeners Chronicle*, 3rd April 1880, p. 433, viewed 20 January 2018 at www.biodiversitylibrary.org, and J. H. Maiden, 1909, 'Maxwell, George (1804–1880)', ADBO, viewed 9 May 2008 at www.anbg.gov.au. In the first four volumes of George Bentham's *Flora Australiensis* reference is made to approximately 854 of Maxwell's specimens (list prepared for Dr Neville Marchant, former Curator, the Western Australian Herbarium and kindly made available to W.G. (Bill) Henderson).

[36] MEL 2112198, no details given, 'Wongerup Creek', undated, AO, MELISR database, 18 November 2005, NHV, RBGM.

[37] Albany District Office, lithograph 446/80, 14 October 1920, J.P. Camm, Surveyor General, Lands and Surveys Department, Western Australia, SROWA. According to Ruth Moir this road has long been referred to by settlers as the Donkey Track, Ruth Moir, personal communication to MHH, 19 October 2010.

[38] Sarah Barrett *et al.*, 2009, p. 14 and Fig. 1, South Coast Ecodistricts.

[39] *Daviesia nudiflora* subsp. *nudiflora*, (MEL 79069), was collected by Oldfield at 'Blackboy Swamp', MELISR database, 18 November 2005, NHV, RBGM. There is a swamp close to Donkey Track, 24 kilometres east of Wongerup Creek, west of Blackboy Hill, Albany District Office, lithograph 446/80, 14 October 1920, J.P. Camm, Surveyor General, Lands and Surveys Department, Western Australia, SROWA. However, this specimen may not have come from this place since, according *to FloraBase*, the southern limit of its range is much further to the north.

Fig. 20.1 East bank of Lower Kalgan River

Banksia attenuata and Christmas Tree/*Moojar* are infilled with shrubs including Scarlet Banksia (*B. coccinea*), Baxter's Banksia (*B. baxteri*), Hood-leaved Hakea (*Hakea cucullata*) and Chittick (*Lambertia inermis*).[40]

The depth of the soil determines other aspects of the *kwongkan*. Sometimes the upper layer is dense with numerous species of Mallees and large shrubs; sometimes it is sparse and reduced to a few scattered Talyerak and Cauliflower Hakea (*Hakea corymbosa*), and sometimes it reduces to heath only in which *Andersonia parvifolia*, Hidden Featherflower (*Verticordia habrantha*), and Plumed Featherflower (*V. plumosa*) are dominant. The steep sides and the floors of valleys cut down into the plain, clothed mainly with Mallee species; *Eucalyptus phaeonophylla*[41] and Hook-leaved Mallee (*E. uncinata*) being the dominant species. Occasional large shrubs of Cayley's Banksia (*Banksia caleyi*), Pincushion Hakea (*Hakea laurina*), Thick-leaved Hakea (*H. crassifolia*) and Cauliflower Hakea (*H. corymbosa*) are seen. A dense understorey of *Melaleuca* spp. covers the ground below. In wet and clayey ground, thickets of Moort (*E. platypus*), Swamp Mallet (*E. spathulata*) and Open-fruited Mallee (*E. annulata*) replace the Mallee.[42]

There is some confusion about the location of the Sandalwood Track. According to Ruth Moir, who, with her husband Keith, frequented Cape Riche over many years from c.1949, the area from Donkey Track into Cape Riche is peppered with bush trails—as they became washed out by heavy rains and impassable, new ones were formed. There were, however, two general routes—one crossed the upper reaches of Eyre River and then ran down to the Cape Riche homestead situated on the west side of the river near its mouth; the other ran a little east of the river past Mount Melville down to Cheyne Bay.[43]

---

[40] J.S. Beard, 1990, pp. 139–144.
[41] Beard referred to *Eucalyptus redunca* as being one of the dominant mallees in the area but according to S. D. Hopper it is *E. phaeonophylla* is dominant along with *E. uncinata*, personal communication to MHH, email attachment, 24 December 2017, HOA.
[42] J.S. Beard, 1990, pp. 139–144.
[43] Ruth Moir, personal communication, 19 October 2010.

Fig. 20.2 Cape Riche homestead

Whichever route they took from the Donkey Track, Oldfield and Maxwell would have travelled for some 16 kilometres through Mallee heath with conspicuous species including Talyerak, Chittick, Hood-leaved Hakea, Albany Blackbutt (*E. staeri*), Bell-fruited Mallee (*E. preissiana*), *Allocasuarina trichodon,* Mouse Featherflower (*Verticordia fastigiata*) and *Acacia sulcata.* As they descended closer to the coast, they passed through coastal heath containing many species, among them *Eucalyptus astringens* subsp. *redacta* (on the lower slopes of Mount Melville/ Konkoberup), *Acacia* spp., White Goodenia (*Goodenia scapigera*), *Leucopogon* spp., *Hakea pandanicarpa* subsp. *crassifolia*, *Kunzea pauciflora*, Hidden Featherflower and Chittick.[44]

The 'Cape Riche' homestead is situated above the east bank of Eyre River, less than one kilometre above its mouth, and a little above the Albany Fraser West Eco-zone along the coast. By the time the naturalists arrived at the Cape, the situation on the property had changed since William Harvey's visit. In 1858, Cheyne sold the property to his nephew, Andrew Moir, who had worked on the station for some years.[45] The homestead and outbuildings, designed and built by Andrew's brother, Alexander[46], had been completed in 1860 with convict labour.[47] Andrew exported wool from the Cape. In 1842 Cheyne had established an independent port, supplying whaling, sealing and some passenger ships with fresh vegetables, meat, flour milled on the property, butter and eggs, and bought up American whalers' cargoes of merchandise. In about 1847 sandalwood cutters, including George Maxwell[48], arrived in the area and it appears that carting the sandalwood to Cape Riche may have been undertaken by bullock teams organised by Cheyne, who exported it to Albany for shipping to China and Ceylon.[49]

[44] This information on the flora of Cape Riche was kindly obtained by Ruth Moir from the records of the Albany Herbarium, a branch of the Western Australian Herbarium, 19 October 2010.
[45] Rica Erickson, 1979, vol. 3, p. 594.
[46] Heritage Council of Western Australia, 'Cape Riche Homestead', Place no. 81, State Heritage Register, viewed 8 November 2013 at www.inherit.stateheritage.wa.gov.au.
[47] Rica Erickson, 1979, vol. 3, p. 594. In this publication Andrew Moir is credited with building the Cape Riche Homestead but the Heritage Council of Western Australia gives Alexander as the architect and builder.
[48] W.G. Henderson, 2011, p. 634.
[49] Amelia Moir and Greg Wardell-Johnson, 1995, *The Continuing Moir Saga: A Genealogical Record of an Albany Family*, Warjam Publishing, Albany, p. ix.

Although Cape Riche had apparently ceased to be an independent port by the early 1860s, Andrew may have continued to operate the trading post.[50]

Bald Island and Cheyne Beach

Map 16 p. 367 - Albany to Cape Riche

In his unpublished manuscript, Oldfield stated that while he was botanising near Doubtful Island Bay, 'the rest of the party started off in the boat, to hunt seals on some small rocky islands in the neighbourhood'.[51] It seems, therefore, that he and Maxwell left the horse and cart or their hacks and pack horses in the care of Andrew Moir and that they obtained passage at Cape Riche on a sealing ship. The crew of this vessel may have been picking up supplies or were using the whaling station located at the end of Cheyne Bay opposite Cheyne Island three kilometres north-west of Cape Riche.[52]

After settlement commenced on the east coast of the continent in 1788, many sealing and whaling vessels visited the south coast of Western Australia. In 1842, William Nairne Clark stated that, according to the 'traditions of the natives' prior to colonisation in Western Australia, the coast was visited by parties in search of Fur Seals[53] (*Arctocephalus pusillusdoriferus*). This species occurs from about Eclipse Island, Western Australia, to a little beyond Kangaroo Island, South Australia.[54] The sealers and whalers, noted Clark,

> frequently made inroads into the territory of the aborigines, and endeavoured to carry off the women, which infringement on their natural rights roused the natives to fury, and several collisions between the blacks and whites. Even now, in talking of these marauders the natives described them with symptoms of loathing and innate hatred.[55]

Among the settlers, sealers had the reputation of being lawless, unscrupulous and violent.[56] They subsisted on kangaroos, small wallabies and birds but chiefly on wild geese and seal flesh, occasionally making a run to King George Sound to purchase flour and other goods, and to sell their seal skins. It is also on record that some sealers, who remained in the Recherche Archipelago with their Aboriginal partners and children during the off-season for seals, hunted kangaroos and small wallabies for domestic purposes and trading with American whalers.[57]

While the Recherche Archipelago appears to have been what Clark referred to as '[t]he great rendezvous of the sealers to the eastward of Doubtful Island Bay'[58], there were, in addition to Cape Riche, other safe anchorages where whaling stations had been established. These were at Cheynes Beach, Cape Riche, Doubtful Island Bay, Middle Island, Recherche Archipelago, and Barrier Anchorage, four kilometres almost due north of Cape Arid. Thomas Fishery is seven kilometres east-north-east of the Cape.[59] Fresh water was available at some, if not all, of these stations.[60]

There must have been an agreement with the sealers regarding the conditions under which Oldfield and Maxwell were transported along the coast, setting out any charges, provisioning arrange-

[50] ibid., p. ix, and Martin Gibbs, 1994, *An Archaeological Conservation and Management Study of 19th Century Shore-Based Whaling Stations in Western Australia*, A Report to the National Trust of Australia for the Australian Heritage Commission, p. 79. According to Gibbs, the operation was 'a bold and astute move which proved highly successful, allowing the masters to avoid paying the excessive harbour fees of Western Australian ports and minimising the risks of desertion and drunkenness that arose on visits to settlements'.
[51] AO, 1865a, fn. to p. 484.
[52] Martin Gibbs, 1994, p. 79.
[53] Clark, W.N., 'Remarks respecting the islands on the coast of S.W. Australia', *Perth Gazette and Western Australian Journal*, 8 October 1842, p.3, DAN, NLA.
[54] Judith E. King, 1964, *Seals of the World*, Trustees of the British Museum (Natural History), London, p. 33.
[55] W.N. Clark, '*Perth Gazette and Western Australian Journal*, 8 October 1842, p.3.
[56] Keith Murray, 2004, *First There Came Ships: Over 200 Years of Ships, Seamen, Wrecks, Pirates, Sealers, Settlers and Miners Along the South-east Coast of Western Australia*, Keith Murray, Esperance, pp. 14–16.
[57] W.N. Clark, *Perth Gazette and Western Australian Journal*, 8 October 1842, p.3.
[58] ibid., pp. 3–4.
[59] Martin Gibbs, 1994, pp. 74–75,79,82-83, 86–87, 90, 94–95, JSBL.
[60] ibid., pp. 74–94, and W.N. Clark, *Perth Gazette and Western Australian Journal*, 8 October 1842, p.3.

ments, where they slept, where they stored their equipment and specimens, and how they were taken ashore to botanise. Unfortunately, they left no details.

Oldfield gathered plants on Bald Island, which lies off the south end of Cheynes Beach where he also collected, but whether they visited this area before or after their trip to the east is not known.[61] The sealers' quest for seals was probably the reason that the naturalists were able to land on Bald Island. While they were there they may have obtained some of the 'wallabies' (Quokka, *Setonix brachyurus*) that abounded there, to augment their meat supplies.[62]

The visit to Cheynes Beach was possibly motivated by the sealers' need to replenish their water supplies from a perennial spring, located in the dunes a little over one kilometre north-west of the headland, or they bivouacked there. Perhaps, however, they used the whaling station, under John Thomas's management in 1861, to prepare seal skins. The station would have been deserted when Oldfield and Maxwell were there as Thomas used it only from June to August as a base from which to hunt Humpback Whales (*Megaptera novaeangliae*).[63]

From a boat, there are many places along the south coast where seals, particularly Fur Seals, can be observed basking in the sun on granite headlands and islands. If the sealers visited Bald Island before travelling east along the coast, it may have been Oldfield's first, possibly hair-raising, experience of landing on an island of granite with no beach to facilitate disembarkation. William Nairne Clark left a colourful description of the method of landing on granite islands on the south coast:

> The sealers approach them [Fur Seals] facing the wind, because the seal is so quick scented that if this precaution is not taken they are sure to feel the approach of man by the organ of sense, and dive into the sea. The crew of a sealing boat is generally composed of 4 men and a boat-steerer, and is built in the whale boat fashion, but much larger than common whale boats, with a lining of canvass of about a foot in breadth all along the gunwale of the boat, to exclude the sea spray in heavy weather. In these boats, the hardy sealer approaches the rocks where the seals repose, and often during a heavy swell. The headsman prepares himself, and standing on the bow of the boat, jumps on the rock armed either with a club or gun. The moment he jumps from the boat, *stern-all* is the cry and it is instantly propelled backwards to prevent it from being dashed to pieces on the rocks... When the work is done the boat is again brought near the rocks and the landsman and his assistant jump in, but it is dangerous work and requires the greatest dexterity and skilful [sic] management of the oars by the [men] in the boat.[64]

Clark did not describe how the seals were transported back to the mainland. If they could be lifted they were probably thrown into the boat but it seems that at times they were towed. When discussing the 'tenacity of life' Oldfield referred to this method of transport in an illustrative example:

> In the early part of 1862, I happened to be botanising in the vicinity of Doubtful Island Bay. One day, while I was pursuing my vocation on shore, the rest of the party started off in the boat, to hunt seals on some small rocky islands in the neighbourhood. Towards evening they returned, bringing with them one of the kind called hair seal, [Australian Sea Lion, *Neophoca cinerea*][65] which animal had been shot through the head, and afterwards beaten with oars and bludgeons, until it was apparently dead, having in fact been towed a distance of ten miles with its head under water. It revived however, the moment that it touched the land, and rising on its hind flippers, boldly faced its captors, who now set to work in earnest to dispatch it. They beat it about the

[61] HAODb.
[62] W.N. Clark, *Perth Gazette and Western Australian Journal*, 8 October 1842, p.3.
[63] Martin Gibbs, 1994, p. 74.
[64] W.N. Clark, letter to the Editor, *The Perth Gazette and Western Australian Journal*, 3 September 1842, p. 3, DAN, NLA.
[65] W.D.L. Ride, 1970, p. 186.

> head, until that member was a perfect jelly, they disembowelled it, and even partially flayed it, yet in this mutilated condition, it exhibited signs of vitality eight hours after its savage captors had abandoned in despair, the task of killing it.[66]

Although Oldfield made no specific comment as to how he felt about this display of brutality, his reference to the seal's 'savage captors' suggests that he may have been appalled at their behaviour. This, then, begs the question as to how he and Maxwell perceived their own safety in the company of the sealers. Both were bush-hardened explorers who from all accounts could handle dangerous situations. Nevertheless, they were probably very careful not to upset the crew by complaining or demanding time on land that did not fit in with the sealing agenda.

Cape Riche to Oldfield River

Map 16 p. 367 and Map 17 p 368

Oldfield gathered at least one plant from Salt River (now Pallinup River).[67] The Pallinup rises in the vicinity of today's Broomehill and flows in a south-easterly direction towards the coast discharging into the Southern Ocean via Beaufort Inlet when the sandbar is open; this occurs every few years. The shoreline is bordered by Swamp Sheoak (*Casuarina obesa*) and possibly Saltwater Paperbark (*Melaleuca cuticularis*) with Mallets (Talyerak, *E. praetermisssa, E. utilis, E. astringens* subsp. *redacta, E. platypus* and *E. newbeyi*)[68] and Mallees (*E. lehmanii, E. tetraptera, E. angulosa, E. notactities, E. macrandra, E. buprestium, E. sporadica, E. uncinata, E. pluricaulis*)[69] on higher ground.[70] The Inlet is about 15 kilometres east of Cape Riche, which could have been a place Oldfield visited with the sealers. However, on their return to Albany after their journey along the coast, he and Maxwell came close to Pallinup River further north as they travelled back to Albany (see below). According to *FloraBase* distribution maps, both specimens are of species Spike Centaury (*Centaurium spicatum*) and Chittick that could have been collected in sandy situations near each locality.

The collectors appear to have spent a few days in the vicinity of Bremer Bay (labels: 'South West Bay') which lies on the eastern boundary of the Albany Fraser West Eco-system but the surrounding area is still well within the Mallee heath of the Marine-Plain Ecodistrict.[71] His specimens from this area seem to have been more numerous than from other places along the south coast, suggesting that he spent more time in this vicinity than at other localities during the expedition. Apart from around the Bay, he obtained plants on the Mallee heath of Point Henry on the western side of the southern end of the granite headland separating Dillon and Bremer Bays.[72] Specimens indicate that he collected on the seaward face of the dune rising from the beach and on the shores of the estuary.[73]

Oldfield must have soon discovered that, while fringing the shores of the estuary were the familiar Saltwater Paperbark, Sea Rush and the samphire *Sarcocornia blackiana*[74], a little further inland the river banks were lined with low outcrops of spongolite, supporting vegetation dominated by Mallees (*Eucalyptus* spp.).[75]

---

[66] AO, 1865a, fn. p. 482.
[67] MEL 709294, *Centaurium spicatum*, 'Salt River', AO, MELISR database, 18 November 2005, NHV, RBGM.
[68] Information regarding the Mallets occurring on the higher ground was kindly provided by S.D. Hopper, personal communication to MHH, email attachment, 24 December 2017, HOA.
[69] Ibid. also for the Mallees.
[70] Anne Brearley, 2005, pp. 447–452.
[71] Sarah Barrett *et al.*, 2009, p. 14 and Fig. 1, South Coast Ecodistricts, p. 13.
[72] HAODb.
[73] For example, MEL 2140688, *Spinifex hirsutus*, AO probably came from the sand dune above the beach and while *Melaleuca cuticularis* from 'near the sea, South West Bay [Bremer Bay]', AO may have been on the shores of the estuary, MELISR database, 18 November 2005, NHV, RBGM.
[74] Water and Rivers Commission Natural Heritage Trust 2001, *Bremer River and Devils Creek, River Action Plan*, Water and Rivers Commission, Water Resource Management Series Report no. WRM 25 June 2001, 23 October 2010, viewed at www.water.wa.gov.au
[75] Anne Brearley, 2005, p.455.

Fig. 20.3 Doubtful Island Bay

There is little doubt that the pair would have called on John Wellstead and his wife Ann (née Crawford) at 'Peppermint Grove' so named for the abundance of WA Peppermint which grows in the area. The homestead is located towards Dillon Bay, on the base of the headland, separating it from Bremer Bay, and close to Cardiminup Swamp, one of the few permanent, freshwater swamps along the south coast. [76] Following tracks created by Wellstead's pastoral activities may well have been Oldfield and Maxwell's preferred option for exploring the area, particularly since the tracks would have enabled them to easily penetrate the dense species-rich coastal heath.

John, Ann and their children had settled at Peppermint Grove in 1857, although John had been running cattle in the area for some time. In 1860, he had applied for 40 acres freehold at Peppermint Grove, where he had built a substantial stone homestead and established a dairy herd and a 'vast' vegetable garden while still conducting his pastoral activities. Around 1858, Wellstead established an outstation, 'Qualup' on the Gairdner River, about eight kilometres north-west of its mouth, and a lesser dwelling at Fitzgerald River.[77] As far as can be ascertained, there was no further pastoral activity east along the coast until a little later in the decade.

From Bremer Bay, the sealers apparently took the naturalists to Doubtful Island Bay. This bay is north-east, and separated from, Bremer Bay by a headland. The granite boss, Point Hood, is the most easterly point of this headland, off which lie the Doubtful Islands ('islands thereabouts'). In Oldfield's account of killing the seal, the sealers returned to Doubtful Island Bay where they landed the animal. This indicates that was where they camped. The Doubtful Island Bay whaling station occupied an area of ten acres around Whale Point (today's Whalebone Point) at the south end of the bay, which may have been where the naturalists and the sealers camped. During the 1860s Hugh McKenzie used the station in conjunction with one at Torbay[78], but whether he was there at the time of Oldfield and Maxwell's visit is not known.

---

[76] Wellstead Heritage Museum, Bremer Bay, 'John Wellstead', viewed 24 October 2010 at www.wellsteadmuseum.com.au.
[77] ibid.
[78] Martin Gibbs, 1994, p. 82.

Oldfield was in the vicinity long enough to collect around the Bay itself, inland near West Mount Barren (about eight kilometres due west from Trigalow Beach and about eight kilometres north of the mouth of the Gairdner River and within today's Fitzgerald River National Park), along the Gairdner River and on Doubtful Islands.[79] From the number of specimen records relating to this expedition in our Oldfield database, his sojourn at Doubtful Island Bay was one of the most fruitful visits he made along the coast. Still within the Marine Plains Ecodistrict that abuts the Albany Fraser Islands Ecodistrict running along the coastline[80], Oldfield would have observed many similarities between floras on the dunes behind Doubtful Island Bay beach, on the heath and Mallee heath lands further inland, along the shores and banks of the estuary and inland along the river, and those he had encountered in corresponding situations near Bremer Bay.

Following their sojourn at Doubtful Island Bay, Oldfield collected a few specimens at various places along the coast that today are within the confines of the Fitzgerald River National Park, which covers an area of about 329,039 hectares between Bremer Bay and Hopetoun, and extends inland for varying distances along its length. It is one of the richest areas for plants in Western Australia, with 1748 identified species; about 75 of these are endemic and some 250 species are either very rare or geographically restricted. Although endemics occur throughout the Park, the highest concentration is in Barren Ranges[81], which, although they have a barren appearance from a distance, are actually covered in *kwongkan*.[82] Oldfield's collecting localities in this area include Fitzgerald River, West and Middle Mount Barren (located about a mile north of the coast), Eyre Range and Phillips River, but his gatherings from each appear to have been few in number and do not seem to reflect the richness of the flora.[83] It may be that he was there too late for the main flowering season and under such circumstances perhaps he focused on gathering seeds for Mueller. East of the Barren Ranges *Banksia speciosa* replaces the other banksias and the conspicuous large shrubs become Chittick, Red Tooth Brushes (*Grevillea hookeriana*) and Christmas Tree.[84]

The party then landed somewhere near Culham Inlet, which receives water from Phillips and Steere Rivers. The Inlet, a coastal salt lagoon surrounded by Saltwater Paperbark, Beaded Samphire (*Sarcocornia quinqueflora*) and Coast Saltbush (*Atriplex isatidea*), is separated from the sea by a high, vegetated sand dune bar, which was unbroken between 1849 and 1993.[85] Behind the Paperbark fringe, the Inlet's eastern banks are backed by Coastal Moort (*Eucalyptus utilis*),[86] and on the western side by Mauw.[87]

The junction of East Mount Barren and the Jerdacuttup Fault at Culham Inlet forms the western boundary of the Esperance Sand-plain West Ecodistrict, which runs east to the eastern divide of the Lort River drainage system. It comprises *kwongkan* heathlands, *Banksia* and Mallee (*Eucalyptus* spp.) shrublands and wetland chains.[88]

Although Oldfield appears to have obtained very few specimens during this sojourn, he must have been there for a few days as he made collections at Phillips River[89], which enters the Inlet from the north-west, and Eyre Range, 12.8 kilometres north-west of the Inlet bar. He and Maxwell also

---

[79] HAODb.

[80] Sarah Barrett *et al.*, 2009, pp. 13 & 14 and Fig. 1, South Coast Ecodistricts.

[81] Sue Moore (Coordinator), Matt Cavana, Kelly Gillen, Chris Hart, Steve Hopper, Kate Orr, Wayne Schmidt, 1991, *Fitzgerald River National Park Management Plan 1991–2001*, Department of Conservation and Land Management (Department of Parks and Wildlife) for the National Parks and Nature Conservation Authority, Perth, Western Australia, p. iii, viewed 24 October 2010 at www.mainroads.wa.gov.au.

[82] J.S. Beard, 1990, p. 150.

[83] HAODb.

[84] J.S. Beard, 1990, pp. 139–144.

[85] Anne Brearley, 2005, pp. 475–478. According to Brearley, the bar was broken artificially in 1920, but for over 70 years since then the Inlet has absorbed river flow without the bar breaking, until 1993 when it broke again as the result of the river flooding, (ibid., p. 475).

[86] Brearley refers to Coastal Moort as *Eucalyptus platypus*, *E. utilis* is correct, S.D. Hopper, personal communication to MHH, email attachment, 24 December 2017.

[87] ibid., pp. 479–480.

[88] Sarah Barrett *et al.*, 2009, p. 13 and Fig. 1, South Coast Ecodistricts, p. 13.

[89] HAODb.

ventured to Mount Desmond (see above) where they jointly collected *Acacia incurva*.[90] Mount Desmond is an elongated whaleback ridge about 35 kilometres north-east from the Inlet bar. On the face of it, the most direct route to the prominence would have been to follow Steere River upstream to its upper reaches. This is probably the course they took. The prominence lies south-east of today's Ravensthorpe and forms the southern part of the Ravensthorpe Range. Overall this range has a diverse flora and within it are areas of high species concentrations and localised endemism[91], which Oldfield may have recognised.

As they moved upstream, their route led them through *kwongkan* heathlands of the Esperance Sandplain West Ecodistrict and into the Greenstone Range Ecodistrict.[92] This is undulating or hilly Mallee country. On the hill slopes the most common species are *E. astringens subsp. redacta, E. gardneri subsp. ravensthorpensis* and Green Mallet, *E. clivicola*). There are at in excess of 50 species in this area. The area has the richest concentration of eucalypts of any local area in Australia. In broad valleys where the soil is at its deepest, are woodlands of Small-fruited Gum (*E. myriadena*) and Salmon Gum (*E. salmonophloia*).[93] The thickets of the summit ridges of Mt Desmond are dominated by low Mallees, Bell-fruited Mallee (*E. preissiana* subsp. *preissiana*), the Bushy Yate (*E. lehmannii* subsp. *parallela*), with Oak-leaved Banksia (*Dryandra quercifolia*).[94] Represented in the botanically diverse understorey are many species of *Acacia, Austrostipa, Boronia, Caladenia, Daviesia, Gastrolobium, Grevillea, Hakea, Hibbertia, Lepidosperma, Leucopogon, Melaleuca, Schoenus, Stylidium* and *Verticordia*.[95]

By this time the summer dry season had set in, but as the collectors trudged through the bush it may not have been hot, since temperatures in this area do not often rise above 30°C.[96] Nevertheless, the terrain is not only hilly in places but also the ground is uneven and covered in loose stone, making for difficult and tiring progress on foot. The water in the river was likely to be too saline to be potable, rendering it necessary for them to carry their supplies.[97]

The collectors next port of call appears to have been near Oldfield River. This is estuarine for eight kilometres and towards its lower end opens to a meandering basin nearly three kilometres long with a maximum width of about 500 metres. If the naturalists had been there following heavy rainfalls in the catchment, the usually wide sandbar across the river mouth may have been broken.[98] The banks of the Oldfield River are rich in 'cultural artefacts and plants' and it has 'clearly has been a trade and ceremonial corridor for Noongars and Ngadju/Bardok people for many generations'.[99]

Their stay there may have been brief because Oldfield apparently confined his collecting to the estuarine end of the river where very few specimens were obtained.[100] Beyond the coastal sand dune facing the ocean, held together by low shrubs and grasses (for example, *Spinifex hirsutus*) is the species-rich coastal scrub of the sheltered, leeward dunes, which overlook the estuary. This is dense low heath characterised by *Melaleuca striata* interspersed with Talyerak (*Eucalyptus pleurocarpa*). Common species include *Calothamnus gracilis, Conothamnus aureus, Adenanthos cuneatus, Xanthorrhoea platyphylla* and *Acacia, Allocasuarina, Leucopogon, Astroloma, Banksia, Hakea* species and the

---

[90] *Acacia incurva*, Mount Desmond, George Maxwell, also A. Oldfield, GB 1864, vol. 2, p. 322.
[91] R.J. Harris, J.D. Majer, C. Williams, A. Buckley & D. Stehlik, 2008, 'Ravensthorpe Range and Overshot Hill – An Overview of Biodiversity Values, Threats and Conservation, ALCOA Foundation's Conservation and Sustainability Fellowship Program, *Sustaining Gondwana*, issue 12, September 2008, p. 20.
[92] Sarah Barrett *et al.*, 2009, p. 14 & Fig. 1, South Coast Ecodistricts.
[93] S.D. Hopper, personal communication, attachment to email, 14 December 2017.
[94] J.S. Beard, 1990, p. 153.
[95] R.J. Harris *et al.*, 2008, p. 22.
[96] According to the Australian Bureau of Meteorology the average daily maximum temperature for Ravensthorpe is 27°C, viewed 26 October 2010 at www.bom.gov.au.
[97] Anne Brearley, 2005, p. 476.
[98] ibid., p. 482.
[99] S. D. Hopper, personal communication, email to MHH, 27 June 2017. Professor Hopper was a member of a survey team (2015) which included Noongar Elders and archaeologists investigating the edge of farmland between the Ravensthorpe and Koomal areas.
[100] HAODb.

sedges *Anarthria laevis, A. scabra* and *Lepidosperma* sp. Fringing the estuary are Saltwater Paperbark (*M. cuticularis*), Swamp Yate (*E. occidentalis*), Bare Twigrush (*Baumea juncea*), *Acacia cyclops* and other species.[101]

Was the Oldfield River named in honour of Augustus?

Although there is no confirmatory official record, it is generally believed that the Oldfield River was named after Augustus. According to a record in the Nomenclature Section of the Department of Lands Administration[102] (WA), the earliest use of the name is in a letter-diary dated 1 June 1866 by William Simon Dempster.[103]

As far as we can ascertain, at the time the only other Oldfields in Western Australia were the ticket-of-leaver John Oldfield, his wife, Sarah, and their six-year-old son.[104] John had arrived in the Colony in 1852 and was a labourer residing in Guildford.[105] No evidence has been found that he was ever involved in any activity on the south coast. It is, therefore, unlikely that the river was named after him.

It has been suggested that it was named by Albert Young Hassell of 'Jerramungup' who crossed this river during an exploration to the east from August to September 1861 and that the 'white man about forty' with Hassell's party may have been Oldfield.[106] However, Albert Hassell's wife Ethel in her account of the journey, based on Albert's diary of the journey, indicates that the 'white man' was 'Old Bob' who was probably one of Albert's employees.[107] Oldfield did not arrive in Western Australia until sometime early in October 1861 (see Chapter 17) and was not on the coast to the east of Oyster Harbour until December of that year. These findings must dispel the idea that Oldfield accompanied Hassell on that expedition. If Hassell did name the stream after Oldfield, it may have been as the result of an earlier encounter. As far as we are aware, Oldfield never visited Jerramungup, north-east of Cape Riche. If the two met it may have been on one of John Hassell's properties closer to Albany, such as 'Takalarup' on the Upper Kalgan River. However, we believe there is a more plausible explanation as to how the river got its name.

There is an 1860 record of *Melaleuca fulgens* with the locality given as 'Cape Arid, Young River' collected in 1860 by Maxwell. This is the only indication that Maxwell had been as far as Cape Arid prior to his and Oldfield's expedition, but it is unfortunate that Young River, which empties into Stokes Inlet, is included on the label.[108] It is, therefore, possible that he was aware of unnamed rivers that fed into the ocean along the south coast. Maxwell may have been responsible for naming the river, possibly having signified his intention to do so to Oldfield while they were there. This could explain why the name appears on specimen labels. If that was the case, Maxwell probably later conveyed his request to the Resident Magistrate at Albany, who, in turn, forwarded it to the Lands Department in Perth. Every effort has been made to locate this correspondence but without success. There is reference to 'Oldfield River' and 'Mount Oldfield' in Edwin Oldfield's handwriting at the foot of the newspaper clipping of Augustus's obituary[109], which suggests that Edwin had been told by Augustus

---

[101] Information kindly provided by Dr Gillian Craig, Consultant Biologist, Ravensthorpe, Western Australia, personal communication, email to MHH, 27 January 2014.
[102] 'Oldfield River', undated official record, Nomenclature File 3664/61, Department of Lands Administration (Western Australia), fol. 610. kindly provided by Mr Brian Goodchild, Nomenclature Section, HOA.
[103] William Simon Dempster, letter-diary, 1 June 1866, Acc. no. 1186A, JSBL.
[104] Rica Erickson (ed), 1979, vol. 2, p. 412.
[105] Rica Erickson (ed), 1979, vol. 3, p. 645. John is included in this volume but was not given his conditional pardon until 1863.
[106] Nomenclature File 3664/61, fol. 610. Department of Lands Administration (Western Australia).
[107] Ethel Hassell, 1975, pp. 193–194.
[108] G. Maxwell, MEL 2161311, *Melaleuca fulgens*, 'South side at the base of Cape Arid, Young River', 1860, MELISR database, 02 April 2009, NHV, RBGM. Young River flows into Stokes Inlet much further to the west of the Cape. It is not clear where he collected this specimen.
[109] Anon.? Edwin Oldfield, *The Argus*, 13 July 1887.

Fig. 20.4 Fitzgerald River estuary

Fig. 20.5 Mid-Mount Barren

Fig. 20.6 From Mount Desmond, looking north

Fig. 20.7 Oldfield River estuary

that these two geographic features were named after him. Mount Oldfield, however, has not been found. Therefore, while extant records tend to support the claim that the river was named after Augustus we cannot be certain.

Stokes Inlet to the western end of the Great Australian Bight

Map 18 p. 369 – Stokes Inlet to the Great Australian Bight

Oldfield and Maxwell visited Stokes Inlet further to the east.[110] They may have agreed with Brearley that it was one of the most attractive estuaries of the south coast. It is set in a deep valley between high, densely-wooded dunes with shady Saltwater Paperbark trees along the shores bordering relatively deep permanent water. There are also sedges along the shore and on low sandy beach ridges, alternating with samphire. The Young and Lort Rivers flow into Stokes Inlet from salt country to the north. Within 20 kilometres of the coast there are spongolite rock outcrops in the valleys. Inland from the coast there is a belt of dune sands five kilometres wide before giving way to the sand plain to the north. The high sand bar, which breaks occasionally after very heavy rains, is covered with coastal dune vegetation.[111] The Lort River forms the boundary of the Esperance Sand-plain West Ecodistrict and Esperance Sand-plain East Ecodistrict.[112] A mosaic of dense scrubland and sand heath dominates the coastal dune system and coastal plain. The dense scrub is mainly a mixture of small *Eucalyptus* spp. and *Acacia* spp., with an understorey of *Spyridium* spp. and *Leucopogon* spp. The sand heath on the coastal plain is dominated by *Banksia* spp., a diverse mix of other proteaceous species and *Lambertia* spp. Further north the bushland is a Mallee of *Eucalyptus* spp. and *Melaleuca* spp.[113]

We have located only one record of an Oldfield specimen which was unequivocally from Young River, and another was stated as being from 'Young River and Esperance Bay'[114]; there are none with the locality given as either Stokes Inlet or Lort River. It will be recalled, however, that Maxwell collected at Stokes Inlet (see above). Again, the visit may have been too brief and too late in the flowering season to allow for more intensive collecting around the inlet, on the dunes, and sand plain.

Their next landing place appears to have been Esperance Bay where, once more, it seems that very few plants were gathered.[115] This area lies within the Albany Fraser Coastal Ecodistrict and to the south of the Esperance Sandplain East Ecodistrict. The former contains vegetation communities similar in structure and often taxa, many of which are endemic, localised or restricted and include Mallet and Mallee systems as well as *kwongkan* heathlands.[116]

We have found no evidence that the pair visited the Recherche Archipelago during this expedition. The Archipelago had long held a reputation among European visitors for being home to many seals, so it seems improbable that, having gone as far as Esperance Bay, the sealers would have bypassed it. Off-shore the plain is submerged and the islands of the Archipelago are formed by granite domes, which continue to protrude above sea level. These islands, which are part of the Albany Fraser Island Ecodistrict, stretch 230 kilometres from near Esperance east as far as Israelite Bay and to 50 kilometres off-shore.[117] Some of the islands are partly encrusted with sand and limestone and their plant cover differs markedly from the adjacent mainland because the islands could not be reached by Aborigines and, thus, were not burnt regularly. Where the soil has developed there are communities of

---

110 G Maxwell, MEL 66199, *Cheiranthera filifolia* var. *filifolia*, river entering Stokes Inlet (Young River), December 1861, MELISR database, 02 April 2009, NHV, RBGM.

111 Anne Brearley, 2005, pp. 485–488.

112 Sarah Barrett *et al.*, 2009, p. 14 and Fig. 1, South Coast Ecodistricts.

113 South Coast Rivercare, 'Young-Lort Catchment', viewed 24 March 2014 at www.rivercare.southcoastwa.org.au.

114 K unnumbered, *Centaurium spicatum*, 'Young River', AO (RBGK); MEL 2296180, *Opercularia hirsuta*, 'Young River and Esperance Bay', AO, MELISR database, 18 November 2005, NHV, RBGM.

115 HAODb.

116 Sarah Barrett *et al.*, 2009, p. 14 and Fig. 1, South Coast Ecodistricts.

117 ibid.

trees and shrubs considerably taller and denser than any found on such outcrops on the mainland. There are distinctive plant communities on the islands and where *kwongkan* occurs on sandy plateau surfaces, the floristic composition is significantly different.[118]

The apparent absence of any specimen records that would bear testimony to Oldfield and Maxwell visiting any of the islands of the Archipelago might be explained by the sealers signifying their intention, before they left, to land only on islands that were devoid of vegetation. Of course, other explanations might have included: one, or both, of the naturalists was ill or nervous about landing on granitic islands, or they thought they would be better employed collecting on the mainland.

From Esperance Bay, the sealers took the naturalists further east but we cannot be certain where on the coast they landed. Based on our Oldfield database records, his collections along this section of the coast were very few. There are only two herbarium specimen records which provide some support for a claim that they landed at Cape Arid and Cape Pasley and possibly places in between; they each bore the locality 'Cape Arid to Cape Pasley'.[119] The Albany Fraser Coastal Ecodistrict lies between these two granitic capes.[120]

It seems that Oldfield made sufficient contact with Aboriginal people in one or more of the territories, possibly facilitated by Maxwell, where *Eucalyptus pleurocarpa* occurred to be able to discern similarities between their languages and customs and those of the tribes inhabiting the Lake Logue area (see Chapter 10). Specimen records for that species in *FloraBase* indicate that its range along the south coast meant that the collectors could have met with members of one or more south coast tribes whose southern mainland territory borders were: the *Minang*, from west of Albany to the west bank of Pallinup (Salt) River; the *Koreng*, from the east bank of Pallinup River to Gairdner River; the *Wadjari* from the east bank of Gairdner River to Point Malcolm whose territory overlapped with the *Njunja*, a breakaway group of the *Wadjari.* The *Njunja*, according to Tindale, had begun to accept circumcision but not sub-incision, whereas the *Wadjari* had steadfastly refused.[121]

From Cape Pasley they travelled further and landed 'Near west end of Great Bight' where Oldfield gathered *Acacia rostellifera*.[122] Oldfield also collected *Templetonia retusa*, which he recorded as being from the 'Great Australian Bight'.[123] Since the Australian Fur Seal occurs along the coast from Eclipse Island to Kangaroo Island, the sealers must have taken the naturalists into the Bight but how far is a question that remains unanswered. It may have been near what, in 1863, was named Israelite Bay by the exploring party comprising [? Charles] Edward and William [Simon] Dempster, G.M. Lanarch and George Maxwell. There is some disagreement as to which of the brothers was a member of the party.[124]

If Oldfield was at Israelite Bay it was probably because the granitic eastern islands of the Recherche Archipelago lay off the Bay and were likely destinations for the sealers in quest of their prey. The mainland linear coastal Nuytsland Ecodistrict surrounds Israelite Bay and extends along the coast of the western end of the Great Australian Bight south of Israelite Bay east to Point Malcolm. It is

---

118 J.S. Beard, 1990, pp. 146–147.

119 MEL 250126, *Kennedia coccinea*, and MEL 2065549, *Poranthera huegelii*, AO, 'Cape Arid to Cape Pasley', MELISR database, 18 November 2005, NHV, RBGM.

120 Sarah Barrett *et al*., 2009, p. 13 and Fig. 1, South Coast Ecodistricts.

121 Norman B. Tindale, 1974, online map of Australian Aboriginal tribal boundaries. There are two specimens in HAODb of *Eucalyptus pleurocarpa* obtained at Kalgan River but where on the river is not known. They each bear the date 1860 but Oldfield was not in Western Australia at that time.

122 NSW 481178, *Acacia rostellifera*, 'Near west end of Great Bight', A. Oldfield, NHNSW.

123 *Templetonia retusa*, 'Great Australian Bight', AO, GB 1864, vol. 2, p. 169.

124 John Rintoul, 1964, *Esperance: Yesterday and Today*, Esperance Shire Council, Esperance, Western Australia, p. 25. Part of the Dempster diary entry for 14 June 1863 reads 'A bay named Israelite Bay in honour of the blacks. They were of the *Dadjunna* [*Ngadjunmaia* of Tindale] tribe.' The *Dadjunna* were circumcised; the name was given to the bay because of the biblical link to the Israelites who were also circumcised, *High Beach*, 'The naming of Israelite Bay', viewed 3 November 2010 at www.hibeach.net. The above list of members of the party is taken from Rica Erickson, 1978, *The Dempsters*, University of Western Australia Press, Nedlands, p. 81. However, according to Wendy Birman it was Andrew Dempster and not William who was part of the party, Wendy Birman, 'Dempster, Charles Edward (1839–1907)', ADBO, viewed 3 November 2010 at www.adb.online.anu.edu.au.

typified by limestone soils, sand dunes and linear saline lake systems with *Acacia* and Mallee shrub-lands and *kwongkan* heathlands.[125]

At each of their stops along the coast, Oldfield would not have been short of material to collect for McCoy, provided, of course, that he had the time and appropriate preservation equipment. From Cape Riche to Cape Arid and beyond the low woodlands, Mallee shrublands, Mallee heath, coastal heath, samphire flats, permanent water bodies, rivers, estuaries, coastal dunes, and granitic headlands and islands are homes to a great variety of terrestrial, riverine, estuarine and marine invertebrates and vertebrates, the more noticeable including many species of birds, mammals, reptiles, fish, insects, crustaceans and molluscs. There are also abundant fossil mollusc deposits in various geological situations.[126]

Return to Albany via the Stirling Range

Map 19 p. 370 – Cape Riche back to Albany

If, as we have suggested, Oldfield and Maxwell went to Cape Riche and boarded the sealers' ship there, the sealers would have returned them to the Cape. The naturalists returned to Albany via the Stirling Range. Oldfield and Maxwell's specimen records indicate that from the Cape, they retraced their path along Sandalwood Track until it crossed Donkey Track and then continued north-west on the same trail across Mallee heath lands of the Marine Plains Eco-system.[127]

About 29 kilometres from where their path crossed Donkey Track, they passed within seven kilometres east of Ellen Peak, one of the largest and most easterly of the prominences in the Stirling Range. From there the track wended its way further north-west to Pallinup River, passing Andrew Moir's 'Sandalwood' property on that river about five kilometres north-east of the eastern end of Stirling Range.[128] There is a specimen of *Conospermum capitatum* at Kew Gardens, which states it collected by Maxwell at 'Woolginup' but was found in 'Herb. Oldfield'[129] and may have been given to him by Maxwell at the time of collection. This place has not been located in any gazetteer but it may have been 'Woolaganup', a place north of, and towards the eastern end of Stirling Range, and about one and a half kilometres west of Pallinup River, close to Woolaganup Creek, a tributary of Pallinup River.[130] In any event, the track traversed this stream. It is not known whether at the time Woolaganup was a property that had been taken up by Europeans for pastoral or farming purposes. The local Aborigines had probably referred to the area by that or a similar name long before colonisation. According to Ethel Hassell[131], the *Wheelman* (*Koreng* of Norman B. Tindale)[132] tribe whose territory was near Jerramungup, referred to the Carnaby's [and Baudin's] White Tailed Black Cockatoo (*Calyptorhynchus latirostris*) as *woolah. Woolaganup* may have referred to the place of that species of cockatoo.[133]

At the time of Oldfield and Maxwell's visit, Sandalwood Track terminated at the junction of the road that ran from Kendenup in the west to Albert Hassell's property Jerramungup in the north-east. Maxwell's old sandalwood camp was nearby on Myerup Pool (now West Myerup). There is a record

[125] Sarah Barrett *et al.*, 2009, p. 15, and Fig. 1, South Coast Ecodistricts.
[126] Anne Brearley, 2005, p. 473.
[127] HAODb.
[128] Peter Rohan, 1979, *Gnowangerup District Heritage Sketchbook*, Westcolor Press, Perth, p. 53.
[129] K unnumbered, *Conospermum capitatum*, G. Maxwell, Herb. Oldfield, RBGK, sighted by A.S. George, 2003, personal communication to MHH, 2005.
[130] National Topographic Map Series, Mount Barker, Sheet SI 50-11 Edition 1, 1: 250,000, Division of National Mapping, Commonwealth of Australia, 1985.
[131] Ethel Hassell, 1975, Map, unnumbered page, and p. 233.
[132] Norman B. Tindale, 1974, map, Australian Aboriginal tribal boundaries.
[133] Among the *Nyungar* (contemporary name for the people of the South-west of Western Australia) the suffix *up* refers to 'place of'.

of a *Darwinia citriodora* with the location '?Yardup'.[134] This species does not occur that far east and was probably obtained at Jardup, which is well within its range.[135] 'Yardup' is a property with its northern boundary close to where Maxwell's sandalwood camp was located.

From Myerup Pool the track ran south-west toward the western end of Stirling Range through Mallee heathland of the Marine Plain Ecodistrict. There are Oldfield specimen records with the locality 'Stirling Range'.[136] 'Monjerrup' and 'Mondrup' were localities on some of Oldfield's specimens, which were probably obtained at Mondurup Peak.[137] An old name for Mondurup Peak was 'Mongerup'.[138] This name was used by Maxwell on four specimens he collected in January 1862.[139] Perhaps because it was damper and colder than the surrounding plain, the pair were motivated to venture into the range on or about 7 January 1862[140], in the hope that uncommon species would still be in flower. On the other hand, Mueller may have been anxious to obtain specimens and seeds from the same locality that Drummond and Maxwell had collected sixteen years previously.[141]

Maxwell and Oldfield gained access to Mondurup Peak from the Jerramungup–Kendenup track via the deep gap along which today the Red Gum Pass Road runs, between Mondurup and Donnelly Peaks. The former is roughly in the centre of and toward the west end of the Range. They probably travelled south along a track forged by sandalwood cutters operating north of the Stirling Range when driving bulky loaded wagons through to the port at Albany, probably via Kendenup.[142]

Travelling along the valley they passed through woodland on colluvium brought down from the mountains, and where the available moisture is highest in the range. Reaching heights of 10–15 metres, Jarrah-Marri occurs on sandier soil and Wandoo on heavier soils. Understorey plants in the former association feature several wattles such as *Acacia drummondii*, *A. myrtifolia* and *A. pulchella* whereas *Bossiaea linophylla* tends to be the most conspicuous beneath Wandoo. If they ascended the prominence they continued to pass through Jarrah–Marri, which declined in height to form low woodland until it merged into a Mallee heath with Jarrah and Marri still the principal tree species. Mallee heath formation that covers all the lower slopes of the mountains through the range includes Christmas Tree, Kingia, the Graceful Grass Tree (*Xanthorrhoea gracilis*), Spearwood Mallee (*Eucalyptus doratoxylon*), Bull Banksia, *Hakea* spp., Heath-leafed Honeysuckle (*Lambertia ericifolia*) and many smaller shrub, herbaceous and rush species. Climbing higher still they were confronted with the dense and difficult-to-penetrate thicket, which covers the top of the mountain in which the most clearly visible species when in flower are Showy Dryandra (*Banksia formosa*), *Isopogon latifolius* and *Gastrolobium leakeanum*.[143]

How they managed the terrain and to what extent the collectors explored Mondurup Peak and other places in the range is not known. When Drummond and Maxwell were in the vicinity of Mondu-

---

[134] MEL 2128116, *Darwinia citriodora*, ? Yardup, undated, coll. AO, MELISR database, 18 November 2005, NHV, RBGM. The query associated with this place name is probably because of its similarity to 'Jardup', an Oldfield collecting locality near Harvey. *FloraBase* specimen records indicate that the species probably occurred in both places.

[135] *FloraBase*.

[136] HAODb.

[137] ibid. There is little doubt that 'Mondrup' refers to Mondurup Peak and it seems likely that 'Monjerrup' also refers to that prominence; an interpretation noted in the MELISR database, 18 November 2005, NHV, RBGM—see for example, MEL 2169980, *Adenanthos filifolius*, 'Monjerrup' (Mondurup Peak [?], Stirling Range), AO. On modern maps, there is a property 'Mungerup' but we have found no evidence that the name was in use at the time of Oldfield's visit. 'Mungerup' has the same location as a property 'Mongup' was established in 1852 by the Barrett Brothers and purchased by George Moir in 1857. It remained in the Moir family's ownership until 1936. It is situated well away from the Sandalwood Track on the Peenebup Creek to the north-east of Stirling Range and the Pallinup River, Peter Rohan, 1979, p. 51.

[138] see Chapter 17.

[139] MEL 2259599, *Stylidium junceum*, January 1862, 'Mongerup', G. Maxwell; MEL 2259740, *Stylidium piliferum*, January 1862, 'Mongerup' G. Maxwell; MEL 2259432, *Stylidium repens*, 7 January 1862, 'Mongerup'; MEL 2261148, *Lobelia rarifolia*, January 1862, 'Mongerup', G. Maxwell, MELISR database, 02 April 2009, NHV, RBGM.

[140] J.S. Beard, 1990, p. 149.

[141] See Chapter 19.

[142] Carolyn Thomson, et al. (eds), 1993, p. 37. Today's Chester Pass Road, which connects Albany with the agricultural country to the north, was originally a track forged by sandalwood cutters, ibid.

[143] J.S. Beard, 1990, pp. 147–149.

rup Peak in 1846, Drummond reported that Maxwell found water 'in a native well about a mile up the water course we were on', so Maxwell and Oldfield may well have headed to the well to replenish their supplies.[144] Drummond's account of his ascent of Mondurup Peak, probably in December 1846, gives some idea of what Oldfield and Maxwell would have experienced had they also climbed the mountain although, at the time of their visit they may have found only the plants mentioned by Drummond in seed:

> Starting at five o'clock I reached the summit of the hill by eleven o'clock. I ascended by the north-east angle, and at about the height of two thousand feet I found first making its appearance a splendid new Banksia with leaves more than nine inches long and about five inches wide, irregularly jagged and serrated like an English Oak [Stirling Range Banksia] (*Banksia solandri*)...I had scarcely time to make myself acquainted with this fine Banksia when I found another exceedingly interesting and beautiful plant, [Mondurup Bell] (*Darwinia macrostegia*) the flowers enclosed by beautiful bracts, white, variegated with crimson veins. These bracts are as elegantly formed as the petals of the finest tulip and they are almost as large, hanging in a bell...at the ends of the slender branches. I thought I could never gather enough of this charming plant and I procured an abundance of perfect seed. As one is obliged to use their hands as well and as often as their feet in ascending or descending these very steep hills I had gone very lightly equipped. I was therefore obliged to make use of my shirt and neck handkerchief, making the shirt into a bag to bring down a supply of Banksia cones, securing the load so as not to impede the use of my hands. I reached our sleeping place at three o'clock much fatigued with my load but highly gratified having this day found at least two plants which will continue to be admired while a taste for the beauties of nature remains to the human race.[145]

Whether from Mondurup Peak Oldfield and Maxwell retraced their footsteps and joined the Jerramungup-Kendenup road north of Stirling Range, or joined it east of the range, it is likely they called on John Frederick Hassell at Kendenup. As noted earlier[146], Oldfield's plants from Kendenup may have been obtained during this visit but it cannot be confirmed.

It is unlikely that the pair returned to Albany by any other route than by travelling the short distance from Kendenup to the Albany-Perth Road and then continuing south past Mount Barker to the Sound. When they arrived, however, is a mystery but Oldfield was there on 27 February 1862 when he collected *Stylidium repens*.[147] Others of his specimens from the area may have been obtained while he was waiting for passage back to Melbourne, but preparing and packing his collections and equipment for transport must have engaged much of his time.

In a letter to Joseph Hooker soon after his arrival in England in 1862, Oldfield said: 'My last W. Aust. collection, although extensive contains little not hitherto known'.[148] After South West Bay (Bremer Bay), Oldfield seems to have collected very few plant specimens at each of the places where he landed. In some cases, this may have been because his visits were too brief to allow for more than a very cursory search or the summer was so far advanced that very few species were in flower or a combination of both factors. As a result, he may have focused his energies on collecting seeds. He offered Joseph Hooker seeds he had collected in Western Australia in March 1862[149] but where he did

---

[144] Rica Erickson, 1969, p. 114.

[145] (ibid., p. 115). This is an excerpt from a letter, probably to William Hooker, but Erickson gives no source details. The names of *Banksia solandri* and *Darwinia macrostegia* in round brackets ( ) appear to have been inserted by Erickson. Words in square brackets [ ] are ours.

[146] See Chapter 13.

[147] MEL 2259437, *Stylidium repens*, coll. AO, 27 February 1862, King George Sound, A. Oldfield, MELISR database, 18 November 2005, NHV, RBGM.

[148] AO to JDH, 4 December 1862, English Letters, vol. 97, 1859–1900, fol. 278 (280), RBGKA.

[149] AO to JDH, 8 December 1862, English Letters, vol. 97, 1859–1900, fol. 274 (281), RBGKA.

not say. Apart from that we have not been able to trace any record of seeds received by Mueller or of natural history specimens sent to Professor McCoy.

We can only imagine Oldfield's frustration if he had arrived too late in the season for the spring flowering, particularly if he recognised that the area was the most botanically diverse, with perhaps the highest degree of endemism of all the places he had visited in Western Australia. There is the possibility that he became very selective and did not collect species that he had already obtained further to the west or elsewhere. If that was the case, it would have gone against his earlier intention[150] to collect in a way that would shed light on species' distribution and variation.

Travelling steerage aboard the steamer *White Swan*[151], Oldfield left Albany on or about 13 March 1862, arriving at Hobsons Bay on 22 March 1862.[152]

[150] See Chapter 8.
[151] 'Shipping Intelligence', *South Australian Register*, 20 March 1862, p. 2, DAN, NLA. Referred to as 'A. Oldfield' in the passenger list. According to this report the *White Swan* left Port Adelaide on 19 March 1862.
[152] 'Shipping Intelligence', *The Argus*, 24 March 1862, p. 4, DAN, NLA.

View from Merton Road looking east, Wandsworth, c. 1880.

Part 5

The end of Oldfield's career and his last years

1862–1887

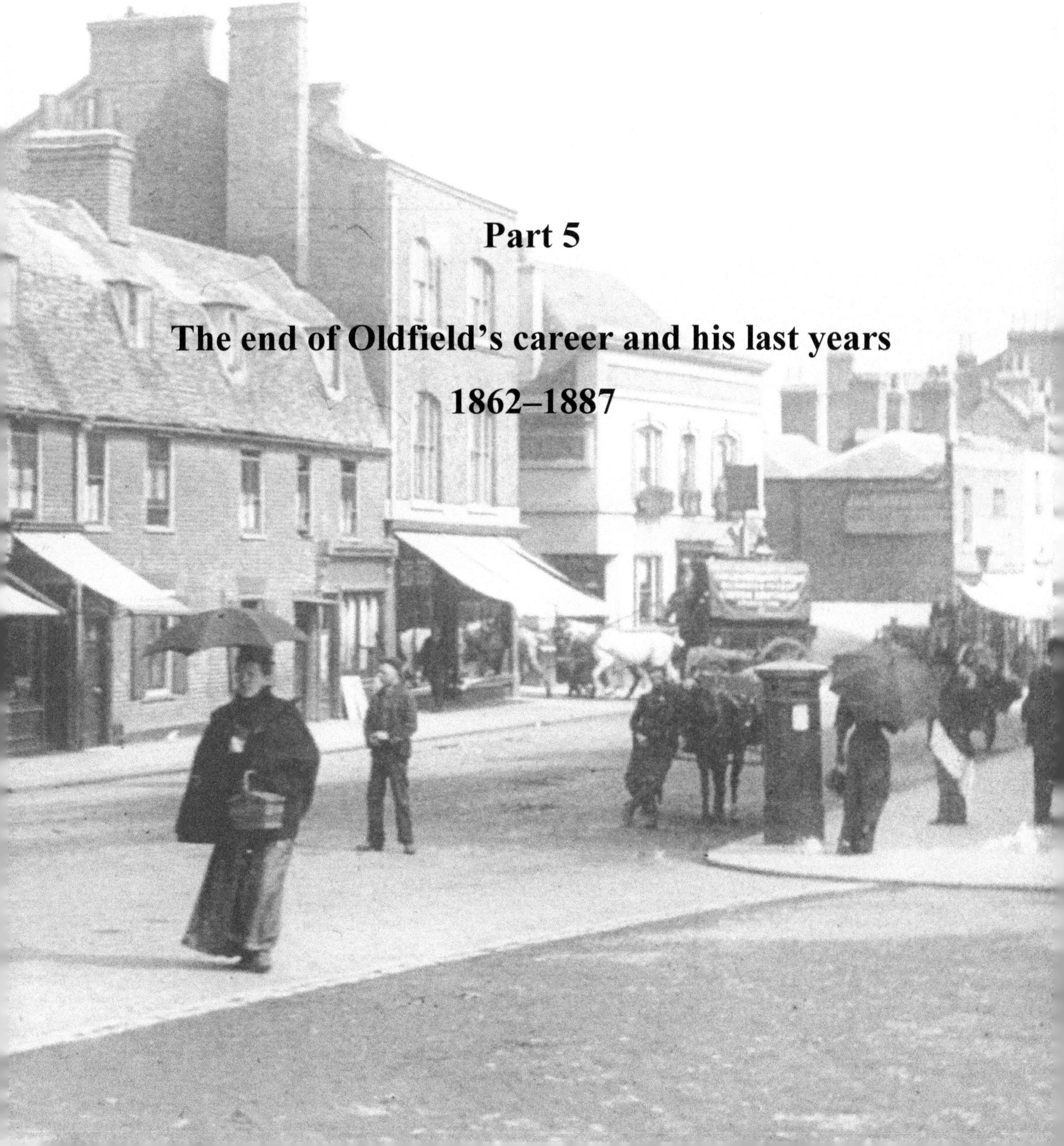

Chapter 21

London 1862–1874

Fig. 21.0 Chthonocephalus oldfieldianus

The demise of Oldfield's botanical career

During his journeys in Western Australia, Oldfield spent several hundred pounds of Mueller's own money.[1] Mueller appears to have used some of this cash honouring his agreement to purchase specimens resulting from Oldfield's 1861–62 expedition. The National Herbarium of Victoria has many of these[2] but, as indicated before, how much of the seed collection remained in Melbourne is unknown.

On 28 April 1862, Oldfield left Melbourne for London on the steamship *Orwell*.[3] He is not mentioned in the published passenger list for this voyage[4] but Mueller stated that he was on board carrying 'plants of Marsilea q[u]adrifolia (L) [M.] hirsuta Br (M. macropus) & of Azolla' destined for Sir William Hooker.[5] Although Oldfield may have worked his voyage back to England, he probably travelled steerage. It was quite common for newspapers not to name passengers undertaking voyages in that class. Four cases of his botanical specimens left Hobart for London on 4 September 1862, onboard the *Isabella Brown*.[6]

How Oldfield occupied the five weeks between leaving Western Australia and departing for England is not known. He may well have spent this period at the Botanic Gardens, Melbourne sorting out the specimens he had obtained for Mueller, McCoy and himself and packing what he intended to keep.

According to one of his obituaries, this journey to London was to obtain treatment for his 'impaired sight'.[7] However, he stated in a letter to Joseph Hooker that his 'sole purpose' for coming back to England was to arrange the publication of his work on 'the philosophy of Natural History'.[8] His intention was probably twofold but perhaps his statement to Hooker was to emphasise the importance he attached to getting the manuscript published.

While he was travelling to England, a letter to the Editor appeared in *The Sydney Morning Herald* dated 27 June 1862 in which 'W.W.' [Reverend William Woolls][9] of George Street, Parramatta reviewed some aspects of the second volume of Mueller's *Fragmenta*. During this discussion, he referred to observations and specimens obtained by a few individual collectors, among them Oldfield and Walcott.[10]

---

[1] See Chapter 17.
[2] See MELISR database, 08 March 2007, NHV, RBGM.
[3] 'Shipping Intelligence', *The Argus*, 29 April 1862, p. 4, DAN, NLA.
[4] ibid.
[5] F. Mueller to W.J. Hooker, 24 May 1862, Director's Letters, vol. 75, Australian and Pacific Letter 1859–1865, RBGKA, excerpt kindly provided by Dr Monika Wells, The Correspondence of Ferdinand Mueller Project, personal communication to MHH, 7 June 2001, HOA.
[6] 'Shipping', *The Mercury*, 4 September 1862, p. 4, DAN, NLA.
[7] Anon. ? Edwin Oldfield, *The Argus*, 13 July 1887, p. 5.
[8] AO to JDH, 1 March 1866, English Letters, vol. 97, 1859–1900, NAT–OXL, fol. 280 (287), RBGKA.
[9] Reverend Wools was a self-taught botanist, and friend of Mueller's, Sara Maroske, personal communication, email 2 October 2017 to MHH, HOA.
[10] 'W.W.', Letter to the Editor, *The Sydney Morning Herald*, 27 June 1862, p. 2, DAN, NLA.

Fig. 21.1 'Dunsford Villas' (foreground) with smaller 'Verandah Cottages' beyond, Wandsworth, 1961

Back with the family

Sometime toward the end of July, or during August, the *Orwell* reached England where Oldfield probably made landfall at one of the many docks that stretched along the River Thames from the Tower of London to Silvertown.[11] On his way up the Thames there is little doubt that he noticed the massive earthworks that were being undertaken by the Metropolitan Board of Works along the banks of the river. A new Thames Embankment was being constructed along the north and south sides of the river under which large underground sewers would run.[12]

No evidence has emerged of any members of the family being at the port to greet Augustus on his arrival. His mother, Mary, and sister, Marian, had continued to live at 1 Dunsford Villas, Wandsworth after Frederick's death. Augustus was to stay with them.[13] Juliana and Vincent's whereabouts at the time was unknown. Vincent may still have had his business in Chancery Lane. It is unlikely that Frederica came to London from Yorkshire as she found such visits 'so expensive'.[14]

Unless he was met by one of his siblings with his or her own or hired conveyance, travelling in those days from the London Docks to Wandsworth by other means was quite an undertaking. Perhaps the most convenient mode of public transport was to go all the way in one of the hansom cabs waiting at the dock but his financial situation may not have allowed him this luxury. Horse-drawn omnibuses operated at the time but to get to Wandsworth he would have suffered the inconvenience of changing vehicles many times along the way. Another option was to take an omnibus or hansom cab from the

[11] Tim Lambert, undated, '19th Century London', viewed 17 November 2010 at www.localhistories.org.
[12] Roy Porter, 1994, pp. 320–321.
[13] AO to JDH, 4 December 1862, English Letters, vol. 97, NAT–OXC, 1859–1900, fol. 278 (280), RBGKA.
[14] Frederica Oldfield to Vincent Oldfield, 5 March 1863, Files 0233–0236, PFA.

dock to Waterloo Station and from there to travel by train on the West London Extension line to Clapham Junction and thence to Wandsworth along the Wimbledon line.[15] Journeys through the streets of London were very slow due to the appalling traffic jams about which Londoners constantly complained—omnibuses, phaetons, hansoms, tilburies, growlers, dogcarts, gigs, broughams, landaus were on the increase, narrow streets and problems with horses and livestock being driven through the streets all contributed to the congestion.[16]

Moving across the suburbs he may have been amazed at the way the city had spread and changed during his absence; row upon row of terrace houses stretched into areas that had been farms and market gardens a decade ago. The extension of the Metropolitan Railway enabled prosperous Londoners to install themselves in new middle-class suburbs further away from the city's centre. The development of the railway also intensified the need for cheap lodgings to house the hosts of people dependent upon West End casual work. The 'respectable' working class had begun an exodus from the streets by the canal and railway, which were now well on the way to becoming 'classic' slums.[17]

By 10 February 1863, Oldfield was residing with Marian at 5 Verandah Cottages in Merton Road, Wandsworth.[18] These small, two-storey cottages were immediately south of Dunsford Villas. Sometime after December 1862, and probably at about the same time as Augustus and Marian moved in together, their mother had relocated to 7 Bass Terrace, Doncaster, where she was living with Frederica who continued to practise there as a music teacher.[19] Under what conditions Augustus and Marian's sharing arrangement operated is not known although it may have been convenient for them both. By then Augustus' sole income was his £35 annuity. Marian, who had been financially dependent on her parents all her life, also seems to have been reliant on an annuity she received from her father's estate. House sharing would have enabled each of them to contribute to the rent and other living expenses to make for a reasonably comfortable middle-class standard of living, particularly if they were able to afford a servant to do the cleaning and laundry.

While Marian and Augustus were at Verandah Cottages, Vincent occasionally took his son, Jonathan (known in the family as Peepsy), to visit. In 1879 Frederica wrote to Jonathan:

> Aunt Marian wishes me to remind you that you used sometimes go to Verandah Cottages (Merton Rd Wandsworth)[20] with your father and you had tea upstairs with her and carried your father's tea downstairs to him while he was talking to Uncle Augustus. She says you wore a little suit of mulberry coloured velvet.[21]

From the description of his suit, these visits were probably c.1863 when Jonathan was about five years old. It is intriguing that Marian and the boy had their 'tea' separately from the two men. Perhaps Marian deliberately stayed away from Vincent as much as possible because of the way he had upset Frederick over the disposal of his property. Nevertheless, there may have been other reasons such as her awareness that Augustus needed to have the opportunity to chat alone with others.

Sometime between 1862 and 1863 news may have come from America that Frederick Jnr was responsible for the musical arrangement of the popular Union soldiers' song 'Abraham's Daughter' as sung by the minstrel, Ben Cotton.[22] It is likely that Frederick and his family were living in Virginia City.[23]

---

[15] ibid., p. 277.
[16] Roy Porter, 1994, p. 273.
[17] ibid., pp. 257–258.
[18] AO to JDH, 10 February 1863, English Letters, vol. 97, NAT–OXC, 1859–1900, fol. 275 (282), RBGKA and Frederica Oldfield to Jonathan Oldfield, 14 January 1879, Files 0116–0122, PFA.
[19] Frederica Oldfield to Vincent Oldfield, 3 March 1863, Files 0295–0297, 0298, PFA.
[20] 'Mr B's' name has not been determined.
[21] Frederica Oldfield to Jonathan Howard Oldfield, 14 January 1879.
[22] F.H.H. Oldfield, undated, 'Abraham's Daughter', M. Gray, San Francisco, Californian Sheet Music Project, viewed 13 January 2006 at www.loc.gov.
[23] *Territorial Enterprise*, Nevada, 20 September 1866, viewed 9 November 2005, at www.ancestry.com.

Oldfield resumes contact with Joseph Hooker 1862–1863

If Mueller's plants survived the voyage, Oldfield must have delivered them to Kew Gardens soon after his arrival. However, a letter to Joseph Hooker from 1 Dunsford Villas, dated 4 December 1862, suggests that the two had not met since his return, apparently due to Oldfield smarting because

> [m]y correspondence with your illustrious father, Sir William, having terminated so abruptly 8 years since (my last collection having been totally unacknowledged by him, although quoted far more than it deserves in your "Flora of Tasmania", [sic]) I feel much diffidence in again writing to you, which I certainly should not do were it not for the following circumstances.

These concerned the possibility of 'Customs house authorities' damaging his collections which he was expecting to land in London at the end of the month. He appealed to Hooker to bring his influence to bear on the authorities to prevent a repeat of the great harm they had caused to his 'last [consignment of] Western Australian specimens' by the 'rigour of the search' and 'careless repacking'.[24]

In a postscript, Oldfield noted that his 'last W.A. collection [1861–62], although extensive, contains little not hitherto known, so of course the proposal made by me [from Hobart 25 August 1861][25] to you with reference to it falls to the ground'. He had received no answer to this communication in which he asked Hooker whether he would purchase a set of any plants and seeds he collected in Western Australia.[26] Even if Hooker had responded care of the collector at *Post Restante,* Perth Post Office, Oldfield would have missed it as he did not go to the city during his visit in 1861–62.

Hooker replied to Oldfield's letter of 4 December 1862 on the following day and from Oldfield's answer, three days later, he was informed that nothing could be done to stop the Customs mishandling his collections. To some further information, Oldfield replied:

> I very much regret that the 'Flora [Tasmaniae]' never reached me, and should very much like to know who has detained it. As I possess the first 9 parts of that splendid work, I do not so much regret the mere loss, but as coming from its Author I should have the more prized it.[27]

At the same time, he accepted an invitation to visit Kew Gardens for the following Wednesday 'if fine'. Oldfield's letter shows that Joseph had approached Sir William regarding the lack of confirmation of the receipt of Oldfield's Tasmanian collections sent from Hobart in 1856 and 1857:

> As I before observed, since the middle of /55 I have had no letter from your Father: as I could impute his silence to nothing but the totally uninteresting nature of my correspondence I was unwilling to trespass on his valuable time by still thrusting myself on his notice. Still, I must own, from Sir William's well-known urbanity and kindness, the non-acknowledgment of any botanical contribution, however, insignificant, sometimes puzzled me. However, your Father's assurance that such acknowledgement was made is ample reward to me.[28]

Apparently, Joseph Hooker had taken issue with Oldfield over a consignment of specimens he had sent to Kew Gardens without authorisation, to which Oldfield replied:

> I feel that some degree of censure is due to me with respect to my after [sic] collections in Tasmania, and can only plead in mitigation that they were sent without my sanction, and that I was quite ignorant of the etiquette to be observed in your case.[29]

[24] AO to JDH, 4 December 1862, English Letters, 1859-1900, vol. 97, NAT–OXL,fol. 278 (280), RBGK.
[25] See Chapter 18.
[26] AO to JDH, 4 December 1862, English Letters, 1859-1900, vol. 97, NAT–OXL, fol. 278 (280), RBGK.
[27] AO to JDH, 8 December 1862, ibid., Fol. 274 (281), RBGKA.
[28] ibid.
[29] ibid.

Fig. 21.2 Dr Joseph Hooker, 1851

There is no indication of who sent the material or from where it had been sent. No evidence has emerged that Oldfield had sent specimens to Kew Gardens from Tasmania after 1857 or from Wandsworth prior to 8 December 1862. Hooker was probably referring to the Mount La Perouse collection.[30] If that was the case it would seem to have been somewhat high-handed and ungrateful of him to have taken Oldfield to task. Hooker must have been aware from his own experience of the great difficulty of botanically exploring the mountainous country of Tasmania. That he knew that specimens from Mount La Perouse were not previously represented in British and European herbaria is evident by their inclusion in an addendum to his *Flora Tasmaniae*.[31] While Hooker's behaviour on this occasion may have been a result of exerting his authority as a metropolitan 'man of science' over what he perceived to be a subordinate colonial collector[32], perhaps the number of specimens being sent to Kew Gardens from all over the world was so great that it had become necessary to control their acceptance.

Undeterred by this, Oldfield offered to donate a collection of his Western Australian seeds, obtained in March 1862, 'if they can be of any service'. Not running the risk of further castigation, he added: 'If you will let me know I can bring them with me.'[33]

[30] See Chapter 14.
[31] JDH, 1860, vol. 1, addendum.
[32] Jim Endersby, 2008, p. 55.
[33] AO to JDH, 8 December 1862, English Letters, 1859-1900, vol. 97, NAT–OXL,fol. 274 (281), RBGk.

Oldfield did, it seems, visit Kew Gardens as indicated in his letter of 8 December.[34] Following this visit, Joseph Hooker wrote, 21 December 1862, to Charles Darwin:

> If you want any information about Australia & its savages, a capital man is now about Kew who has collected plants well & lived amongst & observed natives shrewdly from K.G.S. to Sharks [sic] bay, & W. up to Wide bay [location not known].[35]

Although not named, Hooker was referring to Oldfield and, despite his earlier admonishment, indicated that he held the collector in high regard. Darwin replied three days later:

> If you think of it ask the Australian Traveller whether he ever came across savages starving, & ever knew of their trying various plans to make vegetables eatible [sic]. Ask him whether the Australians blush (Oh oh I forgot they were black); it is an odd theory I have never met in notice by anyone of Expression in Savages; expression is one of my hobby-horses; I have got some funny notions on the subject.[36] 3rd. ask him whether they take any pain in breeding dogs or get them crossed with European dogs.[37]

In his letter of 27 or 28 December, Hooker conveyed Oldfield's reply to Darwin[38] but it has never been found.[39] It seems that Oldfield was not overawed by the author of the recently published and controversial *On The Origin of Species...* He apparently had made a disparaging remark relating to one or more of Darwin's questions, which prompted Darwin to say to Hooker: '...my questions were foolish; but not rarely foolish questions, lead, I find, to good results'.[40] We may never know whether Hooker communicated that remark back to Oldfield. Some inkling of Oldfield's response is given by Darwin in his *Variation of Animals & Plants Under Domestication:*

> Mr Oldfield, who has seen so much of the aborigines of Australia, informs me that "they are all very glad to get a European kangaroo dog, and several instances have been known of the father killing his own infant that the mother might suckle the much-prized puppy".[41]

The collector took delivery of his Western Australian specimens early in February 1863. Thankfully, after he had been 'in an agony of despair' for their safety, they were not damaged. His letter of 10 February asked if it was 'too late' for him to bring the specimens he promised to show Sir William, and 'attend more especially to look out' what he thought 'chiefly interesting to [Joseph], viz. Tasmanian forms'. He went on:

> Should there be any Ferns of which Sir William would like to see more, or better specimens I will attempt to furnish them. As indicated my specimens are large, and in proper states both as regards fruit and variations.

He told Hooker (23 February 1863) that he had not had a letter from Mueller but had received from him the '23rd fascicle' of his *Fragmenta*, the contents of which puzzled him:

> <u>All</u> my Murchison sp. [sic] of Calothamnus <u>new</u> one do. from Drummond's and Preiss' country unseen by them, a new allied genus, new Melaleucae by the score, either

[34] JDH to C.R. Darwin, 21 December 1862, letter 3856, Darwin Correspondence Project, viewed 16 January 2011 at www.darwinproject.ac.uk.
[35] ibid.
[36] According to the Editor's fn. 8 to this letter (ibid.), Charles Darwin, 1872, *The Expression of the Emotions in Man and Animals*, John Murray, London, pp. 316–22, included a discussion of 'Blushing in the various races of man'.
[37] C.R. Darwin to JDH, 24 December 1862, Letter 3875, Darwin Correspondence Project, viewed 16 January 2011 at www.darwinproject.ac.uk.
[38] JDH to C.R. Darwin, 27 or 28 December 1862, Letter 3891, Darwin Correspondence Project, viewed 16 January 2011 at www.darwinproject.ac.uk.
[39] C.R. Darwin to JDH, 24 December 1862, Letter 3875, Darwin Correspondence Project, viewed 16 January 2011 at www.darwinproject.ac.uk. Fn. 9 to this letter states that Oldfield's reply has not been found.
[40] C.R. Darwin to JDH, 3 January 1863, Letter 3898, Darwin Correspondence Project, viewed 16 January 2011 at www.darwinproject.ac.uk.
[41] Charles Darwin, 1888, *Variation of Animals and Plants Under Domestication*, vol. 2, John Murray, London, p. 215.

> the country has been very badly searched previously to my journeys there, or—there is something wrong, that's all.[42]

This may have served to fuel the Hookers' belief that Mueller's taxonomy suffered because of his lack of access to the reference collections in Britain and Europe. Nevertheless, reading between the lines, Oldfield's delight was palpable. According to R.W. Home *et al*., this fascicle was published (or sent to the publishers) on 1 April 1863[43], and the *Calothamnus* and *Melaleuca* species were referred to in fascicle 21[44]; if that was the case, Oldfield appears to have been in error regarding the number of the part he referred to in his letter.

On 3 March 1863, when London was a hive of activity preparing for the wedding of Prince Albert to Princess Alexandra, eldest daughter of King Kristian IX of Denmark, Oldfield wrote informing Hooker that he was sending by 'Parcels Delivery' all his specimens of *Diplarrena,* 'the Donatica Novae Zealandiae' [Donatia novae-zelandiae] and some seeds

> which I at first feared are too old, but as some of them germinated already perhaps with your great facilities at Kew you will be able to raise some of them. If not worth the trial, they will perhaps be useful as fruit specimens.

At the same time, he sent some specimens of the fern *Cyathea australis* 'on speculation', having 'read somewhere that' the Herbarium at Kew Gardens 'did not have proper specimens', and a variety of another fern 'Lomaria procera for Sir William to see; it certainly cannot be your δ'. Oldfield invited Joseph 'to select for Kew from anything' he sent.[45]

Oldfield was ill during March or early April for he wrote to Joseph Hooker on 7 April 1863 that he continued 'too unwell to do much'.[46] He offered to forward to Hooker the whole of his Exogens[47] for [Bentham's] use:

> If you think they will be of service to him and will let me know when he will be ready for them, I will gladly thus far assist in his great work, and am sorry I can do not more towards it.

At the same time, he asked for Hooker's opinion 'on the enclosed Lepidosperma[48] from Tasmania':

> I find one very distinct species of that genus is an addition to the Flor. Tasm. It is one of the largest in that genus, and remarkable for the [very] elevated axis of the culms and leaves. I have it, or one closely allied from S.W. Australia.

He also said he was

> quite in the dark as to L[epidosperma] linearis[49] and L. lateralis:[50] taking La Billardière's figure for a guide. I think I make out L. symanea [?].[51] The plant I had determined as L. linearis could never have been confounded with either of the above, and is probably also an addition to Tasmanian plants.[52]

At an 'exact time' specified by Hooker, Oldfield, in a subsequent communication of 4 May 1863, promised to deliver his collections to Kew Gardens during the last week of April 1863, including the

---

[42] AO to JDH, 10 February 1863, English Letters, vol. 97, 1859–1900, NAT–OXL, fol. 275 (282), RBGK.
[43] FM, 1862–1863, vol. 3 (23), 1 April 1863, cited in R.W. Home *et al*., 1998, vol. 1, p. 587.
[44] ibid., vol. 3 (21), 2 October 1862, pp. 111–119. We are grateful to Alex George for pointing out this error.
[45] AO to JDH, 3 March 1863, English Letters, 1859–1900, vol. 97, NAT–OXL, fol. 276 (283), RBGK.
[46] AO to JDH, 7 April 1863, ibid., Fol. 291 (298), RBGKA.
[47] Exogen refers to a plant belonging to one of the greater part of the vegetable kingdom which are characterised by wood bark, and pith, the wood forming a layer between the other two. The leaves are commonly netted–veined, and the number of cotyledons is two, or, very rarely, several in a whorl, *Biology Online,* viewed 15 January 2011 at www.biology-online.org.
[48] The genus *Lepidosperma* belongs to the family *Cyperaceae*, more commonly known as sedges.
[49] *Lepidosperma linearis* is an orthographic variant of *L. lineare*, APNI.
[50] *Lepidosperma lateralis* is now *L. laterale,* APNI.
[51] *Lepidosperma symanea* does not appear in APNI.
[52] AO to JDH, 7 April 1863, English Letters, 1859–1900, vol. 97, NAT–OXL, fol. 291 (298), RBGKA.

Lepidosperma and *Gleichenia* 'of which he had already written'[53] unless he heard from him to the contrary.

Thanking Hooker 'for the hint about the *Lepidosperma*', he said it was 'another step towards the solution of the puzzle'. He still had had nothing from Mueller except *Fragmenta*.[54] In response to some of Joseph Hooker's queries, Oldfield wrote, 4 May 1863, that he had

> only retained some duplicate specimens of the Australian collections, and the greater part of the Extra Australian: some few of the latter are still among the plants sent, but they can easily be put on one side. I have sent the Endogens[55], so that, whatever befal [sic], you may secure all that may be useful to you. The first selection of course belongs to Kew on the terms you specify as I shall retain nothing myself.[56]

He found 'collections a great incumbrance [sic] in [his] erratic existence' and wanted to get rid of the remainder in bulk, 'so as to be enabled to return to Australia, where I can be of some small use to Science while here I am of none at all'.[57]

He asked if Sir William wanted his 'Extra—as well as Australian Ferns', noting they were all (about 500 specimens) arranged together, as far as his 'scanty means of study' had allowed. He ended the letter by expressing the desire 'that someone would work on his Mosses and Hepaticae' as he thought 'they would find it worth their while'.[58]

It seems from later correspondence that, of those specimens he sent to Kew Gardens for Bentham's use in 1863, Joseph Hooker had agreed to purchase some species not already represented in the Kew herbarium (referred to as '*desiderata*'), and to try and sell the remainder to foreign herbaria. At the same time, he had 'observed that such establishments are very tardy in their payments'.[59]

In addition to the plant material Oldfield sent to Kew Gardens for Bentham, Mueller included over 2,000 of Oldfield's specimens in the parcels of specimens sent on loan from the Botanic Gardens, Melbourne Herbarium, under his agreement to co-operate with Bentham.[60]

Publication of volume 1 of George Bentham's Flora Australiensis

George Bentham wrote in his diary on 30 May 1863, 'Rec'd from Reeve copies of my Flora Australiensis 1s vol. [Ranunculae to Anacardiaceae]'.[61] In this work Bentham cited 189 of Oldfield's specimens. Among the many Australian collectors whose contributions were acknowledged, Bentham said of Oldfield:

> Mr Augustus Oldfield, an acute observer as well as an intelligent collector, who, besides the Tasmanian contributions mentioned in Dr. Hooker's Flora, made large additions to the West Australian plants previously known; in the first instance from the neighbourhood of Murchison river, and afterwards from the south-western districts. Mr. Oldfield is now in this country, and has most generously offered the use of his

[53] *Gleichenia* is a genus of ferns. The specimens sent on this occasion were probably for WJH. However, no mention of the genus has been found in earlier letters.
[54] AO to JDH, 22 April 1863, English Letters, vol. 97, 1859–1900, NAT–OXC, 1859–1900, fol. 292 (299), RBGKA.
[55] Endogen refers to a plant which increases in size by internal growth and elongation at the summit, having the wood in the form of bundles or threads, irregularly distributed throughout the whole diameter, not forming annual layers, and with no distinct pith. The leaves of the endogens have, usually, parallel veins and constitute one of the great primary classes of plants, including all palms, true lilies, grasses, rushes, orchids, the banana, pineapple, *Biology Online*, viewed 17 January 2011 at www.biology-online.org.
[56] AO to JDH, 4 May 1863, English Letters, vol. 97, NAT–OXC, 1859–1900, fol. 277 (284),RBGKA.
[57] ibid.
[58] ibid.
[59] AO to JDH, 8 March 1866, English Letters, vol. 97, NAT–OXL, 1859–1900, fol. 282, 289, RBGKA.
[60] See MELISR database, 08 March 2007, NHV, RBGM.
[61] H.S. Marshall, 1937, 'The dates of publication of Bentham's *Flora Australiensis*', *Journal of the Society for the Bibliography of Natural History*, vol. 1, pp. 69–71.

> own Australian herbarium to the Kew Museum, as a contribution towards the present Flora.[62]

There can be little doubt that the collector was pleased to have had such a favourable mention but he may have been put out over the inaccuracy of Bentham's statements regarding his expeditions in Western Australia. During his three visits to the Colony, he was in the Murchison River area twice and on his way to the Murchison River, on his first visit, he collected in the south-west of the Colony from Cape Naturaliste to Perth. Sometime later Oldfield also took issue on another matter arising from this recognition of his contribution (see below).

The first volume of *Flora Australiensis* was extensively reviewed, anonymously, in *The Sydney Morning Herald* of 13 January 1864. The author referred to the care that Bentham had taken to secure accurate information and listed persons who had contributed in one way or another to the work:

> [W]hilst the collections of recent explorers in Australia have been examined with the strictest attention, even the humble collections of an amateur are duly acknowledged. Thus, for instance, the labours of Drummond, Babbage, Gregory, Fitzalan, Stuart, Walcott, Brown, Landsborough, Leichhardt, Beckler, and Oldfield (who by the way is now in England, and has most generously offered the use of his own Australian Herbarium to the Kew Museum), are honourably recorded.[63]

Oldfield seeks Joseph Hooker's advice on his manuscript on Australian Aborigines

Oldfield may have met with Hooker on at least one occasion in the intervening period, but he does not appear to have had further correspondence with Kew Gardens until 4 November 1863. On this occasion, he wrote requesting Hooker's advice as to where he, Oldfield, might obtain 'the opinion of some competent authority' on the value 'in an ethnographical sense' of the manuscript of his paper on Australia Aborigines. He had 'worked up' the paper from his notes after his return to England.[64] Hooker's reply has not survived, but from subsequent events he apparently sent the manuscript to his colleague, the biologist, Sir John Lubbock, 1st Baron Avebury. Lubbock was the President of the Ethnological Society of London and researcher on 'primitive man'.

Sir William Hooker makes contact

Sir William sent a copy of his 'Tour in Iceland' to Oldfield who responded on 28 November 1863 that he would read it with great interest, 'and, I doubt not fruition'. He returned many thanks for the opportunity of 'adding another to my list of books by an author to whom I owe the chief part of what little I know of Botany, and for whom I entertain in great respect'. Once again, he offered to send fern specimens and drew the botanist's attention to some of his observations:

> In Asplenium bulbifirum as far as I have observed there is no <u>true</u> bulbil[65] formed; the appearance is given by spores germinating while yet attached to the spore case (as I have often seen the seeds of Lotus, Trifolium, Nasturtium germinate while still in the seed-vessels).
>
> In Polystichium vestiutum a true bulbil is formed in the rachis of the frond.[66]

---

[62] GB, 1863, vol. 1, 1863, p. 14.
[63] *The Sydney Morning Herald*, 13 January 1854, p. 3, DAN, NLA.
[64] AO to JDH, 4 November 1863, English Letters, vol. 97, 1859–1900, NAT–OXC, fol. 278 (285). RBGKA.
[65] Bulbil is a small bulb-like structure, one in the axil of a leaf, which may fall to form a new plant, *Oxford Dictionary*, viewed 19 January 2010 at www.oxforddictionaries.com.
[66] AO to WJH, 28 November 1863, English Letters, vol. 72, 1862–65, L–Z, fol. 197, RBGKA.

Sir William appears to have accepted Oldfield's offer of specimens, for on 5 December Oldfield wrote: 'I have looked over my Ferns [etc.] and drawn out a few for you, but I am sure there is nothing to interest you'. There followed what seem to have been responses to some of Sir William's queries:

> With regard to the Isoetes, the most northerly point at which I detected it was on the Murchison, about 80 miles above the mouth.[67] It was growing in shallow pools formed by fresh-water springs (the water of the river is brackish) flowing into basins in the granite soaks. It was of the dwarf nature as in the accompanying specimens, but at Toodyay (Darling Range) I found it attaining a height of 8-10 in. always solitary, and the root of the same bulbous-like form as in the small varieties, and always few-leaved (4 – 6) the leaves for the greater part of their length emerged.
>
> Adiantum Aeth[i]opicum[68] occupies a very limited area in W. Australia, being only found at the southern extremity of the Darling Range — Drummond did not know the plant, but I think that he must have found it in his trip to Port Augusta [? Augusta, Western Australia]. It is curious that after being lost for so great a distance it should appear there, for I believe it has not been found from Victoria to the Darling Range. Gymnogramma leptophylla occupies the same station, but I have found it in South Australia.
>
> Nephrodium unitum[69] first makes its appearance in W. Australia in springy ground at the South Hutt, Port Gregory, and is abundant in similar localities as far north as Freycinets' [sic] Peninsula (Shark's Bay [sic]) to my knowledge. You must have this from Drummond.
>
> I do not recollect a Port Cooper in Australia: one there is in New Zealand.[70]

There is no evidence that Oldfield corresponded further with Sir William although they may have met at Kew Gardens.

A change of Directors at Kew Gardens

On 12 August 1865, the 80-year-old Sir William Hooker died at his home in Kew following a throat infection that was epidemic in London. Officially, his demise was due to 'unspecified causes'. Not long before his death, Sir William, in a memorandum to the Commissioner of Works, pointed out that Joseph would inherit his herbarium. Sir William expressed the hope that the Government would buy the collection at a reasonable valuation decided by six eminent naturalists, as his son did not have the resources to maintain it.[71] The outcome was in accordance with these wishes. The Government paid £5,000 for the herbarium (1,000,000 specimens), £1,000 for the library (4,000 volumes), £1,000 for correspondence (29,000 letters bound in 76 quarto volumes), manuscripts, botanical drawing and other items.[72] This purchase was to be the nation's monument to William Hooker and together with the donations of the herbaria and libraries of George Bentham and William Bromfield[73], the herbarium at Kew Gardens was confirmed as an ongoing resource for botanical taxonomists around the world.

Sir William's published taxonomic output was prodigious and this was greatly assisted by the activities of the global network of collectors he worked to secure. He pursued practical applications of botany and under his direction at Kew Gardens expanded the Museum of Economic Botany (now the

[67] Oldfield's statement that he had detected *Isoetes* on the Murchison, about 80 miles above the mouth has already been referred to in Chapter 11.
[68] *Adiantum aethiopicum* is a maiden hair fern.
[69] Oldfield's statement that *Nephrodium unitum* is abundant as far north as Freycinet Peninsula has already been referred to in Chapter 16.
[70] AO to WJH, 5 December 1863, English Letters, vol. 42, 1862–65, L–Z, fol. 198, RBGKA.
[71] Ray Desmond, 1995, p. 205.
[72] Mea Allan, 1972, p. 494.
[73] Ray Desmond, 1995, p. 198–199.

Fig. 21.3: Herbarium (Hunter House), Royal Botanic Gardens Kew, 2002

Economic Botany Collection). He enlarged the area of the Botanic Gardens from 11 acres to 250 acres, including the arboretum, and erected new glass houses, including the magnificent Palm House.[74] He opened the Gardens to the public. Hooker's improvements to the Gardens, 'coinciding with better travel facilities, transformed Kew into a horticultural mecca for the metropolis'; in the year of his death annual attendance figures exceeded one million. He made botany accessible to the masses through information guides to the contents of the Museum and to the Gardens.[75]

Over the years, Sir William had worked hard to promote Joseph's scientific career, including his appointment as Assistant Director. Joseph gradually became disenchanted with administrative routines and preferred instead to focus on his scientific work. In 1863, he had accepted a commission from the Government of India to write a flora of the sub-continent. He regarded 'succession to [his] father with horror'. Nevertheless, in November 1865 he was appointed Director but he told Charles Darwin:

> I am prepared to improve the Gardens enormously and will do so, but if the scientific character of the establishment is to go down one iota, I shall intimate that I only hold the post with a view to retirement when able.[76]

True to his word, he spent much time and energy on landscaping the gardens. Among his more notable achievements was the network of paths and walks that made the arboretum more accessible to the public.[77]

[74] S.D Hopper, 2015, p. 3.
[75] Ray Desmond, 1995, pp. 221–222.
[76] ibid., p. 225.
[77] ibid., pp. 226–238.

More correspondence with Joseph Hooker

Oldfield's eyesight was worsening and a letter from Frederica to Vincent of 4 March 1865 shows that he was in a poor physical and mental state:

> We heard lately from Aug[ustus] & M[arian] and both were very poorly wh[ich] accounts for A not having been to you—M says he seems very unwell & highly nervous & sometimes does not open his letters for days – but do not mention this at all. I find he opens yours! I fear he is greatly disappointed at his non success and unprofitable mode of life, and I feel deeply for him—but what can be done? On his small means he could not do better without getting employ and that he is not equal to in health I mean.[78]

The Ethnological Society of London accepted the paper on Australian Aborigines and the 'preliminary portion' was read at the meeting of 9 February 1864.[79] This was reported the next day in the *Morning Post* along with a brief overview of the article.[80] Apparently Oldfield was not invited to this meeting. Out of frustration at not hearing from Lubbock regarding the work, Oldfield wrote directly to him in March 1865.[81] This brought forth another castigation, conveyed by Hooker on 4 April 1865, to which the collector, who was ill at the time, replied:

> I exceedingly regret that you have been put to further trouble in the matter [of the paper on Australian Aborigines], but I am not to blame, for in the early part of last month I wrote to Mr Lubbock for his information relative to this affair, but at the same time said that I was induced to apply to him directly, as I wished to spare you further trouble. I should not have taken the liberty of addressing him had it not been that since your communication of the fact that the paper was to be presented (in July last [9 February 1864]), I had received no tidings of any kind as to its destination...I find that I have omitted the most important consideration of your letter. In my last to you, in which I expressed a wish to have presentation copies of my paper, I offered to pay for them being under the impression that such a course is customary, so that if you forwarded that letter to Mr. Lubbock, of course I cannot retract my involuntary [sic] generosity.[82]

Although he was never a member of the Ethnological Society of London, it must have been disappointing not to receive an invitation to be present at the reading of his paper, and how irksome to have been treated by Lubbock as a person of such little consequence that he could not even tolerate Oldfield's direct approach or respond to his letter. Clearly, in Lubbock's mind, naturalists of lower orders, no matter how much they contributed, must be kept in their place. There seems to be no other plausible explanation for his behaviour.

On 2 December 1865 Hooker wrote to Mueller:

> Today's "Times" paper announces the death of Augustus Oldfield, no particulars beyond that he leaves a wife & 2 children. Can you tell me any-thing of his history. I got much interested [here] - he was always gloomy & odd, & when pressed about his nervous condition always darkly hinted at some dreadful scene, the witnessing of which shattered his nerves. We liked what little we saw of him; - he was quiet, unobtrusive, singularly moderate & extremely well informed; his large and most important

[78] Frederica Oldfield to Vincent Oldfield, 4 March 1865, file nos 0157–0159, PFA.
[79] AO, 1865, p. 215.
[80] 'Ethnological Society', *The Morning Post*, 10 February 1864, p. 7, viewed 20 January 2014 at www.explore.bl.uk.
[81] AO to JDH, 4 April 1865, English Letters, vol. 97, NAT–OXL, 1859–1900, fol.279 (286), RBGKA.
[82] ibid.

> Herbarium he had given to me long ago, and would take no price in reward, wished no public notice of it, & only hoped it might be useful. I couldn't help feeling immensely interested in him. I understood that he lived with or near a sister or brother here in England.[83]

Mueller replied, 5 February 1866, regretting hearing of 'Mr Oldfield's death' and said that he 'never knew a word' of 'a dreadful scene' nor have the details emerged in the historical record.[84] To what extent the 'dreadful scene' contributed to his 'gloomy & odd' demeanour we will never know but it must have exacerbated his 'highly nervous' state, of which Frederica wrote, brought about by his failing eyesight and inability to earn a living and to continue his botanical endeavours.

Mueller appears to have misunderstood Hooker's remarks about Oldfield's collections, which pertained to material donated by the collector prior to 1865:

> After his return I allowed [Oldfield] to keep a specimen of the W. Austr. plants he collected on an understanding that they should not be distributed. I am now glad, that after his death they have fallen in your hands; they could not be placed anywhere better. His strength was special knowledge of a number of Tasmanian Mosses & Lichenasters...Oldfield's searches for plants in W. Australia were very creditable to him & I think I have done him full justice in my writings. The 2 or 3 plants bearing our joint authority [i.e. *Eremophila clarkei* and *Halgania bebrana*] were named by myself. I liked to give him every encouragement in my power & did not spare my private means for the purpose.[85]

Mueller also mentioned Oldfield having been a school fellow of Professor George Britton Halford of the University of Melbourne. This is confirmed by his father, 'J. Halford, 14 Down-street, Piccadilly', being one of the signatories to the letter, 1 July 1841, refuting the scurrilous claims which had been made against the Misses and Messrs Liborel's school at Guînes, near Calais.[86] George Britton Halford, who had already distinguished himself as a surgeon, physician and anatomist in England, was the inaugural Professor of Medicine at the newly formed School of Medicine at the University of Melbourne, which had had its first intake of students in 1862. He arrived in Melbourne with his wife Nancy (née Gadd) and family on 23 December 1862[87], about seven months after Oldfield left for England. For Mueller to have known that Halford and Oldfield were at school together, the collector must have learnt of the appointment before he left Melbourne.

Imagine Hooker's surprise when he received a letter (dated 1 March 1866) from Oldfield[88], who later clarified the identity of the deceased: a fourth cousin of his 'of whose existence [he] was not aware until it had terminated'.[89] On 1 March, Oldfield had sought further advice from Hooker:

> It is now more than three years since I returned to England with the sole and express purpose of publishing a work which has engaged my chief thoughts during many years and to complete which I have sacrificed health and fortune. As I am totally unknown, and moreover labour under very great disadvantages, I have considered it useless to offer it to a publisher in the usual manner and have therefore waited thus long hoping that the sale of my collections would enable me to defray the expenses of its publication. Other hope I have none, for considering the state of my health, I dare not

[83] JDH to FM, 2 December 1865, RBGKA, transcription and copy kindly provided by Dr Sara Maroske, 4 March 2012 & 16 March 2012 from the database of The Correspondence of Ferdinand von Mueller Project, Department of History & Philosophy of Science, University of Melbourne.
[84] FM to GB, 5 February 1866, Kew Correspondence, Australia, Mueller, 1858–1870, fols. 196–201., cited in R. W. Home et al., 2002, vol. 2, pp. 337-338.
[85] ibid.
[86] Frederick Oldfield, 'French Schools', Letter to the Editor, *The Times*, London, 8 July 1841, p. 8, TTDA.
[87] K.F. Russell, 1972, 'Halford, George Britton (1824–1910)', ADBO, viewed 18 May 2010 at www.adb.online.anu.edu.au.
[88] AO to JDH, 1 March 1866, English Letters, vol. 97, 1859–1900, NAT–OXL, fol. 280 (287), RBGKA.
[89] AO to JDH, 13 March 1866, English Letters, vol. 97, 1859–1900, NAT–OXL, fol. 284 (291), RBGKA.

> sink the little I hold for [illegible word] in a venture which is sure to fail as a mercantile speculation.
>
> Although of course a firm believer in the truth of my theories, I frankly confess that they will be found so entirely absurd, and the conclusions to which they lead so widely extravagant, that I could not ask any scientific man to implicate himself in their publication, otherwise there are several points on which I would fain seek advice, the which I must however forgo, for I am so unfortunately situated as not to have even a single acquaintance in England, let alone one, who could assist me in the elaboration of my treatise, which, were it more philosophically treated, I should say is on the philosophy of Natural History, while other circumstances prevent me from availing myself of facilities available to others.[90]

Although Oldfield did not explicitly ask Hooker for advice on 'several points', he was probably hoping that the 'scientific man' would offer. There is no further mention of this manuscript in the Oldfield correspondence at Kew Gardens so it seems unlikely that Hooker's advice was forthcoming. If he did read the manuscript, he may have backed away from showing it to any of his colleagues because of Oldfield's criticism of some aspects of Darwin's theory of natural selection and because there were differences between Oldfield's and Hooker's ideas about plant classification.

Hooker was a close friend and colleague of Darwin and supported him, and his theory, through the controversy that raged after *On The Origin of Species...* was published. Oldfield's work was never published and he left no details as to the reason. Financial inadequacy is the most likely explanation. The academic treatise is about 900 handwritten pages plus many, often very lengthy, footnotes on separate sheets. It is not surprising that the writing of it drained his physical resources; he must have been devastated at the outcome. Oldfield continued his letter of 1 March 1866 by enquiring as to whether there existed

> any probability of selling [my] collections (inclusive of the cryptogam[ic] portion which I have here and which is far more perf[ect] than that at Kew) even at a sacrifice, for I feel very anxious to leave England, as I find my health failing me each succeeding winter that I remain here. Indeed, had it not been for the unusual mildness of this season, it is likely that you had not been troubled in this subject.[91]

Despite the clemency of the winter, Oldfield told Hooker on 8 March 1866 that he had been 'prostrate for nearly five months'.[92] Whatever illness afflicted him he seems to have believed it was serious enough to threaten his life; recurrent bouts of pneumonia or bronchitis perhaps?

It seems that any hope of selling his specimens was dashed. In May, Oldfield learned that the terms under which he sent his collections to Kew Gardens three years before, c. April 1863[93], had been 'quite forgotten':

> [W]hile I have been patiently waiting for nearly three years, thinking that the delay was caused by this tardiness [i.e. of foreign herbaria to pay] referred to, I suddenly learn that I am no further advanced than I should have been had I devoted the six months proposed doing what you promised to accomplish in as many weeks, for had I described those plants which I could not in times past identify, I should have had quite as many named (and certainly far more valuable) plants for sale, than I can have under the present arrangement.
>
> Of course, I am glad that the Kew collections should be supplied with desiderata according to our compact, but I really do not perceive what I gain in any manner by the

[90] AO to JDH, 1 March 1866, English Letters, vol. 97, 1859–1900, NAT–OXL, fol. 280 (287), RBGKA.
[91] ibid.
[92] AO to JDH, 8 March 1866, English Letters, vol. 97, NAT–OXL, fol. 282 (289), RBGKA.
[93] AO to JDH, 4 May 1863. English Letters, vol. 97, NAT–OXL, fol. 277 (280), RBGKA. In this letter Oldfield referred to the 'terms' specified by Hooker regarding the disposal of his specimens but there are no details of what these arrangements were.

> transaction, for after the with-drawal [sic] of my normal herbarium (which I regard as the first set); there will remain little except some common K.G. Sound, N.S. Wales and Tasmanian specimens, which by themselves can possess but little value either in a scientific or pecuniary point of view.[94]

He went on to say that, having only Hooker's letter to guide him, he judged 'of the facts of the case'. Oldfield assured Hooker that he placed 'implicit confidence' in his 'honour'

> and that if you consider what has been done for me is at all adequate to the price I pay for it, then am I satisfied. Were I a little stronger I would wait on you, for five minutes conversation and would doubtless place the whole affair in a totally different light.[95]

He then offered to receive back the specimens should they be in Hooker's way, 'inconvenient though it would be just now, but as I still cling to the hope that I shall be able to get rid of them wholesale, I should like one removal to suffice'. He ended by saying that he was in no condition to undertake the labour of identifying and selling the specimens, which he had 'so strenuously sought to evade some three years since'.[96]

Hooker appears to have queried whether Oldfield had sent his collections to Kew Gardens before they had agreed on the terms of their disposal, for in his letter to Hooker of 13 March 1866, Oldfield stated:

> To the best of my recollection (and my memory is pretty good on such points, especially with respect to my rare conversations with scientific men) all arrangements about my plants were made prior to their removal to Kew, and indeed they have only been referred to three times since that event, once by yourself some three months after you had them, when you casually observed ... 'We have done nothing with your plants yet': again about a year after by your late lamented father, Sir William, who spoke to me in commendation of my collections, and afterwards expressed the same on the fly-leaf of a copy of his 'Tour of Iceland' ....and lastly by myself in a letter [4 May 1863] in which I expressed a desire to dispose of my whole collection.
>
> If such be the case, I am quite at a loss to account for your departure from the view taken of the transaction [which appeared] in Mr Bentham's 'Flora Australiensis' of which I was quite ignorant before the receipt of your last letter, for I have not yet seen the 1st vol. of that work, except for a few minutes once at Kew.[97]

On the face of it, the departure from the transaction [between Hooker and Oldfield] appears to have been that Oldfield believed Hooker and Bentham were mistakenly under the impression that the collector had 'most generously offered the use of his own Australian herbarium to the Kew Museum, as a contribution towards the present Flora'. Oldfield's offer of 7 April 1863 to forward the whole of his 'Exogens to Kew for [Bentham's] use' to 'assist in his great work' certainly suggests that he was making them available *gratis*. This, however, does not appear to have been in accord with the terms of his agreement with Hooker although the historical record provides no clarity on the matter. Furthermore, this was not the source of the problem. It seems that Oldfield, by his own admission, had misinterpreted Hooker's words 'according to my own hopes and expectation[s]

> (for I considered the acquisition of a bona fide desideratum by such an Herbarium as that at Kew, as far more important than the receipt of a hundred specimens already possessed by it ... [I]f you consider that I have not acted dishonourably in this affair, I

[94] AO to JDH, 8 March 1866, English Letters, vol. 97, NAT–OXL, fol. 282 (289), RBGKA,
[95] ibid.
[96] ibid.
[97] AO to JDH, 13 March 1866, English Letters, vol. 97, NAT–OXL, fol. 284 (291), RBGKA,

> should like the discussion to stop at this point, and henceforth I will merely refer to my duplicate.[98]

That, however, was not the end of 'the affair'. Within a day or two, Hooker wrote to Oldfield to which letter he replied on 16 March 1866. He thanked the Director for 'the general kind tone' of his letter which had 'exonerated' him from having been 'influenced by any mean motives in having initiated' the correspondence; this had 'removed a load of uneasiness from' his mind. However, he said that Hooker had misunderstood him on 'some points'.[99]

The first seems to have related to one of Hooker's comments that Oldfield's reason for sending his specimens to Kew Gardens was that they were an encumbrance to him. Oldfield took umbrage and reminded Hooker that he had now recollected that there was another time when his collections were mentioned. This was in a letter from Hooker[100] in which he said: 'Are these all the plants we have to overhaul for you?' This, Oldfield said, he had mentioned

> not to prove you are in error but merely to show that I was in some sort justified in the conclusion I arrived at, namely, that the plants were sent to Kew for some other purpose than merely to get them out of my way.[101]

The second point he raised was that Hooker had apparently questioned whether Oldfield had ever said that his entire collection was up for sale. In a response, 8 March 1866[102], he responded that this wish was expressed in a letter to Hooker. The Director appears, then, to have told Oldfield that such an intention should have been conveyed to the then (1863) Director, Sir William Hooker, to which the collector replied:

> I perfectly agree with you that had I boldly mentioned my intentions to your father, he would most cheerfully have given me all the assistance in his power, for I am well assured both by common report and my own personal experience [to] help all—even such undeserving ones as myself—who evince any desire to serve science... as I have observed, the principal part of the misunderstanding has most certainly arisen from my very great reluctance to intrude myself or my affairs on others: for had I fairly stated my wishes to [Sir William] (and let me add in justice to you, or to yourself) all would have gone well.[103]

He went on to recall the occasion, just before his return to Australia in 1850, when he had arranged to meet Sir William to discuss with him the disposal of his future collections; he did not have the 'courage to address' the Director when he arrived to keep the appointment so he left England 'without having entered into arrangements which might have benefited [him] considerably'.[104]

The grounds for Augustus to believe that his circumstances, regarding the sale of his residual collections in 1863, would have been very different had Sir William Hooker lived, were not documented. Perhaps Augustus believed that his collection was commercially valuable enough to realise an amount that, when invested, would have provided him with a substantial increase in his regular income. To achieve this, he may have needed to receive at least £500 for the collection and be able to invest it at 5 per cent per annum to increase his annual income to £60, approximating what he was earning as a school master in Tasmania. Given the historical method by which Sir William acquired specimens for his herbarium, it is highly unlikely that his generosity would have extended to spending that amount himself. On the other hand, Oldfield may have considered that Sir William would have worked more conscientiously than his son to sell his collection. This is doubtful since it is difficult to imagine that

[98] ibid.
[99] ibid.
[100] Oldfield stated he had the letter in his possession but that he was quoting from memory.
[101] AO to JDH, 16 March 1866, English Letters, vol. 97, NAT–OXL, fol. 289 (296), RBGKA.
[102] AO to JDH, 8 March 1866, English Letters, vol. 97, NAT–OXL, fol. 282 (289), RBGKA.
[103] AO to JDH, 16 March 1866, English Letters, vol. 97, NAT–OXL, fol. 289 (296), RBGKA.
[104] ibid.

had he lived, the aging Director had the energy or the will to peddle the specimens to herbaria and private collectors around the world.

Oldfield's 'apparent apathy toward his collections' was, it seems, another matter raised by Hooker to which the collector replied:

> [M]y very much impaired vision is a sufficient explanation, for they have become nearly useless to me, though I still continue to study them mentally, and the notes I made at the time of their gathering have been of material service to me.[105]

What prompted Hooker to say this is a mystery. Perhaps Bentham had complained that the collector was not diligent enough in providing details of his plant observations or that Hooker considered that Oldfield should try to classify his plants or to sell them himself. It would be interesting to know whether Hooker received Oldfield's explanation sympathetically.

Oldfield went on to refer to 'a very generous proposal' put forward by Hooker which resulted in the collector's concurrence with Kew Gardens' entitlement under 'any circumstance' to 'bona fide desiderata' and with the duplicates to be sold.

Once again, Oldfield expressed the hope that Hooker would find time to look over his Cryptogamic collections

> as you would thereby secure authentic specimens of most of Gottsche's [C.M. Gottsche], Hampe's [G.E. Hampe] and C. Müller's Australian Mosses and Hepaticae, beside many of your own in a more satisfactory state than you have hitherto seen them.[106]

The use of 'authentic' in this context appears to have been inappropriate since it is quite clear he is talking about the quality of the specimens for taxonomic purposes.

Hooker wrote again on 28 March 1866, apparently asking whether Oldfield still wanted Kew Gardens to sell his duplicates. Too 'poorly' to reply immediately, he responded on 3 April 1866:

> I should much prefer not quitting the position I assumed in mine of the 16th ult. in which I placed myself unreservedly in your hands. Indeed, after having renounced all claim on your department, I could not conscientiously act the part of a seller.

Nevertheless, he still wanted to know 'whenever there was a fair chance' of finding a purchaser for the sets of his duplicates, so that he could arrange to receive them at his home 'as by working in [sic] some specimens lately received, [he would] somewhat increase its [sic] value'.[107]

On 6 April 1866 Oldfield forwarded to Hooker a receipt for £10. Although he did not specifically state the reason for the payment, there is little doubt that it was to cover the purchase of the *desiderata* specimens. The collector expressed his 'perfect satisfaction' with Hooker's conduct towards him 'throughout this affair'. At the same time, he thought, from what Hooker had said, that it was better for his duplicates 'to be sold from Kew' although he wanted to retain the ferns for 'in the case of finding a purchaser for [his] cryptogamic collection, [he] should like them to be perfect'. Hooker had put a value on the duplicates to which the collector responded that they might go at Hooker's estimation (about 2000 specimens), though he thought 'it low, but that would save trouble'. The Director's estimate was not found.[108] Oldfield assured Hooker that none of the specimens covered by the embargo that Mueller had placed on Oldfield's 1858–1859 Western Australian collections were among those to be sold from Kew Gardens.[109]

---

[105] ibid.
[106] ibid.
[107] AO to JDH, 3 April 1866, English Letters, vol. 97, NAT-OXL, 1859-1900, fol. 286 (293), RBGKA.
[108] The RBGKA holds a collection of volumes entitled 'Goods Inwards' and 'Goods Outwards', containing registers of specimens entering and leaving Kew Gardens. These are not comprehensive and a record of purchase of '*desiderata*' specimens could not be identified. These volumes do not record everything that was received or sent by Kew, Sara Cox, Herbarium, Library, Art and Archives, Kew Gardens, personal communication, email to MHH, 28 January 2011, HOA.
[109] AO to JDH, 2 May 1866, vol. 97, NAT–OXL, 1859–1900, fol. 288(295), RBGKA.

It is not possible to come to any conclusions as to whether Oldfield was treated fairly. In the first place, we do not know what passed between them when Oldfield visited Kew Gardens, which he apparently did on more than one occasion. Secondly, Hooker's letters to Oldfield relating to the matter have disappeared and copies were not kept. Thirdly, it is not known how many specimens were involved and how their value was determined. It seems from the foregoing that the price was very much up to what the purchaser was willing to pay and that very little account was taken of the collector's skill and the time and expense of obtaining them. Still, Oldfield must have been aware, possibly from Charles Stuart before he undertook his last expedition in Western Australia, of the pitfalls associated with selling his specimens.

Hooker does not appear to have been in any hurry to pay for the desiderata, some at least of which had probably been referred to by Bentham in his first two volumes of *Flora Australiensis*, and taken into the Kew Gardens herbarium collection. Nevertheless, if Hooker was being reticent about purchasing Oldfield's *desiderata* it is perhaps understandable. At the time, there was no provision in Kew Gardens' government funding allocated for this purpose. In any event, perhaps unbeknown to Oldfield, the entire collection was being retained in one piece for possible use in the remaining volumes of Bentham's *Flora*.

It is also apparent that Oldfield was his own worst enemy when it came to plan the sale of his specimens. His timidity on the occasion in 1850 when he failed to keep his appointment with Sir William, apparently arising from a deep-rooted shyness about bringing his personal affairs to the attention of others, seems to have set him off on a path of making material freely available to the Hookers. It has been demonstrated that, apart from the letter he wrote in August 1861 asking Hooker if he was interested in purchasing sets of specimens, the purchase of his collections prior to that year appears unmentioned to either of the Hookers. It is to be expected, therefore, that from the beginning of their association with Oldfield until the 1860s, the Hookers would not have considered payment an issue.

In preparing his *Flora Australiensis,* Bentham made significant use of Oldfield's Australian collections. By our count across the seven volumes of this work (1863–1878), Oldfield's specimens, including many that were used to describe species new to science, are cited 1,692 times.[110] Apart from those already in the herbarium at Kew Gardens, including the *desiderata* referred to above, Mueller's consignments on loan to Bentham included over 2,000 specimens collected by Oldfield.[111]

In Hooker's letter to Charles Darwin he suggested that if he wanted any information about Australia & its savages, 'a capital man [namely, Oldfield] is now about Kew[112]'. This poses the question of whether Oldfield ever worked in the Kew herbarium after his return to England. In one or two letters to Hooker he mentioned his intention of visiting the Botanic Gardens.

There are specimens at Kew Gardens and in the National Herbarium of Victoria that have labels with plant habitat and/or form, information that Oldfield appears to have added when he was nearly blind c.1865–66. Thus, comparisons between the handwriting on his original labels and those written in pencil and added later show that his script had changed and deteriorated quite significantly.[113] Using a pencil in these cases probably meant that his failing eyesight rendered the use of pen and ink difficult due to his inability to properly control the flow of ink. We suggest that Oldfield added this information at Kew Gardens at Bentham's request. If this is correct then it tends to confirm that Oldfield kept field notes of plant observations and the specimens he collected.

[110] GB, 1863–1878, vols. 1–7, HAODb.
[111] See MELISR database, 08 March 2007, NHV, RBGM.
[112] JDH to C.R. Darwin, 21 December 1862, letter 3856, Darwin Correspondence Project, viewed 16 January 2011 at www.darwinproject.ac.uk.
[113] See MEL 602747, *Melaleuca oldfieldii*, N. of Yatthoo, Murchison River, AO, lectotype, ibid.

Fig. 21.4 Lectotype of Melaleuca oldfieldii (MEL 602747)

In response to an enquiry from Mueller, Bentham said he had given Oldfield a copy of his *Flora* but had not seen him since 'the last volume was out' (vol. 3, 1866), and he 'did not now know where' to find the collector.[114] Bentham had not bothered to enquire of Hooker for Oldfield's address.

Hooker and Oldfield met at Kew Gardens in late March 1866, and apparently the matter of the embargo Mueller had placed on the distribution of the collector's 1858–1859 Western Australian specimens was raised during their discussion. Oldfield wrote to Hooker on 2 May 1866 'conscientiously' assuring him that none of those specimens were among the plants he had sent to Kew Gardens. At the same time, Oldfield forwarded *Eucalyptus* material according to promise and also subjoined a list of other plants that he wanted identified. He made this request because he did not wish to do the work himself at Kew Gardens as 'the odour of museums (and even of your Herbarium) always renders me ill for several days, so that I avoid such places as much as possible'.[115] This might mean that he provided the additional information regarding his plant observations before his illness which commenced c. November 1865.

On 21st July 1866, Hooker wrote to inform Oldfield that the proofs of his paper on Australian Aborigines were at Kew Gardens and asked if he wanted to look at them; Oldfield replied in the affirmative two days later. He went on to say, however, that he was 'quite unused to correct [sic], and

---

[114] GB to FM, 14 May 1868, RB MSS M4, Library, Royal Botanic Gardens, Melbourne, Correspondence of Ferdinand von Mueller Project, copy kindly provided by Dr Monika Wells, Royal Botanic Gardens, Melbourne to MHH, email, 7 June 2001. Volume 4 of *Flora Australiensis* was not published until 1869.

[115] AO to JDH, 2 May 1866, English Letters, vol. 97, NAT–OXL, 1859–1900, fol. 288 (295), RBGKA.

besides am very busy just now, I should be glad to shirk the duty of correcting if possible'. He restated his request for some copies for which he would bear the expense.[116]

Oldfield was so ill on 26 November 1866 that he found it necessary to enlist the services of one of his sisters, probably Juliana, to pen his reply to a note from Hooker, dated 24 November 1866, in which he requested Oldfield to come to the Gardens. Oldfield, 'being indeed quite prostrate', said he was unable to go in person and that unless the 'affair' could be concluded by letter, an alternative was for his sister to go to Kew to act on his behalf.[117] It appears from Oldfield's communication that the matter was more serious than simply taking delivery of and paying for the off-prints of his paper and that it may have had something to do with his specimens. Unfortunately, no further details have been found.

Mueller enquired of Hooker on 2 March 1868: 'Is Mr Oldfield still in any way active? I fear not on account of his glaucomatous condition'.[118] Apparently Mueller and Oldfield had not corresponded for a lengthy period. Eight years seem to have elapsed before correspondence passed between Augustus and Hooker again.

By about 1869 Oldfield's eyesight was severely impaired. If Hooker had tried to sell Oldfield's specimens he was unsuccessful. On 2 February 1874, Oldfield wrote to the Director enquiring as to whether 'anything had been made' of the remnant of his collections. He wrote the letter himself; his handwriting showing marked signs of his visual disability. Daniel Oliver, Keeper, Kew Herbarium, and Professor of Botany, University College London, replied informing Oldfield that the specimens were still at the Herbarium.[119] Juliana penned to Oliver a reply, dictated on Augustus' behalf (19 February 1874):

> On mature consideration I should prefer to retain the remnant of my collections, and should therefore feel obliged if you or whoever has charge of such matters, would cause that remnant together with the cases in which my collections were originally contained to be forwarded to me here [5 Verandah Cottages, Wandsworth] at your convenience.[120]

This, then, appears to have been the end of Augustus' correspondence with Kew Gardens. It would be understandable if he had felt cast aside by Hooker. His precarious state of health, the disability under which he was labouring, and what that meant for his future might have combined to intensify his emotional reaction; unfortunately, no details have survived.

Oldfield and Joseph Hooker's relationship

After reviewing Joseph Hooker's correspondence with several of his colonial collectors, including Oldfield, Endersby questioned the long-held view that such people were

> marginal to the development of the sciences; they were considered merely passive recipients of knowledge that was created in Europe. In recent decades, the inadequacy of such a view has become increasingly apparent, but there is still considerable debate about what might replace it.[121]

According to Endersby, recent scholarship has begun to recognise that Joseph Hooker's preoccupations—especially taxonomy, botanical distribution and the disciplinary status of botany—are central to understanding the material practices of 19th century natural history, particularly in its imperial context.

[116] AO to JDH, 23 July 1866, English Letters, vol. 97, NAT–OXL, 1859–1900, fol. 294 (301), RBGKA.
[117] AO to JDH, 26 November 1866, English Letters, vol. 97, NAT–OXL, 1859–1900, fol. 295 (302), RBGKA.
[118] FM to JDH, 12 March 1868, copy from The Correspondence of Ferdinand Mueller, provided to MHH, by Dr. Monika Wells, 7 June 2001
[119] AO to JDH, 2 February 1874, English Letters, vol. 97, NAT–OXL, 1859–1900, fol. 296 (303), RBGKA.
[120] ibid.
[121] Jim Endersby, 2008, p. 55.

Endersby has asserted that these practices

> need to be seen as a complex series of negotiations, rather than in terms of straightforward metropolitan dominance; much existing historiography has assumed that those in the colonies were passive servants of imperial science.[122]

Was Oldfield, then, such a servant or did his relationship with the Hookers meet Endersby's claim that, like other colonial collectors, it comprised a 'complex series of negotiations'?

There is no evidence that during the years Oldfield was collecting in Australia or while he was in England from 1849–1850, he and any eminent botanist in England entered into a negotiated agreement as to what he was to receive in return for specimens. In fact, the sentiment expressed in his letters to the Hookers during this period was that his only desire was to serve science in any way he could.

This altruistic approach, as Endersby has asserted, was not uncommon. For example, Ronald Gunn in Tasmania 'considered himself as a gentleman who scorned payment because science must be "followed for its own sake alone."' He did, however, exert pressure on Joseph Hooker by making it clear that he expected taxonomic publications in return for specimens.[123] As far as can be ascertained, Oldfield never gave voice to the same expectation. Nevertheless, he did bemoan to the Hookers that his lack of access to suitable taxonomic reference texts was preventing him from improving his plant classifications.[124] This lament may have been expressed in the hope of extracting taxonomic works from them in return for any collections he sent to Kew Gardens. The Hookers did send him a few of their publications. These gifts, always received with gratitude, may have been intended as inducements to send further material and, as Endersby has suggested, to ensure that the collector's plant classifications were in accordance with the Hookers' requirements.[125]

When Oldfield offered specimens to Kew Gardens he usually requested identifications of some of the plants, but he never explicitly stated that he expected them in return for donating the material. Again, these requests may have expressed a veiled 'gentlemanly' expectation to that effect. He did give voice to one expectation—that his donations be acknowledged. This was an outcome that was not always met, although the tyranny of distance and the unreliable nature of the mails from England meant that he could not be certain in that regard.

As was customary at that time, Oldfield always addressed the Hookers in the respectful format of his day. His letters to them reveal that he greatly respected their advice and botanical expertise and was willing to share with them any of his plant field observations. He had a great thirst for botanical knowledge and a strong desire to ensure that his collections were correctly identified and useful to 'scientific men', although at times he lacked confidence in that regard. Surviving records provide no indication that the Hookers were dissatisfied with the state of the material Oldfield sent. That he did his best to classify his specimens and at no stage attempted to publish plant descriptions would have kept Joseph content.[126]

Although Oldfield considered himself to be a gentleman, no evidence has emerged that, as was the case with Gunn and others, he saw botany as a route to social advancement.[127] Initially, at least, satisfying his desire to be a collector for scientific purposes and discoverer of species new to science, a traveller and possibly a natural history philosopher seem to have dominated his ambition. He may have hoped to achieve a full-time career as a botanist but, as far as can be ascertained, this was never explicitly stated in any documents we located.

---

[122] Jim Endersby, 'Joseph Hooker's place in history', viewed 23 January 2014 at www.kew.org.
[123] Jim Endersby, 2008, pp. 93–94.
[124] See Chapter 7.
[125] Jim Endersby, 2008, p. 17.
[126] ibid., p. 184.
[127] ibid., p. 98.

Later, as Endersby has also concluded[128], forced by poverty and illness Oldfield put his altruism to one side, if somewhat reluctantly, and accepted offers of money or sought it in return for specimens to assist in defraying his expedition expenses. At various times from 1857 to 1862 Mueller, mostly on his own initiative, made available to Oldfield certain monies from his own pocket and from the public purse for this and other purposes. In 1861 the collector sought agreements from Joseph Hooker, Professor Frederick McCoy and Mueller to purchase specimens from his final Western Australian expedition.[129] Many of the specimens from this trip are in the National Herbarium of Victoria[130] so it seems that Mueller kept his promise but if the understanding with McCoy was met it could not be confirmed. Apparently, Hooker came to no such agreement.

Oldfield's eagerness to serve botanical science was evident when, in April 1863, he made available all his Exogens to Bentham to 'assist in his great work, and am sorry I can do not more towards it'. However, as discussed above, it appears that these specimens and the Endogens he subsequently sent to Kew Gardens, and presumably those referred to in Bentham's acknowledgement of Oldfield's contribution to the *Flora Australiensis*, were the subject of an agreement with Hooker. This was that Hooker would select *desiderata* for Kew Gardens for which the collector would be paid with the remainder to be sold through Kew Gardens to other herbaria.

It is our contention that Oldfield was not a passive servant of Imperial Science. He was not backward in expressing his disappointment and loss of enthusiasm for providing further material when his consignments and his letters from Australia to Kew Gardens were not acknowledged and his frustration when promises from Joseph Hooker were not kept. His field activities up to the time Ferdinand Mueller started financially supporting his expeditions were undertaken strictly on his own terms. Oldfield collected in his own time what he thought was new and interesting or extended a species' geographical range. At times, he tried to meet any requests the Hookers might have for particular plant material, but it could not be determined that they ever responded. However, there were occasional requests from Sir William for plant information, which Oldfield provided. His unpublished manuscript on natural philosophy and his apparently scathing response to one of Charles Darwin's queries indicate that he was not afraid to take issue with established scientific authorities. Furthermore, in 1866 he boldly indicated his preference for Hooker to arrange for correcting the proofs of his Aboriginal paper and asked the Director to provide the identifications of a list of plant specimens because the odour of herbaria made him sick and prevented him from doing them himself.

After Oldfield returned to England in 1862, Joseph Hooker castigated him for sending specimens to Kew Gardens without authorisation,[131] for writing directly to Lubbock[132], and criticised him for his 'apparent apathy' towards his collections. It was suggested (see above) that Lubbock and Hooker, even though the latter liked Oldfield and praised his collecting efforts and botanical knowledge, looked upon him as an inferior. Oldfield, however, was not bowed and while he acknowledged that some degree of censure was due to him over sending the specimens without authorisation, he defended his position in each case without adopting the role of a compliant underling. Hooker apparently forgot his 1863 'compact' and, following a spirited correspondence between the two parties, Hooker purchased some *desiderata* for a small sum but took no steps to try and sell the remainder of the collection to other herbaria. There is nothing to suggest that Hooker felt any qualms as a result, indicating perhaps that promises to the likes of Oldfield could be broken easily. Although Oldfield's dealings with Hooker in 1866, over his efforts to sell his residual collection, were unsuccessful, they demonstrate that the collector was anything but a passive player. He was persistent and frank in his letters to the Director but nevertheless was not in a strong bargaining position. Thus, as Mueller was sending

[128] ibid., pp. 99–100.
[129] See Chapters 13, 16 and 17.
[130] See HAODb.
[131] AO to JDH, 8 December 1862, English Letters, vol. 97, 1859–1900, NAT–OXL, fol. 274 (281), RBGKA.
[132] AO to JDH, 4 April 1865, English Letters, vol. 97, 1859–1900, NAT–OXL, fol. 279 (286), RBGKA.

Bentham many of Oldfield's specimens from his Melbourne herbarium, Hooker was probably not overly concerned if the collector withdrew his permission for Bentham to access his collection.

There is no evidence to suggest that the Hookers ever actively courted Oldfield to collect for them. As time went by it became apparent that his plant observations were valuable to the taxonomists, thereby increasing his value to them. The historical record reveals that Oldfield worked hard to get what he wanted from the Hookers—their discoveries in his collections of species new to science, their classifications of many of his specimens and access to their botanical knowledge. In that regard, his donations bore the fruit he so desired. However, was this all he wanted? We can only speculate.

Despite what Edwin said to William Harvey about Augustus' desire to be a collector and traveller, his writings would suggest that his ambition went much deeper than simply gaining recognition as a botanical collector. Plant collecting provided him with opportunities to make a diverse range of observations of the natural environments through which he moved—their respective geological formations, floras and faunas. Always questioning, he attempted to find explanations for what he observed. His unpublished manuscript on natural philosophy and paper on Australian Aborigines bears testimony to that. Scientific theorising might, therefore, have been the underlying reason for him becoming a collector. Furthermore, by gaining the recognition of the eminent taxonomists at Kew Gardens, perhaps he perceived collecting to be a way of providing them with tangible proof of his skills and botanical knowledge with the object of gaining their support for furthering his ambition to become a natural philosopher. His successful efforts to involve Joseph Hooker in getting his Australian Aboriginal paper published, and his apparently veiled attempt to obtain Hooker's advice on his unpublished manuscript on natural philosophy, would provide some support for such a suggestion. While Hooker facilitated the publication of the Australian Aborigines paper, the subject of which was outside his sphere of interest, if he had read the natural philosophy manuscript it is likely he would have distanced himself from it. Thus, as Endersby has pointed out, he would not have agreed with Oldfield's theoretical position on some issues and as a result would not have played any part in promoting its publication.

Available records suggest that Oldfield and Joseph Hooker's relationship was a complex one, which came about because Oldfield actively courted Sir William's interest in his collections and his botanical observations. It was not until after Oldfield returned to England in 1862 that he entered any compacts with an eminent botanist, namely Joseph Hooker, with, in the end, mixed success. Clearly Joseph held Oldfield's collecting skills and plant observations in high regard. However, it seems, to have been more of a relationship where Hooker responded to Oldfield's requests, whether openly expressed or hinted at in veiled terms, with few, if any, coming from Hooker. Nevertheless, it is not possible to assert that their association was one in which the Director of the Royal Botanic Gardens, Kew, exploited his collector. Hooker gave recognition where such was due, and assistance where it suited his interests and within the limits of his resources. On the other hand, Oldfield appears not to have been above trying to exploit his relationship with Hooker. If what we have suggested is true, he set out quite deliberately to achieve Hooker's support in having him recognised as a collector and possibly as a natural philosopher. Perhaps Hooker was mindful of the collector's intentions. In conclusion, we cannot but agree with Endersby that Oldfield and Hooker's relationship was a complex one, but it was more one in which Oldfield worked, with some success, to gain outcomes over and above those which Hooker's other collectors usually expected or received.

Oldfield's relationship with Ferdinand Mueller

Although Mueller sent Oldfield at least one *Fragmenta* fascicle after he returned to England in 1862, as far as can be ascertained there was no correspondence between them after he left Australia. This perhaps is not surprising since it has become obvious that their relationship focussed on discussing botanical matters; a situation which Oldfield would have cherished. Oldfield's reluctance to discuss his personal affairs coupled with Mueller's huge self-imposed work load seems to have kept them from developing a more personal friendship. As far as his limited financial resources permitted, Mueller treated Oldfield generously. He clearly respected Oldfield's collecting skills and botanical observations. This is reflected through Mueller's wide use of his specimens and habitat descriptions across the first nine his volumes of the *Fragmenta* and his references to Oldfield's observations in various articles. However, his generosity was not borne out of charity but out of the realisation that Oldfield could provide him with species new to science, which he could describe and name and, unlike Joseph Hooker, was willing to pay for the service. Perhaps, like other eminent 'scientific men', Mueller deliberately maintained social distance between himself and his collectors because of social class imperatives, but this did not become apparent. Mueller's awareness, during 1857–58, of Oldfield's diminishing eyesight meant that his usefulness to the Government Botanist was also declining. Therefore, it can be inferred that, even had Oldfield wanted it, Mueller could not have appointed him to, or recommended him for a position as a curator or herbarium assistant when on the infrequent occasion such an opportunity arose.

Fig. 22.0 Eremophila oldfieldii

Chapter 22

London 1869–1887

Declining years

In unenviable financial circumstances, Augustus, at the age of 48, was locked in a world of darkness. His ambition, returning to Australia to carry on his botanical explorations, was shattered and he was quite unable to undertake botanical studies fruitfully in England. His desperate situation can be summed up in the words of E.C. Johnson who wrote in 1860:

> Blind men of all ages and from all walks of life are shut out from the ordinary pursuits and excitements of daily life. Their industry is crippled by competition with the seeing; their literature is limited from the very nature of tangible typography and by the paucity of books, while their ordinary tone of thought is fettered by the unconquerable feeling of dependence which besets them at every turn and drives them back to the sad consciousness of their affliction whenever they attempt to go it alone.[1]

Oldfield was widely read in zoology, botany and Australian anthropology. His loss of sight prevented him from following developments in those fields of science without recourse to outside assistance. This must have been a huge additional source of frustration and sadness, even if there were people willing to read scientific works to him. Although, since the late 18th century, writers such as Thomas Blacklock had focused attention on the need to educate the blind in Britain, their plight was still being exacerbated in the 1860s by the low estimation of their intellectual capacity held by many in the wider community. Government neglect of their educational needs prevailed throughout the 19th century.[2] Nevertheless, there had been tactile reading systems in Britain for vision-impaired and blind readers since the late 18th century.[3] By the mid-nineteenth century the most widely used system in England was that of Dr William Moon; for adults, it was simpler and easier to master than previous systems. Moon's first books were published in 1847 but their production was time-consuming. In 1856, a small workshop was built near Brighton where a specially devised typesetting process facilitated the production of books using his system.[4] Louis Braille's system was invented in 1815 and, following the publication of his book, *Method of Writing Words, Music, and Plain Songs by Means of Dots, for Use by the Blind and Arranged for Them*, the system became widely used in Europe and the US by the mid-century.[5] However, it was not until 1868 that it was introduced into England by the British and Foreign Society for Improving the Embossed Literature of the Blind, later the British and Foreign Blind Association. This Association, which eventually became the Royal National Institute for

[1] E.C. Johnson 1860,, 'The Blind of London', Letter to the Editor, TTDA, quoted in John Oliphant, 2008, '"Touching the light": the invention of literacy for the blind', *Pedagogica Historica*, vol. 44, nos. 1–2, p. 77.

[2] John Oliphant, 2008, p. 73.

[3] Holly L. Cooper, 'A Brief History of Tactile Writing Systems for Readers with Blindness and Visual Impairments', *See/Hear*, Spring 2006 vol. 11, no. 2, viewed 12 March 2011 at www.tsbvi.edu.

[4] Omniglot the Language Encyclopedia, 'Moon alphabet', viewed 17 November 2014 at www.omniglot.com.

[5] Wendie Pecharsky, 'The history of teaching the blind', *EHow*, viewed 14 March 2011 at www.ehow.com.

the Blind, was founded by a group of blind men, on the initiative of Thomas Rhodes Armitage, to create and find employment suitable for people who had this disability.[6]

While books using both systems were available in England in the 1870s, and still are today, a question that cannot be answered is whether Oldfield accessed any of them. Training in the use of each system and the publications themselves may have been too expensive or for other reasons unattainable.[7] Furthermore, almost all the books on offer in Britain were religious texts; recreational and scientific reading was not a part of publishing for the blind. It is unlikely, therefore, that he would have made the effort to learn unless, of course, he was able to obtain publications to his taste from France, where a range of secular subjects was available.[8] If there were any scientific works in the 1870s, published in either Braille or Moon systems, it seems they were very few and far between and possibly too expensive for Augustus to purchase. It is, therefore, not surprising that he was depressed. This combined with physical inactivity, possibly a poor diet, climatic conditions that apparently had a deleterious effect on his health, and the knowledge of the misfortunes of his siblings all may have contributed to his recurrent bouts of illness.

Without a wife and children Augustus appears to have been totally reliant on his siblings for support. However, he was probably reluctant to seek their assistance. Nothing has emerged in the historical record, which would offer any indication that he ever had ambitions to marry or that he was a hopeful or experienced lover. The frankness with which he discussed sexual behaviour of the *Watchandies* indicates that he did not suffer from Victorian inhibitions in that regard. He may have deliberately avoided marriage because he wanted to be free of family obligations to satisfy his passion for botany. His collecting and employment activities in Australia would give some weight to such a view. He apparently made no attempt to create for himself a secure financial future, which would have enabled him to provide a standard of living for a family befitting that of the middle-class gentleman he claimed to be. In the absence of other evidence suggesting that he might have remained a bachelor for any other reason—not having found a woman with whom he wished to share his life or who was happy to become his wife or he was a homosexual—at this stage, it seems that it was his passion for botany that kept him from the altar.

Apart from Oldfield's blindness contributing importantly to his impecunious situation, there were other factors that prevented him from making adequate financial provision for times when he could not earn a living. Clearly not the least was his failure to pursue a more lucrative and reliable career. He had the educational background to be indentured to one of the master surveyors in the colonies to qualify for his survey license, thereby fitting him to establish his own business or to be employed as an assistant surveyor by one of the colonial governments or private practitioner. There was a demand for professionally qualified surveyors in Australia at that time, brought about by opening lands for settlement and agricultural, pastoral and mining industries. Such a profession would have created opportunities for some botanical exploration and improved social status. Of course, if his loss of sight was due to one or more factors arising from the lengthy time he spent in the harsh conditions of the Australian bush, following that profession may well have had the same outcome.

From at least 1855, Oldfield had yearned to be a full-time traveler and collector but long before that he may have entertained such an ambition. He said that his retreat from the meeting with Sir William in 1850 meant that he 'left England without having entered into arrangements which might have benefited me considerably'.[9] He left no details as to exactly what he hoped to obtain from Sir William. However, the above statement suggests that he was looking to receive payment for specimens he sent to Kew Gardens, although this is at odds with all his subsequent statements from Austral-

[6] *Wikipedia*, 'Royal National Institute of Blind People', viewed 17 November 2014 at www.en.wikipedia.org.
[7] The elite Worcester College opened in 1866 to cater for blind boys, *Wikipedia*, 'New College Worcester history', viewed 22 December 2014 at www.en.wikipedia.org
[8] John Oliphant, 2008, pp. 67–82.
[9] AO to JDH, 16 March 1866, English Letters, vol. 97, 1859–1900, NAT–OXL, fol. 289 (296), RBGK.

ia, which indicated that his collections were freely given. In any event, it is doubtful he would have been successful in any request for remuneration to Sir William. As noted earlier, Kew Gardens had ceased employing collectors by 1850 and the Hookers relied heavily on specimens donated by interested persons in the colonies. Oldfield may have sought other purchasers in the 1850s but this could not be affirmed.

Oldfield's loss of courage for his 1850 appointment with Sir William, is, perhaps, symptomatic of a personality trait that would have prevented him from conducting a profitable commercial collecting business. Even if he was a hard-headed entrepreneur, the monetary returns from that type of enterprise in Australia were precarious. Distance from the market place, difficulties associated with identifying buyers in England, Europe and the USA, loss and damage to specimens during transport, the apparent lack of a market price based on actual expenses, the time involved and the skill of the collector all combined to make for irregular and uncertain financial rewards.

When employed in Australia, particularly in Tasmania, he collected in his spare time. It seems, however, that when he was between jobs, the pull of botanically unexplored areas on the mainland was so strong he was unable to resist when he could meet the related travel and living expenses. The inevitable result was that his pattern of income-earning employment was erratic.

His stated aim was to be of service to science, opting, particularly prior to 1857, to achieve this by making his privately financed collections available *gratis* to the most eminent botanists in England and Australia. It may not have been until he received financial support from Mueller for the expedition to Mount La Perouse in 1857 and following this, in 1858, for a year as a full-time collector in Western Australia, that he felt able to entertain the possibility of earning a living as a collector or in some other situation as a botanist. By then, however, it would have been too late for him to harbor such a dream.

Support network—family

Although Oldfield's income was inadequate for his needs, he was not indigent; he apparently had enough money to provide himself with food and shelter. Therefore, he did not qualify for financial assistance from the Wandsworth Board of Guardians. Among other welfare services, the Wandsworth Board was responsible for providing financial relief and management of Saint John's Hill Workhouse (also known as the Wandsworth and Clapham Union Workhouse) which was constructed in 1838.[10] Boards of Guardians also had responsibility for self-help groups of blind people, but it is not known whether, in the 1870s, there was such a body in Wandsworth, and if so, whether it catered to the needs of educated middle-class gentlemen who, like Oldfield, were not homeless.[11]

Several personal letters addressed to members of the family by Frederica, Marian and Juliana after 1862 and during the 1870s have survived. The general tone and warmth of this correspondence reveals that the three sisters enjoyed a close relationship; they cared deeply for their mother and were intensely religious.[12] The sisters' poor health, which prompted Frederica to comment to Vincent: 'We are a very shaky set I fear', is a common theme running through the letters. At times referred to as their 'delicacies'—Marian suffered with poor eyesight and was prone to bouts of being 'very poorly', Frederica from a dermatological condition of her hands and recurrent bouts of colds and bronchitis, and Juliana from stress and an unexplained condition.[13] On occasions one or two of the siblings withheld news from others if it was likely to be too upsetting.[14] Some of Frederica's letters contained messages for the addressee from their mother, who while at Doncaster was often ill and bedridden for

[10] *AIM25: Archives in the London and the M25 area,* 'Wandsworth Board of Guardians', viewed 15 March 2011 at www.aim25.ac.uk.
[11] Patrick Phelan, 1984, 'Are we producing the goods?', *The British Journal of Visual Impairment*, vol. 11, pt. 3, p. 70.
[12] See for example, Marian Oldfield to Mary Oldfield, 24 January 1869, Files 0102, 0102, 0103, 0104, PFA.
[13] Frederica Hold to Vincent Oldfield, 4 March 1865, file 0158–0159; c.1867–68, files 0348–0349; 7 July 1872, file 0215–0216; 14 January 1879, files 0116–0122; Marian Oldfield to Mary Oldfield, 24 January 1869, PFA.
[14] See for example, Frederica Hold to Vincent Oldfield, 7 July 1872.

lengthy periods. This correspondence depicted Mary, 'darling mamma', to be an affectionate and caring mother and grandmother but nevertheless one who was not above proffering advice and admonishment when she thought it was needed.[15] There is no inkling in the correspondence of their feelings towards their father. Augustus was mentioned in a few of their letters and his situation was a matter of great concern to them.[16]

On 20 April 1866 Ella Jane Howard Oldfield, Jane and Vincent's second child, was born.[17] Frederica seems to have had a special affection for Vincent although at times she chided him for his failure to keep in touch[18] with her and their mother and to provide adequately for Jane and the children. The sisters' feelings towards Jane are difficult to determine. Depending upon whether their mother or Marian was with her at the time of writing to Vincent, Frederica usually included messages from them for Jane. Until about 1868 she and Juliana wrote of Jane affectionately, and out of concern for her health and wellbeing, and in letters to Vincent, Frederica continued to include her in their loving messages. Nevertheless, at that time they found it difficult to accept that Jane no longer contacted them.[19]

Edwin was kept informed about Augustus' situation (see below), but there is an absence of correspondence in the Pimblett Family Archive between Alfred, Edwin and Frederick and their siblings in England. Any letters from Australia were probably passed around the family and ultimately lost.

Vincent was a worry to his sisters and possibly also to Augustus. By 1868, Vincent no longer had his business in Chancery Lane and was working as a commercial traveller, selling on commission for W. Woodfield & Sons, manufacturers of needles, sewing machine needles and fish hooks in Redditch, north-east Worcestershire.[20] While thus engaged he had negotiated, with a Mr and Mrs Dodd[21], to take over their Cafe de L'Europe at 45-46 Union Street, Plymouth.[22] However, these dealings had not worked to Vincent's advantage and he wrote to Jane (c. August 1868) informing her that he had been 'compelled to leave Dodds' because 'they had behaved diabolically' to him; he did not elaborate in that communication.[23] Understandably Jane, who was seven months pregnant[24] and living at 67 Great Queen Street, Lincolns-in-Fields[25], was upset.[26] A few days later Vincent wrote to her explaining that there had been a dispute with the Dodds concerning their withholding liquor licensing renewal notices because, they claimed, the £600 bill with which Vincent had paid for the business was not honoured.[27] He never denied this to Jane. Probably, therefore, the Dodd's claim was true.[28]

Sometime close to her confinement, Vincent wrote to Jane from Plymouth. He enquired after her pregnancy and added, 'Let me know if it is a boy or girl or both. I will endeavour to keep down my expenses here until you are over your trouble so as to send you all I can'.[29] At that time Jane and the children were residing in Ossulston Street, King's Cross, and by 1871 she was supplementing what little Vincent was sending her by earning income as 'a machinist'.[30]

---

15 See for example, Frederica Hold to Vincent Oldfield, 4 March 1865.
16 See for example, ibid.
17 Christening record for Ella Jane Howard Oldfield, England Births and Christenings, 1538–1975, Index, 20 June 1869, Saint Thomas Regent Street, Westminster, London, England, Birth date: 20 April 1866, *Family Search*, viewed 5 February 2014, www.familysearch.org.
18 See for example, Frederica Hold to Vincent Oldfield, undated ?1871, Files 0348–0349, PFA.
19 Frederica Hold to Vincent Oldfield, [day & month indecipherable] 1875, Files 0225-0226, PFA.
20 W. Woodfield to Vincent Oldfield, 25 April 1868, files 0131–0132, Frederica Hold to Vincent Oldfield, 4 March 1868, PFA.
21 ibid.
22 Vincent Oldfield to Jane Oldfield, undated, File 0338, PFA.
23 Vincent Oldfield to Jane Oldfield, undated (c. August 1868), Files 0344–0345, PFA.
24 William Goodered to Vincent Oldfield, 31 August 1868, Files 0160–0161, PFA.
25 Envelope addressed to Mrs Oldfield, postmarked Plymouth, 17 August 1868, File 0214, PFA.
26 Vincent Oldfield to Jane Oldfield, undated (c. August 1868), Files 0105–0108, PFA.
27 ibid.
28 ibid.
29 Vincent Oldfield to Jane Oldfield, undated (c. October 1868), Files 0432–0343, PFA.
30 Record for Sarah J. Oldfield, 1871 England Census, Registration No: Rg 10/ 227; Folio: 9; Page: 11; Registration District: Pancras: Sub–registration district: Somers Town: 18: County: London, Middlesex, viewed 2 March 2011, www.ancestry.com.

Jane gave birth to a daughter, Louisa Ada Howard, on 8 October 1868 at St Giles, Bloomsbury, presumably in the absence of Vincent.[31] There is evidence throughout the family records that Vincent was very erratic in sending money to Jane and when he did the amounts he sent were totally inadequate. On one occasion Mr F. Pioggio communicated to Vincent that Jane had stated the children were starving and that she was going to make application to the Lord Mayor about Vincent not supporting his children.[32] The couple were not co-habiting during the 1870s-Jane wanted it that way but Vincent could not accept her rejection.[33]

Marian was still residing with Augustus at 5 Verandah Cottages on 24 January 1869 when she wrote to her mother offering her 'special good wishes for many happy or happier returns...of this (now you may be spelling it on the veritable natal day) 25th'. She also expressed concern over Mary's recent illness and her hope that 'now she seem mending' she soon would 'be "Mrs O" again and equal to all your little undertakings'. There is no mention of Augustus in this letter but it was reported that Juliana had cancelled her holiday to Stroud to render assistance during Marian's recent illness. This was of concern to Marian:

> Poor darling, she is so kind and thoughtful, but unhappily her kindnesses distress me—I would much prefer going my usual way and I am convinced it would be best for me in the long run.[34]

Meanwhile, back in Tasmania, Alfred was employed as a master at a private school in Pontville in 1869; how long he was there is not known.[35] Edwin was still running his Commercial Academy in Argyle Street, Hobart.

Fig. 22.1 Edwin and Susanna Oldfield, c. 1880

Edwin raised the matter of Augustus's plight with Mueller when he visited Hobart in January 1869. He expressed the view to Mueller that he thought Augustus's loss of sight 'may be traced to his devotion to the science of Botany, and that because no small portion of his time was spent in investigating the Flora of [Tasmania]' he had 'a legitimate claim upon the consideration of [the Tasmanian] Government'. However, Edwin told Mueller that before he attempted to put Augustus' case to the government, he thought it necessary to 'be armed with proof of his having served' the Colony. On Mueller's advice, Edwin asked Joseph Hooker 'to have the kindness' to send him a few lines 'confirmatory of the assistance which [his brother's] labours have afforded', and also appended Augustus' address in case anything could be done for him in England. Mueller had promised to supplement Hooker's testimonial as far as Augustus' labours in New Holland were concerned.[36]

Edwin informed Hooker that Augustus was quite blind and totally dependent upon the very small annuity he had received under his father's will, and which income was quite inadequate for his wants

> and one which puts it out of his power to render available in the cause of Science the results of a life study associated with a long and close practical observation.[37]

At the same time, Edwin mentioned that his:[38]

---

[31] Birth Record for Louisa Ada Howard Oldfield, 8 October 1868, England Births and Christenings 1538–1975, Page: 529; Vol. 1; Registration quarter: 4, *Family Search*, viewed 5 February 2014 at www.familysearch.org.
[32] See for example, F. Pioggio to Vincent Oldfield, undated but prior to 1874, File 0353 and Henry Martin to J. Oldfield, File 0227, PFA.
[33] Vincent Oldfield to Jane Oldfield, c.1867, File 0341 and undated, File 0224, PFA.
[34] Marian Oldfield to Mary Oldfield, 24 January 1869.
[35] J. Gill to MHH, personal communication, email 21 August 2003, HOA.
[36] E.D. Oldfield to JDH, 30 January 1869, South Australia, Tasmania, Western Australia letters, 1865–1900, fol. 245(246).RBGKA.
[37] ibid.
[38] ibid.

> brother is remarkably sensitive both upon the subject of his affliction and of his pecuniary position and that it is without his cognizance I am moving in the matter. Indeed, I feel sure that, such is his repugnance against what he would term "begging" my intended application on his behalf would not have his sanction. Still, my feelings as his Brother prompt me to do what I feel might be done, in such a manner as in no way to compromise his dignity, but on the contrary to stamp a value upon his gratuitous labors in the botanising of a large portion of New Holland as well as of this Island.[39]

Edwin submitted the application for a pension for Augustus to the Governor-in-Council, Tasmania on 14 June 1871 and 7 August 1871. Although Hooker provided a testimonial it was not until Sir James Milne Wilson, then Premier and Colonial Secretary, wrote on 7 December 1871 that Edwin learnt that his request had been denied:

> I have the honor to acknowledge your communication addressed to His Excellency The Governor, bringing under notice the case of your brother Mr Augustus Oldfield, and seeking from the Government some substantial recognition of his services in the cause of science. I desire to express my great regret that any necessity for such an appeal should exist, and my sincere sympathy in your brother's severe affliction, but while I fully appreciate the testimony of Dr Hooker and Baron Von Mueller to your brother's labour in exploring the Botany of Tasmania I am nevertheless unable, in the present state of the Revenue, to recommend the appropriation of Public Funds to meet his claims, which, in more prosperous times, I should have felt much pleasure in supporting. I return Mr Hooker's and Baron Von Mueller's letters which accompanied your communication.[40]

The application was never considered by the Governor-in-Council[41] and according to Juliana, 'various causes' prevented Edwin from obtaining a hearing but she did not elaborate.[42] Subsequent correspondence explains the reasons for the refusal.

Mary Oldfield died on 16 November 1870 at 33 Cartwright Street, Doncaster, with the certified cause of death 'senile decay'.[43] Her age at the time of her death was given as 79 years, but she must have been at least a year older since she was baptised on 25 January 1790.[44] She was buried in Brompton Cemetery, London, on 23 November 1870, in the same grave as Frederick.[45] Her body may have been transported by train to London from Doncaster, a distance of some 320 kilometres, on the Great Northern Railway line[46] or, less likely, by horse-drawn coach.

By the night of the *1871 England Census* (2 April 1871) Marian and Frederica were together at a dwelling in Potter Street, Harlow, a small market town in the County of Essex about 27 kilometres north-west of London. Marian was recorded as the head of the household but Frederica does not appear to have been a permanent resident since she continued to live in Doncaster through the 1870s.[47] At the time of the *1881 Census* she was at 38 Young Street, Doncaster, where she was recorded as being the household head and an 'annuated [sic] retiree'.[48] Why Marian left Verandah Cottages is not known. She and Augustus may have had a falling out or her health was such that she no longer had the

[39] ibid.
[40] J.M. Wilson to E.D. Oldfield, 7 December 1871, CSD7/1/49, no. 983, TAHO.
[41] E.D. Oldfield to Thomas Reibey, 8 January 1877, CSD10/1/27, File no. 425, TAHO.
[42] Juliana Oldfield to JDH, 2 May 1875, English Letters, vol. 97, NAT–OXL, 1859–1900, fol. 297 (304), RBGKA.
[43] Certified Copy of an Entry of Death, 1870, no. 65, Registration District Doncaster, Sub-district of Doncaster in the County of York, DXZ828296, 9 March 2001, General Register Office, England, HOA.
[44] See Chapter 1.
[45] Register of Burials in the West of London and Westminster Cemetery, Earl's Court, Old Brompton, no 62016, 23 November 1870, copy provided by Laurence Epstein, Cemetery Supervisor, Brompton Cemetery to MHH, 23 October 2002, HOA.
[46] 'History of rail transport in Great Britain 1830–1922', viewed 9 December 2013 at www.wikipedia.org.
[47] Frederica Hold to Jonathan Oldfield, 14 January 1879, Files: 0116–0122, PFA.
[48] Record for Frederica J. Hold, England Census 1881, Registration no.: Rg11; Piece: 4691; Folio: 11; Page: 15; Registration District: Doncaster; Sub District: Doncaster, viewed 28 February 2011, www.familylink.com.

strength to look after him. On the other hand, being in her late fifties, perhaps she decided she had cared for others long enough, and wanted to spend her remaining years free of such burdens.

Meanwhile Juliana was in partnership with Susanna M. Trundle in a school for girls at the White Lodge, Swaffham, in Norfolk. Living at the school on the night of the *1871 England Census* (2nd April 1871) were seven pupils, an assistant mistress and two servants.[49]

Augustus was now the sole occupant of 5 Verandah Cottages. If he had the services of a domestic servant, it must have been from someone who resided elsewhere.[50] Given his disability and poor health, it is difficult to imagine that Marian would have left without ensuring that he had someone to help with the cooking, cleaning and laundry and possibly even shopping. At the time, he was augmenting his income as 'a teacher of languages'[51] so perhaps he could afford to have a woman come in for an hour or two every day. Whether he tutored students at home or taught classes in a school is not known. For a sighted person teaching can be challenging enough, but very difficult for someone who is blind. Perhaps he confined himself to conversation sessions with one or two pupils. The historical record showed his only foreign language was French.[52] Perhaps he secured this work through the British and Foreign Blind Association.

Juliana left Swaffham[53] before the end of May 1875 and established her own girls' seminary, 'Elleray', in Atkins Road, Clapham Park, nearly three kilometres east of Verandah Cottages.[54] By 1881, the school was catering for pupils from the colonies as well as Great Britain. Juliana employed two governesses (one to teach French and the other English), a nurse and three domestic servants.[55]

Augustus and the rest of the family in England experienced a particularly difficult year in 1872.[56] Prior to July of that year, Juliana nursed Augustus through a 'serious' illness, the nature of which is not known. Juliana's 'health and mental powers' had suffered due to the strain she was under while 'looking after poor Augustus' and her emotional reaction to the news of Frederick Jnr's death.[57] He died of hepatitis in San Francisco, on 2 March 1872[58], but the circumstances of his family remain a mystery. Vincent and no doubt Jane, Ella and Jonathan were grief-stricken when, on 1 July 1872, three-year-old Louisa Ada died[59] at 27 Fitzroy Street, Tottenham.[60] At this time, Vincent needed cash, perhaps to pay for Louisa's funeral, and asked for money from his sisters but only managed to glean 'a few shillings' from Frederica.[61]

At the same the time that Frederica learnt of Louisa's death, William Goddered, at Vincent's request, informed her that Vincent had been accused of a very serious offence, the details of which were not divulged in extant letters. Frederica wrote to Vincent expressing her grief and condolences and in the same letter stated that she was highly shocked that

> such foul accusations sh^{d} be brought against you by one who ought to know well whether you are capable of such wickedness –w^{h} for a moment I do not suppose – surely you would not so sin against God & your own soul – besides exposing yourself

---

[49] Record for Juliana Oldfield,1871 England Census, Source information: Rg10/1869; Registration district: Swaffham; Sub-registration district: Swaffham; Folio: 1`67; Page:7, viewed 31 December 2004, www.ancestry.com.

[50] Record for Augustus Oldfield, 1871 England Census Record, Ecclesiastical Parish: All Saints; Town: Wandsworth; Registration District: Wandsworth, Sub-registration district: Wandsworth; Household schedule: 245; Piece:709; Folio 27, p. 48, viewed 12 February 2011, www.ancestry.com.

[51] ibid.

[52] see Chapter 5.

[53] Juliana contemplated leaving Swaffham in March 1868, Frederica Oldfield to Vincent Oldfield, 4 March 1868.

[54] Juliana Oldfield to JDH, 31 May 1875, fol. 298, 305, English Letters, vol. 97, 1859–1900, RBGKA.

[55] Record for Juliana Oldfield, 1881 England and Wales Census, Record, Registration District: Wandsworth; Sub-district: Clapham; Piece: 637; Folio: 14; Page: 21, viewed 12 February 2011 at www.ancestry.com.

[56] Frederica Hold to Vincent Oldfield, 7 July 1872, Files 0215–0216, PFA.

[57] ibid.

[58] Record for Frederick Henry Howard Oldfield, National Probate Calendar from 1861–1941, viewed 1 March 2011 at www.ancestry.com, information provided in personal communication, Jeff Pimblett, email to MHH, 3 March 2011.

[59] Frederica Hold to Vincent Oldfield, 7 July 1872.

[60] Death record for Louisa Ada Oldfield, 1 July 1872, England and Wales, Death Index 1837–1920, Page 19; Vol. 1, Registration quarter: 3; Affiliate Line No: 352, viewed 5 February 2014 at www.familysearch.org.

[61] ibid.

> to the arm of the law – & your family to disgrace – No – I cannot believe it – & the author of such a calumny must be insane.[62]

No evidence of criminal court proceedings against Vincent has emerged[63] but the accusations may have been the reason for the investigation conducted by Coroner Edwin Lankester, on the day after Louisa Ada's death, the cause of which was found to be scarlet fever.[64]

During the year Marian's fortunes 'suffered a severe blow' with the 'smash of the European Office Co.', leaving her with 'a pittance for her maintenance that she had not the bare necessaries of life in a general way'.[65] This was probably because of the banking crisis in Europe, which took hold in 1872 after a mortgage lending boom and which eventually developed into a severe international depression in 1873.[66]

There is little doubt that Augustus was made aware of the matters that concerned his three sisters but how he reacted has not become apparent. He did, however, enjoy the support of his sisters, particularly Juliana, as far as their respective health and limited financial resources would allow.

Sadly, Vincent and Jane's eight-year-old daughter, Ella Jane Howard, succumbed to whooping cough and pneumonia on 21 April 1874 at their home at 107 Euston Road, Gray's Inn.[67] At the time, Vincent was residing with Jane.[68]

Edwin tried once again to obtain a pension for Augustus from the Tasmanian Government. For this application, Juliana requested another letter of support from Joseph Hooker; presumably Edwin had lost the one that had been returned by the Colonial Secretary in 1871. Juliana wrote to Hooker on 2 May 1875:

> I believe you were once good enough to furnish one of my brothers (then living at Hobart Town) with a testimonial on behalf of my brother Augustus Oldfield which was intended to accompany and to enforce an application to the Tasmanian Government for the grant of a small pension to him. Various causes have prevented him from obtaining a hearing hitherto, but he has some hope of succeeding in doing so if he now tries again. Meanwhile he writes to ask me to strengthen his hands [sic] by obtaining from you a re-iteration of your former testimony, as it is so long a period since his first application. I write therefore to beg you to do us this kindness. The necessity for making some attempt to increase my poor brother's income, is very great. It is miserably insufficient, barely enough to provide him with food and shelter and his blindness and general delicacy entirely incapacitate him from doing anything to supplement it himself. You are aware, I believe, that he has now entirely lost his sight.
>
> If, as I trust, you kindly accede to my request, may I beg you to send your testimonial to me, at the above address, at your earliest convenience, as I am instructed to send it to Tasmania by this month's mail.[69]

Hooker replied on the same day requesting 'particulars' about Augustus. Juliana had stressed the need to send his reply to Edwin by the May mail, so it seems strange that she did not respond to Hooker's request until 31 May 1875. She indicated that it would be impossible for her to supply the information for which he had asked because of Augustus's sensitivity regarding his disability and financial situation:

---

[62] ibid.

[63] *The Proceedings of the Old Bailey Online* were searched for evidence of an indictment for a serious criminal offence, viewed 9 March 2011, www.oldbaileyonline.org.

[64] Certified Copy of an Entry of Death for Louisa Ada Oldfield, Registration District: Pancras; Subdistrict Tottenham in the County of Middlesex, No. 279, 3 July 1872, copy no. DYC968988 given at the General Register Office on 10 March 2011, HOA.

[65] Frederica Hold to Vincent Oldfield, 7 July 1872.

[66] Paul Kedrosky, 2008, 'Get your banking crises correct, 1872/73 vs 1929', Viewed 8 March 2011 at www.paul.kedrosky.com.

[67] Certified Copy of an Entry of Death for Ella Jane Howard Oldfield, Registration District: Pancras; Subdistrict Gray's Inn Lane in the County of Middlesex, No. 390, 21 April 1874, copy no. DYC969474 given at the General Register Office on 10 March 2011, HOA.

[68] M. Watson to Vincent Oldfield, 23 April 1874, file 0256–0257, PFA.

[69] Juliana Oldfield to JDH, 2 May 1875, English Letters, vol. 97, NAT to OXL, 1859–1900, fol. 297 (304).

> [I]f he had an idea of the step we are taking, he would at once put his veto upon all further exertion on his behalf, - much as he needs the assistance we are seeking for him. But the particulars that you do remember about him, would I think enable you to do all we contemplated when we applied to you – that is, to bear testimony to his standing as a botanist and to the value of his labours as connected with the Flora of Australia and Tasmania. It may perhaps be within your recollection that Sir William Hooker took a friendly interest in my brother. I have often heard him say, "If Sir William had lived, I should have been in a different position".[70]

Two days later (2 June 1875) Hooker sent the following testimonial:

> I beg to certify that I have known Mr Augustus Oldfield for nearly twenty years as a most active, able, industrious & trustworthy Naturalist, and especially Botanist, whose disinterested labors & collections have thrown great light on the Flora of many distant, and of them previously wholly unexplored districts of Australia.
>
> Mr Oldfield's name stands high as a scientific investigator of Australia. He has great claims not only on the sympathy of the inhabitants of that Continent, but also on their liberality, & I may say their charity, for it is through his disinterested exertions in the promotion of a knowledge of the natural productions of Australia, that he has contracted illnesses and infirmities that have rendered him destitute.
>
> I should add that the liberality with which he has dealt with the materials he collected is beyond all praise. He gave specimens and information of the most valuable description gratis to public institutions, wherever they were likely to be of use, without return of any kind and placed his services & knowledge at the disposal of naturalists in the most enlightened manner. I can truly say that I know no case of modest worth – of heart, hand, and head, - more deserving of public recognition by the Governments of Australia than that of Mr Augustus Oldfield.
>
> J.D. Hooker
> President of the Royal Society,
> Director of the Royal Gardens Kew,
> Author of the Flora of Tasmania and various works on the Botany of Australia71

Writing from Launceston, Edwin put forward his second appeal to the Colonial Secretary on 27 July 1875.[72] After eighteen months (8 January 1877), obviously frustrated by the lack of an official response and smarting over the awarding of a government pension to Ronald Gunn, Edwin wrote to the Colonial Secretary Thomas Reibey:

> That my application has never been under the consideration of the Governor-in Council I have good reason to believe—Sir J. M. Wilson's letter to me, dated 7 Dec 1871, notwithstanding—of which I have great cause of complaint; but, still more so at seeing Dr Hooker's words in favour of my Brother's claim on the consideration of the Colony should have been made of in supporting with success the equal, though not greater one of Mr Gunn, while that of my Brother has been ignored.[73]

In the same letter Edwin requested the return of four enclosures, two of which accompanied the 1869 application and the remaining two, the 1871 application as they were the original documents

[70] Juliana Oldfield to JDH, 31 May 1875, ibid., fol. 298(305).
[71] JDH, AO testimonial, 2 June 1875, apparently the original, HOA. The document was presented to MHH by Lenthall Williams, son of Grace Henrica Oldfield, c. 1985.
[72] Edwin Oldfield to Colonial Secretary, 27 July 1875, CSD10/1/27, File no. 425, TAHO.
[73] Edwin Oldfield to Colonial Secretary, 8 January 1875, ibid., TAHO.

which he wanted back as he was leaving the Colony. Reibey replied on 13 January 1877 addressing Edwin's complaints:

> You state that you have great cause of complaint that your application in 1871 was not submitted for the consideration of the Governor in Council, but you will allow me to point out that Sir James Wilson at that time Colonial Secretary and Premier informed you that he was unable to recommend the appropriation of Public funds to meet your brother's claims, from circumstances mentioned in his letter, which, I regret to say, have never ceased to operate against a more favourable view being taken of the case.
>
> With respect to Mr [R][74] Gunn and Dr Hooker's recommendations, which you complain was 'made use of in' supporting with success the equal 'though not greater claim' of Mr Gunn 'while that of my brother has been ignored,' I am compelled to point out that the cases are not parallel.
>
> While fully appreciating your brother's devotion to the cause of Botanic Science, you must remember that he was not an Officer of the Government, but a private individual, and though I should be glad if it were possible to afford a generous recognition of your brother's labors, I cannot admit that any comparison holds between the cases of these gentlemen.[75]

Why, then, did Edwin try again to obtain a government pension for Augustus in 1875? Perhaps he thought that, as Gunn had been granted a pension, the government had become more prosperous and had made provision for the granting of financial rewards to private citizens who had given exceptional service to the Colony. While it may well be true that Augustus and Gunn's claims were equal with regard to their respective botanical contributions, as an 'Officer of the Government' Gunn's claims for a pension were apparently unquestionable. Between the time of his arrival in Tasmania in 1830 and 1854, when he was elected to the Launceston seat of the Legislative Council and after a short time retired and won the Selby seat in the House of Assembly, Gunn held several important government administrative positions and was firmly located in the Colony's upper echelons. In 1854, he purchased the 'Newstead' estate near Launceston where he built a new home. After he resigned from parliament in 1860, he continued to occupy a series of public service offices until his retirement in 1876. His pension amounted to £250 per annum.[76]

As for Augustus's brothers in Australia, Alfred moved to Sydney sometime during the 1870s but where he lived and what he did for a living is unknown; he was probably a private tutor or teaching in a school. Following the dissolution in 1871 of the Van Diemens Land Mechanics Institute in Melville Street, Hobart, Edwin closed his Argyle Street Commercial Academy and transferred it to the Institute, opening on 17 January 1872[77], where he also offered evening tuition in a range of subjects.[78] A few months later Susanna advertised that she would be happy to 'receive a few little girls to educate with her own' at the Mechanics Institute.[79] She and Edwin remained there for two years until, on 2 April 1873, he reopened the St Andrew's Commercial Academy in Launceston. [80] The previous principal of this school was Susanna's brother, Bernam Swifte.[81] By 1880, Edwin and Susanna had transferred to

---

[74] The transcript of the letter received from the Tasmanian Archive and Heritage Office referred to 'Mr G. Gunn' but this must be an error either on the transcriber's, or Reibey's, part since it was Ronald Campbell Gunn who was one of Hooker's correspondents and who Hooker held in high regard.

[75] Thomas Reibey to E.D. Oldfield, 13 January 1877, CSD10/1/27, File no. 425, TAHO, copy of letter kindly provided by Fiona MacFarlane, Tasmanian Archive and Heritage Office, email to Helen Henderson, 14 April 2011, HOA.

[76] T.E. Burns and J.R. Skemp, 1966, 'Gunn, Ronald Campbell', ADBO, viewed 4 May 2011 at www.adb.online.anu.edu.au.

[77] Advertisement, *The Mercury*, 23 December 1871, pp. 3–4, DAN, NLA.

[78] Advertisement, *The Mercury*, 9 July 1872, p. 3, DAN, NLA.

[79] ibid.

[80] Advertisement, *Launceston Examiner*, 15 March 1873, p. 1, DAN, NLA.

[81] 'Our monthly summary', *Launceston Examiner*, 6 October 1870, p. 2, DAN, NLA.

Melbourne and Edwin was principal of St Jude's Grammar School, Carlton.[82] A Miss Oldfield (probably Grace Henrica)[83] was conducting a preparatory school for boys at 5 Granite Terrace, Gertrude Street, Fitzroy[84]; this was also the family residence.[85]

Extant family records make no mention of Augustus after 1875. It is evident that Augustus could not expect any additional income from his sisters; their financial resources were stretched to the limit in maintaining themselves in very modest lifestyles and any such support from Vincent was out of the question. Nor indeed were his brothers in Australia and America able to help him financially; even if they had been, Augustus may have refused, such was his aversion to accepting charity. Nevertheless, consideration must be given to the support he received from his brothers and sisters during those years.

He probably had very few visits from his elder sisters, who continued through the 1870s and 1880s to reside quite a distance from Wandsworth, with Frederica much further away in Doncaster.

Juliana seems to have been Augustus' main support in England. She lived less than three kilometres from his home. Although still busy with the administration of her school in 1881, she may have visited him frequently and occasionally accompanied him on outings to other parts of London. However, sometime after 1881 and before 1887 she retired, and went to live at Thornwood Lodge, South Penge Park[86], about 14 kilometres east of Verandah Cottages. This may have meant that she did not call on Augustus as often, the occurrence of such visits possibly depending a great deal on her state of health, public transport convenience and on how needy she believed her brother to be.

How long Vincent continued as a commercial salesman after 1871 is not known. By 12 September 1875 he had commenced a career as a translator.[87] Augustus and Vincent appear to have remained on good terms and while he was living in London he probably visited 5 Verandah Cottages from time to time. Vincent, however, took up employment in France and for the remainder of the decade he only returned to London occasionally, for brief visits. Vincent's skills and employment history may not have meshed with those desired by companies in England for positions paying adequate incomes, as sometime after 1876 he obtained a position at La Révolution de Français of 9 Rue d'Aboukir, Paris.[88] There were, at the time, many projects associated with the forthcoming centenary of the French Revolution commencing in 1889.[89] This establishment seemed to be a publishing house associated with one of the projects. A surviving business card of Vincent's shows he was operating out of the same street address as a 'Traducteur [Translator], Interpreter, copiste, etc. (L'Anglais, Francais, Allemande [German])'. In addition, he provided 'redaction [editing] de pieces' and correction of 'Typographique Anglais'.[90]

Vincent's letters to his son, Jonathan, from France indicate that he and his son were on friendly terms. Nevertheless, his son may have been critical when Vincent wrote:

> You must not blame me for your mother's conduct-it has been <u>out of my</u> power to send her anything—in a foreign country—and no immediate employment.[91]

Vincent insisted that their communications be kept from Jane but did not say why.[92] Jonathan's relationship with his mother was strained; Jane's behaviour was disgraceful in the eyes of Frederica but she left no details. Maybe Jane's grief at the deaths of her daughters and her bitterness and anger

[82] Advertisement, *The Argus*, 3 January 1880, p. 8, DAN, NLA.
[83] MHH was told by her father that his mother, Grace Henrica, had been a teacher before she married.
[84] Advertisement, *The Argus*, 22 January 1881, p. 1, DAN, NLA.
[85] *Melbourne Directory, 1887*, Sands and McDougall, State Library of Victoria.
[86] Juliana Oldfield to William Thiselton-Dyer, 12 June 1887, Herbarium Presentations to 1900, vol. 2, fol. 622, RBGKA.
[87] Vincent Oldfield to unknown addressee, 12 September 1875, File 0255, PFA.
[88] Vincent Oldfield to Jonathan Howard Oldfield, undated (c.1877), Files 0123–0124, PFA. This letter is written on notepaper printed with 'La Révolution de Français 9 Rue d'Aboukir, Paris'.
[89] *Musée d'Orsay* 2006, 'Louis-Ernest Lheureux, "Monument to the Glory of the French Revolution"', viewed 5 April 2011 at www.musee-orsay.fr.
[90] Vincent Oldfield, business card, File 0167, PFA.
[91] Vincent Oldfield to Jonathan Oldfield, undated, c.1879, Files 0123–0124, PFA.
[92] See for example: Vincent to Jonathan Oldfield, undated, c.1879, Files 0123–0124; 7 February 1879, Files 0174–0175, PFA.

over her husband's neglect over many years caused her to seek solace in alcohol, to the point where she acted in ways such that she no longer had any standing as a respectable, middle-class woman. Frederica wanted nothing to do with her: 'We can pray for the wanderers but we suffer harm in our own souls by association'.[93]

Interaction with and support from friends and others

What, then, of Augustus' association with people outside the family? Some neighbours may have dropped in on him regularly but this could not be confirmed. His absence from England for nearly twenty years probably meant that the friends of his youth lost contact long ago. However, it is unlikely that he wanted to get in touch with them again since he was so sensitive about his afflictions and his financial circumstances.

Joseph Maiden, Government Botanist, New South Wales, noted that he had received information from Mr H[amilton] Stuart Dove, a naturalist and ornithologist who at the time lived in Cunninghame, Victoria, noting

> some personal reminiscences of Mr Oldfield's declining days, when, quite blind, he used to live in a cottage in Merton-road, Wandsworth, London, consoled by a portion of his herbarium, a specimen of which his visitors would describe as well as they could to the sightless man, and he would then give the botanical name and particulars concerning it.[94]

Dove was born in England in 1864 and migrated to Australia (Tasmania) in 1890, at the age of 26 years, where for a while he worked on his brother's farm at Flowerdale. Later he moved to Devonport where his love of natural history flourished. Like Oldfield he never married. He lived the life of a recluse 'in the company of his numerous natural history specimens, but he was a well-known and well-liked figure in Devonport'. He had a special interest in migratory birds and his many published articles testify to a deep and abiding interest in ornithology.[95] The above quotation suggests that Oldfield had a prodigious memory. Furthermore, the reference to 'visitors' indicates that apart from himself, Dove was aware that Oldfield did receive other callers who were interested in botany or at least indulged his desire to connect with his specimens. Perhaps on these occasions his spirits lifted. His language pupils may have provided some relief from the tedium of day-to-day life, living alone without the ability to undertake any serious biological research.

Oldfield and Dove's friendship may have commenced in Wandsworth during Dove's childhood or adolescence; possibly he was one of Oldfield's language pupils. Perhaps the teacher recognised Dove's interest in natural history and encouraged it through recounting his observations of Australian flora and fauna. Dove's letter to Maiden may have revealed more details of their relationship but, unfortunately, it is missing.[96]

Although no evidence has surfaced that Oldfield ever left his home to visit friends or the local shops and business houses, he may well have done so at times. His obvious resourcefulness and strong desire not to be a burden to others may have driven him to learn how to get around Wandsworth on his own and to seek company at one of the local inns or of friends who lived in the village.

---

[93] Frederica Hold to Jonathan Oldfield, 14 January 1879, Files 0116–0122, PFA.

[94] J.H. Maiden, 1911, pp. 232–234.

[95] Annette Alafaci, 2011, 'Dove, Hamilton Stuart c.1864–1941', *Encyclopedia of Australian Science*, viewed 12 March 2011 at www.eoas.info and C.E.B, 1941, 'Obituary: Hamilton Stuart Dove', *Emu*, vol. 41, pp. 95–96.

[96] The missing letter was, we believe, fol. 8135, Registers of Letters Received, 12 September 1907–23 July 1924, Royal Botanic Gardens & Government Domain, Sydney, SRONSW.

Augustus' death

Sometime c.1881 Augustus retired as a teacher.[97] During the ensuing years his generally poor health deteriorated further until on 22 May 1887, at the age of 66 years, he died at Verandah Cottages with Juliana by his side. His death record gives his occupation as 'Botanist'. As far as can be ascertained he had never officially identified himself as such but clearly the person, probably Juliana, who gave his personal details for the Register of Deaths wanted to accord him the occupation that had most engaged his attention, and for which he had probably sacrificed his health and financial security. Dr W.G. Dickinson certified his cause of death as cancer of the liver and syncope (loss of consciousness from fall in blood pressure).[98] There are many causes of liver cancer, including inherited and lifestyle factors and, therefore, lacking information on Augustus' circumstances, it is not possible to speculate on the cause of his disease.[99]

Following Augustus' death, Vincent wrote to Joseph Hooker on 26 May 1887, addressed from 11 Boulevard Bonne-Nouvelle, Paris:

> It is my painful duty to announce to you the death of my poor brother Augustus Frederick Oldfield who contributed not slightly to the floral exploration of Tasmania and whose hortus siccus was placed at the disposition of the Gardens. He passed away after many years' suffering and 20 years blindness through use of microscope–on Sunday the 22nd...Could you in return for the services rendered to Botany by him send a wreath–or flowers–to adorn his coffin. He is to be buried at Brompton Cemetery on Friday. You might send them direct there. Please also cause a few lines to be inserted in the scientific paper announcing his decease. My sister Juliana will give you his exact age &c. &c. You will excuse my addressing you, but I think it due to him that his memory should be in somewhat [sic] considered by a science to which he fell a voluntary victim.[100]

No evidence has been found that Hooker responded to either of these requests. Of course, if it was posted in France on the day before the funeral it might have been a miracle if the letter arrived at Kew Gardens before the funeral, which took place the next day, 27 May 1887. However, Vincent may have used his usual postal address but wrote from London.

Augustus' funeral was conducted by Reverend A. Veysey, of St. Mary's Anglican Church, West Brompton.[101] Probably in attendance were Juliana and Marian. Frederica and Vincent may have come to London for the ceremony. He was buried in the Brompton Cemetery in his father and mother's grave, atop which stands the broken column placed there for his father. In Augustus' case, there is no doubt that it signified a life unfulfilled.

Vincent was not the only sibling who took steps to have his brother's scientific contribution placed on the public record. The obituaries that appeared in *The Argus* on 13 July 1887 and the *Bairnsdale Advertiser and Tambo and Omeo Chronicle*, 16 July 1887, were undoubtedly submitted by Edwin. A fortnight later, on 27 July, Edwin wrote to Joseph Hooker explaining that all his efforts to obtain 'substantial recognition' by the Tasmanian Government of Augustus' labours had been

[97] England Census 1881, record for Augustus F. Oldfield, Source information: RG11/0656; Registration District: Wandsworth: Sub-registration District: Wandsworth; Folio: 88; Page: 30, viewed 31 December 2004, www.ancestry.com.

[98] Certified copy of death certificate for Augustus Frederick Oldfield, Record No.: 349, Registration District Wandsworth, 1887, Death in the Sub-district of Wandsworth in the County of Surrey, No. DXZ513487, 1 June 1998, General Register Office, England, HOA.

[99] Dr Barry Mendelawitz, personal communication to MHH, 28 April 2011.

[100] Vincent Oldfield to JDH, 26 May 1887, Herbarium Presentations to 1900, vol. 2, fol. 621, RBGKA.

[101] Record for Augustus Frederick Oldfield, No. 136802, Private Grave, 3rd interment, on compartment AC at a measurement of 184'0" x 25'3" with Added Remark '25489BR opened to 12 ft., Register of Burials in the West of London and Westminster Cemetery, Earl's Court, Old Brompton, copy of register entry kindly provided to MHH by Laurence Epstein, Cemetery Supervisor, The Chapel Office, Brompton Cemetery, 23 October 2002, HOA.

Fig. 22.2 The Oldfield grave, Brompton Cemetery

unsuccessful 'owing it was always urged to the low state of the finances there'.[102] This reason was not specifically stated in 1875 records held in Tasmania, but it was given in the Colonial Secretary's response to Edwin's first application (1869). Thomas Reibey's reply to Edwin's second application and his complaint that Gunn had received a substantial pension referenced the fact that the reasons given earlier for the unsuccessful application still held. At the same time, he explained why Gunn qualified for a pension but Oldfield did not. Edwin went on to say to Hooker:

> [I]n justice to him [Augustus] there is one thing it would give me great satisfaction to see done; and if you would kindly take the initiative in effecting it, I shall feel deeply indebted to you. It is that some record of his services might be placed upon the records of the Royal Society. That might not be possible as a late "member" of it; but, as a "Contributor" to it, I should think it might be done. However, I will leave the matter in your hands.[103]

There is no record of Augustus having ever been a member of The Royal Society in London. The organisation to which Edwin was referring was probably the Royal Society of Tasmania, and Edwin wrote to Hooker because he had been, or was, a Fellow of that organisation. No record has been found of Hooker meeting the request, but it seems that Edwin, supported by the testimonials he had retrieved

[102] Edwin Oldfield to Joseph Hooker, JDH, 27 July 1887, Herbarium Presentations to 1900, vol. 2, fol. 627, RBGKA.
[103] ibid.

from the Colonial Office ten years previously, was responsible for the formal recognition of Augustus's contribution being placed in the minutes of the August 1887 meeting of the Society:

A DECEASED SCIENTIST

> Some correspondence was read with reference to Mr Augustus Oldfield, a brother of Mr E.D. Oldfield who kept a commercial school in Hobart for many years. Mr Augustus Oldfield, who was a botanist for many years in Tasmania, and possessed very high testimony of his scientific acquirements and disinterestedness from Sir J. Hooker, Baron Mueller, and other authorities, died in comparative obscurity, and his friends desired that his name should at least be rescued from oblivion, considering his valuable work. The matter came recommended by Dr. Agnew, and the testimonials of the deceased gentleman were read.
>
> MR [Richard Austin] BASTOW said that he had frequently come across the name of Oldfield in his botanical researches, and felt certain that the deceased scientist had done a great deal of useful work, though he knew nothing of him.
>
> MR [Charles Henry] GRANT thought that in the case of a man whose work was so fully acknowledged by eminent authorities, the least the Society could do was to record his name as one who had done valuable work in Tasmania so as to encourage others to follow in his lines.
>
> THE PRESIDENT [Sir Robert George Crookshank Hamilton] said he would assume, from the tone of the meeting, that it was the desire of the Fellows that some formal recognition of the labours of the deceased gentleman should be placed on the records.

This minute appeared in full in *The Mercury* for 16 August 1887.[104] A more detailed account of Augustus' botanical contribution was not published by the Society, which may have been a disappointment to Edwin and to Juliana, Marian, Frederica and Vincent.

Disposal of Augustus' private herbarium

Juliana seems to have undertaken the clearance of Augustus' home. She wrote on 3 June 1887 to 'The Curator Kew Gardens' stating that Augustus had

> left a couple of cases of dried plants which he had collected in Australia, and I think I cannot do better than offer them to you for the Herbarium at Kew Gardens. One case, at any rate, is just as he received it back from Kew, where the specimens were arranged under the superintendence of Dr Hooker, at my brother's cost; for, sad to say, loss of sight had for many years prevented him from pursuing study or research of any kind.[105]

There is no indication in previous correspondence that the specimens kept at Kew Gardens for so many years and subsequently returned to Augustus, had been 'arranged' taxonomically while they were there, but this may have been done to facilitate their sale. Augustus must have over-stretched his finances to pay for this work to be done.

[104] Royal Society of Tasmania, *The Mercury*, 16 August 1887, p. 3.
[105] Juliana Oldfield to Curator, Kew Gardens, 3 June 1887, Herbarium Presentations to 1900, vol. 2, fol. 622, RBGKA.

Joseph Hooker had retired as Director in 1885 and his replacement, William Thiselton-Dyer,[106] replied by telegram six days later (9 June 1887) accepting the collection and agreeing to pay the removal costs.[107] Juliana responded on 12 June:

> The cases of dried plants which I wrote about, will be despatched from Wandsworth by Carter Paterson's van to-morrow, Monday. I was disappointed in not receiving in time the large case wh[ich] I had ordered for those plants which were not already packed, and have had to send them in two smaller ones, with a few packets in a canvas package on the top of the largest case...The other plants I have disturbed as little as possible, endeavouring to leave each package exactly as I found it, so as not to interfere with any classification made by my brother while he still had sight for such work. I shall be glad to know that the cases have reached their destination.[108]

There is no evidence that Juliana knew the value that Augustus' correspondence, diaries and field notes would have been to future botanists, nor, it seems, did Thiselton-Dyer, since he apparently made no effort to obtain them from her. Although these records may still be in existence, we believe that Juliana destroyed them and only retained his unpublished natural philosophical work, which went to Edwin. Perhaps Thiselton-Dyer's oversight was due to his preoccupation with the demands of administering the Royal Botanic Gardens, Kew.[109] Furthermore, both Hooker and Bentham, who had worked on the Australian flora, had long since shifted their research interests away from that continent. Bentham had died in 1884 and Hooker, although still working at Kew Gardens, was probably not informed of Juliana's donation and therefore was unlikely to have suggested the retrieval of Oldfield's field notes. On the other hand, the potential value of such documents may not have been understood at Kew Gardens at that time.

The specimens arrived at Kew Gardens on the afternoon of 13 June 1887, and the carriage fee of 3/- was duly paid.[110] Thiselton-Dyer acknowledged their receipt and at the same time expressed appreciation of the value of Oldfield's labours as a botanical collector. This, said Juliana, was 'very gratifying' for her and his other relatives and she thanked him for making it known to her.[111]

Behind the scenes at Kew Gardens, the specimens were given a somewhat cooler reception than Juliana probably imagined would be the case. Daniel Oliver, Keeper and Professor of Botany, University College London, in a memorandum written on the day after their arrival reported on the collection:

> Phanerogamia
> The Australian specimens, of which a set has been already selected for Kew, are quite worth distributing.
> A general collection probably contains little or nothing [of use] to us. This we shall ascertain later.
> Cryptogamia
> A considerable coll^{n} of Musci. I shall get Mr Wright to report upon & ascertain how [illegible word] already in Herb. Kew.
> A coll^{n} of Filices is prob^{ly} of no use to us. Mr Baker will look into it.[112]

[106] Under Thiselton-Dyer's direction, improvements were made to the Gardens and new features appeared, but he is remembered today not for his rationalisation of Kew's administration nor for the landscaping of its grounds, but rather for emphasising its commitment to colonial prosperity. He retired in December 1905, Ray Desmond, 1995, p. 290.

[107] Note by the Director of Kew, 9 June 1887 on Juliana Oldfield to Curator Kew Gardens, 3 June 1887, Herbarium Presentations to 1900, vol. 2, fol. 622, RBGKA.

[108] Juliana Oldfield to the Director [W. Thiselton-Dyer], 12 June 1887, Herbarium Presentations to 1900, vol. 2, fol. 622, RBGKA.

[109] Ray Desmond, 1995, pp.268–289.

[110] G. Nicholson to Director W. Thiselton-Dyer, 13 June 1837, internal memorandum, Herbarium Presentations to 1900, vol. 2, fol. 623, RBGKA.

[111] Juliana Oldfield to W. Thiselton Dyer, 21 June 1887, Herbarium Presentations to 1900, fol. 626, RBGKA.

[112] D.O. (Daniel Oliver) to unstated (probably W. Thiselton-Dyer), 14 June 1887, internal memorandum, Herbarium Presentations to 1900, vol. 2, fol. 625, RBGKA.

Oliver's memorandum suggests Kew Gardens retained little of Oldfield's private collection. In the case of his *Filices* specimens, this is certainly the case. Thus, in 2002, the authors conducted a thorough search of the Australian and New Zealand *Filices* collection in the Herbarium and only found one of his specimens, *Polypodium pustulatum* collected in New Zealand (AO no. 277) from 'Herbarium Oldfield 6/87'. Enquiries at Kew Gardens were unsuccessful in ascertaining how the remainder of the collection was distributed. At the suggestion of Dr Neville Marchant, former Curator of the Western Australian Herbarium, we asked the Royal Botanic Garden Edinburgh if it held any of Oldfield's plant material. The institution holds 'bryophytes, fungi and vascular plants' that the collector obtained 'from Australia and Tasmania c.1858'. Unfortunately, the collection records of that herbarium have not yet been digitised so details of the plants held could not be provided, nor has any correspondence relating to them survived.[113] While these probably came from Oldfield's private collection after he died, it currently cannot be confirmed.

Some cryptogamic material was sent to taxonomists at other institutions. The algologist, Dr Jacob Georg Agardh, who had been Professor of Botany at Lund University, Sweden, until 1879[114], was loaned Oldfield's algae on 1 November 1887.[115] These were returned toward the end of January 1888. At the time, Oliver commented that Dr Agardh 'from his report, must have taken great pains with them and we sh^{d} treat him liberally'.[116] This may refer to a request to keep specimens that Agardh had made to Kew Gardens.[117] Perhaps some of Oldfield's specimens held in the Agardh Herbarium, Herbarium, Botanical Museum University of Lund in Sweden[118] were the subject of this request.

'A few additional Musci of Oldfield's Herbarium...turned up' and were sent, probably on loan, to the prominent British bryologist, Henry Boswell, in March 1888.[119] Boswell was a trunk-maker in 1861, having inherited the family business at the age of twenty-five from his father. He lived in Oxford for most of his life. He was active botanically in the 1850s, and commenced publishing papers and books on his own and in collaboration with others in the early 1860s. As far as can be ascertained, the University of Oxford never employed him.[120]

At the end of May 1890, Set 2 of Oldfield's Australian and some 'miscellaneous' bryophytes appear to have been donated to the eminent Finnish bryologist, Viktor Ferdinand Brotherus[121], a teacher of natural history and mathematics in the Swedish Girls School of Helsinki. He never held an academic position but obtained a doctorate of Philosophy of international standard in 1884, on the bryogeography of the Caucasus area.[122]

Oldfield's specimens are in at least 32 herbaria in Australia, England, Europe and America (see Appendix 1) but how many came from his private collection is unknown.

Aged 65 years, Augustus' life ended, having suffered for twenty years from his impecunious situation, blindness and poor physical and mental health. His bitter disappointment that he could not continue with his activities as a collector or with his botanical studies dogged his days. However, there

---

[113] Adele Smith, Assistant Curator, Herbarium, Royal Botanic Garden Edinburgh, personal communication to MHH, emails 6 April 2005 and 22 June 2005, HOA.

[114] Gunnar Eriksson, 1970, 'Agardh, Jacob Georg', *Dictionary of Scientific Biography*, vol. 1, Charles Scribner's Sons, New York, pp. 70–71.

[115] D.O. [Daniel Oliver] to unstated [probably W Thiselton-Dyer], 1 November 1887, internal, memorandum, Herbarium Presentations to 1900, fol. 628, RBGKA.

[116] D.O. (Daniel Oliver) to unstated [probably W Thiselton-Dyer since his reply is noted on the memorandum], 16 January 1888, Herbarium Presentations to 1900, fol. 630, RBGKA.

[117] ibid.

[118] Per Lassen, Curator Agardh Herbarium to MHH, email, 8 December 2000, HOA.

[119] D.O. (Daniel Oliver) to unknown (probably W. Thiselton-Dyer), 29 March 1888, Herbarium Presentations, to 1900, vol. 2, fol. 629, RBGKA.

[120] Mark Lawley, undated, 'Henry Boswell (1834/5–1897)' in *A Social and Biographical History of British and Irish Field-bryologists*, viewed 1 May 2011 at www.rbg-web2.rbge.org.uk.

[121] D.O. (Daniel Oliver) to unstated (probably W. Thiselton-Dyer), 30 May 1890, internal memorandum, Herbarium Presentations to 1900, fol. 631, RBGKA.

[122] Lars Soderstrom & Tommy Prestg (eds), 1995, 'State of Nordic bryology today and tomorrow: Abstracts and shorter communications from a meeting in Trondheim December 1995', *Rapport botanisk serie 1996–4*, Norges teknisk-naturvitenskapelige universitet Vitenskapsmuseet, p. 12, viewed 4 May 2011 at www.ntnu.no.

were times when he apparently enjoyed re-counting to visitors his many and varied experiences in the Australian bush and identifying plants from his collection when described by visitors. It was a slow and pitiful decline for Augustus, who, for such a long period, had been so physically and intellectually active, feeding his great thirst for exploring, collecting and understanding the world around him, and who had contributed so much to Australian botanical science. If his eyesight had not failed, the high esteem held by Ferdinand von Mueller, George Bentham, and Joseph Hooker, for his botanical labours and knowledge could have led to a rewarding botanical career in one of the Australian colonies, as a herbarium curator or assistant and perhaps later rising to become a Government Botanist. Any one of these posts would have provided him with opportunities to continue with his philosophical studies and perhaps gain recognition as a leading 19th century botanist of Australian flora. That, however, was not to be.

Epilogue

Despite their respective 'delicacies' Augustus' sisters all lived into old age. Frederica was 87 when she died at her home, 10 Young Street, Doncaster, on 28 November 1901.[1] Juliana died three years later, on 11 October 1904, aged 76[2], at an infirmary at 67 London Road, Tunbridge Wells.[3] Prior to going there she had lived with Marian at her home at 47 Samos Road, Penge since 1891. They enjoyed the services of live-in domestic help.[4] Marian, the eldest of the family, was still living there when she died on 26 March 1906, at the grand age of 94 years.[5]

Vincent's wife, Sarah Jane, died in 1891[6] and Vincent wasted no time in marrying a French woman, Marie Eugenie Gruet, in the same year.[7] They lived in England at 15 Ranelagh Road, Paddington. He was earning his living as a newspaper correspondent.[8]

Augustus' unpublished manuscript came into Edwin's possession, probably sent to him by Juliana. Edwin contacted J.H. Maiden c.1909 and asked his 'acceptance of some mss of his brother'. This request was probably triggered by Edwin having come across Maiden's paper in the 'Records of Tasmanian botanists' in the *Journal of the Royal Society of Tasmania* for 1909, in which he referred to Augustus. According to a note written by Maiden, which is with the manuscript, Edwin

> got the papers from his home in Melbourne [8 Mason Street, Hawthorn][9] some little time afterwards[.] I went to Penshurst [suburb in southern Sydney], where he was staying, and I have a lively recollection of carrying the bundle to the Penshurst Railway Station on a hot Monday afternoon. I have read the whole; it is not all botanical, by any means & contains many philosophical discussions. It should be put in the library & preserved. The outside of the paper should be labelled "Oldfield Mss".[10]

Edwin, after all his unsuccessful endeavours to have Augustus' scientific contribution recognised with a government pension, may have been pleased with these additional remarks by the eminent botanist published in the *Report of the Thirteenth Meeting of the Australasian Association for the Advancement of Science* (1911):

> [Oldfield] was not a mere collector for what he has written shows him to have been a competent linguist, a mathematician, but especially a biologist in both branches of zo-

[1] Certified copy of the entry of death for Frederica Juliana Hold, no. 273 for 1901, Registration District: Doncaster; Sub-district: Doncaster in the County of York, DYC993484, given at the General Register Office, England, HOA.

[2] Census records previously cited indicate Julia's year of birth date as 1828, but her death certificate gave 1830, which would put her age at 74 at the time of her death.

[3] Certified copy of the entry of death for Juliana Maria Hold, no. 40 for 1904, Registration District: Tunbridge; Sub-district: Tunbridge Wells in the County of Kent, DYC828528, given at the General Register Office, England, HOA.

[4] Record for Marian Oldfield, 1891 England Census, Registration district: Croydon; Parish: Penge Registration number: RG12; Piece: 598; Folio: 127; Page number: 29, accessed online 1 April 2011 at www.plus.familylink.com.

[5] Entry for Marian Howard Oldfield, Free BMD, viewed 5 April 2011 at www.findmypast.com.au.

[6] Entry for the death of Sarah Jane Oldfield, Registration District: Pancras, Volume 1b, p. 80, Death Index, General Registry Office, England, accessed 3 April 2011 at www.freebmd.org.uk.

[7] Entry for the marriage of Vincent Leopold Howard Oldfield to Marie Eugenie Gruet, Paris, France & vicinity Marriage Banns 1860–1902, accessed online 10 May 2011, at www.ancestry.com.

[8] 1891 England Census Record for Vincent Oldfield, Registration District: Paddington; Sub-district: St Mary Paddington; Parish: Paddington; Registration number: RG12; Piece: 1; Folio: 87; Page number: 25 accessed online 3April 2011 at www.plus.familylink.com.

[9] Last will and testament of Edwin Douglas Oldfield, 27 August 1911, Register no.139/464, Public Records Office of Victoria.

[10] J.H. Maiden, note on Oldfield's unpublished ms c.1861, 7 May 1924, RBGSL.

> ology and botany. He was evidently a man of very broad reading, and it is a pity he published nothing in regard to the Australian flora.[11]

Edwin died at the home of his daughter, Grace Henrica, 'Headingly', Mason Street, Lower Hawthorn[12], on 3 April 1915 at the age of 92.[13] A brief obituary in *The Mercury* (Hobart) of 7 April 1915 referred to him as 'one of the pioneers of education in Tasmania and Victoria':

> Many citizens of the two States (says the "Argus") will have pleasant recollections of the tuition received at his hands and of his scholastic attainments. The closing years of his life were quietly spent.[14]

It seems that advancing years brought on infirmities that required Edwin and Susanna to live apart. From at least 1911 Edwin had lived with Grace[15], while Susanna was in a nursing home or with one of her other daughters at 106 Bridport Street, Albert Park, where she died on 10 February 1917 in her 90th year.[16] She had 'retained all her faculties, including a brilliant wit, to the last'.[17]

Ten months later, on 16 December 1917, Vincent died at the Kensington Infirmary in London of chronic bronchitis and general debility; no occupation was given. He was 84 years old.[18] Efforts were unsuccessful to find out how he and Eugenie spent the intervening years from 1891, and whether or not she outlived him.

---

[11] J.H. Maiden, 1911, pp. 232–234.

[12] MHH's father built a home in Bridgetown and named it 'Headingly' after his family home in Hawthorn. This is the reason for supposing that it was at Grace Henrica's home that Edwin died.

[13] Death notice, *The Argus*, 5 April 1915, p. 1, DAN, NLA.

[14] *The Mercury* (Hobart), 7 April 1915, p. 4, DAN, NLA.

[15] Last Will and Testament of Edwin Oldfield, no. 139/465, Public Record Office of Victoria.

[16] Death record of Susannah Sara Wilhelmina Henrica Oldfield, 10 February 1917, no. 45 for 1917, Register of Deaths in the District of South Melbourne in the State of Victoria, Victorian Registry of Births, Deaths and Marriages.

[17] *The Mercury* (Hobart), 20 February 1917, p. 3.

[18] Certified copy of the entry of death for Vincent Oldfield, no. 432 for 1917, Registration District: Kensington; Sub-district: Kensington South; in the County of London, DXZ828189, given at the General Register Office, England, HOA.

Appendices

Appendix 1

Herbaria in which Oldfield's specimens are housed[1]

Country	*Herbarium code*	**Herbarium name**
Australia	CANB	Australian National Herbarium, Canberra, Australian Capital Territory. CANB also holds 201 images of Oldfield's K specimens.
	NSW	National Herbarium of New South Wales, Royal Botanic Gardens, Sydney, New South Wales.
	MEL	National Herbarium of Victoria, Royal Botanic Gardens, Melbourne, Victoria.
	BRI	Queensland Herbarium, Toowong, Queensland.
	HO	Tasmanian Herbarium, University of Tasmania, Hobart, Tasmania.
	PERTH	Western Australian Herbarium, Biodiversity Conservation Science Centre. Kensington, Western Australia.
Austria	W	Naturhistorisches Museum Wien, Vienna, Austria.
Belgium	BR	Jardin Botanique National de Belgique, Meise, Belgium.
Finland	H	Viktor Ferdinand Brotherus Herbarium, University of Helsinki, Helsinki, Finland.
France	PC	Muséum Nationale d'Histoire Naturelle, Paris, France.
Germany	HBG	Biozentrum Klein Flottbeck und Botanischer Garten der Universita Hamburg, Germany.
	M	Botanische Staatssammlung, München, Germany.
	B	Botanischer Garten und Botanisches Museum Berlin-Dahlem, Berlin, Germany.
	JE	Friedrich-Schiller Herbarium, University of Jena, Jena, Germany.
Netherlands	U	National Herbarium Nederland, Utrecht University branch, Utrecht, Netherlands.
Republic of Ireland	TCD	School of Botany, Trinity College, Dublin, Ireland.
South Africa	NBG	Compton Herbarium, South African Biodiversity Institute, Cape Town, South Africa.
Sweden	LD	Agardh Herbarium, Botanical Museum, University of Lund, Lund, Sweden.
	S	Naturhistoriska Riksmuseet, Stockholm, Sweden.
Switzerland	G	Conservatoire et Jardin Botaniques de la Ville de Genève, Switzerland.

[1] See Preface for the method used to identify these herbaria. Specimen details in HAODb.

Country	*Herbarium code*	**Herbarium name**
United Kingdom	CGE	Department of Plant Sciences Herbarium, University of Cambridge, Cambridge, England.
	OXF	Fielding-Druce Herbarium and Boswell's Collection, Oxford University, Oxford, England.
	BM	National History Museum, London, England.
	E	Royal Botanic Garden Herbarium, Edinburgh, Scotland.
	K	Royal Botanic Gardens, Kew Herbarium & Economic Botany Collection, Kew, England.
	MANCH	University of Manchester Herbarium, Manchester, England.
USA	GH	Harvard University Herbaria, Cambridge, Massachusetts, USA.
	PH	Philadelphia Herbarium, the Academy of Natural Sciences?[2]
	MO	Missouri Botanical Garden Herbarium, St Louis, Missouri, USA.
	NY	New York Botanical Gardens Herbarium, New York, USA.
	US	Smithsonian Museum of Natural History Herbarium, Washington DC, USA.
	MICH	University of Michigan Herbarium, Ann Arbor, Michigan, USA.

[2] A search of the Academy's online database found no records for 'Oldfield', 'A F Oldfield', 'A Oldfield', 'Augustus Oldfield' or 'Augustus Frederick Oldfield'. However, James A. Mears, 1981, 'Guide to plant collectors represented in the Herbarium of the Academy of Natural Sciences of Philadelphia', *Proceedings of the Academy of Natural Sciences of Philadelphia*, vol. 133, pp. 141–165, includes an Oldfield specimen. Also, W.L. Dix, 1959, 'The lichen collection of the Academy of Natural Sciences of Philadelphia, *The Bryologist*, vol. 62, no. 1, pp. 65–66, states that there are 53 Australian lichen specimens in the herbarium collected by 'E. Oldfield'. An unsuccessful search of the Academy's database was made for this name. It is probably Augustus Oldfield since he is the only collector with that surname listed in Alex S. George, *Australian Botanist's Companion*, Four Gables Press, Kardinya, Western Australia, 2009.

Appendix 2

Bibliography

This bibliography provides a comprehensive list of all the sources referenced in this work

Archives

Australia

Australian Capital Territory
National Library of Australia

Hall, N.V., *English Census Directories Project 1841-1911* (every 10 years), Nick Vine Hall, Albert Park, Victoria, microform, 1991–.

London Directories from the Guildhall Library, 1677-1900, Group 2, 1800-1855 (excluding Post Office London directories), microform.

Pigot & Co.'s Directory of London for 1839, J Pigot & Co., London, vol. 197, microform.

Robson's Commercial Directory, Street Key & Conveyance List for 1828, Part II, Robson, Blades & Co., London, microfilm, vol.179, microform.

Post Office London Directory 1841 & 1850, Frederick Kelly, London, microform.

Post Office London Trades Directory for 1861, W. Kelly, London, microform.

Robson's London Directory: Street Key & Royal Court Guide for 1843, 24th edition, Bowtell & Co.

London Directory and Court Guide Office, Cheapside, vol. 201, microform.

Watkins, F. W., *Watkins Commercial & General London Directory and Court Guide for* 1852-1853, 2nd edn., Longman, Brown, Green, and Longmans, London, vol. 205, microform.

Oldfield, F.H.H., '4 Waltzes [music]: for the Piano Forte, Las & Lyons [18--?], Bib ID: 1457124.

New South Wales
Mitchell Library, New South Wales

Jones, J., *Map of New South Wales and Victoria*, J Jones, Sydney, 1850, **805/1850/1A.**

Royal Botanic Gardens Library, Sydney.

Oldfield, Augustus Frederick 1865a, Untitled ms., 860 pp., Species Manuscripts.

State Records of New South Wales.

The 1862 London International Exhibition records.

Royal Botanic Gardens & Government Domain, Sydney, Registers of Letters Received, 12 September 1907-23 July 1924.

South Australia
South Australian Museum Archives

Tindale, Norman B., Catalogue of Australian Tribes, Collection AA338.

Tasmania
Education Department Tasmania Archives

Education Board Correspondence, 1846.

Southern Board of Education, Report for the Year 1859, Board of Education Reports 1842-85.

Fifth Report of the Southern Board of Education, Tasmania. Report to H.E. the Governor-in-Chief of the Colony of Tasmanian and its Dependencies, Appendix E, Nominal Return of Teachers in the Employ of the Southern Board of Education on 31st December 1861.

Tasmanian Archive and Heritage Office

Annual Official, Financial and Statistical Reports ('Blue Books'), 1841-1847, CSO50/16 – CSO50/22.

Census Van Diemens Land, 1848, Parish No. 2, Microfilm CEN 1/81 79-82.

Colonial Secretary's Office correspondence, CSO 80/1.

Colonial Secretary's Department correspondence, CSD 7/1/49 & CSD10/1/27.

Colonial Tasmanian Family Links database at www.archives.tas.gov.au.

Education Board Correspondence, CSO11/1/160, CSO 24/88/2652, CSO 26/3148, CSO 26/3373.

Lands & Survey Department Correspondence, 1853, LSD 1/1/81/(29), 1854, LSD1/1/81/(37) & LSD1/1/81/(39), microfilm.

Oldfield, E.D., *Commercial Academy*, prospectus, pamphlet, undated, PQ 373.245 Old.

Report of the Inspector of Schools 1857, *Hobart Town Gazette*, no. 1202, p.656.

Southern Board of Education Fourth Report, Report for 1860, *Notes and Proceedings of the Legislative Council*, paper no. 34, Appendix E, Nominal Return of Teachers in the Employ of the Southern Board of Education on 31 December 1860.

Register Book of Marriages for the State of Tasmania and its Dependencies, 1851.

Wood, James (compiler), *Tasmanian Almanack for* 1855, 1857, J W H Walch, Hobart, 1849.

Wood, James (compiler), *Royal Southern Kalendar, Tasmanian Register, and General Australasian & East Indian Official Directory*, J.W.H. Walsh, Hobart Town, 1849, 1850.

Victoria

Desmond and Lyn Williams Family Archive, East Brighton, Melbourne

Oldfield family photographs.

Embroidered sampler by Lavinia Oldfield, 1825.

State Library of Victoria

Melbourne Directory, 1887, Sands and McDougall.

State Records Office of Victoria

Royal Botanic Gardens Melbourne and National Herbarium and Government Botanist, File 84/471, VPRS02873/P001/Unit000034, Type A1.

Correspondence of Lands & Surveys Department, Victoria, VPRS 97/P/001, V/AK/162/03/02.

Chief Secretary's Department, inward registered correspondence, VA475, F58/2731, unit 745, VPRS 1189.

University of Melbourne, Department of History & Philosophy of Science

The Correspondence of Ferdinand Mueller Project database.

Western Australia

Albany History Collection, Albany Public Library

Hassell, CW, 1972, The Hassells of Albany, unpublished ms.

Henderson Oldfield Archive

Research correspondence re Augustus and his family 1995-2014 (to be deposited in the Royal Botanic Gardens, Melbourne, Library Archive), includes copies of Oldfield's unpublished ms, c.1865, and his paper 'On Australian Aborigines', 1865, Joseph Hooker's Augustus Oldfield testimonial, 2 June 1875, Royal Botanic Gardens, Kew, Correspondence regarding Oldfield and Pimblett Family records relating to the Oldfield family.

Henderson Augustus Oldfield Database for Oldfield's specimen records (published and herbarium records) located by the authors.

Grant, Peter, Parks and Wildlife Service, Tasmania, 'Mt La Perouse', unpublished ms, 2008 (with kind permission of the author).

Landgate, Nomenclature Section, Department of Lands Administration (Western Australia)

'Oldfield River', undated official record, Nomenclature File 3664/61.

Royal Western Australian Historical Society Inc. Museum

Alfred Hillman, File of Botanical Specimens—Wildflowers, Accn. no. 2003/44A-Z.

State Library of Western Australia

Towers, Sue & Margie Knight, *Unassisted immigrants & coastal passengers to Hobart Tasmania 1829 - 1865 Index,* microform, Imprint: Macbeth Genealogical Books, Hampton, Victoria, 1993, Microfiche.

J.S. Battye Library of West Australian History

Burges, L.C., Journal, 22 November 1849 to 17 February 1850.

Dempster, William Simon, Diary letter for period 4 December 1865 to 23 January 1866, in Esperance district, sent to his sister Minnie, ms. typescript.

French, William, c.1856, Perth in West-Australien vom Berge Eliza aus, steel engraving, produced by A.H. Payne, Leipzig, Germany.

1995, *A Report on the Aboriginal Ethnohistory of the Warren-Blackwood Region, Southwest Australia*, imprint.

Gaines, C., Bridgetown: One Hundred Years of History, Extracts from thesis, mimeo., c.1970.

Gerritsen, Rupert, 2002, Nhanda villages of the Victoria District, Western Australia, Intellectual Property Publications, Canberra, mimeo.

Gibbs, Martin, 1995, *A Report on the Aboriginal Ethnohistory of the Warren-Blackwood Region, Southwest Australia*, imprint, p. 15.

Gregory, A.C., 'Excursion to the eastward of Kojonup and down the Gordon River April 1846', *Exploration Diaries*, vol. 3, 1836-1846, unpublished transcripts, pp. 764-765.

Hammond, J.E., 1932, 'The Perth water front', *Early Days*, Journal of the Royal Western Australian Historical Society, vol. 2, pt. 11, pp. 16–17.

Lloyd's Register of British and Foreign Shipping, 1840-1845, 1845-1846, 1849-1851.

Perry, N.K., 1956, The History and Present Potentialities of the Old Coast Road from Mandurah to Bunbury, unpublished typed manuscript.

Stokes, J.L., 'Expedition of the Beagle to Port Grey', ***The Inquirer,*** 22 December 1841, p. 2.

Stowe, J.P., 'The Forlorn Hope', ms, 1865.

Turner, Thomas, 1830, 'Augusta, 1830', water colour on linen.

Wolfe, Adam, 1994, *The Albany Maritime Heritage Survey 1627–1994*, prepared for The Heritage Council of Western Australia, The Albany Port Authority and The Albany Maritime Heritage Association, mimeo.

State Records Office of Western Australia (excluding maps)

Burges, W., Resident Magistrate-Champion Bay 1853-1859, CSO Inwards Correspondence, vols. 248, 256, 309, 339, 358, 367, 368, 382, 387, 413, 435.

Harris, J., to Colonial Secretary, 8 April 1856, CSO, Inwards Correspondence, vol. 368.

Wray, H., to Colonial Secretary, 15 October 1856, CSO, Inwards Correspondence, vol. 358.

England

AIM25: Archives in the London and the M25 area,

'Wandsworth Board of Guardians' at www.aim25.ac.uk.

Church Archives

Church of St Mary Rotherhithe Baptisms, Parish Register, 1787.

Church of St Mary Abbots Marriage Register, 1810.

City of Westminster Archives Centre

Register of Baptisms, 1812, 1816, 1817, 1819, 1821, Church of St James Piccadilly, Westminster.

Parish of St James, Westminster, Records Rate Books, 1814-1822, Pall Mall Ward.

Parish of St James, Westminster, Poor Rate book, 1822, Pall Mall Ward.

Parish of St Marylebone, 54 Norton Street, Rate Books, 1828, Microfilm Reel 55, Part 2 & 1829 Microfilm Reel 56, 1829, Part 1.

Register of Baptisms, Church of Saint Mary, St Marylebone Road, 1823.

Jeff and Anne Pimblett Family Archive

Oldfield family correspondence, birth, marriage and death certificates, photographs and ephemera,1828–1916.

General Register Office, England

Birth, marriage and death records 1837–*King's College London*

Calendar for the Year 1840, B. Fellowes, London, 1840, KCLA.

Reports from the Council to the Annual General Court of Governors and Proprietors for the years 1840 -1842.
Statement of the Arrangements for Conducting the Various Departments of King's College London, August, 1838.
Student registers for the period 1828-1848 (incomplete).
King's College calendars for 1829-1844.
London Metropolitan Archives
Middlesex Deeds Registers MDR 1821/3/634 & MDR 1821/3/635.
Holy Trinity Church, Brompton Road, Parish of Kensington, Register of Baptisms, 1839, Arch. P84/TR12, microfilm 089/007.
National Archives, Kew, London
Admiralty and Board of Trade: General Registry and Record Office of Seamen: Alphabetical Index to Registers of Seamen's Tickets: Name of Seaman: MO – OR, 1835-1844, BT 114/15.
Admiralty and Board of Trade General Register and Record Office of Seamen: Agreements and Crew Lists, Series 1: Port of Registry: London Ships Names, JO-K, 1835-1844.
Archer and Green, Architects, Plan drawings for renovations at 28 Bury Street, Westminster, 19 June 1878, LRRO/2348 109860.
Board of Inland Revenue and predecessors: Estate Duty Office and predecessors: Registers of Legacy Duty, Succession Duty and Estate Duty, Frederick Oldfield, IR 26,1860, N-R, fols 1366.
Crime, Prisons & Punishment, Home Office: Criminal Petitions: Series 1, Prison Registers, Great Britain, Series: HO17, Piece no. 49, Item No. HK34.
England and Wales Criminal Registers, 1791-1892, December Sessions, 1797.
King's (Queen's) Bench Prison and Queen's Prison Commitment Books 1719-1862, PRO Pris 4/ 35, November 1822 – November 1823.
The King's Bench Prison and Queen's Prison: Discharges for 1828, PRO PRIS 7/47 1776-1862.
New Zealand Company General Correspondence, 41/4605, PRO CO208/18/1841.
New Zealand Company, Alphabetical List of Surveyors, 1839-1850: PRO CO 208 281.
Middlesex Deeds Registers MDR 1821/3/634, 19 April 1821.
Probate Registry, Court Service, Postal Searches and Copies Department, York, England
Will and testament of Frederick Howard Oldfield, 11 July 1860 and Codicil, 25 October 1860.
Royal Botanic Gardens, Kew Archives
Correspondence with and regarding Oldfield:
Herbarium Presentations to 1900, vol. 2.
Hooker, William Jackson 1856, 'Instructions for collecting and preserving and sending home plants and their usual products', in 'Botany', *Admiralty Manual*, 1856, pp. 1–8, RBGKA.
Kew Correspondence, English Letters, vol. 52, L-Z, 1862-65.
________, Australian Letters, vol. 73, NAT-OXL.
________, Australian Letters, 1851-1858, vol. 74, NAT-OXL.
________, Australia, Mueller, 1858-70.
________, Directors' Letters, Australia Letters, vol. 74, 1851-58.
________, Directors' Letters, Australian and Pacific letters, vol. 75, 1859-65.
________, English Letters, vol. 30, I-Z, 1850.
________, English Letters, vol. 42, L-Z, 1863.
________, English Letters, vol. 97, NAT-OXC, 1859-1900.
________, Supplementary Foreign Letters 1865-1900, vol. 218.
________, South Australia, Tasmania, Western Australia letters, 1865-1900.
Royal Parks, Brompton Cemetery
Copy of Register of Burials in the West of London and Westminster Cemetery, Earl's Court, Old Brompton, for Frederick (entry no. 25489, 24 November 1860), Mary and Augustus Oldfield.
Society of Genealogists
Faculty Office Marriage Licence Allegations (1701-1850) for the year 1810.

France

Guînes Town Museum

Liborel, Louis ?, *Beaulieu House Acadèmy for Young Gentlemen, Under the Direction of M[onsieur] L. Liborel at Guînes, near Calais*, Prospectus, Calais, undated, c.1840. Copy provided by Jacques Louf, 24 November 2005.

Courbot, Chantal, *Education des Jeunes Britanniques: Une Eclosion de Pensionnaits*, Museum of Calais Exhibition, 2005.

Société Historique de Guînes et Des Environs

Chantal Courbot, *Education des Jeunes Britanniques: Une Eclosion de Pensionnaits*, Introduction, Museum of Calais Exhibition, 2005 catalogue.

Herbaria (online sources, including Oldfield specimen record downloads provided on disc or by email)

Australia

Australian Capital Territory

Australian Biological Resources Study, Department of Environment (Australia), Canberra.

Cowan, R.A., 2006, *Australian Marine Algal Name Index* (AMANI), Canberra, and Murdoch University, Perth, at www.anbg.gov.au

McCarthy, P.M 2006a, *Checklist of Australian Liverworts and Hornworts* at www.anbg.gov.au.

_________, 2006b, *Checklist of the Lichens of Australia and its Island Territories*, 2006 at www.anbg.gov.au.

Australian National Herbarium, Australian National Botanic Gardens, Department of Environment and Water Resources (Australia)

Australian Plant Name Index (APNI), part of the Integrated Biodiversity Information System (IBIS) in collaboration with the Centre for Australian National Biodiversity and the Australian Biological Resources Study, Canberra at www.cpbr.gov.au.

Council of Heads of Australasian Herbaria, Australian Government

Australian Plant Census (APC) at www.cpbr.gov.au.

Biography: Biographical notes on plant collectors and illustrators, Australian National Herbarium, Council of Australian Herbaria at www.anbg.gov.au.

Geoscience Australia, Australian Government

Gazetteer of Australia, Place Name Search at www.ga.gov.au.

New South Wales

National Herbarium of New South Wales

Plant Diversity Section, download of Oldfield's specimen records, 23rd October 2006.

PlantNET - New South Wales Flora Online at www.plantnet.rbgsyd.nsw.gov.au.

'The Eucalypts***Icons of the Australian Bush. Eucalink A Web Guide to the Eucalypts' www.plantnet.rbgsyd.nsw.gov.au.

Queensland

Queensland Herbarium, Queensland Department of Environment

Electronic list of 19 Oldfield specimen records provided from HERBRECS, email 14 December, 2000.

Tasmania

Tasmanian Herbarium, Tasmanian Museum and Art Gallery

Oldfield specimen records of *Bryophyta* provided by email, 28 September 2007 and *Angiospermeae*, October, 1999.

Department of Primary Industries, Parks, Water and Environment, *Natural Values Atlas*, www.naturalvaluesatlas.tas.gov.au.

Victoria

National Herbarium of Victoria, Royal Botanic Gardens Melbourne

AusMoss – Catalogue of Australian Mosses, Royal Botanic Gardens, Melbourne, at www.rbg.vic.gov.au.

Maxwell, George, MELISR electronic database download, email 2 April, 2009.

MELISR Data Entry Manual, Royal Botanic Gardens, Melbourne, at www.rbg.vic.gov.au

Oldfield, Augustus, MELISR electronic database downloads, emails to MHH 18 November 2005, 8 March 2007, 1 December 2010, 22 December 2011.

Stuart, Charles, MELISR electronic database download, email 3 September 2007, 3 September 2007.

Western Australia

Western Australian Herbarium, Department of Parks and Wildlife

FloraBase: the Western Australian Flora at www.florabase.dpaw.wa.gov.au.

Oldfield's specimen record download, 9 June 1999.

Assistance with identifying species named in honour of Oldfield.

Albany Herbarium (branch),

Free, Marjorie M.(compiler), Eucalypts in the Albany region, per A.J. Boland, Curator, Forest Research, CSIRO, 11 May 1984.

Kalbarri National Park Herbarium (branch),

Assistance with identifying species named in honour of Oldfield.

Belgium

Jardin Botanique National de Belgique, Brussels, Belgium, at www.br.fgov.be

England, Scotland and Ireland

Fielding-Druce Herbarium, Oxford University Herbaria, Department of Plant Sciences

Oldfield specimen records list provided by email, 27 January 2006.

Natural History Museum, London

Downloads of Oldfield specimen records at http://www.nhm.ac.uk/nature-online.

Royal Botanic Gardens Kew, London

Downloads of Oldfield's specimen records provided by the Trustees of the Royal Botanic Gardens Kew, 22 May 2007; updated via the *Royal Botanic Gardens Kew Herbarium Catalogue* online 30 March 2012 at http://apps.kew.org/herbcat/navigator.do.

Finland

Viktor Ferdinand Brotherus Herbarium, University of Helsinki Herbarium, Helsinki, Finland, Oldfield specimens embrace 3 moss genera (*Zygodon, Dicranum, & Bryum*) but no specimen details provided, email 22 November 2000.

Netherlands

National Herbarium Netherland, Leiden, Netherlands at www.nhn.leidenuniv.nl.

Sweden

Agardh Herbarium, Botanical Museum, University of Lund, Lund, Sweden. Data on five specimens provided, email, 8/12/2000.

Naturhistoriska Riksmuseet, Stockholm, Sweden, data on three Oldfield cryptogamic specimens provided, email 8 December 2000.

Switzerland

Conservatoire et Jardin Botaniques Ville de Genève, Switzerland, online herbarium catalogue at www.ville-ge.ch.

United States of America

Harvard University Herbaria, Cambridge, Massachusetts, at www.kiki.huh.harvard.edu.

New York Botanic Gardens Herbarium, New York.

Herbarium Catalogue online download of Oldfield specimen records, specimen list, 20070531, 31 May 2007 at www.sciweb.nybg.org.

Smithsonian National Museum of Natural History Herbarium, Washington DC at www.collections.nmnh.si.edu.

University of Michigan Herbarium, Ann Arbor, Michigan. The authors sighted a specimen on loan to the Tasmanian Herbarium, October 1999.

Maps

Australia

Australian Capital Territory

Director of Survey, Department of Defence (Army Office), Canberra

Joint Operations Graphic, Ajana, Australia, scale 1:250,000, Series 1501, SH 50-13, Edition 1, Royal Australian Survey Corps, 1983.

________, Geraldton, Australia, scale 1: 250,000, Series 1501, SH 50-1, Edition 1, Royal Australian Survey Corps, 1982.

National Library of Australia

T.L. Mitchell, New South Wales, Surveyor-General, General map of the south eastern portion of Australia: shewing the colony of New South Wales as surveyed and divided by the Sr. Genl. [i.e. Surveyor General] between the years 1827 and 1850, J. Carmichael, sc., North Sydney, [? 1850], online MAP RM 918 (Copy 1), NLA.

National Mapping, Commonwealth of Australia

Australian Surveying & Land Group, Ajana, 1: 250,000, SG50-13, Western Australia, Edition 2, Map 98/047, Commonwealth of Australia, 2000.

National Topographic Map Series, Bunbury, Sheet SI 50-6, 100,000, Edition 1, 1981.

________, Mount Barker, Sheet SI 50-11, 1: 250,000, Edition 1,1985.

New South Wales

Mitchell Library, Sydney, New South Wales

Jones, J., *Map of New South Wales and Victoria Showing the Course of the Mountains, Rivers and Roads, the boundaries of Old, New, and Proposed Counties, the limits of the settled Intermediate and Unsettled Districts and the Situation of the Principal Towns and Villages from a Map by Arrowsmith,* J. Jones, Sydney, 1850.

Tasmania

Allport Library, Tasmanian Archive and Heritage Office

Arrowsmith, J. 1842, *Van Diemen's Land*, map, copied from the original manuscript surveys in the Colonial Office and in the Van Diemen's Land's office, J. Arrowsmith, London.

Department of Environment and Land Management

Huon, 1: 1000,000 Topographic Map, Edition 6, 1997.

South Cape, 1: 1000,000 Topographic Map, 8210, Edition 1, 1994.

Map Supplies, Hobart

Tallis, John, Van Diemen's Island or Tasmania, map 1851, John Tallis & Co., London & New York, reprint, 1991.

Western Australia

Art Gallery of Western Australia

Turner, Thomas, Topographical map County Sussex, Western Australia, 1851, Art Gallery of Western Australia, 1953/ooD6, reproduced in Anne Brearley, *Ernest Hodgkin's Swanland*...2005, p. 310.

Department of Indigenous Affairs, Western Australia

Gascoyne Murchison traditional tribal elders, Gascoyne/Murchison DIA Region Tribal Boundaries, 11 February 1999.

State Records Office of Western Australia

Gregory, F.T., Field Notebook 13, item, Gref/13, Item 253, Cons. 3401.

Public Plan,

________, Albany District Office, lithograph 446/80, 14 October 1920, J.P. Camm, Surveyor General, Lands and Surveys Department, Western Australia.

________, Eye Sketch by A.C. Gregory of part of the Murchison River and Chain Survey of Locn 1, 1850, Cons. 3869, Victoria 0001.

________, Paced Survey of Road between the Wellington & Williams Districts as marked by Acting Surveyor F T Gregory, October 1847, WAS 1861, Cons 5000, Item 245

________, Plantagenet 08 - Plantagenet 15 [80 chain plan, 506630], (1876–1896

________, Plantagenet 16 [80 chain plan, 506635],(1883–).

________,Roads: Bunbury to Blackwood and with branch roads from Balingup to Jayes (Steere's), Wellington and Nelson, District Survey Officer, Bunbury, Survey Department ofWestern Australia 4 September 1860, accession no. 3869, Item No. Roads 127.

________, South West [506065], AU WA S979-cons4900 S42/1 1883-09-24, (Accumulation).

________, Sussex & Blackwood River, undated (c. late 1850s), Cons. 4926, Item Sussex-01, (506223).

________, Sussex, Cons. no. 4926, Item Sussex 3, Series 9789, undated, c. 1860s, (506200).

________, Victoria, 1870, Cons. 5018, (507810).

________, Victoria, 1863, Cons. 5018 (506653).

________, Victoria, 1863, Cons. 5017, Victoria District, Sheet 2, (507475)

________, Victoria, 1863, Cons. 5018 (506654),

________, Victoria, 1863, Cons. 5018, Victoria District, (506653).

________, Victoria, 1865, Cons. 5018 (507508).

________, Victoria, 1853, Cons. 3869, Victoria 076.

________, Victoria 0001, 1850, Cons. 3869.

________, Victoria and North 1860, Cons 5018, Vict.001, (506638).

________, Victoria & North 1, c.1858, Cons 5018, (506637),

________, Victoria 18, Northampton. Cons. 5018 (506666).

________,Wellington and Nelson, Public Plan of Roads: Bunbury to Blackwood and with branch roads from Balingup to Jayes (Steere's), District Survey Officer, Bunbury, Survey Department of Western Australia 4 September 1860, Accession no. 3869, Item No. Roads 127, SROWA.

________,Wellington 4 (506327) & Wellington 5 (506328) Cons. 4929.

________,Wilson's Inlet-Road to Albany and adjacent country compiled from surveys and sketches by F.T. Gregory, 1855, Record series 237 Survey Plans–Roll Plans, 1 January 1841-1 January 1930, Item no. 082, Cons. 3846.

Tasmania

Map Supplies, Hobart

John Tallis 1851, *Van Diemen's Island*, Tasmania, map drawn and engraved by J Rapkin, reprint, 1991.

Online sources

English records:

Ancestry.com.au:

Baptism Index for England, 1780-1837.

England Census, 1841, 1851, 1861, 1871, 1881, 1891.

England and Wales, Criminal Registers, 1791-1892, record for William Oldfield, December Sessions, 1997.

London, England, Births and Baptisms, 1813-1906, 1837.

London, England, Marriages & Banns, 1754-1921.

London, England, Deaths and Burials, 1813-1980.

Free BMD Marriage Index, 1851, 1852.
Death, Burial, Cemetery & Obituaries, England, Wales Free BMD Death Index 1837-1915.
Etchells, Guy, 'A transcript of the 1812 Act for the regulating and preserving Parish and other Registers in England'.
Oldfield, Frederick Henry Howard, 'Four Waltzes for the Piano Forte', 1836, publisher unknown.
British Library at www.explore.bl.uk.
Morning Post, 1864.
Cambridge University Library
Darwin Correspondence Project at www.darwinproject.ac.uk.
National Archives, London at www.nationalarchives.gov.uk
Historical Manuscripts Commission database, Historical Manuscripts Commission National Register of Archives.
Personal name search combining the Personal Index and the Diaries and Papers Index; and the English Strand of the UK.
Archives Network Database at 'St George's Hospital, London', *Hospital Records Database*, (NAL).
Old Bailey online
The Proceedings of the Old Bailey at www.oldbailey online.org.
The Times of London digital archive, 1785 – 2006 viewed at the University of Western Australia Library.
Times of London 1794-1920.

Australian records:
Australian National University
Australian Dictionary of Biography Online at www.adb.anu.edu.au.
National Library of Australia:*Digitised Australian Newspapers online* at www.trove.nla.gov.au.
The Argus, Melbourne, Victoria,1887, 1853, 1858, 1859, 1861, 1862, 1880, 1881, 1887, 1905.
Brisbane Courier, 1864, Queensland.
Bairnsdale Advertiser and Tambo and Omeo Chronicle, 1887, Victoria.
Colonial Times (Hobart, Tasmania), 1842, 1843, 1844, 1854, 1855, 1857.
*The Courier.*Hobart, Tasmania, 1843, 1844, 1845, 1847, 1848, 1849, 1856, 1857, 1858.
Hobart Town Gazette, Tasmania, 1847, 1857.
Hobarton Mercury, Tasmania, 1856, 1857.
The Hobart Town Mercury, Tasmania, 1857.
The Hobart Town Daily Mercury, 1858, 1859, 1860.
Launceston Examiner, Tasmania, 1843, 1870, 1873.
The Maitland Mercury & Hunter River General Advertiser, New South Wales, 1846, 1850.
The Mercury, Hobart, Tasmania, (1860-1854), 1860, 1859, 1871, 1872, 1887, 1917.
South Australian, Adelaide, South Australia,1848.
The South Australian Advertiser, Adelaide, South Australia,1858, 1859, 1861.
South Australian Register, Adelaide, South Australia,1862.
Sydney Morning Herald, New South Wales, 1845, 1854, 1861, 1862, 1913.
Perth Gazette and Western Australian Journal, Western Australia, 1842.
Perth Gazette and Independent Journal of Politics and News, Western Australia, 1851, 1858, 1859, 1861.
Inquirer, 1841
Inquirer and Commercial News, Perth, Western Australia, 1856, 1858.
Tasmanian Archive and Heritage Office at www.archives.tas.gov.au
Colonial Tasmanian Family Links database.
Index of Departures 1817-1867, ref: POL2201.

United States of America records:
Family Search (Church of Jesus Christ of the Latter Day Saints) at www.familysearch.org
English records:
England Births and Christenings, 1538-1975.
England and Wales, Death Index 1837-1920.
Doggett's New York City Directory for 1849-1850, p. 322, New York Genealogical Records, 1675-1920 database.
New York Passenger Lists, 1820-1957.
United States Federal Census, 1850, 1860, National Archives and Records Administration, Washington D.C.
New York Petitions for Naturalization, Common Pleas Court, New York County, US Immigration Records Database.
Territorial Enterprise, Nevada, 20 September 1866.

Books, articles, family tree – online

Alafaci, Annette, 'Dove, Hamilton Stuart (c. 1864–1941)', *Encyclopedia of Australian Science*, at www.eoas.info.

Allen, D.E., 1959, 'The history of the vasculum', *Proceedings of the Botanical Society of the British Isles*, vol. 3, pp. 135–150 at www.archive.bsbi.org.uk.

Anderson, Ross & Rodriquez, Jennifer, 2006, *Pakington Whaling Station*, Report no. 214, Department of Maritime Archaeology, Western Australian Museum at www.museum.wa.gov.au.

Anon., undated, 'Arthur River, Western Australia, history', 'Blackwood River', 'Caroline of Brunswick', 'Curtis's Botanical Magazine', '*Fallopia convolvulus*','Governess', 'History of rail transport in Great Britain 1830–1922', 'Isoetes', 'King's College London', 'Lycopodiopsida', 'Marshalsea', 'New College Worcester history', Nicholas Baudin, 'Portman Estate', 'Royal National Institute of Blind People', 'Reversion Law', 'St George's Square [Pimlico]', *Wikipedia,* at www.wikipedia.org.Anon., 2017,

Anon., undated, 'Bridgetown comparison', Climate Statistics for Australian Locations, Product IDCJCM0030, *Bureau of Meteorology, Australian Government,* at www.bom.gov.au/climate.

Anon., City of Geraldton-Greenough, 'History of Greenough Flats', at www.cgg.wa.gov.au.

Anon., 2008, 'Eaglehawk Neck History' *Parks and Wildlife Service*, Tasmania, at www.parks.tas.gov.au.

Anon., undated, 'Endogens' & 'Exogens', *Biology Online*, at www.biology-online.org.

Anon. 'Ethnological Society', *The Morning Post*, 10 February 1864, p. 7, viewed 20 January 2014 at www.explore.bl.uk. Overview of Augustus Oldfield's Australian Aborigines paper (1865b).

Anon., undated, 'Hatchers Manor', at www.laterooms.com.au.

Anon., undated, 'History', *Murchison House*, at www.murchisonhousestation.com.au.

Anon., 2011, 'History', *St James' Old Cathedral*, at www.sjoc.org.au.

Anon., undated, 'Saltwater people of the coast', *Koori Coast, Living Knowledge,* at www.livingknowledge.anu.edu.au.

Anon., undated, 'Moon alphabet', *Omniglot the Language Encyclopedia*, at www.omniglot.com.

Anon. undated, *Protecting Ramsar wetlands, Priority site 12: Vasse-Wonnerup System, Australian Ramsar Site No. 38, State: Western Australia, Business Plan 2012–2013, Caring for our Country*, Department of Agriculture, Fisheries and Forestry and the Department of Sustainability, Environment, Water, Population and Communities, Australian Government, Canberra, p. 3, at www.nrm.gov.au.

Anon., 'Wandsworth Board of Guardians', *AIM25: Archives in the London and the M25 area* at www.aim25.ac.uk.

Bailey, M.R., Chrimes, M.M., Cox, R.C., Cross-Rudkin, P.S.M., Hurst, B.L., McWilliam, R.C., Rennison, R.W., Ruddock, E.C., Sutherland, R.J.M. & Swailes, T. (eds), 2008, *Biographical Dictionary of Civil Engineers in Great Britain and Ireland*, vol. 2, 1830–1890, The Institution of Civil Engineers, London, p. 21, at www.thomastelford.com/books.

Ballard, T., 1995, *Statement of Travis Ballard, Esq., President of National Congress for Fathers and Children*, At The White House, 27 November 1995, at www.deltabravo.net.
Biosecurity, Queensland Division 2013 *Weeds of Australia,* at www.daff.qld.gov.au.
Birman, Wendy, 1972, 'Dempster, Charles Edward (1839–1907)', *Australian Dictionary of Biography* online, Australian National University, Canberra, at www.adb.online.anu.edu.au.
Blank, Stuart C., undated, 'Royal Navy Service Records', *Military Archive Research,* at www.militaryarchiveresearch.com.
Boulger, George Simonds, c.2011, 'The Botany of the Voyage,*' Dictionary of National Biography, 1885-1900*, vol. 51, viewed 28 December 2017 at www.en.wikisources.org.
Burbidge, N.T., **1969,** 'Bentham, George (1800–1884)' in *Australian Dictionary of Biography Online*, National Centre of Biography, Australian National University at www.adb.anu.edu.au..
Burke & Wills Web undated, 'Diedrich Henne (1825– 1913)', at www.burkeandwills.net.au.
Burns, T.E. & J.R. Skemp, 1966, 'Gunn, Ronald Campbell (1808–1881)', in *Australian Dictionary of Biography Online*, Australian National University, Canberra, at www.adb.anu.edu.au.
Bush, Fiona & Philip Griffiths, 1998, 'Geraldine Lead Mine', Documentary Evidence, Assessment of Cultural Heritage Significance, Register of Heritage Places, Permanent Entry, Heritage Council of Western Australia, at www.register.heritage.wa.gov.au.
Busselton Jetty Environment and Conservation Association undated, 'Busselton Jetty Timeline', Busselton Jetty, at www.d1054021.myweb.westnethosting.com.au.
Calaby, J. H., **1967,** 'Preiss, Johann August Ludwig (1811–1882)', in *Australian Dictionary of Biography Online*, Australian National University, Canberra, at www.adb.anu.edu.au.
Carpenter, Mary Wilson, 2009, *Health, Medicine & Society in Victorian England*, Greenwood Publishing Group, at www.books.google.com.au.
Catholic Archdiocese of Melbourne, undated, 'A Story of St Patrick's Cathedral', at www.cam.org.au.
Chou, B. Ralph, 1997, 'Eye safety during solar eclipses', adapted from NASA RP 1383 *Total Solar Eclipse of 1999 August 11, April 1997*, at www.sunearth.gsfc.nasa.gov.
Clark, Richard, 2007, 'Revegetation of the bank of the Vasse River at the Isaac Street Reserve, Busselton', *Geographe Community Landcare Nursery*, September 2007, at www.geographelandcarenursery.mysouthwest.com.au.
Clifford-Smith, Silas, 2013, 'The reflective gardener', at www.reflectivegardener.blogspot.com.au.
Coffin Bay Brumby Preservation Society Inc., undated, 'South Australia's wild Brumbys', at www.coffinbaybrumby.org.au.
Cooper, Holly L. 2011, 'A Brief History of Tactile Writing Systems for Readers with Blindness and Visual Impairments', *See/Hear*, Spring 2006 vol. 11, no. 2, at www.tsbvi.edu.
Cowan, R.A., 2006, *Australian Marine Algal Name Index* (AMANI), Australian Biological Resources Study, Canberra, and Murdoch University, Perth, at www.anbg.gov.au.
Covent Garden Life, undated, 'Covent Garden: A brief history', at www.coventgardenlife.com.
Crank, Cindy, 2011, 'Transport and carriages in the Victorian era (1837–1901)', *Horses and History Throughout the Ages*, at www.horsesandhistory.wordpress.com.
CSIRO Publishing, 2001, 'Herbarium lends landcarers the vision of Hume', *ECOS: Science for sustainability*, 108, at www.ecosmagazine.com.
Department of the Attorney General, Court and Tribunal Services, 2009, 'Bunbury Court House', at www.courts.dotag.wa.gov.au.
Department of Environment, Australian Government,
undated, 'Muir–Byenup System', *RAMSAR Wetlands information sheet*, at www.environment.gov.au.
2013, 'Vasse-Wonnerup System', *RAMSAR Wetlands information sheet,* at www.environment.gov.au
undated, *Weeds in Australia,* at www.environment.gov.au.
Department of Environment and Conservation (NSW, 2011, 'South Western Slopes – regional history', at www.environment.nsw.gov.au.
Department for Environment and Heritage, South Australia, 2006, *Mount Remarkable National Park Management Plan*, Adelaide, at www.environment.sa.gov.au.

Department of Infrastructure, Energy and Resources, 2010, *Richmond Bridge Conservation Management Plan*, Department of Infrastructure, Energy and Resources, Hobart, at www.transport.tas.gov.au.

Department of Parks and Wildlife, Western Australia (previously Department of Environment and Conservation), 2009, 'Porongurup National Park', information brochure, at www.parks.dpaw.wa.gov.au

Undated, *NatureMap: Mapping Western Australian's Biodiversity,* at www.naturemap.dpaw.wa.gov.au.

Department of Primary Industries, Parks, Water and Environment (Tasmania), 2010, 'Derwent Estuary Habitats', *State of the Derwent Estuary*, Hobart, at www.derwentestuary.org.au.

2008, 'Eaglehawk Neck History' at www.parks.tas.gov.au.

2017, *Natural Values Atlas*, at www.naturalvaluesatlas.tas.gov.au.

Duncan, Adrian, undated, 'The 1862 London Exhibition'. at www.mcgee-flutes.com.

Ducker, Sophie C. 1972, 'Harvey, William Henry (1811–1866)', ADBO, at www.adb.anu.edu.au.

Elrick, C., A. Travers, T. Bond & R. Kay, 2008, *Coastal Zone Climate Change Risk Assessment and Adaptation Action Plan: Issues Paper, Report*, prepared for the City of Mandurah, Coastal Zone Management Pty Ltd, Perth, at www.mandurah.wa.gov.au.

Endersby, Jim, undated, 'Joseph Hooker's place in history', at www.kew.org.

Erickson, Rica, **1966**, 'Drummond, James (c.1787–1863)', in *Australian Dictionary of Biography Online*, Australian National University, Canberra, at www.adb.anu.edu.au.

Forestry Tasmania, 1999, *Derwent Forest District Management Plan*, Forestry Tasmania, Hobart, at www.forestrytas.com.au.

Friends of the Cape to Cape Track, undated, 'Flora', at www.capetocapetrack.com.au.

Geraldton Port Authority, undated, 'Geraldton Port–history and handy facts', at www.midwestports.com.au.

Greening Australia & Alcoa, undated, 'Hutt River: River in profile', *Rivers of Western Australia*, at www.rowa.org.au.

Grieve, M., 1995, *A Modern Herbal* at www.botanical.com.

Gross, Alan, 1972, 'Dallachy, John (1808–1871)', in *Australian Dictionary of Biography Online*, Australian National University, Canberra, at www.adb.anu.edu.au.

Hairston, Creasie Finney, 2002, 'Fathers in prisons: responsible fatherhood and responsible public policies' in Eileen Trzcinski (ed), 2002, *What About Me? Children with Incarcerated Parents*, Michigan Family Impact Seminars, Briefing Report no, 2002–1, at www.hdfs.msu.edu.

Hassell C.W., 1972, 'Hassell, John Frederick Tasman (1839–1919)', in *Australian Dictionary of Biography Online*, Australian National University, Canberra at www.adb.anu.edu.au.

Heath, Jim, 1989, *The Fly in Your Eye*, Viacorp, at www.viacorp.com. This book was originally published by J&E Publishing, Perth.

Heberle, R.L., undated, 'John Septimus Roe 1797–1878 Surveyor Explorer and Botanical Collector, 1786-1867', *History of Orchid Collecting in South Western Australia, 1791–1971*, The Species Orchid Society of Western Australia Inc., at www.members.iinet.net.au.

Henderson, Noel, Henderson Family Tree, at www.trees.ancestry.netscape.com.

Heritage Council of Western Australia

1989, Lynton Convict Hiring Depot (1853–1856), Register of Heritage Places, Place No. 01915, p. 4, at www.register.heritage.wa.gov.au.

1999, *Mount Barker – Settlement and Development of the Mount Barker District*, Heritage Trail pamphlet, H88/021, at www.plantagenet.wa.gov.au.

2009, 'The Cloisters', Place 02119, Assessment Documentation, State Heritage Register, at www.inherit.stateheritage.wa.gov.au.

2012, 'Cape Riche Homestead', Place no. 81, State Heritage Register, at www.register.heritage.wa.gov.au.

Hill, K.D., 2004, 'The Eucalypts***Icons of the Australian Bush. Eucalink: A Web Guide to the Eucalypts', at www.plantnet.rbgsyd.nsw.gov.au.

Hoddinott, W.G., 1967, 'Milligan, Joseph (1807–1884)' in *Australian Dictionary of Biography Online*, Australian National University, Canberra, at www.adb.anu.edu.au.

Hone, J. Anne, 1974, 'Morton, William Lockhart (1820–1898)', in *Australian Dictionary of Biography Online*, Australian National University, Canberra at www.adb.anu.edu.au.

Horner, David, 2003, *Index of River Condition for the Coal River Catchment*, Department of Primary Industries, Environment and Water, Tasmania at www.dpiw.tas.gov.au.

Hopper, Stephen D., April 2015, 'Royal Botanic Gardens Kew', *ELs: Essential Life for Science*, John Wiley & Sons, Chichester, at http://www.els.net.

Humanities Research Institute, University of Sheffield 2012, 'Hospitals', *London Lives, 1690–1800* at www.londonlives.org.

Jangoux, Michel, ? date, 'The Herbarium of the Terres Australes (Baudin Expedition, 1800-1804)', Marine Biology Laboratories, Free University of Brussels (ULB) & University of Mons, at http://sydney.edu.au/research/baudin.

Jones, Frances, 2007, 'Life on the goldfields: living there', *Primefacts*, New South Wales Department of Primary Industries, no. 575, February at www.dpi.nsw.gov.au.Kearey, Terence undated *Cultural Shifts in England*, Chapter IV, Accessed 25 July 2017 at www. en.wikibooks.org.

Kedrosky, Paul 2008, 'Get your banking crises correct, 1872/73 vs 1929' at www.paul.kedrosky.com.

Kingborough Council 2011, 'Early Settlement: Kingston & Kingborough Area', at www.kingborough.tas.gov.au.

Koori Coast, Living Knowledge, undated, 'Saltwater people of the coast' at www.livingknowledge.anu.edu.au.

Lambert, Tim, undated,'19th Century London' at www.localhistories.org.

Landgate (Western Australian Land Information Authority), undated, 'Busselton', 'Williams', *History of Country Town Names*, at www.landgate.wa.gov.au.

Landow, George P., & Skipper, Jim, undated, 'Wages, the Cost of Living, Contemporary Equivalents to Victorian Money', *Wages, the Cost of Living, Contemporary Equivalents to Victorian Money* at www.victorianweb.org.

Larkin, Fred, 2001, 'Survey cairns, James Sprent and the Trigonometrical Survey of Tasmania 1833–1855', *Wildcare Incorporated*, Wildcare News Library at www.wildcaretas.org.au. This article appeared in *Wildtimes* No. 11 on 11th September 2001.

Lawley, Mark, undated, 'Henry Boswell (1834/5–1897)', in *A Social and Biographical History of British and Irish Field-bryologists,* at www.rbg-web2.rbge.org.uk.

Lheureux, Louis-Ernest, 2006, 'Monument to the Glory of the French Revolution', at www.musee-orsay.fr.

Lenoir, Timothy, 1981, 'The Gottingen School and the Development of Transcendental *Naturphilosophie* in the Romantic Era', 1981, *Studies in the History Biology*, vol. 5, 1981, at www.jhfc.duke.edu.

Mackean, D.G., undated, 'The Desert Locust', *Biology Teaching Resources*, at www.biology-resources.com.

Maiden, J. H., 1909, 'Maxwell, George (1804–1880)', viewed 9 May 2008 at *Australian Dictionary of Biography Online*, National Centre of Biography, Australian National University at www.adb.anu.edu.au.

Mann, Horace, undated, 'Notes on Endogenae [Class II: Monocotyledonous or Endogenous Plants], presumably for flora of Hawaii', *Horace Mann Papers*, Library of the Gray Herbarium at www.huh.harvard.edu.

Mastin, Luke, 2009, 'Quantum theory and the uncertainty principle: early development in atomic theory', in *The Physics of the Universe* at www.physicsoftheuniverse.com.

Mayr, Ernst, 1996, 'What is a Species, and What is Not?' originally published in *Philosophy of Science*, vol. 63, pp. 262–277, at www.darwiniana.org.

Mockrin, Miranda, 2001, 'Swan Coastal Plain Scrub and Woodlands (AA1205)', *World Wildlife Fund Full Report,* at www.worldwildlife.org.

Sue Moore (Coordinator), Matt Cavana, Kelly Gillen, Chris Hart, Steve Hopper, Kate Orr, Wayne Schmidt, 1991, *Fitzgerald River National Park Management Plan 1991–2001*, Department of Conservation and Land Management (Department of Parks and Wildlife) for the National Parks and Nature Conservation Authority, Perth, Western Australia, p. iii, at www.mainroads.wa.gov.au.

Morris, Diedre 1874, 'Mueller, Sir Ferdinand Jakob Heinrich von (1825–1896)', *Australian Dictionary of Biography Online*, National Centre of Biography, Australian National University at www.adb.anu.edu.au.
Mueller, F. 1880, letter to the *Gardeners Chronicle,* 3rd April 1880, p. 433, at www.biodiversitylibrary.org,
National Trust of Australia, 'Deeside Homestead Group', *National Trust Register of the Built Heritage of Western Australia,* viewed at www.traces.duit.uwa.edu.au.
O'Donoghue, David James, undated, 'Swifte, Theophilus (1746–1815)', *Dictionary of National Biography, 1885–1900*, at www.wikisource.org.
Oldfield, F.H.H., undated, 'Abraham's Daughter', M. Gray, San Francisco, Californian Sheet Music Project, at www.loc.gov.
Orchard, A.E., 1999, 'Beckler, Hermann (1828–1914)', 'Cunningham, Allan (1791–1839)', 'Collie, Alexander (1793–1835)', 'Drummond, James (1786 - 1863)', 'Fraser, Charles (sometimes 'Frazer') (1788–1831)', 'Gilbert, John (~1810–1845)', 'Labillardière, Jacques J.H. de (1755 - 1834)', 'Milligan, Joseph (1807–1884)', 'Oldfield, Augustus F. (1821—1887)', 'Roe, John S. (1797–1878)', 'Preiss, J.A. Ludwig (1811–1883)', extracted from *A History of Systematic Botany in Australia*, in *Flora of Australia* vol.1, 2nd ed., Australian Biological Resources Studies, Canberra, 1999, sourced *Biography: Biographical notes on plant collectors and illustrators*, Australian National Herbarium, Council of Heads of Australasian Herbaria at www.anbg.gov.au.
Oxford Dictionaries, undated, 'bulbil' at www.oxforddictionaries.com.
Pecharsky, Wendie, undated, 'The history of teaching the blind', *EHow,* at www.ehow.com.
Penrith City Council, 2004, 'Education in NSW in the 19th Century', at www.penrithcity.nsw.gov.au.
Rohan, Peter 1979,, *Gnowangerup District Heritage Sketchbook*, Westcolor Press, Perth.
Russell, K.F., 1972, 'Halford, George Britton (1824–1910)', *Australian Dictionary of Biography* Australian National University, Canberra, at www.adb.online.anu.edu.au.
Sightsavers, undated, 'Causes of blindness' at www.sightsavers.org.
Smith, Russell, 2012, A Flora and Vegetation Survey for the Proposed Busselton-Flinders Bay Rail Trail, ekologica, prepared for ngh environmental, Bunbury at www.environment.gov.au.
Söderström, Lars, & Prestø, Tommy (eds), 1995, 'State of Nordic bryology today and tomorrow: Abstracts and shorter communications from a meeting in Trondheim December 1995', *Rapport botanisk serie 1996–4*, Norges teknisk-naturvitenskapelige universitet Vitenskapsmuseet, p. 12, at www.ntnu.no.
South Coast Rivercare, undated, 'Young-Lort Catchment', South Coast Rivercare at www.rivercare.southcoastwa.org.au.
St James Old Cathedral, 2011, 'History', at www.sjoc.org.au.
Stilwell, G.T., **1969,** 'Archer, William (1820–1874)' in *Australian Dictionary of Biography Online*, Australian National University, Canberra at www.adb.anu.edu.au.
Stockton, Richard, undated, College of New Jersey, 'Art and Architecture of New Jersey', at www.ettc.net.
Success in MICROphth, 2004, 'Glaucoma' in *A Historical Tour of Ophthalmology* at www.mrcophth.com.
Territorial Enterprise, Nevada, 20 September 1866, viewed 9 November 2005, at www.ancestry.com.
Tindale, Norman B., 1974, *Aboriginal Tribes of Australia: Their Terrain, Environmental controls, distribution, limits, and proper names with an Appendix on Tasmanian Tribes by Rhys Jones*, Australian National University Press, Canberra, online reproduction of map of tribal boundaries published in this work at www.archives.samuseum.sa.gov.au.
Tipping, Marjorie J., 1969, 'Becker, Ludwig (c.1808–1861)', ADBO, at www.adb.online.anu.edu.au.
University of Michigan Kellogg Eye Centre, undated, 'Glaucoma' at www.kellogg.umich.edu.
University of Westminster, 'History of the University of Westminster' at www.westminster.ac.uk.
Uren, Malcolm, **1967**, 'Roe, John Septimus (1797–1878)' in *Australian Dictionary of Biography Online*, Australian National University, Canberra at www.adb.anu.edu.au.
Walsh, G. P., 1966, 'Benjamin Boyd (1801–1851)', *Australian Dictionary of Biography Online*, Australian National University, Canberra at www.adb.anu.edu.au.

Waters & Rivers Commission, 1999, 'Albany Waterways Resource Book: Waterway Life' at www.rivercare.southcoastwa.org.au.

2001, *Bremer River and Devils Creek, River Action Plan*, Water and Rivers Commission, Water Resource Management Series Report no. WRM 25 June 2001, at www.water.wa.gov.au.

Wellstead Heritage Museum, undated, Bremer Bay, 'John Wellstead', at www.wellsteadmuseum.com.au.

Willis, J.H., & Cohn, Helen M., 1993, 'Morton, William Lockhart (1820–1898)', *Flora of Victoria*, Vol. 1, Chap 5, 'Botanical Exploration of Victoria', sourced from ANBG, *Biography: Biographical notes on plant collectors and illustrators*, Australian National Herbarium, Council of Heads of Australasian Herbaria, at www.anbg.gov.au.

Wilson, Gwendoline, 1967, 'Wilson, Thomas Braidwood (1792–1843)', *Australian Dictionary of Biography Online*, National Centre of Biography, Australian National University at www.adb.anu.edu.au.

Wong, D., & Blake, G.M., 2009, *Salinity Situation Statement: Tweed River & Gnowergerup Brook*, Water Resources Technical Series, Report no. WRT41, Department of Water, Western Australian Government, October 2009, p. 4, at www.water.wa.gov.au.

Wright, Lois E., & Seymour, Cynthia B., 2002, 'Effects of parental incarceration on children and families' in Eileen Trzcinski (ed) 2002, *What About Me? Children with Incarcerated Parents*, Michigan Family Impact Seminars, Briefing Report no, 2002–1, at www.hdfs.msu.edu.

Wright, R.P., 1974, 'Monger, John Henry (1831–1892)', *Australian Dictionary of Biography Online*, National Centre of Biography, Australian National University, at www.adb.anu.edu.au.

Salzman, L.F. (ed), 1939, 'Schools', *A history of the County of Oxford*, vol. 1, Victoria County History (sponsor), pp. 457–490, at www.british-history.ac.uk.

Books and articles (not online)

Aitken, Richard P., 2006, *Botanical Riches: Stores of Botanical Exploration*, Melbourne University Press at The Miegunyah Press, Carlton with State Library of Victoria.

Alexander, Alison (ed), 2005a, *The Companion to Tasmanian History*, Centre for Tasmanian Historical Studies, University of Tasmania, Hobart.

2005b, 'Hutchins School', in Alison Alexander (ed), 2005a, *The Companion to Tasmanian History*, Centre for Tasmanian Historical Studies, University of Tasmania, Hobart.

2013, *The Ambitions of Jane Franklin, Victorian Lady Adventurer*, Allen & Unwin, Crows Nest, New South Wales.

Allan, Mea, 1972, 'Hooker, William Jackson', in C.C. Gillispie (ed), 1970–1980, *Dictionary of Scientific Biography*, vol. 6, Charles Scribner's Sons, New York, pp. 492–495.

Allen, David Elliston, 1978, *The Naturalist in Britain: A Social History*, Penguin Books, Harmondsworth, England.

Anon., 1952, *Dictionary of Phrase & Fable*, revised edition, Cassell, London.

Anon.? Edwin Oldfield, 'Augustus Oldfield obituary', *The Argus*, 13 July 1887.

Anon. undated, *St James's Church Piccadilly Guide*, St James's Church, Westminster, pamphlet.

Ashton, John, 1898, *The History of Gambling in England,* Duckworth & Co., London, microform, Reid Library, University of Western Australia.

Austin, Robert, 1855, Accompanying Map to *Journal of Assistant-Surveyor R. Austin: Commanding an Expedition Sent by the Government to Explore the Interior of Western Australia, Forth and East of the Settled Districts for Extensive Tracts of Fertile Land Available for Pastoral and Agricultural Purposes...*, Government Printer, Perth.

Baily, Francis, 1812, *Tables for the Purchasing and Renewing of Leases, for Terms of Years Certain and for Lives with Rules for Determining the Value of the Reversion of Estates After Any Such Leases; and for the Solution of Other Useful Problems: Adapted for General Use*., 3rd edn., Royal Exchange, London.

Bain, Sister Mary Albertus, 1975, *Ancient Landmarks: A Social and Economic History of the Victoria District of Western Australia 1839–1894*, University of Western Australia Press,Nedlands, Western Australia.

1996, *A Life of Its Own: A social and economic history of the City of Geraldton and the Shire of Greenough, 1846–1988*, City of Geraldton, Western Australia..

Barker, Anthony J. & Maxine Laurie, 1992, *Excellent Connections: A History of Bunbury, Western Australia 1836–1990*, City of Bunbury, Western Australia.

Barker, Collet, 1830, 'Journal of exploration of the Wilson Inlet area [February 1830]' and 'Log of an expedition west of King George's Sound, February 1830', cited in Joanne Shoobert (ed), 2005, vol. 1, pp. 140–152.

Barker, F. & P. Jackson, 1974, *London: 2000 Years of a City and its People*, Cassell, London, England.

Barker, R.M. & W.R. Barker, 1990, 'Botanical contributions overlooked: the role and recognition of collectors, horticulturists, explorers and others in the early documentation of the Australian flora', in P.S. Short, 1990, *History of Systematic Botany in Australasia: Proceedings of a Symposium Held at the University of Melbourne, 25–27 May 1988*, Australian Systematic Botany Society Inc., South Yarra, Victoria, pp. 37–85.

Barnard, H.C., 1955, *A Short History of English Education from 1760 to 1944*, University of London Press, London, England.

Barrett, Sarah, Sarah Comer, Nathan McQuoid, Meghan Porter, Cameron Tiller & Deon Utber, 2009, 'South coast ecodistricts and their relationship to fire' in *Identification and Conservation of Fire Sensitive Ecosystems and Species of the South Coast Natural Resource Management Region*, South Coast Resource Management, Department of Environment and Conservation (now Department of Parks and Wildlife), Perth, Western Australia, pp. 11-16.

Battye, J.S., 1924, *Western Australia: A History From Its Discovery to the Inauguration of the Commonwealth*, Clarendon Press, Oxford, England.

Beard, J.S., 1990, *Plant Life of Western Australia*, Kangaroo Press, Kenthurst, New South Wales.

Bell, Diane, 1993, *Daughters of the Dreaming*, 2nd edn, University of Minnesota Press, Minneapolis, USA.

Bentham, George, assisted by Ferdinand Mueller, 1863–1878, *Flora Australiensis: A Description of the Plants of the Australian Territory*, 7 vols, Lovell Reeve & Co., London, 1870, reprint, 1967, A. Asher & Co, Amsterdam & L. Reeve & Co., Brook, near Ashford, England.

Berghe, Pierre van den, 1985, 'Social Darwinism', in Adam & Jessica Kuper (eds), 1985, *The Social Science Encyclopaedia*, 2nd edn., Routledge, London, England.

Bernal, J.D., 1953, *Science and Industry in the Nineteenth Century*, Indiana University Press, London, England.

Bernard, H. Russell, 2001, *Research Methods in Anthropology: Qualitative and Quantitative Methods*, 3rd edition, Altamira Press, Oxford, England.

Berndt, Ronald M. & Catherine H., 1965, *The World of the First Australians*, Ure Smith, Sydney.

Bignell, Merle, 1971, *First the Spring: A History of the Shire of Kojonup*, University of Western Australia Press, Nedlands, Western Australia.

Bird, John, 1979, *The Last Coach*, John Bird, Perth, Western Australia.

Blackshaw, Tim, 2003, 'Fathers and childbirth' in Caroline Squire (ed), *The Social Context of Birth*, Radcliffe Publishing, Oxford, England.

Blackwood Environment Society, c.2000, *On the Blackwood: A Guide to the Blackwood River: Bridgetown to Nannup*, Blackwood Environment Society, Bridgetown, Western Australia.

Blainey, Geoffrey, 1991, *Odd Fellows: A History of IOOF Australia*, Allen & Unwin, Sydney, New South Wales. 2003, *Black Kettle and Full Moon*, Penguin Books Australia, Camberwell, Victoria.

Bland, Rivett Henry, 1842, 'R.H. Bland's trip from Albany to the Vasse River, January 1842' in Hercock, Marion, Sheryl Milentis & Phil Bianchi (eds), 2011, *Western Australian Exploration 1836–1845: The Letters, Reports and Journals of Exploration and Discovery in Western Australia*, vol. 2, Hesperian Press, Victoria Park, Western Australia, pp. 294–297.

Blevins, Juliette, 1998, 'A Dutch influence on Nhanda? Wanyjidaga innga!', *Australian Aboriginal Studies*, no. 1, pp. 43-46.

2001, '*Nhanda*: an Aboriginal language of Western Australia', *Oceanic Linguistics Special Publication*, no. 30, University of Hawai'i Press, Hawai'i.

Bolger, Peter, 1973, *Hobart Town*, Australian National University Press, Canberra, Australian Capital Territory.

Bonwick, James, 1870, *The Last of the Tasmanians: or The Black War of Van Diemen's Land*, facsimile edition, 1969, Sampson Low, Son & Marston, London,England.

Boyce, Peter, 2005, 'Britishness' in Alexander, Alison (ed.) 2005, *The Companion to Tasmanian History*, Centre for Tasmanian Historical Studies, University of Tasmania, Hobart, Tasmania, pp. 401–407.

Boyd, William & Edmund J. King, 1975, *The History of Western Education*, A. & C. Black, London, England.

Brandenstein, C.G. von, 1988, *Nyungar Anew: Phonology, text Samples and Etymological and Historical 1500-word vocabulary of an artifially recreated Aboriginal Language in the South-west of Australia*, Department of Linguistics, Research School of Pacific Studies, Australian National University, Canberra,

Brash, J.I. (ed), 1974, *Papers on Scottish Electoral Politics 1832–1854*, Scottish History Society, Edinburgh, Scotland.

Brearley, Anne, 2005, *Ernest Hodgkin's Swanland: Estuaries and Coastal Lagoons of South-western Australia,* University of Western Australia, Nedlands, Western Australia.

Breen, Shayne, 2005 'Class', in Alison Alexander (ed.) 2005a, *The Companion to Tasmanian History*, Centre for Tasmanian Historical Studies, University of Tasmania, Hobart, pp. 408–414.

Brewer's Dictionary of Phrase & Fable, 1952, revised edition, Cassell, London, England..

Bridge, Peter J., Kim Epton, Marion Hercock & Sheryl Milentis (eds), 2009, *'The Finest Goldfields in the World': The Austin Expedition to the Murchison in 1854*, Hesperian Press and the Western Australian Explorers' Diaries Project, Carlisle, Western Australia.

Brock, W.H. & Roy M. Macleod, 2000, 'The Scientists' Declaration: reflection on science and belief in the wake of *Essays and Reviews*, 1864–5', in Roy M. MacLeod (ed) 2000. *The 'Creed of Science' in Victorian England*, Varium Collected Studies Series, Ashgate, Aldershot, III, England.

Brockman, Joan, 1989, 'Captain Pemberton Walcott', *Early Days, Journal of the Royal Western Australian Historical Society*, vol. 10, part 1, 1989, pp. 13–17.

Brooker, M.I.H. & D.A. Kleinig, 1990, *Field Guide to the Eucalypts: South-western and Southern Australia*, Inkata Press, Melbourne.
& S.D. Hopper, 1991, 'A taxonomic revision of *Eucalpytus wandoo*, *E. redunca*, and allied species (E. series *Levispermae* Maiden – Myrtaceae) in Western Australia, *Nuytsia*, vol. 8 (1), pp. 1-189.

Brown, Robert 1810, *Prodromus Florae Novae Hollandiae et Insulae Van-Diemen, Exhibens Characteres Plantarum quas Annis 1802–1805 per Oras utriusque Insulae Collegit et Descripsit Robertus Brown*;..., vol. 1, J. Johnson & Co., London, England.
1830, *Supplementum Primum Prodromi Florae Novae Hollandiae: Exhibens Proteaceas Novas quas in Australia Legerunt DD. Baxter, Caley, Cunningham, Fraser et Sieber; Quarum e Siccis Exemplaribus Characteres Elaboravit Robertus Brown*, Richard Taylor, London.
1832, 'General view of the botany of the vicinity of Swan River', *Journal of the Royal Geographical Society*, vol. 1, pp. 17–21.

Brown-May, Andrew, 1998, *Melbourne Street Life: The Itinerary of Our Days*, Australian Scholarly/Arcadia and Museum Victoria, Kew, Victoria.

Buchanan, A.M., 1990, 'Ronald Campbell Gunn (1808–1881)' in P.S. Short (ed), 1990, *History of Systematic Botany in Australasia*, Australian Systematic Botany Society Inc., South Yarra, Victoria, pp. 179–192.

Burgh, W.J. de, 1986, *The Old North Road*, Western Australian Experience Series, University of Western Australia Press, Nedlands, Western Australia.

Burn, David, 1973, *A Picture of Van Diemen's Land: A Facsimile of a Work Published in 'The Colonial Magazine' 1840–41*, facsimile, Cat & Fiddle Press, Hobart, Tasmania.

Burns, T.E. 1966, 'Gunn, Ronald Campbell (1808–1881)', *Australian Dictionary of Biography Online*, National Centre of Biography, Australian National University at www.adb.anu.edu.au.

Busselton Historical Society, 1978, *Historic Busselton*, 3rd edition, Busselton Historical Society, Busselton, Western Australia.

Cameron, Mary (ed), 1996, *A Guide to Flowers & Plants of Tasmania*, Launceston Field Naturalists Club, revised edition, New Holland Publishers (Australia), Sydney, New South Wales.

Cannon, Michael, 1975, *Life in the Cities: Australia in the Victorian Age*, vol. 3, Thomas Nelson, Melbourne, Victoria.

Carter, Jennie & Bevan, 2011, *Settlement to City: A History of the Armadale Districts and its People*, City of Armadale Western Australia.

C.E.B. 1941, 'Obituary: Hamilton Stuart Dove', *Emu*, vol. 41, 1941, pp. 95–96.

Chauncy, Philip, 1878, 'Notes and anecdotes of the Aborigines of Australia', in R. Brough Smyth, *The Aborigines of Victoria: With Notes Relating to the Habits of the Natives of Other Parts of Australia and Tasmania*, Appendix A, vol. 2, Government Printer, Melbourne, Victoria.

Chessell, Gwen 2008, *Alexander Collie: Colonial Surgeon, Naturalist and Explorer*, University of Western Australia Press, Nedlands, Western Australia.

Christensen, Per, 1992, *The Karri Forest: Its Conservation Significance and Management*, Department of Conservation and Land Management, Como, Western Australia.

Clarke, G.E., 1946, *Early History of Bunbury*, G.E. Clarke, Bunbury, Western Australia.

Clarke, Philip A., 2008, *Aboriginal Plant Collectors: Botanists and Australian Aboriginal People in the Nineteenth Century*, Rosenberg Publishing, Kenthurst, New South Wales.

Colley, Linda, 1992, *Britons: Forging the Nation, 1707–1837,* Yale University Press, New Haven.

Cowin, H.G., c.1971, *The Williams*, Williams Shire Council, Williams, Western Australia.

Cox, Peter, 2005 'Tamar Valley', in Alison Alexander (ed.), 2005a, *The Companion to Tasmanian History*, Centre for Tasmanian Historical Studies, University of Tasmania, Hobart, p. 351.

C. F. Craig, E.J. Hickman, J. Newell, N. McQuoid, A.M. Rick, E.M Sandiford, 2008. *Vegetation of the Ravensthorpe Range Western Australia: Mt Short to Kundip*, Department of Environment & Conservation W.A.. (Department of Parks and Wildlife), Perth, Western Australia.

Crawford, I.M., 1968, *The Art of the Wandjina: Aboriginal Cave Paintings In Kimberley, Western Australia*, Oxford University Press, Melbourne, Victoria.

Cresswell, Gail J., 1990, *The Light of Leeuwin*, Augusta-Margaret River Shire History Group, Margaret River, Western Australia.

Crowley, F.K., 1960, *Australia's Western Third*, Heinemann, Melbourne, Victoria.

Cruden, R.P., 1831, *An Account of the Origin of the Steam-boats in Spain, Great Britain and America and of Their Introduction and Employment Upon the River Thames Between London and Gravesend to the Present Time*, Effingham Wilson, London, England.

Curr, Edward M., 1886–1887, *The Australian Race: Its Origin, Languages, Customs, Place of Landing in Australia, and the Routes by Which it Spread Itself Over that Continent*, vol. 1, Trubner & Co., London, England.

Currie, Mark J., 1829, 'Report of an excursion on the Main' in Joanne Shoobert (ed) 2005, *Western Australian Exploration,* vol. 1 *December 1826–December 1835: The Letters, Reports & Journals of Exploration and Discovery in Western Australia*, Hesperian Press, Carlisle, Western Australia, pp. 76–79.

Curry, Suzanne, Bruce Maslin & John Maslin, 2002, *Allan Cunningham: Australian Collecting Localities*, Australian Biological Resources Study (Flora), Canberra, Australian Capital Territory.

Daley, C., 1935, 'Charles Stuart, An early Australian botanist', *The Victorian Naturalist*, vol. 52. pp. 106–110, 132–137,154–157.

1924, 'Baron Sir Ferdinand von Mueller, KCMG, MD, FRS, botanist, explorer, and geographer', *The Victorian Historical Magazine,* vol. 10, no. 1, May, 1924, pp. 23–25.

Darwin, Charles 1859, *On the Origin of Species by Means of Natural Selection*, John Murray, London, England. 1872 *The Expression of the Emotions in Man and Animals*, John Murray, London, England.

1888, *Variation of Animals and Plants Under Domestication*, vol. 2, John Murray, London.

Davidoff, Leonore & Catherine Hall 1987, *Family Fortunes: Men and Women of the Middle Class 1780–1850*, Hutchinson, London, England.

Desmond, Ray, 1977, *Dictionary of British and Irish Botanists and Horticulturalists including Plant Collectors and Botanical Artists*, Taylor & Francis, London, England.

1995, *The History of the Royal Botanic Gardens Kew*, The Harvill Press with the Royal Botanic Gardens Kew, London.

Dickens, Charles, 1857 edn., *Little Dorrit*, digitally scanned, Amazon kindle, location 1259.

Digby, Anne & Peter Searby, 1981, *Children, School and Society in Nineteenth-Century England*, Macmillan Press, London, England.

Dowson, John, 2008, *Old Albany: Photographs 1850–1950*, National Trust of Western Australia, Perth,

Drummond, James, 1829, 'Journal of Mr James Drummond on an Expedition on the Main under the direction of M. J. Currie Esq.' in Joanne Shoobert (ed) 2005, *Western Australian Exploration*, vol. 1 *December 1826–December 1835: The Letters, Reports & Journals of Exploration and Discovery in Western Australia*, Hesperian Press, Carlisle, Western Australia, pp. 79–80.

1842, *London Journal of Botany*, vol. 3, 1842, p.266.

Ducker, Sophie C. (ed), 1988, *The Contented Botanist: Letters of W. H. Harvey about Australia and the Pacific*, Melbourne University Press at Miegunyah Press, Melbourne.

1990, 'Early Austrian influence on Australian botany', in Short, P. S. (ed.) 1990, History of Systematic Botany in Australasia, Australian Systematic Botany Society, Melbourne, Victoria, pp. 297–304.

Durkheim, Emile, 1915, *The Elementary Forms of the Religious Life*, G. Allen & Unwin, London.

Duyker, Edward, 2003, *Citizen Labillardière: A Naturalist's Life in Revolution and Exploration (1755–1834)*, Miegunyah Press, Melbourne, Victoria.

Eliot, George, 1994, *Middlemarch*, Wordsworth Classics edition, Wordsworth Editions Limited, Ware, Hertfordshire, England.

Elkin, A. P., 1954, *The Australian Aborigines: How to Understand Them*, Angus and Robertson, Sydney, New South Wales.

Elliott, C., 1992, 'Albert Street, Busselton as it was before 1920', *Busselton Historical Society Newsletter*, October, pp. unnumbered.

Elliston Allen, David, 1978, *The Naturalist in Britain: A Social History*, Penguin Books, Harmondsworth, England.

Emery, Merle, 1998, 'Busselton's earliest shops', *Busselton Historical Society Newsletter*, August 1998, pp. unnumbered.

Endersby, Jim, 2008, *Imperial Nature*: *Joseph Hooker and the Practices of Victorian Science*, The University of Chicago Press, London, England.

Endlicher, S. L., 1838, *Stirpium australasicarum herbarii hügeliania decades tres,* J.P. Sollinger, Wien.

Endlicher, S.L., E. Fenzl, G. Bentham & H.W. Schott *et al*., 1837, *Enumeratio plantarum quas in Novae Hollandiae ora austro-occidentali…*, G. Beck, Wien.

Erickson, Rica, 1969, *The Drummonds of Hawthornden*, Lamb Paterson, Osborne Park, Western Australia.

1971, *The Victoria Plains*, Lamb Paterson Pty Ltd., Osborne Park, Western Australia.

1974, *Old Toodyay and Newcastle*, Toodyay Shire Council, Toodyay, Western Australia.

1978, *The Dempsters*, University of Western Australia Press, Nedlands, Western Australia.

1979, *Dictionary of Western Australians 1829–1914*, vol. 3, *Free 1850-1868*, University of Western Australia Press, Nedlands, Western Australia.

Eriksson, Gunnar, 1970, 'Agardh, Jacob Georg'. Dictionary of Scientific Biography, vol. 1, Charles Scribner's Sons, New York, USA, pp. 70–71.

Ewers, John K., 1971, *The Western Gateway: A History of Fremantle*, University of Western Australia Press, Nedlands, Western Australia.

Felmingham, Bruce, 2005 'Economy', in Alison Alexander (ed.) 2005a, *The Companion to Tasmanian History*, Centre for Tasmanian Historical Studies, University of Tasmania, Hobart, Tasmania, p. 421.

Fernie, Lee & Geoff, 1989, *In Praise of a National Park: The Origins and History of the Walpole-Nornalup National Park*, Lee and Geoff Fernie, Walpole, Western Australia,

Fildes, Valerie, 1988, *Wet Nursing, A History from Antiquity to the Present*, Basil Blackwell, Oxford, England.

Finn, Margot C., 2003, *The Character of Credit: Personal Debt in English Culture, 1740-1914*, Cambridge University Press, Cambridge, England.
Finney, Colin, 1993, *Paradise Revealed: Natural History in Nineteenth-century Australia*, Museum of Victoria, Melbourne, Victoria.
Fircks, Tatiana de, 1992, *History of Costume and Style*, Pitman, Sydney, New South Wales.
Flett, James, 1970, *The History of Gold Discovery in Victoria,* 2nd edn, Hawthorn Press, Melbourne, Victoria.
Fornasiero, Jean, Peter Monteath & John West-Sooby, 2004, *Encountering Terra Australis: The Voyages of Nicolas Baudin and Matthew Flinders*, Wakefield Press, Kent Town, South Australia.
Fowke, F., 1867, *Tables of the Results of a Series of Experiments on the Strength of British Colonial and Other Woods Exhibited at the International Exhibition, 1862; Made at the South Kensington Museum with His Report on Similar Experiments in 1855*, Science and Art Department of the Committee of Council on Education, Her Majesty's Stationery Office, London, England.
Fowler, H.W. & F.G., 1964, *The Concise Oxford Dictionary of Current English*, Clarendon Press, Oxford, England.
Fowler, W.S., c.1962, *The Development of Scientific Method*, Pergamon Press, London, England.
Fraser, C., 1830, 'Remarks on the botany, &c of the banks of the Swan River, Isle of Buache, Baie Geographe, and Cape Naturaliste', *Botanical Miscellany*, vol. 1, pp. 221–236.
Frost, A.C., 1979, *Baylya-Balinga: A History of Balingup, W.A.*, Donnybrook-Balingup Shire Council, Western Australia.
Gardam, Julie, c.1984, *The History of Kingston*, [? J. Gardam, Kingston], Tasmanian Archives and Heritage Office, Tasmania.
c. 1988, *Browns River, a History of Kingston and Blackmans Bay,* Rotary Club of Kingston, Kingston, Tasmania.
Garden, Donald S., 1977, *Albany: A Panorama of the Sound from 1827*, Nelson, Melbourne
1978, *Southern Haven: A History of the Port of Albany, Western Australia*, Albany Port Authority, Albany, Western Australia.
1978, *Southern Haven: A History of the Port of Albany, Western Australia*, Albany Port Authority, Albany, Western Australia.
Gardiner, Juliet & Neil Wenborn, 1995, *The History Today Companion to British History*, Collins & Brown, London, England.
Gascoigne, John, 1994, *Joseph Banks and the English Enlightenment: Useful Knowledge and Polite Culture*, Cambridge University Press, Cambridge, England.
1998, *Science in the Service of Empire: Joseph Banks, the British State and the Uses of Science in the Age of Revolution*, Cambridge University Press, Cambridge, England.
Gathe, Janette & Mollie Edgecombe and Bevan Carter, 2001, *Robert Brown: Botanist with Matthew Flinders,* Bevan Carter, Bassendean, Western Australia,
Geerinck, D., 1974, 'Revision of the Australian Iridaceae', *Bulletin du Jardin Botanique de Belgique*, vol. 44, 1974, pp. 29–60.
George, Alex S., 1999, *William Dampier in New Holland: Australia's First Natural Historian*, Bloomings Books, Hawthorn, Victoria.
2009, *Australian Botanist's Companion*, Four Gables Press, Kardinya, Western Australia, p. 583.
2010, 'Australian type material in the Economic Botany Collection, Royal Botanic Garden Kew', *Muelleria*, vol. 28 (2), pp. 163–171.
Geraldton Local Studies Department, 2001, *Geraldton: 150 Years, 150 Lives*, Geraldton Regional Library, Western Australia.
Gerritsen, Rupert, 1994, *And Their Ghosts May Be Heard*, Fremantle Arts Centre Press, Fremantle, Western Australia.
Gibbs, Martin, 1994, *Historic Watering Holes Survey: The Old North Road—Wanneroo to Walkaway Coastal Stock Route Western Australia*, National Trust of Australia (WA), Perth, Western Australia, mimeo, pp. 17–18. May Connor revised and expanded the original report, March 1988.

1994, *An Archaeological Conservation and Management Study of 19th Century Shore-Based Whaling Stations in Western Australia*, A Report to the National Trust of Australia for the Australian Heritage Commission, Canberra, Australian Capital Territory.

2007, 'Lynton: convicts, landscape and colonisation strategies in midwest Western Australia', *Australasian Historical Archaeology*, vol. 25, p. 62.

Gillispie, Charles Coulston (ed), 1970–1980, 'Hooker, William Jackson', *Dictionary of Scientific Biography*, vol. 6, Charles Scribner's Sons, New York, pp. 492–495.

Gioia, Paul & Stephen D. Hopper, 2017, 'A new phytogeographic map for the Southwest Australian Floristic Region after an exceptional decade of collection and discovery', *Botanical Journal of the Linnean Society*, vol. 184, pp. 1–15.

Goode, Brad & Colin Irvine, 2006, *A Survey of Aboriginal Social Water Requirements for the Southern Blackwood Plateau and the Scott Coastal Plain, Southwest, Western Australia*, report prepared for the Department of Environment, Western Australia, Perth.

Goodrick, Joan, 1977, *Life in Old Van Diemens* [sic] *Land*, Rigby, Sydney.

Gottsche, C.M., 1878–1881 'Musci Hepatici (Adans., Hedw.) Sive Lichenastri (Dill., Wallr.) Australiani' in Ferdinand Mueller, *Fragmenta Phytographiae Australiae*, Supplement iv, vol. 11, Auctoritate Guberni, Coloniae Victoriae. ex officina Joannis Ferres, Melbourne.

& J.B.G. Lindenberg, & Nees von Esenbeck, 1844–1847, C.G *Synopsis Hepaticarum*, Hamburg.

Gould, Stephen Jay, 1977, *Ontogeny and Phylogeny*, The Belknap Press, Harvard University Press, Cambridge, Mass.

Gregory, [Augustus], 1845, 'Gregory's excursion down the Blackwood River'. Extract from private letter of Ass. Surveyor Gregory to Surveyor General, giving account of an excursion down the Blackwood River, May 1845' in Marion Hercock. Sheryl Milentis & Phil Bianchi (eds) 2011, *Western Australian Exploration 1836–1845: The Letters, Reports and Journals of Exploration and Discovery in Western Australia*, vol. 2, Hesperian Press, Victoria Park, Western Australia, p.351.

1848, 'The settlers' expedition to the northward from Perth, under Mr Assistant-Surveyor A. C. Gregory, 1848', in *Charles Gregory and Francis Thomas Gregory 2002 Journals of Australian Explorations 1846-1861*, Hesperian Press, Victoria Park, Western Australia, pp. 13–30.

Gregory, F.T., 1857, 'The Murchison River 1857' in *Journals of Australian Explorations by Augustus Charles Gregory and Francis Thomas Gregory*, Hesperian Press, Victoria Park, Western Australia, 2002.

1858, 'Gascoyne River 1858', in *Augustus Charles Gregory and Francis Thomas Gregory*, 2002, pp. 37–51.

1861, 'North-West Coast 1861', in *Journals of Australian Explorations by Augustus Charles Gregory and Francis Thomas Gregory*, Hesperian Press, Victoria Park, Western Australia, 2002, pp. 52–97.

Grey, George, 1838, 'A brief outline of the recent expedition to the north–west coast of Australia…,' *Journal of the Royal Geographical Society*, vol. 8, 1838, pp. 454–459.

1841, *Journals of Two Expeditions of Discovery in North–West and Western Australia during the Years 1837, 38, and 39*, vols 1 & 2, T. & W. Boone, London, facsimile edition, Hesperian Press, Victoria Park, Western Australia, 1984.

Gribbin, John, 2002, *Science: A History*, Penguin Books, London.

Gronow, R.H., 1889, *The Reminiscences and Recollections of Captain Gronow: Being Anecdotes of the Camp, Court, Clubs, and Society, 1810–1860*, Scribner & Welford, New York, 1969 edn., abridged, Bodley Head, London.

Haines, Robin, 2003, *Life and Death in the Age of Sail: The Passage to Australia*, University of New South Wales Press, Sydney.

Hall, Catherine, 1992, *White, Male and Middle Class: Explorations in Feminism and History*, Polity Press, Oxford.

Hall, Norman, 1978, *Botanists of the Eucalypts*, Commonwealth Scientific & Industrial Research Organisation, Melbourne,

Hallam, Sylvia J., 1979, *Fire and Hearth: A Study of Aboriginal Usage and European Usurpation in South-western Australia*, Australian Institution of Aboriginal Studies, Canberra.

Hammond, J.E., 1932,'The Perth water front', *Early Days*, vol. 2, pt. 11, pp. 16–17.
1933, *Winjan's People: The Story of the South-West Australian Aborigines*, Imperial Printing Co., Perth.
1958, *Western pioneers: The Battle Well Fought*, Hesperian Press, Victoria Park, Western Australia
Harris, R.J., J.D. Majer, C. Williams, A. Buckley & D. Stehlik, 2008, 'Ravensthorpe Range and Overshot Hill – An Overview of Biodiversity Values, Threats and Conservation, ALCOA Foundation's Conservation and Sustainability Fellowship Program, *Sustaining Gondwana*, issue 12, September 2008.
Harris, S. & A. Kitchener (eds), 2005, *From Forest to Fjaeldmark: Descriptions of Tasmania's Vegetation*, Department of Primary Industries, Water and Environment, Tasmania.
Harvey, William Henry, 1846, 'Algae of Tasmania', *Tasmanian Journal Natural Science*, vol. 2, pp. 377–384, and vol. 3, pp. 153–159.
1849, 'Algae of Tasmania', *London Journal of Botany*, vol. 3, pp. 428–454.
1858–1863, *Phycologia Australica*, vols. 1–5, Lovell Reeve, London.
Hasluck, Alexandra, 1955, *Portrait with Background: A Life of Georgiana Molloy*, Appendix E, Oxford University Press, Melbourne.
Hassell. Cleve W., 1972, *The Hassells of Albany*, unpublished ms., Albany History Collection, Albany Public Library, Albany.
Hassell, Ethel, 1975, *My Dusky Friends*, C.W. Hassell, Perth, Western Australia.
Henderson, M. Helen 2005, 'Introduction', to Augustus Oldfield 1865b, 'On the Aborigines of Australia', *Transactions of the Ethnological Society of London*, n.s., vol. 3, 1865, pp. 215-298,facsimile edition, Hesperian Press, Victoria Park, Western Australia, pp. iii–lviii.
Henderson, W. G., 2011, 'George Maxwell: bushman and naturalist', *Early Days*, vol. 13, pt. 5, pp. 626–648.
Henn, Percy U. (ed), 1954, *Wollaston's Albany Journals (1848–1856) being volume 2 of the Journals and Diaries (1841–1856) of Revd. John Ramsden Wollaston*, Paterson Brokensha Pty Ltd., Perth, p. 219.
Hercock, Marion, Sheryl Milentis & Phil Bianchi (eds), 2011, *Western Australian Exploration 1836–1845: The Letters, Reports and Journals of Exploration and Discovery in Western Australia*, vol. 2, Hesperian Press, Victoria Park, Western Australia.
Hewson, Helen, 1999, *Australia: 300 Years of Botanical Illustration*, CSIRO Publishing, Collingwood, Victoria.
Hibbert, C., 1987, *The English: A Social History 1066–1945*, Guild Publishing, London, England.
Higham, Geoffrey, 2005. *Gone West: Explorer's Guide to Places from Western Australia's Past*, Geoproject Solutions Pty Ltd., Winthrop, Western Australia,.
Hitchcock, J. K., 1929, *The History of Fremantle: The Front Gate of Australia 1829–1929*, Fremantle City Council, Fremantle, Western Australia.
Home, R.W., A.M. Lucas, Sara Maroske, D.M. Sinkora and J.H. Voigt (eds), 1998 & 2002, *Regardfully Yours: Selected Correspondence of Ferdinand von Mueller* vols. 1, 1840–1859 & 2, 1860–1875, Peter Lang, Bern, Switzerland.
Hooker, Joseph Dalton, 1847-1860, *The Botany of the Antarctic Voyage of H.M. Discovery Ships Erebus and Terror, in the Years 1839–1843under the Command of Captain Sir James Clark Ross, Kt., R.N., F.R.S. & L.S, etc.*, 1844–60, Pt. 1, *Flora Antarctica*, 2 vols, Pt. 2, 1847, *Flora Novae-Zelandiae*, 1855, 2 vols, Pt. 3, *Flora Tasmaniae*, 2 vols, 1860, Reeve, London, England.
1847b, 'Florae Tasmaniae Spicilegium, or contributions towards the Flora of Van Diemen's Land', *The London Journal of Botany*, vol. 6. Reference also made generally to J.D. Hooker's papers in *The London Journal of Botany*, vols 3, 4 & 5, 1844–1846. relating to the flora of Tasmania.
1853, *Introductory Essay to the Flora of New Zealand,* pamphlet, Lovell Reeve, London.
1864, *The Handbook of the New Zealand Flora: a Systematic Description of the Native Plants of New Zealand and the Chatham, Kermadec's, Lord Auckland's, Campbell's, and Macquarie's* [sic.] *Islands*, Lovell Reeve, London.
1878, *The Student's Flora of the British Islands*, 2nd edition, Macmillan & Co., London.

& William Henry Harvey 1845, 'Algae antarcticae' and 'Algae novae-zelandiae', *London Journal of Botany*, vol. 4, pp. 249–276, 293–298 and 521–551.

& William Henry Harvey 1847a, 'Algae tasmanicae', *London Journal of Botany*, vol. 6, 1847, pp. 397–417.

& William Henry Harvey 1847b, 'Algae' in *The Botany of the Antarctic Voyage of H M Discovery Ships Erebus and Terror in the Years 1839–1843*, vol. 1 *Flora Antarctica,* Pt 2, Reeve Brothers, London, pp. 454–519, pls 165–193.

Hooker, William Jackson, 1818–1820, *Musci exotici: containing figures and descriptions of new or little known foreign mosses and other cryptogamic subjects,* 2 vols, London.

1849, *A Manual of Scientific Enquiry: Prepared for the Use of Her Majesty's Navy and Adapted for Travellers in General, London.*

1854, *A Century of Ferns: Being Figures with Brief Descriptions of One Hundred New or Rare,* or his *Imperfectly Known Species of Ferns, from Various Parts of the World; a Selection from the Author's 'Icones Plantarum'*, W Pamplin, London.

1854a, 'Introductory note', J.S. Roe, 'Report of a journey of discovery into the interior of Western Australia, between 8th September 1848 and 3 February 1849', *Hooker's Journal of Botany and Kew Garden Miscellany*, vol. 6, p. 42.

1856 'Instructions for collecting and preserving and sending home plants and their useful products', in 'Botany', *Admiralty Manual*, 1856, London, pp. 1–8.

Hopper, Stephen D., 2004, 'South-western Australia, Cinderella of the World's Temperate Floristic Regions 2', *Curtis's Botanical Magazine*, vol. 21 (2), pp. 132–180.

2010, '660. *Nuytsia floribunda*. Loranthaceae*', Curtis's Botanical Magazine,* vol. 26, pp. 333–368.

2014, 'Sandplain and Kwongkan: historical spellings, meanings, synonyms, geography and definition', in H. Lambers (ed.), 2014, *Plant Life on the Sandplains in Southwest Australia, a Global Biodiversity Hotspot*, pp. 23-33.

Hordern, Marsden, 1997, *King of the Australian Coast: The Work of Phillip Parker King in the Mermaid and Bathurst 1817–1822*, The Miegunyah Press at Melbourne University Press, Carlton South, Victoria.

Horwood, Richard, 1813, *The A to Z of Regency London*: *Plan of the Cities of London & Westminster with the Borough of Southwark',* reduced reprint, H. Margary 1985 in association with Guildhall Library, London

Howe, G. Melvyn, 1997, *People, Environment, Disease and Death: A Medical Geography of Britain Throughout the Ages*, University of Wales Press, Cardiff.

Hudspeth, Audrey, 2005, 'Shipping and ports', in Alison Alexander (ed.) 2005a, *The Companion to Tasmanian History*, Centre for Tasmanian Historical Studies, University of Tasmania, Hobart., p. 332.

Hughes, Kathryn, 1993, *The Victorian Governess*, The Hambledon Press, London.

Inkster, Ian, 1991, *Science and Technology in History: An Approach to Industrial Development*, MacMillan, London.

Jackson, Gavin, 1996, *The Report of an Aboriginal Heritage Study of the Nanda Area Central, Kalbarri Region, Western Australia*: Prepared for the Nanda Aboriginal Corporation, Barrel Well Community, Anthropos Australis, Perth, Western Australia.

Jennings, Rodger, 1983, *Busselton '...outstation on the Vasse' 1830–1850*, The Shire of Busselton, Busselton. 1999, *Busselton: 'A Place to Remember' 1850–1914*, The Shire of Busselton, Busselton.

Johnson, K.A. and J.B. Kirkpatrick, 1996, 'The vegetation and flora of the Coal River Nature Reserve', *Papers and Proceedings of the Royal Society of Tasmania*, vol. 130, pt. 1, pp. 15–23.

Jussieu, Antoine Laurent de, 1789, *Genera Plantarum, secundum ordines naturales disposita juxta methodum in Horto Regio Parisiensi exaratam*, Paris.

Keighery, Greg & Neil Gibson, 2005. 'The Flinders Expedition in Western Australia: Robert Brown, the plants and their influence on W.A. botany', in Juliet Wege, Alex George, Jan Gathe, Kris Lemson and Kath Napier (eds), 2005, *Matthew Flinders and His Scientific Gentlemen*, Western Australian Museum, Perth.

Kennedy, David Jnr, 1876, *Kennedy's Colonial Travel*, Edinburgh, quoted in Michael Cannon 1975, *Life in the Cities, Australia in the Victorian Age*, vol. 3, Thomas Nelson, Melbourne, Victoria.

King, Judith E., 1964, *Seals of the World*, Trustees of the British Museum of Natural History, (now Museum of Natural History), London.

Knight, David, 1986, *The Age of Science: The Scientific World–view in the Nineteenth Century*, Basil Blackwell, Oxford.

Kostoglou, Parry, 1993,'Historic timber-getting between Cockle Creek and Lune River, Block 1', *Archaeology of the Tasmanian Timber Industry*, Report No. 4, mimeo, Tasmanian Forest Research Council and Forestry Commission, Tasmania.

Kynaston, Edward, 1924, *A Man on the Edge: A Life of Baron Sir Ferdinand von Mueller*, Allen Lane, London.

Labillardière, J.J.H., 1804–1806, *Novae Hollandiae Plantarum Specimen.....*, 2 vols, Paris.

Lamb, M.E., 1997, *The Role of the Father in Child Development*, 3rd edition, Wiley, New York.

Lehmann, J.G.C. (ed), 1844–47, *Plantae Preissianae Sive Enumeratio Plantarum Quas in Australasia Occidentali et Meridionale-Occidentali Annis 1838–1841 Collegit Ludovicus Preiss,* vols 1 & 2, Sumptibus Meissneri, Hamburg, Germany.

Levett, John undated, cited in Heather Gaunt, 2010, Identity and nation in the Australian public library: the development of local and national collections 1850s–1940s, PhD thesis, University of Tasmania, Hobart.

Lindley, John, 1839, 'A sketch of the vegetation of the Swan River', Appendix to the first twenty-three volumes of *Edwards's Botanical Register*, James Ridgway, London.

Locke, John, 1690, *Essay Concerning Human Understanding*, London.

1693, *Some Thoughts Concerning Education*, London.

Louf, Jacques 1987, 'L'enseignement à Guînes et les éducateurs due XVIe siècle à nos jours', *Bulletin Historique et artistique du Calais*, no. 111–112, Sept.–Dec.

Lubbock, John 1st Baron Avebury 1865, *Prehistoric Times as Illustrated by Ancient Remains and the Manners and Customs of Modern Savages*, facsimile, 7th edn, 1913, Anthropological Publications, Oosterhout, Netherlands, 1969.

Lucas, Clive, 1992, 'The Architecture of Van Diemen's Land to 1856', in Gillian Winter (ed), 1992, *Tasmanian Insights: Essays in Honour of Geoffrey Thomas Stilwell*, State Library of Tasmania, Hobart.

Lullfitz, Alison, Joe Dortch, Stephen D. Hopper, Carol Pettersen, Ron (Doc) Reynolds, and David Guilfoyle, 2017, 'Human Niche Construction; Noongar Evidence in Pre-colonial Southwestern Australia', *Conservation and Society*, vol. 52 (2), pp. 201-216.

Magnusson, Magnus (ed), 1990, *Chambers Biographical Dictionary*, 5th edn, Chambers, New York.

Maiden, J.H., 1908, 'Records of Australian botanists—(a) general, (b) New South Wales', *Journal of the Royal Society of New South Wales*, vol. 42, 1908, pp. 60–111.

1909a, 'Records of Tasmanian botanists', *Papers and Proceedings of the Royal Society of Tasmania for the year 1909*, 9–29,

1909b, 'Records of Western Australian botanists and Australian botanists who have dealt with the Flora of Western Australia', *Journal of the Western Australian Natural History Society for 1909*, pp. 5–27.

1909–1933, *Critical Revision of the Genus Eucalyptus*, vols. 1–6, William Applegate Gullick, Sydney, 2002, PDF electronic edition, University of Sydney Library, Sydney, New South Wales.

1911, 'Records of Australian botanists', 1st Supplement, Proceedings of Section D, *Report of the Thirteenth Meeting of the Australasian Association for the Advancement of Science, Sydney*, pp. 224–241.

MacLeod, Roy M., 2000, 'The "Bankruptcy of Science" Debate: the Creed of Science and its critics', in Roy M. MacLeod (ed.) 2000. *The 'Creed of Science' in Victorian England*, Varium Collected Studies Series, Ashgate, Aldershot, III, England.

MacKaness, George, 1947, 'Some Private Correspondence of Sir John and Lady Jane Franklin; Tasmania, 1837–1845', Part 2, *Australian Historical Monographs*, vol. 18, New Series, 1947.

Marcet, Jane, 1816, *Conversations in Political Economy; in which the Elements of that Science are Familiarly Explained*, Longman, Hurst, Rees, Orme, Brown and Green, London, England.

1822, *Conversations in Natural Philosophy; in which the Elements of that Science are Familiarly Explained and Adapted to the Comprehension of Young Pupils*, Longman, Hurst, Rees, Orme, Brown and Green, London, England.

1825, *Conversations in Chemistry; in which the Elements of that Science Are Familiarly Explained and Illustrated by Experiments*, 2 vols., Longman, Hurst, Rees, Orme, Brown and Green, London, England.

Marchant, Leslie, 1982, *France Australe: A Study of French Explorations and Attempts to Found a Penal Colony and Strategic Base in South Western Australia 1503–1826*, Artlook Books, Perth, Western Australia.

Marchant, N.G., 1990, 'The Western Australian collecting localities of J.A.L. Preiss', in Short, P. S. (ed.), *History of Systematic Botany in Australasia*, Australian Systematic Botany Society, Melbourne, pp. 131–135.

Marchant James, Ruth, 1977, *Heritage of Pines: A History of Cottesloe*, The Town of Cottesloe Council, Cottesloe, Western Australia.

2001, 'From byway to highway: the linking of port and capital', *Early Days, Journal of the Royal Historical Society of Western Australia*, vol. 12, pt. 1, 2001, pp. 13-26.

Margetson, Stella, 1971, *Regency London*, Cassell, London, England.

Maroske, Sara, 2006. 'Ferdinand Mueller and the Shape of Nature: Nineteenth-century Systems of Plant Classification', *Historical Records of Australian Science*, 2006, vol., 17, pp. 147–168.

2014, 'A taste for botanic science: Ferdinand Mueller's Female Collectors and the History of Australian Botany', *Muelleria*, vol. 32, 2014, pp. 72–91 & Alison Vaughan, 2014, 'Ferdinand Mueller's female plant collectors: biographical register', *Muelleria*, vol. 32, pp. 92–172.

Marshall, Gordon de L, 1994, *Memories of maritime Albany: Les Douglas et al.*, Albany Residency Museum, Albany,

Marshall, H.S., 1937, 'The dates of publication of Bentham's *Flora Australiensis*', *Journal of the Society for the Bibliography of Natural History*, vol. 1, 1937, pp. 69–71.

Marshall, Tony, 2005,'State Library of Tasmania', in Alison Alexander (ed) 2005a, *The Companion to Tasmanian History*, Centre for Tasmanian Historical Studies, University of Tasmania, Hobart, Tasmania

Maslin, B.R. & A.S. George, 2005, 'James Drummond's *Acacia* collections from south-west Western Australia', *Curtin University of Technology, Department of Environmental Biology, Bulletin* 27, pp. 34–39.

Mason-Cox, Margaret, 2005, 'Water', in Alison Alexander (ed.) 2005a, *The Companion to Tasmanian History*, Centre for Tasmanian Historical Studies, University of Tasmania, Hobart, Tasmania, pp. 380–381.

Maxwell-Stewart, Hamish, 2005, 'Maria Island' in (Alison Alexander (ed.) 2005a, *The Companion to Tasmanian History*, Centre for Tasmanian Historical Studies, University of Tasmania, Hobart, Tasmania, p.227.

McConnell, Margaret, 1993, John McGuire & Carrick Moore, *Plateau, Plain and Coast: A History of Dandaragan,* Shire of Dandaragan, Western Australia.

McLaren, Glen, 1996, *Beyond Leichhardt: Bushcraft and the Exploration of Australia*, Fremantle Arts Centre Press, Fremantle, Western Australia.

McNeill, J., F.R. Barrie, W.R. Buck, V. Demoulin, W. Greuter, D.L. Hawkesworth, P.S. Herendeen, S. Knapp, K. Marhold, J. Prado, W.F. Prud'homme van Reine, G.F. Smith, J.H. Wiersema, J.H. Wiersema & N.J. Turland, 2012, *International Code of Nomenclature for algae, fungi, and plants (Melbourne Code) adopted by the Eighteenth International Botanical Congress Melbourne, Australia, July 2011*, Koeltz Scientific Books, Königstein, Germany.

Melville, Henry, 1851, *The present state of Australasia, including New South Wales, Western Australia, South Australia, Victoria and New Zealand, with practical hints on emigration: also, Remarks on prison discipline: with suggestions for obviating the difficulties attending the transportation of convict; to which are added the land regulations, and description of the aborigines and their habit,* G. Willis, London,

Mercer, Alex, 1990, *Disease Mortality and Population in Transition*, Leicester University Press, Leicester, England.

Moir, Amelia & Greg Wardell-Johnson, 1995, *The Continuing Moir Saga: A Genealogical Record of an Albany Family*, Warjam Publishing, Albany, Western Australia.

Molloy, J., 1830, 'Report of Captain Molloy on neighbourhood of the Blackwood River', in Joanne Shoobert (ed) 2005, *Western Australian Exploration,* vol. 1 *December 1826–December 1835: The Letters, Reports & Journals of Exploration and Discovery in Western Australia*, Hesperian Press, Carlisle, Western Australia, pp. 172–173.

Moore, Joanna, 2010, 'Manjimup and the Muirs', *Landscope*, vol. 25, no. 4, p. 26.

Moorehead, Alan, 1963, *Cooper's Creek*, Hamish Hamilton, London, England.

Morgan, David L. & Howard S. Gill, 2004, 'Fish fauna in inland waters of the Pilbara (Indian Ocean) Drainage Division of Western Australia — evidence for three sub-provinces', *Zootaxa*, 636, pp. 1–43.

Morton, A.G., 1981, *History of Botanical Science: An Account of the Development of Botany from Ancient Times to the Present*, Academic Press, London, England.

Mueller, Ferdinand, 1858–1881, *Fragmenta Phytographiae Australiae*, 11 vols., Auctoritate Guberni, Coloniae Victoriae. ex officina Joannis Ferres, Melbourne.

1863, 'A record of the plants collected by Mr Pemberton Walcott and Mr Maitland Brown, in the year 1861, during Mr F.T. Gregory's Exploring Expedition into North-West Australia in the year 1861', *Transactions of the Botanical Society*, vol. 7, 1863, pp. 479–500.

1871, *The Objects of a Botanic Garden in Relation to Industries*, lecture delivered at the Industrial and Technological Museum, Melbourne, Mason, Firth, McCutcheon, Melbourne, Victoria, 23 November 1871 (quoted in R.T.M. Pescott, 1982, *The Royal Botanic Gardens Melbourne: A History from 1845 to 1970*, Oxford University Press, Melbourne. p. 42).

1879–1884, *Eucalyptographia: A descriptive Atlas of the Eucalypts of Australia and the Adjoining Island*, Decades I–X, J. Ferres, Government Printer, Melbourne.

1880, letter to the *Gardeners Chronicle,* 3rd April 1880.

1882a, *Systematic Census of Australian Plants with Chronologic, Literary and Geographic Annotations, Part 1–Vasculares*, Victorian Government, Melbourne, Victoria.

1882b, *General Information Respecting the Present Condition of the Forests and Timber Trade Together with a Report on the Forest Resources of the Colony,* Government Printer, Perth, Western Australia.

Müller, Carolo [Karl], 1849–51, *Synopsis muscorum frondosorum omnium hucusque cognitorum,* A. Foerstner, Berolini [Berlin], Germany.

Mulvaney, D.J. & Neville Green, 1992, *Commandant of Solitude: the Journals of Captain Collet Barker 1828-1831*, Melbourne University Press at the Miegunyar Press, Carlton, Victoria, p. 258.

Mundy, G.C., 1852, *A Record of Observations in Van Diemen's Land*, reprint, Sullivan's Cove, Adelaide, South Australia, 1986.

Munting, Roger, 1996, *An Economic and Social History of Gambling in Britain and the USA*, Manchester University Press, Manchester, England.

Murray, Keith, 2004, *First There Came Ships: Over 200 Years of Ships, Seamen, Wrecks, Pirates, Sealers, Settlers and Miners Along the South-east Coast of Western Australia*, Keith Murray, Esperance, Western Australia.

Nelson, Charles, 1994, 'Nicolaas Witsen's letter of 1698 to Martin Lister about a Dutch expedition to the South Land (Western Australia): the original text and a review of its significance for the history of Australian natural history', *Archives of Natural History*, vol. 21, (2), pp. 154–160.

Newitt, Lyn, 1988, *Convicts and Carriageways*: *Tasmania Road Development Until 1880*, Department of Main Roads, Tasmania, Hobart, Tasmania.

Newman, Terry, 2005, Tasmania, the name' in Alison Alexander (ed.), *The Companion to Tasmanian History*, Centre forTasmanian Historical Studies, University of Tasmania, Hobart, Tasmania, p. 353.

F.J. Odling-Smee. F.J., 2003, *Niche Construction: the Neglected Process in Evolution*, Princeton University Press, Princeton, N.J.

Oldfield, Augustus, 1865b, 'On the Aborigines of Australia', *Transactions of the Ethnological Society of London*, n.s., vol. 3, pp. 215–298.

Oliphant, John, '"Touching the light": the invention of literacy for the blind', *Pedagogica Historica*, vol. 44, nos. 1–2.

Oliphant, G.H.H., 1847, *The Law Concerning Horses, Racing Wagers and Gaming, with an Appendix Containing Recent Cases, Statutes, &c.,* T. & J.W. Johnson, Philadelphia, USA.

Painter, Gwenda, 1987, *The River Trade: Wool and Steamers*, Turton & Armstrong with The Pioneer Settlement Press, Swan Hill, Victoria.

Parker to the Editor, Edward R., 1851, *Perth Gazette and Independent Journal of Politics and News*, 31 January 1851, p. 3, DAN, NLA.

Pashley, A.R., 2000, *Policing Our State: A History of Police Stations and Police Officers in Western Australia 1829–1945*, EDUCANT, Cloverdale, Western Australia.

2002, *A Colonial Pioneer, The Life & Times of John Nicol Drummond, Policeman, Magistrate, Pastoralist*, EDUCANT, Cloverdale, Western Australia.

Pearce, Kim, 2005, 'Public houses', in Alison Alexander (ed.) 2005a, *The Companion to Tasmanian History*, Centre for Tasmanian Historical Studies, University of Tasmania, Hobart, pp. 57–58.

Pearn, John & Kenneth D. Winkel, 2006, 'Toxicology in Australia's colonial era: A chronology and perspective of human envenomation in 19th century Australia', *Toxicon*, vol. 48, pp. 726-737.

Perkin, Harold, 1972, *The Origins of Modern English Society 1780–1880*, Routledge & Kegan Paul, London, England.

Pescott, R.T.M., 1982, *The Royal Botanic Gardens Melbourne: A History from 1845 to 1970*, Oxford University Press, Melbourne, Victoria.

Phelan, Patrick, 1984, 'Are we producing the goods?', *The British Journal of Visual Impairment*, vol. 11, pt. 3, pp. 70–73.

Phillips, Derek, 1985, *Making More Adequate Provision: State Education in Tasmania 1839–1985*, Education Department of Tasmania, Hobart, Tasmania.

Pitt Morison, Margaret & John White, 'Builders and Buildings', in C.T. Stannage (ed) 1979, *A New History of Western Australia*, University of Western Australia Press, Nedlands, Western Australia.

Playford, Phillip, 1996, *Carpet of Silver: The Wreck of the Zuytdorp*, University of Western Australia Press, Nedlands, Western Australia.

Poole, Watkin (ed), 1826, *The Female's Best Friend; or, The Young Woman's Guide to Virtue, Economy, and Happiness: Containing A Complete Modern System of Cookery Formed upon Principles of Economy for Private Families*, New Edition, J. Cleave & Son, Manchester, England.

Pope, Dudley, 1981, *Life in Nelson's Navy*, Chatham Publishing, London, England.

Porter, Roy, 1994, *London: A Social History*, Penguin Books, London, England.

Porter, Roy, 2000, *Enlightenment: Britain and the Creation of the Modern World*, Allen Lane, London, England.

Poulson, Bruce, 'The far south', in Alison Alexander (ed) 2005a, *The Companion to Tasmanian History*, Centre for Tasmanian Historical Studies, University of Tasmania, Hobart, Tasmania, p. 129.

Pugin, A.C., W.H. Leeds & J. Britton 1838, *Illustrations of the Public buildings of London with Historical and Descriptive Accounts of Each Edifice,* 2nd edn, London, England.

Purvis, June, 1989, *Hard Lessons: the Lives and Education of Working-class Women in nineteenth-century England*, Polity, Cambridge, England.

Quennell, Marjorie and C.H.B., 1938, *A History of Everyday Things in England. Done in Four Parts of Which This is the Third. The Rise of Industrialism, 1733–1851*, B.T. Batsford Ltd., London, England.

Quin, ? Robert, cited in Gail J. Cresswell, 1990, *The Light of Leeuwin*, Augusta-Margaret River Shire History Group, Margaret River, Western Australia.

Radcliffe–Brown, A.R., 1931, 'The social organization of Australian tribes', *The Oceania Monographs*, no. 1, The Australian National Research Council, Sydney, Ne3w South Wales.

Rasmussen, S.E. (ed), 1934, *London: The Unique City,* Faber & Faber, London, 1982 edition, MIT Press, Massachusetts, USA.

Reece, R. & R. Pascoe, 1983, *A Place of Consequence: a Pictorial History of Fremantle*, Fremantle Arts Centre Press, Fremantle, Western Australia.

Rich, R.W., 1933, *The Training of Teachers in England and Wales During the Nineteenth Century*, Cambridge at the University Press, England.
Richards, Ronald, 1978, *The Murray District of Western Australia: A History*, Shire of Murray, Pinjarra, Western Australia.
Ride, W.D.L., 1970, *A Guide to the Native Mammals of Australia*, Oxford University Press, Melbourne, Victoria.
Rimon, Wendy, 2005, 'The Depression of the 1840s', 'Regattas', in Alexander, Alison 2005a, *The Companion to Tasmanian History*, Centre for Tasmanian Historical Studies, University of Tasmania, Hobart, Tasmania, pp. 103, 305.
Rintoul, John, 1964, *Esperance: Yesterday and Today*, Esperance Shire Council, Esperance, Western Australia.
Robson, Lloyd, 1983, *A History of Tasmania*, vol. 1, Oxford University Press, Melbourne.(updated by Michael Roe), 1997, *A Short History of Tasmania*, Oxford University Press, Melbourne, Victoria.
Roe, J.S., 1835, 'Journal of an expedition to Moorilup, the Hay and the Sleeman Rivers, & etc. February 1835', cited in Joanne Shoobert (ed), 2005, vol. 1, pp. 410–418.
1854, 'Report of a journey of discovery into the interior of Western Australia, between 8 September, 1848 and 3 February, 1849', *Hooker's Journal of Botany and Kew Garden Miscellany*, vols. 6 & 7, pp. 42–48, 146.
Rogers, L G., & National Landcare Program (Australia),1996, *Geraldton region land resources survey*. Report 13, Department of Agriculture and Food, Western Australia, Perth.
Rohan, Peter, 1979, *Gnowangerup District Heritage Sketchbook*, Westcolor Press, Perth.
Royal Society of Tasmania, *Papers and Proceedings of the Royal Society of Tasmania 1860–1861. Papers and Proceedings of the Royal Society of Van Diemen's Land 1851 –1859.*
Sanders, Theodora, 1975, *Bunbury: Some Early History*, Roebuck Society Publication no. 16, Canberra, Australian Capital Territory.
Sands, Archibald, 1859, 'Description of Perth contained in a letter written by Archibald Sands, Esq., to his son in Melbourne, in 1859', *Early Days*, vol. 2, pt. 13, 1933, pp. 49–50.
Schorer, A., 1968, *History of the Upper Blackwood*, South West Printing and Publishing Co., Bunbury, Western Australia.
Sharr, F.A., 1988, *Western Australian Plant Names and their Meanings: a Glossary*, 2nd edn, University of Western Australia Press, Perth, Western Australia, p. 171.
Sheppard, Francis, 1998, *London: A History*, Oxford University Press, Oxford, England.
Sheppard, F.H.W. (ed), 1960, 'Bury Street', *Survey of London*: *The Parish of St James Westminster,* vols. 29–30, for the London City Council, The Athlone Press, University of London, London, England.
Shoobert, Joanne (ed), 2005, *Western Australian Exploration,* vol. 1 *December 1826–December 1835: The Letters, Reports & Journals of Exploration and Discovery in Western Australia*, Hesperian Press, Carlisle, Western Australia.
Short, P. S. (ed), 1990, *History of Systematic Botany in Australasia*, Australian Systematic Botany Society, Melbourne, Victoria.
Shteir, A.B., 1996, *Cultivating Women Cultivating Science*, The Johns Hopkins University Press, Baltimore, USA.
Singer, Charles, 1997, *A Short History of Science to the Nineteenth Century*, republication of the 1943 edition, Dover Publications, London, England.
Slater, Peter, Pat & Raoul, 1992, *The Slater Field Guide to Australian Birds*, Weldon, Smithfield, New South Wales.
Smith, G.G., 1982, 'Pioneer seashore botanists in Western Australia', *Early Days*, *Journal of The Royal Western Australian Historical Society*, vol. 8 (6), pp. 86-102.
Smith, P.T., 1985, *Policing Victorian London: Political Policing, Public Order, and the London Metropolitan Police*, Contributions in Criminology and Penology, no. 7, Greenwood Press, Westport, England.
Snowden, Dianne, 'Richmond', in Alison Alexander (ed), 2005, *The Companion to Tasmanian History*, Centre for Tasmanian Historical Studies, University of Tasmania, Hobart, p. 308.

South West ARM Region Appraisal Team 2005, *Boyup Brook-Upper Warren Area Catchment Appraisal*, Resource Management Technical Report 319, Department of Agriculture and Food, Perth, Western Australia.

Spencer, Herbert, 1857, 'Progress: its law and causes', *The Westminster Review*, April 1857.
1862, *First Principles*, Williams & Norgate, London, England.

Sprod, Michael, 2005, 'Education', in Alexander, Alison 2005a, *The Companion to Tasmanian History*, Centre for Tasmanian Historical Studies, University of Tasmania, Hobart, pp. 113–115.

Stafleu (ed), **1981**, F.A. *Index Herbariorum: a Guide to the Location and Contents of the World's Public Herbaria,* Pt 2 (5), Scheltema & Holkema, Bohn, Germany.

Stancombe, G. Hawley, 1969, *Highway in Van Diemen's Land*, abridged edition, G. Hawley Stancombe, Glendessary, Western Junction, Tasmania.

Stannage C.T., 1979, *The People of Perth: A Social History of Western Australia's Capital City*, Perth City Council, Perth, Western Australia.

Stanner, W.E.H. 1965, 'Religion, totemism and symbolism', reprinted from Ronald M. Berndt (ed), *Aboriginal Man in Australia: Essays in Honour of Emeritus Professor A.P. Elkin*, Angus & Robertson, Sydney, source Max Charlesworth, Howard Morphy, Diane Bell and Kenneth Maddock (eds), 1984, *Religion in Aboriginal Australia: An Anthology*, University of Queensland Press, St Lucia, Queensland.

Steinmetz, Andrew, 1870, *The Gaming Table: Its Votaries and Victims: In All Times and Countries, Especially in England and in France*, vol. 1, Tinsley Brothers, London, England. Microform, University of Western Australia.

Stephens, W.B., 1998, *Education in Britain*, St Martin's Press, New York, USA.

Stoney, H. Butler, 1856, *A Residence in Tasmania: With a Descriptive Tour Through the Island from Macquarie Harbour to Circular Head*, Smith, Elder & Co., London, Tasmaniana Facsimile Editions no. 6, 1982Melanie Publications, Hobart, Tasmania.

Stokes, J.L., 1841, 'Expedition of the Beagle to Port Grey', ***The Inquirer,*** 22 December 1841, p. 2, DAN, NLA.

Suba, Tanya, 1993 *The Thematic History of the Town of Northampton*, Shire of Northampton, Northampton, Western Australia.

Swifte, Edmund Lenthal, 1860, 'Ghost in the Tower', *Notes and Queries*, vol. 10, 2 September 1860, p. 192.

Syme, J., 1848, *Nine Years in Van Diemen's Land Comprising an Account of its Discovery, Possession, Settlement, Progress, Population, Value of Land, Herds, Flocks etc.; an Essay on Prison Discipline and the Results of the Working of the Probation System; With Anecdotes of Bushrangers*, J Syme, Dundee, Scotland.

Tasmanian Journal of Natural Science, Agriculture, Statistics & etc., vol. 2, no. 6, 1846.

Theißen, Günter, 2006, 'The proper place of hopeful monsters in evolutionary biology', *Theory in Biosciences,* vol. 124, pp. 349-369.

Thomson, Carolyn, Graham Hall & Gordon Friend (eds), 1993, *Mountains of Mystery: A Natural History of the Stirling Range*, Department of Conservation and Land Management, Como, Western Australia.

Tindale, Norman B., 1974, *Aboriginal Tribes of Australia: Their Terrain, Environmental controls, distribution, limits, and proper names with an Appendix on Tasmanian Tribes by Rhys Jones*, Australian National University Press, Canberra, Australian Capital Territory.

Tosh, John, 1991, 'The family of Edward White Benson', in Michael Roper and John Tosh (eds), *Manful Assertions: Masculinities in Britain since 1800*, Routledge, London, England, pp. 44–73.

Tree, Isabella, 1991, *The Bird Man: The Extraordinary Story of John Gould*, Ebury Press, London, England.

Troughton, Ellis, 1957, *Furred Animals of Australia*, 6th edition, Angus and Robertson, Sydney, New South Wales.

Tuckfield, Trevor, 1975, 'Early colonial taverns and hotel', *Early Days, Journal of the Royal Western Australian Historical Society,* vol. 7, pt. 7, 1975, pp. 98–106.

Turner, Thomas, 1834 'Mr Thomas Turner's Journal of an Expedition to trace the Blackwood River to its source-21st to 29th September, 1834' in Shoobert, Joanne (ed), 2005, *Western Australian Exploration,* vol. 1 *December 1826–December 1835: The Letters, Reports & Journals of Exploration and Discovery in Western Australia*, Hesperian Press, Carlisle, Western Australia, pp. 368–374.

Underwood, Roger, 2011, *A Botanical Journey: The Story of the Western Australian Herbarium*, Department of Environment and Conservation, Perth.

Henn, Percy U. (ed), 1954, *Wollaston's Albany Journals (1848–1856) being volume 2 of the Journals and Diaries (1841–1856) of Revd. John Ramsden Wollaston*, Paterson Brokensha Pty Ltd., Perth.

Valentine, P.A.S. & N.J. Enright, 1975, 'The Cape Naturaliste Area, WA – An Environmental Resource Evaluation', *Geowest, Working papers of the Department of Geography, University of Western Australia*, no. 6, November, p. 20.

Vallance, T.G., D.K.T. Moore & E.W. Groves, 2001, *Nature's Investigator: The Diary of Robert Brown in Australia, 1801–1805*, Australian Biological Resources Study, Canberra, Australian Capital Territory.

Vegter, I. Hettie, 1983, *Index herbariorum: a guide to the location and contents of the world's public herbaria*, Part 2(5), Collectors N–R, International Association for Plant Taxonomy, vol. 109.

Vicinus, Martha, 1985, *Independent Women: Work and Community for Single Women 1850–1920*, Virago Press, London, England.

Vickery, Amanda, 1998, *The Gentleman's Daughter: Women's Lives in Georgian England*, Yale University Press, London, England.

Wake, C. Staniland, 1872, The mental characteristics of primitive man, as exemplified by the Australian Aborigines', *Journal of the Royal Anthropological Institute*, vol. 1, pp.74-84.

Wakefield, Priscilla 1807, *An Introduction to Botany in a Series of Familiar Letters*, Darton & Harvey, London, England.

Walvin, J., 1982, *A Child's World: A Social History of English Childhood 1800–1914*, Penguin Books, Harmondsworth, England.

Wege, Juliet & Alex George, Jan Gathe, Kris Lemson and Kath Napier (eds), 2005, *Matthew Flinders and His Scientific Gentlemen*, Western Australian Museum, Perth.

Weinreb, Ben & Christopher Hibbert (eds), 1983, *The London Encyclopaedia*, 2nd edn, 1993 Macmillan London Ltd., London, England.

Wellings, H.P., 1996, *Eden and Twofold Bay: Discovery, Early History and Points of Interest 1797–1965*, Eden Killer Whale Museum, Eden, New South Wales.

Whittell, E.M., 1949, 'The visits of John Gilbert, naturalist to the Swan River Colony, *Early Days,* vol. 4, pt. 1.

Williams, K.J. & B.M. Potts, 1996, 'The natural distribution of *Eucalyptus* species in Tasmania', *Tasforests,* vol. 8, pp. 39-165.

Wilson, A.N., 2005, *London: A Short History*, paperback edn, Phoenix, London, England.

Wilson, W., 1860, in Joseph Dalton Hooker 1847-1860, *The Botany of the Antarctic Voyage of H.M. Discovery Ships Erebus and Terror, in the Years 1839–1843 under the Command of Captain Sir James Clark Ross, Kt., R.N., F.R.S. & L.S, etc.,* 1844–60, Pt. 3, *Flora Tasmaniae*, vol. 2, 1860, Reeve, London, England.

Winter, Gillian, 1993, 'The foundation years of the Royal Society of Tasmania, 1843–86', *Papers and Proceedings of the Royal Society of Tasman*ia, vol. 127.

Woodward, E.L., 1954, *The Age of Reform 1815–1870*, Oxford at Clarendon Press, London. England.

Woolley, Richie & Wayne Smith, 2004, *A History of the Huon and Far South,* vol. 1*, Before the Orchards Grew*, Huon Valley Council, Tasmania.

Unpublished theses

Brown-May, Andrew, 1993, The itinerary of our days: the historical experience of the street in Melbourne, 1837-1923, PhD thesis, Department of History, The University of Melbourne, Melbourne, Victoria.

Carroll, J., c.1972, The Development of Busselton 1832-1872, Thesis, Claremont Teachers College, J.S. Battye Library of Western Australian History, Perth, Western Australia.

Cross, Robert, 1995, The Social Landscape of the Midwest Aboriginal People (1840–1860) as Reflected Through Ethnohistory, Unpublished B.A. (Hons) thesis, Centre for Archaeology, University of Western Australia, Nedlands, Western Australia.

Gaunt, Heather, 2010, Identity and nation in the Australian public library: the development of local and national collections 1850s – 1940s, using the Tasmanian Public Library as case study, PhD Thesis, University of Tasmania, Hobart, Tasmania.

Higgs, Joan, c.1950, Public Transport by Coach, Pt. 1, Thesis, Claremont Teachers' College, J.S. Battye Library of Western Australian History, Perth, Western Australia.

Jones, Jennifer A., 2010, Old Age in a Young Colony: Image and Experience in South Australia in the Nineteenth Century, PhD Thesis, School of History & Politics, University of Adelaide, Adelaide, South Australia.

Maroske, Sara, 2005, Science by correspondence: Ferdinand Mueller and botany in nineteenth-century Australia. Unpublished Dr. Phil. Thesis, Dept. History & Philosophy of Science: University of Melbourne, Melbourne, Victoria.

O'Connor, M.H., 2001, The Historical Ecology of the Greenough Flats, Western Australia, Unpublished PhD, Curtin University of Technology, Bentley, Western Australia.

Staples, A.C., c.1948, Historical Development of the Harvey District up to 1895, unpublished MA thesis, University of Western Australia, J.S. Battye Library of Western Australian History, Perth, Western Australia.

Appendix 3

Images, including Permissions and Acknowledgments

All images are property and copyright of the authors unless where permissions are indicated.

Title Spreads

Part 1 Bury Street, St James's showing Georgian houses, c1850, Thomas Colman Dibdin.
Reproduced with the kind permission of the London Metropolitan Archives (City of London)
Part 2 Mt Wellington, Tasmania, 2007.
Part 3 The Loop, Murchison River, Western Australia, 2010.
Part 4 King George Sound, Western Australia, 2006.
Part 5 View from Merton Road looking east, showing shops on south side of West Hill, Wandsworth, c. 1880.
Reproduced with the kind permission of the Wandsworth Heritage Service, Wandsworth, England.

Chapter 2

Fig 2.1 Herbarium sheets of Oldfield's specimens.
Fig 2.1a *Eucalyptus oldfieldii* TYPE K000347746, K000347747
Fig 2.1b *Pityrodia (=Quoya) oldfieldii* TYPE KOOO975334
Reproduced with kind permission of the Board of Trustees of the Royal Botanic Gardens, Kew.
Fig 2.1c *Melaleuca eremaea*, HOLOTYPE, MEL602165 (=Melaleuca trichophylla)
Fig 2.1d *Caesia dichotoma*, TYPE, MEL107033 (=*Corynotheca micrantha* var. *micrantha*)
Reproduced with kind permission of the Board of Trustees of the Royal Botanic Gardens Victoria.

Chapter 3

Fig. 3.0 *Lasiopetalum oldfieldii*, Kalbarri National Park, Western Australia, 2000.
Fig. 3.1 Court of King's Bench, 1805, Thomas Rowlandson & Augustus Pugin.
Reproduced with the kind permission of Tony Price, The Lord Price Collection, New Maiden, Surrey.
Fig. 3.2 St Mary Abbots, the old church, c. 1810, Edward Edwards.
Reproduced with the kind permission of the Museum of London.
Fig. 3.3 St James's Church, Piccadilly looking from Jermyn St, 1814, John Coney.
Reproduced with the kind permission of the London Metropolitan Archives (City of London).
Fig. 3.4 28 Bury Street, plan for renovations, ground floor, 1878, Architects Archer & Green, London.
Reproduced with kind permission of the National Archives, London.

Chapter 4

Chapter 5

Chapter 6

Chapter 7

Chapter 8

Chapter 9

Chapter 10

Chapter 11

Chapter 12

Chapter 13

Chapter 14

Chapter 15

Chapter 16

Chapter 17

Chapter 18

Chapter 19

Fig. 19.0 *Senecio oldfieldii,* Perup Nature Reserve, Western Australia, 2008.
Fig. 19.1 Albany, King George Sound, Western Australia 1863, Samuel Calvert
Reproduced with the kind permission of the State Library of Victoria, Melbourne.
Fig 19.2 Nr Chester Pass Road at Kamballup old golf course, south of Stirling Range, Western Australia.
Fig. 19.3 Karri (*Eucalyptus diversicolor*), Porongorup Range, Western Australia, 2005.
Fig. 19.4 Near Takalarup Crossing, Upper Kalgan River, Western Australia, 2001.
Fig. 19.5 Vegetation fringing Lake Muir, Western Australia, 2010.
Fig. 19.6 East bank, Tone River, near Tonebridge, Western Australia, 2010.
Fig. 19.7 5 kilometres north-west of Tonebridge, Western Australia, 2010.
Fig. 19.8 Blackwood River near Jayes homestead, Western Australia, 2010.
Fig. 19.9 Cape Leeuwin promontory, looking towards Skippy Rock, 2003.

Chapter 20

Fig. 20.0 *Calytrix oldfieldii,* Kalbarri National Park, Western Australia, Robin Simkin, 2014.
© Robin Simkin, Kalbarri, Western Australia, reproduced with her kind permission.
Fig. 20.1 East bank of Lower Kalgan River, looking north-west, Western Australia, 2001.
Fig. 20.2 Cape Riche homestead from across the Eyre River, Western Australia, 2001.
Fig. 20.3 Doubtful Island Bay from Cape Hood promontory, 2003.
Fig. 20.4 Fitzgerald River estuary from the east bank, 2003.
Fig. 20.5 Mid-Mount Barren from foothills near the coast looking north, 2003.
Fig. 20.6 Oldfield River estuary, looking towards the mouth, 2003.
Fig. 20.7 From Mount Desmond, looking north, 2003.

Chapter 21

Fig. 2.0 *Chthonocephalus oldfieldianus*, Glass Bore, north end of Coolcalalaya Station, Robin Simkin, 2012.
© Robin Simkin, Kalbarri, Western Australia, reproduced with her kind permission.
Fig. 21.1 'Dunsford Villas' (foreground) with smaller 'Verandah Cottages' beyond, Merton Road, Wandsworth, Edward Passmore Olney, 1961.
Reproduced with the kind permission of the London Metropolitan Archives.
© Diligent research failed to identify the copyright holder. Please contact the publisher if you believe you are the copyright holder.
Fig. 21.2 Dr Joseph Hooker, 1851.
Board of Trustees, Royal Botanic Gardens, Kew.
Fig. 21.3 Herbarium (Hunter House), Royal Botanic Gardens Kew, 2002. Part of this building became available in 1852 for the herbarium, the wing was added in 1877.
Fig. 21.4 Lectotype of *Melaleuca oldfieldii* (MEL 602747) showing Oldfield's original label (ink) and two later ones (pencil) demonstrating the effect his failing eyesight was having on his handwriting.
Reproduced with kind permission of the Board of Trustees, Royal Botanic Gardens, Victoria

Chapter 22

Fig. 22.0 *Eremophila oldfieldii,* Eagle Gorge, Kalbarri, Western Australia, 2000
Fig. 22.1 Edwin and Susanna Oldfield, c. 1880, Desmond & Lyn Williams Archive.
Reproduced with the kind permission of Desmond Williams' guardian, Fiona Brooks, Applecross, Western Australia.
Fig. 22.2 The Oldfield grave, Brompton Cemetery, London, where Augustus is interred
© Christine Bassham, Padbury, Western Australia.
Reproduced with the kind permission of Christine Bassham.

INDEX

About the Authors

M. Helen Henderson AM, PhD (Anthropology), FRWAHS

Dr. Helen Henderson AM, PhD (Anthropology), FRWAHS, is the Great Grand-Niece of Augustus Oldfield.

Born in Bridgetown, Western Australia in 1934, she had a distinguished professional career with the WA Museum before joining the WA Health Department where she was the State Co-ordinator of Aboriginal Health Promotion, then a Project Officer for Aboriginal Policy and finally Senior Research Officer of Epidemiology at the Health Department of WA. She eventually retired in 1997 and immediately began work on the autobiography of Augustus Oldfield. She has been a Councillor for the Royal Western Australian Historical Society since 1996 and was Vice-Chairperson from 2001 to 2017. She represents the Society on the Federation of Australian Historical Societies of which she was President from 2004 through to 2008.

She is married to co-author Bill Henderson. They have two children, Noel and Georgina, both of whom are married with children of their own.

William George (Bill) Henderson LS, FISA

Bill Henderson LS, FISA, was born in West Perth on 15 October 1928 and served his survey cadetship under F. G. Medcalf in Albany.

From 1953 he was in the Lands and Surveys Department (WA) as Staff Surveyor and undertook many surveys across the State. A distinguished career culminated in him serving as the Acting Surveyor General of Western Australia, from 1984 to 1986. He retired in 1986 and commenced development of a protea plantation at Toodyay which he operated until 1997. Since then and until 2013, he has focused on mapping Augustus Oldfield's travels and other community activities including being a member of the Royal Western Australian Historical Society Council from 2000 to 2003.

www.ingramcontent.com/pod-product-compliance
Ingram Content Group UK Ltd.
Pitfield, Milton Keynes, MK11 3LW, UK
UKHW061842041026
15591UKWH00014B/882

9 780648 222231